Modern Ferrites, Volume 1

Modern Ferrites, Volume 1

Basic Principles, Processing and Properties

Edited by

Vincent G. Harris
College of Engineering
Northeastern University
Boston, MA, USA

WILEY

IEEE PRESS

This edition first published 2023

Registered Offices
John Wiley & Sons, Inc., 111 River Street, Hoboken, NJ 07030, USA
John Wiley & Sons Ltd, The Atrium, Southern Gate, Chichester, West Sussex, PO19 8SQ, UK

Editorial Office
The Atrium, Southern Gate, Chichester, West Sussex, PO19 8SQ, UK

For details of our global editorial offices, customer services, and more information about Wiley products visit us at www.wiley.com.

Wiley also publishes its books in a variety of electronic formats and by print-on-demand. Some content that appears in standard print versions of this book may not be available in other formats.

A catalogue record for this book is available from the Library of Congress

Hardback ISBN: 9781118971468; ePub ISBN: 9781118971482; ePDF ISBN: 9781118971475; oBook ISBN: 9781118971499

Cover image: Wiley
Cover design by Wiley

Set in 9.5/12.5pt STIXTwoText by Integra Software Services Pvt. Ltd, Pondicherry, India
Printed and bound by CPI Group (UK) Ltd, Croydon, CR0 4YY

C9781118971468_051122

"What one man calls God; another calls the laws of physics."

Contents

Preface

In 1990, I graduated with a Ph.D. degree in electrical engineering having completed a dissertation on the microwave properties of ferromagnetic metallic glassy films. The novelty of my dissertation was the correlation of atomic structure with magnetic and microwave gyromagnetic properties. Soon afterward, I took a National Research Council fellowship at the Naval Research Laboratory (Washington, DC, USA) and eventually was hired in the study of rare earth magnetism and atomic spectroscopy.

In 1994, I informed my then boss, Dr. Norman Koon, of my ambition to study magnetic oxides. He was rather obstinate: "I have studied magnetic metals for 40 years and have just begun to understand them! I am not about to change now!" (Note: Dr. Koon was a giant in magnetism. So his statement is more than a little self-deprecating.) That day marked my excursion into *ferrites*.

Several years ago, after nearly three decades of research in ferrites, I took on the responsibility as the editor of this project, having come to an agreement with the Wiley Publishing Company to complete this assignment in a single year. Although I attribute delays to what I affectionately refer to my *herding* of friends and colleagues, i.e. my coauthors, in truth, the delays were due to my lack of discipline, copious distractions, and duties that weighed heavily upon me at inopportune times. I attribute its completion to the dogged supervision of our efforts by Ms. Sandra Grayson and Ms. Juliet Booker of the Wiley Publishing Company.

The motivation for this undertaking was to create a modern-day text on ferrites. *Modern Ferrites Volume 1: Basic Principles, Processing, and Properties* covers aspects of ferrite structures, chemistry, physics, magnetism, gyromagnetic properties, processing, and properties, among other topics of relevance to students and practitioners.

Modern Ferrites Volume 2: Emerging Technologies and Applications addresses not only conventional ferrites as inductors, permanent magnets, electromagnetic shielding, and absorbers but also their role in emerging fields of magnetoelectrics, multiferroics, medical magnetism, nonlinear microwave devices, and ferrite-based metamaterial devices.

Our aspirations are bold. The contributing authors are experts chosen from the international community who have distinguished themselves throughout their careers and provide here superior products to the reader. Some have invited aspiring younger colleagues as coauthors who show singular potential. Our mutual experiences include an education weaned on tomes of the masters. These include those of Smit and Wijn, von Aulock, Soohoo, Lax and Button, Goldman, and others. Our principal goal is to complement those collections with the addition of a modern text covering both basic principles and emerging applications of particular value to the science and engineering communities while maintaining the standards of excellence embraced by those who came before us.

A project like this is always the result of a group effort, and this text is no different. I am thankful and indebted to my family, students, staff, and friends who assisted in the preparation of this text. A special thanks goes to Ms. Mary Zeng who has assisted me and other authors in editing, formatting, referencing, and submission of chapters to the Wiley Publishing Company. I thank you.

As mentioned, Wiley Publishing Company's Ms. Sandra Grayson and Ms. Juliet Booker, and others on the Wiley editorial staff, were of immense value in guiding this project through to its completion and showing uncommon patience with its editor. I thank you.

Finally, I thank each of the contributing authors and coauthors for their hard work and commitment in sharing with our readers their life lessons and experiences in ferrite research.

In closing, sadly, during the production of this text, our community has lost some of its most valued leaders; persons who were exceptional scholars, mentors, educators, and friends. These individuals include: Dr. Gerald Dione (2020) of MIT

Lincoln Labs, a renowned expert in all aspects of magnetic oxides, including the physics as well as engineering of magnetoceramics; Dr. John Douglas Adam (1943–2018), formally of Westinghouse and Northrop Grumman Corporation, the principal inventor of many important nonlinear ferrite devices and developer of advanced ferrite LPE growth techniques; Prof. Takanori Tsutaoka (1959–2019) of Hiroshima University, an expert and innovator of high-frequency magnetic materials, composites, and metamaterials; and Prof. Boris Kalinikos of St. Petersburg's Electrotechnical University and Colorado State University at Fort Collins, who had made several important contributions to the understanding of nonlinear properties of ferrites and fundamental loss mechanisms and had through his many contributions laid the foundations to the emerging field of *Magnonics* (a growing subfield of Spintronics). I speak for many in saying that we stand on the shoulders of these intellectual giants. May their souls rest in peace, knowing that they leave behind more than technical papers and discoveries, but a legacy of grateful students, valued friendships, and treasured memories.

Alas, we have come to the end of this long road.... the asphalt has turned to sand.

Here, our product is laid bare for your judgment. May you find value in our endeavor.

V.G.H.

List of Contributors

Volume 1

John D. Adam (deceased)
Ph.D., Metamagnetics Inc., Westborough, MA, USA

Everett E. Carpenter
Professor, Virginia Commonwealth University, Richmond, VA, USA

Ogheneyunume Fitchorova
Senior Research Scientist,
Kostas Research Institute at Northeastern University, LLC Burlington, MA, USA

Rongdi Guo
Ph.D., University of Electronic Science and Technology of China, Chengdu, Sichuan, China

Vincent G. Harris
University Distinguished Professor and W. L. Smith Chair Professor, Northeastern University, Boston, MA, USA

Xiaona Jiang
Associate Professor, University of Electronic Science and Technology of China,
Chengdu, Sichuan, China

Pavel Kabos
Ph.D., National Institute of Standards and Technology, Boulder, CO, USA

Boris A. Kalinikos (deceased)
Professor, Saint Petersburg Electrotechnical University, St. Petersburg, Russia

Marina Y. Koledintseva
Technical Lead,
The Boeing Company, St. Louis, MO, USA

David E. Laughlin
ALCOA Professor of Physical Metallurgy, Carnegie Mellon University, Pittsburgh, PA, USA

Yingli Liu
Professor, University of Electronic Science and Technology of China, Chengdu, Sichuan, China

Michael E. McHenry
Professor of Materials Science and Engineering, Carnegie Mellon University, Pittsburgh, PA, USA

Martha Pardavi-Horvath
Professor Emerita, The George Washington University, Washington DC, USA

Sarah E. Smith
Ph.D., Virginia Commonwealth University, Richmond, VA, USA

Alexander S. Sokolov
Research Scientist, Northeastern University, Boston, MA, USA

Ke Sun
Professor, University of Electronic Science and Technology of China, Chengdu, Sichuan, China

Melissa Tsui
Ph.D., Virginia Commonwealth University, Richmond, VA, USA

Takanori Tsutaoka (deceased)
Professor, Hiroshima University, Hiroshima, Chūgoku, Japan

Pieter J. van der Zaag
Professor, University of Groningen, Groningen, The Netherlands

Qiye Wen
Professor, University of Electronic Science and Technology of China, Chengdu, Sichuan, China

Brent Williams
Ph.D., Virginia Commonwealth University, Richmond, VA, USA

Chuanjian Wu
Ph.D., University of Electronic Science and Technology of China, Chengdu, Sichuan, China

Qinghui Yang
Professor, University of Electronic Science and Technology of China, Chengdu, Sichuan, China

Zhong Yu
Professor, University of Electronic Science and Technology of China, Chengdu, Sichuan, China

Huaiwu Zhang
Professor, University of Electronic Science and Technology of China, Chengdu, Sichuan, China

Lan Zhongwen
Professor, University of Electronic Science and Technology of China, Chengdu, Sichuan, China

Xu Zuo
Professor, Nankai University, Tianjin, China

1

Historical Evolution of *Systeme International d'Unites* and Its Application to Electromagnetism and Magnetic Materials

Vincent G. Harris

College of Engineering, Northeastern University, Boston, MA 02115-5000 USA

The International System of Units, popularly known as SI units, was adopted at the 1960 Conférence Générale des Poids et Mesures (CGPM) as the *Systeme International d'Unites*. This system of units is widely accepted by the international science and technology communities and its many disciplines.

Exceptions to its universal acceptance are segments of the magnetics' community that cling doggedly to their historic roots in centimeter–gram–second (CGS).

Interrelationships between these systems are discussed here where we attempt to make clear to the reader units they will encounter in the chapters contained herein that describe the electromagnetic behavior, both static and dynamic, of ferrite materials. The rich history of the *Systeme International d'Unites* will be presented and discussed in the context of their application to magnetism and magnetic materials.

Table 1.1 displays the list of the historical mileposts from the past 150 years marking the evolution and acceptance of SI units.

1.1 Nineteenth-Century Evolution

1.1.1 The CGS Unit System

The precursor to the metric or SI system is the CGS unit system. The foundation of CGS dates back to 1832 and the work of the German mathematician Carl Friedrich Gauss who proposed a system of absolute units based on the fundamental units of length, mass, and time, specifically a system based on millimeter–milligram–second (MMS) [1].

Building on the work of Gauss, in 1873, a committee of the British Association for the Advancement of Science, led by James Clerk Maxwell, William Thomson, and prominent others, proposed the CGS (centimeter, gram and second) system as fundamental units to express all derived electromagnetic units [2]. This CGS system is often referred to as the Gaussian system or the Gaussian–CGS system.

1.1.2 The Lorentz–Heaviside Unit System

A major advance to the Gaussian system of 1873 occurred in 1893 when Hendrik Lorentz and Oliver Heaviside proposed the Lorentz–Heaviside Unit (LHU) system. LHU normalized vacuum permeability and permittivity, i.e. $\varepsilon_0 = 1$ and $\mu_0 = 1$, and revised Maxwell's equations to employ the speed of light c [3].

The LHUs, like SI units, are *rationalized*, meaning there are no 4π factors appearing in Maxwell's equations [4]. As a result, they simplify calculations such as those employed in quantum electrodynamics [5], general relativity [6], atomic physics, and string theory, which have maintained their popularity.

As shown in Figure 1.1, CGS can be further subdivided into electromagnetic units (EMUs) and electrostatic units (ESU). The latter employ less frequently used and unfamiliar units, such as the ESU Franklin, which is also a statCoulomb (electric charge), statWb (magnetic flux), statT (magnetic field), etc., and the EMU abCoulomb (electric charge), Maxwell (magnetic flux), Gilbert (magnetization), etc. [7].

Modern Ferrites: Basic Principles, Processing and Properties, Volume 1, First Edition. Edited by Vincent G. Harris.

Table 1.1 Historical timeline of events contributing to the evolution of the *Systeme International d'Unites.*

Approximate date	Event/contribution
1874	The centimeter gram second (CGS) system of electromagnetic units (EMUs) was formulated by Thomson, Maxwell, and others, through the British Association for the Advancement of Science.
1893	Heaviside–Lorentz Units constitute a system of units within CGS. Lorentz–Heaviside units may be regarded as normalizing $\varepsilon_0 = 1$ and $\mu_0 = 1$ while revising Maxwell's equations to employ the speed of light c. Lorentz–Heaviside units, like SI, are *rationalized*, meaning that there are no factors of 4π appearing explicitly.
1901–1905	1901: Giorgi introduced the rationalized, four-dimensional, meter–kilogram–second (MKS) system in which the fourth electromagnetic unit is not specified. 1902: μ_0 is not numeric; it is a physical quantity having dimensions and measured by experiments 1905: Giorgi recommended Ω as the fourth unit in an "MKS" system. He regarded both μ_0 and the permittivity of vacuum ε_0 as subject to experimental refinement, where $\mu_0 \approx 1.256 \times 10^{-6}$ H/m and $\varepsilon_0 \approx 8.842 \times 10^{-12}$ F/m.
1934	At the International Electrotechnical Commission (IEC) meeting, Giorgi warns "...the evil of 4π and [its] irrationality [will] be perpetuated because permeability of free space would receive the exact value of $4\pi \times 10^{-7}$ instead of being the result of measurement."
1935	IEC recommended the adoption of the "Giorgi system," with the fourth unit to be determined.
1938	The IEC Technical Committee on Electric and Magnetic Magnitudes and Units (EMMU), while recognizing that the fourth unit could be either Ohm, Ampere, Volt, Henry, Farad, Coulomb, or Weber, recommended the permeability of free space of $\mu_0 = 10^{-7}$ in the unrationalized system or $\mu_0 = 4\pi \times 10^{-7}$ in the rationalized system.
1939	The Consultative Committee for Electricity (CCE) formally recommended an absolute four-dimensional MKS system, with the fourth unit being the ampere.
1950, 1956	IEC adopted both the ampere as the fourth unit and rationalization of the Giorgi system in 1950 and affirmed the unit and rationalized value of μ_0 in 1956.
1954	CGPM approved the ampere as a base unit, thereby formalizing the "MKSA" practical system of units.
1960	The General Conference on Weights and Measures established the name Systeme International "d'Unites (SI)" for the system, commonly known today as SI.
1964	US National Bureau of Standards made it a policy to require SI units in its reports.
1966	The Institute of Electrical and Electronics Engineers (IEEE), via its Standards Coordinating Committee on Quantities and Units, recommended the use of SI in all publications.
1974, 1976	Page summarized his philosophy and that of the IEC at the time, which is applicable to all unit systems: "E and B can be defined in vacuo in terms of forces on moving charges. D and H are defined in vacuo in terms of E and B and the magnetic constant (μ_0) of the unit system used."
1983	c became a fixed value in the SI system.
1987	IUPAP recommended "that authors be encouraged to adapt [*sic*] the SI units for data in physics journals."
2014	CGPM stated that the uncertainties in determining the four fundamental constants are not yet sufficiently small to revise the SI. The new values of the constants will be based on determinations that have been published, or at least accepted for publication, by 1 July 2017 and that were used by the International Council for Science's Committee on Data for Science and Technology (CODATA) to recommend values and uncertainties of the four fundamental constants needed to revise the present SI. The four values will later be fixed (i.e. will become exact), and the full set of seven fixed constants will define all units in the revised SI.
2016	The new definition of the ampere fixed the value of e instead of μ_0, and as a result, μ_0 must be determined experimentally. Similarly, the permittivity of vacuum ε_0 must be determined experimentally as it was before c was fixed in 1983.
2017–2018	2017: The International Committee for Weights and Measures (CIPM) formally recommended a major redefinition of the International System of Units (SI). The revised SI will be the most significant change in units of measure since the meter–kilogram–second–ampere (MKSA) system was adopted by the CGPM in 1954. In the revised SI, kilogram, ampere, kelvin, and mole will be redefined in terms of newly fixed values of the Planck constant, the elementary charge, the Boltzmann constant, and the Avogadro constant, respectively. These four constants will join the cesium hyperfine transition frequency, the speed of light, and the luminous efficacy of radiation of frequency 540 THz as the seven defining constants of the SI system.

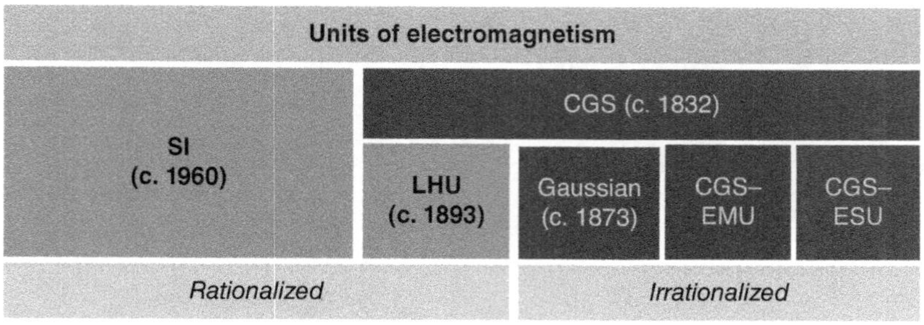

Figure 1.1 Schematic representation of rationalized and irrationalized measurement unit systems, their pedigree, and approximate inception dates.

1.1.3 Rationalized vs. Irrationalized Quantities

Table 1.2 illustrates Maxwell's equations and other important electromagnetic relationships in SI, LHU, and CGS. Here, Maxwell's equations are presented for free space in *differential* form for both macroscopic and microscopic systems. To obtain *integral* forms, one applies the Kelvin–Stokes theorem. This conversion will not be presented here; interested readers are directed to ref. [8] for a comprehensive treatment.

It is clear from Table 1.2 that the SI and LHU systems are *rationalized* quantities, whereas the Gaussian system are *irrationalized* quantities, as evidenced by the presence of 4π in the Gaussian equations. This relationship is depicted visually in Figure 1.1.

Table 1.2 also presents the magnetic and electric polarization fields (i.e. **B**, **M**, **E**, **D**, and **P**) as they are defined in dielectric and magnetic media. These are particularly relevant to the chapter 1. The superscripts "SI," "LH," and "G" indicate the

Table 1.2 Maxwell's equations and other important electromagnetic relationships in SI, LHU, and CGS presented for free space in *differential* form (adapted from ref. [5]).

Maxwell's equations

Name	SI quantities	LH quantities	Gaussian quantities
Gauss's Law (Macroscopic)	$\nabla\cdot\boldsymbol{D}^{\mathrm{SI}}=\rho_{\mathrm{f}}^{\mathrm{SI}}$	$\nabla\cdot\boldsymbol{D}^{\mathrm{LH}}=\rho_{\mathrm{f}}^{\mathrm{LH}}$	$\nabla\cdot\boldsymbol{D}^{\mathrm{G}}=4\pi\rho_{\mathrm{f}}^{\mathrm{G}}$
Gauss's Law (Microscopic)	$\nabla\cdot\boldsymbol{E}^{\mathrm{SI}}=\rho^{\mathrm{SI}}/\epsilon_{\mathrm{o}}$	$\nabla\cdot\boldsymbol{E}^{\mathrm{LH}}=\rho_{\mathrm{f}}^{\mathrm{LH}}$	$\nabla\cdot\boldsymbol{E}^{\mathrm{G}}=4\pi\rho^{\mathrm{G}}$
Gauss's law for magnetism	$\nabla\cdot\boldsymbol{B}^{\mathrm{SI}}=0$	$\nabla\cdot\boldsymbol{B}^{\mathrm{LH}}=0$	$\nabla\cdot\boldsymbol{B}^{\mathrm{G}}=0$
Maxwell–Faraday equation	$\nabla\cdot\boldsymbol{E}^{\mathrm{SI}}=-\dfrac{\partial\boldsymbol{B}^{\mathrm{SI}}}{\partial t}$	$\nabla\cdot\boldsymbol{E}^{\mathrm{LH}}=-\dfrac{1}{c}\dfrac{\partial\boldsymbol{B}^{\mathrm{LH}}}{\partial t}$	$\nabla\cdot\boldsymbol{E}^{\mathrm{G}}=-\dfrac{1}{c}\dfrac{\partial\boldsymbol{B}^{\mathrm{G}}}{\partial t}$
Ampere–Maxwell equation (Macroscopic)	$\nabla\cdot\boldsymbol{H}^{\mathrm{SI}}=\boldsymbol{J}_{\mathrm{f}}^{\ \mathrm{SI}}+\dfrac{\partial\boldsymbol{D}^{SI}}{\partial t}$	$\nabla\cdot\boldsymbol{H}^{\mathrm{LH}}=\dfrac{1}{c}\boldsymbol{J}_{\mathrm{f}}^{\mathrm{LH}}+\dfrac{1}{c}\dfrac{\partial\boldsymbol{D}^{\mathrm{LH}}}{\partial t}$	$\nabla\cdot\boldsymbol{H}^{\mathrm{G}}=\dfrac{4\pi}{c}\boldsymbol{J}_{\mathrm{f}}^{\mathrm{G}}+\dfrac{1}{c}\dfrac{\partial\boldsymbol{D}^{\mathrm{G}}}{\partial t}$
Ampere–Maxwell equation (Microscopic)	$\nabla\cdot\boldsymbol{B}^{\mathrm{SI}}=\mu_{\mathrm{o}}\boldsymbol{J}_{\mathrm{f}}^{\ \mathrm{SI}}+\dfrac{1}{c^2}\dfrac{\partial\boldsymbol{E}^{\mathrm{SI}}}{\partial t}$	$\nabla\cdot\boldsymbol{B}^{\mathrm{LH}}=\dfrac{1}{c}\boldsymbol{J}_{\mathrm{f}}^{\ \mathrm{LH}}+\dfrac{1}{c}\dfrac{\partial\boldsymbol{E}^{\mathrm{LH}}}{\partial t}$	$\nabla\cdot\boldsymbol{B}^{\mathrm{G}}=\dfrac{4\pi}{c}\boldsymbol{J}_{\mathrm{f}}^{\ \mathrm{G}}+\dfrac{1}{c}\dfrac{\partial\boldsymbol{E}^{\mathrm{G}}}{\partial t}$

Fields in a dielectric medium

Name	SI quantities	LH quantities	Gaussian quantities
Displacement field, **D**	$\boldsymbol{D}^{\mathrm{SI}}=\epsilon_{\mathrm{o}}E^{\mathrm{SI}}+P^{\mathrm{SI}}$	$\boldsymbol{D}^{\mathrm{LH}}=\boldsymbol{E}^{\mathrm{LH}}+\boldsymbol{P}^{\mathrm{LH}}$	$\boldsymbol{D}^{\mathrm{G}}=\boldsymbol{E}^{\mathrm{G}}+4\pi\boldsymbol{P}^{\mathrm{G}}$
Polarization density, **P**	$\boldsymbol{P}^{\mathrm{G}}=\chi_{\mathrm{e}}^{\mathrm{G}}\boldsymbol{E}^{\mathrm{G}}$	$\boldsymbol{P}^{\mathrm{LH}}=\chi_{\mathrm{e}}^{\mathrm{LH}}\boldsymbol{E}^{\mathrm{LH}}$	$\boldsymbol{P}^{\mathrm{G}}=\chi_{\mathrm{e}}^{\mathrm{G}}\boldsymbol{E}^{\mathrm{G}}$
Displacement field, **D**	$\boldsymbol{D}^{\mathrm{LH}}=\epsilon\,\boldsymbol{E}^{\mathrm{LH}}$	$\boldsymbol{D}^{\mathrm{LH}}=\epsilon\,\boldsymbol{E}^{\mathrm{LH}}$	$\boldsymbol{D}^{\mathrm{G}}=\epsilon\,\boldsymbol{E}^{\mathrm{G}}$
Permittivity, ϵ	$\dfrac{\epsilon^{\mathrm{SI}}}{\epsilon_{\mathrm{o}}}=1+\chi_{\mathrm{e}}^{\mathrm{SI}}$	$\epsilon^{\mathrm{LH}}=1+\chi_{\mathrm{e}}^{\mathrm{LH}}$	$\epsilon^{\mathrm{G}}=1+4\pi\chi_{\mathrm{e}}^{\mathrm{G}}$

Fields in a magnetic medium

Name	SI quantities	LH quantities	Gaussian quantities
Magnetic induction, **B**	$\boldsymbol{B}^{\mathrm{SI}}=\mu_{\mathrm{o}}(\boldsymbol{H}^{\mathrm{SI}}+\boldsymbol{M}^{\mathrm{SI}})$	$\boldsymbol{B}^{\mathrm{LH}}=\boldsymbol{H}^{\mathrm{LH}}+\boldsymbol{M}^{\mathrm{LH}}$	$\boldsymbol{B}^{\mathrm{G}}=\boldsymbol{H}^{\mathrm{G}}+4\pi\boldsymbol{M}^{\mathrm{G}}$
Magnetization, **M**	$\boldsymbol{M}^{\mathrm{SI}}=\chi_{\mathrm{m}}^{\mathrm{SI}}\boldsymbol{H}^{\mathrm{SI}}$	$\boldsymbol{M}^{\mathrm{LH}}=\chi_{\mathrm{m}}^{\mathrm{LH}}\boldsymbol{H}^{\mathrm{LH}}$	$\boldsymbol{M}^{\mathrm{G}}=\chi_{\mathrm{m}}^{\mathrm{G}}\boldsymbol{H}^{\mathrm{G}}$

(Continued)

Table 1.2 (Continued)

Magnetic induction, **B**	$\boldsymbol{B}^{SI} = \mu^{SI}\boldsymbol{H}^{SI}$	$\boldsymbol{B}^{LH} = \mu^{LH}\boldsymbol{H}^{LH}$	$\boldsymbol{B}^{G} = \mu^{G}\boldsymbol{H}^{G}$
Permeability, μ	$\frac{\mu^{SI}}{\mu_o} = 1 + \chi_m^{SI}$	$\mu^{LH} = 1 + \chi_m^{LH}$	$\mu^{G} = 1 + 4\pi\chi_m^{G}$

Other basic laws

Name	SI quantities	LH quantities	Gaussian quantities
Lorentz force	$\boldsymbol{F} = q(\boldsymbol{E}^{SI} + \mathrm{v} \times \boldsymbol{B}^{SI})$	$\boldsymbol{F} = q(\boldsymbol{E}^{LH} + \frac{1}{c}\mathrm{v} \times \boldsymbol{B}^{LH})$	$\boldsymbol{F} = q(\boldsymbol{E}^{G} + \frac{1}{c}\mathrm{v} \times \boldsymbol{B}^{G})$
Coulomb's law	$\boldsymbol{F} = \frac{1}{4\pi\epsilon_o}\frac{q_1^{SI}q_2^{SI}}{r^2}\hat{\boldsymbol{r}}$	$\boldsymbol{F} = \frac{1}{4\pi}\frac{q_1^{LH}q_2^{LH}}{r^2}\hat{\boldsymbol{r}}$	$\boldsymbol{F} = \frac{q_1^{G}q_2^{G}}{r^2}\hat{\boldsymbol{r}}$
Electric field of stationary point charge	$\boldsymbol{E}^{SI} = \frac{1}{4\pi\epsilon_o}\frac{q^{SI}}{r^2}\hat{\boldsymbol{r}}$	$\boldsymbol{E}^{LH} = \frac{1}{4\pi}\frac{q^{LH}}{r^2}\hat{\boldsymbol{r}}$	$\boldsymbol{E}^{G} = \frac{q^{G}}{r^2}\hat{\boldsymbol{r}}$
Biot–Savart law	$\boldsymbol{B}^{SI} = \frac{\mu_o}{4\pi}\oint\frac{I^{SI}d\boldsymbol{l} \times \hat{\boldsymbol{r}}}{r^2}$	$\boldsymbol{B}^{LH} = \frac{1}{4\pi c}\oint\frac{I^{LH}d\boldsymbol{l} \times \hat{\boldsymbol{r}}}{r^2}$	$\boldsymbol{B}^{G} = \frac{1}{c}\oint\frac{I^{G}d\boldsymbol{l} \times \hat{\boldsymbol{r}}}{r^2}$

system that defines the quantities. For simplicity, the media, be it dielectric or magnetic, are assumed linear, isotropic, and nondispersive so that their permittivity and permeability can be treated as scalars.

In the dielectric media, **E** and **D** are the electric and displacement fields, respectively; **P** is the electric polarization; χ_e is the electric susceptibility, and ϵ is the permittivity, and ϵ_0 is the permittivity of vacuum. In comparison, in magnetic media, **B** and **H** are the magnetic induction and the magnetic field, respectively; **M** is the magnetization; χ_m is the magnetic susceptibility; and μ is the permeability and μ_0 is the permeability of vacuum.

The bottom part of Table 1.2 lists the four basic laws foundational to the electromagnetic theory. These include the Lorentz force, the Coulomb's law, the electric field generated by a stationary point charge, and the Biot–Savart law.

The Lorentz force derives from the combination of electric and magnetic forces on a point charge from electromagnetic fields, whereas the Coulomb's law quantifies these forces (i.e. electrostatic force and magnetostatic force) between two stationary particles (i.e. q_1 and q_2). The Biot–Savart equation describes the magnetic field generated by a constant electric current. It relates the magnetic field to the magnitude and direction of the current, making it fundamental to understanding a system's magnetostatic behavior and is analogous to the role of the Coulomb's law of electrostatics.

Correspondingly, Table 1.3 contains a list of key magnetic units and their conversions from CGS to SI. In the rightmost column, $[\mu_0]$ represents the numerical value of μ_0 that may change over time as instrumentation accuracy and precision improves. In the absence of units, a dimensionless quantity is represented by is associated system of units (i.e. CGS or SI).

Table 1.3 Conversion of units for magnetic quantities from Gaussian (CGS) to SI.

Description	Symbol	Units and conversion from CGS to SI
Magnetic flux	ϕ	G·cm^2; 10^{-8} Wb; 10^{-8} V·s
Magnetic flux density, magnetic induction	B	G; 10^{-4} T; 10^{-4} Wb/m^2
Magnetic field strength	H	Oe; 10^4 A/m
Magnetic moment	m	erg/G (=1 emu); 10^{-3} A·m^2; 10^{-3} J/K
Magnetization (volume)	M	erg/(G·cm^3); emu/cm^3; 10^3 A/m
Specific magnetization (mass)	σ	erg/(G·g) (emu/g); A·m^2/kg
Permeability	μ	Unity (CGS); $[\mu_0]$ H/m; $[\mu_0]$ N/A^2; [μ0] Wb/(A·m)
Volume susceptibility	χ_V	Unity (CGS); 4π (SI)
Specific susceptibility (mass)	χ_m	cm^3/g; $4\pi \times 10^{-3}$ m^3/kg
Demagnetizing factor	N,D	Unity (CGS); $(4\pi)^{-1}$ (SI)

1.2 Early-Twentieth-Century Evolution

At the turn of the twentieth century, Giovanni Giorgi introduced the meter–kilogram–second (MKS) unit system and suggested a fourth electromagnetic unit but did not offer its identity. In a series of papers from 1901 to 1905 [9–11] Giorgi proposed that both the vacuum permeability and permittivity (i.e. μ_0 and ε_0) are numerical, measurable, and dimensioned parameters with the fourth electromagnetic unit being the ohm (Ω).

Circa 1935, the International Electrotechnical Commission (IEC) adopted the Giorgi system with the fourth dimension unresolved [12]. However, soon after (c. 1939), the Consultative Committee for Electricity (CCE) adopted the Giorgi system, now with the fourth dimension being the ampere (i.e. MKSA). This was accepted at the Conférence Générale des Poids et Mesures in 1954 [12].

In 1960, the CGPM formally adopted the *Systeme Internationall d'Unites* (i.e. SI units) as the standard for the international communities. From 1960 to the present, major societies and journals around the world began to require contributing authors to adhere to SI units, formulas, and expressions.

1.2.1 Conflicts Continue to Plague the Magnetics Community

As mentioned, the application of SI and CGS units by magneticians has been a continued source of conflicts and errors.

Although the use of SI units in magnetism does introduce some burdens to practitioners, the benefits outweigh the profound ambiguities introduced by the use of CGS.

The case for adhering to SI remains indisputable [13]. Among its advantages are its unification of magnetic and electrical units, whereas CGS presents conflicts between the rather archaic electromagnetic units (EMUs) and electrostatic units (ESUs). However, a disadvantage arises from competing constitutive conventions [14]. For example, consider the Sommerfeld and Kennelly conventions where the former states:

$$\mathbf{B} = \mu_0(\mathbf{H} + \mathbf{M})$$

where $\mathbf{B}$ is magnetic flux density, μ_0 is the permeability of vacuum, $\mathbf{H}$ is the magnetic field strength, and $\mathbf{M}$ is the magnetization. In comparison, the Kennelly convention states:

$$\mathbf{B} = \mu_0\mathbf{H} + \mathbf{J}$$

where $\mathbf{J}$ is the magnetic polarization. Obviously, the competing conventions confuse students and professors alike. However, being that $\mathbf{M}$ and $\mathbf{J}$ have different units, careful accounting of units can avoid errors.

In contrast, when considering the use of CGS units, the following ambiguous situations arise:

i) The unit for magnetic moment is often expressed in "emu." However, "emu" is simply a measure of the magnitude of magnetic polarization. The proper unit is "erg per gauss" or "erg per oersted."
ii) The volume susceptibility ($\chi_V = \mathbf{M}/\mathbf{H}$) is a dimensionless proportionality constant indicating the degree of magnetic polarization a material experiences in response to an applied magnetic field. Here, $\mathbf{M}$ is the magnetization, or the magnetic moment per unit volume, measured in SI units as A/m, and $\mathbf{H}$ is the magnetic field strength also measured in A/m. However, in CGS (or more specifically CGS-EMU, or Gaussian–CGS units, which in this case are equivalent) $\chi_V^{SI} = 4\pi\chi_V^{G}$, making Gaussian values smaller in magnitude to their SI counterparts even though they are both unitless.
iii) Magnetization, as magnetic moment per unit volume, is commonly expressed as $\mathbf{M}$ in terms of "emu per cubic centimeter" but also as $4\pi M$ in units of gauss (G). These conventions are dimensionally equivalent but differ numerically by 4π. Such competing definitions lead to confusion when one tries to compare results across literature and time.

1.2.2 The "Magnificent 7"

Figure 1.2 presents a logo constructed by the Bureau International des Poids et Mesures (BIPM) relating the seven defining constants and their relationship to the SI base units for the revised SI system of 2019 [15].

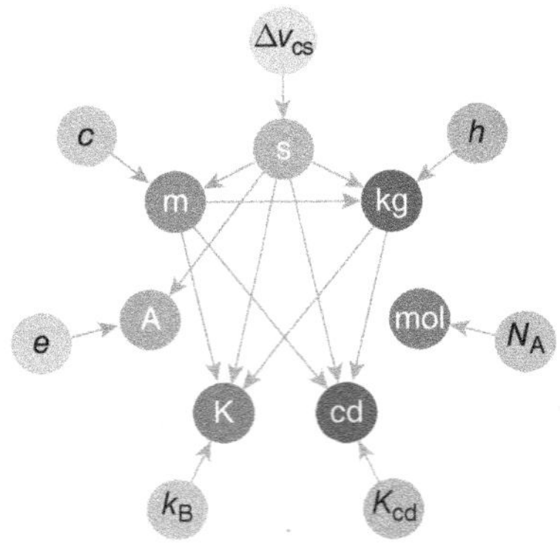

Figure 1.2 The BIPM SI system logo illustrating the dependence of base units on physical constants with fixed numerical values and on other base units. Arrows are shown in the opposite direction compared to those typically depicted in dependency graphs, i.e. a $\longrightarrow$ b , in this chart means *b* depends upon *a*. [15].

Table 1.4 Numerical values for seven defining constants expressed in SI units.

Description	Symbol	Exact value (c. 2019)	SI base units Quantity (unit)
Speed of light	c	299 792 458 m/s	Length (m)
Hyperfine transition frequency of Cs	Δv_{Cs}	9 192 631 770 Hz	Time (s)
Plank's constant	h	$6.626\ 070\ 15 \times 10^{-34}$ J · s	Mass (kg)
Elementary charge	e	$1.602\ 176\ 634 \times 10^{-19}$ C	Electric current (A)
Boltzmann constant	k	$1.380\ 649 \times 10^{-23}$ J/K	Thermodynamic temperature (K)
Avogadro's constant	N_A	$6.022\ 140\ 76 \times 10^{23}$ mol^{-1}	Amount of substance (mole)
Luminous efficacy	K_{Cd}	Photopic luminous efficacy of radiation is 683.002 lm/W for monochromatic light at a wavelength of 555 nm. Scotopic luminous efficacy of radiation is 1700 lm/W for monochromatic light at a wavelength of 507 nm.	Luminous intensity (Cd)

The BIPM was created in 1875 [16] with a mandate to provide a single standard of measurements and units for the international community. Today, that standard is the *Systeme International d'Unites* (i.e. SI). The BIPM logo, depicted in Figure 1.2, is also expressed in numerical detail in Table 1.4, where magnitudes of the seven defining constants are presented with their SI units. The seven constants include the speed of light *in vacuo*, *c*; the hyperfine transition frequency of cesium (Cs), Δv_{Cs}; Planck's constant, *h*; the elementary charge, *e*; the Boltzmann constant, *k*; Avogadro's constant, N_A; and the luminous efficacy, K_{cd} [17].

1.3 Early Twenty-First-Century Evolution: The Revised SI System

In 2017, a major revision was proposed to the *Systeme International d'Unites*, whereupon *h*, *e*, *k*, and N_A, previously precisely measured quantities, were assigned the numerical best estimates from surveyed published values, ensuring continuity with previous definitions of base units.

On 16 November 2018, a group of 60 nations universally approved the revisions to the *Systeme International d'Unites* [18, 19] creating a system that for the first time was exclusively based on constants of nature and not on physical objects stored in vaults and ampoules. This action marked the completion of a journey embarked upon some 150 years ago that endeavored to develop a measurement system based entirely on invariable fundamental properties of nature.

1.3.1 Permeability Revisited: 2019

In the revised SI [12, 20], kilogram, ampere, kelvin, and mole are redefined in terms of newly fixed values of Planck's constant, the elementary charge, the Boltzmann constant, and Avogadro's constant, respectively. These four constants join the hyperfine transition frequency of Cs, the speed of light, and the luminous efficacy of radiation as the seven defining constants of the *Systeme International d'Unites*.

A newly fixed value of elementary charge, in turn, redefines the ampere, making 1 A equivalent to the electric current generated by the flow of $1/(1.602\ 176\ 634 \times 10^{-19})$ elementary charges per second [21].

The definition of the ampere prior to the revised SI system was:

> that constant current which, if maintained in two straight parallel conductors of infinite length, of negligible circular cross section, and placed one meter apart in vacuo, would produce between these conductors a force equal to 2×10^{-7} Newton per meter of length [22].

That definition had the effect of fixing the value of permeability *in vacuo*, μ_0, to be $4\pi \times 10^{-7}$ N/A^2 equivalent to $4\pi \times 10^{-7}$ H/m. The new definition of the ampere derives from the fixed value of e, not that of μ_0, and as a result, μ_0 must be determined experimentally. Similarly, permittivity of vacuum $\varepsilon_0 = 1/(\mu_0 c^2)$ must similarly be determined experimentally as it was before *c* was fixed in Maxwell's equations in 1983. Notwithstanding these revisions, the product $\varepsilon_0\mu_0 = 1/c^2$ remains an exact relationship [23].

In the revised SI of 2019, μ_0 is now a measurable constant proportional to the experimental fine-structure constant α, a.k.a. Sommerfeld's constant. With the advent of new higher precision instrumentation, the modern value of $\alpha, (e^2/4\pi\varepsilon_0 hc)$, is $0.007\,297\,352\,569\,3(11)$ [24] leading to a more exact value of permeability, $\mu_0, (2\alpha/e^2\, h/c)$, of $4\pi \times 1.000\,000\,000\,82\,(20) \times 10^{-7}\,\mathrm{H/m}$ [25, 26].

In summary, at long last, the international science and technology community finally has its best measures and units system in the 2019 revised SI – a system that, for the first time in human history, is entirely based on constants of nature and not upon tangible objects subject to physical transformation or degradation over time.

However, those segments of the magnetics community, who have steadfast embraced CGS units, will likely be slow to adopt new standards. A policy of education and outreach to a new generation of magneticians may pave a path to acceptance.

References

1 Gauss, C.F. (1832). Intensitas vis magneticae terrestris ad mensuram absolutam revocata. *Commentationes Societatis Regiae Scientiarum Gottingensis Recentiores* 8: 3–44.

2 Thomson, W., Foster, G.C., Maxwell, J.C. et al. (1873). First report of the committee for the selection and nomenclature of dynamical and electrical units. Forty-third Meeting of the British Association for the Advancement of Science. Everett, B., Murray, J. edt. 223.

3 Silsbee, F. (1962). Systems of electrical units. *Journal of Research of the National Bureau of Standards Section C* 66C (2): 137–183. doi: 10.6028/jres.066C.014.

4 Kowalski, L. (1986). A short history of the SI units in electricity. *The Physics Teacher* 24 (2): 97–99.

5 Jackson, J.D. (1999). *Classical Electrodynamics*, 3e, 775–784. New York: Wiley.

6 Einstein, A. (2005). *The Meaning of Relativity (1956, 5th Edition)*, 21. Princeton University Press.

7 Cardarelli, F. (2004). *Encyclopaedia of Scientific Units, Weights and Measures: Their SI Equivalences and Origins*, 2e, 20–25. Springer.

8 Katz, V.J. (1979). The history of Stokes's theorem. *Mathematics Magazine* 52 (3): 146–156. doi: 10.2307/2690275.

9 Giovanni, G. (1901), Unità Razionali de Elettromagnetismo. In: *Atti dell' Associazione Elettrotecnica Italiana.*

10 Giorgi, G. (1902). Rational electromagnetic units. In: *Electrical World and Engineer* 40, 355, 368–370. New York: McGraw-Hill.

11 Giovanni, G. (1905). Proposals concerning electrical and physical units. In: *Transactions of the International Electrical Congress*, 136–141. New York: J.B. Lyon Company.

12 Goldfarb, R.B. (2018). Electromagnetic units, the Giorgi System, and the Revised International System of Units. *IEEE Magnetics Letters* 9.

13 Frezza, F., Maddio, S., Pelosi, G. et al. (2015). The life and work of Giovanni Giorgi: The rationalization of the International System of Units. *IEEE Antennas and Propagation Magazine* 57: 152–165. doi: 10.1109/MAP.2015.2486765.

14 Coleman, J.E. (1973). Conventions for magnetic quantities in SI. *American Journal of Physics* 41: 221. doi: 10.1119/1.1987179.

15 BIPM. (2019). https://www.bipm.org/en/measurement-units/si-base-units, Retrieved 6 August 2022.

16 BIPM. (2019). https://www.bipm.org/en/measurement-units/si-base-units, Retrieved 6 August 2022.

17 Newell, D.B., Cabiati, F., Fischer, J. et al. (2017). The CODATA 2017 values of h, e, k, and N_A for the revision of the SI. *Metrologia* 55 (1): L13. doi: 10.1088/1681-7575/aa950a.

18 Historic vote ties kilogram and other units to natural constants. NIST. 16 November 2018. Archived from the original on 18 November 2018. Retrieved 25 February 2021.

19 NIST. (2020). A turning point for humanity: redefining the world's measurement system. https://www.nist.gov/si-redefinition/turning-point-humanity-redefining-worlds-measurement-system

20 BIPM statement: https://www.bipm.org/en/measurement-units/si-base-units, Retrieved 6 August 2022.

21 BIPM. (2016). *The International System of Units, draft 9th ed. Sevres*, France: Bureau Int. `des Poids et Mesures.http://www.bipm.org/utils/common/pdf/sibrochure-draft-2016b.pdf

22 BIPM. (2006). The International System of Units, 8th ed. Sevres, France: Bureau Int. `des Poids et Mesures. http://www.bipm.org/utils/common/pdf/si_brochure_8.pdf

23 Goldfarb, R.B. (2017). The permeability of vacuum and the Revised International System of Units. *IEEE Magnetics Letters* 8: 1110003.

24 Mohr, P.J., Taylor, B.N., and Newell, D.B. (2019). Fine structure constant. CODATA Internationally recommended 2018 values of the fundamental physical constants. National Institute of Standards and Technology.

25 Parker, R.H., Yu, C., Zhong, W., Estey, B., and Müller, H. (13 April 2018). Measurement of the fine-structure constant as a test of the Standard Model. *Science* 360 (6385): 191–195. doi: 10.1126/science.aap7706.

26 Davis, R.S. (2017). Determining the value of the fine-structure constant from a current balance: Getting acquainted with some upcoming changes to the SI. *American Journal of Physics* 85 (5): 364–368. doi: 10.1119/1.4976701.

2

Societal Benefits of Ferrites

Historical, Scientific, and Commercial Breakthroughs

Vincent G. Harris

Center for Microwave Magnetic Materials and Integrated Circuits, Department of Electrical and Computer Engineering, Northeastern University, Boston, MA 02115-5000 USA

2.1 Discovery and Ancient Applications of Lodestone

Ferrite materials have been known to ancient cultures for many centuries. Both the ancient Greek and Chinese cultures make reference to lodestone, a mineral displaying spontaneous magnetic polarization and attraction to iron, as early as the sixth-century BCE [1]. In ancient Greece, Aristotle attributed the first scientific discussion of magnetism to the philosopher Thales of Miletus (c. 625 BCE–545 BCE) [2].

The earliest Chinese reference to magnetism occurs in the fourth-century BC text titled *Book of the Devil Valley Master*. The chronicle *Lüshi Chunqiu*, from the second-century BC, explicitly states that "lodestone makes iron come..." [3,4].

However, astronomer John Carlson discovered an artifact of lodestone from an ancient Mesoamerican civilization, the Olmec society of North America, that may have been used more than a thousand years before the Chinese and Greek cultures [5] as a directional device composed of hematite with titanium lamellas of $Fe_{2-x}Ti_xO_3$ [6]. If true, this artifact would be the first use of a magnetic material for navigation in world history.

2.2 What Is Lodestone?

Lodestone is a historical term that refers to a magnetic iron oxide ore, rich in magnetite but also containing oxidized products of maghemite and hematite. The principal phase, magnetite, having formula Fe_3O_4, is the iron oxide form of the spinel ferrite AB_2O_4, where A and B are Fe^{2+} and Fe^{3+} cations. Fragments of spontaneously magnetized lodestone are only found near the earth's surface and not buried deep below its crust [7]. This observation has fostered the belief that the magnetism of these specimens derives from lightning strikes. When these naturally magnetized lodestone fragments are suspended in a fluid, typically water, they naturally orient toward a magnetic pole thus enabling the development of a magnetic compass [3].

The first suspended magnetic needle compasses were described by a Chinese text of Shen Kuo c. 1088 BCE [8].

Without this invention, global exploration, international commerce, and sharing and development of religions, languages, cultures, and technologies would not have been possible.

Remarkably, the floating needle compass of 1088 BCE has not changed fundamentally in its components and construction in its many present-day manifestations.

Although lodestone find great utility in terrestrial, nautical, and celestial navigation, it was not until the mid-twentieth century that modern variants of lodestone, i.e. ferrites, would be studied for their magnetic and electronic properties for use in power generation, conditioning, and conversion, thus setting society on the course of tremendous technological advance.

Nearly a thousand years after the discovery of the floating lodestone (i.e. ferrite) needle compass, ferrites were rediscovered and developed to address other technological needs, ranging from communication, sensing, transportation, energy, and biomedical applications, among others.

Table 2.1 lists the major development and key inventions that mark the evolution of ferrites. Aspects of the table will be discussed below in a proper sequence.

Modern Ferrites: Basic Principles, Processing and Properties, Volume 1, First Edition. Edited by Vincent G. Harris.

Table 2.1 History of ferrite materials science and technological developments.

Era	Major events	Development and key citations
654–725 BCE	Lodestone, a natural iron ore rich in Fe_3O_4, is reported by Guan Zhong for terrestrial navigation as a "spoon" compass in guiding *feng shui* practices and other forms of divination.	Early discovery and applications of lodestone.
Sixth-century BCE	Greece c. sixth-century BCE reports lodestone is attracted to iron.	
1088 (henceforth AD)	The first suspended lodestone needle compass is reported by Shen Kuo. Today's floating needle compasses remain largely of the same design.	
1909	High-frequency electrical machine applications of ferrites are first studied and patented.	Hilpert [9]
1915	The first study of the structure of spinel ferrite (Fe_3O_4) using newly developed X-ray diffraction.	First determination of spinel ferrite structure by Bragg [26]
1931	The magnetic mineral magnetoplumbite (M-type hexaferrite) is first described.	Guillissen and van Rysselberghe [27]
1932	Reduction in the hysteretic losses in spinel systems are reported by J. L. Snoek.	Invention and development of spinel ferrites and refinement toward commercial engineering applications
1930–1932	Takeshi Takei, professor at Tokyo Tech, discovered that composite oxides containing zinc and iron had valuable magnetic properties. In 1930, Takei applied a patent for his discovery, which was granted in 1932.	
1934	Kramer postulated antiferromagnetism based upon an indirect negative exchange, which eventually became the foundation of superexchange. He also proposed the Kramer's degeneracy theorem of time-reversal symmetric systems with half-integer spin essential for development of RF nonreciprocal devices of the late twentieth century.	First postulation of superexchange theory
1935, 1946	Landau and Lifshitz predicted the existence of ferromagnetic resonance of the Larmor precession essential to understanding the high-frequency properties of ferrite-based microwave materials for device applications. This was independently verified in experiments by Griffiths [7] and Zavoiskij in 1946 [8,9]. Landau won the 1962 Nobel Prize.	Prediction of FMR of the Larmor precession [28–30]
1935	Tokyo Denki Kagaku Kogyo K. K. (now TDK Corp.) was founded to commercialize newly invented ferrite cores. The collaborative work of Tokyo Institute of Technology and TDK led to the development of CuZn ferrite well suited for practical applications at high frequencies up to 1 MHz.	Establishment of TDK Corp. Establishment of TDK Corp. and commercial products of spinel ferrites as "Oxide Core" products
1937	TDK begun mass production of ferrite cores under the product name of "Oxide Core." Most shipped products were used in oscillators, mixers, and transformers in radios.	
1938	Crystal structure of magnetoplumbite (M-type, hexaferrite) was deduced as being hexagonal with the composition $PbFe_{7.5}Mn_{3.5}Al_{0.5}Ti_{0.5}O_{19}$.	Structure of magnetoplumbite hexaferrite determined
1941	Patent and commercialization of *Ferroxcube* by the Philips Co. were carried out.	Development of AB_2O_4-spinel ferrites (A = Zn,Fe, Mn; B = Cu,Mg,Ni) resulting in size reduction of inductor cores
1949	Tensor notation of permeability was proposed as necessary for ferrites as ferrimagnetic materials are shown to become anisotropic in the presence of magnetizing fields.	Introduction of the Polder tensor [31] essential to understanding ferrite gyromagnetic properties

Table 2.1 (Continued)

Era	Major events	Development and key citations
1948	Theoretical studies of magnetic properties of ferrites by Néel, following Kramer's postulate on antiferromagnetism, led to the discovery of ferrimagnetism and a broader understanding of ferrite magnetism based on multiple molecular fields in which net magnetism stems from a negative exchange mediated by the anions' p orbitals.	Superexchange theory developed
1950	Superexchange theory was developed in a robust physical and mathematical framework by Anderson. Anderson later shared in the 1977 Nobel Prize in Physics with van Vleck and Mott.	
1950	Patent and commercialization of *Ferroxdure* by Philips Co.	Development of first hexaferrite commercial products
1950	Industrial activity led to the growth of the power ferrite industry stemming from demand for mass production of large-scale deflection yokes for the burgeoning TV market.	Introduction of power ferrites
1951	First neutron scattering studies of ferrites were carried out. Shull shared in the 1994 Nobel Prize for development of neutron scattering.	Neutron scattering applied to ferrites by Shull et al. [32]
1952	NiZn and MnZn ferrites for use in TV tube deflection yokes and high-voltage flyback transformers.	Philips Co. advanced commercialization of spinel ferrites
1952	$BaFe_{12}O_{19}$, barium hexaferrite, an M-type hexaferrite, was shown to consist of a hexagonal magnetoplumbite structure.	The M-type structure of barium hexaferrite confirmed by Went et al. [33]
1953	Rado described frequency spectra of ferrites and identified resonances with their origins in domain wall and spin dynamics.	Domain wall and spin dynamics resonances explained by Rado [34]
c. 1955	Goodenough–Kanamori rules for superexchange provide phenomenological rules relating local bonding angles formed between triangular ionic arrangements of Me–O–Me to magnetic exchange energy and related properties.	Goodenough and Kanimori [16]
1956	Nonlinearity in ferrites at high microwave power was explored.	Suhl et al. [35]
1956–1957	First development of magnetic garnets was seen.	Bertaut and Forrat [36] Geller and Gilleo [37]
1957	First demonstration of nanoparticle ferrites for cancer remediation by magnetic hyperthermia. Later developed for human trails c. 2010.	Gilchrist et al. [38]
1959	Smit and Wijn published Ferrites [39] and provided descriptions of chemistry, structure, and gyromagnetic properties of garnets, spinels, and M-, X-, Y-, Z-, U-, W-type hexaferrites.	Development of magnetoplumbite and ferroxplana hexaferrites
1960–1961	Single-crystal magnetic garnets.	van Hook [40] and Rudness and Kebler [41]
1967	Mössbauer spectroscopy was employed to describe the magnetic and nuclear structures of spinel $NiFe_2O_4$ as a collinear Néel ferrimagnet. Mössbauer later shared in the 1961 Nobel Prize for development of Mössbauer spectroscopy.	Chappert and Frenkel [42]
1970	Nobel Prize awarded to Louis Eugène Néel "...for fundamental work and discoveries concerning antiferromagnetism and ferrimagnetism..."	Nobel Prize, 1970

(Continued)

Table 2.1 (Continued)

Era	Major events	Development and key citations
c. 1970	Bell Labs introduced nonvolatile "bubble" memory using orthoferrites.	Bobeck and Scovil [43] Scovil, Michaelis, and Bobeck [44]
1970–1972	Understanding of microwave relaxation mechanisms in ferrites was advanced.	Patton [45]
c. 1972–1978	Ferrite-based magnetoelectric heterostructures/composites of magnetostrictive ferrites and piezoelectrics later made popular by Srinivas, Viehland, Harris, and others.	Magnetoelectric heterostructures [46]
1971–1972	Liquid phase epitaxy (LPE) growth of garnet thick films demonstrated [47].	LPE growth by Glass et al. and others demonstrated
1976–1977	LPE growth of spinel ferrite thick films was demonstrated [48].	
1978	LPE growth of hexaferrite thick films was demonstrated [49].	
1975	Goldman [26] advanced NiZn ferrites for RF applications at Spang & Company.	Microwave applications of spinels demonstrated and advanced
1977	Nobel Prize in Physics awarded to Phillip W. Anderson (shared with Mott and Van Vleck) "...for their fundamental theoretical investigations of the electronic structure of magnetic and disordered systems..." Anderson played a major role in transforming Néel's superexchange theory to its modern form.	Nobel Prize [50]
1981–1982	Aqueous coprecipitation of ferrite nanoparticles was demonstrated.	Massart [51]
1982	Barium hexaferrite thin film–based perpendicular recording media proposed.	Ferrite-based perpendicular, high-density storage media proposed [52]
1987	Low-temperature, spin-spray ferrite film plating was demonstrated.	Abe [53]
c. 1989	Barium hexaferrite thin film–based longitudinal recording media proposed.	Longitudinal, high-density storage media proposed by Speliotis [54]
1990	First single-phase, room temperature, multiferroic– Bismuth ferrite was reported.	Kubel and Schmid [55]
1992	Pulsed laser deposition (PLD) of thin-film ferrites was demonstrated.	Vittoria [56,57]
1994	Nobel Prize awarded to Clifford G. Shull and Bertram N. Brockhouse for "...neutron scattering to better explore the atomic structure of matter."	Nobel Prize [58]
c. 1990–1995	Ferrites were used as bead line filters for many commercial electronics.	Bead line filters commercialized
1980–2000	Research into the synthesis and characterization of nanostructured ferrite systems and nanoparticles grew rapidly.	Nanostructured ferrite systems developed
1980–present	Development of nonlinear effects in ferrites and related devices by the Kalinikos et al.	Nonlinear effects of ferrites and devices
2000–present	Room temperature, low field magnetoelectric response reported in long wavelength hexaferrite magnetic structures.	Magnetoelectric effect in hexaferrites reported by Kimura [59]
[262004–present	Ferrite-based solitons and related devices were developed.	Soliton-based technologies by Wu et al. [60]
2000–present	Research and applications focus on microwave ferrites, magneto-optical properties, ferrite films, nanostructured ferrites and low-temperature co-fired ceramics (LTCC) multilayer chip inductors.	Microwave and magneto-optical applications pursued

Table 2.1 (Continued)

Era	Major events	Development and key citations
2000–present	Self-biased hexaferrites were developed.	Self-biased hexaferrite ceramics demonstrated [61–66]
c. 2005–present	Superparamagnetic ferrites (as SPIONS) developed as T_2 magnetic resonance imaging (MRI) contrast agents.	Medical diagnostic applications developed [67]
c. 2008–present	Harris demonstrated successful ferrite integration with semiconductor substrates (SiC and GaN) for integrated system-on-a-wafer technology.	Underlying materials science for integrated system-on-a-wafer technology demonstrated
2010–present	Metamagnetics Inc. commercialized first self-biased microstrip circulators.	Self-biased devices demonstrated and commercialized
2016–present	Metamagnetics Inc. commercialized first broadband passive frequency selective limiters as *Autotune filters*.	*Autotune Filters* product line introduced
2012, 2020	Ultrahigh coercive fields measured in hexaferrites with FMR frequencies appearing beyond 240 GHz.	Namai et al. [68] – ε-Fe_2O_3 Gorbachev et al. [69] – hexaferrites
Twenty-first-century advances	Advances in the synthesis and characterization of ferrites continues. Research focuses on crystal growth; thin films; nanoparticles; nanostructured ferrites and ferrite-based composites; and multilayer chips inductors, ferrofluids, Spintronics, soft magnetic materials, magneto-optics, EMI suppression and shielding, as well as energy, biomedical, environmental applications, and RF technologies.	Materials for RF, sub-millimeter wavelength, and optical applications continue to grow. New fields of Spintronics and magnetoelectronics emerge and evolve. Medical application of multifunctional NPs continues to develop

2.3 Hilpert, Takei, Tokyo Institute of Technology and TDK

At the beginning of the twentieth century, c. 1909, Hilpert filed the first patent detailing the processing and application of ferrites as insulating magnetic materials for "Electric and Magnetic Apparatus" [9]. However, this early attempt to develop ferrites for commercialization and industrialization did not meet with success. The principal reason was that their intrinsic magnetic properties paled in comparison to existing ferromagnetic materials for low-frequency technologies popular in the day. The need for insulating magnets had yet to be made clear at that time. In essence, Hilpert's invention was far ahead of its time.

In 1934 and 1935, approximately 25 years after Hilpert, Takeshi Takei of Tokyo Institute of Technology, discovered spinel oxides of zinc and iron had unique and attractive high-frequency magnetic properties [10].

Tokyo Denki Kagaku Kogyo, now known as the TDK Corp., was founded in 1935 to commercialize these ferrite materials. The collaborative works of Tokyo Institute of Technology and TDK led to the development of CuZn ferrite cores well suited for practical applications at frequencies up to ~1 MHz [11].

TDK began the mass production of ferrite cores in 1937 under the product name "Oxide Core." Most of the shipped product were used in oscillators, mixers, and transformers for radios to meet the needs of the rapidly escalating Pacific Rim war that would become World War II. These were the world's first practical uses of ferrite materials [11].

These ferrites contributed to reduced volume and weight of radio receivers compared with the air cores commonly used at the time. Until the end of World War II, TDK was the only company that supplied ferrite cores, shipping a total of nearly five million units during the war [11].

For comparison, it is projected that by 2024, more than 1.5 million *metric tons* of ferrite cores will be produced per year, reflecting the critical role that ferrites play in modern technologies [12].

2.4 Kramer, Néel, and Anderson and the Development of the Superexchange Theory

In the 1940s to 1950s, much of the existing theory that explains the origins of magnetism in ferrimagnetic and antiferromagnetic oxides was developed.

Key contributions by Kramers, Néel (Nobel Prize, Physics, 1970), and Anderson (Nobel Prize, Physics, 1977), led to the theory of superexchange that explained the origins of antiferromagnetism and ferrimagnetism in magnetic oxides [13–15]. Goodenough and Kanimori [16] provided phenological rules that greatly aided experimentalists in refining compositions and structures for greater commercial utility.

Although a successful quantum mechanical model of magnetism was first put forth in 1927 by Heitler and London [17], it is van Vleck [18] who shared the 1970 Nobel Prize in Physics (with Anderson and Mott) and credited as the "Father of Modern Magnetism." Building upon these theories, Sham and Kohn (and others), developed the first density functionals as an *ab initio* quantum mechanical theory [19]. The application of density functional theory (DFT) to metals and oxide systems has proven extremely successful and resulted in the awarding of the 1998 Nobel Prize in Chemistry to Walter Kohn. Kohn shared the Prize with Pople who made major contributions to molecular orbital theory [20]. Due to challenges stemming from the incorporation of finite temperature spin–orbit coupling, a full *ab initio* DFT of magnetic oxides such as ferrites remains elusive.

With an improved theoretical understanding, engineering applications of ferrites rapidly progressed in the latter half of the twentieth and early twenty-first century.

The fundamental physics of these theories will be presented in great detail in the following chapters of this text.

2.5 Snoek and the Philips Company

Ferrite materials are a unique class of insulating magnetic oxides that possess moderate values of magnetization, high permeability, high permittivity, and low losses at frequencies up to and beyond millimeter wavelengths.

These properties afford great value to high-frequency devices that require strong coupling to electromagnetic signals while experiencing low losses. In addition, due to their intrinsic magnetism, these materials also provide time-reversal symmetry breaking that leads to nonreciprocal behavior essential to many device applications in radar and communications systems as isolators, circulators, phase shifters, etc.

The value of ferrites as materials for high-frequency applications continues to grow.

In the 1940s to 1950s, ferrites for RF applications were systematically studied by Snoek [21,22] of Philips for applications in devices that send, receive, and manipulate electromagnetic signals at RF frequencies as well as for power conversion and conditioning applications.

Snoek worked primarily on refining the process conditions for optimized performance of high-frequency spinels, developed products under the tradename *Ferroxcube* (c. 1941, type I CuZn ferrite, type II MgZn ferrite, type III MnZn ferrite, type IV NiZn ferrite, and type V Mn ferrite). These product lines allowed enhanced performance and reduced form factors for communication coils, transformers, flyback transformers, and deflection yokes. MnZn ferrites, operating from 1 kHz to 1 MHz, profoundly influenced international markets of these technologies and emerging commercial electronics.

As valuable as MnZn ferrites were for applications up to 1 MHz, at frequencies above 1 MHz, their low electrical resistivities caused them to become too lossy, requiring alternative materials to be developed.

Cation substitutions of Ni^{2+} for Mn^{2+}, which preferentially reside on the octahedral sublattice, led to large increases in electrical resistivities and a reduction in RF losses. This provides NiZn ferrites with unique properties for high-frequency operations, including those well above 1 MHz. Other substitutions included Mg and Cu that are both divalent cations that strongly prefer octahedral site occupancy and, therefore, play a similar role as that of Ni^{2+}. Cu cations were also found to provide enhanced densities, and therefore, NiCuZn ferrites have become very popular as a commercial composition.

With the introduction of NiCuZn ferrites, new applications at higher frequencies such as RF antennas; RF transformers; and RF isolators, circulators, and switches up to and beyond 1 GHz have become available.

However, NiCuZn-ferrite saturation magnetization is typically 30% less than that of MnZn ferrite, causing an abrupt increase in the form factor near ~1 MHz when one transitions from one ferrite composition to the other.

Although he lived a relatively short life, dying as a result of an automobile crash in 1950 at the age of 48 years [23], perhaps no one had advanced ferrite processing and product development more in fewer years of active research than J. L. Snoek.

From 1970 to 2000, led by experts such as Alex Goldman of Spang & Company, new spinel ferrite materials, having permeabilities as high as 20 000 to 30 000, were developed to be used at frequencies up to and exceeding 1 MHz [24].

2.6 Hexagonal Ferrites

Circa 1950, barium hexaferrite ($BaO{\bullet}6Fe_2O_3$) was discovered somewhat serendipitously at Philips Physics Laboratory when engineers preparing samples of hexagonal lanthanum ferrite for use as a semiconductor material produced barium hexaferrite of the magnetoplumbite structure. Barium hexaferrite, advertised under the trade name *Ferroxdure*, was shown to have both a high coercivity (i.e. ~170 kA/m) and relatively low raw materials cost. Soon after, Philips developed strontium hexaferrite (i.e. $SrO{\bullet}6Fe_2O_3$), with superior properties to barium hexaferrite. Barium and strontium hexaferrites have since dominated the market for hard ferrites due their excellent performance and low costs.

Hexaferrites exist in several structural variants, i.e. M, Y, Z, X, U, and W, in which the metal and oxygen stoichiometry vary [25]. Due to their highly symmetric structures and low intrinsic magnetocrystalline anisotropy energy, cubic garnets and spinel ferrites tend to have low magnetic anisotropy fields. In comparison, crystal symmetry breaking in the hexaferrites provide them with extremely high magnetic anisotropy fields. As will be discussed in greater detail, the ferromagnetic resonance (FMR) frequency shifts with changes in magnetic anisotropy field, together with the demagnetizing field, and the magnetization and applied magnetic fields (i.e. bias fields), leading to the operation of hexaferrites at microwave, millimeter-wave, and quasi-optical frequencies.

2.7 Crystals, Films, and Nanoparticles

In recent years, advances in the synthesis and characterization of ferrite films, nanostructured ferrites, and multilayer chips inductors have ushered in new technologies from self-biased circulators, integrated microinductors, integrated transmit and receive modules, and Spintronics, to MRI contrast agents and cancer remediation treatments, including magnetohyperthermic and drug delivery agents.

Today, ferrites play an essential role in modern society. Nearly every consumer product has one or more ferrites as embedded components. And nearly all future products considered today are anticipated to contain ferrites.

Major technological market growth of ferrites is enabled by the development of high-quality crystals (by several techniques), polycrystalline compacts (including LTCC), films (as PLD, spin-spray films, LPE, sputtering, and MBE), and nanoparticles (by a large variety of techniques reviewed by Willard et al. [70]).

Many of these breakthroughs and major developments in these areas are given in Table 2.1. Technologies range from magnetic storage; medical (as MRI contrast agents and magnetic hyperthermia), power generation, conditioning, and conversion; RF and millimeter-wave devices for the sending, receiving, and manipulating electromagnetic signals for communication, radar, and sensing.

Recent developments in power generation, condition, and conversion, including miniaturization and higher operation frequency for applications of massive scales such as cloud computing centers, the retail high-efficiency magnetic core market sets all-time high benchmarks every year.

2.8 Global Ferrite Research Groups

Table 2.2 presents a partial list of university and national laboratories active in ferrite research and their specific research areas. Certainly, programs have been unintentionally omitted with the author's apology. Additionally, there are many programs that have made profound contributions to ferrite technologies that have been left off this table due to inactivity in recent years. Industrial players are not listed due to space constraints. These are listed in the market discussion in Section 2.9.

Table 2.2 Global Ferrite Research Groups (not industry)[a].

Active University and National Lab Ferrite Groups	Country	Specialty
Northeastern University	USA	Self-biased magnetoceramics and devices, PLD and LPE films, nanoparticles, metamaterials, impedance tuning of magnetoceramics, and medical magnetism
Colorado State University, Ft. Collins	USA	Soliton devices, spin dynamics, microwave losses, and Spintronics
Morgan State University	USA	PLD thin films and characterization
University of Alabama	USA	DFT and hexaferrites
University of Colorado, Colorado Springs	USA	Magnetoceramics, millimeter-wave measurements, medical magnetism, and electromagnetic theory
NIST	USA	Nonlinear dynamics and related RF devices, Spintronics
Virginia Commonwealth University	USA	Ferrite nanochemistry and medical magnetism
Oakland University	USA	Magnetoelectric heterostructures, nonlinear dynamics, electromagnetic theory, Spintronics
University of Minnesota	USA	Spin dynamics, RF devices, electromagnetic theory
MIT	USA	Spin dynamics, RF devices
Tufts University	USA	Hexaferrites and millimeter-wave characterization
Virginia Tech	USA	Inductor cores, magnetoelectric heterostructures
Tokyo Institute of Technology	Japan	Sputtering and spin-sprayed films, RF devices
University of Tokyo	Japan	Microwave/millimeter-wave absorption materials
University of Osaka	Japan	Magnetoelectrics
Toyohashi University of Technology	Japan	Multiferroics
Teikyo Heisei University	Japan	Ferrite-based magnetic recording
Kyoto University	Japan	Hexaferrite magnets
Meiji University	Japan	Nano-ferrite magnets
University of Tsukuba	Japan	Ferrite thin films
Tohoku University	Japan	On-chip inductors
Chiba Institute of Technology	Japan	Spintronics
University Applied Sciences, Jena	Germany	Magnetoceramic design and processing
Technische Universität Kaiserslautern	Germany	Linear and nonlinear spin dynamics, electromagnetic theory, magnons, and Spintronics
University of Aveiro	Portugal	Magnetoceramic design & processing, hexaferrites
University of Electronic Science and Technology of China (UESTC)	China	LPE and LTCC magnetoceramics, Spintronics, and impedance tuning of magnetoceramics
Huazhong University of Science and Technology (HUST)	China	Microwave magnetics and devices
Chinese Academy of Science, Institute of Physics	China	Magnetoelectrics
Tsinghua University	China	Multilayer chip components
Beijing University	China	Microwave absorption materials
Southwest Jiaotong University	China	Magnetodielectric materials and devices
Xian University	China	Magnetoelectrics and devices
Scientific Research Company "Electron-Carat"	Ukraine	Liquid phase epitaxy of ferrites
Moscow Institute of Physics and Technology	Russia	Multiferroics and magnetoelectrics
St. Petersburg Electrotechnical University	Russia	Nonlinear spin dynamics of ferrites, electromagnetic theory, and RF devices
Moscow State University	Russia	Magnetoceramics and giant coercive field ferrites
India Institute of Technology – Bombay	India	Sputtering and PLD ferrite films
India Institute of Technology – New Delhi	India	LTCC, magnetoceramics

[a]Table 2.2 was assembled with the assistance of Dr. Yajie Chen.

2.9 State of Global Markets (c. 2019)

Based on the Micromarket Monitor [71], the major driver for the global ferrite market (both hard and soft ferrite magnets) is the rapid growth of the consumer electronics and automotive industries that have high demands for ferrite cores. The global ferrite market value estimation at 2021 was US$6.4 billion with a compound annual growth rate (CAGR) of 3.57%, which is expected to reach US$7.9 billion by 2027. Based on Markets Media [72], soft ferrites make up US$1.7 billion of this market in 2017 and is expected to hit US$1.87 billion by 2025 with a CAGR of 1%.

Among end users, the automotive market demand for ferrites has been experiencing the highest growth rate as this industry has been shifting toward electrical power generation. The automotive industry is one of the largest end user industries for ferrite magnets, and the growth in the automobile sector has led to an increase in the demand for ferrite magnets.

In addition to automotive, the consumer electronics market has been experiencing a large wave of market demand, much more since the rise of personal computers and its trend toward miniaturization and lightweight laptops, pads, and cell phones.

Ferrite inductors are compact and lightweight and suitable for use in consumer electronics. Consumer electronics are also increasingly becoming more compact and lighter in weight due to the miniaturization of components. The increasing demand for smart mobile devices is likely to drive an increase in inductor market for consumer electronics applications.

The major regional contributions to this market have been Asia-Pacific with a market of US$2.93 billion, holding about 75% of the market share, while the rest belongs to USA, Europe, the Middle East, and Africa.

South America's ferrite market is estimated to experience high growth and is projected to reach a CAGR of 12.9% for the period under consideration.

The key players in the ferrite market include Hitachi; Shin-Etsu Chemical Co., Ltd.; Lynas Corp. Ltd.; Molycorp Chemicals and Oxides; Hangzhou Permanent Magnet Group; Ningbo Yunsheng High-Tech Magnetics Co., Ltd.; Daido Steel Co., Ltd.; and JPMF Guangdong Co. Ltd.

The key soft ferrite core manufacturers include TDK, DMEGC, Vacuumschmelze, Magnetics, TDG, Acme Electronics, Ferroxcube, Nanjing New Conda, Haining Lianfeng Magnet, Hec Group, JPMF, Kaiyuan Magnetism, NBTM New Materials, Samwha Electronics, Toshiba Materials, etc. [72].

In summary, although ferrites are perhaps the oldest functional materials in our history, having been used by ancient civilizations for more than ~3000 years, today they remain among the most important enabling materials for conventional and emerging engineering fields. From Spintronics to cancer remediation, no other material displays the wide-ranging positive society impact. There appears no end in sight to their value and potential.

Reference

1 Keithley, J.F. (1999). *The Story of Electrical and Magnetic Measurements: From 500 B.C. to the 1940s.* John Wiley and Sons, p. 2.
2 Fowler, M. (1997). Historical beginnings of theories of electricity and magnetism. Retrieved 2008- 04-02.
3 Dillon, M. (2017). *Encyclopedia of Chinese History*. Routledge, p. 98.
4 Shu-hua, L. (1954). Origine de la Boussole II. Aimant et Boussole. *Isis* (in French). 45(2): 175–196. doi: 10.1086/348315. JSTOR 227361. un passage dans le *Liu-che-tch'ouen-ts'ieou* [...]: "La pierre d'aimant fait venir le fer ou elle l'attire." From the section "*Jingtong*" of the "Almanac of the Last Autumn Month".
5 Carlson, J.B. (1975). Lodestone compass: Chinese or Olmec primacy?: Multidisciplinary analysis of an Olmec hematite artifact from San Lorenzo, Veracruz, Mexico. *Science* 189 (4205): 753–760.
6 Evans, B.J. (1977). Magnetism and archaeology: Magnetic oxides in the first American civilization, p. 1097, Elsevier. *Physica B+C* 86–88: S. 1091–1099.
7 Livingston, J.D. (1996). *Driving Force: The Natural Magic of Magnets.* Harvard University Press, p. 14–20.
8 Needham, J. (1986). *Science and Civilization in China: Volume 4, Physics and Physical Technology, Part 1, Physics.* Taipei: Caves Books Ltd.
9 Hilpert, S. (1909). Manufacturing method of magnetic materials having small electric conductivity for electric and magnetic apparatus. German Pat. Nos. 226,347 and 227,787 (1909).
10 Kato, Y., Ebara-Ku, T., and Takei, T. (1934). Permanent magnet and method of manufacturing the same. Patent No. 1976,230; Kato, Y., Ebara-Ku, T., and Takei, T. (1935). Permanent magnet and method of manufacturing the same. Patent No. 1,997,193.
11 *"TDK: Our History"*, https://en.wikipedia.org/wiki/TDK. Accessed July 1, 2022.

12 https://www.researchandmarkets.com/reports/5633319/ferrite-magnet-market-global-industry-trends%23utm_source=GNOM&utm_medium=PressRelease&utm_code=nlfm8c&utm_campaign=1553212+-+Global+Ferrite+Magnet+Market+(2021+to+2026)+-+Industry+Trends%2C+Share%2C+Size%2C+Growth%2C+Opportunity+and+Forecasts&utm_exec=jamu273prd. Accessed August 17, 2022.

13 Kramers, H.A. (1934). L'interaction Entre les Atomes Magnétogènes dans un Cristal Paramagnétique. *Physica* 1 (1–6): 182.

14 Néel, L. (1948). Magnetic properties of ferrites: Ferrimagnetism and antiferromagnetism. *Annals of Physics* (Paris) 3: 137–198.

15 Anderson, P.W. (1950). Antiferromagnetism. Theory of superexchange interaction. *Physical Review* 79 (2): 350.

16 Goodenough, J.B. (1955). Theory of the role of covalence in the perovskite-type manganites [La, M(II)]MnO_3. *Physical Review* 100 (2): 564. Goodenough, J.B. (1958). An interpretation of the magnetic properties of the perovskite-type mixed crystals $La_{1-x}Sr_xCoO_{3-\lambda}$. *Journal of Physics and Chemistry of Solids* 6 (2–3): 287; Kanamori, J. (1959). Superexchange interaction and symmetry properties of electron orbitals. *Journal of Physics and Chemistry of Solids* 10 (2–3): 87.

17 Heitler, W. and London, F. (1927). Wechselwirkung neutraler Atome und homöopolare Bindung nach der Quantenmechanik. *Zeitschrift für Physik* 44: 455–472.

18 Van Vleck, J.H. (1928). The correspondence principle in the statistical interpretation of quantum mechanics. *Proceedings of the National Academy of Sciences of USA* 14: 178–188.

19 Kohn, W. and Sham, L.J. (1965). Self-consistent equations including exchange and correlation effects. *Physical Review* 140 (4A): A1133–A1138.

20 Pople, J.A. (1951). Molecular association in liquids: II. A theory of the structure of water. *Proceedings of the Royal Society A* 205 (1081): 163–178.

21 Snoek, J.L. (June 1936). Magnetic and electrical properties of the binary systems MO. Fe_2O_3. *Physica* 3 (6): 463–483.

22 Snoek, J.L. (1946). Non-metallic magnetic material for high frequencies. *Philips Technical Review* 8: 353.

23 Koiwa, M. (15 April 2004). A note on Dr. J.L. Snoek. *Materials Science and Engineering: A* 370 (1–2): 9–11.

24 Goldman, A. (2006). *Modern Ferrite Technology*. Springer US, p. 438.

25 Pullar, R.C. (2012). Hexagonal ferrites: A review of the synthesis, properties and applications of hexaferrite ceramics. *Progress in Materials Science* 57: 1191.

26 Bragg, W.H. (1915). The structure of magnetite and the spinels. *Nature* 95: 561.

27 Guillissen, J. and Van Rysselberghe, P.J. (1931). Studies on Zinc and Barium Ferrites. *Transactions of The Electrochemical Society* 59: 95.

28 Landau, L.D. and Lifshitz, E.M. (1935). Theory of the dispersion of magnetic permeability in ferromagnetic bodies. *Physikalische Zeitschrift der Sowjetunion* 8: 153.

29 Griffiths, J.H.E. (1946). Anomalous high-frequency resistance of ferromagnetic metals. *Nature* 158 (4019).

30 Zavoisky, E. (1946). Spin magnetic resonance in the decimeter-wave region. Zhurnal, F. and Zavoisky, E. (1946). Paramagnetic absorption in some salts in perpendicular magnetic fields. *Zhurnal Eksperimental'noi i Teoreticheskoi Fiziki* 16 (7): 603–606.

31 Polder, D. (1949). On the theory of ferromagnetic resonance. *The London, Edinburgh, and Dublin Philosophical Magazine and Journal of Science* 40.

32 Shull, C.G., Wollan, E.O., and Koehler, W.C. (1951). Neutron scattering and polarization by ferromagnetic materials. *Physical Review* 84: 912.

33 Went, J.J., Rathenau, G.W., Gorter, E.W., and van Oosterhout, G.W. (1952). Hexagonal iron-oxide compounds as permanent-magnet materials. *Physical Review* 86: 424.

34 Rado, G.T. (1953). Magnetic spectra of ferrites. *Reviews of Modern Physics* 25: 81.

35 Suhl, H. (1956). The nonlinear behavior of ferrites at high microwave signal levels. *Proceedings of the Ire* 44: 1270–1284. Suhl, H. (1957). The theory of ferromagnetic resonance at high signal powers. *Journal of Physics and Chemistry of Solids* 1: 209–227.

36 Bertaut, F. and Forrat, F. (1956). Structure des ferrites ferrimagnetiques des terres rares. *Comptes Rendus* 242: 382.

37 Geller, S. and Gilleo, M.A. (1957). Structure and ferrimagnetism of yttrium and rare-earth-iron garnets. *Acta Crystallographica* 10: 239.

38 Gilchrist, R.K., Medal, R., Shorey, W.D., Hanselman, R.C., Parrott, J.C., and Taylor, C.B. (1957). Selective inductive heating of Lymph Nodes. *Annals of Surgery* 146 (4): 596–606.

39 Smit, J. and Wijn, H.P.J. (1959). *Ferrites*. New York: John Wiley & Sons.

40 Van Hook, H.J. (1961). Phase relations in the system Fe_2O_3–Fe_3O_4–$YFeO_3$ in Air. *Journal of the American Ceramic Society* 44: 208.

41 Rudness, R.G. and Kebler, R.W. (1960). Growth of single crystals of incongruently melting yttrium iron garnet by flame fusion process. *Journal of the American Ceramic Society* 43.

42 Chappert, J. and Frankel, R.B. (1967). Mössbauer study of ferrimagnetic ordering in nickel ferrite and chromium-substituted nickel ferrite. *Physical Review Letters* 19: 570.

43 Bobeck, A.H. and Scovil, H.E.D. (June 1971). Magnetic bubbles. *Scientific American* 78: 224.

44 H.E.D. Scovil, P.C. Michaelis and Bobeck were awarded the IEEE Morris N. Liebmann Memorial Award by the IEEE with the following citation: For the concept and development of single-walled magnetic domains (magnetic bubbles), and for recognition of their importance to memory technology.

45 Patton, C. (1972). A review of microwave relaxation in polycrystalline ferrites. *IEEE Transactions on Magnetics* 8 (3): 433–439.

46 Van Suchetelene, J. (1972). Product properties: a new application of composite materials. *Philips Research Reports* 27: 28–37; Van den Boomgaard, J. and Born, R.A.J. (1978). A sintered magnetoelectric composite material BaTiO3 -Ni(Co,Mn)Fe3O4. *Journal of Materials Science* 13: 1538–1548.

47 Levinstein, H.J., Licht, S., Landorf, R.W., and Blank, S.L. (December 1971). Growth of high-quality garnet thin films from supercooled melts. *Applied Physics Letters* 19: 486–488.

48 Stearns, F.S. and Glass, H.L. (1976). Liquid phase epitaxy of hexagonal ferrites and spinel ferrites on non-magnetic spinel substrates. *Materials Research Bulletin* 11 (10): 1319–1325.

49 Glass, H.L. and Liaw, J.H.W. (1978). Growth and characterization of LPE hexagonal ferrites. *Journal of Applied Physics* 49: 1578.

50 The Nobel Prize in Physics 1977. NobelPrize.org. *Nobel Prize Outreach AB 2022. Friday 19 Aug 2022.* https://www.nobelprize.org/prizes/physics/1977/summary/

51 Massart, R. (1982). Magnetic fluids and process for obtaining them. US Patent 4329241; Massart, R. (1981). Preparation of aqueous magnetic liquids in alkaline and acidic media. *IEEE Transactions on Magnetics* 17: 1247–1248.

52 Fujiwara, T., Isshiki, M., Koike, Y., and Oguchi, T. (1982). Recording performances of Ba-ferrite coated perpendicular magnetic tapes. *IEEE Transactions on Magnetics* MAG-18 (6): 1200.

53 Abe, M. and Tamaura, Y. (1984). Ferrite plating in aqueous solution: New technique for preparing magnetic thin film. *Journal of Applied Physics* 55: 2614.

54 Speliotis, D.E. (1989). High density recording on particulate and thin film rigid disks. *IEEE Transactions on Magnetics* 25 (5): 4048–4050.

55 Kubel, F. and Schmid, H. (1990). Structure of a ferroelectric and ferroelastic monodomain crystal of the perovskite BiFeO3. *Acta Crystallographica* B46 (6): 698–702.

56 Carosella, C.A., Chrisey, D.B., Lubitz, P., Horwitz, J.S., Dorsey, P., Seed, R., and Vittoria, C. (1992). Pulsed laser deposition of epitaxial BaFe12O19 thin films. *Journal of Applied Physics* 71: 5107.

57 Vittoria, C. (1994). Fabrication of ferrite films using laser deposition. USPTO, Patent number: 5320881, Date of Issue: June 14.

58 The Nobel Prize in Physics 1994. NobelPrize.org. *Nobel Prize Outreach AB 2022. Thu. 18 Aug 2022.* https://www.nobelprize.org/prizes/physics/1994/summary/

59 Kimura, T. (2012). Magnetoelectric hexaferrites. *Annual Review of Condensed Matter Physics* 3: 93–110.

60 Wu, M., Kalinikos, B.A., and Patton, C.E. (2004). Generation of dark and bright spin wave envelope soliton trains through self-modulational instability in magnetic films. *Physical Review Letters* 93: 157207.

61 Chen, Y.J., Sakai, T., Chen, T.Y. et al. (2006). Screen printed thick self-biased, low-loss, barium hexaferrite films by hot-press sintering. *Journal of Applied Physics* 100: 043907.

62 Chen, Y., Geiler, A.L., Sakai, T., Yoon, S.D., Vittoria, C., and Harris, V.G. (2006). Microwave and magnetic properties of self-biased barium hexaferrite screen printed thick films. *Journal of Applied Physics* 99: 08M904.

63 Harris, V.G. and Sokolov, A.S. (2019). The self biased circulator: Materials considerations and processing. *Journal of Superconductivity and Novel Magnetism* 32: 97–108; 30th Jubilee Edition.

64 Wu, C., Yu, Z., Sokolov, A.S., Yu, C., Sun, K., Jiang, X., Lan, Z., and Harris, V.G. (2018). Tailoring magnetic properties of self-biased hexaferrites using an alternative copolymer of isobutylene and maleic anhydride. *AIP Advances* 8 (5): 56221.

65 Zhang, X., Meng, S., Song, D., Zhang, Y., Yue, Z., and Harris, V.G. (2017). Epitaxially grown BaM hexaferrite films having uniaxial axis in the film plane for self-biased devices. *Scientific Reports* 7: 44193.

66 Su, Z., Chen, Y., Hu, B., Sokolov, A.S., Bennett, S., Burns, L., Xing, X., and Harris, V.G. (2013). Crystallographically textured self-biased W-type hexaferrites for X band microwave applications. *Journal of Applied Physics* 113 (17): 17B305.

67 Jun, Y.W., Lee, J.H., and Cheon, J. (2007). Nanoparticle contrast agents for molecular magnetic resonance imaging. In: *Nanobiotechnology II: More Concepts and Applications* (ed. C.A. Mirkin and M.C. Niemeyer). Weinheim: Wiley-VCH.

68 Namai, A., Yoshikiyo, M., Yamada, K. et al. (2012). Hard magnetic ferrite with a gigantic coercivity and high frequency millimetre wave rotation. *Nature Communications* 3: Article number: 1035.

69 Gorbachev, E.A., Trusov, L.A., Sleptsova, A.E. et al. (January–February 2020). Hexaferrite materials displaying ultra-high coercivity and sub-terahertz ferromagnetic resonance frequencies. *Materials Today* 32: 13–18.

70 Willard, M.A., Kurihara, L.K., Carpenter, E.E., Calvin, S., and Harris, V.G. (2004). Chemically prepared magnetic nanoparticles. *International Materials Reviews* 49: 125–170.

71 Micromarketmonitor.com

72 MarketsMedia.com

3

Structure and Chemistry of Ferrites and Related Oxide Systems

Michael E. McHenry and David E. Laughlin

3.1 Introduction

This chapter discusses *ferrite* structures and related phases, which will be related in a subsequent chapter 4 to magnetic properties. William Gilbert's 1600 work [1], *De Magnete*, gives a historical discussion of magnetism and efforts to understand the earth's magnetism. The first permanent magnets date to 600 BC when lodestone, Fe_3O_4, was used in compass needles. A loadstone's south pole points to the earth's north pole and vice versa, giving evidence of the earth's global magnetic field. The discovery of lodestone magnets and their orientation in the earth's magnetic field was followed by identification of iron ores and their crystalline structures. 2-3 ferrites are double oxides of trivalent Fe and a divalent cation. 4-2 ferrites have divalent Fe and a tetravalent metal. Ferrites knowledge is intertwined with the oxides of iron: magnetite, Fe_3O_4; hematite, Fe_2O_3; and wustite, FeO. Early studies of ferrites date back to Snoek [2] and Verwey [3] of the Phillips Research Lab, who from 1933 to 1948 studied conductivity mechanisms in ferrites involving electron hopping from ferric, Fe^{2+}, to ferrous, Fe^{3+}, cations.

The theory of ferrimagnetism was developed by Louis **Néel** in 1948 [4] and explained T-dependent magnetic response in ferrites. Charles Kittel [5], also in 1948, developed the theory of ferromagnetic resonance that was central to understanding the high-frequency magnetic response of these materials. He also extended the **Néel** theory of ferrimagnetism to noncollinear dipole moments [6]. These magnetic properties are discussed in more detail in a subsequent chapter 4.

Magnetite is the prototype ferrite with the formula:

$$(FeO)Fe_2O_3 \tag{3.1}$$

making apparent the distinction between the ferric, Fe^{2+}, cation and the ferrous, Fe^{3+}, cation. 2-3 cubic ferrites based on magnetite have the formula:

$$(MO)Fe_2O_3 \tag{3.2}$$

where M is a divalent transition metal. These adopt a spinel structure [7]. In a *normal spinel*, the M^{2+} cations occupy tetrahedrally coordinated A-sites and the Fe^{3+} cations occupy octahedrally coordinated B-sites surrounded by O^{2-} anions. Magnetite is an *inverse spinel* where an Fe^{3+} cation occupy a tetrahedral A-site and Fe^{3+} and Fe^{2+} cations occupy octahedral B-sites. In 4-2 inverse spinels, Fe^{2+} cations occupy octahedral B-sites and a tetravalent ion occupies the tetrahedral A-site. The *cubic ferrites* are soft magnets [8] (with the exception of cobalt ferrite, $CoFe_2O_4$, which is relatively hard). The *hexagonal ferrites* are hard magnets important for high-frequency microwave responses and motor applications.

The *garnets* are related cubic materials of interest for their magneto-optical properties. The cubic *garnets* are silicates with the chemical formula:

$$(MO)_3(Fe_2O_3)(SiO_2)_3 \tag{3.3}$$

with M^{2+} cations dodecahedrally coordinated.

Hexaferrites are low-symmetry ferrites based on hematite, Fe_2O_3, with the formula:

$$(MO)(Fe_2O_3)_6 \tag{3.4}$$

Modern Ferrites: Basic Principles, Processing and Properties, Volume 1, First Edition. Edited by Vincent G. Harris.

with M^{2+} cations typically larger (Ba^{2+}, Sr^{2+}, etc.) and dodecahedrally coordinated.

Ferrites and related oxides are also important for extraterrestrial magnetism [9]. Mars is called the red planet because of its hue that results from iron-containing minerals, in particular, hematite, which gives the red color to rust. The magnetic properties of oxides and oxyhydroxides, which form in regions with water, are of fundamental interest in identifying the composition of the Martian soil. These minerals carry implications for the broader question of life on Mars. The water-containing or forming minerals have implications for astrobiology, and many of these minerals are those studied on Earth in many current biomedical applications [10].

This chapter introduces crystallography of spinel ferrites and relationships with other magnetic oxides. Magnetic properties that derive from the structure are discussed in subsequent chapters. Here, we revive and analyze the crystal structures of ferrites and related oxide materials considering:

i) Oxides and oxyhydroxides of iron, Fe
ii) Magnetic dipole moments, magnetization, exchange coupling, anisotropy, and magnetostriction
iii) 2-3 and 4-2 cubic spinel ferrites
iv) titanomagnetites
v) Garnets
vi) Low-symmetry hematite and hexaferrites

3.2 Oxides and Hydroxides of Iron

The study of Fe-containing minerals (1) gives evidence to Fe as an abundant chemical species in terrestrial environments, (ii) identifies Fe oxidation states in these minerals, and (iii) identifies water or hydroxide groups in the minerals, which give clues to the history of water on a planet. These minerals often form the basis for functional materials and are particularly important in biomedical applications, for example.

The identification of Fe-containing minerals in biological processes on Earth is striking. The changing of the valence state of Fe is important to its function in metabolism and energetic processes in animals. Fe in *hemoglobin* captures and transports molecular oxygen. Fe is found in *ferritin* in the liver and the human brain, where in both cases it is present as a ferrihydrite core surrounded by a ferritin shell. In the brain, this Fe is necessary for neurotransmitter synthesis and myelination [10]. The *biomineralization* process of Fe plays an important role in the storing of Fe in organs. A *biomineralization* process producing biogenic magnetite and/or maghemite, as opposed to ferritin cores, has recently been suggested as giving rise to aberrant peptide accumulation in Alzheimer's disease [10]. Oxides, hydroxides, and oxyhydroxides of Fe of interest in terrestrial and extraterrestrial samples are investigated in products of biomineralization. Many synthetic biomaterials are examples of *biomimetic materials*. Biomimetic materials are developed using inspirations from nature for achieving functionality for other biological or nonbiological functions on Earth.

3.2.1 Oxides of Iron

Iron, the earth's most abundant metal, is an important structural and magnetic material. Iron oxides are important as (i) magnetic materials, (ii) corrosion products of iron and ferrous alloys, (iii) terrestrial and extraterrestrial minerals, and (iv) functional biomaterials. Figure 3.1 shows the iron–oxygen phase diagram [11]. The inset shows the structures of three stable iron oxides – wustite, hematite, and magnetite – in polyhedral and ball-and-stick representations showing tetrahedral and octahedral coordination polyhedra. Magnetite and hematite are discussed in the following sections.

Iron oxides include FeO, wustite; Fe_3O_4, magnetite; and Fe_2O_3. Fe_2O_3 exists in four allotropic forms: α-Fe_2O_3, hematite, and γ-Fe_2O_3, maghemite, which are magnetic oxides. β- and δ-Fe_2O_3 are the other two allotropes of Fe_2O_3. Wustite adopts a rocksalt crystal structure with octahedral coordination of Fe^{2+} cations. Magnetite crystallizes in a spinel structure [7]. Magnetite has Fe^{2+} and Fe^{3+} cations occupying tetrahedrally and octahedrally coordinated polyhedra. At low temperatures, Magnetite is nearly a line compound, pointing to difficulties in synthesizing it as a single phase. Maghemite is a spinel ferrite and is also ferrimagnetic. It is a Fe^{3+}-deficient magnetite. Its formula can be expressed $\left(Fe_8^{3+}\right)\left[Fe_{\frac{40}{3}}^{3+}Vac_8\right]O_{32}$, where Vac represents a vacancy. Hematite has Fe^{3+} cations in planes of octahedra that are a basic structural unit in the lower symmetry *hexaferrites* that possess more complicated superlattice structures.

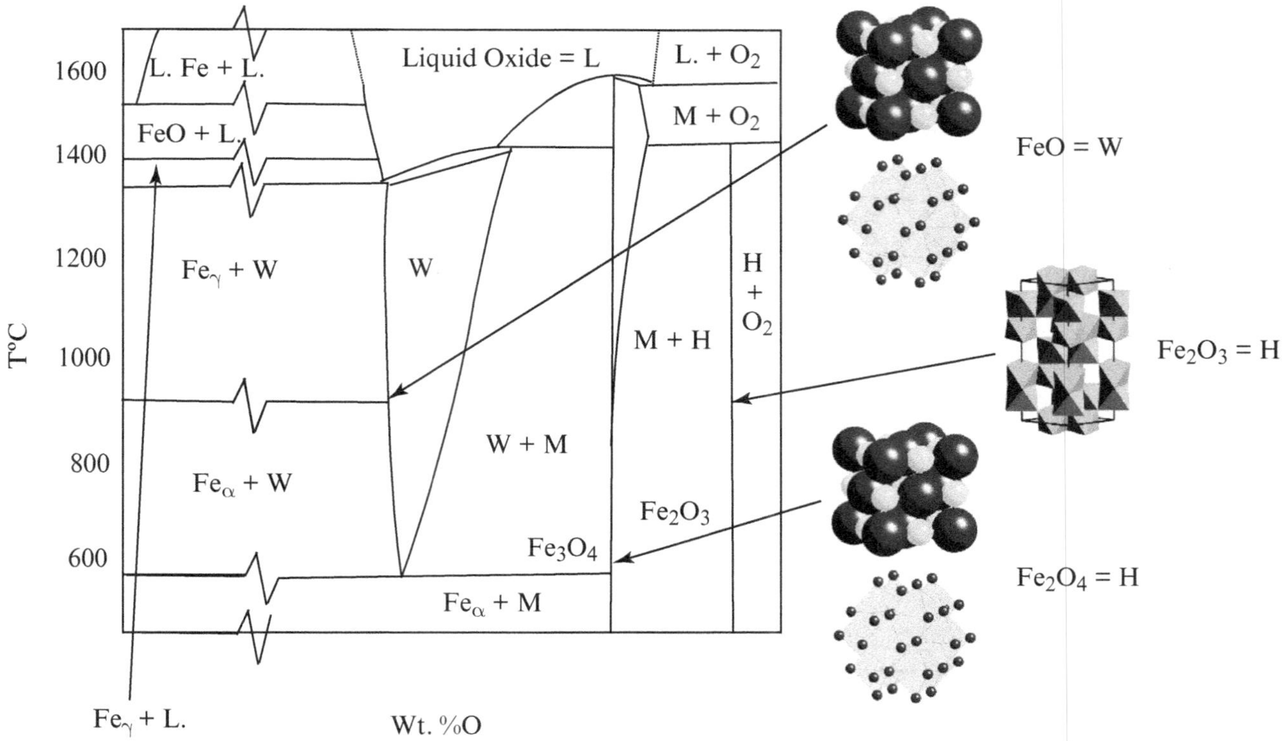

Figure 3.1 Iron–oxygen phase diagram and crystal structures of wustite, hematite, and magnetite.

3.2.2 Hydroxides of Iron

Important hydroxides of iron include Goethite (α-FeO(OH)), Akaganite (β-FeO(OH)), Lepidocrocite (γ-FeO(OH)), Feroxyhyte (δ-FeO(OH)), and ferrihydrite ($5Fe_2O_3{\cdot}9H_2O$).The three allotropic forms of FeO(OH) are (i) α-FeO(OH) is Goethite; (ii) β-FeO(OH) is Akageneite, a Cl-containing hydroxide associated with Fe corrosion in acidic water; (iii) γ-FeO(OH) is Lepidochrosite. All are associated with Fe corrosion in water. Minerals like ferrihydrite, $5Fe_2O_3{\cdot}9H_2O$ have water directly incorporated into an Fe oxide (hematite) structure. Hydroxides have hydroxyl, $(OH)^{-1}$, groups or water molecules. The weaker ionic or hydrogen bonds to the hydroxyl groups often cause bond strengths to be smaller than in the oxides [12].

3.2.2.1 Goethite

Goethite, α-FeO(OH), named after the German poet J. W. Goethe (1749–1832), is an iron oxyhydroxide isostructural with diaspore α-AlO(OH). Both minerals have related allotropes. γ-AlO(OH) is Boehmite, and γ-FeO(OH) is Lepidocrocite. Goethite adopts the *Pbnm* space group also described in an alternate setting of the *Pnma* space group (#62). In this setting, Goethite has orthorhombic lattice constants $a = 0.995$, $b = 0.301$, and $c = 0.462$ nm. Fe, O, and H ions occupy 4c special positions. O^{2-} and $(OH)^{-1}$ anions follow *hcp* stacking with Fe^{3+}.

Octahedrally coordinated with edge sharing octahedra. Double chains of edge sharing octahedral (Figure 3.2a) run out of the plane (along the *b*-axis), with chains joined through vertex sharing octahedra. Goethite is an antiferromagnet with a 400 K **Néel** temperature.

Goethite formed by weathering of iron-rich minerals is common in soils, precipitated by groundwater, and seen in hydrothermal deposits. Goethite nanoparticles are the most common iron oxyhydroxide formed by local sedimentation in marine and lake sediments on Earth [13]. Goethite observed by NASA's Spirit rover in the Martian crater Gusev suggests liquid water's presence in the developing stages of Mars.

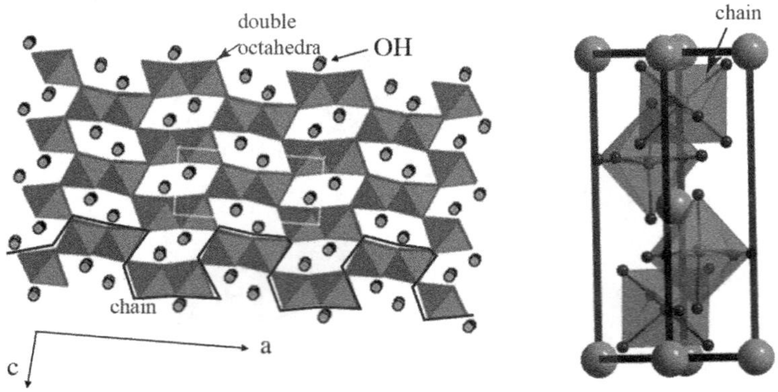

Figure 3.2 Double chains of Fe^{3+} octahedra in goethite.

3.2.2.2 Akaganeite

Akaganeite is an Fe^{3+} oxide–hydroxide (with Cl substitutions possible for (OH) and Ni for Fe) mineral with the formula: $Fe^{3+}O(OH,Cl)$. $FeO_{0.833}(OH)_{1.167}Cl_{0.167}$ is reported with Cl driven from the structure in a transformation to hematite [14]. It is studied in the corrosion of iron in salt water and other acidic environments. It also forms from *pyrrhotite* in ferric-sulfate-bearing solutions. Akaganeite, β-FeO(OH), adopts a monoclinic crystal structure and transforms to hematite at 290–310 °C. It is named after the Akagane mine in Iwate, Japan, where it was first discovered. It has been observed in rocks from the moon and meteorites. β-FeO(OH) products in meteorite corrosion is attributed to Cl^- in a more open structure [15, 16]. Akaganeite's space group has been disputed [17]. X-ray diffraction studies [18, 19] suggested an *I4/m* space group. Crystal structure refinement [20] shows a monoclinic *I2/m* space group to better fit X-ray diffraction (XRD) data. Akaganeite, adopting the *I2/m* space group (#12), has lattice constants $a = 1.0587$, $b = 0.30311$, $c = 1.0515$ nm, and $\beta = 90.03$ [21]. Figure 3.2b shows cation coordination polyhedral in monoclinic Akageneite.

Surface areas and energies of Fe-containing oxides and minerals influence phase stability in small particles. In nanoparticles, hematite is the most stable crystal structure at large particle sizes; at intermediate particle sizes, Goethite is the more stable structure; and at the smallest sizes, Akageneite is most stable.

3.2.2.3 Lepidocrocite

Lepidocrocite, γ-FeO(OH), is isostructural with Boehmite, γ-AlO(OH). O^{2-} and $(OH)^{-1}$ anions form an *hcp* structure, and Fe^{3+} is octahedrally coordinated. Vertex sharing octahedra link to form chains. The chains are connected by edge sharing of the octahedra, resulting in corrugated chains parallel to [0 1 0]. Sheets are weakly coupled through (OH)– bonds, resulting in a micaceous nature. Lepidocrocite is an antiferromagnet with a 50 K **Néel** temperature. The assignment of a space group for Lepidocrocite has been made in several settings. The compound was first described by Goldsztaub [22]. Ewing [23] reported a *Bbmm* space group (#63), with lattice constants $a = 1.24$, $b = 0.387$, $c = 0.306$ nm, while later Rooksby [24] assigned *Amam* (#63) with lattice constants $a = 0.388$, $b = 1.254$, and $c = 0.307$ nm. These are the same space groups but in different settings. Figure 3.3 shows a ball-and-stick representation of the ions in a single monoclinic unit cell of Lepidocrocite, (b) polyhedral representation, and (c) corrugated chains of Fe^{3+} octahedra in the same. Planes are weakly coupled through hydrogen bonds.

3.2.2.4 Ferrihydrite

Ferrihydrite with chemical composition: $5Fe_2O_3{\cdot}9H_2O$, is a mineral found on Mars that is of biogenic origin and that is often grown as magnetic nanoparticles [25]. Ferrihydrite is also a precursor for magnetite *biomineralization* in magnetosomes [26, 27] such as *magnetotactic bacteria* [28] and chitons [29]. Ferrihydrite forms the core component (micelle) of the iron storage protein, ferritin, in bacteria and higher organisms, including the human liver. On Earth, biochemically precipitated magnetite is found in teeth of marine mollusks called chitons. Metabolic iron in chiton teeth, deposited as ferrihydrite, is subsequently converted to magnetite.

Biomineralization of magnetite is important in the evolutionary processes in life forms [30] where we distinguish between *prokaryotes* and *eukaryotes*. Prokaryotes are bacteria and other organisms without nuclei, whereas eukaryotes are organisms with nuclei. The evolutionary descent of eukaryotes to prokaryotes is the basis of the biochemical similarity of life on Earth [31]. Eukaryotes and prokaryotes both sequester iron for metabolic processes through the use of *ferritins*.

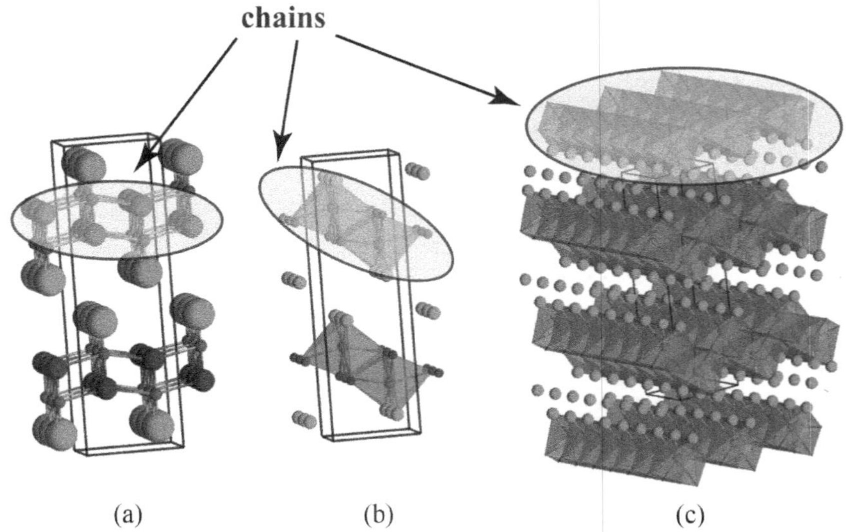

Figure 3.3 (a) Ball-and-stick representation of ions in a monoclinic unit cell of Lepidocrocite, (b) polyhedral representation, and (c) corrugated chains of Fe^{3+} octahedra in the same.

Ferritins are multisubunit proteins that store iron as a mineral particle. The role of ferritins in iron metabolism in organisms is important in understanding life forms.

The protein ferritin is used to synthesize organic–inorganic nanocomposites of iron oxides, called *magnetoferritin* [32]. The protein is first depleted of a weakly paramagnetic ferrihydrite core, followed by addition of ferrous iron under anaerobic conditions, yielding a superparamagnetic core (~7 nm in diameter) [33, 34]. In the human iron cycle, excess iron is temporarily stored in the liver as ferritin molecules. Recent work characterizing human liver ferritin cores by using analytical electron microscopy has shown that the ferritin core crystallizes predominantly in the ferrihydrite structure [35].

3.3 Cubic Spinel Ferrites

$MgAl_2O_4$ is the prototypical spinel structure with Pearson symbol $H1_1$. It is a *normal spinel* for which Mg^{2+} occupies tetrahedral sites, A-sites, and Al^{3+} occupies octahedral sites, B-sites. Magnetite, Fe_3O_4, is an example of an *inverse* spinel. In this structure, octahedra share edges with each other and octahedra and tetrahedra share vertices with each other. The cations are bridged by O^{2-} anions. The normal and inverse spinels are distinguished by different occupations of tetrahedral (A) and octahedral (B) sites. A cartoon representation of the spinel structure is shown in Figure 3.4.

The spinel structure consists of eight *fcc* octants. For a single *fcc* octant, there are a total of four O^{2-} anions. For each O^{2-} anion, there are one octahedral and two tetrahedra sites.

Thus, there are a total of eight filled tetrahedra and four filled octahedra per octant and per formula unit. In the spinel structure, one-eighth of the 64 tetrahedral sites are occupied by cations and one-half of the 32 octahedral sites are occupied by cations. In a *normal spinel*, divalent ions fill the tetrahedral sites while trivalent cations occupy the octahedral sites. In an *inverse spinel*, the tetrahedral sites are occupied by trivalent and half of the octahedral sites are occupied by divalent cations and half by trivalent cations.

Ferrites are magnetic oxides with multivalent cation species with a spinel structure AB_2O_4, where A designates a tetrahedral site and B an octahedral site defined as special positions in a cubic-close packed O^{2-} sublattice. 2-3 ferrites are double oxides of Fe^{3+} and M^{2+}. 4-2 ferrites are double oxides of Fe^{2+} and M^{4+}. We discuss the crystallography of each of these structures later. The prototype of the cubic spinel structure is $MgAl_2O_4$. The space group for spinel is $Fd\bar{3}m$ #227. O^{2-} anions occupy the *32e* special position with coordinates (*x*,*x*,*x*); A-sites are the *8a* special position with coordinates $\left(\frac{1}{2},\frac{1}{2},\frac{1}{2}\right)$, and B-sites are at the *16d* special position with coordinates $\left(\frac{1}{8},\frac{1}{8},\frac{1}{8}\right)$ for a cubic unit cell centered at $\left(-\frac{1}{8},-\frac{1}{8},-\frac{1}{8}\right)$. In an ideal spinel structure, the position of the O^{2-} anion would be at $\left(\frac{3}{8},\frac{3}{8},\frac{3}{8}\right)$, i.e. $x = 3/8$, or in terms of an *oxygen u*-parameter $\left(\frac{1}{8}+u,\frac{1}{8}+u,\frac{1}{8}\right)+u$, where $u = 1/4$. The oxygen u-parameter is illustrated in Figure 3.5. Deviations from ideality depend on ions' sizes and charge balance.

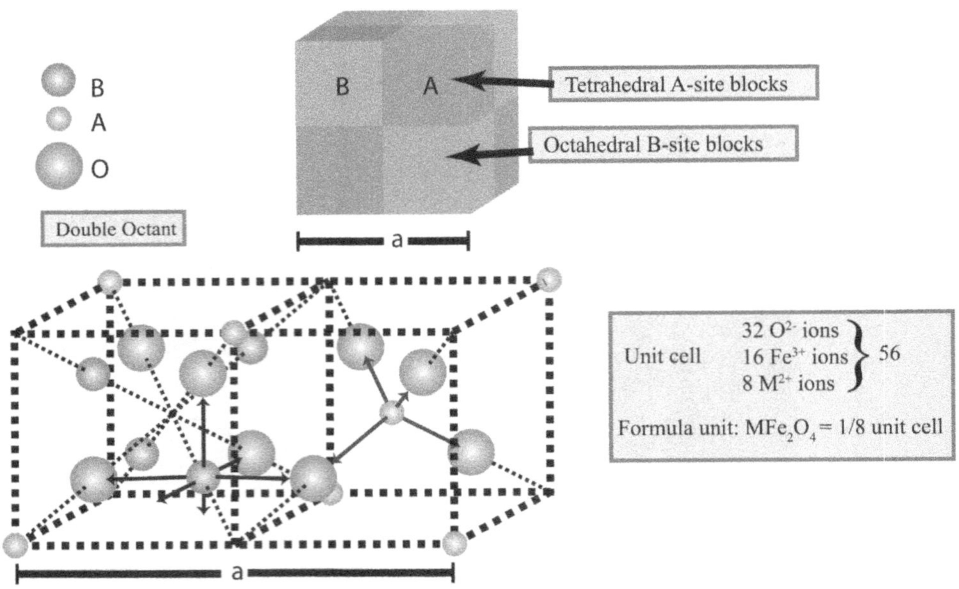

Figure 3.4 Spinel structure illustrating octahedral and tetrahedral blocks accounting for the eight formula units per cubic unit cell and locations of ions in octahedral/tetrahedral di-blocks.

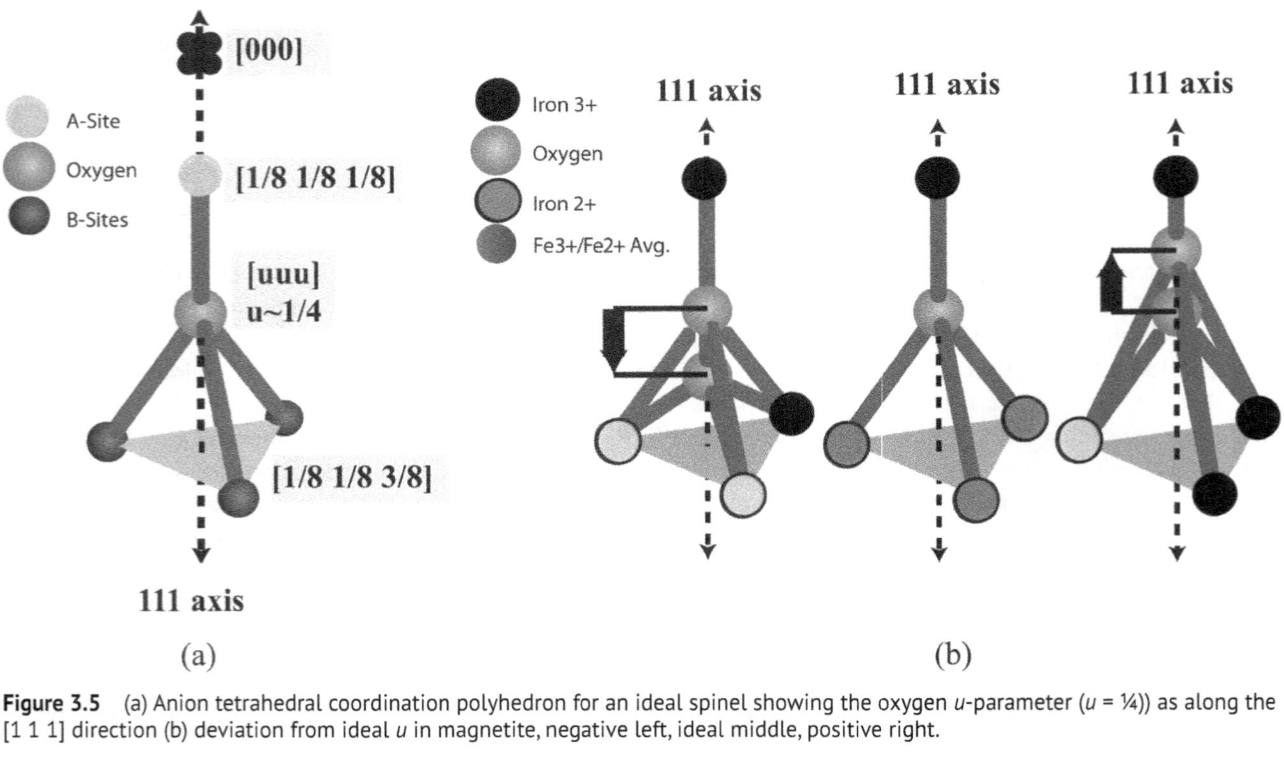

Figure 3.5 (a) Anion tetrahedral coordination polyhedron for an ideal spinel showing the oxygen *u*-parameter (u = ¼)) as along the [1 1 1] direction (b) deviation from ideal *u* in magnetite, negative left, ideal middle, positive right.

Table 3.1 summarizes selected Wyckoff positions of space group #227 $(Fd\bar{3}m)$ with origin at $3m$ for future reference. Notice that the B-site *16d* special position has a vacant partner *16c* special position with the same multiplicity. Similarly, the A-site *8a* special position has a vacant partner *8b* special position with the same multiplicity.

These vacant sites are useful for illustrating certain motifs in the spinel structure. A convenient representation of the spinel structure considers the stacking of polyhedral planes. By inference, these polyhedral planes are also those exposed at particular crystallographic surfaces. Figure 3.6 shows a unit cell of the spinel structure and oxygen polyhedral

Table 3.1 Selected Wyckoff positions of space group #227 ($Fd\bar{3}m$) with origin at $3m$.

Multiplicity	Wyckoff letter	Site symmetry	Coordinate
32	e	$3m$	(x,x,x)
16	d	$3m$	$\left(\frac{1}{2},\frac{1}{2},\frac{1}{2}\right)$
16	c	$3\,m$	$(0,0,0)$
8	b	$4m3m$	$\left(\frac{3}{8},\frac{3}{8},\frac{3}{8}\right)$
8	a	43	$\left(\frac{1}{8},\frac{1}{8},\frac{1}{8}\right)$

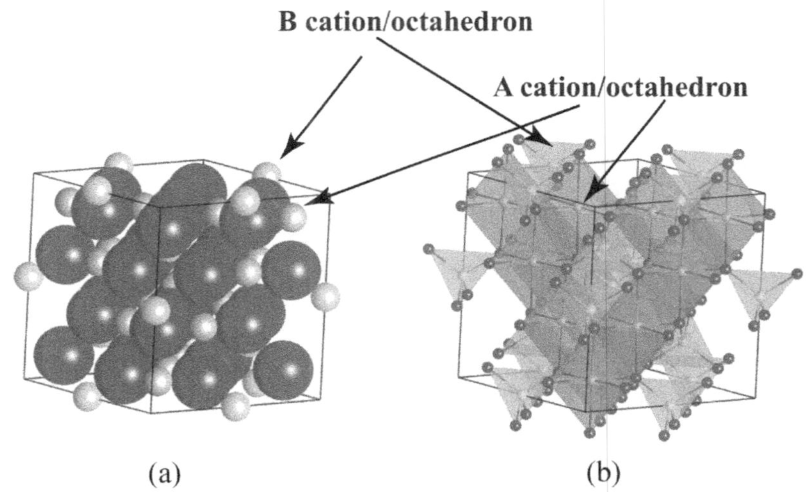

Figure 3.6 (a) Unit cell of the cubic spinel crystal structure in a space filling model (oxygen = large ion, small dark gray = B-site cations, and small light gray = A-site cations) and (b) coordination polyhedral representation showing tetrahedral A-sites and octahedral B-sites.

environments of the cations. Note that alternating (0 0 4) planes have exclusively Mg tetrahedra and Al octahedra. The Al octahedra form chains along the [1 1 0] and [1 1 0] directions in every other plane with Al cations as a consequence of a 41-screw axes that is a symmetry operation in this space group. The structure can be understood in terms of the alternating chains of edges sharing octahedra that also share vertices with Mg tetrahedra in intervening planes. A feature of *fcc* stacking is that if we choose any [111] direction, half of the tetrahedra have apices pointing parallel to this direction and the other half antiparallel. These are designated as T+ *and* T− *tetrahedra*, respectively. Although only a fraction of the tetrahedral interstices is occupied by cations in the spinel structure, the T^+ and T^- tetrahedra have equal occupation. The spinel structure and the stoichiometry are illuminated further in our discussion of the important ferrite magnetic materials that adopt this structure in a subsequent chapter. When the *A* and *B* cations are magnetic ions, this gives rise to technologically important magnetic properties.

3.3.1 2-3 Cubic Spinel Ferrites

2-3 ferrites are double oxides of Fe^{3+}, and $M^{2+}Fe_3O_4$ is the compound for which M is Fe^{2+} that was introduced earlier as an important oxide of iron. A key property of a spinel ferrite is the antiparallel coupling of A- and B-site Fe atomic *magnetic dipole moments*. This antiparallel coupling results from *superexchange* interactions mediated by O^{2-} anions between the two Fe cation sites, as developed more fully in a subsequent chapter 4 [36]. As the cation distribution can deviate from the normal spinel structure, it is instructive to express the spinel formula more generally as:

$$\left(M_{\delta}^{2+}M_{1-\delta}^{3+}\right)\left[M_{1-\delta}^{2+}M_{1+\delta}^{3+}\right]O_4 \tag{3.5}$$

where the cations in parentheses, ( ), are on A-sites and those in brackets, [ ], are on B-sites. The case of $\delta = 1$ corresponds to the *normal spinel structure*, $\delta = 0$ corresponds to the *inverse spinel structure*, and $\delta = 2/3$ corresponds to a *random distribution* of cations. The inversion process can be written as a quasi-chemical reaction [37] exchanging cations on A- and B-sites, respectively:

$$(A)+[B]=(B)+[A] \tag{3.6}$$

with a reaction constant:

$$K=\frac{(A)[B]}{(B)[A]}=\frac{\delta^2}{(1-\delta)(2-\delta)} \tag{3.7}$$

and a configurational entropy:

$$S_C=-R\left[\delta\ln\,\delta+(1-\delta)\ln(1-\delta)+\delta\ln\frac{\delta}{2}+(2-\delta)\ln\left(1-\frac{\delta}{2}\right)\right] \tag{3.8}$$

which is maximized for $\delta = 2/3$ for which SC = 2 Rln 2, which is the case of random inversion. 2/3 reflects 2 B-sites for every A-site.

3.3.1.1 2-3 Cubic Spinel Ferrites – Substituents

Manganese aluminate spinels, $(Mn_xAl_{1-x})_3O_4$ [38], are examples of spinels in which nonmagnetic Al^{3+} cations preferentially occupy octahedral B-sites [39]. Manganese aluminate spinels are also a spinel system with only one magnetic cationic species Mn (that can assume more than one valence state – 2+, 3+, and 4+). The end member of the manganese aluminate series is the mineral *hausmannite*, Mn^3O^4. Hausmannite has a tetragonally distorted spinel structure [40] caused by the Jahn–Teller effect associated d^4 electronic configuration of Mn^{3+} [41]. Aluminum substitution into Huasmannite can also give rise to Yafet–Kittel magnetic spin states, as discussed in a later chapter 4.

For an inverse spinel that mixes $M_1Fe_2O_4$ and $M_2Fe_2O_4$, the spinel formula is:

$$\left(Fe^{3+}\right)\left[\left(M_1^{2+}\right)_x\left(M_2^{2+}\right)_{1-x}Fe^{3+}\right]O_4 \tag{3.9}$$

The spinel formula for a system, which mixes a normal, $M_1Fe_2O_4$, spinel and an inverse, $M_2Fe_2O_4$, spinel as:

$$\left(Fe_{1-x}^{3+}\left(M_1^{2+}\right)_x\right)\left[Fe_{1+x}^{3+}\left(M_2^{2+}\right)_{1-x}\right]O_4. \tag{3.10}$$

In a spinel system, which mixes a normal spinel, $M_1Fe_2O_4$, and another normal spinel, $M_2Fe_2O_4$, we can express the spinel formula as:

$$\left(\left(M_1^{2+}\right)_{1-x}\left(M_2^{2+}\right)_x\right)\left[Fe^{3+}\right]_2O_4 \tag{3.11}$$

A cation disorder may influence degrees of inversion in cubic ferrites. A disorder can result from processing conditions [38, 42, 43] or from radiation-induced damage. A cation disorder also significantly influences the magnetic properties in spinels with nonmagnetic cations [38, 39].

3.3.1.2 2-3 Cubic Spinel Ferrites – Bond Valence Analysis

Ionic crystals (MX) can be analyzed in terms of *Pauling's rules* [44], summarized in Box 3.1. Typical ionic structures have a close-packed anion (X) lattice, with cations (M) occupying a fraction of the octahedral and tetrahedral interstices. The composition of an ionic material is determined by fractional cation occupation on interstitial sites. This occupancy is dictated by size considerations, charge balance, and electrostatic interactions. Typically, $R_X > R_M$, the lattice constant is determined by summing neighboring anion–cation ionic radii and multiplying by a geometric constant.

Box 3.1 Pauling's Rules for Ionic Structures [11]

Pauling's rules for ionic structures are:

i) An anion coordination polyhedron is formed about each cation. Cation–anion distances are determined by the sum of their radii. The *coordination number* is determined by the radius ratio.
ii) The *bond strength*, *s*, is the ratio of the cation charge to the cation coordination number. In a stable structure, the total strength of bonds reaching an anion from neighboring cations in a coordination polyhedron equals the anion charge.
iii) The linkages of coordination polyhedra are such that of edges, and especially faces, tend not to be shared. If edges are shared, they tend to be shortened.
iv) Cations with high charge and low coordination especially do not share edges.
v) The number of crystallographically or chemically distinct atoms in a structure is small.

Bond strength arguments are illustrative for the 2-3 spinels. If we consider a spinel structure for magnetite, Fe_3O_4, O^{2-} is coordinated by an A cation and three B cations, respectively, as shown in Figure 3.5a. In Box 3.2, we show a bond valence analysis for magnetite in a normal and an inverse spinel structure, respectively.

Note that the bond valence analysis of Box 3.2 requires an averaging over the ferric and ferrous ion charge states. If the Fe^{2+} and Fe^{3+} ions are localized on the triangular B-sites coordinating the O^{2-} anion, we would require two configurations – (Fe^{2+}, Fe^{2+}, Fe^{3+}) and (Fe^{3+}, Fe^{3+}, Fe^{2+}) (ignoring the possibility of (Fe^{2+}, Fe^{2+}, Fe^{2+}) and (Fe^{3+}, Fe^{3+}, Fe^{3+}) configurations as having too large cation–cation repulsion energies) – to maintain the charge balance. In fact, magnetite is a reasonable conductor so that electron hopping to maintain an average $Fe^{2.5+}$ charge state also maintains the charge balance. As electron hopping is thermally activated, it is frozen out at low temperatures and charge localization occurs.

Figure 3.5b illustrates the consequences of charge localization on the oxygen *u*-parameter. The O^{2-} anion experiences a stronger Coulombic attraction to the (Fe^{3+}, Fe^{3+}, Fe^{2+}) triangular configuration than the (Fe^{2+}, Fe^{2+}, Fe^{3+}), as a result the oxygen *u*-parameter is expected to be different in the two configurations. More than one *u*-parameter is not allowed in the spinel $Fd\bar{3}m$ space group. Local distortions would thus lead to the broadening of spinel Bragg peaks. In the context of Jahn–Teller [45] activity, such distortions can be accommodated dynamically through vibrational modes or statically through a symmetry lowering structural transformation.

Box 3.2 Bond Valence Analysis for Fe_3O_4

For Fe_3O_4 in a normal spinel structure, the bond strength at A-sites (Fe^{2+}) is s = 2/4 = ½ and at B-sites (Fe^{3+}) is s = 3/6 = ½ and the bond sum is given by:

$$\sum s_i = \frac{1}{2} + \left(\frac{1}{2} + \frac{1}{2} + \frac{1}{2}\right) = 2.$$

This charge cancels out the O^{2-} charge, stabilizing the structure.

For an Fe_3O_4 inverse spinel, the bond strength at A-sites (Fe^{3+}) is s = ¾; at B-sites with Fe^{3+}, the bond strength is s = 3/6 = ½; and at B-sites with Fe^{2+}, it is s = 2/6 = 1/3. An average B-site charge is:

(Continued)

Box 3.2 (Continued)

$$s = \frac{1}{2}\left(\frac{1}{2} + \frac{1}{3}\right) = \frac{5}{12}$$

and the bond sum is:

$$\sum s_i = \frac{3}{4} + \left(\frac{5}{12} + \frac{5}{12} + \frac{5}{12}\right) = \frac{3}{4} + \frac{5}{4} = 2$$

again indicating structural stability.

Another avenue for accommodating Fe^{2+} and Fe^{3+} on B-sites would be through an ordering reaction. A $(Fe^{2+})[Fe^{3+}Fe^{2+}]_2O_4$ spinel is of the type $AB'B''X_4$ discussed by Talanov [46] who used group theoretical techniques to identify three possible symmetry lowering ordering possibilities on B-sites in the $Fd\bar{3}m$ space group, called *1:1 octahedral order*. These can lead to one of the three thermodynamically stable structures with space groups: (i) *Imma*, (ii) $P4_122$ (or its enantiomorph $P4_322$), and (iii) $F\bar{4}3m$ for which there are known examples for (i) and (ii) but not for (iii). It is interesting that examples for (ii) include M_2TiO_4 with (M = Mg, Zn, Mg). Figure 3.7 shows ball-and-stick representations of the (a) $P4_122$ and (b) *Imma* space groups for *1:1 octahedral order* in spinels. These are the two space groups that have been experimentally observed. B-site chirality is apparent in the $P4_122$ ordered structure.

It is also possible to have ordering on A-sites, called 1:1 *tetrahedral order* [47], yielding thermodynamically stable structures with space groups (i) $F43m$ and (ii) $R3m$. The second is a superlattice in which $A'B_2X_4$ and $A''B_2X_4$ layers alternate along the trigonal axis in the rhombohedral space group, i.e. A′ occupying T^+ and A″ occupying T^- tetrahedra. This is the analog to layering of Fe^{2+} octahedral and Ti^{4+} octahedral planes in ilmenite solid solutions with hematite.

This configurational charge imbalance in Fe_3O_4 can also be accommodated by a structural phase transition with the tilting of octahedral cation coordination polyhedra. This gives rise to a slight rhombohedral distortion of the lattice along the [1 1 1] direction that corresponds to the easy magnetization direction (EMD) in the parent spinel phase. This rhombohedral distortion manifests below 122 K in a *Verwey transition* to a monoclinic structure [48, 49]. Verwey showed that resistivity increases by two orders of magnitude on cooling through the Verwey transition, with the transition temperature decreasing with increasing oxidation. Much work has focused on the possibility of observing *charge ordering* accompanying the Verwey

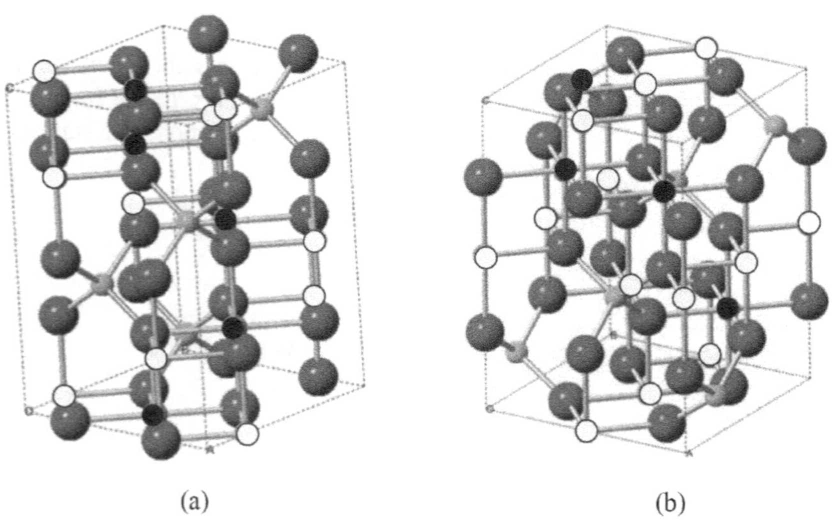

Figure 3.7 Ball-and-stick representations of (a) $P4_122$ and (b) *Imma* space groups for 1:1 octahedral order in spinels (*Source:* showing black/white ordering on octahedral sites and gray tetrahedral sites).

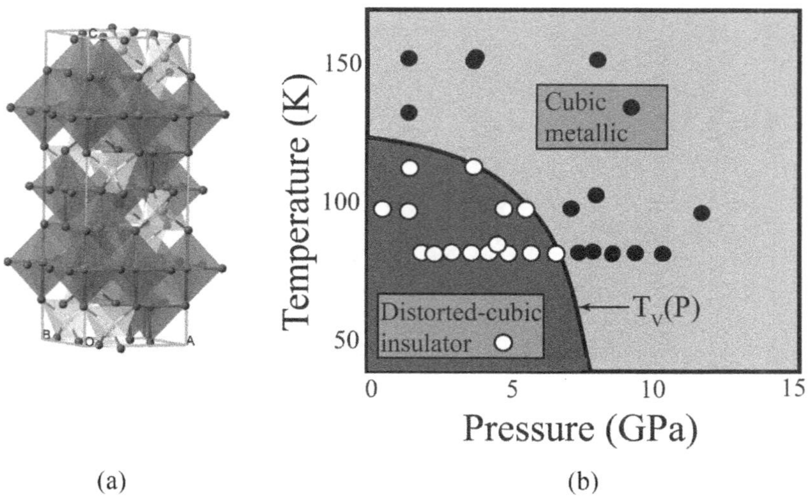

Figure 3.8 (a) Polyhedral representation of a unit cell for the monoclinic crystal structure of magnetite below its Verwey transition and (b) P–T phase diagram for magnetite.

transition. Figure 3.8a shows a polyhedral representation of a unit cell for the monoclinic crystal structure of magnetite below its Verwey transition. As described by Rozenberg et al. [50], the quasi-chemical solid state reaction:

$$\left(Fe^{3+}\right)\left[2Fe^{3+}+e^{-}\right]\rightarrow\left(Fe^{3+}\right)\left[Fe^{2+}Fe^{3+}\right] \tag{3.12}$$

describes charge ordering on cooling through the Verwey transition. Charge ordering is often seen in interesting *multiferroic materials* and has led some to postulate ferroelectricity in charge ordered Fe_3O_4 [51].

The Verwey transition temperature is known to be very sensitive to impurities [52], stoichiometry [53], and oxidation [54]. Of recent interest has been the strain and pressure dependence of the Verwey transition [55]. In $Fe_{3-x}Ti_xO_4$, the Verwey transition is suppressed completely by $x = 0.04$. In fact, for $Fe_{3(1-\delta)}O_4$ and $Fe_{3x}M_xO_4$, with M = Zn or Ti, a universal depression of the Verwey transition as a function of $x = 3\delta$ is observed. With complete suppression of the transition at $x = 0.04$ [56]. Figure 3.8b illustrates the P–T phase diagram for Fe_3O_4 [57]. At temperatures below 122 K at atmospheric pressure and below ~8 GPa, Fe_3O_4 has a structure (*mP*) with a monoclinic space group [58], its Pearson symbol is *mP*56 and has lattice constants $a = 0.5944$, $b = 0.5925$, and $c = 1.6775$ nm, with $\beta = 90.236$ 3, as illustrated in Figure 3.8a. The cubic phase has a slightly larger molar volume than the monoclinic. This indicates entropy to decrease from monoclinic to cubic, perhaps because of a magnetic phase change.

3.3.1.3 Spinel Surfaces

Finite size and surface effects in ferrite nanoparticles can include changes in the degree of inversion (cation preference for A- and B-sites), preferential breaking of like (B–B) or unlike (A–B) bonds at terminating surfaces, symmetry breaking, and chemistry changes at the surface. Chemical and structural changes influence such magnetic properties as surface superexchange interactions, dipole moments, and magnetic anisotropy. Oxygen-mediated B–B and A–B superexchange bonds determine the temperature dependence of the magnetization and whether cation magnetic dipole moment alignment is collinear or noncollinear [59].

We consider the stacking sequence in the spinel structure in terms of cation coordination polyhedra to predict the structures of terminating surfaces in spinel crystals, as illustrated in Figure 3.9. The crystallography of terminating faces determines the symmetry of the atomic environments at the surface. For oxides, we find it more profitable to consider cation coordination polyhedral units, rather than atoms, as the basic structural units. The changes in polyhedral environments at the surface, compared with the bulk, can significantly influence the magnetic properties [59].

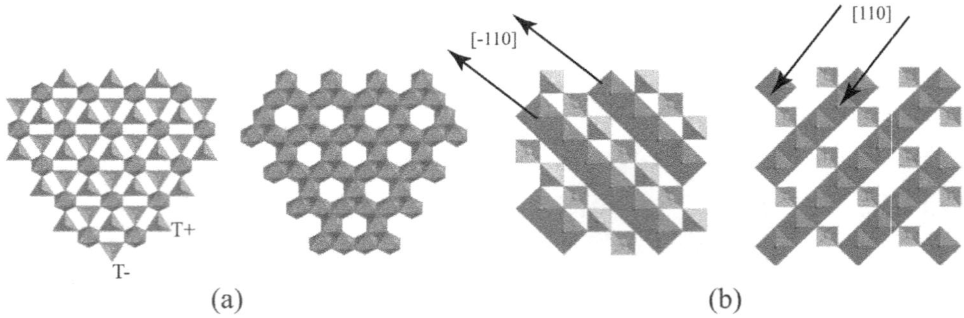

Figure 3.9 Polyhedral representation showing tetrahedral A-sites and octahedral B-sites in proposed terminating (a) [1 1 1] mixed and octahedral and (b) [1 0 0] (octahedral chain) surfaces. *Source*: (showing black/white ordering on octahedral sites and gray tetrahedral sites).

Figure 3.9 shows the stacking of polyhedral unit planes in the spinel structure in (a) (1 1 1) and (b) (1 0 0) terminating surfaces [59]. For each, two possibilities can be thought of as the surfaces on either side of a crystal cleavage plane. Broken fourfold symmetry at the surfaces reflects the fact that the #227 space group does has a 4_1 screw axis, which can have important implications for magnetic surface magnetic anisotropy in (1 0 0) terminated ferrites. Note that for these surfaces, the octahedra align in chains along the [1 1 0] or [1 $\underline{1}$ 0] direction. Both of these planes have the AB_2 stoichiometry of the parent spinel phase. The uniaxial chains will cause uniaxial magnetic surface anisotropy. The (1 1 1) terminations can be called octahedral and mixed. The first of these has a pseudo-hexagonal arrays of octahedra and a stoichiometry B_3. The mixed surface has octahedra coordinated by six tetrahedra and a stoichiometry A_2B. The two average to give the AB_2 stoichiometry of the parent spinel.

3.3.2 4-2 Cubic Spinel Ferrites

Ulvospinel, Fe_2TiO_4, is an example of 4-2 spinel in which Ti^{4+} ions have no d electrons, and therefore, they have no magnetic dipole moment. Like the manganese aluminate spinels, nonmagnetic Ti^{4+} cations substitute exclusively on the octahedral B-sites. In the pseudo-binary titanomagnetite system discussed in a subsequent chapter 4, this causes a monotonic decrease in magnetization and ferrimagnetic **Néel** temperature. On reaching the end member ulvospinel, Fe_2TiO_4, all of the iron is in the form of divalent cations, Fe^{2+}.

Like magnetite, for which there is a cation-deficient maghemite, Fe_2O_3, with a spinel structure, there is a *Titanomaghemite* phase [60], and therefore, synthesizing stoichiometric ulvospinel requires careful control of process atmosphere. Several atmosphere-dependent transformations of ulvospinel have been observed with increasing temperature [61]. Oxidation of ulvospinel to form metastable Titanomaghemite has been shown to occur at 300 °C under atmospheric conditions [61] as inferred from peak broadening in XRD and changes in magnetization.

To monitor phase transformations, high T and P energy dispersive XRD experiments have been performed at the HPCAT 16BM-B beamline Argonne National Laboratory, Advanced Photon Source (APS). Figure 3.10 shows data taken on an unannealed ulvospinel powder sample. The waterfall plot shows Fe_2TiO_4 ulvospinel peaks (dark blue), α-Fe (purple), hexagonal BN (orange), and MgO (gray) peaks. The latter two are artifacts from the pressure cell. *Titanomaghematitzation* was shown to be reduced in *in-situ* heating experiments under high pressures. Formation of ilmenite and hematite from ulvospinel decomposition can be observed at high T with decomposition Ts as a function of P.

At 650 MPa, transformation of the original ulvospinel solid solution with heating was observed and broadening of the spinel peaks above 300 °C, which at atmospheric pressure were attributed to *titanomaghematization*, was observed to be substantially suppressed. The suppression of titanomaghemitization was argued to be a result of high pressures. Oxidation is known to take place by a vacancy-assisted diffusive mechanism, and the activation energy barrier for vacancy formation typically increases under pressure, giving a plausible explanation for this observation.

XRD peaks in pristine ulvospinel samples are generally quite broad. To understand peak broadening in ulvospinel, bond strength arguments are again illustrative for 4-2 spinels. Consider a spinel structure for ulvospinel, Fe_2TiO_4, O^{2-}, coordinated by an A cation and three B cations, respectively, analogous to that shown in Figure 3.5a now with Ti^{4+} cations exclusively on B-sites. In Box 3.3, we perform a bond valence analysis for ulvospinel in both a normal structure and an inverse spinel structure, respectively.

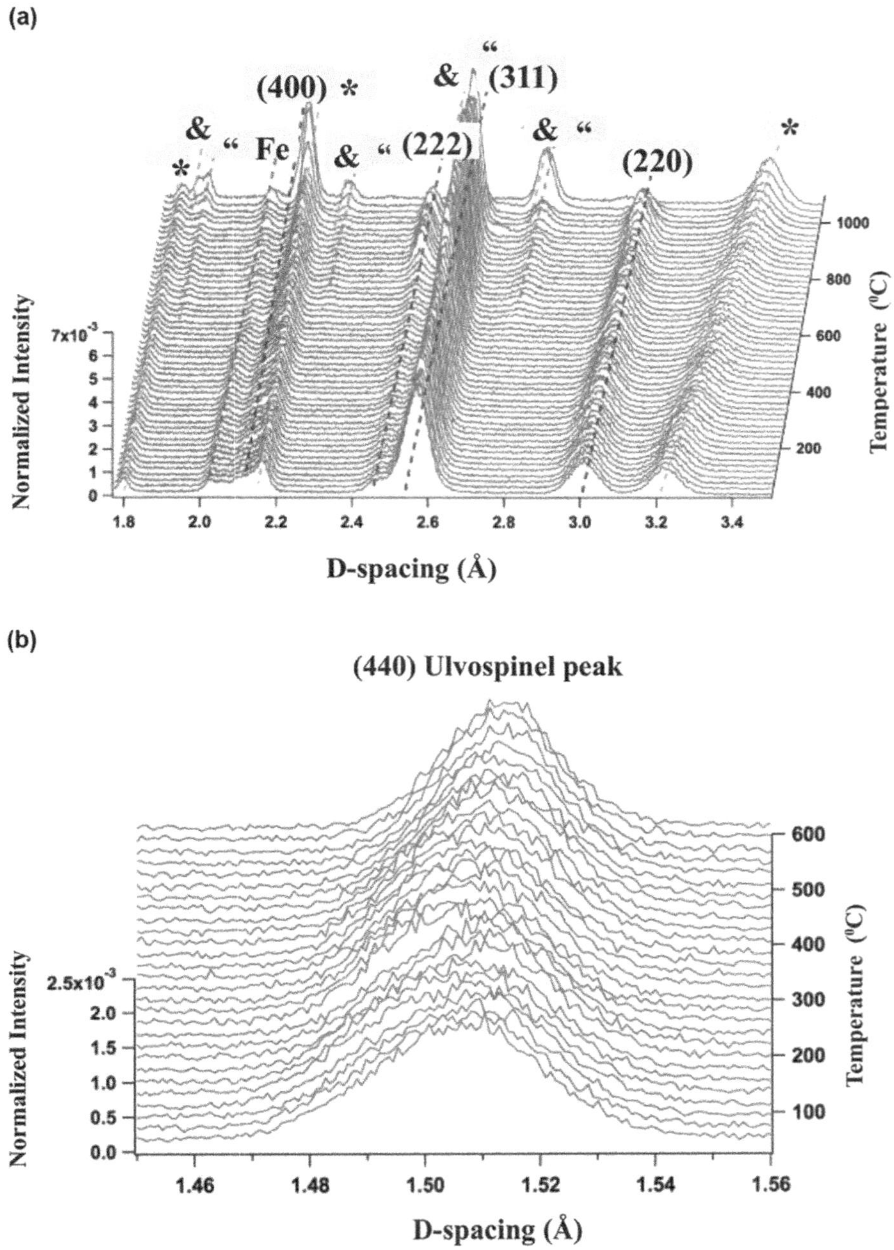

Figure 3.10 (a) High T and P energy dispersive XRD results for ulvospinel. Waterfall plots shows Fe_2TiO_4 (indexed), bcc α-Fe (purple) and BN (*) from the pressure cell. Above 700°C, $FeTiO_3$, ilmenite (") and 1000 °C hematite (&) are observed. (b) Ulvospinel (4 4 0) plane peak evolution from 300 to 600 °C.

Box 3.3 Bond Valence Analysis for Fe_2TiO_4

For Fe_2TiO_4 in a normal spinel structure, we can express the bond strength at A-sites (Fe^{2+}) as $s = 2/4 = 1/2$; at B-sites with (Fe^{2+}) $s = 2/6 = 1/3$ and $s = 4/6 = 2/3$ and at B-sites with (Ti^{4+}) as $s = 4/6 = 2/3$; yielding an average B-site charge of ½ with the bond sum:

$$\sum s_i = \frac{1}{2} + \left(\frac{1}{2} + \frac{1}{2} + \frac{1}{2}\right) = 2$$

which exactly cancels the O^{2-} charge, indicating structural stability.

For Fe_2TiO_4 in an inverse spinel structure, we can express the bond strength at A-sites (Ti^{4+}) as $s = 4/4 = 1$ and at B-sites with Fe^{2+} as $s = 2/6 = 1/3$, with the bond sum:

(Continued)

Box 3.3 (Continued)

$$\sum s_i = \frac{1}{3} + \left(\frac{1}{3} + \frac{1}{3} + \frac{1}{3}\right) = 2$$

indicating structural stability; however, it is known that Ti^{4+} has a preference for octahedral B-sites in ulvospinel and in the pseudo-binary titanomagnetite system.

Note that the bond valence analysis of Box 3.3 requires an averaging over the $Fe^{2+}$and Ti^{4+}cation charge states. If the Fe^{2+} and Fe^{3+} ions are localized on the triangular B-sites coordinating the O^{2-} anion, we would require two configurations – (Fe^{2+}, Fe^{2+}, Ti^{4+}) and (Ti^{4+}, Ti^{4+}, Fe^{2+}) to maintain the charge balance. Ulvospinel is a very good insulator so that electron hopping to maintain an average 3+ charge state ($QO = 6$) that maintains the charge balance is not possible. As electron hopping is not probable, charge localization must occur.

Figure 3.5b illustrates the consequences of charge localization on the oxygen *u*-parameter for magnetite. In ulvospinel, the O^{2-} anion will feel a comparable Coulombic attraction in the (Fe^{2+}, Fe^{2+}, Ti^{4+}) to the (Fe^{3+}, Fe^{3+}, Fe^{2+}) triangular configurations in magnetite. However, the O^{2-} anion will feel a much stronger Coulomb attraction to the (Ti^{4+}, Ti^{4+}, Fe^{2+}) configuration than the (Fe^{2+}, Fe^{2+}, Fe^{3+}) configuration in magnetite. As a result, the oxygen *u*-parameter would be expected to be different in the two configurations and more strained than in magnetite.

Chemically distinct charge states on a single crystallographic site are more problematic in ulvospinel as it has a larger resistivity [62] of 7-8 orders of magnitude. If charge is localized, and averaging accommodated by having (Fe^{2+}, Fe^{2+}, Ti^{4+}) and (Ti^{4+}, Ti^{4+}, Fe^{2+}) B-site cation configurations about oxygen anions that average to ensure charge balance, neither configuration is locally stable. (Fe^{2+}, Fe^{2+}, Fe^{2+}) and (Ti^{4+}, Ti^{4+}, Ti^{4+}) configurations are unlikely as local stability is worse [63]. If charge is localized in B-site cation configurations about oxygen anions (Fe^{2+}, Fe^{2+}, Ti^{4+}) and (Ti^{4+}, Ti^{4+}, Fe^{2+}), one would expect a diffraction peak broadening [47] from distributed bond strains reflecting the relative stability of the configurations. These strains could be detectable in low-temperature Mossbauer spectra. Such measurements have not been carried out.

An avenue for accommodating Fe^{2+} and Ti^{4+} on B-sites is through ordering, *1:1 octahedral order.* A (Fe^{2+}) [$Ti^{4+}Fe^{2+}$]$_2O_4$ spinel is also of the type $AB'B''X_4$ discussed by Talanov [46] [M.91a], with one of the three thermodynamically stable structures with space groups: (i) *Imma*, (ii) $P4_122$ (or its enantiomorph $P4_322$) [65], and (iii) $F43m$. However, to date, no reports of ordering in this compound have been carried out.

3.4 Pseudo-binary Cuts Through the Ternary Ti–Fe–O Phase Diagram

Figure 3.11 is a section of the ternary Fe–Ti–O phase diagram at 700 °C, including magnetically interesting Pseudobrookite–Ferropseudobrookite, hematite–ilmenite, magnetite–ulvospinel (titanomagnetite) pseudo-binaries, and wustite [64], listed

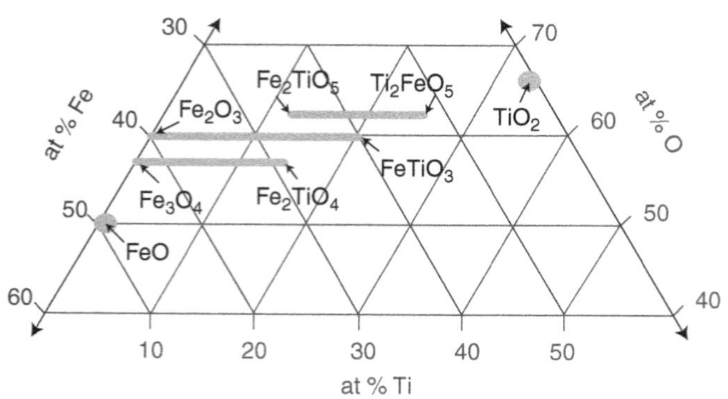

Figure 3.11 Section of a ternary Fe–Ti–O phase diagram at 700 °C, showing Pseudobrookite–Ferropseudobrookite, hematite–ilmenite, magnetite–ulvospinel (titanomagnetite) pseudo-binaries [Rag87] and wustite.

in order of decreasing oxidation. Lamellar structures exsolved from cooled ilmenite–hematite solid solutions are postulated for remanence dominated planetary magnetic field anomalies [66], while others suggest spinodal decomposition in magnetite-rich titanomagnetites as responsible [67].

The titanomagnetites are a pseudo-binary oxide system, between a 2-3 cubic ferrite and a 4-2 cubic ferrite, with compositions:

$$x\left(Fe_2TiO_4\right) - (1-x)(Fe_3O_4). \tag{3.13}$$

The pseudo-binary Fe_2TiO_4–Fe_3O_4 system is reported to have a miscibility gap upon spinodal decomposition [68, 69]. Early reports suggest two phases appearing in the decomposition are a strongly magnetic magnetite phase and a weaker magnetic Ti-rich ulvospinel. The fine microstructure resulting from spinodal decomposition may be important to explain a magnetic remanent state for these minerals on Mars. Certain compounds are of particular extraterrestrial interest because they have magnetic transitions in the day to night T swing on Mars (and the moon). This can be detected with miniaturized magnetic sensors and identified on the basis of **Néel** temperatures.

Compounds in the pseudo-binary Fe_2TiO_4–Fe_3O_4 system are spinels. Figure 3.12a and b shows a proposed phase diagram for pseudo-binary titanomagnetites. The phase diagram assumes a regular solution. Miscibility gap asymmetry is an open question. A Vegards Rule estimation of the magnetic **Néel** temperature of the pseudo-binary system, ranges from 575°C for Fe_3O_4 to 0 K for antiferromagnetic Fe_2TiO_4, the end members of the pseudo-binary system. The T-dependence of the magnetic transition for the solid solution series between magnetite and ulvospinel shows the magnetite phase to be magnetic at all temperatures in the day to night temperature swing on Mars.

Ti-rich phases developed in exsolution may go through magnetic transformations in the day to night temperature swings, depending on their state of decomposition. The decomposition microstructures could be cuboidal magnetite surrounded by ulvospinel and or magnetite/ulvospinel lamella [70, 71]. Both magnetite and ulvospinel are cubic inverse spinels, and the lattice constant of solid solutions varies linearly with composition in the pseudo-binary system.

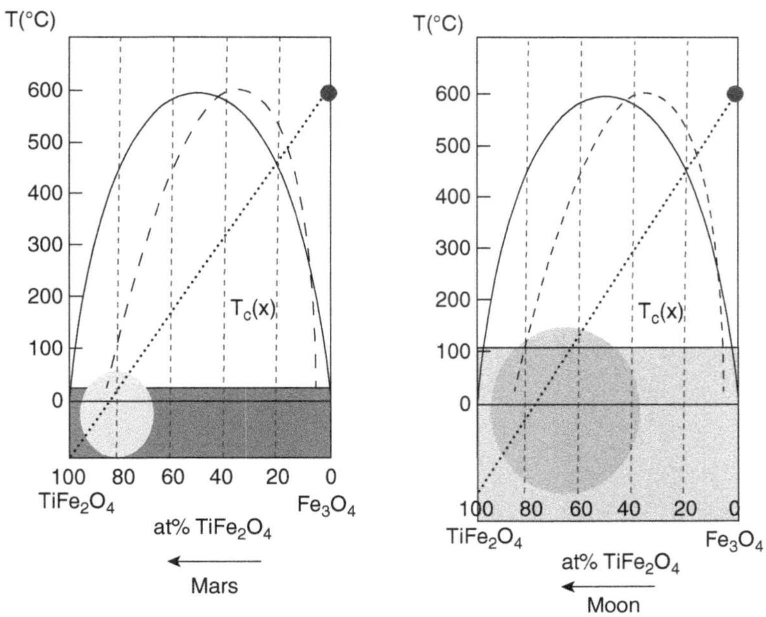

Figure 3.12 Magnetite–ulvospinel phase diagram. Solid line shows predicted the regular solution theory miscibility gap, and the dashed curve is experimentally determined. The dotted line is the magnetic (T_N) transition, red/green areas represents a day to night T swing on the surface of Mars/Earth.

3.4.1 Titanomagnetites – Cation Distributions

With the exception of the antiferromagnetic ulvospinel [72], all of the solid solutions in the pseudo-binary titanomagnetite system are ferrimagnetic. Cation distributions reflect the fact that each Ti^{4+} ion substituted for a Fe^{3+} ion requires the conversion of a second Fe^{3+} ion to Fe^{2+}. Titanomagnetites can deviate from ideal stoichiometries due to (i) exsolution as discussed [70], (ii) nonstoichiometry at high and low temperatures, and (iii) cation impurities [73]. Cation impurities are a consideration in extraterresterial minerals but not in compounds synthesized from relatively pure species. While a thermally activated degree of inversion in titanomagnetite spinels has been described, its activation energy [73] barrier is similar to that for atomic diffusion by a Frenkel defect mechanism and therefore also consistent with exsolution. Given a small difference in the spinel lattice constants across the series [74], it is difficult to detect and therefore rule out early exsolution through spinodal decomposition. A framework for considering both chemical and charge order in the titanomagnetites was proposed by Callen [75]. We begin with a quasi-chemical formula for mixing the end members in this pseudo-binary system:

$$(1-x)\mathrm{Fe_3O_4} = (x)\mathrm{Fe_2TiO_4} = \mathrm{Ti}_x\mathrm{Fe}_{3-x}\mathrm{O}_4. \tag{3.14}$$

Now, consider the valence states of all the cations in the system, i.e. Ti^{4+}, Fe^{3+}, and Fe^{2+}. Given that in the pseudo-binary Fe_2TiO_4–Fe_3O_4 system, if every substituted Ti^{4+} is for a Fe^{3+} cation, which, in turn, causes an additional Fe^{3+} to change valence to Fe^{2+}, then we write the chemical formula as:

$$\mathrm{Ti}_x^{4+}\mathrm{Fe}_{1-2x}^{3+}\mathrm{Fe}_{1+x}^{2+}\mathrm{O}_4^{2-} \tag{3.15}$$

For a general system, we can define partitioning of the ions between the octahedral B-sites and tetrahedral A-sites in the spinel structure in terms of two order parameters η and ξ where (i) η is taken as the fraction of Ti^{4+} cations on the A-sites and (ii) ξ is taken as the fraction of Fe^{2+} cations on the A-sites, respectively. The crystal chemical formula can be written as:

$$\left(\mathrm{Ti}_{\xi x}^{4+}\mathrm{Fe}_{\eta(1+x)}^{2+}\mathrm{Fe}_{1-\eta-\xi x-\eta x}^{3+}\right)\left[\mathrm{Ti}_{(1-\xi)x}^{4+}\mathrm{Fe}_{(1-\eta+(1-\eta)x)}^{3+}\mathrm{Fe}_{((1+\eta)-(2-\eta-\xi)x)}^{2+}\right]\mathrm{O}_4^{2-} \tag{3.16}$$

This general formula is complex, but fortunately, it can be simplified for titanomagnetites where it is observed that the Ti^{4+} cations reside exclusively on the B-sites [76] so that $\xi = 0$ and the formula can be simplified, using a single order parameter, to:

$$\left(\mathrm{Fe}_{\eta(1+x)}^{2+}\mathrm{Fe}_{1-\eta-\eta x}^{3+}\right)\left[\mathrm{Ti}_x^{4+}\mathrm{Fe}_{(1-\eta+(1-\eta)x)}^{3+}\mathrm{Fe}_{((1+\eta)-(2-\eta)x)}^{2+}\right]\mathrm{O}_4^{2-} \tag{3.17}$$

It is this model that is used in a later chapter 4 to define two prevailing models for the compositional dependence of net magnetic dipole moment per formula unit and by in reference the magnetization of the titanomagnetites. There are several models proposed for cation distributions in the titanomagnetites. The most widely cited models have been proposed by Akimoto [77] and **Néel** [78]. In the Akimoto model, the titanomagnetite formula is written as:

$$\left(\mathrm{Fe}_x^{2+}\mathrm{Fe}_{1-x}^{3+}\right)\left[\mathrm{Ti}_x^{4+}\mathrm{Fe}_{(1-x)}^{3+}\mathrm{Fe}_1^{2+}\right]\mathrm{O}_4^{2-} \tag{3.18}$$

This model assumes magnetite to be an *inverse spinel* and that every Ti^{4+} substitution changes *both* an A-site and B-site Fe^{3+} cation to Fe^{2+} cations to balance the charge. This model has $\eta = 0$ for $x = 0$ and changing continuously to 1 at $x = 1$. Another way of expressing this is that the Ti^{4+} additions are partially compensated by changing the degree of inversion of the spinel and partially by charge compensation on the B-sites.

The **Néel** model forces an inverse spinel structure as long as charge balance is compensated for on B-sites, i.e. the A-site cations remain Fe^{3+} until there is no possibility of compensating for the extra positive charge of Ti^{4+} on B-sites. In the formula for titanomagnetite solid solutions, the cation distribution is:

$$\left(Fe^{3+}_1\right)\left[Ti^{4+}_x Fe^{3+}_{(1-2x)} Fe^{2+}_{1+x}\right]O^{2-}_4 \tag{3.19}$$

for compositions $0 \leq x \leq 0.5$. The **Néel** model indicates that the tetrahedral sites are exclusively occupied by Fe^{3+} when $0 \leq x \leq 0.5$. This indicates that every Ti^{4+} replaces a Fe^{3+} on the B-site as well as transforms another Fe^{3+} on B-sites to Fe^{2+}. This can obviously only continue until all of the Fe^{3+} on the B-sites are depleted when $x = 0.5$. When the composition of the titanomagnetite is in the range $0.5 \leq x \leq 1.0$, the **Néel** model cation distribution is expressed:

$$\left(Fe^{2+}_{2x-1} Fe^{3+}_{2-2x}\right)\left[Ti^{4+}_x Fe^{2+}_1\right]O^{2-}_4 \tag{3.20}$$

and the Fe^{2+} cations are distributed randomly on the tetrahedral A-sites.

3.5 Titanomagnetite – Bond Valence Analysis

Bond valence arguments are illustrative for the titanomagnetites. The details, while straightforward, are lengthier because of the need to consider three cations, Fe^{2+}, Fe^{3+}, and Ti^{4+}, leading to more configurations possible in the nodal analysis. We forgo such an analysis here. Given the possibility of different possible local oxygen u-parameters for one composition, a distribution of anion–cation bond lengths and angles will cause a distribution of super exchange interactions. There are multiple possible cations at the B-sites, and if the cations are randomly distributed on these sites, an average oxygen u-parameter can be chosen. This average will change with composition [74, 79]. Distribution in the oxygen u-parameter position is akin to local strains and results in XRD peak broadening.

In some spinels, cations are not randomly distributed on their special positions, but instead form ordered structures. Such ordering is entropy lowering, but even if the ordered phase is stable, it may not be observed because of sluggish kinetics of ordering. It is worth noting that ordering reactions that apply to the $x = 0.25$ and $x = 0.75$ compositions have been enumerated [47]. Titanomagnetites of these compositions are of the type $AB'_{\frac{1}{2}}B''_{\frac{3}{2}}X_4$. If B' is Ti^{4+}, we have the $x = 0.25$ compound, and if B'' is Ti^{4+}, we have the $x = 0.75$ compound.

Talanov identified [47] $P4_122$ and its enantiomorph $P4_322$ as the possible space groups for symmetry lowering ordering possible on the B-sites in the $Fd\overline{3}m$ space group for 1:3 *octahedral order* [80]. Of particular interest in the enantiomorphs is the fact that both the anions and cations form helical spirals in each enantiomorph. The choice of chirality of the spirals depends on the choice of the screw axis, 4_1 or 4_3, in the enantiomorphic space group pairs. For either choice, the anion and cation helix chirality is opposite so that the two are intertwined in a double helix [47]. Figure 3.13 shows the structure of an ordered $AB'_{\frac{1}{2}}B''_{\frac{3}{2}}X_4$ spinel where $x = 0.25$ and blue is Ti^{4+}.

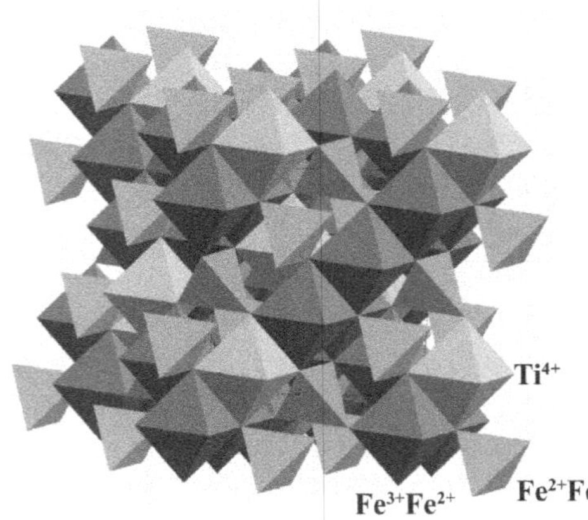

Figure 3.13 The structure of an ordered (1:3) $AB'_{\frac{1}{2}}B''_{\frac{3}{2}}O_4$ spinel, where $x = 0.25$, light gray octahedron = Ti^{4+}.

3.6 Garnets

Garnet is an example of an orthosilicates and a gem stone mineral. The chemical formula for garnets can be written as:

$$C_3A_2D_3O_{12} \tag{3.21}$$

where the *C* cations occupy dodecahedral sites, the *A* cations occupy octahedral sites, and the *D* cations occupy tetrahedral sites in the structure. An example of a garnet is the mineral *Andradite* ($Ca_3Fe_2Si_3O_{12}$, Figure 3.14, which has Fe^{3+} cations in octahedral, Si^{4+} cations in tetrahedral and large Ca^{2+} cations in dodecahedral sites). Figure 3.14a shows all atoms in

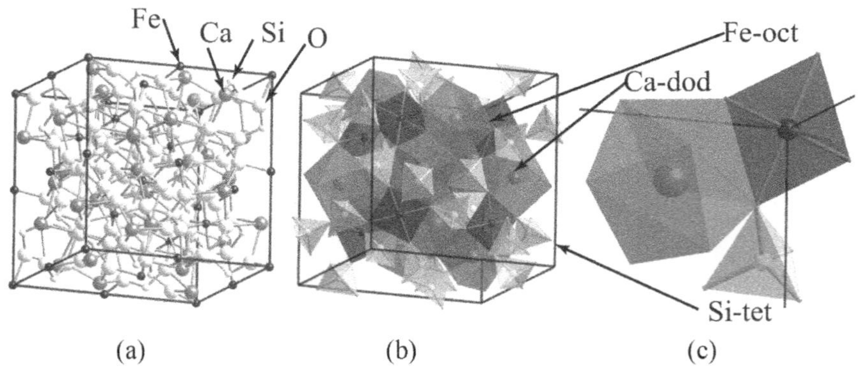

Figure 3.14 Unit cell of the andradite garnet: (a) ball-and-stick depiction; (b) cation coordination polyhedra; and (c) connectivity of octahedra, tetrahedra, and dodecahedra.

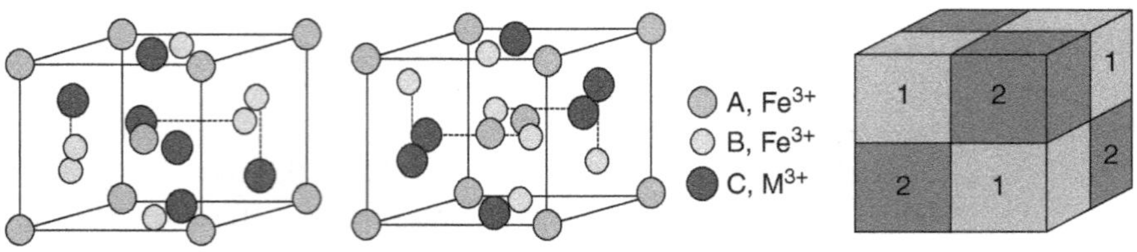

Figure 3.15 Cations in an octant (1) of the garnet structure of $M_3Fe_5O_{12}$ (left), a second octant (2) with mirror symmetry (middle), and the stacking of these units in a 2 × 2 × 2 supercell (right).

a single cubic unit cell in a ball-and-stick depiction; Figure 3.14b shows the cation coordination polyhedral. The connectivity of the three types of cation coordination polyhedra is shown in Figure 3.14c. The cubic unit cell of garnet contains eight formula units for a total of 160 atoms.

Magnetic garnets include the compound $Gd_3Fe_5O_{12}$ having Gd^{3+} occupying the dodecahedral sites and Fe^{3+} on both octahedral and tetrahedral sites. A wide variety of cations in different valence states can reside on these different sites (especially on the large dodecahedral sites). Ionic size is the primary consideration for site occupancy. The similarity in the ionic radii of the rare earth ions causes rare earth iron garnets to readily form solid solutions. Y (ionic radius = 0.090 nm) and Gd (ionic radius = 0.0938 nm) form solid solutions with the formula: $Y_xGd_{3-x}Fe_5O_{12}$ ($0 \leq x \leq 3$); substitution of Y for Gd occurs on the dodecahedral site. Figure 3.15 shows the unit cell of the $M_3Fe_5O_{12}$ garnet as a stacking of two different building blocks 1 and 2.

All cations in an octant cell are drawn (O^{2-} anions are omitted for clarity). The second octant is the inverse of the first (inversion with respect to the octant center). Both octants are used to decorate a 2 × 2 × 2 supercell, which forms the unit cell of the garnet structure. This is reminiscent of cubic ferrite octants.

The similarity in the ionic radii of the rare earth ions causes rare earth iron garnets to readily form solid solutions. Y (ionic radius = 0.090 nm) and Gd (ionic radius = 0.0938 nm) form solid solutions with the formula: $Y_xGd_{3-x}Fe_5O_{12}$ ($0 \leq x \leq 3$). The end member in this series $Y_3Fe_5O_{12}$ is called *yttrium iron garnet* (YIG). YIG was called the "fruit fly" of ferromagnetic resonance studies by Charles Kittel and Marshal Sparks because of the ability to grow single crystal YIG with extremely small FMR linewidths.

YIG is also an important magneto-optical material. YIG has a large optical transparency and magneto-optic Kerr effect coefficient that can be enhanced with Bi substitution for Y [81]. Although garnets are cubic materials, it is desirable to induce magnetic anisotropy in the materials by application of a magnetic field. Anisotropy fields in these materials make their directional properties useful in microwave devices such as isolators, circulators, and gyrators.

3.7 Low-Symmetry Iron Oxides – Hematite

Hematite is of particular interest as it is the most stable form of iron oxide with unusual magnetic behavior. Hematite derives its name from the Greek word *haimatos*, meaning blood, for its red color. This is the color of rust and hematite is an important oxide of iron for it role in oxidation of ferrous alloys. Historically, hematite, Fe_2O_3, has been extensively studied because of its fundamental, as well as its technological important properties.

Hematite has a corundum-type crystal structure designated by the $R\bar{3}c(D_{3d}^6)$ rhombohedral space group [82]. The crystal structure of hematite can be visualized as stacking of nearly hexagonal close-packed oxygen sheets, with two-thirds of the

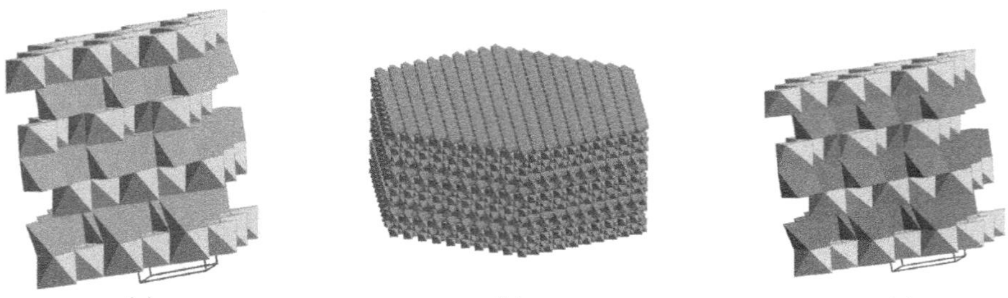

Figure 3.16 (a) Polyhedra representation of Fe_2O_3, (b) hematite and stacking of octahedra sheets in a hematite nanoparticle, and (c) coordination polyhedra in ilmenite (Ti^{4+} = brown).

almost octahedral interstices between the sheets being occupied by Fe^{3+} in a regular manner. Thus, there are alternate iron and oxygen layers stacked along the *c*-axis [0 0 1] of the hexagonal unit cell. Figure 3.16a illustrates the crystal structure of α-Fe_2O_3 in a polyhedra representation. Figure 3.16b shows the stacking of octahedral sheets in a hematite nanoparticle.

A hematite–ilmenite, α-Fe_2O_3–$FeTiO_4$, pseudo-binary in the Fe–Ti–O ternary system is also of importance for geomagnetic response [83]. Figure 3.16c shows the crystal structures of (b) ilmenite. While hematite adopts space group *R*3*c*, ilmenite has space group *R*3. Figure 3.16c shows that in ilmentite, planes of Ti^{4+} and Fe^{2+}octahedra have alternated stacking along the trigonal axis, yielding a stratified structure.

3.8 Low-Symmetry Ferrites: Hexaferrites

Hexaferrites are important layered oxide compounds with stoichiometry:

$$(MO)(Fe_2O_3)_6 \tag{3.21}$$

M is a divalent cation like Ca^{2+}, Ba^{2+}, Sr^{2+}, ... These structures are based on stacking of α-Fe_2O_3 spinel-like (S) layers with rhombohedral (R) MO layers. The structure of α-Fe_2O_3 is the rhombohedral structure of corundum, α-Al_2O_3. The prefix *hexa* refers to not only the six Fe_2O_3 units in each compound but also the hexagonal β-alumina or *magnetoplumbite structure*. Compounds with these structures exhibit a wide variety of technically important properties and are important *microwaves ferrites*.

M-type *barium hexaferrite*, i.e. $BaFe_{12}O_{19}$ (BaM), as a parent compound, is among the most versatile of hexagonal ferrites and finds applications as permanent magnet materials for varied applications from motors to microwave, millimeter, and quasi-optical wave devices. The substitution of Sr for all or some of the Ba ions provide enhanced magnetic properties leading to higher magnetic energy products and value as superior permanent motor materials. The introduction of the large Ba (or Sr) ions break the cubic crystal symmetry of the spinel structure, leading to the stabilization of larger hexaferrite unit cell structures.

Figure 3.17 shows the (a) ball-and-stick, (b) space filling, and (c) polyhedral representations of the hexagonal structures of $BaFe_{12}O_{19}$. Figure 3.17(d) shows the connectivity of cation coordination polyhedra where the polyhedra from several neighboring cells are illustrated. Large Ba ions are (dodecahedrally) 12-fold coordinated. There are five crystallographically distinct Fe cation sites, three are octahedrally coordinated and one tetrahedrally coordinated. The fifth Fe cation site has fivefold triangular prismatic coordination. The structure has layers with BaO and trigonal prismatic coordinated Fe^{3+}, separated by layers containing the Fe^{3+} octahedra and tetrahedra. This, and other hexaferrites, are superlattice structures.

The superlattice structure of BaM can be altered to accommodate other stacking sequences of S and R layers, yielding different hexaferrite stoichiometries, as depicted in the ternary diagram of Figure 3.18. Compounds include the M-, X-, Y-, W-, U-, and Z-type hexaferrites whose unit cells are illustrated in Figure 3.19. M-type barium hexaferrite ($BaO{\cdot}6Fe_2O_3$), having the magnetoplumbite crystal structure (see Figure 3.19a) adopts a $P6_3/mmc$ space group with lattice parameters $a = 0.59$ nm and $c = 2.32$ nm. The cell alternates R and S blocks. An S block contains two O^{2-} layers and six Fe^{3+} ions, whereas an R block consists of three oxygen layers; the center layer has one O^{2-} anion replaced by a barium cation.

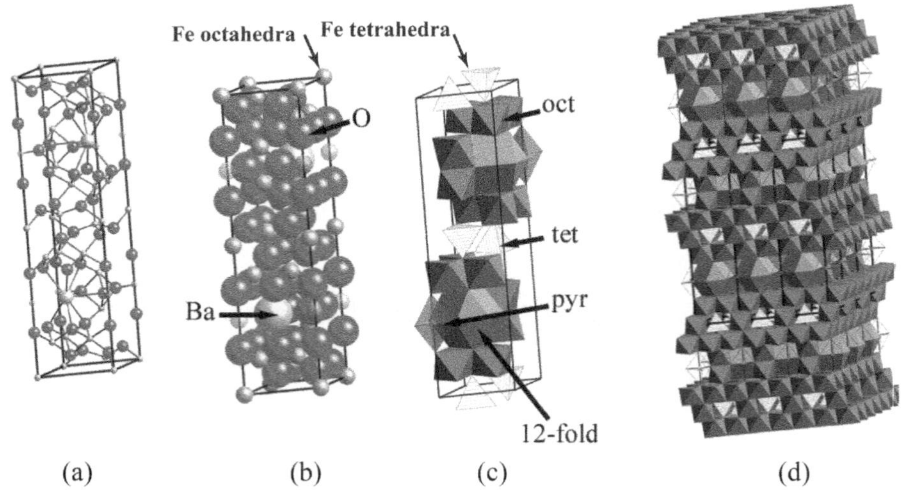

Figure 3.17 (a) Ball-and-stick, (b) space filling, and (c) polyhedral representations of the structure of barium hexaferrite, $BaFe_{12}O_{19}$. (d) Connectivity of the cation coordination polyhedra, showing polyhedra from several neighboring cells.

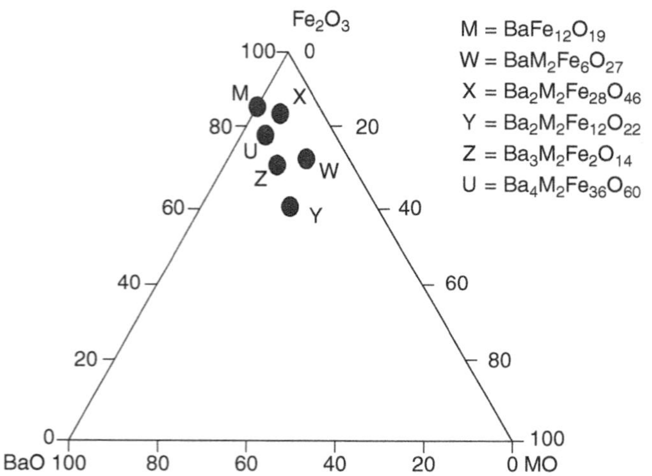

Figure 3.18 Ternary Fe_2O_3–BaO–MO phase diagram showing known hexaferrite compounds.

Figure 3.19 contains unit cell depictions as ball-and-stick models of all the known variants of the hexaferrite compounds known by their designations X, Y, Z, U, W, and X. Table 3.2 lists their formula and stacking sequence as well as some of their intrinsic magnetic properties. The value of these hexaferrites lie in their flexibility to be tailored by cation substitution allowing for adjusting their magnetic anisotropy fields and as such their operational frequencies and performance. These compounds exist as either ferroxplana or magnetoplumbite structures. Although all possess hexagonal crystal symmetries, the spontaneous alignment of the magnetic easy axis varies from along the crystallographic *c*-axis (i.e. magnetoplumbite structure) or within the *a*–*b* basal plane (i.e. ferroxplana structures). The strength of the magnetic anisotropy field that defines the magnetic easy direction is sensitively defined by cation substitution and the resulting distortions to the crystal lattice. As we will see in the coming chapters, such distortions affect the bonding angles between cation–anion–cation, altering the magnetic exchange and increasing or decreasing the exchange energy (J) and related intrinsic magnetic properties such as K, T_C, and M.

The alignment of the magnetic easy axis along the crystallographic *c*-axis in the magnetoplumbite structure allows one to routinely process M-type hexaferrites through ceramic methods to align the crystallographic c-axis, and as such the magnetic easy axis, out the sample plane – a geometry that is required for applications such as isolators and circulators.

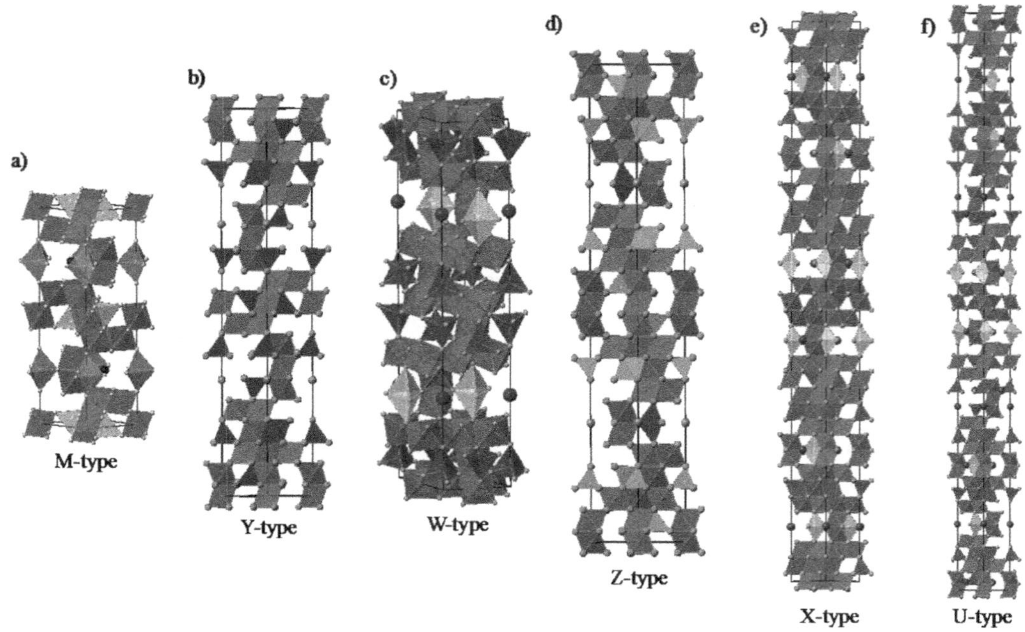

Figure 3.19 Ball-and-stick models for all known variants of hexaferrites: (a) M, (b) Y, (c) W, (d) Z, (e) X, and (f) U.

Table 3.2 Hexaferrite compound variants and their magnetic properties.

Ferrite	Designation	Stacking sequence	Ms ($A{\cdot}m^2/kg$)	H_{AO} (kA/m)	$H_{A\Phi}$ (kA/m)	Tc (K)
Magnetoplumblte						
$BaFe_{12}O_{19}$	BaM	SRS*R*	72	1353		450
$SrFe_{12}O_{19}$	SrM	SRS*R*	92	1692		460
Ferroxplana						
$Ba_2Co_2Fe_{12}O_{22}$	Co_2Y	3(ST)	34	2228	12	340
$Ba3Co_2Fe_{24}O_{41}$	Co_2Z	STSRS*T*S*R*	50	1035	1	410
$BaCo_2Fe_{16}O_{27}$	Co_2W	SSRS*S*R*	50	1687		490
$Ba_2Co_2Fe_28\theta4_6$	Co_2X	SRS*S*R* RSR*R*S*	57	756		467
$Ba4Co_2Fe3_6O_60$	Co_2U	SRS*R*S*T	50	637		434

Note: * denotes the rotation of the unit block 180° about the c-axis.

With anisotropy fields tunable by chemical substitution from 1000s to 10 000s Oe, the FMR frequency can be adjusted from <X-band to >W-band, allowing great utility for microwave to millimeter-wave communication, sensing, and radar applications.

Ferroxplana hexaferrites can similarly be processed in a manner with the easy magnetic axis aligned along the a–b plane. This alignment is advantageous for applications ranging from phase shifters, filters, delay lines, couplers, and antennas to other applications.

Acknowledgments

The authors would like to thank the National Science Foundation (NSF) through Materials World Network (Grants DMR1106943 and DMR083903) for providing support. In addition, they would like to thank for many useful discussions with M. Sorescu, M. Willard, and M. DeGraef and their students, including A. Velazquez, W. Schoenthal, X. Liu, M. Saenko,

S. Long, C. Grothschner, J. Runco, A. Wise, A. Leary, and M. Lucas; and M. Diaz-Michelena, M. Maicas, C. Aroca, and their many students, including A. B. Fernandez, R. Sanz, and M. F. Cerdan.

Bibliography

1 Gilbert, W. (1958). *De magnete*. Courier Corporation.

2 Snoek, J.L. (1936). Magnetic and electrical properties of the binary systems MO. Fe2O3. *Physica* 3 (6): 463–483.

3 Verwey, E.J.W. and Heilmann, E.L. (1947). Physical properties and cation arrangement of oxides with spinel structures I. Cation arrangement in spinels. *The Journal of Chemical Physics* 15 (4): 174–180.

4 Louis, N. (1948). Propértiés Magnétique des Ferrites; Ferrimagnetism et Antiferromagnetism. *Annales de Physique* 3: 137–198.

5 Kittel, C. (1948). On the theory of ferromagnetic resonance absorption. *Physical Review* 73 (2): 155.

6 Yafet, Y. and Kittel, C. (1952). Antiferromagnetic arrangements in ferrites. *Physical Review* 87 (2): 290–294.

7 Bragg, W.H. (1915). The structure of magnetite and the spinels. *Nature* 95 (2386): 561.

8 Snelling, E.C. (1988). Soft ferrites: Properties and applications.

9 Fernández, A.B., McHenry, M.E., Dı, M., Aroca, C., and Maicas, M. (2013). Data base of extraterrestrial magnetic minerals, test and magnetic simulation. *IEEE Transactions on Magnetics* 49 (7): 3533–3536.

10 Collingwood, J.F., Chong, R.K., Kasama, T., Cervera-Gontard, L., Dunin-Borkowski, R.E., Perry, G., ... Dobson, J. (2008). Three-dimensional tomographic imaging and characterization of iron compounds within Alzheimer's plaque core material. *Journal of Alzheimer's Disease* 14 (2): 235–245.

11 De Graef, M. and McHenry, M.E. (2012). *Structure of Materials: An Introduction to Crystallography, Diffraction and Symmetry*. Cambridge University Press.

12 Cornelius, K. (1985). *Manual of Mineralogy*. John Wiley & Sons.

13 van der Zee, C., Roberts, D.R., Rancourt, D.G., and Slomp, C.P. (2003). Nanogoethite is the dominant reactive oxyhydroxide phase in lake and marine sediments. *Geology* 31 (11): 993–996.

14 Ståhl, K., Nielsen, K., Jiang, J., Lebech, B., Hanson, J.C., Norby, P., and van Lanschot, J. (2003). On the akaganéite crystal structure, phase transformations and possible role in post-excavational corrosion of iron artifacts. *Corrosion Science* 45 (11): 2563–2575.

15 Buchwald, V.F. (1977). The mineralogy of iron meteorites. *Philosophical Transactions of the Royal Society of London A: Mathematical, Physical and Engineering Sciences* 286 (1336): 453–491.

16 Bucnwar-o, Y.F. (1989). Corrosion of Fe-Ni alloys by Cl-containing akagan6ite (p-FeOOH): The Antarctic meteorite case. *American Mineralogist* 74: 656467.

17 Garcia, K.E., Morales, A.L., Barrero, C.A., Arroyave, C.E., and Greneche, J.M. (2004). Magnetic and crystal structure refinement in akaganeite nanoparticle. *Physica B: Condensed Matter* 354 (1): 187–190.

18 Mackay, A.L. (1960). β-Ferric oxyhydroxide. *Mineral Magazine* 32: 545.

19 Murad, E. (1979). Mössbauer and X-ray data on β-FeOOH (akaganéite). *Clay Minerals* 14: 273.

20 Post, J.E. and Buchwald, V.F. (1991). Crystal structure refinement of akaganeite. *American Mineralogist* 76 (1–2): 272–277.

21 Post, J.E., Heaney, P.J., Dreele, R.B.V., and Hanson, J.C. (2003). Neutron and temperature-resolved synchrotron X-ray powder diffraction study of akaganéite. *American Mineralogist* 88 (5–6): 782–788.

22 Goldsztaub, S. (1931). Dehydration of natural ferric hydrates. *Comptes Rendus* 193: 533.

23 Ewing, F.J. (1935). The crystal structure of lepidocrocite. *The Journal of Chemical Physics* 3 (7): 420–424.

24 Rooksby, H.P. (1951). *X-ray Identification and Crystal Structures of Clay Minerals* (ed. W.F. Bradley), 264. London: Mineralogical Society.

25 Šafařík, I. and Šafaříková, M. (2002). Magnetic nanoparticles and biosciences. *Monatshefte für Chemie/Chemical Monthly* 133 (6): 737–759.

26 Lang, C. and Schüler, D. (2006). Biogenic nanoparticles: Production, characterization, and application of bacterial magnetosomes. *Journal of Physics: Condensed Matter* 18 (38): S2815.

27 Kirschvink, J.L., Jones, D.S., and MacFadden, B.J. (Eds.) (2013). *Magnetite Biomineralization and Magnetoreception in Organisms: A New Biomagnetism*, Vol. 5. Springer Science & Business Media.

28 Frankel, R.B., Papaefthymiou, G.C., and Blakemore, R.P. (1985). Mössbauer spectroscopy of iron biomineralization products in magnetotactic bacteria. In: Kirschvink, J.L., Jones, D.S., and MacFadden, B.J. (eds) *Magnetite Biomineralization and Magnetoreception in Organisms*. Topics in Geobiology, vol 5. Springer, Boston, MA.

29 Lowenstam, H.A. (1962). Magnetite in denticle capping in recent chitons (Polyplacophora). *Bulletin Geological Society America* 73: 435.

30 Chang, S.B.R. and Kirschvink, J.L. (1989). Magnetofossils, the magnetization of sediments, and the evolution of magnetite biomineralization. *Annual Review of Earth and Planetary Sciences* 17 (1): 169–195.

31 Margulis, L. and Bermudes, D. (1985). Symbiosis as a mechanism of evolution: Status of cell symbiosis theory. *Symbiosis (Philadelphia, Pa.)* 1: 101–124.

32 Meldrum, F.C., Heywood, B.R., and Mann, S. (1992). *Science* 257: 1522523.

33 Mornet, S., Vasseur, S., Grasset, F., and Duguet, E. (2004). Magnetic nanoparticle design for medical diagnosis and therapy. *Journal of Materials Chemistry* 14 (14): 2161–2175.

34 Tartaj, P., Morales, M.P., Gonzalez-Carreno, T., Veintemillas-Verdaguer, S., and Serna, C.J. (2005). Advances in magnetic nanoparticles for biotechnology applications. *Journal of Magnetism and Magnetic Materials* 290: 28–34.

35 Pan, Y., Brown, A., Brydson, R., Warley, A., Li, A., Powell, J., ... Falke, M. (2005). TEM and STEM studies of the atomic structure and chemistry of ferritin in human liver biopsies. *Microscopy and Microanalysis* 11 (S02): 1102.

36 Gorter, E.W. (1954). Saturation magnetization and crystal chemistry of ferrimagnetic oxides. *Philips Research Reports* 9 (4): 295–320.

37 O'Neill, H.S.C. and Navrotsky, A. (1983). Simple spinels; crystallographic parameters, cation radii, lattice energies, and cation distribution. *American Mineralogist* 68 (1–2): 181–194.

38 Nakamura, Y., Smith, P.A., Laughlin, D.E., DeGraef, M.J., and McHenry, M.E. (1995). Structure and magnetic properties of quenched (MnxAl1-x)3O4 Spinels and Hausmannites. *IEEE Transactions on Magnetics* 31: 4154–4156.

39 Willard, M.A., Nakamura, Y., Laughlin, D.E., and McHenry, M.E. (1999). Magnetic properties of ordered and disordered spinel-phase ferrimagnets. *Journal of the American Ceramic Society* 82 (12): 3342–3346.

40 Pollert, E. (1980). Tetragonal-to-cubic transformation of hausmannite. *Journal of Solid State Chemistry* 33 (3): 305–308.

41 Goodenough, J.B. and Loeb, A.L. (1955). Theory of ionic ordering, crystal distortion, and magnetic exchange due to covalent forces in spinels. *Physical Review* 98 (2): 391.

42 Guire, M.R., Kalonji, G., and O'Handley, R.C. (1990). Cation distributions in rapidly solidified cobalt ferrite. *Journal of the American Ceramic Society* 73 (10): 3002–3006.

43 De Guire, M.R., O'Handley, R.C., and Kalonji, G. (1989). The cooling rate dependence of cation distributions in CoFe2O4. *Journal of Applied Physics* 65 (8): 3167–3172.

44 Pauling, L. (1939). *Nature of the Chemical Bond*, 1e. Ithaca, NY: Cornell University.

45 Jahn, H.A. and Teller, E. (1937, July). Stability of polyatomic molecules in degenerate electronic states. I. Orbital degeneracy. *Proceedings of the Royal Society of London A: Mathematical, Physical and Engineering Sciences* 161 (905). 220–235. The Royal Society.

46 Talanov, V.M. (1990). Structural modelling of low-symmetry phases of Spinels. I. Phases with 1: 1 octahedral order. *Physica Status Solidi (B)* 162 (1): 61–73.

47 Talanov, V.M. (1990). Structural modelling of low-symmetry phases of Spinels. II. Phases with 1: 3 octahedral order. *Physica Status Solidi (B)* 162 (2): 339–346.

48 Verwey, E.J.W. (1939). Physical properties and cation arrangement of oxides with Spinel structures, Part I: cation arrangements in Spinels. *Nature* 144: 327.

49 Verwey, E.J.W. and Haayman, P.W. (1941). Physical properties and cation arrangement of oxides with Spinel structures, Part I: cation arrangements in Spinels. *Physica* 8: 979.

50 Rozenberg, G.K., Pasternak, M.P., Xu, W.M., Amiel, Y., Hanfland, M., Amboage, M., ... Jeanloz, R. (2006). Origin of the Verwey transition in magnetite. *Physical Review Letters* 96 (4): 045705.

51 Van Den Brink, J. and Khomskii, D.I. (2008). Multiferroicity due to charge ordering. *Journal of Physics: Condensed Matter* 20 (43): 434217.

52 Brabers, V.A.M., Walz, F., and Kronmüller, H. (1998). Impurity effects upon the Verwey transition in magnetite. *Physical Review B* 58 (21): 14163.

53 Aragón, R., Buttrey, D.J., Shepherd, J.P., and Honig, J.M. (1985). Influence of nonstoichiometry on the Verwey transition. *Physical Review B* 31 (1): 430.

54 Özdemir, Ö., Dunlop, D.J., and Moskowitz, B.M. (1993). The effect of oxidation on the Verwey transition in magnetite. *Geophysical Research Letters* 20 (16): 1671–1674.

55 Carporzen, L. and Gilder, S.A. (2010). Strain memory of the Verwey transition. *Journal of Geophysical Research: Solid Earth* 115 (B5).

56 Kakol, Z., Kozlowski, A., Sabol, J., Metcalf, P., and Honig, J.M. (1994). The crystal structure of the sigma-phase in the Co-Cr system. *Acta Physica Polonica A* 85: 223–227.

57 Rozenberg, G.K., Hearne, G.R., Pasternak, M.P., Metcalf, P.A., and Honig, J.M. (1996). Nature of the Verwey transition in magnetite Fe_3O_4 to pressures of 16 GPa. *Physical Review B* 53: 6482–6487.

58 Wright, J.P., Attfield, J.P., and Radaelli, P.G. (2002). Charge ordered structure of magnetite Fe_3O_4 below the Verwey transition. *Physical Review B* 66 (21): 214422.

59 Swaminathan, R., McHenry, M.E., Poddar, P., and Srikanth, H. (2005). Magnetic properties of polydisperse and monodisperse NiZn ferrite nanoparticles interpreted in a surface structure model. *Journal of Applied Physics* 97 (10): 10G104.

60 Readman, P.W. and O'reilly, W. (1972). Magnetic properties of oxidized (cation-deficient) Titanomagnetites $(Fe, Ti, \square)_3O_4$. *Journal of Geomagnetism and Geoelectricity* 24 (1): 69–90.

61 Groschner, C., Lan, S., Wise, A., Leary, A., Lucas, M.S., Park, C., ... McHenry, M.E. (2013). The role of atmosphere on phase transformations and magnetic properties of ulvospinel. *IEEE Transactions on Magnetics* 49 (7): 4273–4276.

62 O'reilly, W. and Banerjee, S.K. (1965). Cation distribution in titanomagnetites (1− x) Fe3O4− xFe2TiO4. *Physics Letters* 17 (3): 237–238.

63 Liu, X., Sorescu, M., Wang, J., Schoenthal, W., Dong, Z., Ramanujan, R.V., ... McHenry, M.E. (2014). Structural and Magnetic Properties of xFe (2) TiO4 center dot (1-x) Fe3O4 (0.75<= x<= 1). *IEEE Transactions On Magnetics* 50 (11).

64 Raghavan, V. (1987). Phase diagrams of ternary iron alloys. I. ASM International 1987, 226.

65 Darul, J., Nowicki, W., Piszora, P., Baehtz, C., and Wolska, E. (2005). Synchrotron X-ray powder diffraction studies on the order–disorder phase transition in lithium ferrites. *Journal of Alloys and Compounds* 401 (1): 60–63.

66 Robinson, P., Harrison, R.J., McEnroe, S.A., and Hargraves, R.B. (2002). Lamellar magnetism in the haematite–ilmenite series as an explanation for strong remanent magnetization. *Nature* 418 (6897): 517–520.

67 Buddington, A.F. and Lindsley, D.H. (1964). Iron-titanium oxide minerals and synthetic equivalents. *Journal of Petrology* 5 (2): 310–357.

68 Smith, P.P. (1980). Spinodal decomposition in a titanomagnetite. *American Mineralogist* 65 (9–10): 1038–1043.

69 Lindsley, D.H. (1981). Some experiments pertaining to the magnetite-ulvospinel miscibility gap. *American Mineralogist* 66 (7–8): 759–762.

70 Wise, A., Saenko, M., Velázquez, A.M., Laughlin, D.E., Díaz-Michelena, M., and McHenry, M.E. (2011). Phase evolution in the Fe3 O4 – Fe2TiO4 pseudo-binary system and its implications for remanent magnetization in Martian minerals. *IEEE Transactions on Magnetics* 47 (10): 4124–4127.

71 Sanz, R., Cerdán, M.F., Wise, A., McHenry, M.E., and Diaz-Michelina, M. (2011). Temperature dependent magnetization and remanent magnetization in pseudo-binary x (Fe2TiO4)-(1-x)(Fe3O4)(0.30). *IEEE Transactions on Magnetics* 47 (10): 4128.

72 Readman, R.W. (1978). Magnetic properties of ulvöspinel. *Physics of the Earth and Planetary Interiors* 16 (3): 196–199.

73 Bleil, U. (1976). An experimental study of the titanomagnetite solid solution series. *Pure and Applied Geophysics* 114 (2): 165–175.

74 Wechsler, B.A., Lindsley, D.H., and Prewitt, C.T. (1984). Crystal structure and cation distribution in titanomagnetites (Fe (sub 3-x) Ti x O 4). *American Mineralogist* 69 (7–8): 754–770.

75 Callen, H.B., Harrison, S.E., and Kriessman, C.J. (1956). Cation distributions in ferrospinels. Theoretical. *Physical Review* 103 (4): 851.

76 Blasse, G. (1964). *Crystal Chemistry and Some Magnetic Properties of Mixed Metal Oxides with Spinel Structure*. Philips Research Laboratories.

77 Akimoto, S.I. (1954). Thermo-magnetic study of ferromagnetic minerals contained in igneous rocks. *Journal of Geomagnetism and Geoelectricity* 6 (1): 1–14.

78 Néel, L. (1955). Some theoretical aspects of rock-magnetism. *Advances in Physics* 4 (14): 191–243.

79 Bosi, F., Hålenius, U., and Skogby, H. (2009). Crystal chemistry of the magnetite-ulvöspinel series. *American Mineralogist* 94 (1): 181–189.

80 Braun, P.B. (1952). A superstructure in spinels. *Nature* 170 (4339): 1123–1123.

81 Kim, T., Nasu, S., and Shima, M. (2007). Growth and magnetic behavior of bismuth substituted yttrium iron garnet nanoparticles. *Journal of Nanoparticle Research* 9 (5): 737–743.

82 Prewitt, C.T., Shannon, R.D., Rogers, D.B., and Sleight, A.W. (1969). C rare earth oxide-corundum transition and crystal chemistry of oxides having the corundum structure. *Inorganic Chemistry* 8 (9): 1985–1993.

83 Robinson, P. and Putnis, A. (1999). The magnetic properties and crystal chemistry of oxide spinel solid solutions. *Surveys in Geophysics* 19 (6): 461–520.

4

Ferrite Magnetism

Fundamentals of Néel's Molecular Field Theory

Michael E. McHenry and David E. Laughlin

4.1 Introduction

This chapter discusses magnetism in ferrites in the context of magnetic dipole moments, molecular field theory, and magnetic exchange interactions. Chapter 4 builds on the material presented in Chapter 3.

4.2 Magnetic Dipole Moments, Magnetization, Exchange, and Anisotropy

This section summarizes the magnetic properties that will be discussed in reviewing structure–properties relationships in ferrites and related oxides.

4.2.1 Magnetic Dipole Moments – Definitions

A *magnetic dipole moment*, $\sim\mu$, is a vector, which points normal to the circulating current, with dimensions of current times area. An atomic magnetic dipole moment originates from a circulating charge, $\sim\mu = IA\sim u$, with $\sim\boldsymbol{r}$, the position vector of a circulating charged particle; I, the current; $A = \pi r^2$ the area swept out by the charge; and $\sim\boldsymbol{u}$, a unit vector normal to the area, A. The magnetic dipole moment is proportional to the *angular momentum vector*, $\sim\boldsymbol{\Pi}$.$\sim\boldsymbol{\Pi}$ represents the total angular momentum, which has contributions from spin and orbital angular moment, as discussed later.

The magnetic dipole moment and the general angular momentum vector, $\sim\boldsymbol{\Pi}$, are related through (MKSA and CGS):

$$\mu = ge/2m\boldsymbol{\Pi} \quad \mu = ge/2mc\boldsymbol{\Pi} \tag{4.1}$$

where g is the Lande *g-factor*. For an orbiting electron, the constant g = 1. The dipole moment *spin angular momentum* has g = 2. Angular momentum is quantized in units of **Planck's constant** divided by $2\pi(\text{h}^{-} = \text{h}/2\pi = 1.05\times10^{-34}$ $\mathbf{J}-\text{s} = 1.05\times10^{-27}\,\text{erg}-\text{s})$. The fundamental unit of magnetic moment is the *Bohr magneton*:

$$\mu B = (\text{e}/2\text{m})\text{h}^{-} \quad \mu B = (\text{e}/2\text{mc})\text{h}^{-} \tag{4.2}$$

with: $\mu B = 9.27\times10^{-24}\,Am^2(\text{J}/\text{T}) = 9.27\times10^{-21}\,\text{erg/G}$. The atomic dipole moment, μ_{atom}, is the sum of all electron dipole moments for an atom in accordance with Hund's rules [1]. For identical atoms the magnetization, M, is:

$$M = N_a\mu_{\text{atom}} \tag{4.3}$$

where N_a is the number of dipole moments per unit volume.

Modern Ferrites: Basic Principles, Processing and Properties, Volume 1, First Edition. Edited by Vincent G. Harris.

4.2.2 Magnetization and Dipolar Interactions

Magnetization is an extrinsic materials property [2] that depends on the total number of atoms in a system, their individual dipole moments, and the vectorial addition of the dipole moments. Magnetization, *M*, is formally defined as the net dipole moment per unit volume:

$$M = \Sigma_{atoms} \times \mu_{atom} / V \tag{4.4}$$

with *V* the volume of the material. Magnetization is often also reported as a *specific magnetization*, which is the net dipole moment per unit weight. Individual atomic dipole moments can be aligned randomly, in a collinear arrangement, or in some other ordered but noncollinear fashion. Considering *collinear dipoles* that may add or subtract, depending on whether they are parallel or antiparallel, allows for the definition of interesting types of *collective magnetism*. The case of *noncollinear magnetic dipole moments* is also of interest in ferrite systems leading to richer magnetic response as described later.

In a *paramagnet*, atomic dipole moments are randomly oriented. The paramagnetic magnetization is zero in the absence of an applied field, H_a, as the "net" sum of random vectors is zero. Only strong coupling can align dipoles in $H_a = 0$ to yield a *spontaneous magnetization*. Such interactions are called *exchange interactions*, as discussed later. Atomic dipolar interactions are too weak to couple at appreciable temperatures, but they do give rise to ordered states at sufficiently low temperatures. For two collinear dipoles pointing perpendicular to $\sim r_{12}$, the potential energy between (parallel/antiparallel) dipoles is expressed by Coulomb's law:

$$E_p = \pm \mu_1 \mu_2 / 4\pi\mu_0 r_{12}^3 \quad E_p = \pm \mu_1 \mu_2 / r_{12}^3 \tag{4.5}$$

4.2.3 Magnetic States

In ionic compounds, electrons localized on specific atomic sites determine atomic dipole moments and discrete magnetic states are calculated using *Hund's rules*. Systems with delocalized electrons states are treated in the band theory of solids. The general angular momentum vector, $\sim\boldsymbol{\Pi}$, defined earlier, has contributions from *orbital*, $\sim\boldsymbol{L}$ and *spin*, $\sim\boldsymbol{S}$, angular momentum both of which quantized in units of ^{-}h. The orbital, spin, and total angular momenta are summed using *Hund's rules* shown in Figure 4.1.

The *ground state multiplet* with m_l and m_s eigenstates are used to calculate components of orbital, $\boldsymbol{L}$; spin, $\boldsymbol{S}$; and the total angular momentum, *J*. The magnitudes of orbital and spin angular momenta are summed over a multielectron shell:

$$L = n\sum i = 1(m_l)_i h \quad S = n\sum i = 12(m_s)_i h \tag{4.6}$$

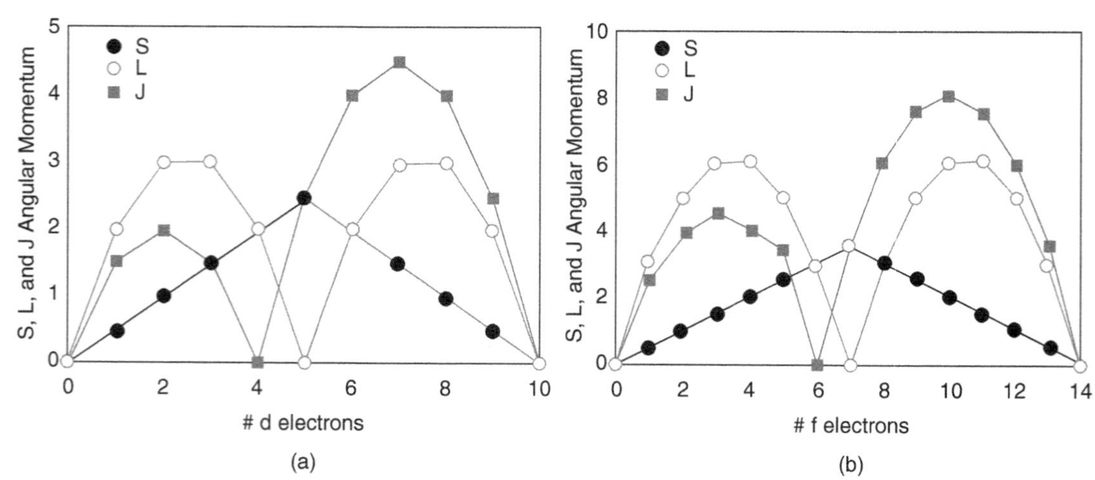

Figure 4.1 J, L, and S quantum numbers for (a) TM^{2+} and (b) RE^{3+} ions.

The projection of the total angular momentum vector, $J = L + S$ along a field direction is subject to quantization conditions requiring J = |L − S| for less than half-filled shells and $J = |L + S|$ for greater than half-filled shells. Occupation of eigenstates of S, L, and J are predicted by Hund's rules:

i) i We fill m_l states (which are $2l + 1$-fold degenerate) in such a way as to first maximize total spin in accordance with the Pauli exclusion principle.
ii) ii We maximize the total orbital angular momentum, consistent with rule 1.
iii) iii The total angular momentum [3] is taken as the sum of spin and orbital angular momentum and with = – for less than half-filled shells and = + for more than half-filled shells (Table 4.1).

A ground state multiplet can be written compactly in the spectroscopic term symbol:

$$^{2S+1}L_J \tag{4.7}$$

where L is the symbol for orbital angular momentum ($L = 0 = S, L = 1 = P, L = 2 = D, L = 3 = F$, etc.) and $2S + 1$ and J are the numerical values. Fe^{3+}, with $S = 5/2$, $L = 0$, and $J = 5/2$, is assigned the term symbol $^6S_{5/2}$. The atomic moment vector is related to the total angular momentum vector by:

$$\mu = \gamma^{-} hJ = -g(J, L, S)\mu_B J \tag{4.8a}$$

where γ is the *gyromagnetic factor* and g = g(J, L, S) is the *Lande g-factor*:

$$g(J, L, S) = 3/2 + 1/2 \frac{[S(S+1) - L(L+1)]}{J(J+1)} \tag{4.8b}$$

The Lande g-factor accounts for angular momentum precession and quantum mechanical rules for projection onto the field axis. For identical ions with angular momentum J, an effective magnetic moment in units of μB is:

$$P_{\text{eff}} = g(J, L, S)[J(J+1)]^{1/2} \tag{4.9}$$

The term *quenched orbital angular momentum* refers to the fact that often the orbital angular moment vector is strongly tied to a crystalline *easy magnetization direction* (EMD). If this is true, we can take $L = 0$ and $J = S$ so that $g = 2$ and $P_{\text{eff}} = 2[S(S + 1)]^{1/2}$. This is true for transition metals and their simple oxides.

4.2.4 Illustration 1: 2-3 Cubic Spinel Ferrites – Dipole Moments and Substituents

Moments in ferrites are well described by *Hund's rules*, with quenched angular momentum from crystal fields. Factors that determine site preference for cations include the following: (i) A-sites are smaller and thus prefer smaller cations, (ii) cations with larger valence are smaller, and (iii) the electronic configuration (bonding) and the crystal-field effects give preference to certain sites for certain cations. Table 4.2 summarizes important fundamental and technical magnetic properties for some scientifically and commercially important ferrite materials [4, 5].

A sample calculation of the dipole moments and magnetization [6] for magnetite, $FeFe_2O_4$, in Box 4.1 to illustrate the important of degree of inversion on the magnetization in ferrites. It is instructive to consider variation of dipole moments

Table 4.1 Transition metal ion spin and dipole movements (L = 0).

d electrons	1	2	3	4	5
Cations	Ti^{3+},V^{4+}	V^{3+}	V^{2+},Cr^{3+}	Cr^{2+},Mn^{3+}	Mn^{2+},Fe^{3+}
S	$\frac{1}{2}$	1	$\frac{3}{2}$	2	$\frac{5}{2}$
$\mu(\mu_B)$	1	2	3	4	5
d electrons	6	7	8	9	10
Cations	Fe^{2+}	Co^{2+}	Ni^{2+}	Cu^{2+}	Cu^{+},Zn^{2+}
S	2	$\frac{3}{2}$	1	$\frac{1}{2}$	0
$\mu(\mu_B)$	4	3	2	1	0

Table 4.2 Physical and technical magnetic properties for some scientifically and commercially important ferrite materials.

Compound	B_s (T)	Néel temp.	Density	Resistivity	Lattice	K_1	λ
	$4\pi M_s$	T_N (K)	$\frac{g}{cm^3}$	$\Omega-cm$	constant	$10^2\frac{erg}{cm^3}$	ppm
	$\mu_0 M_s$		$10^3\frac{kg}{m^3}$	$10^{-2}\Omega-m$	a_0 (nm)	$10^3\frac{J}{m^3}$	
$MnFe_2O_4$	0.50	573	5.00	10^4	0.85	−4	−5
$FeFe_2O_4$	0.60	858	5.24	4×10^{-3}	0.839	−12	40
$CoFe_2O_4$	0.53	793	5.29	10^7	0.838	200	−110
$NiFe_2O_4$	0.34	858	5.38	10^8	0.834	−7	−26
$CuFe_2O_4$	0.17	728	5.42	10^5	0.8445	—	—
$MgFe_2O_4$	0.14	713	4.52	10^7	0.836	−4	−6
$ZnFe_2O_4$	0	—	5.40	10^2	0.84	—	—
$CdFe_2O_4$	0	—	5.76	—	0.873	—	—
$Li_{0.5}Fe_{2.5}O_4$	0.39	943	4.75	—	0.833	—	—
γ-Fe_2O_3	0.52	848	4.89	—	0.834	−5	—

(and thus magnetization) with composition [7] in ferrites with nonmagnetic substituents. Figure 4.2a shows the variation in net dipole moment per formula unit, n_B, for the Mn–Zn, Co–Zn, and Ni–Zn ferrites, $M_{1-x}Zn_xFe_2O_4$ as a function of composition. Substitution of Zn increases the net dipole moment by substituting a zero moment atom on a site that typically has a ferromagnetic cancellation. This does not continue past $x \sim 0.6$ due to a *Yafet–Kittel triangular spin configuration* [8].

Yafet–Kittel structures in $M_{1-x}Zn_xFe_2O_4$ (M = Mn, Ni) ferrites result from weakening ferromagnetic exchange upon substitution of the nonmagnetic Zn^{2+} cations on the tetrahedral A-sites. As such, these substituents destroy exclusively A–B superexchange bonds. The canted spin problem is known to lead to nonsaturating behavior in systems where weakening of the A–B exchange bonds in preference to B–B exchange bonds is caused by substitution of nonmagnetic cations [9]. It is similarly possible to have nonmagnetic substituents on the B-sites of a spinel where both A–B and B–B exchange bonds are broken. It is the weakening of the A–B exchange, which causes deviation from the original Néel two-sublattice model [10] in favor of the Yafet–Kittel [11] noncollinear triangular spin configurations.

Engineering application of ferrites involves the considerations of the intrinsic properties including the induction, Néel temperature [12, 13], magnetic anisotropy, magnetostriction coefficients, and electrical resistivity. Figure 4.2 shows the compositional variation of (b) the magnetocrystalline anisotropy and (c) magnetostriction coefficients in the ternary MnO–ZnO–Fe_2O_3 systems. Optimization of these properties as well as the maximum induction is the basis for achieving attractive properties in Mn–Zn ferrites.

Box 4.1 Magnetization and specific magnetization of magnetite

Let us consider $FeFe_2O_4$; the atomic dipole moments of Fe^{2+} and Fe^{3+} are $\mu_{Fe}{}^{2+} = 4\mu_B$ and $\mu_{Fe}{}^{3+} = 5\mu_B$. In a normal spinel, Fe^{2+} sits on the A-site and its moment is antiparallel to Fe^{3+} on the two B-sites. The net dipole moment per formula unit is

$$-\mu_{Fe}{}^{2+} + 2\mu_{Fe}{}^{3+} = (-4 + 2(5))\mu_B = 6\frac{\mu B}{fu}$$

In the prototype $MgAl_2O_4$ normal spinel, the space group is $Fd\bar{3}m$, #227, with the Pearson symbol *cF*56. There are 56 atoms and eight formula units in a unit cell. The cubic lattice constant is 0.85 nm. The magnetization (net dipole per unit volume) is calculated as:

Box 4.1 (Continued)

$$M = \frac{6\frac{\mu_B}{fu}\times\frac{9.27\times10^{-24}\,\mathrm{A-m^2}}{\mu_B}}{\left(\frac{\left(0.85\times10^{-9}\right)^3}{8}\frac{\mathrm{m^3}}{fu}\right)} = 724\times10^3\,\frac{\mathrm{A}}{\mathrm{m}}\left(=724\,\frac{\mathrm{emu}}{\mathrm{cm^3}}\right)$$

and its induction:

$$B=\mu_0 M = 724\times10^3\,\frac{\mathrm{A}}{\mathrm{m}}\times4\pi\times10^{-7}\,\frac{Wb}{\mathrm{A-m}} = 0.9098\,\frac{Wb}{\mathrm{m^2}} = 0.9098T(=9098G)$$

Molecular weight per formula unit for magnetite is: $3(55.983)+4(16)=231.541\ \mathrm{g/mole}=0.23154\ \mathrm{kg/mole}$ and the density, p, and the specific magnetization, $\sigma = M/P$

$$\rho=\frac{0.23154\frac{\mathrm{kg}}{\mathrm{mole}}}{\left(6.023\times10^{23}\frac{fu}{\mathrm{mole}}\right)\left(\frac{\left(0.85\times10^{-9}\right)^3}{8}\frac{\mathrm{m^3}}{\mathrm{fu}}\right)} = 5008\,\frac{\mathrm{kg}}{\mathrm{m^3}}\left(=5.008\,\frac{\mathrm{g}}{\mathrm{cm^3}}\right)$$

$$\sigma=\frac{724\times10^3\,\frac{\mathrm{A}}{\mathrm{m}}}{5008\,\frac{\mathrm{kg}}{\mathrm{m^3}}} = 144\,\frac{\mathrm{A-m^2}}{\mathrm{kg}}\left(=144\,\frac{\mathrm{emu}}{\mathrm{g}}\right)$$

In an inverse spinel, Fe^{3+} sits on the B-site and has its dipole moment antiparallel to Fe^{2+} and Fe^{3+} dipoles on the B-sites. The net dipole moment per formula unit is:

$$-\mu_{Fe^{3+}}+\mu_{Fe^{2+}}+\mu_{Fe^{3+}} = (-5+(4+5))\mu_B = 4\,\frac{\mu_B}{fu}$$

Repeating the calculations yields $M = 483\times10^3\,\mathrm{A/m}(=483\ \mathrm{emu/cm^3})$.

These are closer to those observed for magnetite, which is not a completely inverse (the degree of inversion is ~ 80%) spinel.

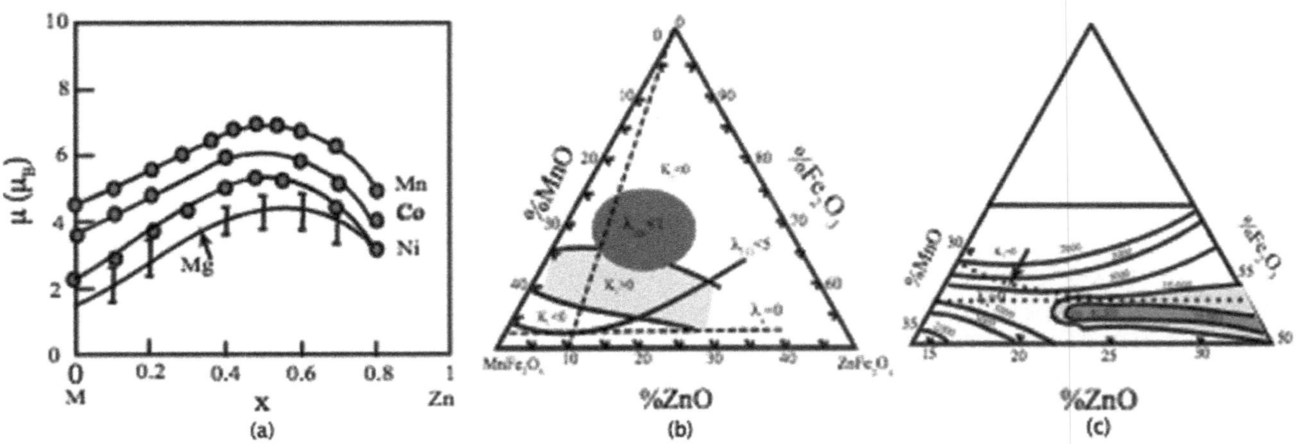

Figure 4.2 Variation of the dipole moment per formula unit with x in $M_{1-x}Zn_xFe_2O_4$ ferrites.

Manganese aluminate spinels, $(Mn_xAl_{1-x})_3O_4$ [14], are examples of spinels in which nonmagnetic Al^{3+} cations preferentially occupy octahedral B-sites [15] and also give rise to Yafet–Kittel structures. Manganese aluminate spinels are also a spinel system with only one magnetic species, Mn (that can assume more than one valence state: 2+, 3+, and 4+). This offers a reduced number of exchange parameters needed to describe exchange couplings and ferromagnetic ordering temperatures if we consider only J_{AB}, J_{BB}, and J_{AA} interaction between the same species. However, there is another superexchange mechanism the *double-exchange mechanism* [16] that can be operative between ions in different oxidation states. Zener's [17] theory predicts electron exchange between two species by a hopping mechanism. The end-member of the manganese aluminate series is the mineral *hausmannite*, Mn_3O_4. Hausmannite [18] has a tetragonally distorted spinel structure caused by the Jahn–Teller effect associated d^4 electronic configuration of Mn^{3+}. Aluminum substation into hausmannites can also give rise to Yafet–Kittel spin structures.

Box 4.2 considers the formula for the net dipole moments for several pseudo-binary ferrite compound systems. These are (i) mixtures of two inverse spinels; (ii) mixtures of an inverse and normal spinel; and (iii) mixtures of two normal spinels.

Cation disorder may influence the degrees of inversion in cubic ferrites. The disorder can result from processing conditions [18] or from radiation-induced damage. Cation disorder also significantly influences the magnetic properties in spinels with nonmagnetic cations[15, 19, 20].

Box 4.2 Net dipole moments in ferrites

Net dipole moment per formula unit for pseudo-binary spinels are of interest. For an inverse spinel that mixes $M_1Fe_2O_4$ and $M_2Fe_2O_4$, the spinel formula is: $(Fe^{3+})[(M_1^{2+})_{1-x}(M_2^{2+})_{1-x}Fe^{3+}]\,O_4$. If M_1^{2+} has a dipole moment, n_1, and M_2^{2+}, n_2, then as Fe^{3+} dipoles on the A- and B-sites cancel, the net dipole per formula unit is:

$$\mu = \left[(1-x)n_1 + xn_2\right]\mu_B$$

We can express the spinel formula for a system, which mixes a normal, $M_1Fe_2O_4$, and an inverse, $M_2Fe_2O_4$, spinel as:

$$\left(Fe_{1-x}^{3+}\left(M_1^{2+}\right)_x\right)\left[Fe_{1+x}^{3+}(M_2^{2+})_{1-x}\right]O_4$$

If M_1^{2+} has a dipole moment, n_1, and M_2^{2+}, n_2, the net dipole per formula unit is:

$$\mu = \left[5(1+x) + n_2(1-x) - 5(1-x) - n_1x\right]\mu_B = \left[(10 - n_2 - n_1)x + n_2\right]\mu_B$$

In the Zn ferrites, $n_2 = 0$ and $\mu = \left[(10 - n_2)x + n_2\right]\mu_B$.

In a spinel system, which mixes a normal spinel $M_1Fe_2O_4$ and another normal spinel $M_2Fe_2O_4$, we can express the spinel formula as: $\left(\left(M_1^{2+}\right)_{1-x}\left(M_1^{2+}\right)_x\right)\left[Fe^{3+}\right]O_4$. If M_1^{2+} has a dipole moment, n_1, and M_2^{2+}, n_2, the net dipole per formula is:

$$\mu = \left[2\times5 - n_1(1-x) - n_2x\right]\mu_B = \left[(10 - n_1) - (n_2 - n_1)x\right]\mu_B$$

The Mn–Zn-ferrites for which $n_1 = 5$, $n_2 = 0$ yield $\mu = [5(1+x)]\mu_B$.

4.2.5 Illustration 2: Titanomagneties – Dipole Moments

We consider the titanomagnetites as an example of a multication pseudo-binary spinel system for which the study of the atomic dipole moments is rich due to charge transfer effects. There are several models for cation distributions in the titanomagnetites based on the arguments in Chapter 1. The most widely cited models have been proposed by Akimoto [21] and Neel [22], respectively. In the Akimoto model, the titanomagnetite formula is written as:

$$(Fe_x^{2+}Fe_{1-x}^{3+})\left[Ti_x^{4+}Fe_{(1-x)}^{3+}Fe_1^{2+}\right]O_4^{2-} \qquad (4.10)$$

Note that this model assumes magnetite to be an *inverse spinel* and that every Ti^{4+} substitutes changes *both* an A-site and a B-site Fe^{3+} cation to Fe^{2+} cations to balance the charge.

This model then has $\eta = 0$ (as defined in Chapter 3) for $x = 0$ and changing continuously to 1 at $x = 1$. Another way of saying this is that the Ti^{4+} additions are partially compensated for by changing the degree of inversion of the spinel and partially by charge compensation on the B-sites. If we consider the dipole moments for Fe^{3+} and Fe^{2+} predicted from their Hund's rule ground states, with quenched orbital angular momentum, this particular formula predicts that in solid solutions, as the amount of Ti^{4+} is increased, the net dipole moment per formula unit changes linearly from 4 μ_B per formula unit for magnetite to 0 μ_B per formula unit for ulvospinel.

The Néel model forces an inverse spinel as long as charge balance can be compensated for on B-sites, i.e. the A-site cations remain Fe^{3+} until there is no possibility of compensating for the extra positive charge of Ti^{4+} on B-sites. The formula for titanomagnetite solid solutions, the cation distribution is written as:

$$\left(Fe_1^{3+}\right)\left[Ti_x^{4+}Fe_{(1-2x)}^{3+}Fe_{1+x}^{2+}\right]O_4^{2-} \tag{4.11}$$

for composition $0 \leq x \leq 0.5$. The Néel model indicates that the tetrahedral sites are exclusively occupied by Fe^{3+} when $0 \leq x \leq 0.5$. This indicates that every Ti^{4+} replaces an Fe^{3+} on the B-site as well as transforming another Fe^{3+} on the B-sites to Fe^{2+}. This can obviously only continue until all of the Fe^{3+} on the B-sites are depleted when $x = 0.5$.

When the composition of the titanomagnetite is in the range $0 \leq x \leq 0.5$, the Néel model cation distribution is expressed:

$$\left(Fe_{2x-1}^{2+}Fe_{2-2x}^{3+}\right)\left[Ti_x^{4+}Fe_1^{2+}\right]O_4^{2-} \tag{4.12}$$

and the Fe^{2+} cations are distributed randomly on the tetrahedral A-sites. The Néel model predicts an abrupt change in slope of the composition dependence of the dipole moment occurring at $x = 0.5$, thus magnetic measurements can be used to distinguish between the two models. The fact that the magnetite phase is more strongly magnetic than the ulvospinel phase suggest that magnetic energies may contribute to the asymmetry in the miscibility gap [23].

Figure 4.3 plots cation distributions predicted by the Néel model and summarizes the predicted dipole moments, μ, and average charge on the tetrahedral, Q_T, and octahedral, Q_O, cations, respectively, as a function of the composition variable, x. Note that for $x \leq 0.5$, the average charge on the tetrahedral and two octahedral sites remains constant at 3 and 5, respectively. The net dipole moment per formula unit (in μ_B) varies linearly with composition with a slope of −6. For $x \leq 0.5$, the average charge on the tetrahedral site, Q_T, varies linearly from 3 to 2 and on the octahedral sites, Q_O, varies linearly from 4 to 6. The net dipole moment per formula unit (in μ_B) varies linearly with composition with a slope of −2. Experimental determination of cation distributions to support magnetization measurements is discussed later in the context of Mössbauer spectroscopy.

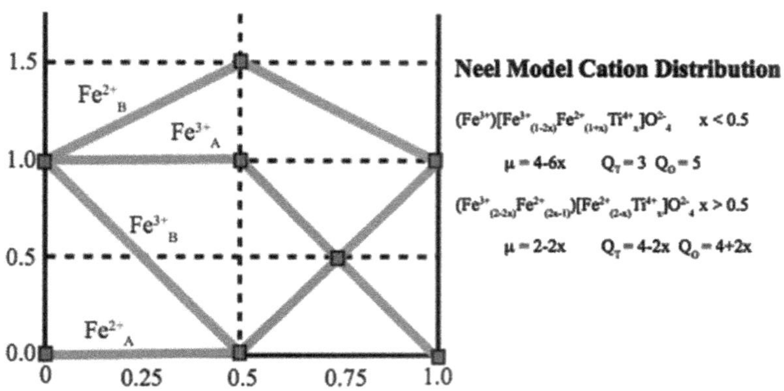

Figure 4.3 Cation distributions predicted by the Néel model and predicted dipole moments, μ, and charge on the tetrahedral, Q_T, and octahedral, Q_O, cations, respectively, as a function of the composition variable, x.

In Ti^{4+}-rich titanomagnetics, precise lattice constant changes with composition were absent in earlier literature. This is presumably due to the problems with achieving precise stoichiometries. The exclusion of data at large values of x [24] reflects the fact that oxygen fugacity must be controlled to keep the titanomagnetites from chemically decomposing. New synthesis of titanomagnetites has required advanced solid-state synthesis techniques and the precise control of oxygen fugacity to yield reliable lattice constants. Single-phase TMX solid solutions of compositions $x = 0.75$–1.0, i.e. TM75 to TM100, were successfully synthesized.

Figure 4.4a shows the X-ray diffraction (XRD) patterns for TM75–95. Indexed to spinel peaks ([1 1 0] peaks of excess Fe left over from the process described by [WSV+11] [LGW+13]) [25] [26] are starred. Lattice constants of these solid solution samples [27] were fit from XRD data employing the method of Nelson and Riley [28]. Lattice constants as a function of composition ($x > 0.75$) were fit to a straight line and the extension of the data (Figure 4.4b) follow published literature values for $x \leq 0.75$.

Transmission electron microscopy (TEM) characterization of TMX solid solutions of compositions $x = 0.75$–1.0 titanomagnetite samples was performed [29] , and lattice constants measured by TEM were shown to agree with those in Figure 4.4. High-resolution TEM (HRTEM) simulation was used to determine atomic site occupancies. HRTEM images were collected on a JEOL-2010 microscope with a double-tilt holder. Several zone axis diffraction patterns with corresponding high-resolution images were acquired from different grains by tilting the specimen. The images were further filtered and processed via Fourier filtering methods by CRISP software. To examine the structure of a single grain, multislice simulations were conducted by using JEMS software developed by Pierre Stadelmann. Starting with XRD-refined results for the structure model in multislice calculations and imputing microscope and specimen parameters 1, stimulated micrographs with varying thickness and defocus were collected. Atom occupancy information was confirmed by comparing processed and stimulated pictures.

Figure 4.5a (adapted from [25]) shows an HRTEM image taken from ulvospinel single solid solution, the center black dots represent Fe cations at tetrahedral sites; (b) shows a software simulated image of the same; and (c) indicates the projected spinel structure. HRTEM observations were made on single-phase ulvospinel samples viewing crystals along the (0 0 1) zone axis. By comparing the simulated and real atom sites, it was concluded that 8a sites are occupied only by Fe^{2+} cations and that 16d sites are occupied by both Fe^{2+} and Ti^{4+} cations at equal concentration predicted probability. O^{2-} anions occupy the 32e sites. These data confirm the preference for cations for the octahedral B-sites in the titanomagnetics as a random solid solution on this sublattice. Figure 4.5d shows an HRTEM images of a TM90 sample, on a (0 0 1) zone axis as seen from the diffraction pattern in (e). The HRTEM image allows an atomic resolution of (22) planes from which their spacing was measured and lattice constant confirmed. The absence of second phases further confirms the homogeneity of the sample. HRTEM pictures of TM85 to 100 compositions further confirmed that the samples were homogenous single solid solutions with spinel structure and Ti^{4+} cations occupying exclusively octahedral B-sites.

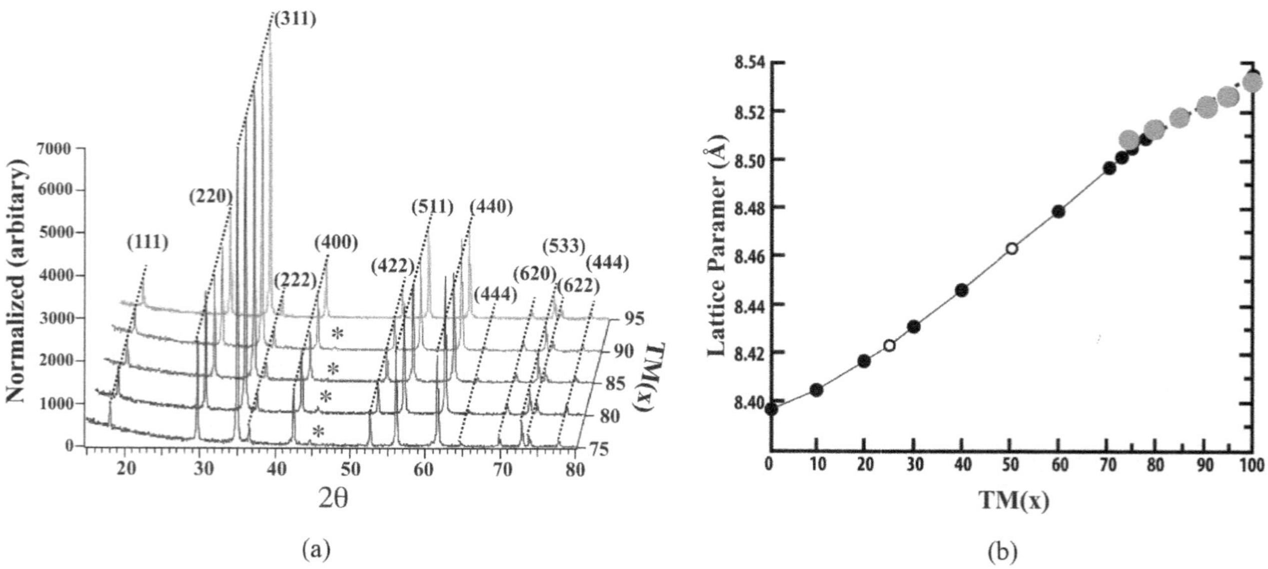

Figure 4.4 (a) XRD patterns for TM75-95 as indexed to spinel peaks; excess Fe (110) peaks are starred. (b) Lattice parameter-based XRD data (gray) as determined from data fitting. Gray shown as extensions to published data (black, white).

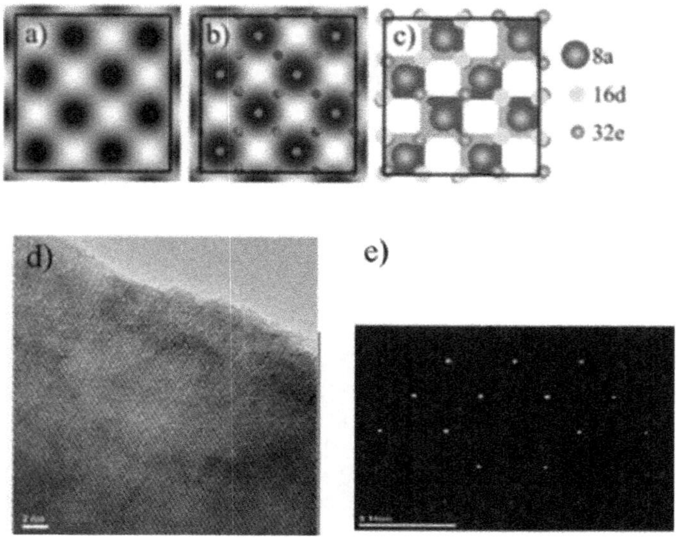

Figure 4.5 Black outline in (a), (b), and (c) marks the (2 × 2) spinel unit cell: (a) an HRTEM image taken from ulvospinel single solid solution and the center black dots stand for Fe atoms at tetrahedral sites; (b) stimulated image; and (c) spinel structure. Adapted from X. Liu, M. Sorescu, J. Wang, W. Schoenthal, Z. Dong, R.V. Ramanujan, D.E. Laughlin and M.E. McHenry, "Structural and magnetic properties of xFe2TiO4(1-x)Fe3O4 (0.75<x<1)", *IEEE Transactions on Magnetics* 50, 6101504, (2014).

4.3 Exchange, Molecular, and Mean Field Theory

To explain spontaneous magnetization, we identify the origin of an internal field that aligns dipoles in the absence of an applied field. In a *ferromagnet*, an *internal field* or equivalent *exchange interaction* aligns atomic dipole moments parallel in the absence of an applied field ($H = 0$). Interactions supporting ferromagnetism are explained by two historical models:

i) *Mean Field Theory*, which considers a nonlocal *Weiss field* that aligns magnetic dipole moments in the absence of an applied field, $H_a = 0$
ii) *Heisenberg Exchange Theory*, which considers interactions through direct or indirect overlap of atomic orbitals responsible for dipole moments

Both theories allow convenient representations of other collective magnetic phenomena: *antiferromagnetism*, *ferrimagnetism*, *helimagnetism*, etc. **An *antiferromagnet* is a material for which dipoles of equal magnitude on adjacent nearest-neighbor atomic sites (or planes) are arranged in an antiparallel fashion in $H_a = 0$.** Antiferromagnets have zero magnetization in $H_a = 0$ because of the vector cancellation of moments. They exhibit temperature-dependent collective magnetism because the dipole moment arrangement is not random but precisely ordered.

A *ferrimagnet* is a material having two (or more) sublattices, where the magnetic dipole moments of *unequal* magnitude on adjacent nearest-neighbor atomic sites (or planes) are arranged in an antiparallel fashion. Ferrimagnets have nonzero magnetization in $H_a = 0$ as dipole moments do not cancel. Ferromagnets, antiferromagnets, and ferromagnets are *collinear magnets*, where the dipole moments are either parallel or antiparallel.

It is also possible to have ordered magnets where the dipole moments are not parallel or antiparallel. The *helimagnet* is an example of a *noncollinear magnets*. Other examples of noncollinear ordered magnetic states include the triangular spin arrangements in some ferrites.

The temperature dependence of magnetization in a system with collective magnetic response is calculated in statistical mechanics, considering the ordering effect of an applied and internal field and the disordering effect of thermal fluctuations. Mean field theory was introduced by Pierre Weiss in 1907. Weiss postulated an *internal magnetic Weiss field*, H_{int}, acting to align atomic dipole moments in the absence of an external applied field, H_a. The *Weiss field* is directly proportional to magnetization:

$$H_{int} = \lambda M \tag{4.13}$$

The constant of proportionality, λ, is the *Weiss molecular field constant.*
The quantum theory of paramagnetism predicts magnetization as a function of H/T in terms of a *Brillouin function*, $B_J(x)$. For $J = S$, it reduces to a hyperbolic tangent function, and in the classical limit, $J = \infty$, the *Langevin function*, $L(x)$:

$$M = M_s(\coth(x) - 1/x) = M_s L(x) \qquad x = \frac{\mu_{atom}\mu_0 H}{k_B T}$$

In the classical limit, with superposition of applied and internal magnetic fields:

$$M = M_s(\coth(x') - 1/x') = M_s L(x') \qquad x' = \frac{\mu_{atom}\mu_0 H}{k_B T}\left[H_a + \lambda M\right] \tag{4.14}$$

for a collection of classical dipole moments. Similarly, $M = N_m < \mu_{atom} >$ and:

$$\frac{M}{N_m \mu_{atom}} = \frac{M}{M_s} = L(\frac{\mu_0 \mu_{atom}}{k_B T}\left[H_a + \lambda M\right]) \tag{4.15}$$

This transcendental equation has solutions with a *spontaneous magnetization* in the absence of an applied field that can be solved for considering M ($H = 0$) the variables:

$$b = \frac{\mu_0 \mu_{atom}}{k_B T}[\lambda M] \quad T_c = \frac{N_m \mu_0 \left(\mu_{atom}\right)^2 \lambda}{3k_B} \tag{4.16}$$

The *Curie temperature,* T_c, has units of temperature and:

$$\frac{b}{T_c} = \frac{3}{T}\left[\frac{M(0)}{M_s}\right] \quad \frac{M(0)}{M_s} = \frac{bT}{3T_c} = L(b) \tag{4.17}$$

Two equations for reduced magnetization (i.e. $M(0)/M_a = bT/3T_c$ and $M(0)/M_a = L(b)$) can be solved graphically, for any choice of T by considering the intersection of the two functions $b/3$ (T/T_c), and $L(b)$. For $T \geq T_c$, the only solution is for $M = 0$, i.e. paramagnetic response. For $T < T_c$, solutions with a nonzero, spontaneous magnetization are determined. From $T = 0$ T_c, the intersection of the two functions $b/3$ (T/T_c) and $L(b)$ allows us to determine the zero-field magnetization, $M(0)$, as a fraction of the spontaneous magnetization at a given T. M $(0, T)/M_a$ decreases monotomically from 1, at 0 K, to 0 at $T = T_c$. At $T = T_c$, a higher order ferromagnetic phase to paramagnetic phase transition occurs. In summary, mean field theory for ferromagnets predicts:

i) For $T < T_c$, a spontaneous magnetization occurs even in the absence of a field.
ii) For $T > T_c$, the misaligning effects of T randomize the direction of atomic dipole moments in the absence of a field, defining the paramagnetic response.

Mean field theory for antiferromagnets considers an A sublattice for which the spin moment is up and a B sublattice for which a spin moment of equal magnitude is down. Internal fields on the A and B sites are:

$$\vec{H}_A^{int} = -\lambda_{BA}\vec{M}_B \quad \vec{H}_B^{int} = -\lambda_{AB}\vec{M}_A \tag{4.18}$$

where M_A and M_B are A and B sublattice magnetizations, and by symmetry, $\lambda_{BA} = \lambda_{BA}$.

4.3.1 Ferrimagnetic Exchange and the Temperature Dependence of Magnetization

For ferrites and other oxides, we distinguish between direct and indirect exchanges. A *superexchange* interaction is an indirect exchange mechanism transmitter though O^{2-} anions. The exchange results from overlap between magnetic d orbitals on the cation sites and the p orbitals of oxygen. This indirect exchange interaction is illustrated in Figure 4.6. The magnitude of superexchange interactions depends on the distance and angles of d–p–d bonds in the crystal as determined by crystallography.

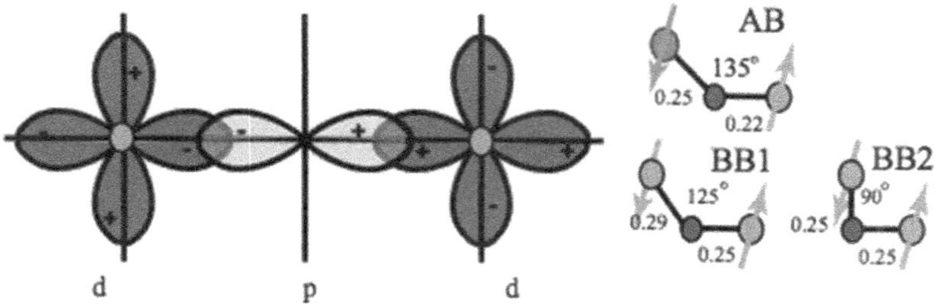

Figure 4.6 Superexchange interaction of magnetic cation d orbitals mediated by an O p orbital and the geometry of the superexchange interactions in the spinel structure.

Figure 4.6 also illustrates the geometry of the nearest-neighbor superexchange interactions in the spinel structure. The strongest superexchange interactions occur for bond angles approaching 180° and shorter bond distances. Superexchange is proportional to the overlap between the p and d orbitals. While it is possible to draw superexchange bonds with cations at large distance from O^{2-} anions, as overlap decays exponentially with separation, these interactions are much weaker than the interactions illustrated in Figure 4.6. This implies that as there is not a direct hop between A-site cations through a single O^{2-} anion, A–A exchange is very weak. Only superexchange interactions with bond distances ≤ 0.35 nm are considered.

The phenomenon of ferrimagnetism as described using Néel's two-sublattice theory extend the theory of antiferromagnetism to internal Weiss fields to the A and B sublattices, having different dipole moments. For ferrites, sublattice magnetizations are associated with cation moments in octahedral and tetrahedral sites. Originally, Néel chose to consider only the AOB interactions (J_{AB}). This model fails to describe properties of mixed ferrite structures. In Néel's two-sublattice theory, the simple ferromagnetic mean field model can be complicated in several ways:

i) In the inverse spinel structure, the octahedral B-sites are occupied by both divalent and trivalent ions, which may have different moments.
ii) The moments may not be collinear.
iii) Disorder may lead to different degrees of inversion for a given composition.

These three superexchange interactions needed to describe spinel ferrites complicates analysis of M(T). Néel's mean field theory (1948) was extended to predict six types of temperature dependences of the magnetization, as illustrated in Figure 4.7a. The more complicated cases consider splitting of the B sublattice into B′ and B″ sublattices, respectively. Here A–A, B′–B′, B″- B″, A–B′, A–B″ and B′–B″ interactions lead to the possibility of six T-dependences of M_s(T) (associated with different T-dependences of the mean fields). The sublattice magnetizations, M_A and M_B, associated with moments of ions in the octahedral and tetrahedral sites, give a net saturation magnetization: $M_s = |M_A - M_B|$. The six possibilities are denoted as M, P, Q, and R (exhibiting a single ferromagnetic Néel temperature) and N and V (exhibiting a compensation point where the sign of the net magnetization changes). Compensation temperatures are observed in *garnet* magnetic oxides. The P- and N-type magnetizations are the most commonly observed. The temperature dependences are decomposed into the magnetization of the A and B sublattices in Figure 4.7b.

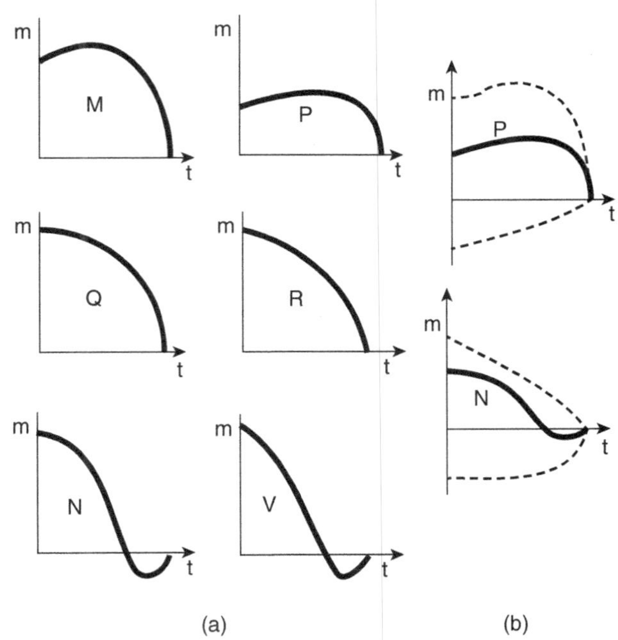

Figure 4.7 (a) Reduces magnetizations, $m(T)$, of the Néel mean field theory for cubic ferrites and (b) decomposition of P- and N-type magnetizations into $m(T)$'s of the A and B sublattices.

A Yafet–Kittel mean field theory model considers that for the B-site cation coordination, the four B sublattices are pairwise degenerate, and thus, only two B sublattices are needed, i.e. B′ and B″. The mean field theory is expressed as:

$$\vec{H}'_A = n(\alpha_1 \vec{M}'_A + \alpha_2 \vec{M}''_A - \vec{M}'_B - \vec{M}''_B) \tag{4.19a}$$

$$\vec{H}''_A = n(\alpha_2 \vec{M}'_A + \alpha_1 \vec{M}''_A - \vec{M}'_B - \vec{M}''_B) \tag{4.19b}$$

$$\vec{H}'_B = n(-\vec{M}'_A - \vec{M}''_A + \gamma_1 \vec{M}'_B - \gamma_2 \vec{M}''_B) \tag{4.19c}$$

$$\vec{H}''_B = n(-\vec{M}'_A - \vec{M}''_A - \gamma_1 \vec{M}'_B - \gamma_2 \vec{M}''_B) \tag{4.19d}$$

where n sets the scale of the systems and the mean field constants are $\alpha_1, \alpha_2, \gamma_1, \gamma_2 < 0$; $\alpha_1 + \alpha_2 = 2\alpha$; and $\gamma_1 + \gamma_2 = 2\beta$.

The degree of inversion is important in determining the net dipole moment per formula unit. The magnitudes of the J_{AA}, J_{BB}, and J_{AB} superexchange interactions depend on cation–cation distances and bridging oxygen bond angles. As A-sites are not directly bridged by a single O^{2-} anion, these interactions are particularly weak. In *ferrite nanoparticles*, the mean field theory is further complicated by finite size effects where the relative number of *A–A*, *A–B*, and *B–B* exchange bonds is different on different crystallographic surfaces [30].

The *Yafet–Kittel theory* postulated noncollinear magnetization vectors M_A, M'_B, and M''_B, as shown in Figure 4.8a, leading to the so-called triangular spin structures. *Hausmanite*, Mn_3O_4, is an example of a Yafet–Kittel spinel. The anion coordination polyhedra show triangular nets of B cations with (1 1 1) plane normals passing through the anion and the A-site cation. For triangular Yafet–Kittel spinel structures, the magnetization may cant from the [111] directions with projections in (111) planes.

4.3.2 Angularly Dependent Magnetic Properties: Crystal Fields and Magnetic Anisotropy

4.3.2.1 Angularly Dependent Magnetic Properties: Crystal Fields

We now consider anisotropy in the charge distributions of atoms or ions. We view closed-shell atoms as spherical with spherical charge distributions. For atoms with open shells, the Hund's rules occupation of angular momentum states leads to anisotropic charge distributions. An illustrative example is rare earths [31] shown in Figure 4.9, which shows contours of constant charge density for 4f electrons in tripositive (RE^{3+}) ions in the rare earth series. Depending on the details of the filling of the 4f states, the resulting charge density can be oblate, prolate, or nearly spherical (i.e. for Gd^{3+}).

For a material to exhibit *magnetic anisotropy*, both the orbital angular momentum (L_z) and the *crystalline electric field or crystal field* must have a less-than-spherical symmetry. The crystal field causes orbital in a crystal lattice. This is also the criterion for *quenched orbital angular momentum*. As the value of the orbital angular momentum is small, when quenched, this does not give a complete explanation of magnetic anisotropy. We must further consider the spin angular momentum and its coupling with the orbital angular momentum. This results from *spin–orbit interaction*. Spin–orbit interactions give rise to energy; the *spin–orbit energy* is expressed as:

$$E_{so} = -\lambda L \cdot S \tag{4.20}$$

where λ is a phenomenological parameter called the *spin–orbit coupling constant*. For $\lambda > 0$, the lowest spin–orbit energy occurs for spins aligned parallel to the orbits, and for $\lambda < 0$, for spins aligned antiparallel to the orbits. We can thus see that

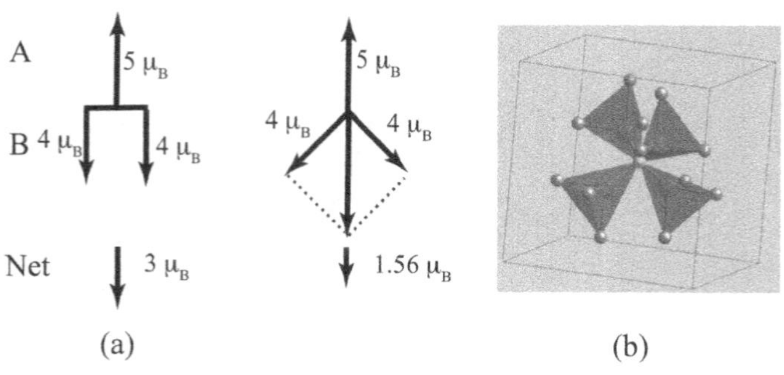

Figure 4.8 (a) Hypothetical collinear and observed noncollinear spin arrangements on Mn^{2+} and Mn^{3+} in the Mn_3O_4 spinel tetrahedra; (b) connectivity of O^{2-} anion coordination polyhedral with reference to the spinel tetrahedra unit cell.

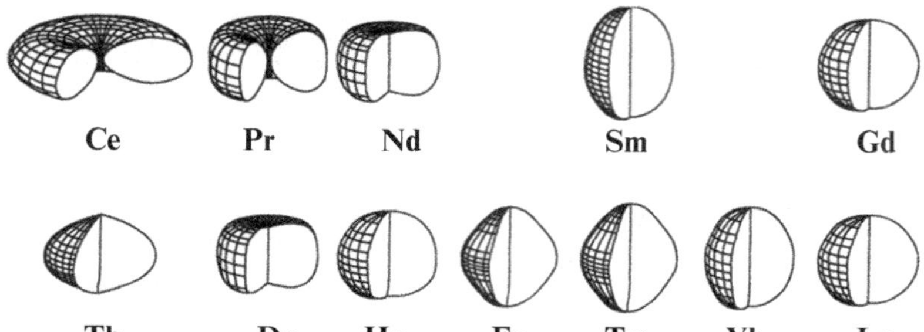

Figure 4.9 Tripositive (RE^{3+}) ion 4f electron charge distributions of the rare earth series. Adapted from: Rare Earth Permanent Magnets Edited by J. M. D. Coey Oxford Science Publications.

the spin–orbit interactions, even though quenched orbital angular moment is small, can lead to an energy lowering for spin dipole moments aligned with the orbitals and consequently in specific crystallographic directions through crystal-field effects. This lower energy for dipoles in particular crystallographic directions in the origin of *magnetocrystalline anisotropy.*

Crystalline electric fields [32, 33] and their symmetries are also important in determining the optical and magnetic properties of many systems, most importantly in ionic solids with rare earth or transition metal cations (including ferrites). We will now use a simple illustration of crystal field theory to show how the symmetry of a potential influences the eigenstates of a quantum mechanical system.

If we consider an ion of interest in an ionic solid, much of the energy of the electrons(s) localized on the ion depends on nearest-neighbor *Coulomb interactions*. Coulomb interactions between electrons on the site of interest and charges external to the ion are described by an *electrostatic potential*, $V(\mathrm{r})$. For transition metals, the outermost d electrons are most strongly perturbed by the crystalline potential. For rare earths the crystalline potential influences eigenstates of the f electrons but less so due to shielding by the outermost s and p electrons. In some cases, it is a good approximation to treat charge associated with neighboring ions as point charges and this is also convenient for illustrating crystal-field effects. The electrostatic potential for point charges obeys the *Laplace equation*:

$$\nabla^2 V(r,\theta,\varphi)=0 \tag{4.21a}$$

and therefore $V(r,\theta,\varphi)$ is conveniently expanded in spherical harmonics:

$$V(r,\theta,\varphi)=A_{lm}r^{1}Y_{lm}(\theta,\varphi) \tag{4.21b}$$

In the simple point charge approximation for the crystalline electric field, it is common to consider only nearest-neighbor interactions. The $l=0$ term is typically ignored as it contributes only to the zero state of energy, and many higher order terms disappear due to the symmetry of the crystal field. The precise expansion of the crystal field depends on the choice of the axis of quantization (i.e. the orientation of the z-component of the total angular momentum vector with respect to the symmetry axes of the point charge distribution). A simplification of the expansion of the potential arises if instead it is expressed in *tesseral harmonics*, $Z_{n\alpha}$, as:

$$V(r,\theta,\phi)=\sum_{l}\sum_{m}\gamma_{lm}r^{l}Z_{n\alpha}(\theta,\phi) \tag{4.21c}$$

where r is the radial coordinate and the *gamma coefficients*, γ_{lm}, are given by:

$$\gamma_{lm}=\sum_{j}\frac{4\pi}{2n+1}q_j\frac{Z_{lm}\left(\theta_j,\phi_j\right)}{Rj^{l+1}} \tag{4.21d}$$

where R_j is the position of the jth atom and q_j is its charge. This results in an expansion in terms of sines and cosines. Tesseral harmonics are simply constructed from spherical harmonics ($Z_{10} = Y_n^0$, $Z_{nm}^c = Re[Y_{lm}]$, and $Z_{nm}^s = Im[Y_{lm}]$, where s and c denote sine and cosine expansions, respectively. A further simplification for point groups with inversion symmetry has sine terms disappear and only even order ls appear in the expansion of the harmonics.

A *crystal-field Hamiltonian* is constructed by multiplying the potential by the charge of the ion. A perturbing potential partially lifts the (2l+1) degeneracy of the ground state multiplet of the free atom or ion. The symmetry-derived splittings that arise is similar to those that are the origin of *Jahn–Teller effects*. Figure 4.10 illustrates the orientation of the d orbitals with respect to ions in a cubic crystal field. Symmetry dictates that the equivalent the d_{xy}, d_{xz}, and d_{yz} orbitals have the same energy but different from the $d_{x^2-y^2}$ and d_{x^2} pair of orbitals. Note the energies are different because the first three orbitals are oriented toward the nearest-neighbor ion sites and the second toward the second nearest-neighbor sites. Spins aligned with these orbits would necessarily have different energies when oriented in [1 1 1] and [1 0 0] directions, respectively. The notation used for labeling our states is a group theoretical notation, designating *irreplaceable representations* of the particular point group of interest with the conventions in Table 4.3.

Subscripts g or u which stand for the German words *gerade* and *ungerade*, denoting even and odd symmetries, respectively, with respect to inversion. Note that it is symmetry that dictates the splitting of the originally fivefold degenerate atomic d states into the three- and two fold degenerate T_2 and E irreducible representations of the cubic group. The arrangement of orbitals with respect to the point charges determine the ordering of these two states. For negative ions (point charges) surrounding an ion with d orbitals at the origin, the octahedral environment shown in Figure 4.11a yields the splitting $E > T_2$, and in the cubic environment (Figure 4.11b), the splitting $T_2 > E$.

4.3.3 Crystal-Field Hamiltonians

In a quantum mechanical framework, the energy of electronic states for atoms experiencing a crystal field can be summarized in terms of a *crystal-field Hamiltonian*:

$$H = H_0 - eV_{el} + \lambda L \cdot S \tag{4.22}$$

where H_0 is the Hamiltonian giving rise to electronic energy levels in the absence of the crystal field. V_{el} is the electrostatic Coulomb potential due to ions (viewed as point charges) surrounding the ion for which the crystal-field energy levels are

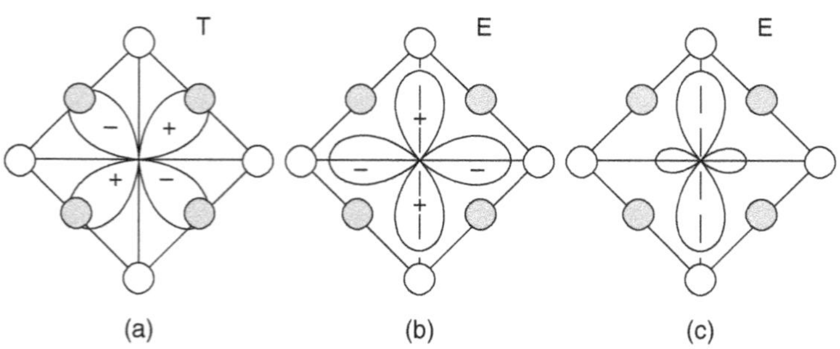

Figure 4.10 Interaction of the magnetic ion d orbitals with the charge on neighboring ions in a cubic environment: (a) d_{xy}, d_{zy}, and d_{yz}; (b) $d_{x^2-y^2}$; and (c) the d_{x^2} orbitals.

Table 4.3 Labeling of irreducible representations (Schoenflies).

A	1-d (one-fold degenerate) representation
B	1-d (one-fold degenerate) representation
E	2-d (twofold degenerate) representation
T_1	3-d (threefold degenerate) representation
T_2	3-d (threefold degenerate) representation
G	4-d (fourfold degenerate) representation
H	5-d (fivefold degenerate) representation

to be calculated and $\lambda \boldsymbol{L}\cdot\boldsymbol{S}$ is the aforementioned *spin–orbit coupling*. The electrostatic potential of an ion of interest and the resulting total electrostatic energy is expanded:

$$V(\mathrm{r},\theta,\phi)=\sum_j \frac{q_j}{\left|\left(\vec{R}_j-\vec{\boldsymbol{r}}\right)\right|} \quad V(\mathrm{r},\theta,\phi)=\sum_i\sum_j \frac{q_i q_j}{\left|\left(\vec{R}_j-\vec{\boldsymbol{r}}\right)\right|} \tag{4.23a}$$

where $\boldsymbol{r}$ is the position vector with respect to the ion of interest, R_j is the position of the jth ion surrounding the ion of interest, and q_j is the charge of the jth surrounding ion. The total energy involves summing over all ions in the structure $\left(\sum_i\right)$, where q_i is the charge of the ith reference ion. The solution to problems with such electrostatic potentials is usually aided by expanding in terms of powers of $\boldsymbol{r}$ and *spherical harmonics* or *tesseral harmonics*, as the potential must be a solution to the *Laplace's equation*.

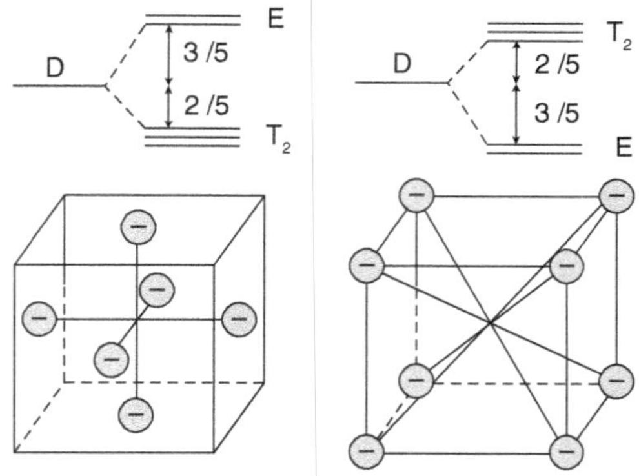

Figure 4.11 Crystal-field splitting of d orbitals due to the charge on neighboring ions in (a) an octahedral and (b) a cubic environment.

The expansion in spherical harmonics is of the form:

$$V(r,\theta,\phi)=\sum_{l,m} C r^l Y_l^m(\theta,\phi) \tag{4.23b}$$

where C is a constant and $Y_l^m(\theta,\varphi)$ are spherical harmonics defined as follows:

$$Y_l^m(\theta,\phi)=(-1)^{\frac{m+|m|}{2}}\left[\frac{(2l+1)(l-|m|)!)}{2(l+|m|)!}\right]^{\frac{1}{2}}\frac{P_l^{|m|}(\cos\theta)\exp(im\phi)}{(2\pi)^{\frac{1}{2}}} \tag{4.23c}$$

$$P_l^{|m|}(x)=(1-x^2)^{\frac{|m|}{2}}\frac{d^{|m|}P_l^0(x)}{dx^{|m|}} \quad P_l^0(xa)=\frac{1}{2^l l!}\frac{d^l\left(x^2-1\right)^l}{dx^l} \tag{4.23d}$$

Examples of the first few spherical harmonics of interest are:

$$Y_2^0=\frac{1}{4}\left(\frac{5}{\pi}\right)^{\frac{1}{2}}(3\cos^2\theta-1) \quad Y_2^{\pm 2}=\frac{1}{4}\left(\frac{15}{2\pi}\right)^{\frac{1}{2}}\sin^2\theta\exp(\pm 2i\phi)$$

and

$$Y_4^0=\frac{3}{16}\left(\frac{1}{\pi}\right)^{\frac{1}{2}}(35\cos^4\theta-30\cos^2\theta+3) \quad Y_4^{\pm 2}=\frac{3}{8}\left(\frac{5}{2\pi}\right)^{\frac{1}{2}}\sin^2\theta(7\cos^2\theta-1)\exp(\pm 2i\phi)$$

$$Y_4^{\pm 3}=\frac{\mp 3}{8}\left(\frac{35}{\pi}\right)^{\frac{1}{2}}\sin^3\theta\cos\theta\exp(\pm 3i\phi) \quad Y_4^{\pm 4}=\frac{3}{16}\left(\frac{35}{2\pi}\right)^{\frac{1}{2}}\sin^4\theta\exp(\pm 4i\phi)$$

The expansion of the electrostatic potential in tesseral harmonics is of the form:

$$V(r,\theta,\phi)=\sum_{l,m}\gamma_l^m r^l Z_l^m(\theta,\phi) \quad \gamma_l^m=\sum_i \frac{4\pi q_i}{2l+1}\frac{Z_l^m\left(\theta_j,\phi_j\right)}{R^{l+1}} \tag{4.24a}$$

where the terms $Z_l^m(\theta_j,\varphi_j)$ represent the tesseral harmonics evaluated at the positions of the surrounding ions. Tesseral harmonics offer simplifications of the crystal-field problem in that they represent real functions as opposed to the imaginary functions ($\exp(im\varphi)$) in the spherical harmonics. The tesseral harmonics are defined:

$$Z_{l0} = Y_l^0 \tag{4.25a}$$

$$Z_{l0}^c = \frac{1}{\sqrt{2}}(Y_l^{-m} + (-1)^m Y_l^{-m}) \tag{4.25b}$$

$$Z_{l0}^s = \frac{i}{\sqrt{2}}(Y_l^{-m} - (-1)^m Y_l^{-m}) \tag{4.25c}$$

where $m > 0$ and c refers to cosine and s to sin expansions, respectively. Comparing with the definition of spherical harmonics it can be seen that:

$$Z_{l0} = Y_l^0 \tag{4.25d}$$

$$Z_{lm}^c = \left[\frac{(2l+1)(l-m)!)}{2(l+m)!}\right]^{\frac{1}{2}} P_l^m(\cos\theta)\frac{\cos(m\phi)}{\sqrt{\pi}} \tag{4.25e}$$

$$Z_{lm}^s = \left[\frac{(2l+1)(l-m)!)}{2(l+m)!}\right]^{\frac{1}{2}} P_l^m(\cos\theta)\frac{\sin(m\phi)}{\sqrt{\pi}} \tag{4.25f}$$

As an illustration, we consider an example of an expansion of a crystal-field potential ions coordinated in a simple cubic lattice in Box 4.3. We note the similarity in this expansion and that of the magnetocrystalline anisotropy discussed later.

Box 4.3 Crystal-field calculation for a simple cubic lattice

Consider ions in a simple cubic lattice with nearest neighbors at ($\pm a$,0,0), (0,$\pm a$,0) and (0,0,$\pm a$). We use the following useful expansion:

$$|(r - R_j)|^2 = \left[(x \pm a)^2 + y^2 + z^2\right] = (x^2 + y^2 + z^2) \pm 2ax + a^2 = r^2 \pm 2ax + a^2$$

The crystal-field expansions for x-, y-, and z-pairs, respectively, are:

$$V_x = q\left[\left(r^2 + 2ax + a^2\right)^{\frac{1}{2}} + \left(r^2 - 2ax + a^2\right)^{\frac{1}{2}}\right]$$

$$V_y = q\left[\left(r^2 + 2ay + a^2\right)^{\frac{1}{2}} + \left(r^2 - 2ay + a^2\right)^{\frac{1}{2}}\right]$$

$$V_z = q\left[\left(r^2 + 2az + a^2\right)^{\frac{1}{2}} + \left(r^2 - 2az + a^2\right)^{\frac{1}{2}}\right]$$

where q is the charge of the ion at each of the cubic sites. We now use the following substitutions: $A = (r^2 + a^2)$, $B = r^2 / a^2$, $X = 2ax / A$, $Y = 2ay / A$, and $Z = 2az / A$ to write the total potential:

$$V = \frac{q}{\sqrt{A}}\left[(1+X)^{\frac{-1}{2}} + (1+Y)^{\frac{-1}{2}} + (1+Z)^{\frac{-1}{2}} + (1-X)^{\frac{-1}{2}} + (1-Y)^{\frac{-1}{2}} + (1-Z)^{\frac{-1}{2}}\right]$$

(Continued)

Box 4.3 (Continued)

Now, use the Taylor series expansion of the type:

$$(1+X)^{\frac{-1}{2}}+(1-X)^{\frac{-1}{2}}=2+\frac{3X^2}{4}+\frac{35X^4}{64}+\frac{693X^6}{64\cdot 24}+\dots$$

to yield:

$$V=\frac{q}{\sqrt{A}}\left[6+\frac{3\left(X^2+Y^2+Z^2\right)}{4}+\frac{35\left(X^4+Y^4+Z^4\right)}{64}+\frac{693\left(X^6+Y^6+Z^6\right)}{64\cdot 24}+\dots\right]$$

or back-substituting x, y, and z:

$$V=\frac{6q}{a}+\frac{35q}{4a^5}\left[\left(x^4+y^4+z^4\right)-\frac{3r^4}{5}\right]-\frac{21q}{2a^7}\left[\left(x^6+y^6+z^6\right)\right.$$
$$\left.+\frac{15}{4}\left(x^2y^4+x^2y^4+y^2x^4+y^2z^4+z^2x^4+z^2y^4\right)+\frac{15r^6}{14}\right]+\dots$$

4.3.4 Method of Operator Equivalents

The crystal-field theory discussed earlier is developed further using the *method of operator equivalents* in Hutchings. In this section, we show its application schematically. The reader is referred to Hutchings for further details. Hutchings presents a point charge model for Coulombic interactions between an ion with magnetic states and its neighboring ions in a crystalline lattice. Energy levels for electrons on a given ion (magnetic ion) are calculated based on the nearest-neighbor coordination by other ions in the crystal. The energy levels of interest are (i) d electrons for transition metals and (ii) f electrons for rare earths. Treated are atomic-like energy states which are split by the symmetry of the crystal field. The model is presented in the language of quantum mechanical operator equivalents.

Consider another example of a Coulombic point charge potential for another lattice of interest. Box 4.4 summarizes the Coulomb potential for a C_s Cl structure.

Box 4.4 Crystal-field calculation for a C_s Cl structure

Consider ions in a simple cubic lattice with nearest neighbors at: $(\pm a/2, \pm a/2, \pm a/2)$. Define the quantity $d=\sqrt{3}a$ and use the expansion of Box 4.3 to determine that:

$$V=\frac{8q}{d}-\frac{70q}{9d^5}\left[\left(x^4+y^4+z^4\right)-\frac{3r^4}{5}\right]-\frac{224q}{9d^7}\left[\left(x^6+y^6+z^6\right)\right.$$
$$\left.+\frac{15}{4}\left(x^2y^4+x^2y^4+y^2x^4+y^2z^4+z^2x^4+z^2y^4\right)+\frac{15r^6}{14}\right]+\dots$$

Comparing Boxes 4.3 and 4.4 we see a pattern that in general for any cubic material of the Coulomb energy can be expanded to 6th order terms as:

$$E_{\text{coul}}=C_4\left[\left(x^4+y^4+z^4\right)-\frac{3r^4}{5}\right]+D_6\left[\left(x^6+y^6+z^6\right)\right.$$
$$\left.+\frac{15}{4}\left(x^2y^4+x^2y^4+x^2y^4+z^2x^4+z^2x^4\right)+\frac{15r^6}{14}\right] \tag{4.26}$$

For common cubic coordinations the coefficients C_4 and D_6 are shown in Table 4.4.

Up to this point, the discussion and that of the last section are similar. We now continue by considering another useful expansion of the kernel of the Coulomb potential, $\frac{1}{|R-r|}$ in terms of *Legendre polynomials*:

$$\frac{1}{|R-r|}=\sum_{l=0}^{\infty}\frac{r^{l}}{R^{l+1}}P_{l}^{0}(\cos\,\omega) \tag{4.27a}$$

where $P_l^0(\cos\,\omega)$ is the Legendre polynomial:

$$P_l^0(\cos\,\omega)=\frac{4\pi}{2l+1}\sum_{m=-l}^{m=l}Y_l^{-m}\left(\theta_j,\phi_j\right)Y_l^{m}\left(\theta_i,\phi_i\right) \tag{4.27b}$$

where Y_l^m is a *spherical harmonic*, which are complex functions. It is often more convenient to express the Legendre polynomials and expansion in tesseral harmonics:

$$P_l^0(\cos\,\omega)=\frac{4\pi}{2l+1}\sum_{\alpha}Z_l^{\alpha}(\theta_j,\phi_j)Z_l^{\alpha}(\theta_i,\phi_i) \tag{4.27c}$$

where Z_l^{α} are the tesseral harmonics. Within this framework, we can again consider the general expansion for any cubic material to sixth-order terms as:

$$E_{\text{coul}}=D_4'\left[Z_{40}(\theta,\phi)+\sqrt{\frac{5}{7}}Z_{44}^{c}(\theta,\phi)\right]+D_6'\left[Z_{40}(\theta,\phi)-\sqrt{7}Z_{64}^{c}(\theta,\phi)\right] \tag{4.27d}$$

For common cubic coordinations, the coefficients D_4' and D_6' are shown in Table 4.4.

Table 4.4 Crystal field coefficients for common cubic coordinations: D_4' and D_6' are for expansion in tesseral harmonics

	D'_4	D'_6
8-fold	$\frac{-56\sqrt{\pi}qq'r^4}{27d^5}$	$\frac{32\sqrt{\pi}qq'r^6}{9\sqrt{13}d^7}$
6-fold	$\frac{7\sqrt{\pi}qq'r^4}{3d^5}$	$\frac{3\sqrt{\pi}qq'r^6}{2\sqrt{13}d^7}$
4-fold	$\frac{-28\sqrt{\pi}qq'r^4}{27d^5}$	$\frac{16\sqrt{\pi}qq'r^6}{9\sqrt{13}d^7}$

At this point, we consider the *method of operator equivalents*. The premise of this model is that spherical harmonics (tesseral harmonics) are the eigenfunctions of angular momentum operators. They can, therefore, be converted into operator equivalents written in terms of components of the total angular momentum vectors J_x, J_y, J_z, and J, and the so-called raising and lowering operators:

$$J_+=J_x+iJ_y\;J_+=J_x-iJ_y \tag{4.28a}$$

The Coulomb energy expansions described earlier can then be turned into operator equivalents substituting J_x for x, J_y for y, J_z for z, etc. This involves the following definitions:

$$Z_l^m=(\text{constant})\theta_l\langle r^l\rangle O_l^m \tag{4.28b}$$

$$H_{\text{CEF}}=eV(r,\theta,\phi)=\sum_{l,m}B_l^m O_l^m \tag{4.28c}$$

$$B_l^m = -e(constant)\theta_l\langle r^l\rangle\gamma_l^m = \theta_l\langle r^l\rangle A_l^m \quad (4.28d)$$

where the quantity $\langle r^1\rangle$ are the expectation values of powers of the radius of orbits in question (i.e. r^2, r^4, r^6, etc.) as calculated from quantum mechanical calculations. The crystal-field Hamiltonian is obtained by compiling the values of θ_l (where θ_l is the *Steven's coefficient* for a given J) that are tabulated in Hutchings, and $O_1{}^m$ (where $O_1{}^m$ is the operator equivalent of $Z_1{}^m$) also given in Hutchings. $\gamma_1{}^m$ (defined as the constant in the potential is only dependent on the configuration of neighboring ions) can easily be evaluated. In many cases, $\gamma_1{}^m$ is zero due to symmetry.

The crystal-field Hamiltonian expansion depends on the choice of the quantization axis. The J_z-axis is typically chosen to be the highest symmetry axis. The matrices are diagonalized to give crystal-field splitting of ground states of different magnetic ions. The matrix elements of the operator equivalents give the crystal-field Hamiltonians. The anisotropy constants to calculate magnetic anisotropy can be determined from crystal field.

4.3.5 Spin–Orbit Interactions

We can now further amplify the origin of energy level splittings, which arise from *spin–orbit interactions*. If we consider an electron orbiting a nucleus (in the electron reference from this will appear as the nucleus orbiting the electron [at rest]). Invoking the Biot–Savart form of Amperes Law, we can calculate $\vec{B}$ due to this relative motion:

$$\vec{B} = \frac{\mu_0}{4\pi}\frac{\vec{J}\times\vec{r}}{r^3} = \frac{-Ze\mu_0}{4\pi}\frac{\vec{v}\times\vec{r}}{r^3} \quad (4.29a)$$

where Z is the atomic number, J the current density, r the position, and v is the relative velocity. The electric field is expressed as:

$$\vec{E} = \frac{-Ze}{4\pi\epsilon_0}\frac{\vec{r}}{r^3} = \frac{1}{e}\frac{\partial V}{\partial r}\frac{\vec{r}}{r} \quad (4.29b)$$

Therefore:

$$\vec{B} = -\mu_0\epsilon_0(\vec{v}\times\vec{E}) = \frac{-1}{c^2}(\vec{v}\times\vec{E}) = \frac{-1}{ec^2 r}(\vec{v}\times\vec{r})\frac{\partial V}{\partial r} \quad (4.29c)$$

relating induction to orbital angular momentum using $L = r \times mv$:

$$\vec{B} = \frac{-1}{emc^2}\left(\frac{1}{r}\right)\frac{\partial V}{\partial r}\vec{L} \quad (4.29d)$$

Finally, the origin of the spin–orbit interaction lies in the potential energy of the spin dipole moment interacting with this field:

$$U_p = -\overrightarrow{\mu_{spin}}\cdot\vec{B} = \frac{-g\mu_B}{emc^2\hbar}\left(\frac{1}{r}\right)\frac{\partial V}{\partial r}\vec{L}\cdot\vec{S} \quad (4.29e)$$

finally incorporating a relativistic correction of $\frac{1}{2}$:

$$\Delta E_{so} = \frac{1}{2mc^2}\left(\frac{1}{r}\right)\frac{\partial V}{\partial r}\vec{L}\cdot\vec{S} \quad (4.29f)$$

4.3.6 Mössbauer Spectroscopy

Mössbauer spectroscopy [34] is a powerful tool for studying local structure and magnetic states in solids. The *Mössbauer effect* is recoil-free γ-ray resonant absorption. Mössbauer spectroscopy probes the local environments and magnetic structure in solids and gives information as per the local symmetries and chemical environments of atoms and directly probes the crystalline electric field. It is also an important tool in probing local symmetries in nanocrystalline and amorphous materials.

The Mössbauer effect involves emission of γ-rays with the decay of an atomic nucleus from an excited state to its ground state. In a crystal, the entire lattice takes up the recoil energy required for momentum conservation. The large mass of the lattice results in the recoil energy ($p^2/2m$) being negligible. Thus, the γ-ray energy is the difference in energy between the excited and ground states of the emitting nucleus. An emitted γ-ray can be absorbed, to pump a nucleus into an excited state; after a mean lifetime τ, this excited nucleus will fall back into the ground state, reemitting the γ-ray isotropically. This γ-ray resonance fluorescence is used as a spectroscopic tool by modulating the emitted γ-ray energy by a Doppler shift, associated with the relative velocity of the γ-ray source and the emitting nuclei. The Doppler energy shift is given by $E = \frac{v}{c} E_\gamma$, where v is the source velocity, c is the speed of light, and E_3 is the γ-ray energy. The most common Mössbauer experiment involves a radioactive source containing the Mössbauer isotope in an excited state (^{57}Fe) and an absorber, consisting of the material under investigation, that contains the same isotope in the ground state.

A Mössbauer spectrum consists of a γ-ray detector and collected as a function of the source/absorber relative velocity (and, therefore, the modulation energy, E). A resonance occurs for $E = \Gamma = h/2\pi\tau$. For a ^{57}Fe Mössbauer active nucleus, Γ is 4.6×10^{-12} keV. Figure 4.12a shows a typical unsplit Mössbauer resonance line, characterized by a position, δ; a line width, Γ; and area $A = c_\varepsilon \Gamma$. Mössbauer spectra splittings are useful to obtain chemical and field signatures. A Mössbauer active nucleus in a noncubic environment is subject to a quadrupole splitting, Δ, which is proportional to the electric field gradient at the nucleus resulting from the chemical environment of the surrounding nuclei (Figure 4.12b). This is, therefore, a measure of the crystal-field strength. For a magnetic environment, the hyperfine field, H_0, splits the Mössbauer signal into a hextet (Figure 4.12c).

Mössbauer spectroscopy is a unique probe of cation distributions in the titanomagnetities. As a result of a linear relationship between the oxidation state of Fe cations and the Mössbauer-derived isomer shift, the average oxidation state of Fe cations in titanomagnetite solid solutions can be inferred from may the isomer shift of Fe cations in the room temperature Mössbauer spectra. Sorescu et al. [36] examined a series of titanomagnetite solid solutions $Ti_xFe_{3-x}O_4$, $0 \leq x \leq 1$, produced by the solid-state reaction of nanopowders.

Magnetic circular dichroism (MCD) [37] has also been used to probe cation distributions in the titanomagnetites [35]. Figure 4.13 shows the reported cation distributions for titanomagnetites as determined by (a) Mössbauer spectroscopy and (b) MCD for samples prepared by two different methods. The Mössbauer data suggest charge localization on the A-sites in ulvospinel-rich titanomagnetites, reaching $Q_T = 2$ by $x = 0.75$, while the MCD data have a linear variation extrapolated to reach reaching $Q_T = 2$ at $x = 1.0$. Like many previous observations [38] of the compositional variation of the cation distribution, the data points lie between the Néel and Akimoto model predictions discussed earlier.

Substituting Ti^{4+} in the titanomagnetite series is known to influence charge localization with resistivity changing by 7–8 orders of magnitude across the series. The increase is not monotonic, with jumps near $x = 0.2$, 0.5, and 0.8. the jump at $x = 0.5$ can be associated with the discontinuity in cation distributions in the Néel model. The Néel model prediction that Ti^{4+} substitutes for Fe^{3+} on B-sites, simultaneously causing Fe^{3+} to Fe^{3+} charge transfer at a different B-site. At $x = 0.5$, only Fe^{3+} remains on the B-sites. For further Ti^{4+} substitution (on the B-sites), two A-site Fe^{3+} are converted to Fe^{2+}, to maintain the charge balance. At $x = 0.75$, an average charge of 2.5 is reached corresponding to a local maximum in the A-site configurational entropy. As we consider a binary mixture of Fe^{3+} and Fe^{2+} on A-sites, the configurational entropy is $-R(X_{Fe2+} \ln X_{Fe2+}) + (1 - X_{Fe2+}) \ln(1 - X_{Fe2+})$, where X_{Fe2+} is the fraction of Fe^{2+} on the A-sites. This entropy is maximized when $X_{Fe2+} = 0.5$, giving $S_c = R \ln 2$.

4.3.6.1 Angularly Dependent Magnetic Properties: Magnetic Anisotropy

Anisotropy refers to the dependence of properties on the direction in which they are measured. Only *magnetocrystalline anisotropy* is intrinsic, depending on the material's symmetry. Most ferromagnetic materials derive from cubic or uniaxial materials, and therefore, the discussion will also be limited to these anisotropies. The previous discussion of crystal fields showed that electrostatic interactions can explain the physical origin of magnetocrystalline anisotropy. This phenomenon is also

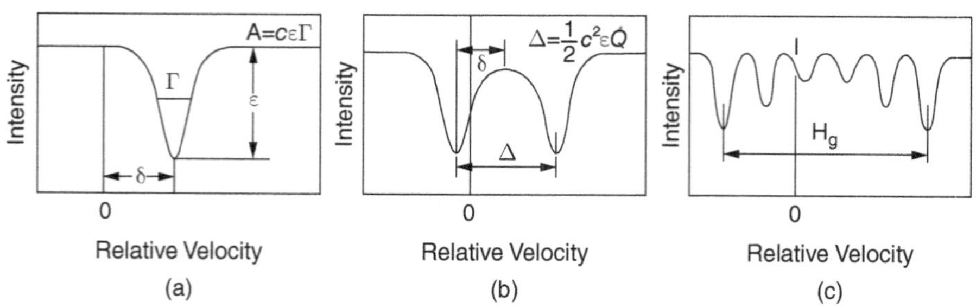

Figure 4.12 Mössbauer spectroscopy phenomena: (a) resonance peak, (b) quadruple splitting, and (c) hyperfine field splitting.

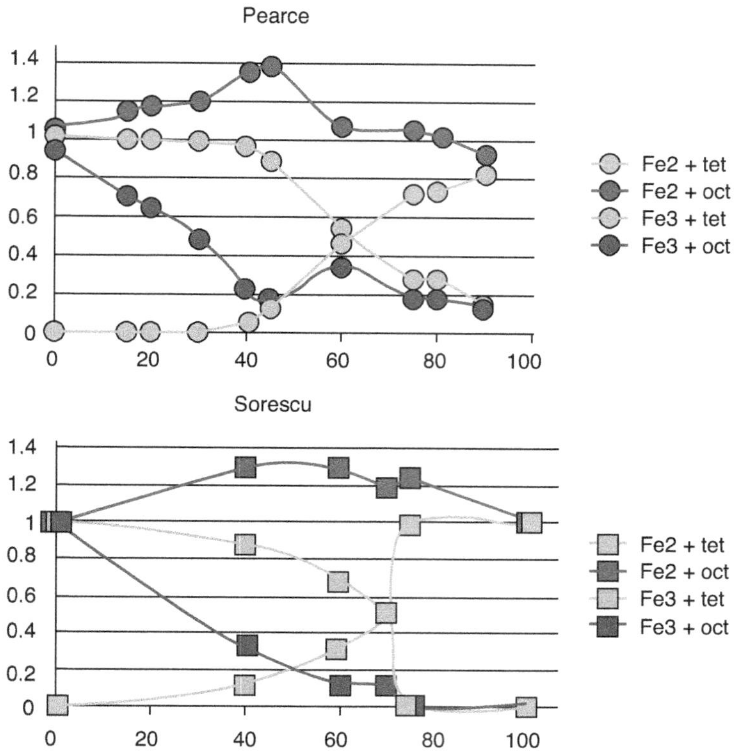

Figure 4.13 Cation distributions as determined by (a) Mössbauer spectroscopy and (b) MCD for titanomagnetite samples redrawn from literature data.

rooted in the spin–orbit interactions. As a result, the first principle predictions of magnetocrystalline anisotropies require quantum mechanics. The phenomenology of magnetocrystalline anisotropy, however, seeks to describe the angular dependence of the internal energy as a function of the orientation of the magnetization vector. To this end, simple models can be very descriptive. We begin this a discussion by introducing one of the earliest such models, proposed by Nobel laureate Louis Néel.

The *Néel model* for magnetic anisotropy is a simple pair model considering the direction cosines of the magnetization vector as *order parameters*. This model has been applied to a variety of important magnetic anisotropy issues, including bulk magnetocrystalline anisotropy, bond anisotropy in A–B substitutional solid solutions, order–disorder transitions, surface and interfacial anisotropy, directional ordering, and field annealing. Figure 4.14 illustrates the geometry for a pair of adjacent spins in the Néel model rotation into an orientation at an angle φ with respect to the bond axis and the same where the rotation is accompanied by strain in a magnetoelastic material. The phenomenological Néel model explains the angular dependence of magnetic anisotropy in crystals.

It has also been used to explain surface anisotropy based on symmetry-breaking ideas [39, 40]. The premise of this model is that anisotropy can be expanded in simple pair potentials. The elementary internal energy of a spin pair is expanded in a series of *Legendre polynomials* as:

$$W(r,\cos\phi) = g(r) + l(r)\left(\cos^2\phi - \frac{1}{3}\right) + q(r)\left(\cos^4\phi - \frac{6}{7}\cos^2\phi - \frac{3}{35}\right) = \ldots \tag{4.30a}$$

where r is the distance between atoms and φ is the polar angle between the bond and spin direction S:

$$\cos\phi = \frac{\vec{M}\cdot\vec{r}_{ij}}{|\vec{M}||\vec{r}_{ij}|} = \vec{m}\cdot u\vec{r} \tag{4.30b}$$

where m is the reduced magnetization vector and $u\vec{r}$ is a unit vector in the direction of the bond axis.

The common direction of the spins is determined by exchange interactions. The first term, g(r), which is spatially isotropic, corresponds to the *exchange interaction energy*. The second term is called the *dipole–dipole interaction* term. The magnitude of $l(r)$ is much larger than the third term $q(r)$, which is called *quadrupolar interaction* and therefore, we neglect $q(r)$ in the future discussion. For classical dipole–dipole interactions, predictions of the coefficients fail by a factor of 100

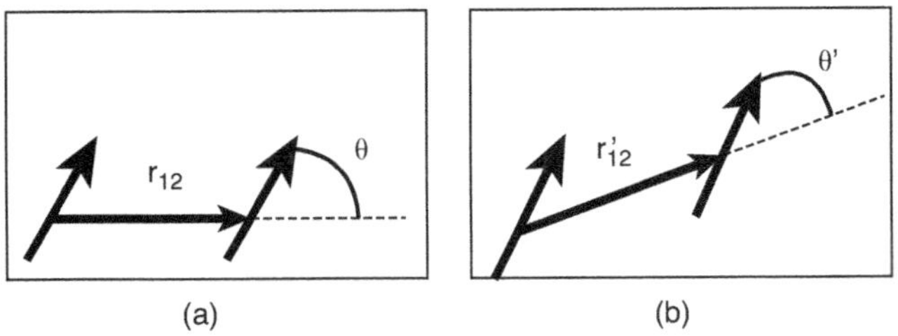

Figure 4.14 Geometry for two adjacent spins in the Néel model (a) rotation into an orientation at an angle φ with respect to the bond axis and (b) rotation accompanied by strain in a magnetoelastic material.

1000. A more sophisticated approach considers spin–orbit interactions more rigorously.

The magnetic anisotropy in the Néel model is calculated summing over all spin pairs in a unit volume of the crystal. To calculate the total anisotropy energy, K_A, sum over all pairs of atoms, but typically restrict to the nearest-neighbor pair interactions:

$$E_{\mathrm{me}} = \sum_{NN} \frac{\partial_1(r)}{\partial r} e r_{\mathrm{eq}} \tag{4.30c}$$

where $\in$ is the strain and r_{eq} is the equilibrium spacing between the dipoles (atoms). Note that discussing magnetoelastic response requires a higher order tensorial description and strain can be tensile, compressive, or shear [41–43].

Polynomials in cos φ describe the directional dependence of the pair energy. Summing over coordination shells symmetrizes these functions, and terms disappear in accordance with the symmetry of the coordination shells. For example, in cubic symmetry, angular terms multiplied by 1(R) sum to zero, and thus, only fourth-order terms in cos φ persist. In uniaxial systems (hexagonal and tetragonal), the second-order terms in cos φ remain. We now consider an example of such an expansion for a simple cubic lattice of spins in Box 4.5.

It is easily shown that:

$$K_1 = -2Nq, \quad K_1 = \frac{16}{9} N_q, \quad K_1 = N_q \tag{4.30d}$$

For the simple cubic, body centered cubic, *bcc*, and face centered cubic, *fcc*, lattices respectively.

Box 4.5 Néel model circulation of the magnetocrystalline anisotropy angular dependence for a simple cubic lattice

Consider atoms in a simple cubic lattice with nearest neighbors at $(\pm a,0,0)$, $(0,\pm a,0)$, and $(0,0,\pm a)$. The Néel expansions for *x*-, *y*-, and *z*-pairs, respectively, are:

$$(E_A)_x = N\left[l(r)\left(\alpha_1^2 - \frac{1}{3}\right) + q(r)\left(\alpha_1^4 - \frac{6}{7}\alpha_1^2 + \frac{3}{35}\right) + \ldots\right]$$

$$(E_A)_y = N\left[l(r)\left(\alpha_2^2 - \frac{1}{3}\right) + q(r)\left(\alpha_2^4 - \frac{6}{7}\alpha_2^2 + \frac{3}{35}\right) + \ldots\right]$$

$$(E_A)_z = N\left[l(r)\left(\alpha_3^2 - \frac{1}{3}\right) + q(r)\left(\alpha_3^4 - \frac{6}{7}\alpha_3^2 + \frac{3}{35}\right) + \ldots\right]$$

where $\alpha_1, \alpha_2, \alpha_3$ are the direction cosines of a spin with respect to the cube axes and *N* is the number of atoms per unit volume on an SC lattice. These terms may now be summed to yield:

$$(E_A)_{\mathrm{tot}} = Nq(\alpha_1^4 + \alpha_2^4 + \alpha_3^4) = -2Nq(\alpha_1^2\alpha_2^2 + \alpha_2^2\alpha_3^2 + \alpha_3^2\alpha_1^2) \tag{4.31}$$

where we recognize that the identities ${\alpha_1}^2 + {\alpha_2}^2 + {\alpha_3}^2 = 1$ and $({\alpha_1}^2 + {\alpha_2}^2 + {\alpha_3}^2)^2 = 1$ can be used to simplify the expressions. The term $-2N_q$ is called K_1. Furthermore, we can recognize that the direction cosines are equivalent to reduced magnetization components, $m_x = \alpha_1$, $m_y = \alpha_2$, and $m_z = \alpha_3$ to write:

$$(E_A)_{\mathrm{tot}} = -2Nq(m_x^2 m_y^2 + m_y^2 m_z^2 + m_z^2 m_x^2) \tag{4.32}$$

If the magnetic anisotropy is expanded to sixth-order terms, it takes the form:

$$(E_A)_{tot} = K_1(m_x^2 m_y^2 + m_y^2 m_z^2 + m_z^2 m_x^2) + K_2(m_x^6 + m_y^6 + m_z^6) \tag{4.33}$$

The easy directions of magnetization can be determined by minimizing this energy to find the equilibrium values of the direction cosines or equivalently the components of the reduced magnetization vector.

Extending the phenomenology of the Néel model further, it is common to expand *the cubic magnetocrystalline anisotropy energy* to sixth order as:

$$E_a = K_1(\alpha_1^2\alpha_2^2 + \alpha_2^2\alpha_3^2 + \alpha_3^2\alpha_1^2) + K_2(\alpha_1^2\alpha_2^2\alpha_3^2) \tag{4.34}$$

where αs are the direction cosines of angles between M and the $\langle 100\rangle$ axes. $K_2 = 0$ minima are along the $\langle 100\rangle$ axes if $K_1 > 0$ and along the $\langle 111\rangle$ axes if $K_1 < 0$. The second case is true for magnetite, which has a $\langle 111\rangle$ *EMD*.

In hexagonal, or any *uniaxial material*, we need to include second-order terms in the magnetocrystalline anisotropy energy expansion:

$$E_\alpha = K_0 + K_1 \sin^2\theta \tag{4.35}$$

Here, $\theta = \alpha_3$ is the common symbol used to denote the one unique angle (i.e. with respect to the c-axis). The *EMD* can be parallel to the c-axis, as in the hexaferrites, or lie in the basal planes, depending on the sign of K_1. When the fourth-order terms are included in the expansion, the anisotropy is:

$$E_\alpha = K_0 + K_1 \sin^2\theta + K_2 \sin^4\theta \tag{4.36}$$

and conical EMDs are possible. The field associated with magnetic anisotropy is called the *anisotropy field*, H_K:

$$H_K = \frac{2K_1}{M_s} \tag{4.37}$$

If $K_2 = 0, H_K$, is the field necessary to saturate the material along the hard direction and serves as an upper bound for the coercivity.

The coupling of magnetic and mechanical properties is termed *magnetostriction*. Magnetoelastic interactions result from coupling magnetic anisotropy and elastic response. Magnetostriction can cause dissipative energy loss in magnetic materials. Coupling of residual stress in alloys with nonzero *magnetostriction coefficient*, λ (units of strain), results in stress-induced anisotropy that can reduce the ease of magnetization, lower μ, and increase H_c.

A change in dimension during magnetization by an applied field is *anisotropic magnetostriction*. Along the direction of the applied magnetic field, the sample elongates or shrinks and is perpendicular to the field in which it shrinks or elongates. An *isotropic magnetostriction* or *exchange magnetostriction* arises from coupling of exchange energy to interatomic spacing. This occurs during cooling through T_c in $H_a = 0$. As a tensor property, magnetostriction is tied to crystallography.

4.3.7 Titanomagnetite – Spin-Flop and Spin Glass/Liquid Response

In ferrite materials, it is possible to rotate against antiparallel, or noncollinear, superexchange interactions and fully saturate the materials by *spin-flop transitions*. As this rotation against superexchange is much more difficult than the rotation against magnetocrystalline anisotropy, full collinear parallel saturation will not occur until very high fields. An exchange field can be taken as the measure of a saturating field for a spin-flop transition in the same way that an anisotropy field is taken as a saturating field when considering magnetocrystalline anisotropy.

In spinel systems, where nonmagnetic substituents are added, the weakening of superexchange can allow for observation of nonsaturating $M(H)$ curves to fields that are now accessible in high-field magnet lab facilities [44]. Figure 4.15a shows typical $M(H)$ curves for manganese aluminate spinels and hausmannites where nonsaturating response at fields as high as 5 T is observed. This has been attributed to the rotation against superexchange interactions in a noncollinear Yafet–Kittel state for these compounds. Figure 4.15b shows high-field $M(H)$ response of titanomagnetite, TM90–TM100, powdered samples at low T, taken at the pulsed field facility at the National High Field Magnet Facility (NHFML) at Los Alamos National Lab. Note both plateaus in the data and no saturating response to fields of 50 T.

In the titanomagnetites, if we assume the Néel model of cation distribution discussed earlier, whereas Fe is added to ulvospinel, Fe_2TiO_4, it replaces Ti^{4+} on a B-site as Fe^{2+}, and two A-site Fe^{2+} cations are converted to Fe^{3+} cations, to

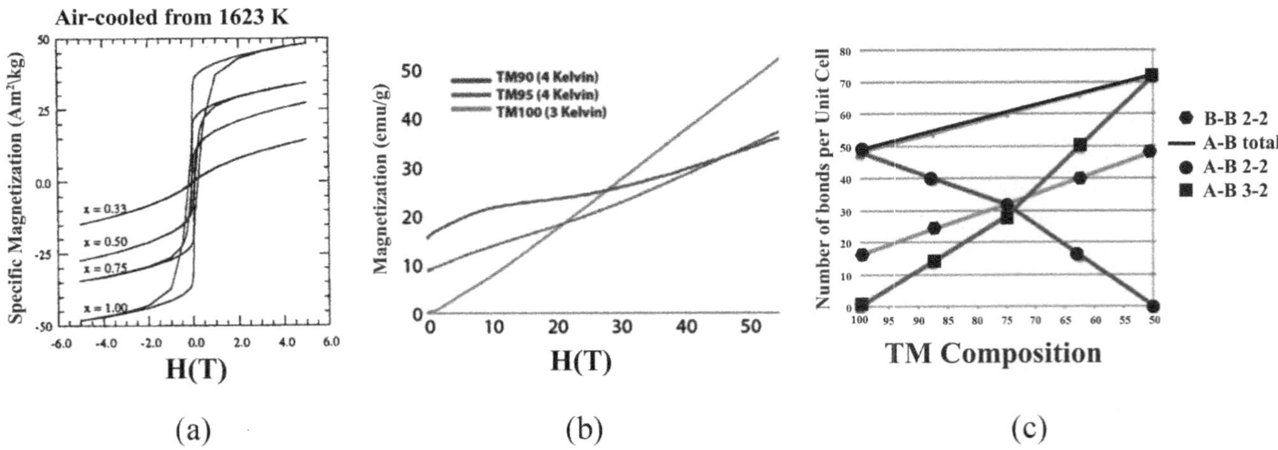

Figure 4.15 (a) Typical *M*(*H*) curves for manganese aluminate spinels and hausmannites; (b) high-field magnetization curves of titanomagnetite, TM90–TM100, powdered samples at low *T*; and (c) number of different superexchange bonds as the function of composition in a TM unit cell, with 2–2 and 2–3, referring to Fe^{2+}–Fe^{2+} bonds and Fe^{2+}–Fe^{3+} bonds, respectively. Reproduced from: Y. Nakamura, P.A. Smith, D. E Laughlin, M. J. De Graef, and M. E. McHenry. "Structure and Magnetic Properties of Quenched (MnxAl1-x)3O4 Spinels and Hausmannites". *IEEE Trans. Mag.* 31, 4154–56, (1995).

maintain the charge balance. As these substitutions of a magnetic cation form a nonmagnetic cation increases the number of superexchange bonds, the spin-flop saturation of TMs will increase with increased iron concentration. Figure 4.15c shows the count of the number of different superexchange bonds as the function of composition in a TM unit cell, with 2–2 and 2–3, referring to Fe^{2+}–Fe^{2+} bonds and Fe^{2+}–Fe^{3+} bonds respectively.

When spin-flop is achieved, the magnetization of the material is the sum of all the magnetic moments of its constituents, aligned in parallel. If we take the quenched orbital angular momentum magnetic moments of the ions as Fe^{3+}, $5\mu_B$; Fe^{3+}, $4\mu_B$; and Ti^{4+}, $0\mu_B$, the cations in TM compositions would give rise to a spin-flop fully aligned state with a saturation magnetization for Ti_x^{4+}, $Fe^{3+}{}_{2-2x}Fe^{2+}{}_{1+x}$ given by:

$$0+5(2-2x)+4(1+x)=14-6x\frac{\mu_B}{fu} \tag{4.38}$$

for $0 \leq x \leq 1$. This value is independent of cation distribution. If the high-field *M*(*H*) curves of Figure 4.15b are extrapolated to intersection with the fully aligned state's saturation magnetization, the exchange field required for a full spin-flop can be estimated. While for magnetite at low temperatures this field is in excess of 500 T, and not experimentally accessible, for ulvospinel, it has been estimated to be as low as 200 T, a field that can be reached in the NHFML-pulsed field facility.

Cation charge and positional disorder in spinel ferrimagnets are known to have remarkable effects on the magnetic properties [45]. In Ti^{4+}-rich titanomagnetites, magnetic frustration has recently been attributed to changes in ferromagnetic superexchange bonds as Fe^{3+} is substituted for Ti^{4+} ion octahedral B-sites in ulvospinel. Distributed exchange bonds, known to influence the T-dependent magnetization in systems coupled by direct exchange [46], were investigated in their role in T-dependent magnetization and spin frustration in systems coupled by superexchange. The plateau in Figure 4.15b is similar to response observed in magnetically frustrated systems [47]. Other spinel systems have exhibited magnetic frustration as B-sites in spinels form a *Kagome lattice*, for which frustrated noncollinear spin arrangements are predicted [48]. The curved transitions of the plateaus of Figure 4.15b can partially be attributed to the samples being polycrystalline, while flat response has been observed in single-crystal spinels.

Magnetic interactions are frustrated if an atomic dipole moment cannot arrange its orientation such that it lowers its exchange energy through the interaction with its neighbors. A simple example considers antiferromagnetic interactions on a triangle, where if two antiparallel collinear atomic dipoles occupy two vertices of a triangle, a third collinear atomic dipoles cannot lower its exchange energy because it must be parallel to one and antiparallel to the other of the first two and thus is frustrated.

To consider frustrated dipole moments in the context of the spinel structure, we can examine 2-d motifs consistent with frustrated magnetic interactions. Figure 4.16a shows degenerate spin configurations for equal atomic dipole moments on a triangular lattice. Figure 4.17a shows a network of vertex-sharing B-site (green) tetrahedra centered around the vacant 8b special position (yellow) for the spinel structure. Figure 4.17b orients this network to see the Kagome lattice for which frustrated spin arrangements are also possible.

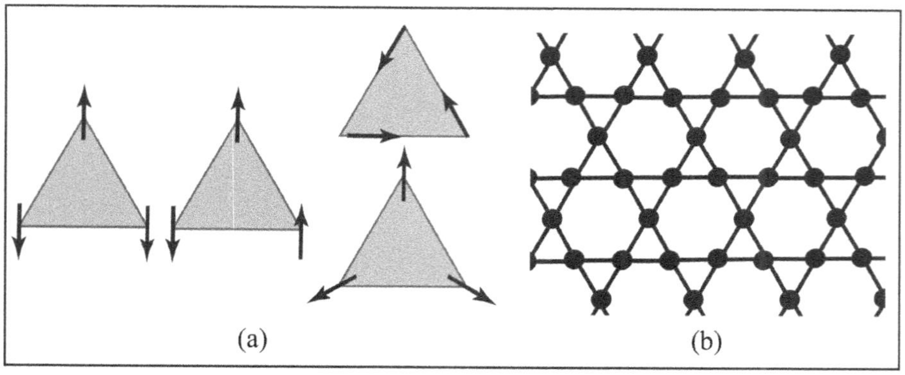

Figure 4.16 (a) Two degenerate spin configurations for equal atomic dipole moments on a triangular lattice; (b) Yafet–Kittel triangular spin structures; and (c) Kagome lattice for which frustrated spin arrangements are also possible.

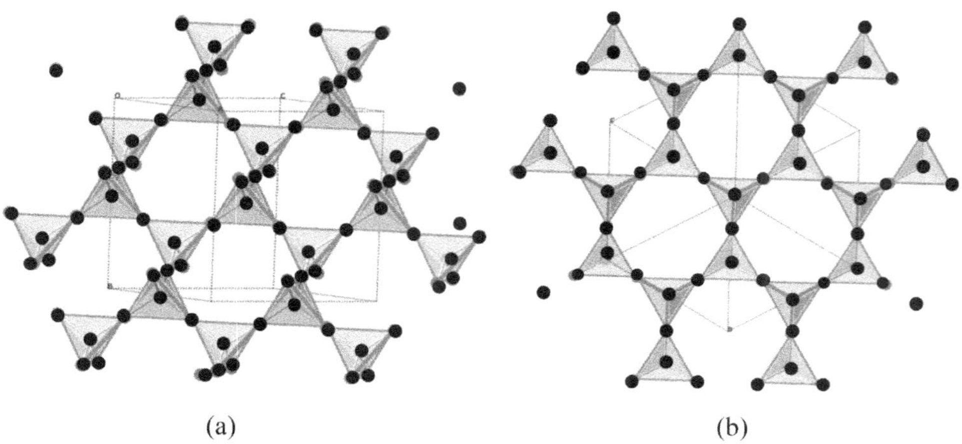

Figure 4.17 (a) Network of vertex-sharing B-site (green) tetrahedra centered around the vacant 8b special position (yellow) for the spinel structure and (b) projection of the same into a (1 1 1) plane viewed along the trigonal axis of the plane normal.

In the spinel lattice, B-site cations form a pyrochlore sublattice [49] consisting of a network of vertex-sharing tetrahedra. These tetrahedra are centered about the vacant 8b special position of the *Fd*3*m* space group of the spinel lattice. Figure 4.17a shows the pyrochlore sublattice of B cations and the connectivity of the tetrahedra. Note that the tetrahedra are centered by a vacancy as depicted in yellow. Figure 4.17b shows a projection of the same into a (1 1 1) plane viewed along the trigonal axis of the plane normal. Note the analogy with the 2-d Kagome lattice for which frustrated spin arrangements are possible.

In substituting Fe into Fe_2TiO_4, changes in cation distribution bring the system from having only Fe^{2+}–Fe^{2+} A–B superexchange bonds to adding Fe^{3+}–Fe^{2+} double-exchange bonds. Adding iron also increases the number of B–B Fe^{2+}–Fe^{2+}superexchange bonds. For Fe_2TiO_4, $M(H)$, as shown in Figure 4.15b, reflects the rotation of the dipole moments against the antiferromagnetic exchange in ulvospinel. As the sample T is increased, the rotation becomes easier and the field eventually saturates the antiferromagnet. Beyond that temperature, the magnetization diminishes as in accordance with the mean field theory. This causes the cusp in the susceptibility observed in Figure 4.18. The cusp is broadened both because of the polycrystallinity of the sample and distribution of the exchange interactions. On increasing the Fe content, samples saturated at high temperatures in an 8 T field and field-cooled values suggest a T-dependent remnant [50] state. This is reminiscent of unfreezing of a spin glass [51]. The reappearance of random Fe^{3+}–Fe^{2+} and Fe^{2+}–Fe^{2+} A–B superexchange bonds on a pseudo-Kagome lattice is hypothesized to cause magnetic frustration.

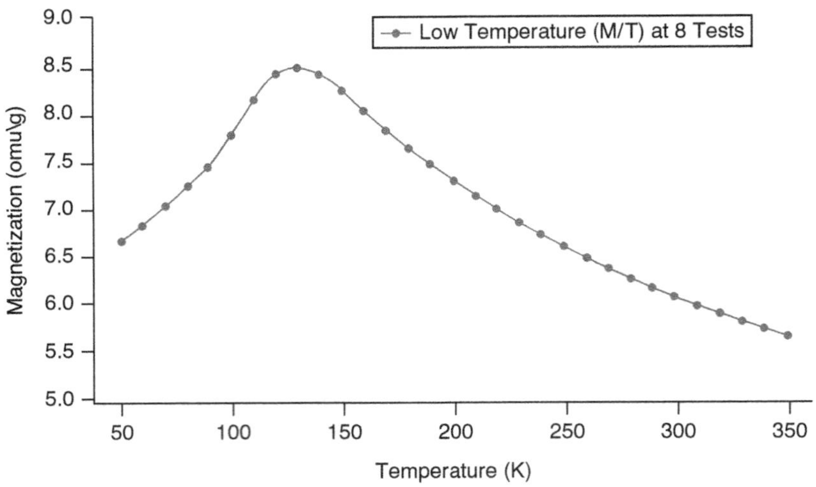

Figure 4.18 (a) Cusp in the susceptibility vs. T data for ulvospinel.

Acknowledgments

Authors would like to acknowledge the support from the National Science Foundation (NSF) through Materials World Network under Grants DMR1106942 and DMR093903 and their students. They would also like to thank M. Díaz-Michelena, M. Maicas, C. Aroca, and many of their students for useful discussions.

Bibliography

1 Hund, F. (2013). *Linienspektren: Und Periodisches System der Elemente*, Vol. 4. Springer-Verlag.
2 McHenry, M.E. and Laughlin, D.E. (2015). Magnetic properties of metals and alloys. In: David E. Laughlin and Kazuhiro Hono (eds), *Physical Metallurgy (Fifth Edition)*, 1881–2008. Elsevier.
3 Russell, H.N. and Saunders, F.A. (1925). New regularities in the spectra of the alkaline earths. *The Astrophysical Journal* 61: 38.
4 McCurrie, R.A. (1982). *Ferromagnetic Materials Structure and Properties*. San Diego: Academic Press.
5 Goldman, A. (2006). Ferrite processing. *Modern Ferrite Technology*, Springer, New York 151–216.
6 Néel, L. (1950). Aimantation a Saturation de Certains Ferrites. *Comptes R. Ac. Sc. Paris* 220: 190–192.
7 Guillaud, C. (1951). Propriétés magnétiques des ferrites. *Journal de Physique et le Radium* 12 (3): 239–248.
8 Murthy, N.S., Natera, M.G., Youssef, S.I., Begum, R.J., and Srivastava, C.M. (1969). Yafet–Kittel angles in zinc-nickel ferrites. *Physical Review* 181 (2): 969.
9 Dwight, K. and Menyuk, N. (1960). Magnetic properties of Mn 3 O 4 and the canted spin problem. *Physical Review* 119 (5): 1470.
10 Néel, M.L. (1948). Propriétés magnétiques des ferrites; ferrimagnétisme et antiferromagnétisme. *Annales de Physique* 12: 137–198.
11 Yafet, Y. and Kittel, C. (1952). Antiferromagnetic arrangements in ferrites. *Physical Review* 87 (2): 290.
12 Weiss, P. (1907). L'hypothèse du champ moléculaire et la propriété ferromagnétique. *Journal of Physics: Theories and Applications* 6 (1): 661–690.
13 Néel, L. (1932). *Influence des fluctuations des champs moléculaires sur les propriétés magnétiques des corps* (Doctoral dissertation, Universite).
14 Nakamura, Y., Smith, P.A., Laughlin, D.E., De Graef, M., and McHenry, M.E. (1995). Structure and magnetic properties of quenched (Mn/sub x/Al/sub 1-x/)/sub 3/O/sub 4/spinels and Hausmannites. *IEEE Transactions on Magnetics* 31 (6): 4154–4156.
15 Willard, M.A., Nakamura, Y., Laughlin, D.E., and McHenry, M.E. (1999). Magnetic properties of ordered and disordered spinel-phase ferrimagnets. *Journal of the American Ceramic Society* 82 (12): 3342–3346.

16 Goodenough, J.B. and Loeb, A.L. (1955). Theory of ionic ordering, crystal distortion, and magnetic exchange due to covalent forces in spinels. *Physical Review* 98 (2): 391.

17 Zener, C. (1951). Interaction between the d-shells in the transition metals. II. Ferromagnetic compounds of manganese with perovskite structure. *Physical Review* 82 (3): 403.

18 Pollert, E. (1980). Tetragonal-to-cubic transformation of hausmannite. *Journal of Solid State Chemistry* 33 (3): 305–308.

19 De Guire, M.R., Kalonji, G., and O'Handley, R.C. (1989). Cation distributions in rapidly solidified cobalt ferrite. *Journal of Applied Physics* 65: 316772.

20 De Guire, M.R., O'Handley, R.C., and Kalonji, G. (1990). The cooling rate dependence of cation distributions in CoFe2O4. *Journal of American Ceramics Society* 73: 3202206.

21 Akimoto, S.I. (1954). Thermo-magnetic study of ferromagnetic minerals contained in igneous rocks. *Journal of Geomagnetism and Geoelectricity* 6 (1): 1–14.

22 Néel, L. (1955). Some theoretical aspects of rock- magnetism. *Advances in Physics* 4 (14): 191–243.

23 Miodownik, A.P. (1982). The effect of magnetic transformations on phase diagrams. *Journal of Phase Equilibria* 2 (4): 406.

24 Wechsler, B.A., Lindsley, D.H., and Prewitt, C.T. (1984). Crystal structure and cation distribution in titanomagnetites (Fe (sub 3-x) Ti x O 4). *American Mineralogist* 69 (7–8): 754–770.

25 Lan, S., Groschner, C., Runco, J., Wise, A., Díaz-Michelena, M., Laughlin, D., and McHenry, M.E. (2013). Phase identification and temperature-dependent magnetization of Ti-Rich Titanomagnetite ($0.5 \leq x \leq 1$) in different atmospheres. *IEEE Transactions on Magnetics* 49 (7): 4314–4318.

26 Schoenthal, W., Liu, X., Cox, T., Mesa, J.L., Maicas, M., Díaz-Michelena, M., ... McHenry, M.E. (2014). Synthesis and magnetic properties of single phase titanomagnetites. *Journal of Applied Physics* 115 (17): 17A934.

27 Wise, A., Saenko, M., Velázquez, A.M., Laughlin, D.E., Díaz-Michelena, M., and McHenry, M.E. (2011). Phase evolution in the Fe_3O_4–Fe_2TiO_4 pseudo-binary system and its implications for remanent magnetization in Martian minerals. *IEEE Transactions on Magnetics* 47 (10): 4124–4127.

28 Nelson, J.B. and Riley, D.P. (1945). An experimental investigation of extrapolation methods in the derivation of accurate unit-cell dimensions of crystals. *Proceedings of the Physical Society* 57 (3): 160.

29 Liu, X., Sorescu, M., Schoenthal, W., Dong, Z., Ramanujan, R.V., Laughlin, D.E., and McHenry, M.E. (2014). Structural and magnetic properties of xFe2TiO4(1-x) Fe3O4 (0.75¡x¡1). *IEEE Transactions on Magnetics* 50: 101504–101506.

30 Swaminathan, R., McHenry, M.E., Poddar, P., and Srikanth, H. (2005). Magnetic properties of polydisperse and monodisperse NiZn ferrite nanoparticles interpreted in a surface structure model. *Journal of Applied Physics* 97 (10): 10G104.

31 Coehorn, R.L. (1991). *Supermagnets: Hard Magnetic Materials*. Long, G.J. and Grandjean, F., editors. Chapter 8. Kluwer.

32 Hutchings, M.T. (1964). Crystal field theory. *Solid State Physics* 16.

33 Freeman, A.J. and Watson, R.E. (1962). Theoretical investigation of some magnetic and spectroscopic properties of rare-earth ions. *Physical Review* 127 (6): 2058.

34 Herber, R.H. (1982). Mössbauer effect. In: Parker, Sybil P. (ed), *The McGraw Hill Encyclopedia of Physics*. New York: McGraw Hill.

35 Hamdeh, H.H., Barghout, K., Ho, J.C., Shand, P.M., and Miller, L.L. (1999). A Mössbauer evaluation of cation distribution in titanomagnetites. *Journal of Magnetism and Magnetic Materials* 191 (1–2): 72–78.

36 Sorescu, M., Xu, T., Wise, A., Díaz-Michelena, M., and McHenry, M.E. (2012). Studies on structural, magnetic and thermal properties of xFe2TiO4-(1− x) Fe3O4 (0≤ x≤ 1) Pseudo-binary System. *Journal of Magnetism and Magnetic Materials* 324 (7): 1453–1462.

37 Pearce, C.I., Henderson, C.M.B., Telling, N.D., Pattrick, R.A., Charnock, J.M., Coker, V.S., ... van der Laan, G. (2010). Fe site occupancy in magnetite-ulvospinel solid solutions: A new approach using X-ray magnetic circular dichroism. *American Mineralogist* 95 (4): 425–439.

38 O'reilly, W. and Banerjee, S.K. (1965). Cation distribution in titanomagnetites (1−x) Fe3O4− xFe2TiO4. *Physics Letters* 17 (3): 237–238.

39 Bruno, P. (1988). Magnetic surface anisotropy of cobalt and surface roughness effects within Néel's model. *Journal of Physics F: Metal Physics* 18 (6): 1291.

40 Chuang, D.S., Ballentine, C.A., and O'Handley, R.C. (1994). Surface and step magnetic anisotropy. *Physical Review B* 49 (21): 15084.

41 Chikazumi, S. and Graham, C.D., Jr. (1997). *Physics of Ferromagnetism*. Oxford: Clarendon Press.

42 Victora, R.H. and MacLaren, J.M. (1993). Theory of anisotropy in strained superlattices. *Journal of Applied Physics* 73 (10): 6415–6417.

43 Victora, R.H. and MacLaren, J.M. (1993). Theory of magnetic interface anisotropy. *Physical Review B* 47 (17): 11583.

44 Schoenthal, W. and Jaime, M. and Diaz-Michelena, M. and Laughlin, D.E. and McHenry, M.E. (2014). Magnetic frustration dependence on superexchange bonds in titanomagnetites. J. *Appl. Phys.* xxx.

45 Laughlin, D.E., Willard, M.A., and McHenry, M.E. (2000). Magnetic ordering: Some structural aspects. In: Turchi, P., Gonis, A., (Eds.), *Phase Transformations and Evolution in Materials*, 121–137. The Minerals, Metals and Materials Society.

46 Gallagher, K.A., Willard, M.A., Laughlin, D.E., and McHenry, M.E. (1999). Distributed magnetic exchange interactions and mean field theory description of temperature dependent magnetization in amorphous Fe Zr B Cu alloys. *Journal of Applied Physics* 85: 5130–5132.

47 Zapf, V., Jaime, M., and Batista, C.D. (2014). Bose-Einstein condensation in quantum magnets. *Reviews of Modern Physics* 86 (2): 563.

48 Diaz, S., De Brion, S., Chouteau, G., Canals, B., Simonet, V., and Strobel, P. (2006). Magnetic frustration in the spinel compounds Ge Co 2 O 4 and Ge Ni 2 O 4. *Physical Review B* 74 (9): 092404.

49 Van Den Brink, J. and Khomskii, D.I. (2008). Multiferroicity due to charge ordering. *Journal of Physics: Condensed Matter* 20 (43): 434217.

50 Groschner, C., Lan, S., Wise, A., Leary, A., Lucas, M.S., Park, C., ... McHenry, M.E. (2013). The role of atmosphere on phase transformations and magnetic properties of ulvospinel. *IEEE Transactions on Magnetics* 49 (7): 4273–4276.

51 Anderson, P.W. (1956). Ordering and antiferromagnetism in ferrites. *Physical Review* 102 (4): 1008.

5

Molecular Field and Exchange-Interaction Theories Applied to Ferrimagnetic Systems

Ke Sun, Chuanjian Wu, Zhong Yu, Rongdi Guo, Xiaona Jiang, and Zhongwen Lan

School of Materials and Energy, University of Electronic Science and Technology of China, Chengdu 610054, China

5.1 Introduction

Ferromagnetism and ferrimagnetism are typically magnetic materials with strong magnetization at room temperature. The atomic magnetic moments of ferromagnetics/ferrimagnetics can align parallel/antiparallel to one another within magnetic cells called magnetic domains. Their magnetization disappears reversibly upon heating above a given temperature. The fact that these affects signal long-range order and occurs at high temperatures can be explained in the context of molecular field theory. In the following sections, we focus on the calculation of molecular field coefficients in ferrite systems. Further, the response to the nature of molecular fields will be introduced. Heisenberg direct-exchange and Anderson indirect-exchange models are essential elements. According to exchange-interaction theory, we will then provide a description of calculations of exchange integrals based upon density functional theory (DFT). Chapter 3 deals with first-principles calculations of ferrite systems in greater detail and hence Section 3 of Chapter 2 provides a necessary primer for Chapter 3.

5.2 The Néel Molecular Field Theory

The celebrated theory of interactions between atomic magnetic moments in ferromagnetic systems was established based on the seminal studies performed by Pierre Weiss. It is assumed that in addition to any externally applied field, H, there is an internal "molecular field" H_{mf} in a ferromagnet proportional to its magnetization M [1]:

$$H_{mf} = \omega M \tag{5.1}$$

where ω is the molecular field coefficient. Here, H_{mf} (~10^9A/m) becomes very large and manifests as a induced spontaneous magnetization at room temperature. Weiss's theory, explained by a quantum mechanical treatment of the many-body problem, proposes the existence of long-range magnetic order in which a molecular field defines the interaction between spins. This theory successfully addressed many of the experimental observations of ferromagnetic behavior in metals at the time, including the temperature-dependent behavior of magnetization and the existence of magnetic domains. However, it failed when applied to ferrites and other magnetic oxides that we now know are ferrimagnetic or antiferromagnetic. Néel [2] put forward a molecular field theory as an expansion of Weiss's theory.

In the Néel theory, multiple molecular fields were introduced to describe coupled aligned and anti-aligned spin lattices to explain complex thermomagnetic behavior in ferrite and other magnetic oxide systems. In order to explore this theory, we consider here a simplified mode composed of two sublattices (see Figure 5.1).

From the figure, it is obvious that a given sublattice is subject to the effect from both the neighboring and same sublattice. Therefore, the molecular field H_{Amf} and H_{Bmf} in A and B sublattice can be defined as

$$\begin{cases} H_{Amf} = \omega_{AA}\lambda M_A - \omega_{AB}\mu M_B \\ H_{Bmf} = \omega_{BB}\mu M_B - \omega_{BA}\lambda M_A \end{cases} \tag{5.2}$$

Modern Ferrites: Basic Principles, Processing and Properties, Volume 1, First Edition. Edited by Vincent G. Harris.

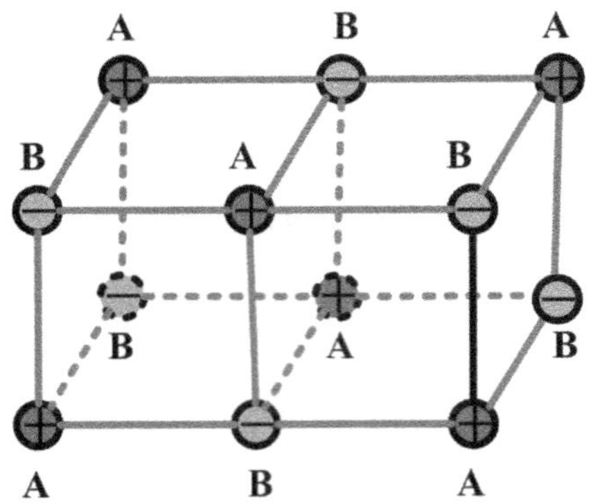

Figure 5.1 The schematic diagram of two sublattices: A and B sublattice.

Note that $\omega_{AB} = \omega_{BA}$, $\omega_{ij}(i, j = \text{A and B})$ is the molecular field coefficient between i and j sublattices, and λ and μ are the ratio of magnetic ions in A and B sublattices. The spontaneous magnetization $M_i(T)$ in the i sublattice are established on the basis of the Weiss molecular field:

$$M_i(T) = M(0)B_s(y_i) \tag{5.3}$$

with

$$B_s(y_i) = \frac{2J+1}{2J}\coth\left(\frac{2J+1}{2J}y_i\right) - \frac{1}{2J}\coth\left(\frac{1}{2J}y_i\right),\quad M(0) = NgJ\mu_B \tag{5.4}$$

where $M(0)$ is magnetization at $T = 0$ K, $B_s(y_i)$ the Brillouin function, J the angular quantum number, N the number of magnetic ions, g the Lande factor, and μ_B is the Bohr magneton, respectively. However, the Néel theory gives an additional expression for H_{mf}. The distinctive coefficient y_i is thereby rewritten as:

$$y_i = \frac{\mu_B g J}{k_B T}(\omega_{ii}M_i(T) - \omega_{ij}M_j(T)) \tag{5.5}$$

In Equation (5.5), ω, k_B, and T are the molecular field coefficient, Bohr magneton, and temperature, respectively. The solution of Equations (5.2)–(5.5) can be divided into two cases: spontaneous and paramagnetic magnetization.

5.3 Spontaneous Magnetization

Assuming that $\alpha = \omega_{AA}/\omega_{AB}$ and $\beta = \omega_{BB}/\omega_{AB}$, the total energy E in a ferrimagnetic system between two sublattices is expressed by:

$$E = -\frac{1}{2}\lambda M_A H_{Amf} + \mu M_B H_{Bmf} = -\frac{\omega_{AB}}{2}(\alpha\lambda^2 M_A^2 - 2\lambda\mu M_A M_B + \beta\mu^2 M_B^2) \tag{5.6}$$

In ferrimagnetic systems, as $\omega_{AA} < \omega_{AB}$ and $\omega_{BB} < \omega_{AB}$, $\alpha < 1$ and $\beta < 1$. According to the minimum energy principle, the A-site and B-site spins must align antiparallel with one another. Thus, the variation of spontaneous magnetization $M_s(T)$ as a function of temperature might be established. The corresponding regions of the (α, β) plane are shown in Figure 5.2. Considering $M/R/V$-type $M_s(T)$ curves require explanation not necessarily from the Néel molecular field theory, but by Yafet–Kittel-type magnetic order [3], we provide the analytical description of three other types of relationships as follows:

i) ***P/Q*-type $M_s(T)$ curve:** At $T \to 0$ K, the rate of change of $M_s \sim T$ (dM_s/dT) is greater than zero for the P-type model, while $dM_s/dT < 0$ for the Q-type model. In the range between 0 K and T_c, dM_A/dT and dM_B/dT for the Q-type curve have similar values; however, for the P-type curve, the declining rate in M_B tends to lag that of M_A. The majority of spinels and hexaferrites have been confirmed experimentally to be contained within these material models [4].
ii) ***N*-type $M_s(T)$ curve:** In most garnet ferrites ($R_3Fe_5O_{12}$), which are introduced in the next section, R occupies the (24c) site and has stronger magnetization at $T = 0$ K. As magnetization on the (24c) site falls rapidly with increasing temperature, $|M_{24c}| = |M_{24d} - M_{16a}| \neq 0$ and the Néel offset point T_d, which differs from the Curie temperature T_c, are then observable. At higher temperatures, the total magnetization $M_s \neq 0$ is mainly ascribed to the magnetization $|M_{24d} - M_{16a}|$. It is emphasized that the Curie temperature, T_c, for garnet ferrites results from the strong coupling effect from Fe^{3+}–Fe^{3+} interactions, and hence, almost all garnet ferrites demonstrate a similar Curie temperature. Table 5.1 shows the typical Néel offset point, T_d, and Curie temperature, T_c, of garnet ferrites.

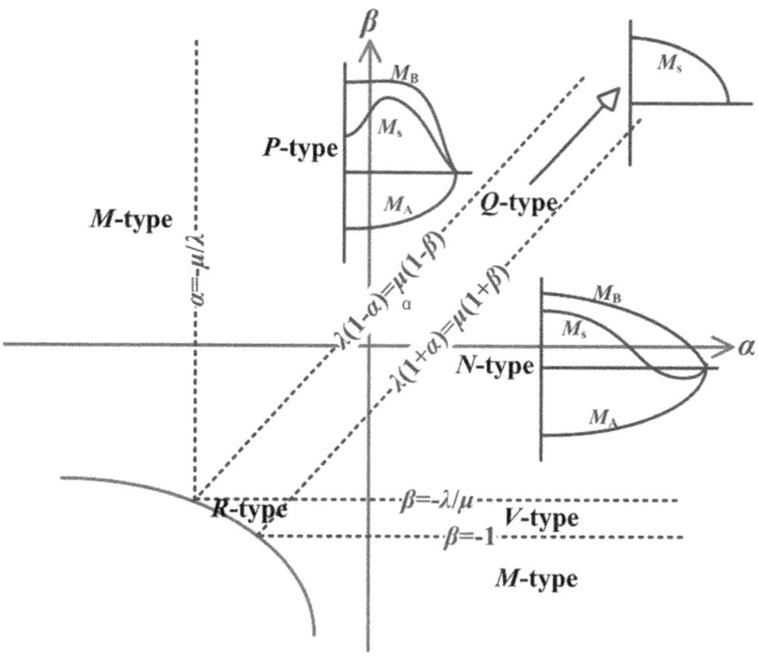

Figure 5.2 Spontaneous magnetization as a function of T in the regions of the (α, β) plane.

Table 5.1 Typical T_d and T_c in garnet ferrites ($R_3Fe_5O_{12}$).

R	Y	Gd	Tb	Dy	Ho	Er	Tm	Yb	Lu
T_d (K)	560	564	568	563	567	556	549	548	549
T_c (K)	—	290	246	220	136	84	—	~0	—

5.4 Paramagnetic Magnetization

At $T > T_c$, the ferrimagnetism transforms to paramagnetism. Due to $y_i << 1$, the Brillouin functions can be simplified to be:

$$B_S(y_i) = \frac{(J+1)y_i}{3J} \tag{5.7}$$

Combining Equations (5.7) and (5.3), the magnetization $M_A(T)$ and $M_B(T)$ in A and B sublattices are obtained by:

$$\begin{aligned} M_A(T) &= \frac{C}{T}[H + \omega_{AB}(\alpha\lambda M_A(T)) - \mu M_B(T)] \\ M_B(T) &= \frac{C}{T}[H + \omega_{AB}(-\lambda M_A(T)) + \beta M_B(T)] \end{aligned} \tag{5.8}$$

with

$$C = \frac{N\mu_B^2 g^2 J(J+1)}{3k_B} \tag{5.9}$$

For the externally applied field H, both $M_A(T)$ and $M_B(T)$ align parallel to H, and the total magnetization M_s is:

$$M_s = \lambda M_A + \mu M_B \tag{5.10}$$

According to Equations (5.8)–(5.10), we get:

$$\begin{aligned}\frac{M_A}{H} &= \frac{\frac{T}{C} - \omega\mu(\beta+1)}{\left(\frac{T}{C} - \omega\beta\mu\right)\left(\frac{T}{C} - \omega\alpha\mu\right) - \omega^2\lambda\mu} \\ \frac{M_B}{H} &= \frac{\frac{T}{C} - \omega\lambda(\alpha+1)}{\left(\frac{T}{C} - \omega\beta\mu\right)\left(\frac{T}{C} - \omega\alpha\lambda\right) - \omega^2\lambda\mu}\end{aligned} \tag{5.11}$$

Then the relationship between the reciprocal of magnetic susceptibility $1/\chi_m$ and temperature T can be deduced as:

$$\frac{1}{\chi_m} = \frac{T}{C} + \frac{1}{\chi_0} - \frac{\rho}{T-\theta} \tag{5.12}$$

where:

$$\begin{aligned}\frac{1}{\chi_0} &= \omega(2\lambda\mu - \lambda^2\alpha - \mu^2\beta) \\ \rho &= C\omega^2\lambda\mu[\lambda(\alpha+1) - \mu(\beta+1)]^2 \\ \theta &= C\omega\lambda\mu(\alpha+\beta+2)\end{aligned} \tag{5.13}$$

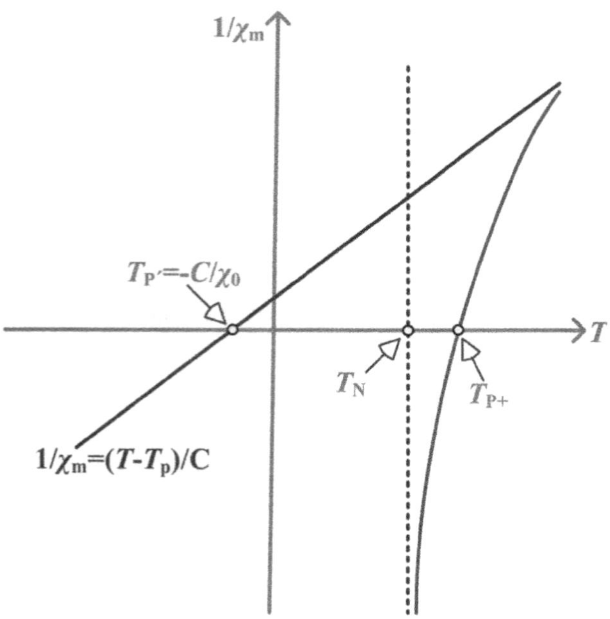

Figure 5.3 $1/\chi_m$–T theoretic curve in ferrimagnetic system.

Hence, Equation (5.12) demonstrates a hyperbola with the asymptote $1/\chi_m = (T - T_{p'})/C$ (see Figure 5.3). At $T = T_{p+}$, the reciprocal of magnetic susceptibility $1/\chi_m = 0$. T_{p+} is regarded as the paramagnetic Curie point.

Hence, we obtain:

$$T_p^2 - C\omega(\alpha\lambda + \beta\mu)T_p + C^2\omega^2\lambda\mu(\alpha\beta - 1) = 0 \tag{5.14}$$

with the solution (i.e. paramagnetic Curie point T_p)

$$T_p = \frac{1}{2}C\omega[(\lambda\alpha + \mu\beta) - \sqrt{(\alpha\lambda - \beta\mu)^2 + 4\lambda\mu}] \tag{5.15}$$

Néel's molecular field theory provides a simplified model that accounts for Coulomb interactions based on the mimic field, and a general conceptual framework that connects and clarifies the previous work on systems with both short-range and long-range magnetic order. Especially at higher T_c, the estimation of Néel molecular field theory coincides well with the experiment. Therefore, it has been used widely in the ferromagnetic and ferrimagnetic systems. We will next give an introduction for calculation of molecular field coefficients in ferrite systems (spinels, garnets, and hexaferrites) according to this theory.

5.5 Molecular Field Coefficients in Ferrites

5.5.1 Spinels

Considering the spinel structure, space group *Fd3m*, in its typical form $[A]\{B\}_2O_4$ (where [] indicates divalent cations occupying fourfold tetrahedral lattice sites, and {} the trivalent cations occupying sixfold octahedral lattice sites) [5]. $[A]\{B\}_2O_4$ is a single formula unit that constitutes one-eighth of the spinel unit cell (see Figure 5.4). Cations have either fourfold or sixfold coordination forming tetrahedral (A) and octahedral (B) sublattices that are arranged in a close packed arrangement with respect to each other. A cations reside on 8 of 64 available tetrahedral sites whereas B cations reside on 16 of 32 available octahedral sites. Oxygen anions form a close packed structure and contribute 32 ions that electrically balance the unit cell [6, 7].

In LiZnTi ferrite systems, Zn^{2+} prefers to enter [A] sublattices, while Ti^{4+} the {B} sublattices [8]. However, Gorter [9] found that as the Zn contents were increased, the resultant magnetic moment departed from the behavior expected for collinear spins with Zn^{2+} in [A] sites. In 1974, Dionne [10] suggested modifying Equation (5.4):

$$\begin{aligned} M_A(0) &= gS\mu N_B(1-k_A) \\ M_B(0) &= 2gS\mu N_B(1-k_B)(1-f_A) \end{aligned} \tag{5.16}$$

where k_A, k_B, and f_A represent the fractions of diamagnetic ions substituted for Fe^{3+} ions in the respective sublattices. From the experimental work of Blasse and Gorter, the empirical relations for k_A, k_B, and f_A have been defined as

$$\begin{cases} k_A = 0.45t, \quad k_B = 0.525t + 0.25 & \left(0 \le t \le \dfrac{2}{3}\right) \\ k_A = 0.584t - 0.092, \quad k_B = 0.462t + 0.292 & \left(t > \dfrac{2}{3}\right) \end{cases} \tag{5.17}$$

$$\begin{cases} f_A = 0.8(z^2 - t^2) & \left(0 \le t \le \dfrac{2}{3}\right) \\ f_A = 0 & \left(t \ge \dfrac{2}{3}\right) \end{cases} \tag{5.18}$$

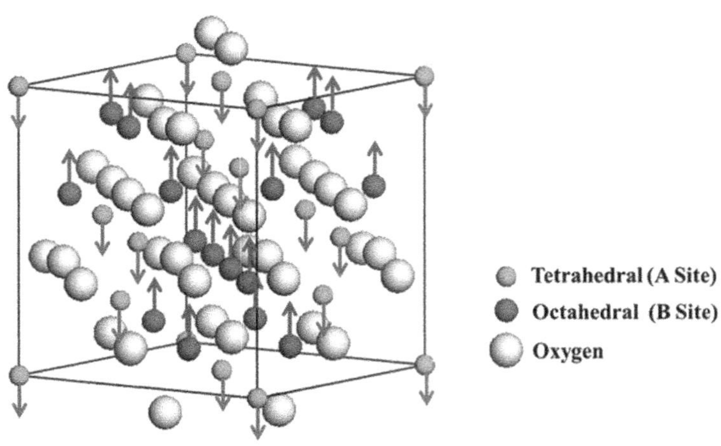

Figure 5.4 Schematic representation of a spinel ferrite structure.

In Equations (5.17)–(5.18), z and t are the contents of Zn^{2+} and Ti^{4+} in LiZnTi ferrite systems, respectively. According to Néel's model of collinear-spin ferrimagnetism, Dionne fitted the experimental $M_s \sim T$ and obtained the relationship between molecular field coefficients (ω_{AA}, ω_{AB}, and ω_{BB}) and the amounts of Zn^{2+} and Ti^{4+} (z and t) in the LiZnTi ferrites:

$$\begin{cases} \omega_{AA} = -150\left(1 - \frac{4}{3}k_B^t - \frac{4}{3}k_B^l\right) \\ \omega_{AB} = 273\left[1 - 0.8k_B^t\left(k_A^z + k_A^l\right) - 0.2k_A^z\right] \\ \omega_{AA} = -60\left(1 - k_A^l - \frac{1}{3}k_A^z\right) \end{cases} \tag{5.19}$$

For $t \le z$, it could be determined from the aforementioned cation-distribution model that $k_A{}^l = 0$, $k_A{}^z = z$, $k_B{}^t = 0.5t$, and $k_B{}^l = 0.25t–0.25z$. While for $t > z$, the contributions are $k_A{}^l = 0.\ 5t–0.5z$, $k_A{}^z = z$, $k_B{}^t = 0.5t$, and $k_B{}^l = 0$.

According to the Néel molecular field theory, Sun et al. [11] gave detailed calculations of the molecular fields in the Ni-substituted MnZn ferrites ($Mn_{0.506-x}Zn_{0.244}Ni_xFe_{2.250}O_{4.0}$, $x = 0.066$, 0.084, 0.103, and 0.122). Assuming that orbit angular momentums are completely frozen in MnZn ferrites, the equivalent spin quantum numbers S_i of [A] and {B} sublattices are approximately equivalent to be:

$$S_i = \frac{\sum x_i S_{x_i}}{\sum x_i} \tag{5.20}$$

where S_{xi} is the spin quantum number of one magnetic ion in the i sublattice and x_i is the corresponding proportion of magnetic ions. Hence, the magnetization for a given sublattice at 0 K can be written as:

$$\begin{aligned} M_A(0) &= gS_A\mu_B N(1 - k_A) \\ M_B(0) &= 2gS_A\mu_B N(1 - k_B) \end{aligned} \tag{5.21}$$

where k_A and k_B represent the fractions of nonmagnetic ions substituted for Fe^{3+} ions in the [A] and {B} sublattice, respectively. By the nonlinear fitting of the experimental M_s–T curve (see Figure 5.5), the molecular field coefficients (ω_{AA}, ω_{AB}, and ω_{BB}) were determined and subsequently presented in Table 5.2.

With regard to NiZn ferrite systems, Guo et al. [12] investigated the Curie temperature, T_c, and paramagnetic Curie temperature, T_p, according to the cation distribution $(Zn^{2+}{}_tFe^{3+}{}_{1-t})_A(Ni^{2+}{}_{0.9146-t}Cu^{2+}{}_{0.1005}Fe^{3+}{}_{1.9899-(1-t)})_BO_4$, where t is the Zn^{2+} content. Table 5.3 illustrates the Curie temperature and molecular field coefficients for select NiCuZn ferrites. As $T \rightarrow T_c^-$, the magnetization $M(T)$ of Equation (5.3) would be zero, and consequently, the Curie temperature T_c could be deduced from:

$$T_c = \frac{\omega_{AB} g\mu_B M_B(0)(\beta + \varphi_c)(S_B + 1)}{3k_B} \tag{5.22}$$

with

$$\varphi_c = \frac{\left[\alpha - \beta\frac{(S_B+1)M_B(0)}{(S_A+1)M_A(0)}\right] + \sqrt{\left[\alpha - \beta\frac{(S_B+1)M_B(0)}{(S_A+1)M_A(0)}\right]^2 + 4\frac{(S_B+1)M_B(0)}{(S_A+1)M_A(0)}}}{2\frac{(S_B+1)M_B(0)}{(S_A+1)M_A(0)}} \tag{5.23}$$

At higher T_c, the modified reciprocal of magnetic susceptibility $1/\chi_m$ (see Equation (5.12)) is derived to be:

$$\frac{1}{\chi_m} = \frac{T - \theta}{\lambda C_A + \mu C_B} - \frac{\varsigma}{T - T_c'} \tag{5.24}$$

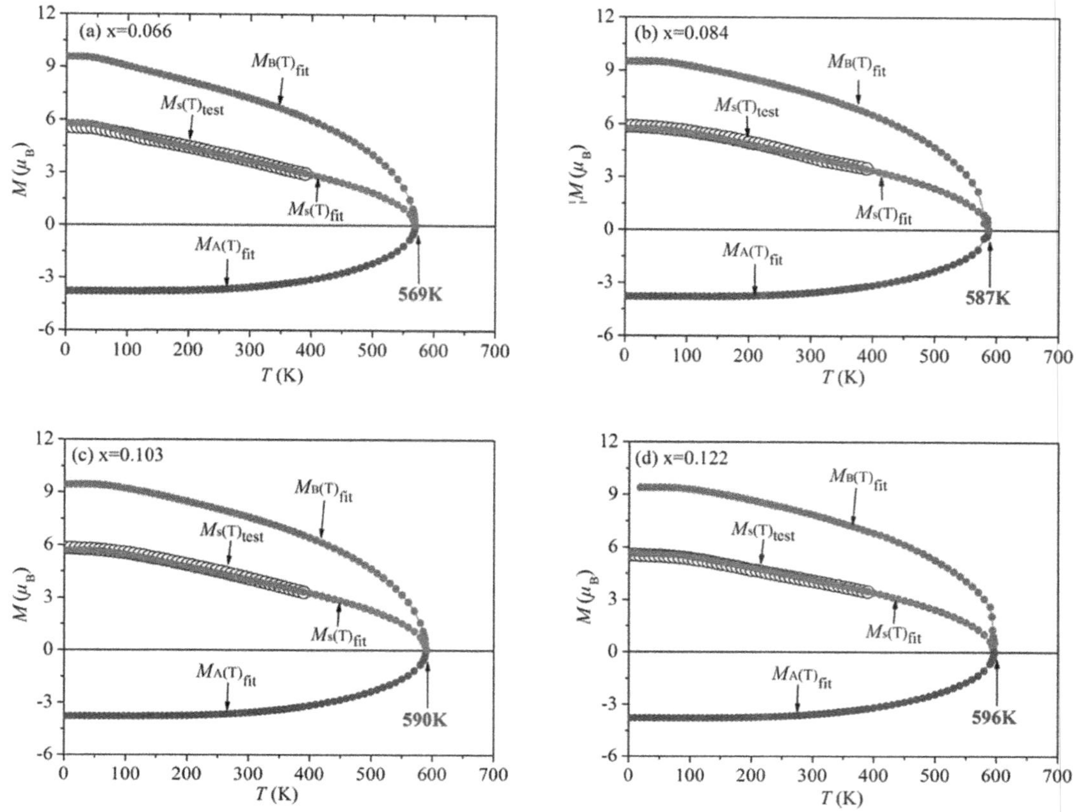

Figure 5.5 Experimental and fitting molecule magnetic moment versus temperature for $Mn_{0.506-x}Zn_{0.244}Ni_xFe_{2.250}O_{4.0}$ ferrites.

Table 5.2 The molecular field coefficients and Curie temperature, T_c, for $Mn_{0.506-x}Zn_{0.244}Ni_xFe_{2.250}O_{4.0}$ ferrites: $T_{c(test.)}$ and $T_{c(fit.)}$ represent the measuring and fitting Curie temperature, respectively.

x	ω_{AA}	ω_{AB}	ω_{BB}	$T_{c(test.)}$(K)	$T_{c(fit.)}$(K)
0.066	-359	292	-90	582	569
0.084	-387	300	-83	595	587
0.103	-401	303	-81	599	590
0.122	-417	309	-78	606	596

Table 5.3 Fitting $T_{c(fit.)}$, calculated $T_{c(cal.)}$, experimental $T_{c(test.)}$, and molecular field coefficients for selected NiCuZn ferrites.

				T_c (K)		
t	ω_{AA}	ω_{AB}	ω_{BB}	Fit.	Cal.	Test.
0.1854	−199	275	-45	745	745	730
0.2641	-196	271	-45	725	724	706
0.2976	-194	267	-45	706	705	683
0.3221	-192	261	-45	679	678	660
0.3526	-190	257	-45	662	658	637

with

$$\theta = \frac{\omega_{AB}\left(\alpha\lambda^2 C_A^2 + \beta\mu^2 C_B^2 - 2\lambda\mu C_A C_B\right)}{\lambda C_A + \mu C_B}$$

$$T_C' = \frac{\omega_{AB}\lambda\mu C_A C_B(\alpha+\beta+2)}{\lambda C_A + \mu C_B}$$

$$\zeta = \frac{\omega_{AB}^2\lambda\mu C_A C_B\left[(\alpha+1)\lambda C_A - (\beta+1)\mu C_B\right]^2}{\left(\lambda C_A + \mu C_B\right)^3} \tag{5.25}$$

$$C_A = \frac{N\mu_B^2 g^2 S_A(S_A+1)}{3k_B}$$

$$C_B = \frac{N\mu_B^2 g^2 S_B(S_B+1)}{3k_B}$$

where the T_c' can be obtain from Equation (5.25) and T_p deduced from $1/\chi_m = 0$. Table 5.4 shows the calculated T_c' and paramagnetism T_p derived from the reciprocal of $1/\chi_m$ in select NiCuZn ferrites. It is, hence, found that T_c' is very similar to $T_{c(cal.)}$ and $T_{c(fit.)}$ (see Table 5.3).

5.5.2 Garnets

Garnet ferrites are represented by the general formula $R_3Fe_5O_{12}$, where R is a rare earth trivalent cation, commonly the yttrium cation (see Figure 5.6). A unit cell of garnet consists of eight formula units, where the oxygen sublattice forms three types of polyhedra sites: these are the dodecahedral (eightfold), octahedral (sixfold), and tetrahedral (fourfold) sites. All the polyhedral are distorted, where the cation distribution is defined as $\{R_3\}[Fe_2](Fe_3)O_{12}$, where {}, [], and () represent dodecahedral (24*c*), octahedral (16*a*), and tetrahedral (24*d*) sites, respectively [13,14].

In yttrium iron garnets (YIG), the magnetic sublattices are [16*a*] and (24*d*) sublattices due to the diamagnetic Y^{3+} completely occupying the {24*c*} sublattice. When Fe^{3+} ions are replaced by diamagnetic substitutions at higher levels, its thermodynamic properties departed from the Néel molecular field model. In 1964, Geller [15] indicated that the tendency toward antiferromagnetic alignment with lower Curie temperature caused the canting effects more readily measured in (24*d*) sublattices. Dionne then proposed the approximation that for substitutions in one sublattice, the molecular field coefficient of the opposite sublattice is reduced in magnitude while its own coefficient remains unchanged, and for substitutions in either sublattice, the intersublattice coefficient ω_{dd} is also reduced. By comparing

Table 5.4 The calculated T_c' and paramagnetism T_p in selected NiCuZn ferrites.

T	C_A	C_B	T_c'(K)	T_p(K)
0.1854	4.3747	2.6460	742.43	753.38
0.2641	4.3747	2.7864	718.93	720.31
0.2976	4.3747	2.8472	700.95	701.03
0.3221	4.3747	2.8921	680.60	680.73
0.3526	4.3747	2.9484	660.79	662.14

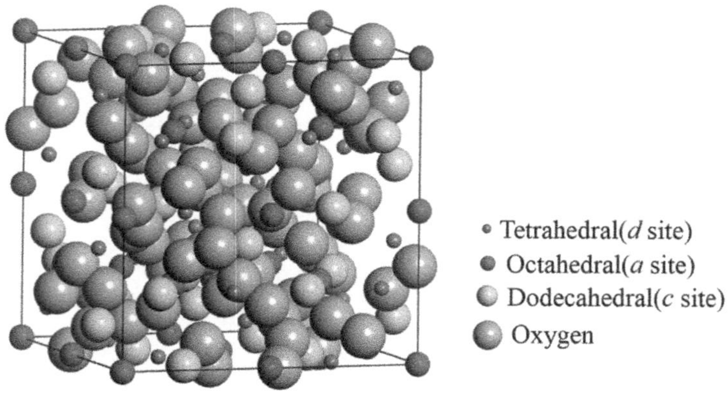

Figure 5.6 Schematic representation of garnet structure.

the magnitudes of the coefficients with the changes in sublattice moment from canting, the empirical functions of $M_d(0)$ and $M_a(0)$ are given by:

$$\begin{cases} M_d(0) = 3gS\mu_B N(1-k_d)(1-0.1k_a) \\ M_a(0) = 2gS\mu_B N(1-k_a)(1-k_d^{5.4}) \end{cases} \tag{5.26}$$

where k_a and k_d represent the fractions of diamagnetic ions substituted for Fe^{3+} ions in the respective sublattices. Combining Equation (5.26) with Equations (5.3)–(5.5), the linear relations of the molecular field coefficients for the $\{Y_3\}[Fe_{2-x}P_x](Fe_{3-y}Q_y)O_{12}$ system with diamagnetic substitutions P and Q for $x \leq 0.7$ and $y \leq 1.95$ are given by:

$$\begin{cases} \omega_{dd} = -30.4(1-0.43x) \\ \omega_{aa} = -65(1-0.42x) \\ \omega_{ad} = 97(1-0.125x-0.127y) \end{cases} \tag{5.27}$$

It is emphasized that Dionne suggested the antiferromagnetic transition occurs when ω_{aa} and ω_{dd} are around –20 mole/cm^3.

In rare earth iron garnet systems, the magnetic interactions associated with {24*c*} sublattices are generally thought to be very weak, Dionne [16] assumed that spin canting appears to cause a significant reduction in the net moment of {24*c*} sublattices. Hence, the total angular momentum J' had been redefined as:

$$J' = g\mu_B(L+S)\cos\varphi \tag{5.28}$$

where L, S, and φ represent the orbital number, the spin quantum number, and the canting angle, respectively. By contrast with the molecular field coefficients derived from the paramagnetic susceptibility measurement and with other values determined without taking into account the spin canting, the molecular field coefficients of rare earth garnets were calculated and presented in Table 5.5.

5.5.3 Hexaferrites

In contrast to the spinel and garnet ferrites, hexaferrites have more complicated crystal structures with hexagonal symmetry. Among the most popular of hexaferrites (M-, Y-, U-, X-, and Z-) are those derived from M-type hexaferrites ($BaFe_{12}O_{19}$, BaM). As shown in Figure 5.7, the magnetoplumbite structure of BaM is described as RSR^*S^*, where $R = (Ba^{2+}Fe_6^{3+}O^{2-}_{11})^{2-}$ is a barium-containing hexagonal block with three oxygen layers, and $S = (Fe_6^{3+}O^{2-}_8)^{2+}$ is a spinel block with only two layers: R^* and S^* are obtained from R and S blocks, respectively, by a rotation of 180° about the c-axis. And its cation distribution is given by

$$\begin{array}{lllll} (Ba^{2+}Fe_2^{3+} & \left[Fe_2^{3+}\right]O_3^{2-} & Fe_2^{3+} & \left[Fe_2^{3+}\right] & \left[Fe_2^{3+}\right]O_3^{2-})_2 \\ 4f_2(\downarrow\downarrow) & 2b(\uparrow) & 4f_1(\downarrow\downarrow) & 2a(\uparrow) & 12k(6\uparrow) \end{array}$$

Table 5.5 The molecular field coefficients in moles/cm^3 of rare earth garnets.

{24c}	J'	Φ (°)	ω_{ac}	ω_{cd}	ω_{ac}[a]	ω_{cd}[a]	ω_{ac}[b]	ω_{cd}[b]
Gd^{3+}	3.5	0	−3.44	6.0	−1.2	3.4	–	–
Tb^{3+}	4.6	40	−4.5	6.5	−4.4	4.6	−1.8	3.4
Dy^{3+}	5.3	45	−5.2	6.0	−3.2	3.6	−3.35	3.95
Ho^{3+}	4.98	51.5	−1.5	4.0	−4.0	2.4	−0.75	1.5
Er^{3+}	4.62	52	−0.5	2.2	−0.6	1.0	−0.75	1.25
Yb^{3+}	1.49	64.8	−6.0	8.0	−1.0	8.8	−1.7	2.0

a) Data derived from paramagnetic susceptibility measurements.
b) Results from other calculations without considering spin canting.

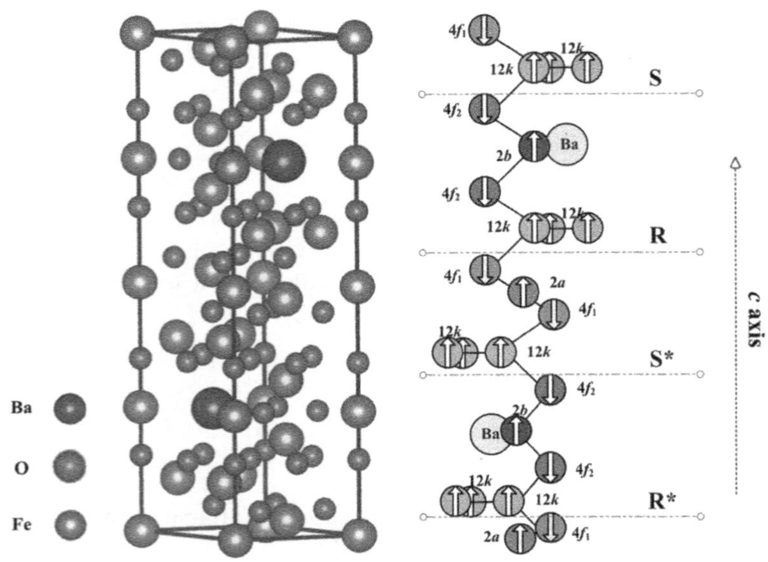

Figure 5.7 The unit cell and spin configurations for $BaFe_{12}O_{19}$: The arrows represent the local magnetic moment at each atom site.

where 2a, 2b, $4f_1$, $4f_2$, and 12k denote five different sublattices. The 24 Fe^{3+} ions are distributed over five different sublattices: three parallel (2a, 2b, 12k) sites and two antiparallel ($4f_1$ and $4f_2$) sites [17, 18].

Recently, cation distributions within the substituted BaM have been investigated by various techniques, such as Mössbauer, nuclear magnetic resonance (NMR), extended X-ray absorption fine structure (EXAFS), Rietveld refinement, and predictive DFTs [19, 20]. The Alkali metal and rare earth ions mainly occupy Ba^{2+} sites, with 3d transition metal ions residing on the Fe^{3+} sites. La–Co substitution have become typical examples of cation doping in BaM systems. The calculation of molecular field coefficients of substituted BaM pursue is [21]:

$$M_s(T) = 6M_{12k}(T) - 2M_{4f_1}(T) - 2M_{4f_2}(T) + M_{2b}(T) + M_{2a}(T)$$

where $M_i\,(T)$ is the magnetization of one Fe^{3+} in site i at the temperature T and $M_s\,(T)$ is saturation magnetization per molecular formula of barium hexaferrites. As the exchange parameters J_{bf2} (35.96 K/μ_B), J_{kf1} (19.63 K/μ_B), J_{af1} (18.15 K/μ_B), J_{kf2} (4.08 K/μ_B), and J_{bk} (3.69 K/μ_B) are far greater than the others (less than 0.1 K/μ_B) in BaM, the molecular field coefficients ω_{bf2}, ω_{kf1}, ω_{af1}, ω_{kf2}, and ω_{bk} could be taken into account. According to the Néel model, the Brillouin functions is:

$$M_i(T) = M_i(0)B_{J_i}(y_i) \tag{5.29}$$

$$B_{J_i}(y_i)=\frac{2J_i+1}{2J_i}\coth\left(\frac{2J_i+1}{2J_i}y_i\right)-\frac{1}{2J_i}\coth\left(\frac{1}{2J_i}y_i\right)$$

with a series of simplified y_i

$$\begin{cases} y_{2a}=\dfrac{gJ_{2a}\mu_B}{k_BT}\omega_{af1}\cdot M_{4f_1}(T) \\ y_{2b}=\dfrac{gJ_{2b}\mu_B}{k_BT}[\omega_{bf2}\cdot M_{4f_2}(T)+\omega_{bk}\cdot M_{12k}(T)] \\ y_{4f_1}=\dfrac{gJ_{4f_1}\mu_B}{k_BT}[\omega_{af1}\cdot M_{2a}(T)+\omega_{kf1}\cdot M_{12k}(T)] \\ y_{4f_2}=\dfrac{gJ_{4f_2}\mu_B}{k_BT}[\omega_{bf2}\cdot M_{2b}(T)+\omega_{kf2}\cdot M_{12k}(T)] \\ y_{12k}=\dfrac{gJ_{12k}\mu_B}{k_BT}[\omega_{bk}\cdot M_{2b}(T)+\omega_{kf1}\cdot M_{4f_1}(T)+\omega_{kf2}\cdot M_{4f_2}(T)] \end{cases} \tag{5.30}$$

To calculate the molecular field coefficients, simultaneous equations have been fitted by a least square method to determine the experimental values of the saturation magnetization. Figure 5.8 demonstrates the fitting results of $M_s \sim T$ in $Ba_{1-x}La_xFe_{12-x}Co_xO_{19}$, and the fitted molecular field coefficients, and Curie temperature ($T_{c(test.)}$ and $T_{c(fit.)}$) are listed in Table 5.6. For BaM, the molecular field coefficient ω_{bk} shows an opposite sign to the ω_{bf2}, ω_{kf1}, ω_{af1}, and ω_{kf2}.

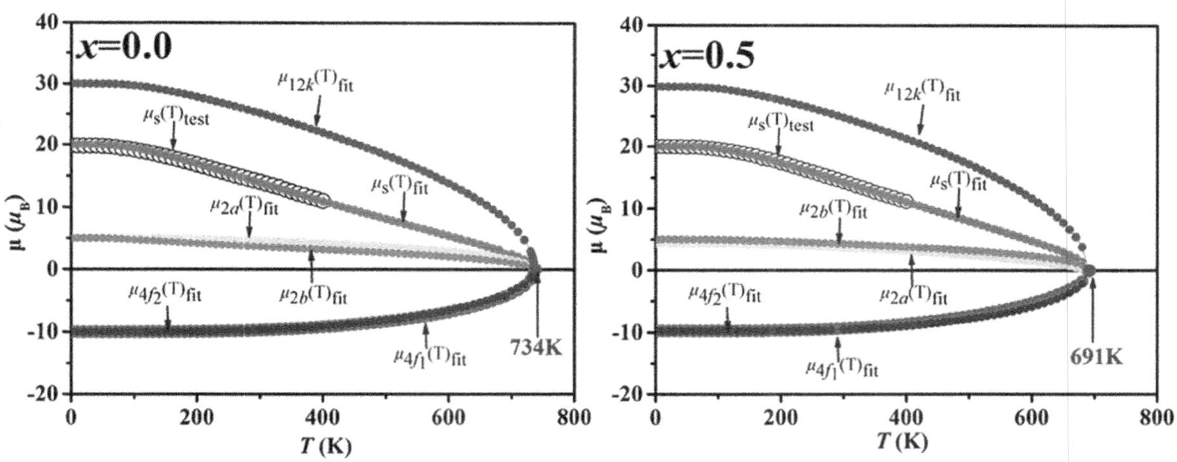

Figure 5.8 Magnetic moment μ_s vs. temperature T for $Ba_{1-x}La_xFe_{12-x}Co_xO_{19}$ samples.

Table 5.6 Molecular field coefficients and Curie temperature ($T_{c(test.)}$ and $T_{c(fit.)}$) for $Ba_{1-x}La_xFe_{12-x}Co_xO_{19}$ samples.

Samples	ω_{bf2}	ω_{kf1}	ω_{af1}	ω_{kf2}	ω_{bk}	$T_{c(test.)}$	$T_{c(fit.)}$
$x = 0.0$	90	30	71	29	−19	712	734
$x = 0.5$	81	28	62	27	−17	685	694

5.6 Exchange Interactions

The Néel theory provides one model that successfully explains the relationships between spontaneous magnetization and temperature in ferrimagnetic and ferromagnetic systems. However, the nature of molecular fields was not deeply explored at the time. In 1928, Frenkel suggested the spontaneous magnetization originates from the specific electrostatic electron–electron interaction (which is subsequently regarded as an exchange interaction) within a quantum mechanical framework. Heisenberg [22] then developed a systematic ferromagnetic exchange theory based on hydrogen exchange interaction. After that the Anderson theory [23] describing ferromagnetic/ferrimagnetic systems and the Ruderman–Kittel–Kasuya–Yosid (RKKY) theory [24] describing rare earth metal systems were proposed. All these theories focus on the cognition that exchange interaction between the neighboring magnetic ions will force the individual moments into parallel or antiparallel alignment with their neighbors as the lowest energy configuration. According to the exchange types between magnetic ions, exchange interaction could then be divided into direct and indirect exchanges.

5.6.1 Direct Exchange

Direct exchange operates between moments, which are close enough to have sufficient overlap of their wavefunctions. It gives a strong but short-range coupling, which decreases rapidly as the ions are separated. Direct exchange was originally proposed by Heisenberg, and he gave these assumptions in a system with N magnetic sublattices:

i) Considering the distance between each other is very large, only one electron spin magnetic moment of one given atom possesses the contribution to ferromagnetism.
ii) Assuming the polarization state of atoms, only the electron–electron interactions between different atoms would be taken into account.

According to description (i), the exchange interaction in this system can be written as:

$$E_{\text{ex}} = -2A\sum_{i,j}\sigma_i\sigma_j \tag{5.31}$$

where σ_i and σ_j represent the spin-angle quantum number of i and j electrons and A is the exchange integral in this system. For i electron, as its effect from each neighboring electron can be thought to be equivalent, the neighboring interaction is defined by

$$E_{\text{ex}}^i = -(2A\sum_{i,j}\sigma_i)\sigma_j = -g\mu_B S_i \cdot H_{\text{mf}} \tag{5.32}$$

In Equation (5.32), H_{mf} is the molecular field that was introduced in the preceding section. Thus, Heisenberg established the close relationship between molecular fields and exchange interactions. According to Hund's rule and description (ii), the final exchange interaction was obtained as:

$$E_{\text{ex}} = -2A\sum_{i<j}S_iS_j \tag{5.33}$$

where S_i and S_j is the total spin-angle quantum number of i and j electrons.

5.6.2 Indirect Exchange

Indirect exchange couples moments over relatively large distances through nonmagnetic conduction often via cation–anion–cation connectivity. In rare earth metals whose magnetic electrons in the 4f shell are shielded by 5s and 5p electrons, direct exchange is rather weak and indirect exchange via the conduction electrons gave rise to magnetic order. This exchange is also known as the RKKY interaction. While in ferromagnetic and ferrimagnetic systems (e.g. ferrites), their

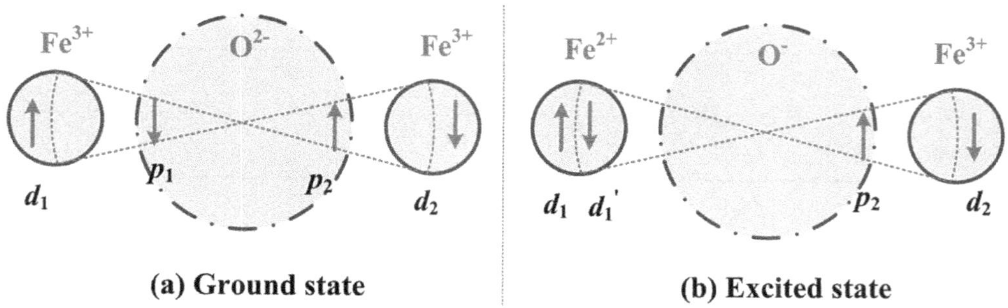

Figure 5.9 The typical Fe^{3+}-O^{2-}-Fe^{3+} interactions in ferrite systems: In order to introduce conveniently, two Fe^{3+} ions are assumed to be antiparallel.

indirect interactions (i.e. superexchange interactions) between metal ions are mediated by nonmagnetic anions. As shown in Figure 5.9, it evolves from the typical Fe^{3+}–O^{2-}–Fe^{3+} interactions in ferrites. In the ground state (see Figure 5.9a), the Fe^{3+} electron configuration of the valence shell is $3d^5$, and it is $2p^6$ in O^{2-}. Although spontaneous magnetization could not occur due to nonmagnetic O^{2-} ions residing between the two Fe^{3+}cations, both Fe^{3+} and O^{2-} ions could experience the overlap of their wavefunctions. When it occurs, one electron transfers from O^{2-} to Fe^{3+}, and the systems would develop into an excited state (see Figure 5.9b) with Fe^{2+} and O^- ions. The magnetic O- thereby produces direct interaction with other Fe^{3+}. In 1950, the description of superexchange interaction was proposed by Anderson:

$$E_{ex} = -\sum_{R_l \neq R_m} A_{in}(R_l R_m) S(R_l) \cdot S(R_l) \tag{5.34}$$

where A_{in} is the indirect-exchange integral, R_l and R_m are the locations of two electrons. Equation (5.34) is similar in form to that of the Heisenberg model.

So far, the first-principles calculation had been proven to be the only valid way of obtaining estimations of the exchange-interaction strength in the limit of long wavelength magnons including ferrites [25, 26]. This calculation derives from the condition that spin-wave energies should be small as compared to the band with exchange splitting (i.e. adiabatic approximation). Therefore, at this time, we introduce the calculations of exchange interaction in ferrites by way of a first-principles DFT approach. The reader is encouraged to read Chapter 4 of this text for a more in-depth treatment of DFT as applied to ferrite systems.

5.7 Exchange Interactions in Ferrites

5.7.1 Spinels

In order to obtain an accurate calculation of the exchange integrals and the energy gap, values of the transfer integral (t), on-site Coulomb repulsion (U), and charge-transfer energy (D) are required. These could not be ascertained in a self-consistent fashion using the Hubbard model or the molecular field theory. In 2002, Zuo and Vittoria [27] introduced a mixture of Fock and Becke exchange to improve the estimations of the transfer integral, band gap, and exchange correlation:

$$E_X = (1-\omega)(E_X^{\mathrm{LSDA}} + 0.9\Delta E_X^B) + \omega E_X^{\mathrm{exact}} \tag{5.35}$$

where E_X^{exact}, E_X^{LSDA}, and ΔE_X^B are the exact exchange, local spin-density approximation (LSDA) exchange, and the Becke gradient correction. The weight ω controls the balance between LSDA-based and Fock exchanges. This approach had been applied to calculate the exchange integrals of Mn_2FeO_4 [27], $NiFe_2O_4$ [28], and $CuFe_2O_4$ [29] systems by energy differences within the framework of the DFT. As shown in Table 5.7, the spin structure and their energy differences from ferrimagnetic structure in $NiFe_2O_4$ systems were listed by Xu et al. calculated exchanged integrals as a function of the weight ω are

Table 5.7 Spin structure and their energy differences from ferrimagnetic structures.

Spin structure				**Energy difference**
Designation	**Fe(A)**	**Ni(B)**	**Fe(B)**	
Ferrimagnetic	↑↑	↑↑	↑↑	—
Ferrimagnetic-1	↓↓	↑↑	↑↑	$E_1 = 48J_{NiFe}^{(AB)}S_{Ni}S_{Fe} + 48J_{FeFe}^{(AB)}S_{Fe}S_{Fe}$
Ferrimagnetic-2	↑↓	↑↑	↑↑	$E_2 = 16J_{FeFe}^{(AA)}S_{Fe}S_{Fe} + 24J_{NiFe}^{(AB)}S_{Ni}S_{Fe} + 24J_{FeFe}^{(AB)}S_{Fe}S_{Fe}$
Ferrimagnetic-3	↑↑	↑↑	↓↓	$E_3 = 48J_{FeFe}^{(AB)}S_{Fe}S_{Fe} + 32J_{NiFe}^{(BB)}S_{Ni}S_{Fe}$
Ferrimagnetic-4	↑↑	↓↓	↑↑	$E_4 = 48J_{NiFe}^{(AB)}S_{Ni}S_{Fe} + 32J_{NiFe}^{(BB)}S_{Ni}S_{Fe}$
Ferrimagnetic-5	↑↑	↑↑	↑↓	$E_5 = 24J_{FeFe}^{(AB)}S_{Fe}S_{Fe} + 16J_{NiFe}^{(BB)}S_{Ni}S_{Fe} + 8J_{FeFe}^{(BB)}S_{Fe}S_{Fe}$
Ferrimagnetic-6	↑↑	↑↓	↑↑	$E_6 = 24J_{NiFe}^{(AB)}S_{Ni}S_{Fe} + 8J_{NiNi}^{(BB)}S_{Ni}S_{Ni} + 16J_{NiFe}^{(BB)}S_{Ni}S_{Fe\ e}$

represented in Figure 5.10. Zuo et al. indicated that the best fit to the experimental results at $\omega = 0.3$ in Mn_2FeO_4 systems, and $\omega = 0.4$ in $NiFe_2O_4$ and $CuFe_2O_4$ systems.

5.7.2 Garnets

Owing to nonmagnetic Y^{3+}, the superexchange interactions in YIG arise from [16a]–[16a], [16a]–(24d), and (24d)–(24d) interactions. Applying the first-principles calculations, the next nearest-neighbor exchange interactions J_{ij} are usually neglected. Therefore, Xie et al. [30] divided the calculated J_{ij} into two different cases: (1) only nearest-neighbor exchange interactions (NN case), (2) both nearest-neighbor and next-neighbor exchange interactions (NNN case). Ten different spin configurations (SCs) analogy to Zuo's methods were used to determine the exchange integrals (see Table 5.8). From Table 5.8, the difference between the total energies and the spin model $|\Delta E_{NNN}|$ ($<<|\Delta E_{NNN}|$) has a significantly smaller value of 0.66%, which might be acceptable, that is, Xie et al. suggested minimal reliance on the experimental data must put the justification for both the nearest and the next-neighbor exchange interactions on a more solid footing in YIG systems. By substituting the four equations $E_{aa} = J_{aa}S_aS_a$, $E_{dd} = J_{dd}S_dS_d$, and $E_{ad} = J_{ad}S_aS_d$ into the expression of E'_{tot} in Table 5.8, the exchange integrals of NN and NNN models can be obtained in Table 5.9.

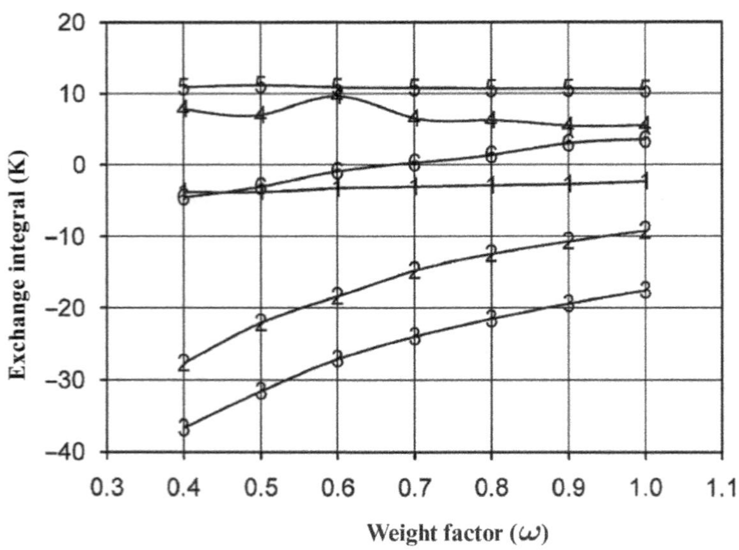

Figure 5.10 Calculated exchange integrals as a function of weight ω: (1) $J_{FeFe}^{(AA)}$, (2) $J_{NiFe}^{(AB)}$, (3) $J_{FeFe}^{(AB)}$, (4) $J_{NiNi}^{(BB)}$, (5) $J_{NiFe}^{(BB)}$, and (6) $J_{FeFe}^{(BB)}$.

Table 5.8 Total energies (meV) for different spin configurations: E_{cal} is the total energies for $U - J = 4.7$ eV relative to the ground state, ΔE_{NNN} and ΔE_{NN} represent the differences between the fitted total energies from the NNN and NN cases.

SC	E'_{tot}	E_{cal}	ΔE_{NNN}	ΔE_{NN}
1	$E_{tot} + 24E_{aa} + 48E_{dd} + 48E_{ad}$	0.00	0.37	−59.97
2	$E_{tot} + 24E_{aa} + 48E_{dd} - 48E_{ad}$	4225.32	−0.31	−3.69
3	$E_{tot} + 24E_{aa} + 16E_{dd} + 32E_{ad}$	1907.02	0.39	−58.19
4	$E_{tot} + 24E_{aa} + 16E_{dd} + 16E_{ad}$	566.01	0.38	44.42
5	$E_{tot} + 12E_{aa} + 32E_{dd} + 32E_{ad}$	778.86	0.23	5.97
6	$E_{tot} + 24E_{aa} + 48E_{dd}$	1987.42	−0.21	−36.19
7	$E_{tot} + 24E_{aa} + 48E_{dd}$	1228.54	0.24	52.29
8	$E_{tot} + 24E_{aa} + 16E_{dd}$	1848.59	−3.04	43.44
9	$E_{tot} + 24E_{aa} + 16E_{dd}$	1885.68	−7.62	49.55
10	$E_{tot} + 24E_{aa} + 48E_{dd}$	2018.23	−13.40	−39.80

Table 5.9 Exchange integrals (meV) in NN and NNN cases.

J_{ad}	J_{dd}	J_{aa}	J'_{ad}	J'_{dd}	J'_{aa}	methods
3.176	0.223	0.112	—	—	—	NN case (GGA + U, $U - J = 3.7$ eV)
2.917	0.213	0.09	0.218	0.228	0.005	NNN case (GGA + U, $U - J = 3.7$ eV)
2.584	0.160	0.091	—	—	—	NN case (GGA + U, $U - J = 4.7$ eV)
2.387	0.154	0.072	0.163	0.179	0.04	NNN case (GGA + U, $U - J = 4.7$ eV)

5.8 Hexaferrites

Novák and Rusz [31] investigated the total density of states (DOSs) for the ground state spin arrangement in BaM calculated with LSDA, GGA, and GGA + U employing the WIEN2k program and suggested that an insulating state was only obtained with the GGA + U. Wu et al. [32] thereby calculated the exchange integral and Curie temperature in La-substituted BaM systems, which is mentioned next.

When the spin of one single sublattice is inverted relative to the ground state, the difference $\Delta(\alpha_i - \alpha_0)$ between the exchange energies of ground state α_i and excited state α_0 is obtained based on Heisenberg Hamiltonian:

$$\Delta(\alpha_i - \alpha_0) = -2S_i n_i \sum_{j \neq i} z_{ij} J_{ij} S_j \sigma_i^{(\alpha_0)} \sigma_j^{(\alpha_0)} \tag{5.36}$$

where i and j represent two sublattices, n_i the number of ith sublattice, z_{ij} the number of neighboring sites in the jth sublattice to the ith sublattice, J_{ij} the exchange integrals, and S_i and S_j are the spins in the ith and jth sublattices, respectively. For the spins of two sublattices inverted, the difference $\Delta(\alpha_{ij} - \alpha_0)$ could be also given by:

$$\Delta(\alpha_{ij} - \alpha_0) = \Delta(\alpha_i - \alpha_0) + \Delta(\alpha_j - \alpha_0) + 4n_i z_{ij} J_{ij} S_i S_j \sigma_i^{(\alpha_0)} \sigma_j^{(\alpha_0)} \tag{5.37}$$

Hence, they developed the relationship between total energies in these systems and exchange integrals:

$$J_{ij} = \frac{\Delta(\alpha_{ij} - \alpha_0) - \Delta(\alpha_i - \alpha_0) - \Delta(\alpha_j - \alpha_0)}{4n_i z_{ij} S_i S_j \sigma_i^{(0)} \sigma_j^{(0)}} \tag{5.38}$$

Figure 5.11 shows their calculations of exchange integrals as a function of the effective U_{eff} employing Vienna *Ab initio* Simulation Package (VASP). It was observed that the strength of superexchange interaction in BaM systems mainly depends on the 2b–$4f_1$, $4f_2$–12k, 2a–$4f_1$, and $4f_1$–12k interactions. In the following, the Curie temperature T_c derived from the aforementioned exchange integrals had been then estimated by the mean-field approximation (MFA) and random-phase approximation (RPA) methods. Table 5.10 lists their final results of the calculated Curie temperature T_c. The MFA and RPA estimations provide an upper and lower bound of the Curie temperature. As the arithmetic average takes all the magnon values with the equal weight in MFA, and in RPA, it is the harmonic average, the T_c determined by the RPA is in better agreement with the experimental T_c.

In a related work, Geiler et al. [20] used a novel alternating target laser ablation deposition technique and showed that Mn-dopant cations were placed in specific interstitial sites of $BaFe_{12}O_{19}$ thin films as opposed to being distributed throughout the unit cell as in conventional bulk materials. The distribution of Mn cations has been confirmed experimentally and

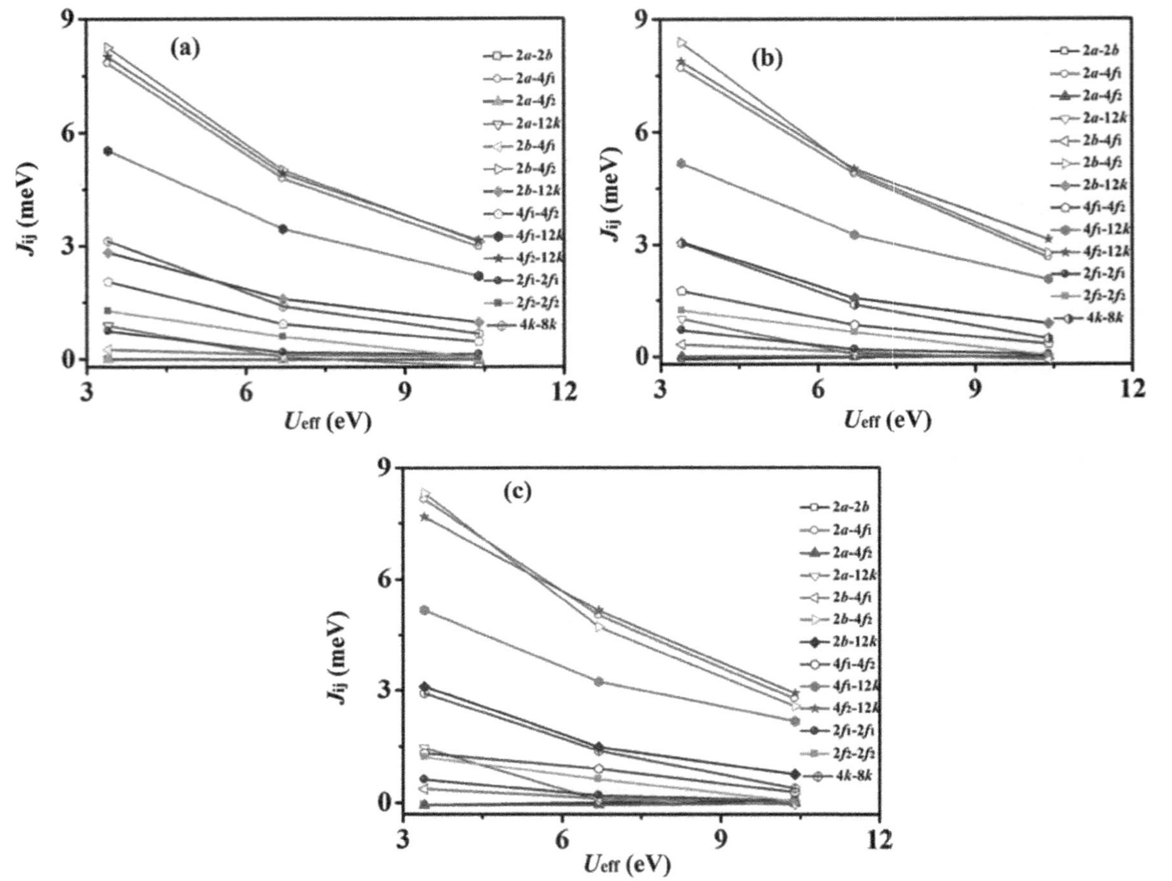

Figure 5.11 The exchange integrals as a function of U_{eff} for $Ba_{1-x}La_xFe_{12}O_{19}$ systems: (a) $x = 0.0$, (b) $x = 0.5$, and (c) $x = 1.0$.

Table 5.10 The experimental (EV) and calculated Curie temperatures (K) for $Ba_{1-x}La_xFe_{12}O_{19}$ systems.

	$x = 0.0$			$x = 0.5$			$x = 1.0$		
U_{eff} (eV)	3.4	6.7	10.4	3.4	6.7	10.4	3.4	6.7	10.4
EV	723	723	723	705	705	705	695	695	695
MFA	1514	980	629	1450	951	581	1419	928	579
RPA	1009	653	419	967	634	387	946	619	386

predicted by LSDA + GGA. As a result of site selection, the saturation magnetization increased by 12–22% and the Néel temperature increased by 40–60 K compared with bulk materials.

5.9 Summary

In this chapter, we introduced and applied two essential theories of magnetism: molecular field and exchange-interaction theories. In molecular field theory, an internal suppositional field analogy to external applied field had been proposed by Weiss. Néel subsequently expanded this notion to ferrimagnetisms and ferromagnetisms. He suggested the valence electrons would be localized at the atoms and contribute a magnetic moment. Néel molecular field theory successfully explained the relationships between spontaneous magnetization and temperature. Unfortunately, he could not give a concrete description of the nature of this internal suppositional field. The establishment of exchange-interaction theory gave a final response: spontaneous magnetization originates from electrostatic electron–electron interactions. By far, the most successful method currently available for calculating exchange integrals is the DFT. No assumptions in DFT are made as to whether the electrons are localized or itinerant, rather the electrons choose the arrangement that will give them the lowest possible total energy. However, the first-principles calculations are both computationally intensive and difficult because of the fact that the exact form of the exchange correlation of interelectronic interaction energy is unknown, especially for large clusters of atoms (i.e. supercells).

References

1 Weiss, P. (1907). L'hypothèse du champ moléculaire et la propriété ferromagnétique. *Journal de Physique* 6: 661–690.
2 Néel, L. (1970). Magnetism and local molecular field. *Science* 174: 985–992.
3 Yafet, Y. and Kittel, C. (1952). Antiferromagnetic arrangement in ferrites. *Physical Review* 87: 290–294.
4 Smart, J.S. (1955). The Néel theory of ferrimagnetism. *American Journal of Physics* 6: 356–370.
5 Sickafus, K.E., Wills, J.M., and Grimes, N.W. (1999). Structure of spinel. *Journal of the American Ceramic Society* 82: 3279–3292.
6 Goodenough, J.B. and Loeb, A.L. (1955). Theory of ionic ordering, crystal distortion, and magnetic exchange due to covalent forces in spinels. *Physical Review* 98: 391–408.
7 Harris, V.G. (2012). Modern microwave ferrites. *IEEE Transactions on Magnetics* 48: 1075–1104.
8 Sláma, J., Grusková, A., Šoka, M. et al. (2012). Analysis of magnetic properties of substituted Li ferrites. *IEEE Transactions on Magnetics* 46: 455–458.
9 Gorter, E.W. (1954). Saturation magnetization and crystal chemistry of ferromagnetic oxide. *Philips Research Reports* 9: 321–365.
10 Dionne, G.F. (1974). Molecular-field coefficients of Ti^{4+}- and Zn^{2+}- substituted lithium ferrites. *Journal of Applied Physics* 45: 3621–3626.
11 Sun, K., Wu, C., Yang, Y. et al. (2015). Cation distribution and the temperature dependence of Brillouin function for nickel-substituted MnZn ferrites. *IEEE Transactions on Magnetics* 51: 1–4.
12 Guo, R., Yu, Z., Yang, Y. et al. (2015). Relationship between Curie temperature and Brillouin function characteristics of NiCuZn ferrites. *Journal of Applied Physics* 117: 073905 1–6.
13 Geller, S. and Gilleo, M.A. (1957). The crystal structure and ferrimagnetism of yttrium-iron garnet, $Y_3Fe_2(FeO_4)_3$. *Journal of Physics and Chemistry of Solids* 3: 30–36.
14 Xu, Y.N. and Ching, W.Y. (1999). Electronic structure of yttrium aluminum garnet ($Y_3Al_5O_{12}$). *Physical Review B* 59: 10530–10535.
15 Dionne, G.F. (1970). Molecular field coefficients of substituted yttrium iron garnets. *Journal of Applied Physics* 41: 4874–4881.
16 Dionne, G.F. (1976). Molecular-field coefficients of rare-earth iron garnets. *Journal of Applied Physics* 47: 4220–4221.
17 Obradors, X., Solans, X., Collomb, A. et al. (1988). Crystal structure of strontium hexaferrite $SrFe_{12}O_{19}$. *Journal of Solid State Chemistry* 72: 218–224.
18 Dixit, V., Nandadasa, C.N., Kim, S.G. et al. (2015). Site occupancy and magnetic properties of Al-substituted M-type strontium hexaferrite. *Journal of Applied Physics* 117: 243904 1–7.

19 Wu, C., Yu, Z., Yang, Y. et al. (2016). Computational and experimental study on the cation distribution of La-Cu substituted barium hexaferrites. *Journal of Alloys and Compounds* 664: 406–410.

20 Geiler, A.L., Yang, A., Zuo, X. et al. (2008). Atomic scale design and control of cation distribution in hexagonal ferrites. *Physical Review Letters* 101: 067201 1–4.

21 Wu, C., Yu, Z., Yang, Y. et al. (2015). Brillouin function characteristics for La-Co substituted barium hexaferrites. *Journal of Applied Physics* 118: 103907 1–5.

22 Nesbet, R.K. (1961). Molecular model of the Heisenberg exchange interaction. *Physical Review* 122: 1497–1508.

23 Anderson, P.W. (1959). New approach to the theory of superexchange interactions. *Physical Review* 115: 2–12.

24 Sherafati, M. and Satpathy, S. (2011). RKKY interaction in graphene from the lattice Green's function. *Physical Review B* 83: 165425 1–8.

25 Katsnelson, M.I. and Lichtenstein, A.I. (2000). First-principles calculations of magnetic interactions in correlated systems. *Physical Review B* 61: 8906–8912.

26 Rusz, J., Bergqvist, L., Kudrnovský, J. et al. (2006). Exchange interactions and Curie temperatures in Ni2−*x*MnSb alloys: First-principles study. *Physical Review B* 73: 214412 1–10.

27 Zuo, X. and Vittoria, C. (2002). Calculation of exchange integrals and electronic structure for manganese ferrite. *Physical Review B* 66: 184420 1–8.

28 Zuo, X., Yang, S., Baibiellini, B. et al. (2006). A computational study of nickel ferrite. 303: e432–e435.

29 Zuo, X., Yang, A., Vittoria, C. et al. (2006). Computational study of copper ferrite ($CuFe_2O_4$). *Journal of Applied Physics* 99: 08M909 1–3.

30 Xie, L.S., Jin, G.X., He, L. et al. (2016). First-Principles study of exchange interactions of yttrium iron garnet. arXiv preprint arXiv. 170.00110.

31 Novák, P. and Rusz, J. (2005). Exchange interactions in barium hexaferrite. *Physical Review B* 71: 184433 1–6.

32 Wu, C., Yu, Z., Sun, K. et al. (2016). Calculation of exchange integrals and Curie temperature for La-substituted barium hexaferrites. *Scientific Reports* 6: 36200 1–8.

6

Ferrite Magnetism

A First-Principles Approach

Xu Zuo

College of Electronic Information and Optical Engineering, Nankai University, Tianjin 300071, P. R. China

6.1 A Brief Intorduction to First-Principles Calculation

Although nowadays first-principles calculations can be simply done by several clicks on a personal computer, it might not be trivial to interpret and justify the calculated results. In fact, first-principles calculations can be understood as a subtle theoretical framework that guarantees its correctness and accuracy in certain cases. Therefore, it is necessary for anyone who cares about the results by first-principles calculations to know the fundamentals. In this section, we will sketch out first-principles calculations. The readers who want more details are referred to other volumes [1, 2].

6.1.1 The Fundamental Problem

The problem of electrons in solids can never be trivial, as long as its Hamiltonian is written down:

$$\mathcal{H} = T_e + V_{eI} + V_{ee} + T_I + V_{II} \tag{6.1}$$

where the terms from left to right are the kinetic energy of electrons, electron–ion Coulomb interactions, electron–electron Coulomb interactions, kinetic energy of ions, and ion–ion Coulomb interactions. Hamiltonian (6.1) is a many-body problem [3], to which there is usually no exact solution. Thus, many (approximate) approaches have been developed, including the most successful density functional theory (DFT) [4].

An immediate simplification of Hamiltonian (6.2) is the *adiabatic* or *Bohn–Oppenheimer approximation* [5] by which the ionic kinetic energy term is ignored. Ion mass is about 1800 times that of electron. Thus, the ionic kinetic energy term that is inversely proportional to ion mass should be small and usually negligible compared with the electronic kinetic energy term. The Bohn–Oppenheimer approximation implies that, when the electrons are considered, the ions can be treated as fixed. This separates the motions of electrons and ions and makes calculations a lot easier to be formulated in theory and implemented by computer codes.

6.1.2 Single-Electron Approximation

After the Bohn–Oppenheimer approximation, Hamiltonian (6.1) is slightly simplified; the ionic kinetic energy in is neglected, and the ion–ion Coulomb interactions are constant. It can be rewritten as:

$$\mathcal{H} = T_e + V_{ext} + V_{ee} + E_{II} \tag{6.2}$$

where we rename V_{eI} as V_{ext} for the convenience of further derivations. We rewrite V_{II} as E_{II} and keep it in the Hamiltonian to preserve the charge neutrality. Otherwise, the system will be charged and, more critically, the total energy will diverse.

It is obvious that Hamiltonian (6.2) is still a many-body problem, which is usually impossible to solve even numerically. Thus, further approximations should be taken to make it at least numerically solvable. Most first-principles approaches take the *single-electron approximation* [6] by which the difficult many-body problem is simplified to the well-studied single-electron problem and, at the same time, the fundamental physics of the problem is preserved.

Modern Ferrites: Basic Principles, Processing and Properties, Volume 1, First Edition. Edited by Vincent G. Harris.

One way to take single-electron approximation is to formally ignore the electron–electron Coulomb interaction in Hamiltonian (6.1). This is the so-called *noninteracting approximation*, by which the electrons move in an effective "central field" [7]. In fact, this approximation is the central step in most first-principles approaches, such as DFT [8] and Hartree–Fock (HF) method [9]. The single-electron Hamiltonian after noninteracting approximation can be simply written,

$$h_{\text{eff}} = -\frac{1}{2}\nabla^2 + v_{\text{eff}} \tag{6.3}$$

where we adopt the atomic units. It should be noted that Hamiltonian (6.3) is for a single electron and the total Hamiltonian is simply the summation of the single ones. The effective potential v_{eff} should include the periodic potential from the positively charged ions in solids and certain corrections from the electron–electron Coulomb interactions.

Without the corrections, the kinetic energy and periodic potential of Hamiltonian (6.2) will lead to the well-known concept of *band structure* [6] – the dispersion of electron wave in a solid, that is, the energy (or frequency) as a function of the wave number. In solid, wave number is a good quantum number according to the *Bloch theorem* [5]. The band structure, or quasiparticle spectrum, is the core result of any first-principles calculation based on the single-electron approximation. The corrections in v_{eff} are crucial to align the calculated results with the experimental observations. There are many ways to formulate the corrections. For example, it can be formulated by HF approximation, where only the *exchange* from the electron–electron Coulomb interactions is included. More sophisticated corrections can include *correlation* and give a closer description to the *true* effective potential. It should be noted that the corrections are from the electron–electron Coulomb interactions and usually too large to be treated by perturbation method.

6.1.3 Hartree–Fock Approximation

HF method is the most successful first-principles approach widely applied to atoms and molecules, although it might not be suitable for solids. HF method is very similar to the noninteracting approximation and a good prototype to introduce the concept of exchange and correlation.

6.1.3.1 Basic Formulation and Introduction to Exchange

To introduce the HF method, we calculate the total energy by taking the quantum average of Hamiltonian (6.2) over the antisymmetric multi-electron wave function, the Salter determinant of single-electron spin orbitals:

$$\begin{aligned} E = \langle \Phi | \mathcal{H} | \Phi \rangle &= \sum_{i\sigma} \langle i\sigma | -\frac{1}{2}\nabla^2 + V_{\text{ext}} | i\sigma \rangle + E_{\text{II}} \\ &+ \frac{1}{2} \sum_{i\sigma_i, j\sigma_j} \langle i\sigma_i, j\sigma_j | v_{\text{ee}} | i\sigma_i, j\sigma_j \rangle - \frac{1}{2} \sum_{i\sigma, j\sigma} \langle i\sigma, j\sigma | v_{\text{ee}} | j\sigma, i\sigma \rangle \end{aligned} \tag{6.4}$$

where

$$\Phi = \frac{1}{\sqrt{N!}} \begin{vmatrix} \phi_1(\mathbf{x}_1) & \phi_1(\mathbf{x}_2) & \cdots \\ \phi_2(\mathbf{x}_1) & \phi_2(\mathbf{x}_2) & \cdots \\ \cdots & \cdots & \cdots \end{vmatrix} \tag{6.5}$$

is the multi-electron wave function, including both orbital and spin degrees of freedom in the single-electron spin orbitals ϕs by defining $\mathbf{x} \equiv \mathbf{r}\sigma$, and

$$v_{\text{ee}} = \frac{1}{|\mathbf{r} - \mathbf{r}'|} \tag{6.6}$$

is the Coulomb interaction. The first term on the right hand side of Equation (6.4) involves only one spin–orbital, and V_{ext} here is the potential from ions. The third and fourth terms are from the electron–electron Coulomb interactions and involve two-spin orbitals. The third term is called *Hartree energy*, which is, in fact, the Coulomb interaction between two electron clouds (or densities) defined by $\left|\phi_{i\sigma_i}\right|^2$ and $\left|\phi_{j\sigma_j}\right|^2$, respectively.

The fourth term is called *exchange energy*, because the order of i and j is exchanged in the Dirac ket, compared with the order in the bra. It can be shown that the (direct) exchange integral is always positive [10]. Thus, the exchange will always lower down the total energy. In addition, it is obvious that the exchange only exists between the orbitals with the same spin. In other words, there is no exchange between the orbitals with different spins. The exchange energy is indeed a result of the *Pauli exclusion principle* [11] that requires the multi-electron wave function should be antisymmetric.

The equations of single-electron spin orbitals can be derived by minimizing the total energy (6.4) with respect to all degrees of freedom. This can be accomplished by variational method, and the resulting equation of a single-electron orbital is:

$$\left(-\frac{1}{2}\nabla^2 + V_{\text{ext}} + \sum_{j\sigma_j}\langle j\sigma_j \,|\, v_{\text{ee}} \,|\, j\sigma_j\rangle\right)\phi_{i\sigma} + \left(-\sum_{j}\langle j\sigma \,|\, v_{\text{ee}} \,|\, i\sigma\rangle\phi_{j\sigma}\right) = \varepsilon_{i\sigma}\phi_{i\sigma} \tag{6.7}$$

where the first parenthesis includes the kinetic energy, the external potential, and the Hartree potential from the Hartree energy, and the second parenthesis is the exchange potential from the exchange energy. Equation (6.4) can be fit into the form of Hamiltonian (6.3), if we define that:

$$V_{\text{eff}} = V_{\text{ext}} + V_{\text{H}} + V_{\text{X}i\sigma} \tag{6.8}$$

where V_{ext} and V_{H} are the external potential and the Hartree potentials, respectively, and:

$$V_{\text{X}i\sigma} = -\sum_{j}\langle j\sigma \,|\, v_{\text{ee}} \,|\, i\sigma\rangle\frac{\phi_{j\sigma}}{\phi_{i\sigma}} \tag{6.9}$$

is the exchange potential exerted on the single-electron orbital $\phi_{i\sigma}$. It should be noted that $V_{\text{X}i\sigma}$ depends on the ith orbital $\phi_{i\sigma}$ itself, which is the fundamental difference between the Hartree potential and exchange by HF.

Equation (6.7) is a set of equations, where both Hartree and exchange potentials depend on the orbitals. It is usually solved in a self-consistent way, where the potentials are constructed from the initial orbitals; the orbitals are updated by solving Equation (6.7), the potentials are then updated from the new orbitals, and a new self-consistent cycle starts. In general, a breaking condition should be defined to end the iteration, which is usually the total energy difference between two consecutive cycles. In practice, we can select a basis set and write the orbitals as linear combinations of the basic functions. This leads to a set of eigenvalue problems in the matrix form that can be numerically solved by computers.

6.1.3.2 Koopmans' Theorem and Single-Electron Spectrum

All approaches based on single-electron approximation face a subtle but fundamental question, "What is the meaning of the single-electron energies?" The answer to this question is nontrivial and depends on the specific approach. For the HF method, Koopmans' theorem clearly shows that a single-electron energy given by Equation (6.7) is equal to the change in total energy induced by removing an electron for a filled orbital but preserving all the other orbitals [2].

In Koopmans' theorem, the single-electron energy of a filled orbital can be interpreted as the ionization energy. Extending this theorem to an empty orbital and calculating affinity energy involve the concept of self-interaction. The Hartree potential of Equation (6.7) includes the term $\langle i\sigma_i \,|\, v_{\text{ee}} \,|\, i\sigma_i\rangle$ that is the Coulomb potential from the ith orbital itself, which is called *self-interaction*. This unphysical self-interaction term is perfectly canceled by the same term in exchange potential when $i = j$.

However, the cancellation of self-interaction only happens to the filled orbitals. To consider an empty (virtual) orbital, we can explicitly extend the HF method to include occupation numbers of orbitals, and the Hartree and exchange potentials can be rewritten as follows, respectively,

$$\begin{aligned} V_{\mathrm{H}} &= \sum_{j\sigma_j} f_j \langle j\sigma_j | v_{\mathrm{ee}} | j\sigma_j \rangle \\ V_{\mathrm{X}i\sigma} &= -\sum_{j} f_j \langle j\sigma | v_{\mathrm{ee}} | i\sigma \rangle \frac{\phi_{j\sigma}}{\phi_{i\sigma}} \end{aligned} \tag{6.10}$$

where f_j is the occupation number of the jth orbital. If f_j's are either 0 or 1, Equation (6.10) is equivalent to limit the summations to all filled orbitals, and thus, will result in the exactly same Equation (6.7) and the cancellation of self-interaction for a filled orbital. However, if the ith orbital is empty, it will be excluded from the summations in Equation (6.10), and thus, there will be no self-interaction and cancellation.

In fact, a filled orbital will experience the Hartree and exchange potentials of all the other filled ones. An empty orbital will, however, experience those potentials of all filled ones. The asymmetry between the filled and empty orbitals will induce a large (and overestimated) gap between them in the band structure. Thus, the single-electron energy of the lowest empty orbital by HF should *not* be interpreted as the affinity energy. The affinity is usually calculated by the total energy difference between the systems with N and $N+1$ electrons.

In solids, the asymmetry results in an overestimated bandgap between the valence and conduction bands. It should be noted that the bandgap problem is rooted in the single-electron approximation and affects any first-principles approach based on this approximation. For example, local spin-density approximation (LSDA) usually leads to an underestimated bandgap. Intuitively, a mixture of HF and LSDA may correct the bandgap, which is so-called hybrid DFT.

6.1.3.3 Exchange and Correlation

The exchange energy in Equation (6.4) or the potential in Equation (6.7) is always negative, which can be intuitively understood as if a spin would avoid another spin in the same direction. In fact, the Pauli exclusion principle prohibits two parallel spins appear at the same location. This implies that the distributions of two spins cannot be independent to each other but correlated. In tradition, the dependency included in HF is called *exchange* and that not included in HF is called *correlation*.

To show the exchange between two parallel spins in HF, the pair distribution function is handily introduced from the Slater determinant (6.5):

$$n(\mathbf{r},\mathbf{r}') = |\Phi(\mathbf{r},\mathbf{r}')|^2 + |\Phi(\mathbf{r}',\mathbf{r})|^2 \tag{6.11}$$

where the spin coordinates are omitted. The physical meaning of Equation (6.11) is that the possibility of one spin at position $\mathbf{r}$ and the other at $\mathbf{r}'$ should be equal to the possibility of the first spin at $\mathbf{r}$ and the second spin at $\mathbf{r}'$ plus that of two spins exchanging positions. Inserting the Slating determinant of two spins, we can explicitly write Equation (6.11) in terms of single-electron orbitals:

$$n(\mathbf{r},\mathbf{r}') = n(\mathbf{r})n(\mathbf{r}') - \left|\sum_i \bar{\phi}_i(\mathbf{r})\phi_i(\mathbf{r}')\right|^2 \tag{6.12}$$

where the summation is over all occupied orbitals, and:

$$n(\mathbf{r}) = |\phi_1(\mathbf{r})|^2 + |\phi_2(\mathbf{r})|^2 \tag{6.13}$$

is the single-spin distribution function. It is obvious in Equation (6.12) that, in general, the exchange distribution:

$$n_{\mathrm{X}}(\mathbf{r},\mathbf{r}') \equiv -\left|\sum_i \bar{\phi}_i(\mathbf{r})\phi_i(\mathbf{r}')\right|^2 \neq 0 \tag{6.14}$$

and obviously:

$$n(\mathbf{r},\mathbf{r}') \neq n(\mathbf{r})n(\mathbf{r}'), \tag{6.15}$$

which clearly shows that the distributions of two parallel spins are *not* independent. Moreover, Equation (6.12) shows that the exchange energy is always negative and lowers down the possibility to find a pair of parallel spins at positions **r** and **r′**, respectively. In fact, we can calculate the energy corresponding to the exchange distribution of Equation (6.12):

$$\begin{aligned} E_X &= \frac{1}{2'}\iint \mathrm{d}\mathbf{r}\mathrm{d}\mathbf{r}' \frac{n_X(\mathbf{r},\mathbf{r}')}{|\mathbf{r}-\mathbf{r}'|} \\ &= \frac{1}{2}\{\langle 11 | v_{ee} | 11\rangle + \langle 22 | v_{ee} | 22\rangle + \langle 12 | v_{ee} | 21\rangle + \langle 21 | v_{ee} | 12\rangle\} \end{aligned} \tag{6.16}$$

which is exactly the exchange energy between two parallel spins.

The exchange distribution n_X can be interpreted as a hole distribution superposed on the independent distribution. In fact, it is easy to show that

$$\int \mathrm{d}\mathbf{r}' \left|\sum_i \bar{\phi}_i(\mathbf{r})\phi_i(\mathbf{r}')\right|^2 = n(\mathbf{r}) \tag{6.17}$$

and thus, for N parallel spins:

$$\int \mathrm{d}\mathbf{r}' n(\mathbf{r},\mathbf{r}') = n(\mathbf{r})(N-1) \tag{6.18}$$

which implies that the number of holes is exactly 1. The hole induced by exchange is called "exchange hole."

For two antiparallel spins, it is unnecessary to write a two-spin wave function as a Slater determinant. In fact, two antiparallel spins are identifiable, and thus, the two-spin wave function is simply the product of two single-spin orbitals:

$$\Phi = \phi_1(\mathbf{r}_1)\phi_2(\mathbf{r}_2) \tag{6.19}$$

which is the wave function in the so-called *Hartree* method, and leads to the formulation very similar to that of HF, except the exchange energy and potential. In additional, the pair distribution function satisfies:

$$n(\mathbf{r},\mathbf{r}') = n(\mathbf{r})n(\mathbf{r}') \tag{6.20}$$

which shows that two antiparallel spins are uncorrelated in the HF method.

It should be noted that the Salter determinant is not the unique way to write down the multi-electron wave function. For example, in configuration interaction (CI) method, the wave function is defined as:

$$\Phi = c_0\Phi_0 + c_1\Phi_1 + \cdots \tag{6.21}$$

which is a linear combination of Salter determinants Φ defined by the electron configurations included in the calculation. Wave function (6.21) makes it possible to take into account excited states. For example, in the calculation of NiO solid, the ground configuration of Ni^{2+}, $3d^8$, can be included in Φ_0, and the low excited configuration, $3d^7 4s^1$, can be included in Φ_1.

The total energy given by CI should be lower than that by HF because the degrees of freedom increase in CI by including additional coefficients (c)'s and orbitals in the excited configurations. Thus, there should be an extra negative term besides the exchange term defined in total energy (6.4). This is the so-called "correlation energy." Accordingly, there should be an extra term (n_C) besides n_X in the pair distribution function of two spins, which can be interpreted as an adjustment to the exchange hole distribution to further lower down the total energy. It should be noted that, in general, correlation energy is not limited to two parallel spins and does *not* vanish for two antiparallel spins. Thus, the pair distribution of two antiparallel spins is no longer independent, and the correlation term will redistribute the spins to lower down the total energy. In addition, the correlation term should integrate to zero, because a spin should interact with exactly $N-1$ other spins.

6.1.4 Density Functional Theory

The HF method is a typical *orbital approach*, in which the potentials are constructed directly from the orbitals without any interim quantity. The computational complexity of HF method is estimated about N^4, where N is the dimension of the numerical problem that can be estimated by the number of electrons and that of the basic functions. It is obvious that the complexity of HF method will fast increase when the number of electrons increases. This difficulty leads to the DFT, where the electron (or spin) density is hired as the interim quantity to construct the potential and thus to significantly reduce the complexity.

6.1.4.1 Thomas–Fermi Approximation

The complexity of the HF method is from the calculation of the interorbital integrals when updating the Hartree and exchange potentials in Equation (6.7). It is, however, obvious that the Hartree energy can be rewritten in terms of density:

$$E_{\mathrm{H}} = \frac{1}{2}\iint \mathrm{d}\mathbf{r}\mathrm{d}\mathbf{r}' \frac{n(\mathbf{r})n(\mathbf{r}')}{|\mathbf{r}-\mathbf{r}'|} \tag{6.22}$$

where,

$$n(\mathbf{r}) = \sum_i |\phi_i(\mathbf{r})|^2 \tag{6.23}$$

In general, it is impossible to rewrite the exchange energy by HF approximation in terms of density. However, for an important special case, the uniform electron gas, the exchange energy per spin can be rewritten as:

$$\varepsilon_{X\sigma} = \frac{E_{X\sigma}}{N_\sigma} = -\frac{3}{4}\left(\frac{6}{\pi}n_\sigma\right)^{1/3} \tag{6.24}$$

Further, by using the kinetic energy of uniform electron gas, the total energy can be rewritten in terms of density as:

$$\begin{aligned} E[n] = {} & c_1\int \mathrm{d}\mathbf{r}n(\mathrm{r})^{5/3} + \int \mathrm{d}\mathbf{r}V_{\mathrm{ext}}(\mathbf{r})n(\mathbf{r}) \\ & + \frac{1}{2}\iint \mathrm{d}\mathrm{r}\mathrm{d}\mathrm{r}' \frac{n(\mathrm{r})n(\mathrm{r}')}{|\mathrm{r}-\mathrm{r}'|} + c_2\int \mathrm{d}\mathbf{r}n(\mathbf{r})^{4/3} \end{aligned} \tag{6.25}$$

which is the so-called "Thomas–Fermi approximation" [12]. The important is that each term in Equation (6.25) is a functional of density $n(\mathbf{r})$. Thus, the ground state energy can be calculated by minimizing the total energy functional versus the density, instead of calculating the orbitals, which will tremendously reduce the computational complexity.

However, the Thomas–Fermi approximation fails to predict shell structures of atoms, chemical bonds of molecules, and bandgaps of insulators, all of which are the direct results of atomic orbitals. In fact, the "orbitals" in the uniform gas model are plane waves that are so different from atomic orbitals. Thus, the Thomas–Fermi approximation, the original DFT, is too crude to apply to most materials.

6.1.4.2 Hohenberg–Kohn Theorems

The Thomas–Fermi approximation is crude and need to be improved. However, there is a fundamental question, "Can a density functional be exact, and at the same time, universal for any many-body system?" If the answer is negative, it would not be necessary to improve the Thomas–Fermi approximation or propose anything similar to it as an accurate and universal theoretical approach guiding calculations for a broad spectrum of materials.

Hamiltonian (6.2)[1] will determine the multi-electron ground state wave function Φ, whose integration will determine the ground state density as:

$$n(\mathbf{r}) = \langle \Phi | \psi^* \psi | \Phi \rangle \tag{6.26}$$

---

1 We can ignore the ion–ion energy E_{II} that is a constant under Bohn–Oppenheimer approximation.

where ψ is the one-body field operator [11]. Thus, it is trivial that Hamiltonian (6.2) will determine the ground state density n. The first Hohenberg–Kohn theorem [4], however, states conversely that the external potential V_{ext} in Hamiltonian (6.2) can be uniquely determined, up to a constant, by the ground state density n for any system. Because the kinetic energy and the interelectron Coulomb interaction terms are fixed in form, the first theorem in fact claims that Hamiltonian (6.2) can be determined up to a constant by the ground state density.

To prove the first theorem, we can assume another potential V'_{ext} that results in the ground state Φ', different from Φ, but the same density n. Thus, we can obtain an inequality:

$$\begin{aligned} E' &= \langle\Phi'|\mathcal{H}'|\Phi'\rangle < \langle\Phi|\mathcal{H}'|\Phi\rangle = \langle\Phi|\mathcal{H} - V_{ext} + V'_{ext}|\Phi\rangle \\ &= E + \langle\Phi|V'_{ext} - V_{ext}|\Phi\rangle = E + \int d\mathbf{r}\left(V'_{ext} - V_{ext}\right)n(\mathbf{r}) \end{aligned} \tag{6.27}$$

where we assume that the ground state is nondegenerate. Noting that the primed quantities can be interchanged with the unprimed ones, we can obtain another inequality:

$$E < E' + \int d\mathbf{r}(V_{ext} - V'_{ext})n(\mathbf{r}) \tag{6.28}$$

Summing inequalities (6.7) and (6.8) leads to the absurdity:

$$E' + E < E + E' \tag{6.29}$$

Because Hamiltonian (6.2) that sets up the multi-electron Schrödinger equation can be fully determined from the ground state density, the multi-electron wave functions of all states and, consequently, all properties of the system should also be determined from it. Thus, it is possible to reformulate the many-body problem defined by Hamiltonian (6.2) in terms of the density. In fact, this is the first part of the second Hohenberg and Kohn theorem [4], i.e. it is possible to define a universal total energy functional $E[n]$ in terms of the density n for any external potential V_{ext}. Moreover, the second part of the theorem states that the functional $E[n]$ for a given V_{ext} assumes its global minimal at the ground state density n_0.[2]

To prove the second theorem, we simply calculate the ground state energy as:

$$\begin{aligned} E &= \langle\Phi|\left(T + V_{ee} + V_{ext}\right)|\Phi\rangle \\ &= \langle\Phi|\left(T + V_{ee}\right)|\Phi\rangle + \int d\mathbf{r}v(\mathbf{r})n(\mathbf{r}) = E[n]. \end{aligned} \tag{6.30}$$

From the first theorem, it is obvious that the multi-electron wave function of the ground state is a functional of its own density. Thus, the first term of the second line in Equation (6.30) is a functional of the density, and Equation (6.30), in fact, defines the universal total energy functional in terms of the density. For any excited state Φ', we can also write:

$$\begin{aligned} E' &= \langle\Phi'|\left(T + V_{ee} + V_{ext}\right)|\Phi'\rangle \\ &= \langle\Phi'|\left(T + V_{ee}\right)|\Phi'\rangle + \int d\mathbf{r}v(\mathbf{r})n'(\mathbf{r}) = E[n'] \end{aligned} \tag{6.31}$$

which is the same functional as that defined by Equation (6.10) but assuming its value at n'. It is obvious that $E' > E$ and the functional $E[n]$ assumes its minimum at the ground state density n_0.

It will be helpful to write the total energy functional in more detail. To do this, we write Hamiltonian (6.2) in terms of field operator ψ as:

$$\begin{aligned} \mathcal{H} &= \int d\mathbf{r}\left[\frac{1}{2}\nabla\bar{\psi}(\mathbf{r})\nabla\psi(\mathbf{r}) + v(\mathbf{r})\bar{\psi}(\mathbf{r})\psi(\mathbf{r})\right] \\ &+ \frac{1}{2}\iint d\mathbf{r}d\mathbf{r}'\frac{1}{|\mathbf{r}-\mathbf{r}'|}\bar{\psi}(\mathbf{r})\bar{\psi}(\mathbf{r}')\psi(\mathbf{r}')\psi(\mathbf{r}) \end{aligned} \tag{6.32}$$

where the first line is the one-body terms, including the kinetic energy and external potential, and the second line is the two-body term, from which the Hartree and exchange correlation can be derived.

2 We emphasize the ground state here by subscript 0.

Here, we only consider the two-body term and take its average over the ground state Φ:

$$\langle\Phi|V_{ee}|\Phi\rangle=\frac{1}{2}\iint d\mathbf{r}d\mathbf{r}'\frac{1}{|\mathbf{r}-\mathbf{r}'|}\langle\Phi|\bar{\psi}(\mathbf{r})\bar{\psi}(\mathbf{r}')\psi(\mathbf{r}')\psi(\mathbf{r})|\Phi\rangle \quad (6.33)$$

where the quantum average in the Dirac bracket is exactly the pair distribution function $n(\mathbf{r},\mathbf{r}')$. Using the commutation relations of ψ and $\bar{\psi}$, we can expand $n(\mathbf{r},\mathbf{r}')$ as:

$$n(\mathbf{r},\mathbf{r}')=n(\mathbf{r})n(\mathbf{r}')-\langle\Phi|\bar{\psi}(\mathbf{r})\psi(\mathbf{r}')\delta(\mathbf{r}-\mathbf{r}')|\Phi\rangle \quad (6.34)$$

where the first term leads to the Hartree energy and the second to exchange correlation. Expanding the field operator and the delta function into a complete set of single-body states, for example, the single-electron spin orbitals, we can rewrite the second term:

$$E_{XC}=\frac{1}{2}\sum_{ijk}\langle\Phi|a_j a_k|\Phi\rangle\int d\mathbf{r}d\mathbf{r}'\frac{1}{|\mathbf{r}-\mathbf{r}'|}\bar{\phi}_j(\mathbf{r})\bar{\phi}_i(\mathbf{r}')\phi_i(\mathbf{r})\phi_k(\mathbf{r}') \quad (6.35)$$

where the diagonal terms ($j=k$) restore the exact exchange and the off-diagonal terms ($j\neq k$) implies the correlation.

6.1.4.3 Kohn–Sham Ansatz

Although the Hohenberg–Kohn theorems established the crucial role of the (ground state) density in many-body problems, our understanding on atoms, molecules, and solids still relies on the concepts and quantities, well established in history and single body in nature, such as orbital, electron configuration, chemical bond, band structure, etc. At the same time, the benefit implied in the two Hohenberg–Kohn theorems is obvious – the computational complexity can be remarkably reduced by introducing the density as the targeting variable of the variational problem that is exactly equivalent to the original many-body Schrödinger equation. In a practical point of view, the computational complexity of Hartree–Fock is substantially from updating the potentials in Equation (6.7) by directly calculating them from the single-electron orbitals. Thus, if the potentials of a set of equations similar to Equation (6.7) can be written in terms of the density, the complexity will be remarkably reduced, because it is trivial to calculate the density from the orbitals. This is the basic idea of the Kohn–Sham ansatz.

The Kohn–Sham ansatz assumes that there is a noninteracting system whose ground state density is equal to that of the interacting system under investigation. Thus, the density shared by the both interacting and noninteracting systems will uniquely determine the external potential and total energy functional of the noninteracting system, according to the two Hohenberg–Kohn theorems, and will lead to a noninteracting many-body problem that is ready to be solved. The single-body equation of a noninteracting system is simply defined by Hamiltonian (6.3) and the Kohn–Sham ansatz indeed writes the effective potential in terms of the density.

To formulate the Kohn–Shame scheme, we start from the total energy functional defined in Equation (6.30) that can be rewritten:

$$E[n]=T_s+\int d\mathbf{r}v(\mathbf{r})n(\mathbf{r})+\frac{1}{2}\iint d\mathbf{r}d\mathbf{r}'\frac{1}{|\mathbf{r}-\mathbf{r}'|}n(\mathbf{r})n(\mathbf{r}')+E_{XC} \quad (6.36)$$

where the first term T_s is the kinetic energy of the noninteracting system that can be readily and exactly formulated from the single-electron orbitals as:

$$T_s=\sum_i\int d\mathbf{r}\bar{\phi}_i(\mathbf{r})(-\frac{1}{2}\nabla^2)\phi_i(\mathbf{r}) \quad (6.37)$$

where the summation is over all occupied orbitals. In Equation (6.36), the third term (Hartree energy) and the fourth term (exchange correlation) are derived from the two-body term (6.33) using Equation (6.34). Further, the exact formula of exchange correlation is unknown but can be approximated in terms of the density as well. For example, the local (spin) density approximation assumes:

$$E_{XC}=\int d\mathbf{r}n(\mathbf{r})\varepsilon_{XC}[n(\mathbf{r})] \quad (6.38)$$

where $\varepsilon_{XC}[n(\mathbf{r})]$ is the exchange-correlation energy density of homogeneous electron gas. The second Hohenberg–Kohn theorem states that the total energy functional assumes its minimum at the ground state density. Thus, the variation of total energy (6.36), under the constrain that the total number of electrons is conserved, leads to:

$$\int d\mathbf{r}\frac{\delta T_s[n]}{\delta n}\delta n+\int d\mathbf{r}\delta n\{v(\mathbf{r})+v_H[n]+v_{XC}[n]\}=\lambda\int d\mathbf{r}\delta n \tag{6.39}$$

where $v_{XC}[n]$ is the exchange-correlation potential:

$$v_{XC}[n]=\frac{\delta E_{XC}[n]}{\delta n} \tag{6.40}$$

$v_H[n]$ is the Hartree potential:

$$v_H[n]=\int d\mathbf{r}'\frac{n(\mathbf{r}')}{|\mathbf{r}-\mathbf{r}'|} \tag{6.41}$$

and the Lagrange multiplier λ is, in fact, the chemical potential μ:

$$\frac{\delta T_s}{\delta n}+v+v_H+v_{XC}=\mu \tag{6.42}$$

The density can be simply written in term of single-electron orbitals:

$$n(\mathbf{r})=\sum_i\bar{\phi}_i(\mathbf{r})\phi_i(\mathbf{r}) \tag{6.43}$$

where the summation is over all occupied orbitals. Thus, the variation δn can be written in terms of the variation of $\bar{\phi}_i(\mathbf{r})$, keeping all other orbitals stationary, and we obtain a set of single-electron equations very similar to Equation (6.7):

$$\left\{-\frac{1}{2}\nabla^2+v(\mathbf{r})+v_H[n]+v_{XC}[n]\right\}\phi_i(\mathbf{r})=\varepsilon_i\phi_i(\mathbf{r}) \tag{6.44}$$

where we, in fact, utilize the kinetic energy in terms of the orbitals [see Equation (6.37)] and rename the Lagrange multiplier as single-electron energy levels ε_i. In fact, utilizing the exact form of kinetic energy in terms of the orbitals avoids the severe problems, such as missing the shell structure of atoms, caused by approximating the kinetic energy by an analytical function of the density in the Thomas–Fermi approximation.

Comparing Equation (6.44) with Equation (6.7), it is obvious that the only difference is the exchange correlation that is a functional of the density in Equation (6.44) instead of a functional of the orbitals in Equation (6.7). Moreover, an orbital $\phi_i(\mathbf{r})$ can only "experience" the exchange-correlation potential v_{XC} determined by the local density $n(\mathbf{r})$, whereas the exact exchange derived from Hartree–Fock is an integration, including the orbitals that is nonlocal. To solve Equation (6.44), one can start from an initial density, solve the orbitals from Equation (6.44), calculate the density using Equation (6.43), and update the potentials of Equation (6.44) and continue the next cycle.

6.1.4.4 Slater–Janak Theorem and Derivative Discontinuity

In Hartree–Fock method, the Koopmans' theorem connects the single-electron energy level to the energy of removing an electron from a many-body system without disturbing the other electrons. It can be shown that there is a similar theorem for DFT, the Slater–Janak theorem [13]:

$$\frac{\partial \tilde{E}}{\partial n_i}=\varepsilon_i \tag{6.45}$$

where $\tilde{E}$ is the extended total energy functional that includes the occupation numbers n_i, which can be fractional, of single-electron orbital ϕ_i. It should be noted that Equation (6.45) is derived under the assumptions that the occupation numbers follow the Fermi–Dirac distribution of fermions and that all the other orbitals do *not* relax. Applying theorem (6.45) to the highest occupied orbital (HO) of metals, we obtain the chemical potential:

$$\mu = \varepsilon_{\text{HO}} = \frac{\partial \tilde{E}}{\partial n_{\text{HO}}} \tag{6.46}$$

However, when applied to insulators, theorem (6.45) leads to the *derivative discontinuity* [14–16]. Suppose there is an insulator with M ($M \to \infty$) electrons in total. The electrons fill up to the valence band maximum (VBM). When the total electron number, N, as a continuous variable, approaches M from both sides, the extended total energy functional $\tilde{E}$ approaches the total energy functional E by its construction [13]. Thus, the extended total energy functional is a continuous function of the total electron number. On the contrary, its derivative is discontinuous, that is, the left derivative is unequal to the right one. The left derivative can be calculated by removing an infinitesimal amount of electron, and gives the energy of VBM ε_{v}, equal to the negative of ionization energy I:

$$-I = \varepsilon_{\text{v}} = \left(\frac{\partial \tilde{E}}{\partial N}\right)_{-} \tag{6.47}$$

Similarly, the right derivative gives the energy of conduction band minimum ε_{c}, equal to the negative of electron affinity A:

$$-A = \varepsilon_{\text{c}} = \left(\frac{\partial \tilde{E}}{\partial N}\right)_{+} \tag{6.48}$$

It is obvious that the left and right derivatives of the extended total energy functional with respect to the total electron number are unequal and that the bandgap is given by:

$$E_{\text{g}} = I - A = \left(\frac{\partial \tilde{E}}{\partial N}\right)_{+} - \left(\frac{\partial \tilde{E}}{\partial N}\right)_{-} \tag{6.49}$$

Equation (6.49) connects the bandgap of insulators and the derivative discontinuity of the total energy with respect to the total electron number.

When an infinitesimal amount of electron is removed or added, it induces an infinitesimal variation of the density. Thus, the derivative discontinuity can also be expressed as the discontinuity of the functional derivative of the total energy functional with respect to the density. In fact, it can be shown that:

$$\frac{\delta E[n]}{\delta n} = \frac{\partial \tilde{E}}{\partial N} = \mu \tag{6.50}$$

keeping the external potential v of Equation (6.36) fixed [17]. From Equation (6.42), both external potential and Hartree potential are continuous to an infinitesimal variation of the density, and thus, the kinetic energy and exchange correlation imply the derivative discontinuity. Thus, the bandgap of a real crystal is given by two parts:

$$E_{\text{g}} = \varepsilon_{\text{g}} + v_{\text{XC}}{}^{(+)} - v_{\text{XC}}{}^{(-)} \tag{6.51}$$

where ε_{g} is the bandgap of the noninteracting system given by the derivative discontinuity of kinetic energy and the $v_{\text{XC}}{}^{(+)} - v_{\text{XC}}{}^{(-)}$ is the correction induced by the derivative discontinuity of exchange correlation.

6.1.4.5 Exchange Correlations

The Kohn–Sham ansatz makes it possible to conduct practical calculations for materials based on the Hohenberg–Kohn theorems. It, however, encapsulates all the difficulties of a many-body problem in the exchange correlation, because all the other terms in the total energy functional are exactly known.

As the exact formulation of exchange correlation is unknown, many efforts have been made to develop the approximations. In general, more accurate the approximation is, more computationally expensive the calculation.

6.1.4.5.1 Local (Spin) Density Approximation

In Equation (6.38), we have introduced LSDA that can be rewritten in terms of spin densities:

$$E_{XC}^{LSDA}[n_\uparrow, n_\downarrow] = \int d\mathbf{r} n(\mathbf{r}) \varepsilon_{XC}^{hom}(n_\uparrow(\mathbf{r}), n_\downarrow(\mathbf{r})) \tag{6.52}$$

where ε_{XC}^{hom} is the exchange-correlation energy density of homogeneous gas. It should be noted that ε_{XC}^{hom} is a function, instead of functional, of the spin densities, $n_\uparrow$ and $n_\downarrow$. The ε_{XC}^{hom} can be separated into the exchange and correlation parts, ε_X^{hom} and ε_C^{hom}, where ε_X^{hom} can be analytically and exactly formulated by applying the Hartree–Fock approximation to homogeneous electrons gas (see Equation (6.24)).

The correlation part is more complicated than the exchange, because there is no analytical exact result for correlation hole or energy. However, correlation hole and energy can be numerically calculated by more accurate techniques, such as quantum Monte-Carlo simulation [18], and the analytical formulation can be generated by fitting the numerical result [19, 20].

LSDA is satisfactory in predicting most ground state properties of many real materials, at least better than expected. Note that the density in real materials is inhomogeneous, whereas LSDA simply utilize the exchange correlation of homogeneous gas. LSDA has shown systematic errors in structural and elastic properties in solids. The lattice constants of *sp* materials, such as Si and Al, are underestimated by about 2% and the cohesive energies are overestimated by 15% or more [21]. This problem is directly related to the overbinding problem that LSDA systematically overestimates the bond energies of simple dimmers [22].

The most serious problem is that LSDA systematically underestimate bandgaps of semiconductors/insulators by about 40% or more. For example, the LSDA bandgap of silicon is 0.54 eV at experimental lattice constant that is narrower than the experimental value of 1.17 eV by about 54% [21]. This problem should be worse in ferrites, where the open *d* shells of transition metal ions determine the electronic structure and consequently the magnetic properties.

In spite of those deficiencies, LSDA is the starting point to improve the formulation of exchange correlation for better consistency between first-principles calculations and experiments.

6.1.4.5.2 Gradient Approximations

In local density approximation (LDA), the exchange-correlation potential is strictly local, that is, the potential, $v_{XC}(\mathbf{r})$, depends only on the local charge density, $n(\mathbf{r})$, which leads to the deficiencies discussed earlier and, on the other hand, significantly reduces the computational complexity. In fact, LDA is exact only for homogeneous gas, where the charge density is uniform. On the contrary, the charge density[3] of any real material cannot be uniform, which should be included in the construction of exchange correlation but is neglected by LDA.

A handy remedy is gradient approximation, a natural extension of LDA. As suggested by its name, a gradient approximation includes the gradient of the density, ∇n, into the formulation of exchange correlation, which is equivalent to bring in more information about the density, $n(\mathbf{r})$, not only exactly at position $\mathbf{r}$ but in an infinitesimal proximity of $\mathbf{r}$. It is the semi-local nature of gradient approximation that makes it possible to partially cure the deficiencies of LDA.

The first gradient approximation was proposed in the original papers of DFT [4, 8] as a gradient expansion approximation (GEA) of exchange-correlation functional, assuming a slow variation of the density:

$$E_{XC} = \int d\mathbf{r} \left\{ \varepsilon_{XC}^{(0)}(n) n + \varepsilon_{XC}^{(2)}(n) |\nabla n|^2 + \cdots \right\} \tag{6.53}$$

where the first coefficient $\varepsilon_{XC}^{(0)}$ should be the exchange-correlation density of homogeneous gas and the second coefficient $\varepsilon_{XC}^{(2)}(n)$ can be obtained either by dimensional argument with an empirical parameter [23] or by inverse-space perturbation method [24]. At first glance, GEA as an improvement based on LDA, should cure the deficiencies of LDA at least partly. However, testing GEA over atoms and others show that GEA is not only insufficient to correct the errors by LDA but also inappropriate for certain systems such as metallic surfaces [25].

The failure of GEA can be attributed to the large density gradient in atoms and other real systems that break the slow-varying condition:

$$|\nabla n| / n \ll k_F(n) \tag{6.54}$$

3 Because the exchange energy follows the spin-scaling rule, it is only necessary to focus on the fully spin-polarized case, where the spin can be dropped.

where

$$k_F(n) = (3\pi^2 n)^{1/3} \tag{6.55}$$

is the local Fermi wave number of density n. In fact, if considering only the exchange part and rewriting Equation (6.53):

$$E_X = \int d\mathbf{r} n \varepsilon_X^{\mathrm{LDA}}(n) F_X(n, \nabla n) \tag{6.56}$$

it is trivial to obtain the GEA enhancement factor:

$$F_X(n, \nabla n) = 1 + \beta\left(|\nabla n| / n^{4/3}\right)^2 + \cdots \tag{6.57}$$

where β is a dimensionless constant. Thus, the slow-varying condition (6.54) is equivalent to the condition that the gradient terms in the enhancement factor (6.57) should be much less than 1. For convenience, the normalized gradient can be defined:

$$s \equiv |\nabla n| / (2k_F n) \propto |\nabla n| / n^{3/4} \tag{6.58}$$

and the GEA enhancement factor is simply:

$$F_X(s) = 1 + \tilde{\beta} s^2 + \cdots \tag{6.59}$$

However, the shell structure of atoms is so peculiar that at the boundary of each shell defined by the principal quantum number (or subshell defined by both principal and angular quantum number), the density rapidly becomes zero and the density itself is low. Thus, the second term in the enhancement factor (6.57) can be unbounded for the outer shells and, consequently, violate the slow-varying condition. For solids, the GEA is also not optimistic, because the density of a solid can be approximated by superposition of atomic densities. Moreover, most properties of a solid are determined by the valence electrons in the interstitial region characterized of low density and large gradient.

The failure of GEA can also be attributed to violating some important rules posed on the exchange correlation by the many-body theory; for example, the exchange hole has to be summed to exactly –1. In fact, we have introduced the concept of exchange hole in Section 4.3.3, for which we can explicitly define the exchange-correlation hole distribution,

$$h_{XC}(\mathbf{r}, \mathbf{r}') = \frac{n(\mathbf{r}, \mathbf{r}')}{n(\mathbf{r})} - n(\mathbf{r}') \tag{6.60}$$

which is the difference between the conditional distribution $n(\mathbf{r}, \mathbf{r}') / n(\mathbf{r})$ and the single-electron distribution, i.e. the density, $n(\mathbf{r}')$. Now, following Equation (6.12), it is trivial to obtain the exchange hole:

$$h_X(\mathbf{r}, \mathbf{r}') = -|\rho(\mathbf{r}, \mathbf{r}')|^2 / n(\mathbf{r}) \tag{6.61}$$

where

$$\rho(\mathbf{r}, \mathbf{r}') = \sum_i^{\mathrm{occ}} \bar{\phi}_i(\mathbf{r}) \phi_i(\mathbf{r}') \tag{6.62}$$

is the single-electron density matrix defined in terms of the Kohn–Sham noninteracting orbitals that can be orthogonal to each other and normalized. It is trivial to prove that:

$$\int d\mathbf{r}' h_X(\mathbf{r}, \mathbf{r}') = -1 \tag{6.63}$$

and that the exchange energy is exactly:

$$E_X = \int d\mathbf{r} n(\mathbf{r}) \left\{ \frac{1}{2} \int d\mathbf{r}' \frac{1}{|\mathbf{r} - \mathbf{r}'|} h_X(\mathbf{r}, \mathbf{r}') \right\} \tag{6.64}$$

where the exact exchange density ε_X is the integral in the braces involving the exchange hole distribution and the Coulomb interaction kernel. A gradient expansion of the hole distribution leads to a more general form of GEA and, more importantly, reveals that expanded hole distribution violates the sum rule (6.63) because of the undamped oscillation at large $|\mathbf{r}-\mathbf{r}'|$ [26]. Moreover, the undamped oscillation also violates another rule of any interacting fermion system:

$$h_X(\mathbf{r},\mathbf{r}') \leq 0 \tag{6.65}$$

It should be noted that, although LDA is the lowest order approximation, its exchange hole is exactly summed to −1 and is negative anywhere [27]. Thus, although LDA is unnecessarily accurate for real systems, it is more physically reasonable, in the sense of exchange hole distribution, than GEA.

To cure the pathologic hole distribution induced by GEA, a real-space cutoff technique is introduced [26], and a generalized gradient approximation (GGA) can be constructed by assuming an analytic form of enhancement factor with parameters and numerically fitting it to the corrected hole behavior, which leads to the PW86 and PW91 parameterizations [28, 29]. Moreover, it is more direct but empirical to numerically fit the analytic form to the exchange data of atoms and molecules, which leads to the B86 and B88 parameterizations [22, 30]. It should be noted that the choice of analytic form is not arbitrary but predefined by the known properties, such as asymptotic behavior, scaling rules, and bounds, which, in fact, lead to the PBE parameterization without any numerically fitting [31]. In addition, PBE can be obtained from the real-space cutoff technique with a proper damping factor [32]. It is interesting that the PBE exchange enhancement factor is in the same form as that of B86, except that the two parameters in B86 are determined by fitting the data of atoms and molecules.

The GGA of correlation is more complicated, because the correlation hole cannot be explicitly defined like the exchange hole (6.61) that can be expanded in terms of the density gradient. However, it is possible to obtain the lowest order correlation enhancement factor similar to that of Equation (6.57) at high-density limits using the many-body perturbation theory [33]. Moreover, the PW91 and PBE correlations are deduced from the asymptotic behaviors and scaling rules [29, 31]. It should be noted that unlike exchange that follows the spin-scaling rule:

$$E_X(n_\uparrow,n_\downarrow) = \frac{1}{2}\{E_X(2n_\uparrow) + E_X(2n_\downarrow)\} \tag{6.66}$$

by which the partly spin-polarized exchange is simply the average of fully spin-polarized ones, correlation does not show that property and, thus, is generally a complicated function of both charge density and spin-polarization rate:

$$n = n_\uparrow + n_\downarrow, \text{ and } \zeta = (n_\uparrow - n_\downarrow)/n \tag{6.67}$$

and the charge density gradient.

Both LYP correlation [34] and its precursor, the Colle–Salvetti correlation [35] can be understood as the gradient improvements of the original Wigner correlation [36]. This is more obvious for LYP, where the gradient correction that is proportional to the square of normalized gradient (s^2), as expected, is inserted into the nominator of the Wigner correlation.

GGA is successful on correcting the overbinding problem of LDA. In fact, the bond energies of simple molecules and lattice constants of *sp* metals predicted by GGA are consistent with the experimental results with relative errors usually less than 1% [29]. However, GGA improvement on bandgap is limited. For example, the GGA bandgap of Si is 0.65 eV, slightly more closer to the experimental value of 1.17 eV than the LSDA value of 0.49 eV.

6.1.4.5.3 GW Approximation and LDA+U Approach

The bandgap problem of LSDA/GGA can be attributed to the fact that the derivative discontinuity of LSDA/GGA is mostly missing as an approximation to the *true or exact* exchange correlation and, thus, is too weak to open the bandgap as wide as the experimental value for an insulator/semiconductor, according to Equation (6.51). In fact, Green's functional theory shows that the quasiparticle orbitals are solutions of equation:

$$\left\{-\frac{1}{2}\nabla^2 + v(\mathbf{r}) + v_H[n]\right\}\phi(\mathbf{r}) + \int d\mathbf{r}''\Sigma(\mathbf{r},\mathbf{r}'';\varepsilon)\phi(\mathbf{r}'') = \varepsilon\phi(\mathbf{r}) \tag{6.68}$$

where Σ is the self-energy operator, of which the lowest order expansion leads to the HF exchange potential [37]. If we *believe* that the Kohn–Sham orbitals are the quasiparticle orbitals, both HF exchange potential in Equation (6.7) and the

Kohn–Sham exchange-correlation potential in Equation (6.44) are indeed approximations to the self-energy term in Equation (6.68) that is nonlocal and energy dependent. HF exchange potential, as the lowest order expansion of the self-energy term, is nonlocal but independent of energy. LDA exchange potential, as a simplification of the HF one, is both local and energy independent. Self-energy Σ can be expanded in terms of Green's function G and screened potential W:

$$\Sigma = iGW + \cdots \tag{6.69}$$

and leads to the GW approximation (GWA) that can be performed either as a post correction to LDA [38] or in a self-consistent way [39]. Although GWA can successfully correct the bandgap problem of insulators/semiconductors, including transition metal compounds, such as ferrites, without any empirical parameter, its high computational cost makes it impractical to apply it to complex systems, such as hexagonal ferrites, especially in a self-consistent way.

The bandgap problem is fatal when LDA/GGA is applied to antiferromagnetic transition metal monoxides that share the same superexchange mechanism with ferrites. For example, the bandgap of NiO is 0.9 eV by LSDA [40] and 1.2 eV by GGA [41], both of which are much narrower than the experimental value of 4.3 eV [42]. In fact, most transition metal oxides are *not* band insulator, such as diamond, in which the bandgap can be substantially attributed to the derivative discontinuity of kinetic energy in a language of DFT. They are Mott insulator [43], in which the bandgap should be attributed to the Coulomb repulsion in the open d shell of the transition metal ions.

The single-band Hubbard model includes the Coulomb repulsion by amending the tight-banding model:

$$\mathcal{H} = -t\sum_{\langle i,j\rangle} a_j^{\dagger} a_i + U\sum_{i} n_{i\uparrow} n_{i\downarrow} \tag{6.70}$$

where the first term is the tight binding, whose Fourier transformation is simply the energy band, and the second term is the on-site Coulomb repulsion. Intuitively, the repulsion term raises the energy of the configuration that two electrons are on the same site and, thus, impedes electron transfer from one site to another. Thus, if $U \gg t$, the electrons will be localized on each site and the ground state is insulator. It should be noted that, this is true, even if the band is half-filled, such as the e_g band of NiO. In fact, in a band structure picture, the half-filled band will be split by the repulsion terms into fully occupied lower Hubbard band and empty upper one, and a bandgap size of U will be open between them.

Inspired by the Hubbard model, LDA+U approach introduces addition terms to the LDA total energy [44, 45]:

$$E = E_{\text{LDA}} + \frac{1}{2}U\sum_{i\sigma}(n_{i\sigma} - n_0)(n_{i-\sigma} - n_0) \tag{6.71}$$

where $0 \le n_i \le 1$ is the occupation number of spin σ on the ith orbital (the localized orbital of the ith site), and n_0 is the average occupation number of that localized orbital. Now, applying the Janak theorem (6.45) to total energy (6.71), we obtain the single-spin energy level:

$$\varepsilon_{i\sigma} = \varepsilon_{i\sigma}^{\text{LDA}} + U(n_{i-\sigma} - n_0) \tag{6.72}$$

where the correction induced by "+U" depends on spin, i.e. the correction to the ith orbital of spin σ depends on the occupation number of the ith orbital of spin $-\sigma$. The spin-dependent correction (6.72) results in the splitting of the ith orbital into the lower orbital with spin σ and the upper with spin $-\sigma$, like the Hubbard model (6.70). For the half-filled case, $n_0 = 1/2$, the lower orbital is full and the upper is empty and the gap size of U between the lower and upper orbitals, which implies an insulating ground state. The correction (6.72) implies a correction on the exchange-correlation potential:

$$v_{i\sigma} = v_{i\sigma}^{\text{LDA}} + U(n_{i-\sigma} - n_0) \tag{6.73}$$

which indeed restores the derivative discontinuity of the *true* exchange correlation at the bandgap.

In an open d or f shell, there are usually more than one localized orbitals, degenerated or nearly degenerated, and indexed by the magnetic quantum number, m. Thus, there are on-site exchange interactions J between the localized orbitals with the same spin in addition to on-site Coulomb repulsion U, while J is usually about tenth of U. Moreover, it is usually assumed that the Coulomb repulsions between a pair of orbitals with the same m is approximately equal to that between those with different values of m. Thus, a typical LDA+U calculation usually includes two parameters, U and J, for each species of transition metal ion, which can be chosen as *ad hoc* parameters or determined by constrained LSDA calculations based on the linear response theory of DFT [46].

In fact, the LDA+U approach can be considered as an approximation to GWA [44]. Assuming that most LDA orbitals are accurate except those localized d or f orbitals, only the GWA correction to those localized orbitals should be considered and the correction to Σ from the correlation part of W results in the Coulomb repulsion terms of LDA+U.

The LDA+U approach has been widely applied to transition metal compounds, including ferrites, to correct the problems related to the underestimated correlation of the open d or f shell by LDA, with a low computational complexity [45]. For example, it enhances the bandgap of NiO from the LDA value of 0.2 to 3.1 eV (cf. experimental value of 4.3 eV), with U and J determined by constrained LDA calculations and, at the same time, enhances the local magnetic moment of Ni^{2+} ion from the LDA value of 1.0 μ_B to 1.59 μ_B (cf. experimental value of about 1.77 μ_B). It also restores the antiferromagnetic ground state of $CaCuO_2$, one of the parent materials of high-T_c superconductors. In fact, LDA underestimates the localization of the d orbitals of transition metal ions and overestimates the covalency between those d orbitals and the sp orbitals of the nearby anions, which will lead to a crystal field stronger than the real situation and probably a low spin state. Moreover, it improves the calculated structural parameters over LDA for NiO, such as lattice constant, cohesive energy, and elastic constants [47].

6.1.4.5.4 Hybrid Density Functional

The fundamental difficulty of an interacting many-body system is the interaction term between the particles, which is encapsulated into the exchange correlation, the exact of which is unknown. However, we can imagine that the interaction term is scaled by a coupling strength λ that continuously decreases from 1 to 0 but keeps the density unperturbed. It is obvious that λ equal to 0 and 1 correspond to noninteracting and fully interacting situations, respectively. Thus, we can write the Hamiltonian of the many-body system:

$$H_\lambda = T + V_{\text{ext}}\left[n;\lambda\right] + \lambda V_{\text{ee}} \tag{6.74}$$

where V_{ee} is the electron–electron interaction and $V_{\text{ext}}\left[n;\lambda\right]$ is the external field that is uniquely determined by the density according to the first Hohenberg–Kohn theorem for each coupling strength λ. Equation (6.74) is the so-called *adiabatic connection* of DFT, which leads to the exact exchange correlation [48, 49]. Now, let's write the general expression of exchange-correlation energy:

$$E_{\text{XC}} = \frac{1}{2}\iint n(\mathbf{r})\frac{1}{\left|\mathbf{r}-\mathbf{r}'\right|}h_{\text{XC}}(\mathbf{r},\mathbf{r}')\mathrm{d}\mathbf{r}\mathrm{d}\mathbf{r}' \tag{6.75}$$

where $h_{\text{XC}}(\mathbf{r},\mathbf{r}')$ is the exchange-correlation hole defined by Equation (6.60) that is an integration over λ:

$$h_{\text{XC}}(\mathbf{r},\mathbf{r}') = \int_0^1 h_{\text{XC}}(\mathbf{r},\mathbf{r}';\lambda)\mathrm{d}\lambda \tag{6.76}$$

The coupling-strength-dependent exchange-correlation hole $h_{\text{XC}}(\mathbf{r},\mathbf{r}';\lambda)$ can be defined following Equation (6.60), but replacing only the pair distribution function $n(\mathbf{r},\mathbf{r}')$ by its scaled counterpart and following Equation (6.33):

$$n(\mathbf{r},\mathbf{r}';\lambda) = \langle\Phi_\lambda | \bar{\psi}(\mathbf{r})\bar{\psi}(\mathbf{r}')\psi(\mathbf{r}')\psi(\mathbf{r}) | \Phi_\lambda\rangle \tag{6.77}$$

where Φ_λ is the many-body wave function at coupling strength λ.

Although the adiabatic connection gives the exact formula of exchange correlation, it is far away from a practical calculation, because there is coupling-strength-dependent pair distribution function in the formula that is usually unknown. However, if we know exactly or approximately the coupling-strength-dependent pair distribution functions at some points of λ, we can interpolate integral (6.76) and obtain an approximation of the *exact* exchange correlation. This leads to the hybrid density functionals.

The adiabatic connection (6.74) is, in fact, partially replacing the electron–electron interaction by the external (mean) field. At the end point $\lambda = 0$, there is only mean field, which is equivalent to Hartree–Fock approximation. Thus, the exchange correlation at $\lambda = 0$ is nothing more or less than the exact exchange defined by Equation (6.61). Moreover, it was proposed that LDA is a reasonable approximation to the other end point $\lambda = 1$ [50]. Thus, an interpolation of integral (6.76) is simply:

$$\tilde{h}_{\mathrm{XC}} = \frac{1}{2} h_{\mathrm{X}}^{\mathrm{HF}} + \frac{1}{2} h_{\mathrm{XC}}^{\mathrm{LDA}} \tag{6.78}$$

and the corresponding exchange-correlation energy is given by:

$$\tilde{E}_{\mathrm{XC}} = \frac{1}{2} E_{\mathrm{X}}^{\mathrm{HF}} + \frac{1}{2} E_{\mathrm{XC}}^{\mathrm{LDA}} \tag{6.79}$$

which is simply an average of HF and LDA.

In the point of view of GWA, the HF approach, as the lowest order expansion of the self-energy Σ, keeps only the bare Coulomb interaction and ignores any correction to it implied in the electron–electron interaction, i.e. the screening effect encapsulated in W. Thus, a possible way to correct HF is to replenish those corrections from a "good source," i.e the LDA, which is exact for homogeneous gas, and overestimates those corrections for real materials, especially insulators/semiconductors. Thus, a mixture of HF and LDA should be closer to the *true* exchange correlation than either of them for real materials, in the sense of taking into the screening effect.

In practice, the half–half mixture (6.79) might be insufficient to take into the screening effect, which can be observed from its binding energy results of simple molecules that are usually lower than the experimental but better than HF. Thus, more sophisticated form of hybrid density functionals were proposed to numerically fit the experimental data, where the half–half formula is generalized by introducing semiempirical parameters, such as the B3PW91 [51]:

$$E_{\mathrm{XC}} = E_{\mathrm{XC}}^{\mathrm{LDA}} + a_0 (E_{\mathrm{X}}^{\mathrm{HF}} - E_{\mathrm{X}}^{\mathrm{LDA}}) + a_{\mathrm{X}} \Delta E_{\mathrm{X}}^{\mathrm{B88}} + a_{\mathrm{C}} \Delta E_{\mathrm{C}}^{\mathrm{PW91}} \tag{6.80}$$

where the first term is the LDA exchange correlation, the second is the HF correction to the LDA exchange weighted by the parameter a_0, and the third and fourth terms are the B88 and PW91 gradient corrections to exchange and correlation, respectively, regulated by the parameters a_{X} and a_{C}. The parameters in Equation (6.80) can be determined by numerically fitting the experimental results of fundamental data of atoms and simple molecules. Moreover, it is possible to choose other exchange or correlation gradient corrections to improve Equation (6.80). For example, the B3LYP exchange correlation replaces the correlation of Equation (6.80) with a mixture of VWN and LYP [52]. It should be noted that, although the parameters in hybrid density functionals are usually semiempirical, it is possible to justify that the major mixture parameter a_0 is the reciprocal of an index n, which describes how fast the exact exchange correction to the density functional approximation becomes zero when $\lambda \to 1$. Because n also defines the order of expansion of partially interaction exchange correlation as a polynomial of coupling strength λ, it should generally be integer and thus limits a_0 to values 1/2, 1/3, 1/4, and so on. This leads to parameterless hybrid density functionals, such as PBE0:

$$E_{\mathrm{XC}}^{\mathrm{PBE0}} = \frac{1}{4} E_{\mathrm{X}}^{\mathrm{HF}} + \frac{3}{4} E_{\mathrm{X}}^{\mathrm{PBE}} + E_{\mathrm{C}}^{\mathrm{PBE}} \tag{6.81}$$

It is interesting that, although most hybrid density functionals are generated by numerically fitting the experimental data of atoms and small molecules, they also successfully correct the bandgap problems of solids, even including the transition metal compounds. For example, the B3LYP bandgap of NiO is 3.9 eV [53], closer to the experimental value of 4.3 eV and the result (3.1 eV) given by the LSDA+U approach, with U from constrained first-principles calculations.

6.1.5 Remarks

In this section, we briefly introduced the first-principles calculations. Our focus is clearly on the exchange correlation, because it will directly affect the calculated magnetic properties, such as exchange integrals, as will be seen in the following sections. What we missed is how to implement the calculations, which includes a lot of details of first-principles calculations. For example, to perform first-principles calculations for solids, we should construct the one-body Kohn–Sham wave function for self-consistent calculations, which usually follows one of three basic approaches: plane waves, localized orbitals, and augmented functions. Moreover, because the valence electrons determine most physical and chemical properties of solids, the nuclear and core electrons are usually approximated by an effective potential, i.e. pseudo potential. As we point out at the beginning of this section, the readers who want more details are referred to other volumes [1, 2].

6.2 First-Principles Calculations of Ferrites

In this section, we will apply first-principles calculations to ferrites, a family of practical magnetic materials. The structure and chemistry of ferrites have been introduced in the first chapter. We will focus on the first-principles calculations of exchange integral and magnetic anisotropy, both of which are important magnetic properties impacting the applications of ferrites.

Before further discussion, we put a brief remark on local magnetic moment by first-principles calculations. It should be noted that magnetic moment or spin is generally a vector field dispersive in real space like charge density that is a scalar field. In fact, we can define the spin-dependent density by including spins in Equation (6.23):

$$n^{\alpha\beta}(\mathbf{r}) = \sum_i \bar{\phi}_i^{\alpha}(\mathbf{r})\phi_i^{\beta}(\mathbf{r}) \tag{6.82}$$

where α and β denote majority and minority spins, respectively, and the summation is over all occupied orbitals. Equation (6.82) can be rewritten in the matrix form as:

$$\mathbf{n}(\mathbf{r}) = \begin{pmatrix} \sum_i \bar{\phi}_i^{\uparrow}(\mathbf{r})\phi_i^{\uparrow}(\mathbf{r}) & \sum_i \bar{\phi}_i^{\uparrow}(\mathbf{r})\phi_i^{\downarrow}(\mathbf{r}) \\ \sum_i \bar{\phi}_i^{\downarrow}(\mathbf{r})\phi_i^{\uparrow}(\mathbf{r}) & \sum_i \bar{\phi}_i^{\downarrow}(\mathbf{r})\phi_i^{\downarrow}(\mathbf{r}) \end{pmatrix} \tag{6.83}$$

which is obviously a 2×2 Hermit matrix that can be expanded in terms of the tree Pauli matrices and the identity matrix:

$$\mathbf{n}(\mathbf{r}) = \rho(\mathbf{r})I + \mathbf{m}(\mathbf{r})\cdot\sigma \tag{6.84}$$

where ρ and $\mathbf{m}$ are charge and magnetization densities, respectively, and it is obvious that the magnetization is a vector field. In noncollinear magnetism, the spin quantization axis depends on the position, and there is no uniform spin axis. However, in collinear magnetism, there is a uniform spin axis, i.e. Equation (6.84) can be globally diagonalized by a unitary matrix, and thus, Kohn–Sham formalism can be handily extended to the magnetic system by reorganizing the spin-up and spin-down densities into charge density and magnetization density (or spin-polarization rate; see Equation (6.67)).

6.2.1 Exchange Integrals

In spite of the particular definition of magnetization in DFT, a fact about ferrites is that most of them are good insulators, which implies that magnetization densities are localized around magnetic ions. Thus, the local magnetic moment, or local spin, is well defined, and the Heisenberg model is valid. In fact, our first-principles calculations of exchange integrals are based on the Heisenberg model:

$$\mathcal{H} = -\frac{1}{2}\sum_{i,j} J_{ij}\mathbf{S}_i \cdot \mathbf{S}_j \tag{6.85}$$

where J_{ij} is the exchange integral between the local spins on sites i and j. It should be note that, as a spin Hamiltonian, Equation (6.85) may include the contributions to exchange integrals from different mechanisms. For example, it may contain the direct exchange between two spins on two ions, respectively:

$$J_{ij}^{d} = \iint \mathrm{d}\mathbf{r}\mathrm{d}\mathbf{r}' \frac{1}{|\mathbf{r}-\mathbf{r}'|}\bar{\phi}_i(\mathbf{r})\bar{\phi}_j(\mathbf{r}')\phi_j(\mathbf{r})\phi_i(\mathbf{r}') \tag{6.86}$$

which can be proven to be *always* positive by Fourier transformation [10].

It is obvious that the always positive direct exchange cannot explain the antiferromagnetic order of transition metal monoxides, such as NiO, that implies a negative exchange integral in the Heisenberg model (6.85). Moreover, the magnetic ions in those antiferromagnets are separated by oxygen ions, and thus, the direct exchange integral (6.86) should be much lower than the value inferred from the critical temperature of paramagnetic–antiferromagnetic phase transition. The superexchange was proposed as an indirect exchange mechanism in those antiferromagnetic oxides [54]. In fact, we can consider an example of well-separated H_2 molecule, which can be described by the single-orbital Hubbard model (6.70):

$$\mathcal{H} = t\left(c_{1+}^{*}c_{2+} + c_{1-}^{*}c_{2-}\right) + t\left(c_{2+}^{*}c_{1+} + c_{2-}^{*}c_{1-}\right) + Un_{1+}n_{1-} + Un_{2+}n_{2-} \tag{6.87}$$

where + and – denote the up and down spins, respectively. To diagonalize Hamiltonian (6.87), we first write down all possible two-body states:

$$|1_+, 2_-\rangle,\ |1_-, 2_+\rangle,\ |1_\pm, 2_0\rangle,\ |1_0, 2_\pm\rangle,\ |1_+, 2_+\rangle,\ |1_-, 2_-\rangle \tag{6.88}$$

where the first four states are triplet and the last two are singlet. We then write the matrix form of the Hamiltonian for the singlets:

$$\mathcal{H}_{\text{sng}} = \begin{pmatrix} 0 & 0 & t & t \\ 0 & 0 & t & t \\ t & t & U & 0 \\ t & t & 0 & U \end{pmatrix} \tag{6.89}$$

the eigenvalues of which are:

$$E = \frac{U - \sqrt{16t^2 + U^2}}{2}, 0,\ U,\ \frac{U + \sqrt{16t^2 + U^2}}{2} \tag{6.90}$$

and if $t << U$, the ground state energy is:

$$E_0 \approx -\frac{4t^2}{U} \tag{6.91}$$

and the ground state is approximately a mixture of the first two singlets of (6.88) that is the antiferromagnetic state. Noticing that the triplet states are ferromagnetic and their energy is zero, we obtain the exchange integral in the corresponding spin Hamiltonian:

$$J \approx -\frac{4t^2}{U} \tag{6.92}$$

In spinel ferrites, there are two sublattices of cations, namely, A- and B-sites (or *a* and *d* Wyckoff positions). Thus, if considering only the exchange couplings between nearest neighbors inter- and intra-sublattices, we rewrite the Heisenberg model as:

$$\mathcal{H} = -J_{\text{AB}} \sum_{\langle i,j \rangle} \mathbf{S}_i^{\text{A}} \cdot \mathbf{S}_j^{\text{B}} - J_{\text{BB}} \sum_{\langle i,j \rangle} \mathbf{S}_i^{\text{B}} \cdot \mathbf{S}_j^{\text{B}} - J_{\text{AA}} \sum_{\langle i,j \rangle} \mathbf{S}_i^{\text{A}} \cdot \mathbf{S}_j^{\text{A}} \tag{6.93}$$

For most ferrites, where both A- and B-sites are occupied by transition metal ions, J_{AB} is usually much stronger than J_{BB} and J_{AA} and, thus, leads to the so-called *Néel configuration* – the spins on one sublattice are parallel to each other but antiparallel to those on the other one.

To calculate the exchange integrals of ferrites from first principles, we need to connect Equation (6.93) to the quantities calculated by first principles. We first ignore J_{AA} that is usually very weak in most ferrites. We then replace the quantum operators with their quantum-averaged values, in which the spins can be approximately calculated from the first-principles local magnetic moments. To map the energy of spin Hamiltonian to first-principles ones, we choose the first-principles total energy of ferromagnetic state as the reference point. We then calculate the total energies of selected spin configurations and finally obtain the exchange integral by solving a set of linear equations.

Equation (6.92) shows that the superexchange integral is inversely proportional to the on-site Coulomb repulsion U and proportional to the square of hopping integral t. The repulsion U should be the value with proper screening, instead of a bare value given by the HF approach or overscreened by LDA. In addition, the hopping integral t can be either indirect or direct. The indirect hopping is usually via the anions (mostly oxygens) between transition metal ions, which, in fact, depends on the weak p–d covalency bond between the transition metal ions and anions. It should be noted that HF and LDA are well known for the underbinding problem and the overbinding problem, respectively. The hopping integral t can be related to the crystal field parameters that determine the split between manifolds with different symmetries. The direct hopping is usually between the near-neighbor B-sites in spinel ferrites and depends on the d–d covalency bond between transition metal ions.

6.2.1.1 Manganese Ferrite ($MnFe_2O_4$)

6.2.1.1.1 Introduction and Approach

We calculated the electronic structure and exchange integrals in manganese ferrite [55]. Manganese ferrite is a well-known microwave material. More importantly, it is simple in the sense of cation electron configuration, i.e. both Mn^{2+} and Fe^{3+} are of half-filled open d shell that is ideal for exploring the impacts of strong correlation on exchange integral in practical materials. As shown in Equation (6.92) and mentioned before, the superexchange coupling strength is quantitatively determined by two electronic parameters, U and t, both of which rely on a proper treatment of the screening effect of the open d shells of Mn^{2+} and Fe^{3+} in manganese ferrite. Previous calculation using GGA gave the exchange coupling J_{AB} much stronger than the experimental result [56]. Thus, we chose B3PW91 exchange correlation (6.80) in the calculation. Because the parameter a_0 that controls the mixture of HF and LDA exchanges is determined by numerically fitting experimental data of atoms and molecules *without* the d shell, it was selected to be a variable parameter (w) in the calculation. We chose Gaussian basis functions, using which the HF exchange can be handily calculated. The combination of B3PW91 and Gaussian basis functions is implemented in CRSTAL98 code.

To calculate the exchange integrals, J_{AB} and J_{BB}, we designate three spin configurations, namely, ferromagnetism (FM), ferrimagnetism-1 (FI-1), and ferrimagnetism-3 (FI-3), to the primitive unit cell, where FI-1 is the experimental Néel configuration and FI-3 is the configuration with the A-site and one B-site spin-up and another B-site spin-down. We investigated both normal and inverse spinel structures, where Mn^{2+} ions are on A-sites in the normal structure and on B-sites in the inverse structure. The formula to calculate the exchange integrals from the first-principles total energies with different spin configurations are:

$$
\begin{aligned}
J_{AB} &= \frac{1}{96S^2}\left(E_{FI\text{-}1} - E_{FM}\right) \\
J_{BB} &= \frac{1}{32S^2}\left[E_{FI\text{-}3} - \left(E_{FI} + E_{FM}\right)/2\right].
\end{aligned}
\tag{6.94}
$$

In addition, the calculation diverged when a_0 is less than 0.4, because of the mismatch between the HF-optimized basis functions and the exchange correlation close to LSDA. We thus introduced a scaling factor λ to the radial part of the Fe^{3+} basis functions to make its d wave functions more dispersive in real space as:

$$
\tilde{R}(r) = \lambda^{3/2} R(\lambda r) \tag{6.95}
$$

which conserves the normalization of wave functions.

6.2.1.1.2 Exchange Integrals

We plot the calculated J_{AB} and J_{BB} in Figure 6.1, which clearly shows the effects of the screening effect on the calculated exchange integrals. As discussed before, HF underestimates t and overestimates U and consequently underestimates the superexchange contribution to J_{AB}. As the distance between an A-site and its nearest-neighbor B-site is 3.5 Å and the bonding angle of A–O–B is 122°, the direct exchange contribution to the exchange integral is negligible. Thus, we obtain a negative J_{AB} even using only the Fock exchange, although the value of J_{AB} is about 20% of the experimental value. On the other hand, as the distance between the nearest-neighbor B-sites is 3 Å and the bonding angle of B–O–B is 94.5°, the direct exchange exceeds the superexchange contribution underestimated by HF. Thus, a pure Fock exchange leads to a qualitatively incorrect positive J_{BB} for a normal structure.

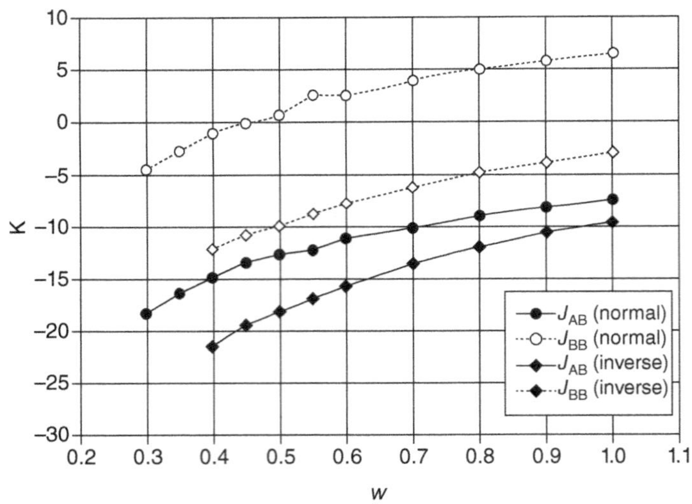

Figure 6.1 J_{AB} and J_{BB} of normal and inverse structures using the basis set in which the $3d$ wave functions of Fe^{3+} is scaled to 130% in the radial direction.

When w decreases, the underestimation of superexchange is compensated by the LSDA contribution in the B3PW91 exchange correlation, as pure LSDA usually overestimates t and underestimates U, and consequently, overestimates the superexchange contribution to the exchange integrals. Therefore, in the case of J_{AB}, as the direct exchange is negligible, the negative J_{AB} becomes stronger when w decreases. The J_{BB} also drops when w decreases, and the positive J_{BB} of normal structure turns to be negative at $w \sim 0.45$. Compared to the experimental data, $J_{AB} \sim -20$ K and $J_{BB} \sim -3$ K, the best fit to experiment is achieved at $w \sim 0.35$.

6.2.1.1.3 Electronic Structure

We plot the calculated density of states (DOSs) of $MnFe_2O_4$ in Figure 6.2. Compared with the DOS given by GGA, the DOS given by our calculation is quite different. First, the DOS given by GGA implied that $MnFe_2O_4$ is half metallic, where the spin-down Mn^{2+} and the spin-down Fe^{3+} d bands overlapped each other and crossed over the Fermi level. However, in our calculation, the DOS shows that $MnFe_2O_4$ is an insulator with a bandgap between the spin-down Mn^{2+} and spin-down Fe^{3+} d bands. Second, the DOS given by GGA implied that $MnFe_2O_4$ is a typical Mott insulator for both Fe^{3+} and Mn^{2+}, in which the ligand p band is located at a lower energy than the d bands of transition-metal ions in the valence band. However, in our calculation, the Fe^{3+} and O^{2-} system is a typical charge-transfer insulator and the Mn^{2+} and O^{2-} system is an interim state between the Mott insulator and the charge-transfer insulator.

We also extracted the screened on-site Coulomb repulsions of Fe^{3+} and Mn^{2+} as a function of w from the calculated DOS (see Figure 6.3). The U parameters linearly decrease, when w decreases. This is expected because the B3PW91 exchange correlation is simply the linear interpolation of HF and LSDA weighted by parameter w. Extrapolating the linear dependence of U on w at $w \sim 0.35$, which is the value given by the best fit to experimental exchange integrals, we obtain U of Fe^{3+} and Mn^{2+} are about 14 and 10 eV, respectively, which are close to the experimental data. We also obtained the bandgap from DOS as a function of w, which assumes its experimental value of about 4.0 eV at $w \sim 0.40$.

6.2.1.1.4 Conclusion

It should be noted that, although w is a variable parameter in the calculation, it quantitatively improves the calculated results of exchange integrals, J_{AB} and J_{BB}, screened on-site Coulomb repulsions of both Fe^{3+} and Mn^{2+} and the bandgap. This shows the potential of hybrid density functionals, such as B3PW91, as a low-computational-cost approach to improve the first-principles calculations on practical ferrites.

6.2.1.2 Lithium Ferrite ($Li_{0.5}Fe_{2.5}O_4$) and Magnesium Ferrite ($MgFe_2O_4$)

In previous calculations of exchange integrals in $MnFe_2O_4$, we introduce a scaling factor to make the basic functions and the Fe^{3+} d wavefunctions more dispersive in real space. This approach did solve the divergence problem when $w < 0.4$ in the calculations and can be justified that the optimized wave function for B3PW91 exchange correlation should be

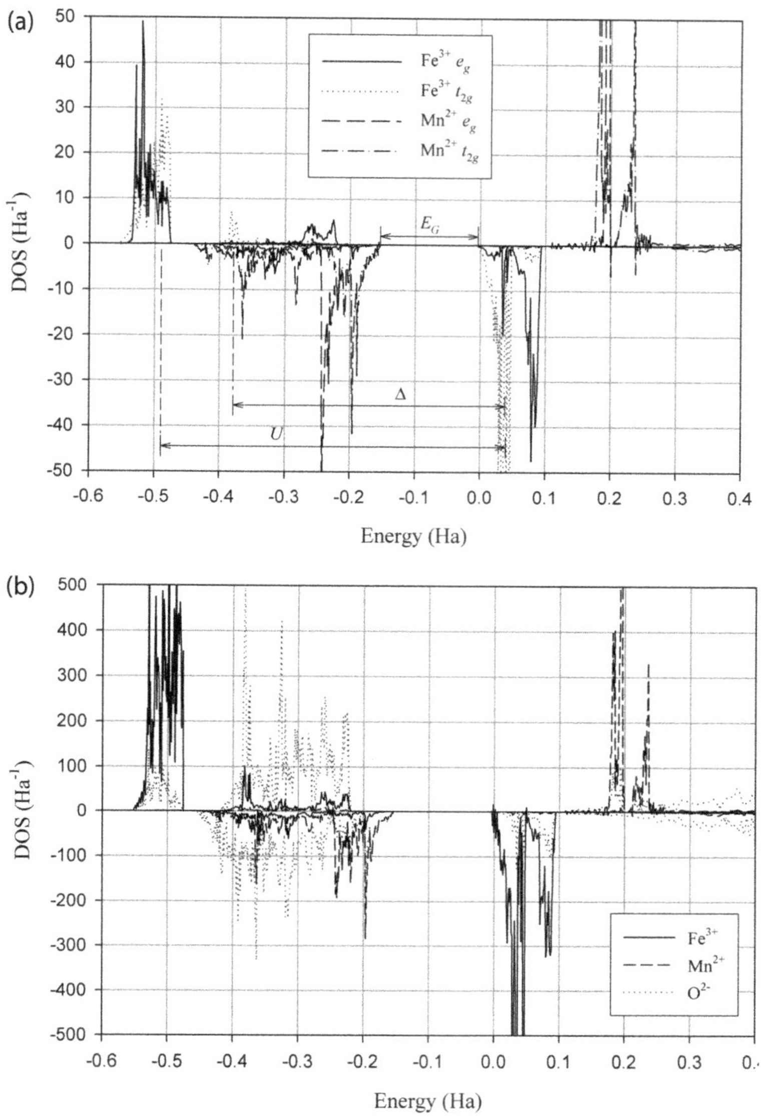

Figure 6.2 Calculated DOS of $MnFe_2O_4$ in the normal structure and FI-1 magnetic structure at $w = 0.4$: (a) projected on symmetries and (b) projected on atoms.

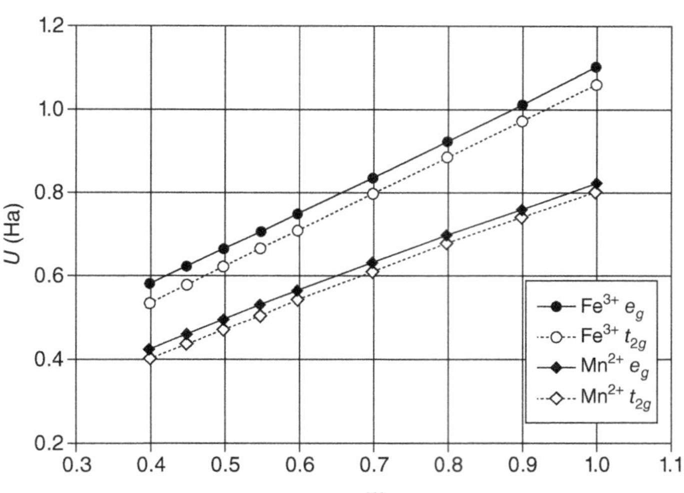

Figure 6.3 U of Fe^{3+} and Mn^{2+} extracted from the calculated density of states.

more dispersive than that optimized for HF because of the LSDA component. A fundamental question, however, was, *whether the parameters* $w \sim 0.35$ *and* $\lambda \sim 1.3$ *are universal*, although a computational study of nickel (II) oxide (NiO) using similar density functional suggested a close percentage of HF exchange (~30%) in total exchange correlation [57].

To answer that question, we chose lithium ferrite ($Li_{0.5}Fe_{2.5}O_4$) and magnesium ferrite ($MgFe_2O_4$), both of which contain only Fe^{3+} ions [58]. We followed the same approach to calculating the exchange integrals in $MnFe_2O_4$. In addition to J_{AB} and J_{BB}, we calculated J_{AA} in this study. We then considered four spin configurations, namely, FM, FI-1, FI-2, and FI-5. The linear equations derived from Equation (6.93) for those spin configurations are listed in Table 6.1. Both lithium and magnesium ferrites are inverse spinel, where the Li^+ and Mg^{2+} ions are on B-sites.

The calculated J_{AB} in $MgFe_2O_4$ and $Li_{0.5}Fe_{2.5}O_4$ shows similar dependence on parameters w and λ in $MnFe_2O_4$; J_{AB} becomes more negative as w decreases or λ increased (see Figure 6.4a and b). This behavior can be attributed to the definition of w and λ. As w (the percentage of HF exchange in total exchange) decreases, more LSDA enters the total exchange, which decreases the effective Coulomb repulsion (U) energy and increases the hopping integral (t). As a result, the superexchange interaction is enhanced and J_{AB} becomes more negative. Increasing λ shows similar effects such as decreasing w, but directly operates on the d wavefunctions of the magnetic ions (Fe^{3+}).

Table 6.1 Spin configurations and total energies derived from the Heisenberg model.

(a) $MgFe_2O_4$			
Spin configuration			
Designation	**A-sites**	**B-sites**	**Energy**
FM	↑↑	↑↑	$-E_{FM} = 8J_{AA}S_A^2 + 24J_{AB}S_AS_B + 4J_{BB}S_B^2$
FI-1	↓↓	↑↑	$-E_{FI\text{-}1} = 8J_{AA}S_A^2 + 24J_{AB}S_AS_B + 4J_{BB}S_B^2$
FI-2	↑↓	↑↑	$-E_{FI\text{-}2} = -8J_{AA}S_A^2 + 4J_{BB}S_B^2$
FI-5	↑↑	↑↓	$-E_{FI\text{-}5} = 8J_{AA}S_A^2 + 4J_{BB}S_B^2$
(b) $Li_{0.5}Fe_{2.5}O_4$			
Spin configuration			
Designation	**A-sites**	**B-sites**	**Energy**
FM	↑↑	↑↑↑	$-E_{FM} = 8J_{AA}S_A^2 + 36J_{AB}S_AS_B + 12J_{BB}S_B^2$
FI-1	↓↓	↑↑↑	$-E_{FI\text{-}1} = 8J_{AA}S_A^2 - 36J_{AB}S_AS_B + 12J_{BB}S_B^2$
FI-2	↑↓	↑↑↑	$-E_{FI\text{-}5} = -8J_{AA}S_A^2 + 12J_{BB}S_B^2$
FI-5	↑↑	↑↑↓	$-E_{FI\text{-}2} = 8J_{AA}S_A^2 + 12J_{BB}S_AS_B - 4J_{BB}S_B^2$

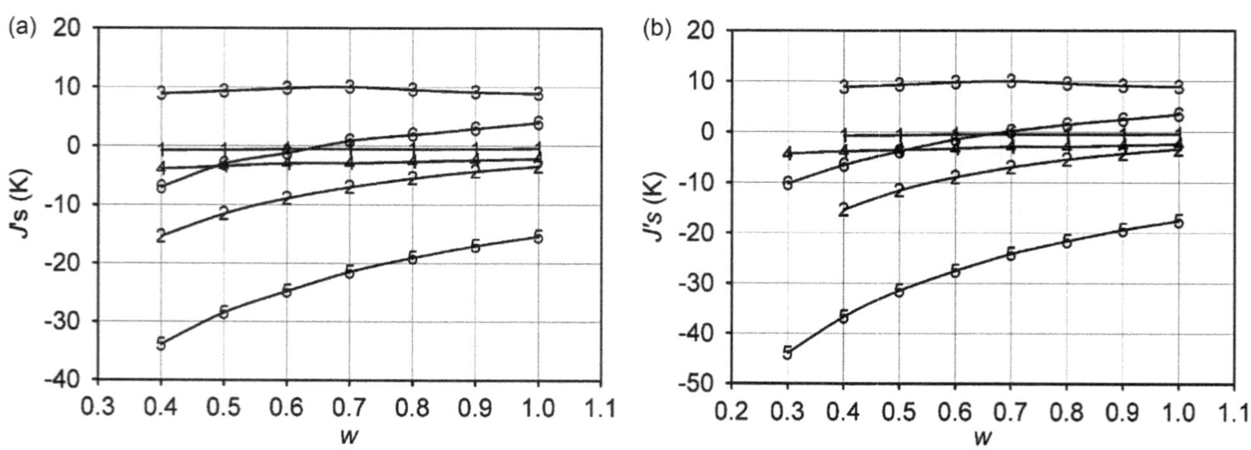

Figure 6.4 Calculated exchange constants of (a) $MgFe_2O_4$ and (b) $Li_{0.5}Fe_{2.5}O_4$ as a function of parameters w and λ. In both, curves (1), (2), and (3) represent J_{AA}, J_{AB}, and J_{BB} calculated with $\lambda = 1.0$, respectively, and curves (4), (5), and (6) represent those calculated with $\lambda = 1.3$, respectively.

With $w = 0.7$ and $\lambda = 1.3$, the calculated J_{AB} in $MgFe_2O_4$ matches the experimental result (see Table 6.2). For $Li_{0.5}Fe_{2.5}O_4$, matching the experimental result of J_{AB} leads to $w = 0.6$ and $\lambda = 1.3$ (see Table 6.2). Although these optimal values are different from those ($w \sim 0.35$ and $\lambda = 1.3$) deduced from $MnFe_2O_4$, they are concentrated in a small parameter space. This implies that there is at least a set of values of parameters w and λ, which can possibly be universal for *Fe^{3+}-only* spinel ferrite.

6.2.1.3 Nickel Ferrite ($NiFe_2O_4$) and Copper Ferrite ($CuFe_2O_4$)

The ferrites discussed in the previous sections contain only Fe^{3+} and Mn^{2+}, whose electronic configuration is d^5, the half-filled d shell. In this section, we extend our discussion to ferrites containing more than half-filled d shell, namely, nickel ferrite ($NiFe_2O_4$) and copper ferrite ($CuFe_2O_4$) [59, 60]. Cobalt ferrite is excluded, because it is a well-known magnetic anisotropic material.

6.2.1.3.1 Nickel Ferrite ($NiFe_2O_4$)

Nickel ferrite is an inverse spinel, where Ni^{2+} ions are on B-sites. The electronic configuration of Ni^{2+} ion is d^8, of which the multi-electron ground state in the octahedral crystal field is nondegenerate Γ_2, where there is no Jahn–Teller distortion and the spin-orbit coupling effect is the only second-order perturbation. In fact, the e_g manifold of Ni^{2+} on the B-site is half-filled like Fe^{3+} and Mn^{2+}.

To calculate the exchange integrals, we followed the same approach as applied before. In addition, because Ni^{2+} is also magnetic, there are six exchange integrals, namely, $J^{AB}{}_{NiFe}$, $J^{AB}{}_{FeFe}$, $J^{BB}{}_{NiNi}$, $J^{BB}{}_{NiFe}$, $J^{BB}{}_{FeFe}$, and J^{AA}. The spin configurations and corresponding linear equations of J have been listed in Table 6.3.

According to the interacting sites, the exchange constants can be classified into three types: J^{AA} (between A-sites), J^{AB} (between A- and B-sites), and J^{BB} (between B-sites). The calculation shows that J^{AA} is always negative and weak (curve 1 in Figure 6.5), as expected according to the empirical rule of superexchange [61], given the bonding angle of A–O–A is smaller than 90°.

Table 6.2 Comparison between experimental and calculated exchange constants in $MgFe_2O_4$ and $Li_{0.5}Fe_{2.5}O_4$.

		J_{AB}	J_{BB}	J_{AA}
$MgFe_2O_4$	Exp.	−22.5	−11.5	−7.0
	w=0.7, λ=1.3	−21.4	0.9	−2.9
$Li_{0.5}Fe_{2.5}O_4$	Exp.	−29.0	−10.0	−20
	w=0.6, λ=1.3	−27.5	−1.5	−3.3

Table 6.3 Spin structures and energy differences from the ferromagnetic structure.

Spin structures				**Energy difference**
Designation	Fe(A)	Ni(B)	Fe(B)	
Ferromagnetic	↑↑	↑↑	↑↑	–
Ferromagnetic-1	↓↓	↑↑	↑↑	$\Delta = 148J^{AB}_{NiFe}S_{Ni}S_{Fe} + 48J^{AB}_{FeFe}S_{Fe}S_{Fe}$
Ferromagnetic-2	↑↓	↑↑	↑↑	$\Delta_2 = 16J^{AA}S_{Fe}S_{Fe} + 24J^{AB}_{NiFe}S_{Ni}S_{Fe} + 24J^{AB}_{FeFe}S_{Fe}S_{Fe}$
Ferromagnetic-3	↑↑	↑↑	↓↓	$\Delta_3 = 48J^{AB}_{FeFe}S_{Fe}S_{Fe} + 32J^{BB}_{NiFe}S_{Ni}S_{Fe}$
Ferromagnetic-4	↑↑	↓↓	↑↑	$\Delta_4 = 48J^{AB}_{NiFe}S_{Ni}S_{Fe} + 32J^{BB}_{NiFe}S_{Ni}S_{Fe}$
Ferromagnetic-5	↑↑	↑↑	↑↓	$\Delta_5 = 24J^{AB}_{FeFe}S_{Fe}S_{Fe} + 16J^{BB}_{NiFe}S_{Ni}S_{Fe} + 8J^{BB}_{FeFe}S_{Fe}S_{Fe}$
Ferromagnetic-6	↑↑	↑↓	↑↑	$\Delta_6 = 24J^{AB}_{NiFe}S_{Ni}S_{Fe} + 8J^{BB}_{NiNi}S_{Ni}S_{Ni} + 16J^{BB}_{NiFe}S_{Ni}S_{Fe}$

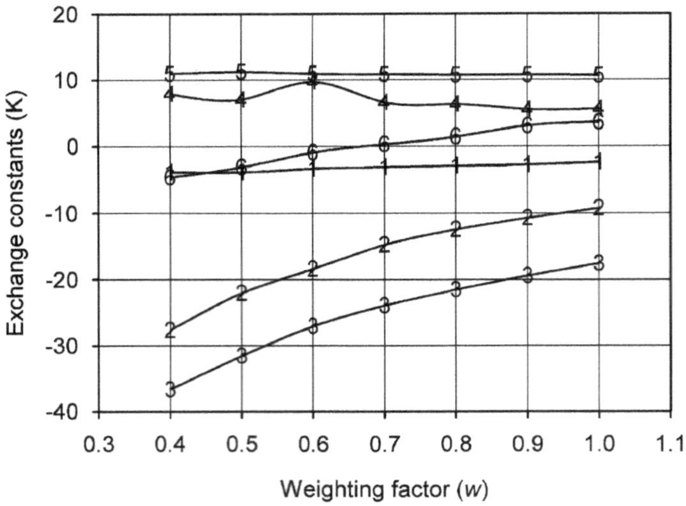

Figure 6.5 Calculated exchange constants as a function of w at $\lambda = 1.3$, where curve 1 is J^{AA}, curve 2 is J^{AB}_{NiFe}, curve 3 is J^{AB}_{FeFe}, curve 4 is J^{BB}_{NiNi}, curve 5 is J^{BB}_{NiFe}, and curve 6 is J^{BB}_{FeFe}.

J^{AB} can be classified into two subtypes: J^{AB}_{NiFe} (between Ni^{2+} and Fe^{3+}) and J^{AB}_{FeFe} (between Fe^{3+}). The calculation shows that, at a fixed value of w, J^{AB}_{NiFe} is weaker than J^{AB}_{FeFe} (curves 2 and 3 in Figure 6.5). According to the simplified formula of superexchange (6.92), the result of J^{AB}_{NiFe} and J^{AB}_{FeFe} implies that the hopping term (t) between Ni^{2+} and Fe^{3+} is smaller than that between Fe^{3+} ions, if we assume that the on-site repulsions (U) of Ni^{2+} and Fe^{3+} are about equal to each other. As the bonding angle of A–O–B is ~ 120º, the major contribution to hopping term of the Hubbard model is from the covalency between transition metal ions and anions. Thus, our result on J^{AB}_{NiFe} and J^{AB}_{FeFe} implies that the covalency of the Ni–O bond is weaker than that of Fe–O bond. Both J^{AB}_{NiFe} and J^{AB}_{FeFe} are negative and become stronger as w decreases. In fact, as w decreases, the LSDA weight in the exchange correlation increases; this enhances the superexchange. Fitting the average of J^{AB}_{NiFe} and J^{AB}_{FeFe} to the experimental results of J_{AB}, we find that the best fit is at $w \sim 0.4$, which is close to the previously determined value for manganese ferrite.

J^{BB} can be classified into three subtypes: J^{BB}_{NiNi} (between Ni^{2+}), J^{BB}_{NiFe} (between Ni^{2+} and Fe^{3+}), and J^{AB}_{FeFe} (between Fe^{3+}). As several competing exchange mechanisms contribute to J^{BB}, the behavior of J^{BB} is more complex than that of J_{AB}. In these competing mechanisms, the superexchange via the direct hopping between B-sites (J^{BB}_{sd}), the direct exchange between B-sites (J^{BB}_{d}), and the superexchange via oxygen (J^{BB}_{O}) are the three major contributions, where J^{BB}_{sd} is negative and J^{BB}_{d} and J^{BB}_{O} are positive. In the case of J^{BB}_{NiNi} and J^{BB}_{NiFe}, J^{BB}_{sd} vanishes, as all t_{2g} orbitals of Ni^{2+} are fully occupied. As a result, the calculation shows that J^{BB}_{NiNi} and J^{BB}_{NiFe} are positive (curves 4 and 5 in Figure 6.5). In addition, J^{BB}_{NiFe} is slightly stronger than J^{BB}_{NiNi}. However, in the case of J^{BB}_{FeFe}, J^{BB}_{sd} is nonzero and competing with J^{BB}_{d} and J^{BB}_{O}. As a result, the calculation shows that J^{BB}_{FeFe} monotonically decreases as w decreases and change its sign at $w \sim 0.65$ (curve 6 in Figure 6.5).

The DOS in nickel ferrite is calculated (see Figure 6.6). The calculated DOS at $w = 0.4$ and $\lambda = 1.3$ clearly shows that nickel ferrite is a good insulator, where the bandgap is about 4 eV. From the DOS, the on-site Coulomb repulsion of Fe^{3+} (U_{Fe3+}) is about 10 eV and that of Ni^{2+} (U_{Ni2+}) is about 12 eV, which is close to experimental results [62]. From the DOS, the lower Hubbard bands of Ni^{2+} are lower than oxygen bands. This shows the strong charge-transfer nature of Ni^{2+}. However, the lower Hubbard bands of Fe^{3+} strongly interact with the oxygen bands. In addition, the DOS shows the crystal field splits between e_g and t_{2g} bands. From the split of Fe^{3+}, it is obvious that the crystal field of B-site is stronger than A-site, as predicted by the crystal field theory.

The spin-density map (SDM) is a visual presentation of the spins in the magnetic solid state. In the calculated SDM, we clearly observe different spin states of A and B-sites and A–O–B bonds (see Figure 6.7). In more detail, the A–O and B–O bonds are asymmetric and the B–O bond is stronger. This picture is in agreement with the results of DOS and exchange integral. In addition to the A–O–B bond, the SDM of the (1 1 0) plane shows the B–B direct hopping between Fe^{3+} on *B-sites* (see Figure 6.7a). However, the SDM of the $(1\bar{1}0)$ plane only shows the $d_{3z2\text{-}r2}$ orbital, as the spin density of fully occupied t_{2g} orbitals of Ni^{2+} is zero and that of $d_{x2\text{-}y2}$ on $(1\bar{1}0)$ plane is also zero (see Figure 6.7b).

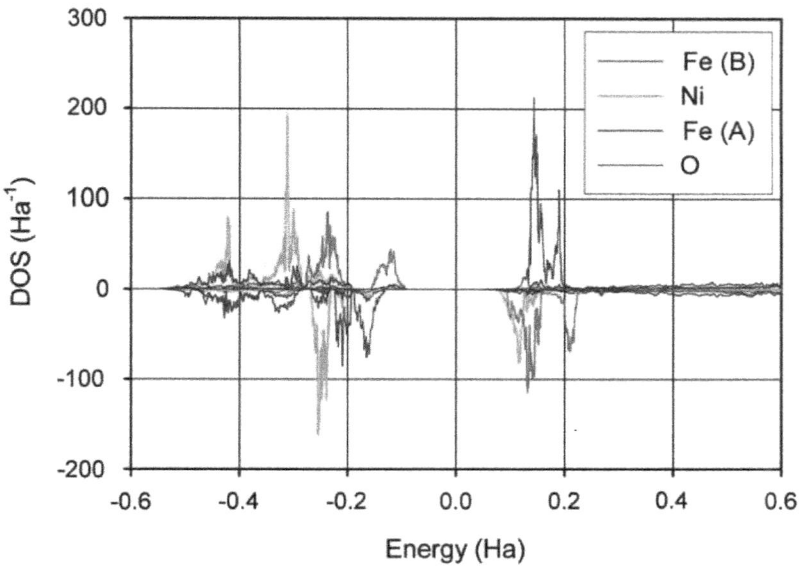

Figure 6.6 Calculated DOS of nickel ferrite at $w = 0.4$ and $\lambda = 1.3$.

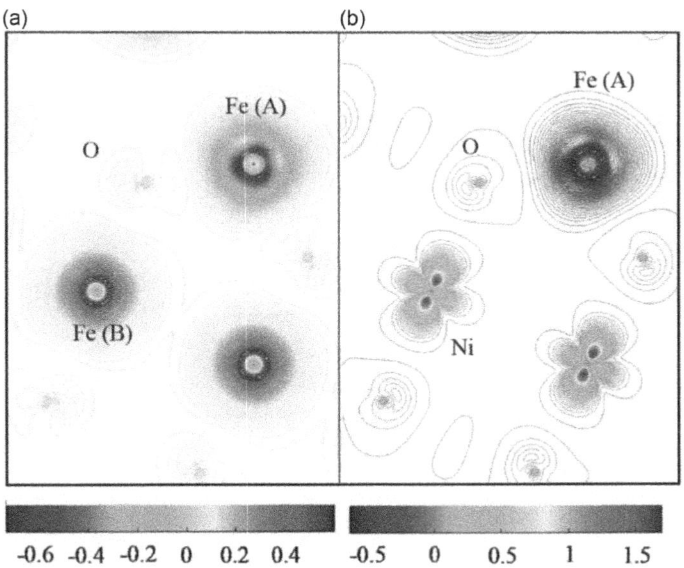

Figure 6.7 Calculated SDM of nickel ferrite at $w = 0.4$ and $\lambda = 1.3$, where (a) is on the (1 1 0) plane and (b) is on the $(1\bar{1}0)$ plane.

6.2.1.3.2 Copper Ferrite ($CuFe_2O_4$)

Copper ferrite is a partially inverse spinel, where 6–24% Cu^{2+} are on A-sites, depending on the sample preparation, and the corresponding magnetization is 1.48–2.92 μ_B. This implies that it is possible to enhance the magnetization of copper ferrite by pushing more Cu^{2+} ions on A-sites. The electronic configuration of Cu^{2+} is d^9, which implies a strong Jahn–Teller effect. In fact, a tetragonal distortion with $c/a > 1$ was observed in the experiment.

To calculate the exchange integrals, we followed the same approach as applied before. We consider both inverse and normal structures. We adopted the experimental lattice constants and atomic positions in the calculation of the inverse structure and assume $c/a = 0.9$ referring to that value in $CuCr_2O_4$ and $CuRh_2O_4$.

The calculated exchange constants of the inverse structure are listed in Table 6.4. As observed in the previous computational results of S-state ferrites (manganese, magnesium, and lithium ferrites), J^{AA} is weak and antiferromagnetic (negative) and J^{BB} are strong and antiferromagnetic. Both J^{AA} and J^{AB} become stronger as w decreases. In fact, as w decreases, the

Table 6.4 Calculated exchange constants of the inverse structure using $\lambda = 1.3$.

w	J^{AA}_{FeFe}	J^{AB}_{FeCu}	J^{AB}_{FeFe}	J^{BB}_{CuCu}	J^{BB}_{CuFe}	J^{BB}_{FeFe}
0.40	−9.1	−16.9	−39.4	−508.6	−88.4	−41.2
1.00	−2.3	−5.9	−18.0	22.5	8.0	3.7

Table 6.5 Calculated exchange constants of the normal structure using $\lambda = 1.3$.

w	J^{AA}_{CuCu}	$J^{AB_1}_{CuFe}$	$J^{AB_2}_{CuFe}$	$J^{B_1B_1}_{FeFe}$	$J^{B_1B_2}_{FeFe}$	$J^{B_2B_2}_{FeFe}$
0.40	−84.3	−60.0	−12.0	−3365.3	−109.4	−132.7

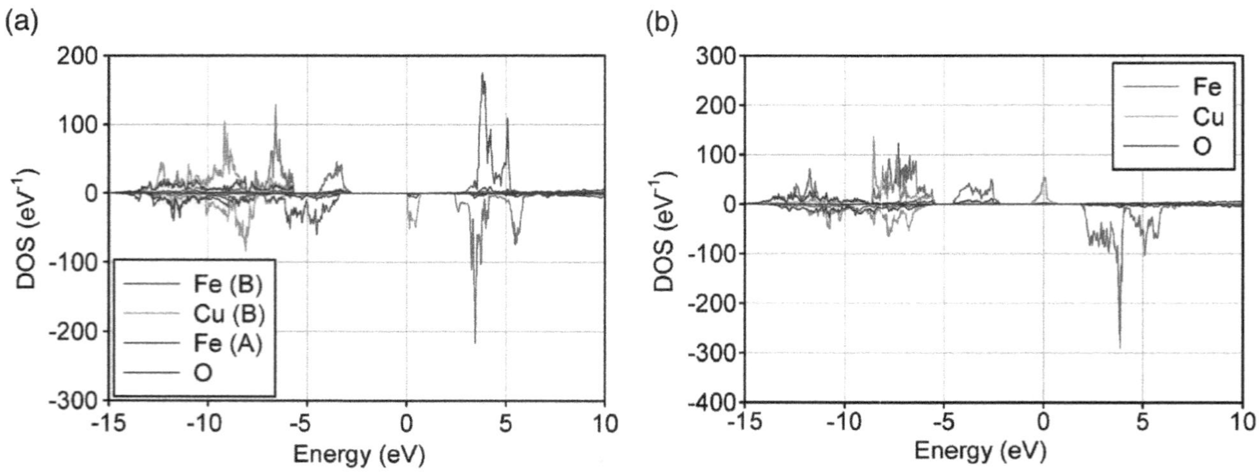

Figure 6.8 Calculated DOS of the (a) inverse and (b) normal structures with Néel configuration at $w = 0.4$.

covalency between magnetic ions and oxygen increases and the on-site Coulomb repulsion of magnetic ions decreases. This results in a stronger antiferromagnetic superexchange contribution to the exchange constant. In contrast to the S-state ferrites, the J^{BB} of the inverse structure show a significant change as w decreases. At $w = 1.0$, J^{BB} are ferromagnetic (positive) and $J^{BB}{}_{CuCu}$ is the strongest interaction in all of the exchange interactions, which implies that Néel configuration may be stable. However, at $w = 0.4$, J^{BB} are antiferromagnetic and J^{BB}s are stronger than J^{AB}, which implies that Néel configuration is unstable. More calculations on the other values of w between 0.4 and 1.0 show that $J^{BB}{}_{CuCu}$ oscillates with the amplitude of 4500 K at $w = 0.7$. This phenomenon may be related to orbital degeneracy of Cu^{2+}. As the e_g doublet is split by weak next nearest-neighbor crystal field, the occupied orbital may change as the exchange correlation is changing from HF ($w = 1.0$) toward LSDA ($w = 0.4$). The calculated exchange constants of the normal structure are listed in Table 6.5. At $w = 0.4$, the antiferromagnetic J^{BB} are much stronger than J^{AB}, which suggests that Néel configuration is unstable.

DOS is calculated as the key electronic structure result for both inverse and normal structures with Néel configuration at $w = 0.4$. In the calculated DOS of the inverse structure (see Figure 6.8), the spin-down Cu^{2+} d band is right above the Fermi level in energy. As we adopt the level shift technique to guarantee the insulating states, which shifts the conduction bands up by 0.3 Ha in a self-consistent cycle and shifts it back after the cycle, the Fermi level is very likely to fall into the spin-down Cu^{2+} d band, in fact. If this is true, the inverse structure may be half metallic. However, in the calculated DOS of the normal structure (see Figure 6.8), the Fermi level directly falls into the spin-up Cu^{2+} d band. This suggests that the normal structure may also be half metallic.

6.2.1.4 Cadmium–manganese Ferrite ($Cd_{1-x}Mn_xFe_2O_4$)

In this section, we turn to the mixed cadmium–manganese ferrite ($Cd_xMn_{1-x}Fe_2O_4$) [63]. The mixed ferrite, $M_xMn_{1-x}Fe_2O_4$ (M = Zn, Cd) was developed to enhance the magnetization of manganese ferrite ($MnFe_2O_4$), where the magnetization of A

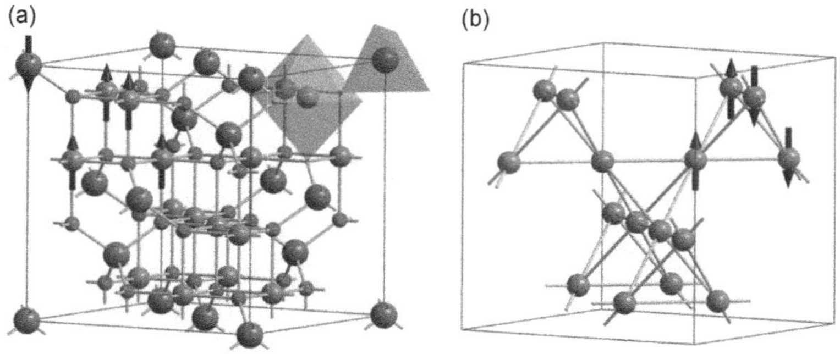

Figure 6.9 (a) Spinel structure with Néel ferrimagnetic configuration. (b) Geometric frustrated B sublattice formed by corner-sharing tetrahedrons.

sublattice cancels a half of that of B sublattice for the well-known Néel configuration (see Figure 6.9). Divalent nonmagnetic ions, such Zn^{2+} and Cd^{2+}, strongly prefer A-sites, because of the sp^3 hybridization of the empty s and p orbitals and weak convalency bond with coordinating oxygen ions under tetrahedral local symmetry. Thus, when these nonmagnetic ions are doped into manganese ferrite, they will occupy A-sites, and thus, the total magnetization will hopefully increase. However, because they are nonmagnetic and not magnetically coupled to the B-site Fe^{3+} ions, the average J_{AB} will decrease by oping. When it is weaker than J_{BB}, the Néel configuration of spin will be unstable and the ground magnetic is the Yafet–Kittel spin canting state [64].

Moreover, the B sublattice is a network of tetrahedrons sharing corners (see Figure 6.9), where a spin cannot find an equal number of nearest neighbors in parallel and antiparallel directions. This is a spin-frustration structure in three dimensional. In a spin-frustration structure, simple antiferromagnetism that requires an equal number of nearest neighbors in parallel and antiparallel directions for any spin, is unstable, and the magnetic ground state is usually complex antiferromagnetic state, where the exchange couplings beyond nearest neighbors may involve.

In the previous sections, we applied the hybrid density functional, B3PW91, to the first-principles calculations of exchange integrals of ferrites. In fact, LSDA+U is an approach developed for strong correlated systems, specially transition metal oxides. In this section, we apply the LSDA+U approach to calculate the exchange integrals of cadmium–manganese ferrite.

In the calculation, we considered five concentrations of Mn, i.e. $x = 0$, 0.25, 0.5, 0.75, and 1. To calculate the exchange integrals, we followed the approach applied before, except that we replaced the hybrid density functional with the GGA + U approach and applied U to the d orbitals of Mn and Fe. Instead of determining U from constrained LSDA calculations, we let U vary as an empirical parameter to show its effect on the calculated exchange integrals. Because the magnetic ground state of the frustrated B sublattice involves the exchange interactions beyond nearest neighbors, we extended our calculation to the exchange couplings between the second and third nearest neighbors. The wave functions were constructed by the projector augmented-wave method. The calculation was implemented by VASP code [65].

Our first concern is the calculated exchange integral of $MnFe_2O_4$, which can be directly compared to the previous calculation using hybrid density functional. The results calculated with the optimized structure show that the exchange interaction between A- and B-sites (J_{AB}) is 90 K and dominant, and its magnitude reaches maximum at $U \sim 2$ eV. The exchange interaction between the nearest-neighbor B-sites (J_{BB1}) is positive and increases when U increases, while the second nearest-neighbor (J_{BB2}) is negative. However, the experimental value of J_{AB} is ~ -20 K and that of J_{BB} is low. As the superexchange is sensitive to the local geometry and the optimized structure that are different from the experimental ones, we can expect the deviation between the J values calculated with the optimized structure and the experimental one. Therefore, we also calculated J using the experimental structure. The value of J_{AB} calculated with the experimental structure is lower than that calculated with the optimized structure and is closer to the experimental value (see Figure 6.10). J_{AB} is calculated with the experimental structure and is about −50 K. In addition, J_{BB1} calculated with the experimental structure is negative when $U < 3.5$ eV and becomes positive when $U > 3.5$ eV. When $2.5 < U < 3.5$ eV, the value of is low and quantitatively in agreement with the experimental results.

We calculated the dependence of exchange integrals, J_{AB} and J_{BB1}, on manganese concentration x, which enable us to predict the concentration at which Néel configuration transits to Yafet–Kittel configuration. Experiments show that in zinc manganese ferrite ($Zn_{1-x}Mn_xFe_2O_4$), the spins on B-sites are canting with respect to the direction of spins on A-sites, when the concentration of Mn is below a critical value ($x < 0.5$). In fact, A (or B)-sites of the spinel structure further split into A′ and A″ (or B′ and B″) sites, and spin canting will happen, if the molecular field on one B′-site from the nearest neighbor

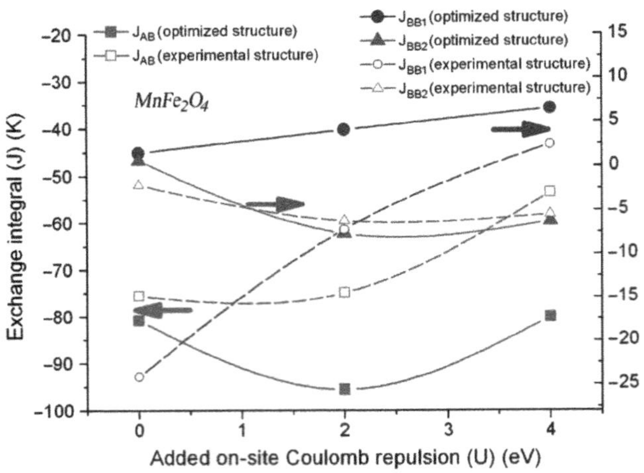

Figure 6.10 Dependences of exchange integrals in $MnFe_2O_4$ on the added on-site Coulomb repulsion U. The solid and hollow symbols represent the results calculated with the optimized and experimental structures, respectively.

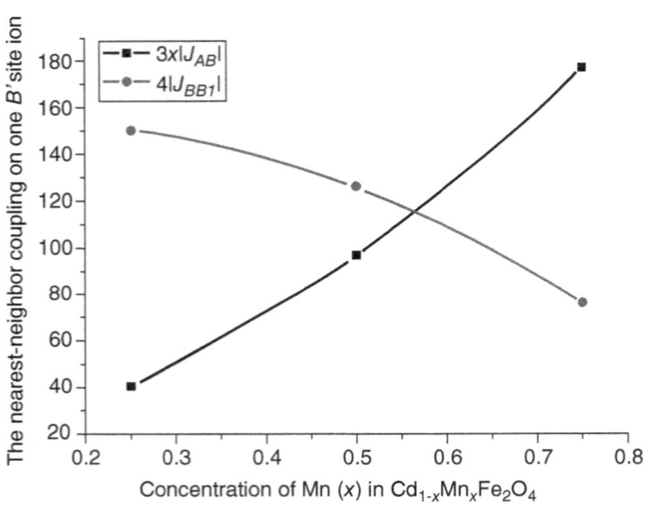

Figure 6.11 Exchange couplings on one B′ site ion from the nearest neighbor A′ and B″ sites.

B″ sites is stronger than that from the nearest neighbor A′ (or A″)-sites. Each B′site has four nearest-neighbor B″ sites and three nearest-neighbor A′-sites. It should be noted that only part of A-sites are magnetic and contribute to exchange interaction depending on x. As the spins of all magnetic ions, in our cases, are identical in model ($S = 5/2$), we can estimate the competition between the molecular field on one B′-site from the nearest neighbor A′ (or A″) and that from the nearest-neighbor B″-sites by comparing $3xJ_{AB}$ and $4J_{BB1}$. The results based on the exchange integrals given by GGA calculations are shown in Figure 6.11, which predict that spin canting may happen when $x < 0.6$.

6.2.1.5 Hexagonal Ferrite ($SrCo_2Ti_2Fe_8O_{19}$)

In this section, we turn to calculation of exchange integrals of hexagonal ferrites. We calculated the exchange integrals of strontium M-type hexagonal ferrite ($SrFe_{12}O_{19}$, in abbreviation, SFO) and Co–Ti codoped derivation ($SrCo_2Ti_2Fe_8O_{19}$, in abbreviation, SCTFO). SFO is characterized by a strong magnetic anisotropy and was applied to microwave ferrite devices at Ka band (26.5–40 GHz), such as a circulator [66]. SFO is doped with cobalt to adjust its magnetic anisotropy. Titanium is codoped to compensate the charge state of Co ions. Recently, it was discovered that SCTFO is a magnetoelectric material, in which magnetization can be modulated by electric field [67]. The magnetoelectric effect can be attributed to the noncollinear spin configuration by the spin current model [68]. Thus, a better understanding of magnetoelectric effect in this material does depend on its spin configuration, to which cation distribution, exchange integrals, and magnetic anisotropy are important.

In fact, Co^{2+} and Ti^{4+} pair in the lattice and prefer the tetrahedral 4f1 and octahedral 12k positions, respectively. It should be noted that the ground state of Co^{2+} under tetrahedral crystal field is nondegenerate Γ_2, which implies low magnetic anisotropy. Moreover, Ti^{4+} is nonmagnetic, which interrupts the exchange chain along the crystallographic c-axis that couples the layers of magnetic ions. In fact, like Néel configuration in spinel ferrite, the experimental Gorter spin configuration here is established on the antiferromagnetic exchange couplings between the magnetic sublattices, and leads to the unbalanced total spins of difference sublattices. In the magnetic sublattices of M-type hexagonal ferrite, namely, 2a, 4f1, 12k, 2b, and 4f2, 12k is unique as the interchange to magnetically connecting the S-block sites (2a and 4f1) and the R-block sites (2b and 4f2). Thus, the exchange chain will be interrupted, if the nonmagnetic Ti^{4+} ions are doped on the 12k sites. The combination of low magnetic anisotropy and interrupted exchange chain may lead to noncollinear spin configurations in SCTFO that further leads to magnetoelectric effects.

To calculate the exchange integrals, we followed the approach applied before, i.e. to solve a set of linear equations derived for different spin configurations, in which the coefficients are determined by structure and the inhomogeneous terms are first-principles total energies. We applied the GGA+U approach to enhance localization of d orbitals of Co^{2+} and Fe^{3+} and let both U_{Fe} and U_{Co} vary as empirical parameters. To limit the number of exchange integrals to be calculated, we set a cutoff radius of 4 Å and ignore the exchange coupling beyond this radius.

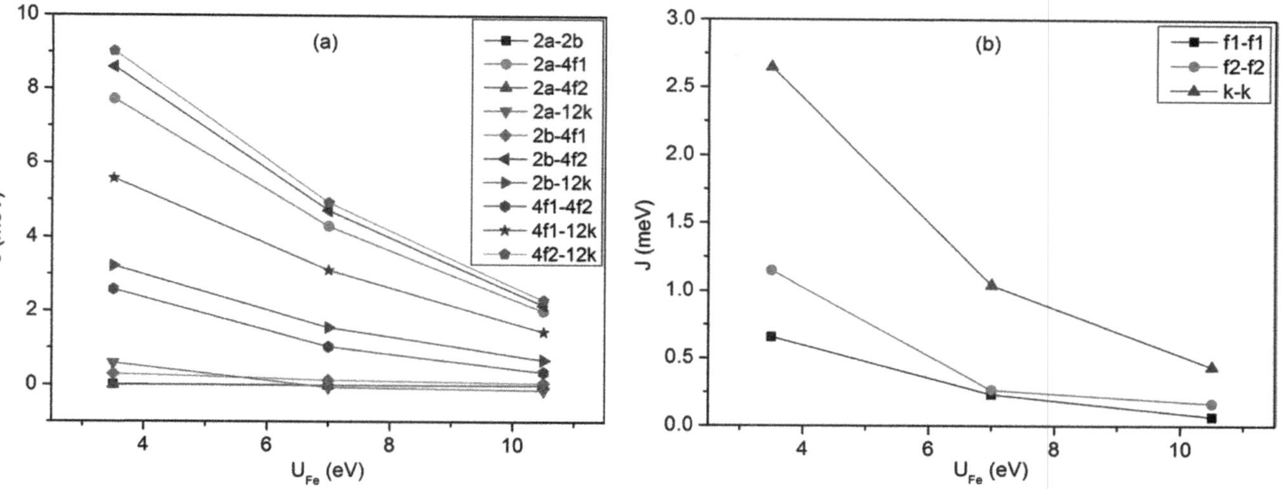

Figure 6.12 The dependence of exchange integrals on U_{Fe} in $SrFe_{12}O_{19}$. (a) Intersublattice exchange integrals and (b) intrasublattice exchange integrals.

We first calculated the exchange integrals of SFO (see Figure 6.12). All interactions are antiferromagnetic, and the magnitude of exchange integrals decreases as U increases. The decrease of exchange integrals with increasing U is in agreement with the superexchange theory, in which interaction is proportional to b^2/U, where b is the hopping integral. The dominant exchange interactions in SFO are the intersublattice exchange integrals 4f2-12k, 2b-4f2, 2a-4f1, and 4f1-12k, which are at least twice stronger than the intersublattices. The first-principle energies calculated with different spin configurations show that Gorter configuration is the magnetic ground state. In fact, from the dominant exchange integrals given by our calculations, we rediscover the exchange chain that leads to Gorter configuration, 12k-(4f2-2b-4f2)-12k-(4f1-2a-4f1)-12k-..., where the exchange couplings in the parentheses are inside the R and S blocks (or R* and S*) respectively. It is obvious that the 12k sites uniquely connecting the R and S blocks.

We also calculated the exchange integrals of SCTFO (see Figure 6.13). The calculations with U_{Co} fixed to 3.3 eV show that the dependence of exchange integrals on U_{Fe} in SCTFO is similar to the case in SFO. All exchange interactions are antiferromagnetic. The magnitude of exchange integrals decreases with increase in U_{Fe}, but 2a-4f1 and 4f1-12k decrease slower than others. This is because Co atoms occupy the 4f1 sites, and the exchange interactions involving the 4f1 sites do not solely depend on U_{Fe}. Figure 6.13 shows the dependence of exchange integrals on U_{Co}, with U_{Fe} fixed to 7 eV. In the figure, we can see that those interactions involving the 4f1 sites are impacted by U_{Co}, but the variations of those exchange integrals are relatively slow. The other interactions are almost invariant as U_{Co} changes, except that 12k-12k decreases slowly as U_{Co} increases.

We choose $U_{Fe} = 7$ eV and $U_{Co} = 3.3$ eV as the appropriate values for the GGA+U calculation of SCTFO. With these U values, the comparison of exchange integrals in SFO and SCTFO is shown in Figure 6.14, where it is obvious that the exchange integrals are separated by a level $J = 2.5$ meV into two groups and that the dominant ones are 2a-4f1, 2b-f2, 4f1-12k, and 4f2-12k. The CoTi-codoping induce large changes to the exchange integrals 4f1-12k and 4f2-12k that magnetically connect R and S blocks. The strength of 4f1-12k increases from ~ 3 to ~ 4 meV and that of 4f2-12k decreases from ~ 5 to ~ 4 meV. At the same time, the strength of 4f1-4f2 is almost unaffected by the codoping. Thus, the large drop of 4f2-12k causes the ferromagnetic alignment of 4f1-4f2 unstable and may result in the noncollinear spin cone structure. Moreover, the local magnetic anisotropy of 4f1 is lowered by the codoping, which further favors the noncollinear structure.

6.2.2 Magnetic Anisotropy

Magnetic anisotropy is a macroscopic magnetic phenomenon that free energy depends on the direction of magnetization. In history, it was attributed to the magnetic dipole–dipole interaction between magnetic ions that was later proven to be too weak to predict the experimental magnetic anisotropic energy (MAE) for ferromagnetic metals, such as Ni. It was then proposed that MAE was dominantly from the spin–orbital coupling interaction in transition metal ions, i.e. the single-ion magnetic anisotropy [69]. In fact, the spin–orbital coupling interaction is implied in the Dirac equation, the relativistic

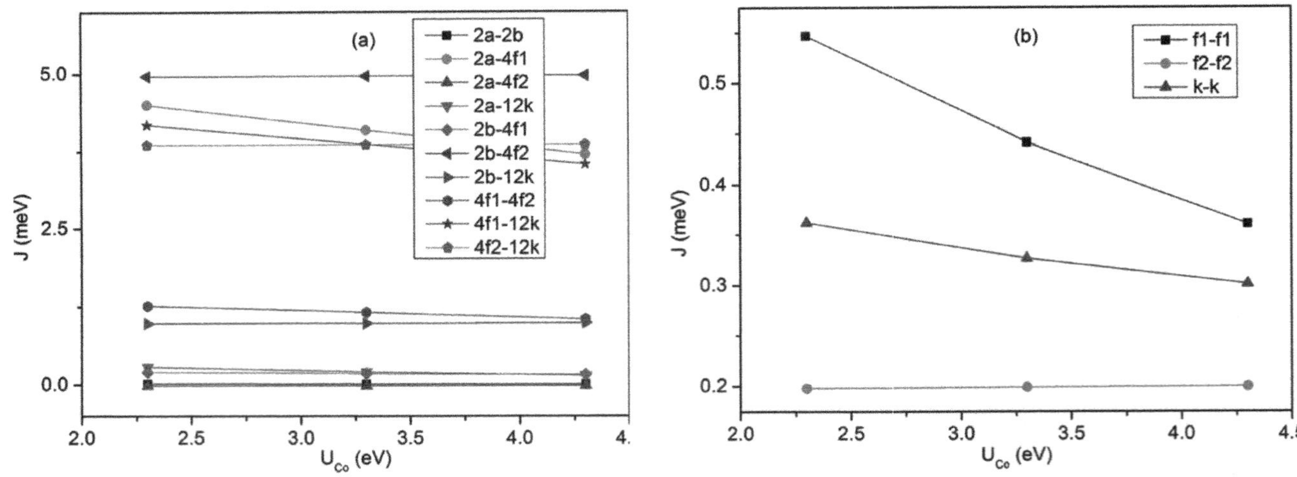

Figure 6.13 The dependence of exchange integrals on U_{Co} in $SrCo_2Ti_2Fe_8O_{19}$. (a) Intersublattice exchange integrals and (b) intrasublattice exchange integrals.

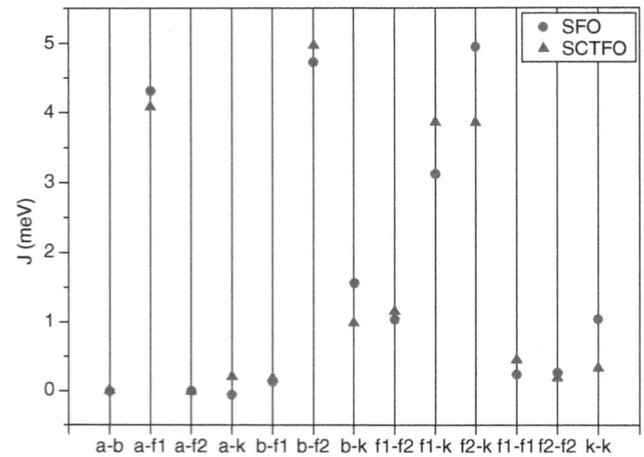

Figure 6.14 The comparison of exchange integrals in $SrFe_{12}O_{19}$ and $SrCo_2Ti_2Fe_8O_{19}$.

equation of electron. From the Dirac equation, there are several relativistic corrections to Schrödinger equation, where the spin–orbital coupling interaction is the first-order vector correction, usually written as

$$\mathcal{H}_{SO} = \lambda \mathbf{L} \cdot \mathbf{S} \tag{6.96}$$

where λ is a coefficient depending on the potential of the ion core and radial parts of the electron wave function. The physical picture of spin–orbital coupling interaction is simple. The orbital motion of electron generates an effective magnetic field on its own spin degree of freedom. In fact, we can imagine that the nuclear is equivalently rotating around the electron and generating a magnetic field on the electron.

It is possible to extend the concept of spin–orbital coupling interaction to multi-electron atoms or ions, for which the **L** and **S** in Hamiltonian (6.96) are hence multi-electron operators. For 3*d* transition metal ions in solids, the crystal field effect is usually stronger than the spin–orbital coupling interaction. Thus, the spin–orbital Hamiltonian (6.96) can be treated as a perturbation [10]. For nondegenerate multi-electron ground state, the orbital angular momentum is quenched and only the second-order perturbation of spin–orbital is nonzero, which implies a week magnetic anisotropy. On the contrary, for the triply degenerate ground state, the orbital angular moment can be partially restored as the pseudoangular momentum, and the first-order perturbation is nonzero, which implies a strong or moderate magnetic anisotropy. The multi-electron approach results in the spin Hamiltonian:

$$\mathcal{H}_{SO} = DS_z^2 + E(S_x^2 - S_y^2) \tag{6.97}$$

where D is the so-called "zero-field split," a measure of the anisotropy along the z-axis that will split the spin states of different magnetic quantum numbers, and E is the measure of the anisotropy inside the x–y plane.

Band structure calculation can also include the spin–orbital coupling interaction and calculate the magnetic anisotropy by a perturbation approach [70, 71]. In this approach, the spin-polarized band structure is first calculated without the spin–orbital coupling interaction, and then the spin–orbital coupling interaction is calculated as a perturbation for each band. The direction of magnetization will involve in the calculation when calculating the matrix elements of $\mathbf{L}\cdot\mathbf{S}$, and the magnetic anisotropy energy is given by the difference of total energies perturbed along different magnetization directions, as suggested by the force theorem [72]. In this approach, it is assumed that the spin densities do not depend on magnetization direction. Moreover, the degenerate (or almost) states near Fermi energy may require a special treatment [73].

More generally, it is possible to calculate the MAE by noncollinear band structure calculations [74, 75]. In this approach, the total energy with a magnetization direction is not calculated by perturbing the collinear spin-polarized result, but directly and self-consistently by noncollinear calculation, in which the magnetization direction is an input. In this approach, the spin densities with different magnetization directions can be different, which violates the force theorem that requires the densities are conserved. However, when the spin–orbital coupling interaction is strong enough, it is reasonable that changing the spin direction will affect the orbital motion, thereby violating the force theorem.

6.2.2.1 Magnetic Anisotropy of $SrFe_{12}O_{19}$ and $SrCo_xTi_xFe_{12-2x}O_{19}$

As an example of first-principles calculation of magnetic anisotropy of ferrites, we discuss the calculation of MAE of Sr M-type hexagonal ferrite (SFO) and its Co–Ti codoped derivation (SCTFO) [76]. In the previous section, we have introduced the background of the calculation and calculated exchange integrals of this material.

We follow the force theorem approach. We first conduct a collinear spin-polarized calculation and then do the perturbation calculations with different magnetization directions. We let the angle between the magnetization and the crystallographic c-axis vary from 0° to 90°. We also assume the Gorter spin configuration. We apply GGA+U approach and choose $U_{Fe} = 7$ eV and $U_{Co} = 3$ eV. It is shown that the correlation effect can affect the calculated MAE and orbital momentum [77].

We first consider the calculated MAE of SFO. Figure 6.15 shows the dependence of calculated E_{MCA} on θ, from which we can see that our results fit very well to $K_1 \sin^2\theta$ curve, with positive K_1 indicating that the hexagonal c-axis is the magnetic preferential axis. K_1 for $SrFe_{12}O_{19}$ can be determined by curve fitting to be 0.718 meV per unit cell (about 1.67×10^6 erg/cm^3), which is in the same order of magnitude of the experimental value (about 4.5×10^6 erg/cm^3 at temperature near 0 K). The quantitative match of our results to the theoretical description and experimental value demonstrates that force theorem works well in calculating E_{MCA} for M-type hexagonal ferrites.

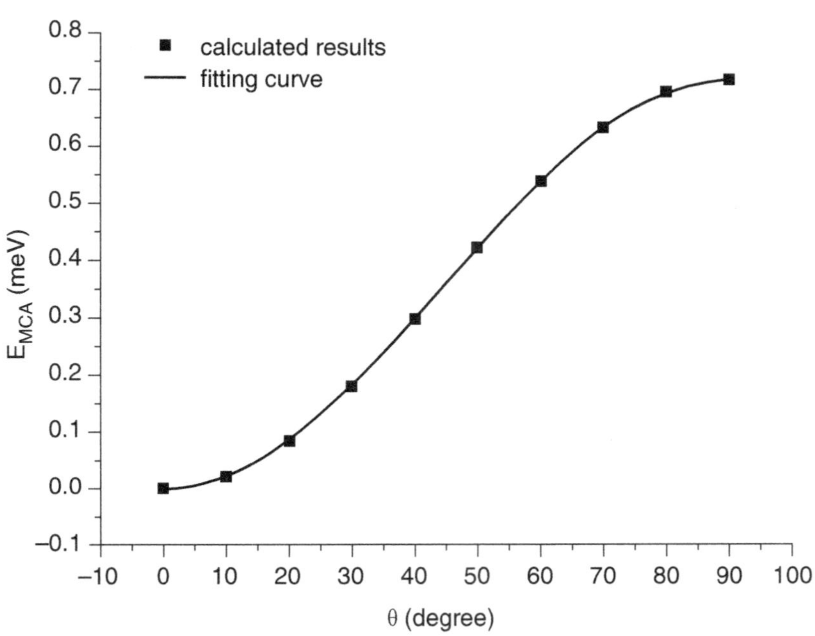

Figure 6.15 The calculated magnetic anisotropy energy $E_{MCA} = E(\theta) - E_{001}$ and the fitting curve of $K_1 \sin^2\theta$.

We then consider the Co–Ti codoped SCFO. When Co and Ti are doped, they will prefer certain sites. It was reported that Co prefers 4f1 sites and that Ti prefers 12k and 4f2. Thus, we calculated and compared the total energies of certain combinations of these locations (see Table 6.6). The lowest total energy is achieved when Co ions occupy 4f1 sites and Ti ions occupy 12k sites. In addition, moving Ti ions from 12k sites to 4f2 sites slightly increases the total energy. This result confirms that Ti prefers 12k and 4f2 sites.

It is obvious from Table 6.6 that the location of Co ion impacts the MAE. When Co ions are on 4f1 sites, the MAE is positive or weak negative. On the contrary, when they are on 4f2 sites, the MAE is negative and strong. This is due to the different local symmetry of 4f1 and 4f2 sites. In fact, the local symmetry of 4f1 is tetrahedral, and the Co^{2+} ground state is nondegenerate Γ_2, which implies quenching of orbital angular momentum and weak magnetic anisotropy from second-order perturbation. However, the local symmetry of 4f2 is octahedral, and the Co^{2+} ground state is triply degenerate Γ_4, which implies partial restoration of orbital angular momentum and strong (or moderate) magnetic anisotropy from first-order perturbation. Moreover, comparing the results of Co(4f1)Ti(4f2) and Co(4f1)Ti(12k) in Table 6.6, we find that the Ti location also impacts the MAE.

Assuming that Co ions occupy 4f1 sites and Ti occupy 12k sites, we calculated the dependent of MAE on Co–Ti codoping concentration (see Figure 6.16). It is obvious that the Co–Ti codoping converts the uniaxial magnetic anisotropy of SFO to the in-plane magnetic anisotropy. It should be noted that the in-plane magnetic anisotropy will assist the formation of noncollinear spin configuration if the exchange chain that aligns the layers of magnetic ions is interrupted by Ti doping on 12k sites.

Table 6.6 Total energy difference relative to the lowest energy configuration (meV) and magnetic anisotropy energy ($E_{MCA} = E_{100} - E_{001}$ in meV) per unit cell with different locations of Co^{2+} and Ti^{4+} in $SrCo_2Ti_2Fe_8O_{19}$.

Configuration	Total energy	E_{MCA}
Co(4f1)Ti(4f2)	0.518	0.216
Co(4f1)Ti(12k)	0	−0.025
Co(4f2)Ti(12k)	4.472	−0.383
Co(4f14f2)Ti(12k)	3.695	−0.8
Co(4f14f2)Ti(12k4f2)	2.385	−0.606

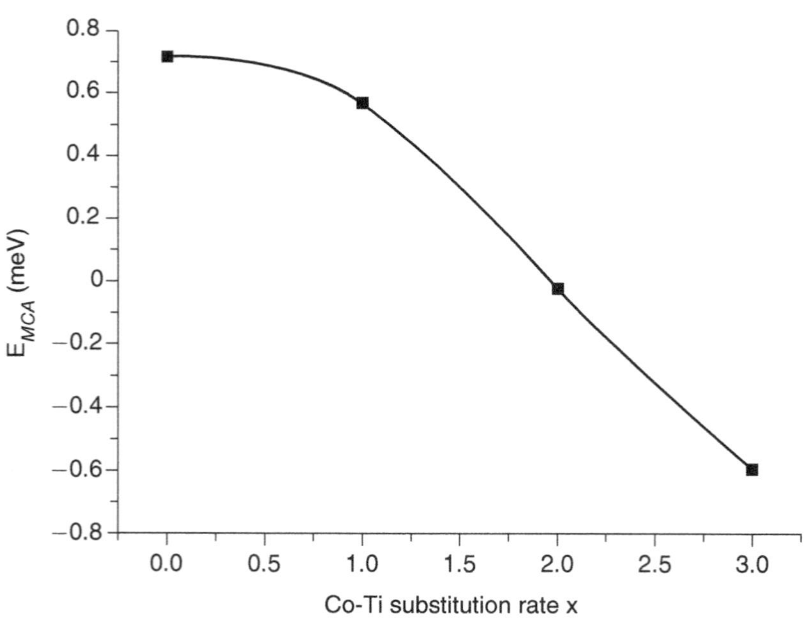

Figure 6.16 The dependence of magnetic anisotropy energy on substitution rate of Co–Ti.

References

1 Martin, M.R. (2004). *Electronic Structure: Basic Theory and Practical Methods*. Cambridge: Cambridge University Press.

2 Kohanoff, J. (2006). *Electronic Structure Calculations for Solids and Molecules: Theory and Computational Methods*. Cambridge: Cambridge University Press.

3 Mahan, G.D. (2000). *Many-Particle Physics*, 3e. New York: Kluwer Academic.

4 Hohenberg, P. and Kohn, W. (1964). Inhomogeneous electron gas. *Physical Review* 136: B864.

5 Ashcroft, N.W. and Mermin, N.D. (1976). *Solid State Physics*. Fort Worth: Harcourt.

6 Harrison, W.A. (1999). *Elementary Electronic Structure*. Singapore: World Scientific.

7 Hartree, D.R. (1928). The wave mechanics of an atom with non-coulombic central field. part I. Theory and methods. *Mathematical Proceedings of the Cambridge Philosophical Society* 24: 89.

8 Kohn, W. and Sham, L.J. (1965). Self-consistent equations including exchange and correlation effects. *Physical Review* 140: A1133.

9 Fock, V. (1930). Näherungsmethode zur losung des quantenmechanichanischen mehrkörperprobleme. *Zeitschrift für Physik* 61: 126.

10 Yosida, K. (1996). *Theory of Magnetism*. Berlin: Springer-Verlag.

11 Landau, L.D. and Lifshitz, E.M. (1977). *Quantum Mechanics (Mon-relativistic Theory)*, 3e. Oxford: Pergamon.

12 Dirac, P.A. (1930). Note on exchange phenomena in the Thomas atom. *Mathematical Proceedings of the Cambridge Philosophical Society* 26: 376.

13 Janak, J.F. (1978). Proof that $\frac{\partial E}{\partial n_i}=\epsilon_i$ in density-functional theory. *Physical Review B* 18: 7165.

14 Perdew, J.P., Parr, R.G. and Levy, M. (1982). Density-functional theory for fractional particle number: Derivative discontinuity of the energy. *Physical Review B* 49: 1691.

15 Sham, L.J. and Schlüter, M. (1983). Density-functional theory of the energy gap. *Physical Review B* 51: 1888.

16 Perdew, J.P. and Levy, M. (1983). Physical content of the exact Kohn–Sham orbital energies: Band gap and derivative discontinuities. *Physical Review B* 51: 1884.

17 Parr, R.G. and Bartolotti, L. (1983). Some remarks on the density functional theory of few-electron systems. *The Journal of Chemical Physics* 87: 2810.

18 Ceperley, D.M. and Alder, B.J. (1980). Ground state of the electron gas by a stochastic method. *Physical Review Letters* 45: 566.

19 Perdew, J.P. and Zunger, A. (1981). Self-interaction correction to density-functional approximaition for many-electron systems. *Physical Review B* 23: 5048.

20 Vosko, S., Wilk, L. and Nusair, M. (1983). Accurate spin-dependent electron liquid correlation energies. *Canadian Journal of Physics* 58: 1200.

21 Juan, Y.M., Kaxiras, E. and Gordon, R.G. (1995). Use of the generalized gradient approximation in pseudopotential calculations of solids. *Physical Review B* 51: 9521.

22 Becke, A.D. (1986). Density functional calculations of molecular bond energies. *The Journal of Chemical Physics* 84: 4524.

23 Herman, F., Van Dyke, J.P. and Ortenburger, I.B. (1969). Improved statistical exchange approximation for inhomogeneous many-electron systems. *Physical Review Letters* 22: 807.

24 Kleiman, L. (1984). Exchange density-functional gradient expansion. *Physical Review B* 30: 2223.

25 Perdew, J.P., Langreth, D.C. and Sahni, V. (1977). Corrections to the local density approximation: Gradient expansion versus wave-vector analysis for the metallic surface problem. *Physical Review Letters* 38: 1030.

26 Perdew, J.P. (1985). Accurate density functional for the energy: Real-space cutoff of the gradient expansion for the exchange hole. *Physical Review Letters* 55: 1665.

27 Gunnarsson, O. and Lundqvist, B.I. (1976). Exchange and correlation in atoms, molecules, and solids by the spin-density-functional formalism. *Physical Review B* 13: 4274.

28 Perdew, J.P. and Wang, Y. (1986). Accurate and simple density functional for the electron exchange energy: Generalized gradient approximation. *Physical Review B* 33: 8800.

29 Perdew, J.P., Chevary, J.A., Vosko, S.H., Jackson, K.A., Pederson, M.R., Singh, D.J. and Fiolhais, C. (1992). Atoms, molecules, solids, and surface: Application of the generalized gradient approximation for exchange and correlation. *Physical Review B* 46: 6671.

30 Becke, A.D. (1988). Density-functional exchange-energy approximation with correct asymptotic behavior. *Physical Review A* 38: 3098.

31 Perdew, J.P., Burke, K. and Ernzerhof, M. (1996). Generalized gradient approximation made simple. *Physical Review Letters* 77: 3865.

32 Perdew, J.P., Burke, K. and Wang, Y. (1996). Generalized gradient approximation for the exchange-correlation hole of a many-electron system. *Physical Review B* 54: 16533.

33 Ma, S.K. and Brueckner, K.A. (1968). Correlation energy of an electron gas with a slowly varying high density. *Physical Review* 165: 18.

34 Lee, C., Yang, W. and Parr, R.G. (1988). Development of the Colle-Salvetti correlation-energy formula into a functional of the electron density. *Physical Review B* 37: 785.

35 Colle, R. and Salvetti, D. (1983). A general method for approximating the electronic correlation energy in molecules and solids. *The Journal of Chemical Physics* 79: 1404.

36 Wigner, E.P. (1938). Effects of the electron interaction on the energy levels of electrons in metals. *Transactions of the Faraday Society* 34: 678.

37 Hedin, L. (1965). New method for calculating the one-particle Green's function with application to the electron-gas problem. *Physical Review* 139: A796.

38 Hybertsen, M.S. and Louie, S.G. (1985). First-principles theory of quasiparticles: Calculation of band gaps in semiconductors and insulators. *Physical Review Letters* 55: 1418.

39 Faleev, S.V., van Schilgaard, M. and Kotani, T. (2004). All-electron self-consistent GW approximation: Application to Si, MnO, and NiO. *Physical Review Letters* 93: 126406.

40 Terakura, T., Williams, A.R., Oguchi, T. and Kübler, J. (1984). Transition-metal monoxides: Band or Mott insulators. *Physical Review Letters* 52: 1830.

41 Dufek, P., Blaha, P., Sliwko, V. and Schwarz, K. (1994). Generalized-gradient-approximation description of band splittings in transition-metal oxides and fluorides. *Physical Review B* 49: 10170.

42 Sawatzky, G.A. and Allen, J. (1984). Magnitude and origin of the band gap in NiO. *Physical Review Letters* 53: 2339.

43 Mott, N.F. (1937). Discussion of the paper by de Boer and Verwey. *Proceedings of the Physical Society* 49: 72.

44 Anisimov, V.I., Aryasetiawan, F. and Lichtenstein, A.I. (1997). First-principles calculations of the electronic structure and spectra of strongly correlated systems: The LDA+U method. *Journal of Physics: Condensed Matter* 9: 767.

45 Anisimov, V.I., Zaanen, J. and Anderson, O.K. (1991). Band theory and Mott insulators: Hubbard U instead of Stoner I. *Physical Review B* 44: 943.

46 Cococcioni, M. and de Gironcoli, S. (2005). Linear response approach to the calculation of the effective interaction parameters in the LDA+U method. *Physical Review B* 71: 035105.

47 Dudarev, S.L., Botton, G.A., Savrasov, S.Y., Humphreys, C.J. and Sutton, A.P. (1998). Electron-energy-loss spectra and the structural stability of nickel oxide: An LSDA1U study. *Physical Review B* 57: 1505.

48 Becke, A.D. (1988). Correlation energy of an inhomogeneous electron gas: A coordinate space model. *The Journal of Chemical Physics* 88: 1053.

49 Harris, J. (1984). Adiabatic-connection approach to Kohn–Sham theory. *Physical Review B* 29: 1648.

50 Becke, A.D. (1993). A new mixing of Hartree–Fock and local density functional theories. *The Journal of Chemical Physics* 98: 1372.

51 Becke, A.D. (1993). Density-functional thermochemistry. III. The role of exact exchange. *The Journal of Chemical Physics* 98: 5648.

52 Stephens, P.J. (1994). Ab initio calculation of vibrational absorption and circular dichroism spectra using density functional force Fields. *Journal of Physical Chemistry* 98: 11623.

53 Muscat, J., Wander, A. and Harrison, N.M. (2001). On the prediction of band gaps from hybrid functional theory. *Chemical Physics Letters* 342: 397.

54 Anderson, P.W. (1959). New approach to the theory of superexchange interactions. *Physical Review* 115: 2.

55 Zuo, X. and Vittoria, C. (2002). Calculation of exchange integrals and electronic structures for manganese ferrite. *Physical Review B* 66: 184420.

56 Singha, D.J., Gupta, M. and Gupta, R. (2002). First-principles investigation of MnFe2O4. *Physical Review B* 64: 064432.

57 de P. R. Moreira, I., Illas, F. and Martin, R.M. (2002). Effect of Fock exchange on the electronic structure and magnetic coupling in NiO. *Physical Review B* 65: 155102.

58 Zuo, X., He, Y., Yang, A., Bernardo, B., Harris, V.G. and Vittoria, C. (2005). Calculation of exchange constants in spinel ferrites with magnetic S-state ions. *Journal of Applied Physics* 97: 10F104.

59 Zuo, X., Yan, S., Barbiellini, B., Harris, V.G. and Vittoria, C. (2006). A computational study of nickel ferrite. *Journal of Magnetism and Magnetic Materials* 303: e432.

60 Zuo, X., Yang, A., Vittoria, C. and Harris, V.G. (2006). Computational study of copper ferrite (CuFe2O4). *Journal of Applied Physics* 99: 08M909.

61 Goodenough, J.B. (1958). An interpretation of the magnetic properties of the perovskite-type mixed crystals La1−xSrxCoO3−λ. *Journal of Physics and Chemistry of Solids* 6: 287.

62 Fujimori, A. and Minami, F. (1984). Valence-band photoemission and optical absorption in nickel compounds. *Physical Review B* 30: 957.

63 Feng, M., Zuo, X., Vittoria, C., Harris, V.G. and Wu, J. (2011). Ab initio study on manganese doped cadmium ferrite (Cd1-xMnxFe2O4). *IEEE Transactions on Magnetics* 47: 324.

64 Yafet, K. and Kittel, C. (1952). Antiferromagnetic arrangements in ferrites. *Physical Review* 87: 290.

65 Kresse, G., Marsman, M. and Furthmüller, J. (2014). *VASP the GUIDE. Computational Materials Physics*. Wien, Austria: Faculty of Physics, Universität Wien,Sensengasse 8/12, A-1090.

66 Zuo, X., How, H., Somu, S. and Vittoria, C. (2003). Self-biased circulator/isolator at millimeter wavelengths using magnetically oriented polycrystalline strontium M-type hexaferrite. *IEEE Transactions on Magnetics* 39: 3160.

67 Wang, L., Wang, D., Cao, Q., Zheng, Y., Xuan, H., Gao, J. and Du, Y. (2012). Electric control of magnetism at room temperature. *Scientific Reports* 2: 223.

68 Katsura, H., Nagaosa, N. and Balatsky, A.V. (2005). Spin current and magnetoelectric effect in noncollinear magnets. *Physical Review Letters* 95: 057205.

69 van Vleck, J.H. (1937). On the anisotropy of cubic ferromagnetic crystals. *Physical Review* 52: 1178.

70 Brooks, H. (1940). Ferromagnetic anisotropy and the itinerant electron model. *Physical Review* 58: 909.

71 Bruno, P. (1989). Tight-binding approach to the orbital magnetic moment and magnetocrystalline anisotropy of transition-metal monolayers. *Physical Review B* 39: 865.

72 Daalderop, G., Kelly, P.J. and Schuurmans, M.F. (1990). First-principles calculation of the magnetocrystalline anisotropy energy of iron, cobalt, and nickel. *Physical Review B* 41: 11919.

73 Wang, D.-S., Wu, R. and Freeman, A.J. (1993). State-tracking first-principles determination of magnetocrystalline anisotropy. *Physical Review Letters* 70: 869.

74 Hobbs, D., Kresse, G. and Hafner, J. (2000). Fully unconstrained noncollinear magnetism within the projector augmented-wave method. *Physical Review B* 62: 11556.

75 Błoński, P., Lehnert, A., Dennler, S., Rusponi, S., Etzkorn, M., Moulas, G. ... Hafner, J. (2010). Magnetocrystalline anisotropy energy of Co and Fe adatoms on the (111) surfaces of Pd and Rh. *Physical Review B* 81: 104426.

76 Feng, M., Shao, B., Wu, J. and Zuo, X. (2013). Ab initio study on magnetic anisotropy change of SrCoxTixFe12-xO19. *Journal of Applied Physics* 113: 17D909.

77 Yang, I., Savrasov, S.Y. and Kotliar, G. (2001). Importance of correlation effects on magnetic anisotropy in Fe and Ni. *Physical Review Letters* 87: 216405.

7

Gyromagnetic Properties of Ferrites

Marina Koledintseva[1] *and Takanori Tsutaoka*[2]

[1] *The Boeing Company, Saint Louis, MO, U.S.A.*
[2] *(Deceased) University of Hiroshima, Hiroshima, Japan*

7.1 Introduction

Radio frequency (RF)/microwave ferrites find applications over a significant portion of electromagnetic (EM) spectrum – from very low-frequency (VLF) RF band (3–30 kHz) to extremely high frequencies (EHFs, 30–300 GHz), which lies in the millimeter-wave band.

Ferrites are essentially insulating ceramics sintered of various oxides of metals possessing significant magnetic properties. The magnetic properties of ferrites arise due to the overexpressed spin magnetic moments of electrons on the external shell of the atoms of these metals. The examples of these metals are iron, cobalt, nickel, and some rear earth elements. Ferrites are formed by specific types of crystalline lattices, and below some temperature, they possess unique magnetic, or *ferrimagnetic*, properties. L. Néel (France) was the first who introduced the term "*ferrimagnetism*" (1948) and provided the theoretical basis to understand physics of ferrites [1].

Before that, the fundamentals of classical theory of ferromagnetism in magnetic metals were developed, and it could be credited to many outstanding scientists [2]. Herein we will name just a few physicists, including P.-E. Weiss (France) who discovered domains in ferromagnetic metals in 1906 and introduced the concept of a "molecular" field acting on an electron of an atom [3]; V. K. Arkadiev (Russia) who was the first to observe ferromagnetic resonance (FMR) at UHF in metals in 1911 [4]; Y. G. Dorfman (USSR) in 1923 who explained FMR by the Zeeman effect (1896) of splitting in magnetic field and transitions between the split components [5]; L. Landau and E. Lifshitz (USSR) in 1935 proposed the phenomenological model of FMR based on Larmor precession of magnetization vector) [6], and then this model was verified experimentally independently by E. K. Zavoisky (USSR) [7] and J. H. Griffiths (UK) [8] in 1945–1946.

Ferrimagnetism is a complex form of magnetic order in crystal structures of such ionic compounds as magnetic oxides. The magnetic structure is composed of two magnetic sublattices (called A and B) separated by oxygen. The exchange interactions are mediated by the oxygen anions. When this happens, the interactions are called indirect, or superexchange interactions. The strongest superexchange interactions result in an antiparallel alignment of spins between the A and B sublattices.

In ferrimagnets, the magnetic moments of the A and B sublattices *are not equal* and result in a net magnetic moment. Ferrimagnetism is, therefore, similar to ferromagnetism. It exhibits all the hallmarks of ferromagnetic behavior, such as *spontaneous magnetization*, *Curie temperatures*, *hysteresis*, and *remanence*. However, ferro- and ferrimagnets have very different magnetic order. Magnetite (Fe_3O_4) is a well-known ferromagnetic material. Indeed, magnetite was considered a ferromagnet until Néel in the 1940s provided the theoretical framework for understanding ferrimagnetism. Other examples of ferrimagnets are maghemite (γ-Fe_2O_3), jacobsite ($MnFe_2O_4$), trevorite ($NiFe_2O_4$), and magnesioferrite ($MgFe_2O_4$).

If the A and B sublattice moments *are exactly equal* but opposite, the net magnetic moment is zero. This type of magnetic order is called *antiferromagnetism*. Examples of antiferromagnetic materials are ulvospinel Fe_2TiO_2, hematite α-Fe_2O_3, ilmenite $FeTiO_2$, and orthoferrite $LaFeO_6$. The clue to antiferromagnetism is the behavior of susceptibility above a critical temperature, called the Néel temperature (T_N). Above T_N, the susceptibility obeys the Curie–Weiss law for paramagnets but with a negative intercept indicating negative exchange interactions.

Ferrites are dense, homogeneous ceramic structures made by mixing iron oxide (Fe_2O_3) with oxides or carbonates of one or more metals such as manganese, zinc, nickel, or magnesium. They are formed by sintering at high temperatures (of about 1000 °C and higher). Oxide-type ferrites are defined as ferrimagnetic semiconductors with the general chemical structure $MeO \cdot Fe_2O_3$, where Me is the bivalent metal ion (e.g. Cu^{2+}, Zn^{2+}, Ni^{2+}, Mg^{2+}, etc.). More complex compounds such

Modern Ferrites: Basic Principles, Processing and Properties, Volume 1, First Edition. Edited by Vincent G. Harris.

as $x(Me_aO_b)\cdot y(Fe_2O_3)$ are also called ferrites. If Fe_2O_3 is substituted by an oxide of any other transitional element of the Fe group or by combination of oxides, this will be the class of allied substances, as they exhibit ferrimagnetism. Some of the ferrites, such as orthoferrites, exhibit antiferromagnetic properties. Ferrites can be monocrystalline and polycrystalline. Ferrites have very high resistivity and low loss for eddy currents, so they can be used at very high frequencies (MHz to GHz).

Historically, the first sintered ferrites were with the cubic crystal structure of *spinel* type; their composition was $MeO\cdot Fe_2O_6$. The world's first ferrite magnet, the spinel-type ceramic magnetic core material was discovered in 1930 by two professors of Tokyo Institute of Technology, Y. Kato and T. Takei, and their invention laid the basis to the establishing the company TDK (Tokyo Denki Kagaku Kogyo K.K.) in 1935 [9]. However, their product's low magnetic force limited its scope of application. The first commercially useful ferrites (including their "Ferroxcube," Ni–Zn and Mn–Zn ferrites) were developed at Philips Research Laboratories in the Netherlands by J. L. Snoek and his associates, mainly during 1933–1945 [10]. Some more examples of spinel ferrites are Li, Mg–Mn, Mg–Al, Mg–Cr, and Ni–Zn–Cr. Advances in ferrite material science and sintering technology by 1960s paved the road to broad spectrum of applications of ferrites ever as. Even nowadays, over 80 years after the invention of soft ferrites, the two most common ferrites belong to the spinel-type group. These are the ferrites which are widely used for various RF filters and rod antennas and for the wide range of electromagnetic compatibility (EMC) applications (see Chapter 6 of Volume 2).

Ferrites of the other magnetic crystalline structures also find their important "ecological niches" in RF electronics and microwave engineering.

The second group after spinels is called *garnets*, which have a general composition $3(Me_2O_3)\cdot 5(Fe_2O_3)$, where Me is one of the rare earth metals – Y, Gd, Dy, Ho, Er, Tm, Lu, Yb, Sm, Eu, or Tb. Garnets possess a complex cubic crystalline structure. The most widespread is yttrium iron garnet (YIG), where Me is the yttrium (Y) ion. YIG material is often used in tiny high-quality (high-Q) resonators to build various microwave devices, e.g. very narrowband tunable filters for preselectors of heterodyne receivers, spectrum analyzers, and frequency-selective power meters. Other applications include various frequency-selective nonreciprocal microwave devices – circulators, phase shifters, and isolators, which could be designed either for a specific frequency, or could be electrically controlled (tunable) over some operating frequency range [11]. However, YIGs for their performance need to be magnetized to saturation, and as microwave frequencies increase, their application needs more and more intense bias magnetic field, which makes a magnetic system too bulky. Therefore, YIGs are used up to 30–40 GHz at most. Another important application of ferrite garnets is nonvolatile computer memory and processor systems based on movable bubble domains.

The third important magnetic crystalline group is *magnetoplumbites*, which have hexagonal structure. Barium ferrite, $BaO\cdot 6\,Fe_2O_3$, is a representative of this group. *Hexagonal ferrites*, or hexaferrites, unlike spinels and garnets, possess high intrinsic magnetization due to the internal magnetocrystalline anisotropy field. Their usage in various functional devices of microwave electronics results in significant miniaturization of the devices, as they can perform over microwave band, at frequencies from a few gigahertz to short-wave part of millimeter-wave band (~300 GHz) without applying strong bias magnetic fields [12].

The fourth group is *perovskites*, such as strontium–manganese ferrites, Sr_2MnGaO_5 or Sr_2MnAlO_6. Perovskite ferrites have comparatively low resistivity and Curie temperature. Orthoferrrites, which are antiferromagnetic materials, also belong to the group of perovskites. An example of orthoferrite is $LaFeO_3$ with Néel temperature $T_N = 750$ K.

One more example of a practically important ferrite structure is *maghemite* γ-Fe_2O_3, which has already been mentioned earlier. This is an unstable compound, which transforms to an antiferromagnetic hematite α-Fe_2O_3 on heating over 400 °C. γ-Fe_2O_3 is used as a recording medium for data storage. Ferrimagnetic intermediate phases also occur in some metallic alloy systems, e.g. Mn_2Sb, Mn_2Sn, Mn_2Ga, Mn_2Ge, Mn_3In, $FeGe_2$, FeSe, Cr_3As_2, and $CrPt_6$. Many binary alloys contain rear earth elements. They have been recognized as ferrimagnetic substances rather than ferromagnetic due to the presence of specific sublattices, which are typical for ferrimagnets.

RF and microwave engineering mainly use *magnetically soft ferrites*, which have low coercivity $H_c \sim 10$ Oe; high resistivity (up to $10^{12}\,\Omega\cdot$m); high static permeability μ_s; low loss at remagnetization; low magnetostriction; and practically no eddy currents due to high resistivity. Magnetically hard ferrites are also employed but mainly for producing bias magnetic fields as permanent magnets. Their coercivity is $H_c \sim 10^2\ldots10^3$ Oe, high remagnetization loss (or hysteresis curve area is high). Microwave ferrites usually have a rectangular hysteresis curve and are employed in the magnetically saturated state near their FMR. In this chapter, we will consider *gyromagnetic phenomena in microwave ferrites* near their FMR. Gyromagnetic phenomena are associated with magnetization motion of the spin magnetic moment of electrons in the magnetic atoms of ferrites. The name "gyromagnetic" comes from the phenomenological model, in which magnetization moment moves as an axis of a gyroscope, i.e. follows precessional motion.

Ferrites are anisotropic materials that rotate the polarization plane of electromagnetic waves (EMWs) propagating in a ferrite medium, i.e. they exhibit *gyrotropic effects*. Their permeability is described by a tensor. They exhibit nonreciprocal interaction with EMWs of different directions of propagation and polarization. These gyromagnetic properties of ferrites and gyromagnetic effects will be considered in this chapter.

The overall structure of this chapter is mentioned next.

Section 7.2 presents the physical phenomena of ferrites near FMR and peculiarities of ferrimagnetic resonance are considered. Magnetization motion equation (MME) is presented with different relaxation terms. The linearized MME in small-signal approximation is solved to obtain the magnetic susceptibility tensor for an unbounded ferrite medium and for a ferrite ellipsoid. Demagnetization form factors and crystallographic magnetic anisotropy in ferrites are taken into account. Also, the grain structure in polycrystalline ferrites and presence of magnetic domains are considered as factors affecting FMR in ferrites.

Section 7.3 describes and analyzes various techniques to measure magnetic and dielectric spectra of ferrites as gyromagnetic media are overviewed. The examples of the typical measures microwave magnetic spectra of spinel, garnet, and hexagonal ferrites are given.

Section 7.4 provides an overview of some electrodynamic problems of wave propagation in extended ferrite media and interaction of EMWs with localized ferrite samples. A special attention is paid to small perturbation technique, widely used to analyze waveguides, transmission lines, and cavities containing ferrite samples with at least one dimension, which allows for magnetostatic approximation. Self-matched field technique to describe interaction of a small ferrite ellipsoid in a waveguide is shown. This lays the basis to design of various nonreciprocal and resonance devices for microwave engineering.

Section 7.5 presents a gyromodel based on gyrovector formalism developed by a Russian physicist L. K. Mikhailovsky (1928–2014). This gyromodel has laid the basis for developing a new area of electronics – currentless spin electronics. Numerous devices and absorbing coatings find application in microwave engineering and electromagnetic compatibility.

Section 7.6 discusses stable nonlinear phenomena, which take place in the vicinity of FMR in ferrites, at the levels of microwave power below spin-wave instability. Principles of operation of some devices of gyromagnetic electronics based on these phenomena are discussed.

Section 7.7 draws the conclusions on the "ecological niche" and perspectives of gyromagnetic electronics and electrodynamics.

7.2 Physics of Gyromagnetism in Ferrites

Gyromagnetic medium is a material where the *gyromagnetic effect* takes place. Gyromagnetic effect is the phenomenon, when the magnetization vector after being excited comes back to its equilibrium position as the damping precession around the direction of the bias magnetic field. This is the consequence of gyromagnetic properties of elementary magnetic moments of a ferrite. The existence of the magnetic moment is always related to the nonzero angular momentum, which means that an elementary magnet has the properties of a gyroscope. This behavior of magnetic moment is the reason of gyromagnetic effects and gyromagnetic (or ferromagnetic, which is the same) resonance.

Note that a ferrite is a *ferrimagnetic* material. The peculiarity of its magnetic order is the presence of two sublattices. However, the magnetic moment of an elementary cell (either a crystallite or the entire sample if there is no domain structure, i.e. all this sample is a single domain) is large. This means that the magnetic moments of sublattices are of opposite signs. They subtract but are not equal and do not completely compensate each other. Therefore, the magnetic susceptibility values, both static and dynamic, are high; this makes them similar to ferromagnetic materials, and the phenomenological description of ferrites and ferromagnetic materials are practically identical. We will briefly discuss some differences, though, too.

7.2.1 Microscopic Nature of Gyromagnetism

Ferrites are gyromagnetic materials; their resultant spontaneous magnetization is nonzero. Their crystals are "hard" systems, characterized by the resultant nonzero angular momentum $\boldsymbol{J}$ and the corresponding quantum numbers. The external magnetic field $\boldsymbol{H}$ typically cannot change the internal electron structure of an atom; the atom behaves as a single particle; and $\boldsymbol{J}$ plays the part of the spin moment. If the orbital moments of atoms are "frozen," then the total spin $\boldsymbol{S}$ equals to the sum of spin moments of all the atoms. The allowed states are determined by the spatial quantization of the vector $\boldsymbol{S}$ in the direction of $\boldsymbol{H}$. There are $(2s+1)$ states, and the energy between two neighboring states is $\Delta E = ge\hbar\,/\,(2mc)H$, where g is the splitting Lande factor (for self-spinning electrons it is close to 2); e is absolute value of the charge of electron; m is the electron mass; c is the speed of light; and $\hbar = h/(2\pi) = 6.62\times10^{-34}[\mathrm{J\cdot s}]/(2\pi) = 1.055\cdot10^{-27}[\mathrm{erg\cdot s}]$ is the Planck's constant; the latter is expressed in Gauss unit system.

Let us consider an electron as an isolated particle, with only magnetic field acting on it. The electron energy can have two discrete values, corresponding to two various spin orientations. The difference of energies between these two states equals to work needed to turn an electron spin over.

$$\Delta E = 2\mu_B H = \frac{ge\hbar}{2mc}H\bigg|_{g_{\mathrm{spin}}=2} = \frac{e\hbar}{mc}H. \tag{7.1}$$

The transition between these two energy levels is accompanied by a quantum absorption or radiation, and its frequency (frequency of Larmor's precession) in Gauss unit system is:

$$\omega = \frac{\Delta E}{\hbar} = \frac{eH}{mc} = \gamma H, \tag{7.2}$$

γ is the magnetomechanical, or gyromagnetic ratio.

In the SI system, the gyromagnetic ratio is $\gamma = \frac{e}{m} = 1.761\times10^{11}\,[\mathrm{C/kg} = \mathrm{rad/(s\cdot T)}]$, and the relation between the Larmor resonance frequency and bias field is:

$$\omega = \mu_0 \gamma H, \tag{7.3}$$

where $\mu_0 = 4\pi\cdot10^{-7}\mathrm{H/m}$ is the permeability of vacuum. Note that using the SI system is preferable. However, specialists in magnetism and magnetic materials often use Gauss system of magnetic units. In this system, the unit for magnetic induction B and magnetization vector M is 1 G=10^{-4} T; and the unit for magnetic field intensity H is 1 Oe=79.58 A/m.

If there is a crystal with the total spin $\boldsymbol{S}$ and there is a microwave field with frequency ω, then the crystal will absorb quanta of energy of microwave field. The magnetization and the total spin $\boldsymbol{S}$ in small jumps diverge from the direction of the external field $\boldsymbol{H}$. As $\boldsymbol{S}$ and $\boldsymbol{M}$ rotate around $\boldsymbol{H}$, an absorption of microwave field energy can be represented as the resonance process, where the angle of precession and the energy of the precessional motion increase until reaching some steady-state condition, when the energy supply from microwave field and dissipation at the motion are balanced. If the spin system keeps all the stored energy, the angle of precession would increase until the direction of $\boldsymbol{M}$ would change to the opposite, and the resonance would be saturated. But there is energy loss, and at some angle of precession, there is a balance between the absorbed and the lost energies. Loss of energy is associated with relaxation (dissipation), which turns into heat and radiation.

7.2.2 Magnetization Motion Equation

Spin magnetic moment is unambiguously related with the mechanical angular momentum:

$$\boldsymbol{m} = -\gamma \boldsymbol{J}, \tag{7.4}$$

where γ is the aforementioned gyromagnetic ratio.

From classical mechanics, it is known that the torque $\boldsymbol{T}$, or the rotating moment, is associated with the derivative of the mechanical angular momentum (see Figure 7.1):

$$\boldsymbol{T} = \frac{d\boldsymbol{J}}{dt}. \tag{7.5}$$

At the same time:

$$\boldsymbol{T} = \boldsymbol{\omega}_0 \times \boldsymbol{J}, \tag{7.6}$$

so that

$$\frac{d\boldsymbol{J}}{dt} = \boldsymbol{\omega}_0 \times \boldsymbol{J}. \tag{7.7}$$

If the magnetic moment $\boldsymbol{m}$ is affected by the external bias magnetic field $\boldsymbol{H}$, at the equilibrium, or stationary condition, the torque (5) affects this magnetic moment. Then

$$\frac{d\boldsymbol{J}}{dt} = \mu_0 \boldsymbol{m} \times \boldsymbol{H}. \tag{7.8}$$

As:

$$\mu_0 \boldsymbol{m} \times \boldsymbol{H} = \boldsymbol{\omega}_0 \times \boldsymbol{J} \Rightarrow \quad \mu_0 \boldsymbol{m} \times \boldsymbol{H} = -\gamma \boldsymbol{\omega}_0 \times \boldsymbol{m}, \tag{7.9}$$

the Larmor precession frequency is:

$$\omega_0 = \mu_0 \gamma H. \tag{7.10}$$

Differentiating (7.4) by time, we get:

$$\frac{d\boldsymbol{m}}{dt} = -\gamma \frac{d\boldsymbol{J}}{dt}, \tag{7.11}$$

and then the MME for one magnetic moment (spin) is obtained as:

$$\frac{d\boldsymbol{m}}{dt} = -\mu_0 \gamma \boldsymbol{m} \times \boldsymbol{H}. \tag{7.12}$$

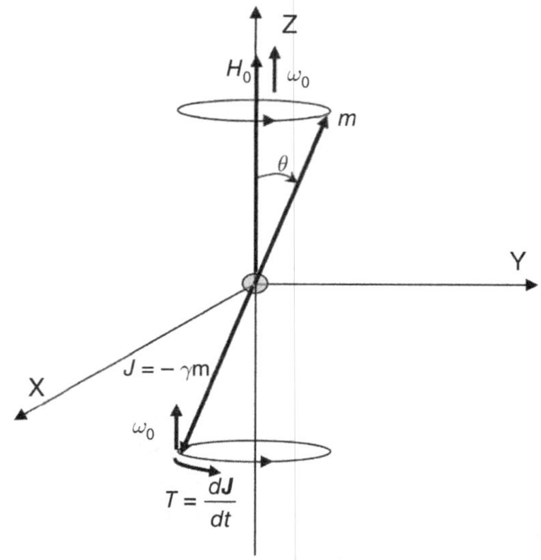

Figure 7.1 Precession of magnetization (magnetic moment) vector.

The average (in space) magnetic moment of a magnetic medium, or magnetization vector, is:

$$\boldsymbol{M} = \lim_{\substack{N \to \infty \\ \Delta V \to 0}} \frac{\sum_{i=1}^{N} \boldsymbol{m}_i}{\Delta V} \tag{7.13}$$

Therefore, the MME for the average magnetic moment is written as:

$$\frac{d\boldsymbol{M}}{dt} = -\mu_0 \gamma \boldsymbol{M} \times \boldsymbol{H}. \tag{7.14}$$

This equation was first proposed by L. D. Landau and E. M. Lifshitz in 1935 [6] and is often named after them as Landau–Lifshitz equation. According to this equation, the vector $\boldsymbol{M}$ precesses around the vector of field $\boldsymbol{H}$ with the frequency $\omega_0 = \mu_0 \gamma H$ (in SI units), where $\gamma = 1.76 \times 10^{11}$ C/kg; or $\omega_0 = \gamma_g H$ (in Gauss units), where $\gamma_g \approx 2.8$ GHz/kOe.

7.2.3 MME with Relaxation Terms

Note that the MME (7.14) does not include any dissipation, so it describes nondamping precession around the equilibrium position, i.e. speed of precession and magnitude do not change as precession proceeds in time. It can be easily shown in coordinate form that the first integral corresponding to Equation (7.14) is $|\boldsymbol{M}| = \text{const}$. In the coordinate form, Equation (7.14) can be written as:

$$\begin{aligned} \frac{dM_x}{dt} &= -\mu_0 \gamma \left(M_y H_z - H_y M_z \right); \\ \frac{dM_y}{dt} &= -\mu_0 \gamma \left(M_x H_z - H_x M_z \right); \\ \frac{dM_z}{dt} &= -\mu_0 \gamma \left(M_x H_y - H_x M_y \right). \end{aligned} \tag{7.15}$$

Then:

$$\begin{aligned} \frac{d|\boldsymbol{M}|}{dt} &= \frac{d}{dt} \sqrt{M_x^2 + M_y^2 + M_z^2} = \\ & \frac{1}{\sqrt{M_x^2 + M_y^2 + M_z^2}} \left(M_x \frac{dM_x}{dt} + M_y \frac{dM_y}{dt} + M_z \frac{dM_z}{dt} \right) = 0. \end{aligned} \tag{7.16}$$

This means that in the equilibrium position $\boldsymbol{H} = \boldsymbol{H}_0 = \text{const}$ and $\boldsymbol{M} = \boldsymbol{M}_0 = \text{const}$, then $\boldsymbol{M}_0 \times \boldsymbol{H}_0 = 0$, and therefore, the vectors are parallel: $\boldsymbol{M}_0 \parallel \boldsymbol{H}_0$.

To take into account the loss, a relaxation (damping, dissipation) term $\boldsymbol{R}$ should be introduced in the MME:

$$\frac{d\boldsymbol{M}}{dt} = -\mu_0\gamma\boldsymbol{M}\times\boldsymbol{H} + \boldsymbol{R}. \tag{7.17}$$

Formally, the relaxation term is associated with some effective relaxation field $\boldsymbol{H}_r$ as:

$$\boldsymbol{R} = \mu_0\gamma\boldsymbol{M}\times\boldsymbol{H}_r. \tag{7.18}$$

There are various phenomenological forms of the relaxation term $\boldsymbol{R}$, including Landau–Lifshitz (1936), Bloch (1950), Codrington-Olds-Torrey (1954), Gilbert (1955), modified Bloch–Bloembergen (1956), Callen (1958), and Fletcher–LeCrow–Spenser (1960) forms [13–15]. All these relaxation terms describe different mechanisms of losses in precessional dynamics, i.e. "drive to equilibrium" for the magnetization vector. Some of the most widely used forms are summarized in Table 7.1.

It is important to note that Landau–Lifshitz and Gilbert forms are practically coinciding; therefore, they are often combined as LLG – Landau–Lifshitz–Gilbert – relaxation term [6, 16].

The magnetization or the magnetic moment vector precession affected by dissipation is illustrated in Figure 7.2.

Table 7.1 Relaxation terms in magnetization motion equation.

No.	Dissipative term	Precessional velocity ω and angle θ	Drive for decay
1	**Landau–Lifshitz (1936)** $\boldsymbol{R} = -\lambda_{LL}\left(\boldsymbol{H} - \frac{(\boldsymbol{H}\cdot\boldsymbol{M})\cdot\boldsymbol{M}}{\lvert\boldsymbol{M}\rvert^2}\right)$ – the first term is $\perp$ to $\boldsymbol{M}$, the second is $\parallel$ to $\boldsymbol{M}$, or $\boldsymbol{R} = -\frac{\lambda_{LL}}{\lvert\boldsymbol{M}\rvert^2}\boldsymbol{M}\times(\boldsymbol{M}\times\boldsymbol{H}) = -\frac{\alpha_{LL}\mu_0\gamma}{\lvert\boldsymbol{M}\rvert}\boldsymbol{M}\times(\boldsymbol{M}\times\boldsymbol{H})$ $\lambda_{LL}, \alpha_{LL}$ are the LL damping constants; $\lambda_{LL} = \mu_0\gamma\alpha_{LL}\lvert\boldsymbol{M}\rvert$ $-\gamma\mu_0(\boldsymbol{M}\times\boldsymbol{H})$ by $\frac{\partial\boldsymbol{M}}{\partial t}$; but the dynamics of decay is different. $\alpha = \frac{\alpha_{LL}}{1-\alpha_{LL}}$; $\lambda_{LL} = \mu_0\gamma\lvert\boldsymbol{M}\rvert\frac{\alpha}{1+\alpha^2}$	$\omega = \mu_0\gamma H_Z(1+\alpha_{LL}^2)$ $\theta = \text{Arcsin}\left(\frac{\gamma\mu_0 H_r}{\alpha\omega}(1+\alpha_{LL}^2)\right)$	$\boldsymbol{H}_{\text{conpous}} \perp \boldsymbol{M}$; $\lvert\boldsymbol{M}\rvert = const$ Tip trajectory driving toward internal $\boldsymbol{H}(t)$; damping from precession torque only.
2	**Gilbert (1955)** [15], [16] $\boldsymbol{R} = \frac{\alpha}{\lvert\boldsymbol{M}\rvert}\boldsymbol{M}\times\frac{\partial\boldsymbol{M}}{\partial t}$ - exact algebraic equivalence with LL term, replacing $-\gamma\mu_0(\boldsymbol{M}\times\boldsymbol{H})$ by $\frac{\partial\boldsymbol{M}}{\partial t}$; but the dynamics of decay is different; $\alpha = \frac{\alpha_{LL}}{1-\alpha_{LL}}$; $\lambda_{LL} = \mu_0\gamma\lvert\boldsymbol{M}\rvert\frac{\alpha}{1+\alpha^2}$	$\omega = \mu_0\gamma H_z/\sqrt{1+\alpha^2}$ $\theta = \arcsin(\frac{\mu_0\gamma H_\tau}{\alpha\omega})$	$\lvert\boldsymbol{M}\rvert = const$; $\left\lvert\frac{\partial\boldsymbol{M}}{\partial t}\right\rvert \perp \boldsymbol{M}$; $\boldsymbol{M}(t)\times\frac{\partial\boldsymbol{M}}{\partial t}$ magnetization tip trajectory driving towards internal $\boldsymbol{H}(t)$; damping from precession torque only.

Table 7.1 (Continued)

No.	Dissipative term	Precessional velocity ω and angle θ	Drive for decay
3	**Bloch (1950)** [13] $R_{x,y} = -\frac{1}{T_2}M_{x,y}$- normal to the bias H_z field $R_z = -\frac{1}{T_1}(M_z - M_0)$ - along bias H_z field M_0- equilibrium magnetization; T_1 is spin-lattice relaxation time; T_2 is relaxation time for precession damping, including spin-spin relaxation time	$\omega = \mu_0\gamma H_z$ $\theta = \text{Arctan}\,(\mu_0\gamma T_{1,2}H_\tau)$	$\boldsymbol{M}$ relaxes only toward z-direction of bias field (false equilibrium!). Magnetization $\boldsymbol{M}$ is NOT conserved! Valid in the small-signal limit; $\boldsymbol{M}$ is conserved at setting $T_1 = 2T_2$.
4	**Modified Bloch, or Bloch–Bloembergen (1956)** [13], [15] $R_{x,y} = -\omega_{rel}(M_{x,y} - \chi_0 H_{x,y})$- transverse components; $\frac{1}{T_2} = \frac{1}{T_1} + \frac{1}{T_{spin-spin}}$. $\omega_{rel} = 1/T_2 = \mu_0\gamma H_{rel}$ is the relaxation frequency; $\chi_0 = \frac{M_0}{H_{x,y}}$ is static susceptibility	$\omega = \mu_0\gamma H_z$ $\theta = \text{Arctan}\,(\gamma\mu_0 T_2 H_\tau)$	$\lvert\boldsymbol{M}\rvert \neq const$ Driven toward instantaneous equilibrium for the difference $m_{x,y} - m_{x,y}^{eq}$; where $m_{x,y}^{eq} = \frac{M_0}{H_{x,y}}h_{x,y}$
5	**Codrington-Olds-Torrey (1954)** [14] $\boldsymbol{R}_{\parallel H} = -\frac{1}{T_1}\left[\left(\frac{M_0\lvert\boldsymbol{H}\rvert - \boldsymbol{M}\cdot\boldsymbol{H}}{\lvert\boldsymbol{H}\rvert}\right)\cdot\frac{\boldsymbol{H}}{\lvert\boldsymbol{H}\rvert}\right]$ - along the total instantaneous H-field $\boldsymbol{R}_{\perp H} = -\frac{1}{T_2}\left[\left(\frac{(\boldsymbol{M}\times\boldsymbol{H})\times\boldsymbol{H}}{\lvert\boldsymbol{H}\rvert^2}\right)\right]$- perpendicular to the total instantaneous H-field	$\omega = \mu_0\gamma H_z$	$\boldsymbol{M}_{compens} \perp \boldsymbol{H}$ Valid in the small-signal limit.
6	**Callen (1958)** [17] $\boldsymbol{R} = \alpha_C\boldsymbol{M} - \lambda_C\boldsymbol{M}\times[\boldsymbol{M}\times\boldsymbol{H}]$	$\omega = \mu_0\gamma H_z\left(1+\left(\frac{\lambda_C M}{\gamma\mu_0}\right)^2\right)$ $\theta = \text{Arcsin}\left(\frac{\mu_0^2\gamma^2 H_\tau}{\lambda_C M\omega}\times\left[1+\left(\left(\frac{\lambda_C M}{\gamma\mu_0}\right)^2\right)\right]\right)$	The direct relaxation process is a function of the magnetization angle relative to the crystalline anisotropy axis.

It can be shown that $\lvert H_r\rvert \approx H_r^{\max}\cdot\Phi(\theta)$, where $\Phi(\theta)$ is a function of the precession angle θ, and $H_r^{\max} = \Delta H_{0.5}/2$, or resonance line at the level of 0.5 from maximum loss (the line width will be discussed later). The expressions for the relaxation field in the cases of a few most widely used dissipation terms are the following:

i) For Landau–Lifshitz, the relaxation term is $\lvert\boldsymbol{H}_r^{LL}\rvert \approx \alpha_{LL}\lvert\boldsymbol{H}_{DC}\rvert\sin\theta$.

ii) For Gilbert, the relaxation term is $\lvert\boldsymbol{H}_r^{LL}\rvert \approx \alpha\lvert\boldsymbol{H}_{DC}\rvert\sin\theta$.

iii) For Bloch–Bloembergen (BB), or modified Bloch, the term is $\lvert\boldsymbol{H}_r^{B}\rvert \approx \frac{\lvert\boldsymbol{H}_{DC}\rvert}{\gamma T_2}\tan\theta$.

iv) For Callen, the term is $\lvert\boldsymbol{H}_r^{C}\rvert \approx \frac{\lambda_C\lvert\boldsymbol{M}\rvert}{\gamma}\lvert\boldsymbol{H}_{DC}\rvert\sin\theta$.

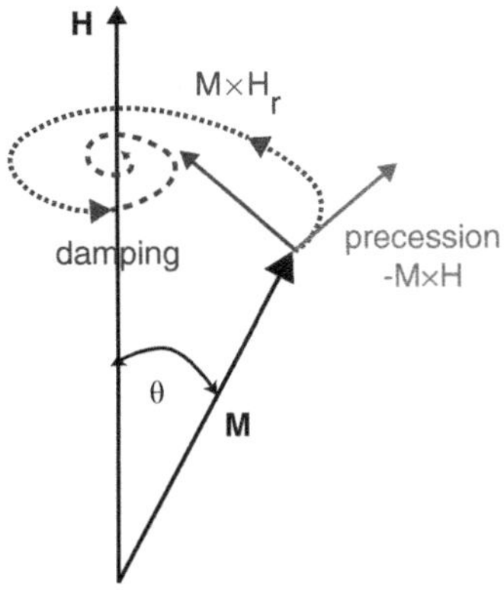

Figure 7.2 Illustration of damping during magnetization vector dynamics.

where $\boldsymbol{H}_{DC}$ is the total DC field acting on the magnetic moment (the sum of the bias field $\boldsymbol{H}_0$, the crystallographic anisotropy field $\boldsymbol{H}_A$, and the demagnetization field $\boldsymbol{H}_D$).

7.2.4 Linearized MME for Small-Signal Approximation

Imagine that the magnetization moment vector $\boldsymbol{M}$ is disturbed, i.e. deflected by the angle $\theta > 0$ from the direction of the DC bias field $\boldsymbol{H}_{DC}$ by some force. This may be a strong short pulse of magnetic field perpendicular to the bias field. The precession of the vector $\boldsymbol{M}$ will be associated with the damping free oscillations, and after some relaxation time, the vector $\boldsymbol{M}$ will return to its initial position at $\theta = 0$ along the bias field. However, in the case of the continuous forced oscillations, the angle of precession will be kept nonzero, as energy supplied to the system will be sufficient to sustain the precession.

In this section, we will consider the case when there is a high-frequency (RF/microwave) magnetic field acting on a ferrite in addition to the DC bias magnetic field, as is illustrated in Figure 7.3.

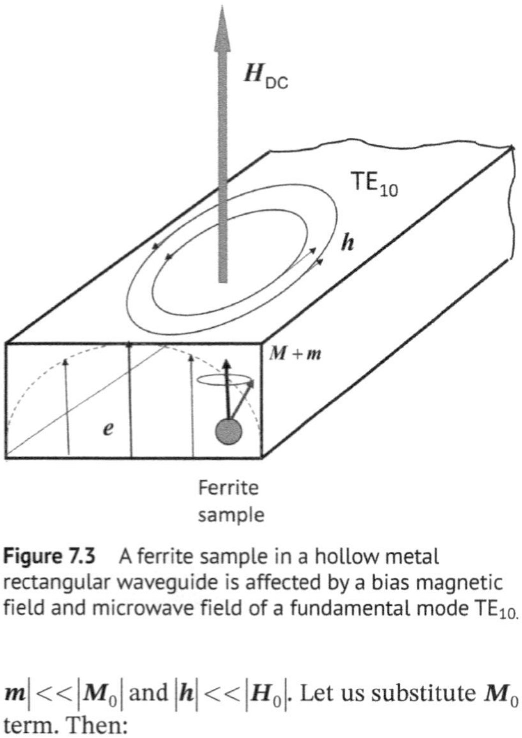

Figure 7.3 A ferrite sample in a hollow metal rectangular waveguide is affected by a bias magnetic field and microwave field of a fundamental mode TE_{10}.

Let us assume that the total field acting on ferrite is

$$\boldsymbol{H}_{\Sigma} = \boldsymbol{H}_{DC} + \mathrm{Re}(\boldsymbol{h}e^{j\omega t}), \tag{7.19}$$

where $\boldsymbol{H}_{DC} = \boldsymbol{H}_i + \boldsymbol{H}_e$; $\boldsymbol{H}_i$ is the internal static field inside the ferrite (if it is an ellipsoid, then the internal field includes demagnetization factors; it also includes the crystallographic anisotropy field); $\boldsymbol{H}_e = \boldsymbol{H}_0$ is the external bias magnetizing field; and $\boldsymbol{h}$ is the complex amplitude of the microwave (dynamic) field. Herein and throughout this chapter, j is the imaginary unit. The total magnetization can be written as:

$$\boldsymbol{M}_{\Sigma} = \boldsymbol{M}_0 + \mathrm{Re}(\boldsymbol{m}e^{j\omega t}), \tag{7.20}$$

where $\boldsymbol{m}$ is the complex amplitude of the microwave (dynamic) magnetization vector; $\boldsymbol{M}_0\big|_{\max} = M_S$ is the maximum magnetization equal to the saturation magnetization.

The MME is easily solved in the linearized form, i.e. in the *small-signal approximation*. Let us assume that the DC bias magnetic field is directed along the z-direction and that the high-frequency magnetic field has only $h_{x,y}$ components in the Cartesian coordinates. Let us also assume that the magnitudes of high-frequency magnetization and magnetic field are much less than their DC counterparts, $|\boldsymbol{m}| << |\boldsymbol{M}_0|$ and $|\boldsymbol{h}| << |\boldsymbol{H}_0|$. Let us substitute $\boldsymbol{M}_0 + \boldsymbol{m}e^{j\omega t}$ and $\boldsymbol{H}_{DC} + \boldsymbol{h}e^{j\omega t}$ in the MME (7.14), first omitting the relaxation term. Then:

$$\frac{d(\boldsymbol{M}_0 + \boldsymbol{m}e^{j\omega t})}{dt} = -\mu_0\gamma\left[(\boldsymbol{M}_0 + \boldsymbol{m}e^{j\omega t}) \times (\boldsymbol{H}_{DC} + \boldsymbol{h}e^{j\omega t})\right]. \tag{7.21}$$

Keeping only the terms of the first order (assuming the second-order terms as infinitesimally small), the linearized equation is obtained:

$$j\omega\boldsymbol{m} = -\mu_0\gamma\boldsymbol{m} \times \boldsymbol{H}_{DC} - \mu_0\gamma\boldsymbol{M}_0 \times \boldsymbol{h}. \tag{7.22}$$

The same equation in the coordinate form is:

$$\begin{aligned} &j\omega m_x + \omega_0 m_y = \mu_0\gamma M_0 h_y;\\ &-\omega_H m_x + j\omega_0 m_y = -\mu_0\gamma M_0 h_x;\\ &j\omega m_z = 0, \end{aligned} \tag{7.23}$$

where $\omega_0 = \mu_0\gamma H_{DC}$ is the Larmor frequency.

The solutions for the microwave components of magnetization vector are:

$$m_x = \frac{\mu_0 \gamma M_0 \omega_0}{\omega_0^2 - \omega^2} h_x + j \frac{\mu_0 \gamma M_0 \omega}{\omega_0^2 - \omega^2} h_y;$$
$$m_y = -j \frac{\mu_0 \gamma M_0 \omega}{\omega_0^2 - \omega^2} h_x + \frac{\mu_0 \gamma M_0 \omega_0}{\omega_0^2 - \omega^2} h_y; \tag{7.24}$$
$$m_z = 0.$$

The same system for the microwave magnetization can be written in the tensor form:

$$\boldsymbol{m} = \ddot{\chi} \cdot \boldsymbol{h}, \tag{7.25}$$

where $\ddot{\chi}$ the magnetic susceptibility tensor, which is a 3×3 matrix:

$$\ddot{\boldsymbol{\chi}} = \begin{vmatrix} \chi & j\kappa & 0 \\ -j\kappa & \chi & 0 \\ 0 & 0 & 0 \end{vmatrix}. \tag{7.26}$$

The diagonal and off-diagonal susceptibility terms in Equation (7.26) are calculated as:

$$\chi = \frac{\omega_M \omega_0}{\omega_0^2 - \omega^2} \quad \text{and} \quad \kappa = \frac{\omega_M \omega}{\omega_0^2 - \omega^2}, \tag{7.27}$$

where $\omega_M = \mu_0 \gamma M_0$. The frequency $\omega = \omega_0$ is the frequency of FMR, which is also called the frequency of *gyromagnetic resonance*, as it is associated with the spin magnetic moment precession.

Note that the third component of the tensor (7.26) is $\chi_{\|} = 0$. This case corresponds to a ferrite in the saturated state, when $M_0 \approx M_S$, and $\omega_M = \mu_0 \gamma M_S$. The more general case of the susceptibility tensor is with the nonzero component $\chi_{\|}$, e.g. for unsaturated ferrites, and when relaxation term is taken into account:

$$\ddot{\chi} = \begin{vmatrix} \chi & j\kappa & 0 \\ -j\kappa & \chi & 0 \\ 0 & 0 & \chi_{\|} \end{vmatrix}. \tag{7.28}$$

The corresponding magnetic permeability tensor, which relates the magnetic induction and the magnetic field, $= \ddot{\mu} \cdot \boldsymbol{H}$, for the case of (7.26), is:

$$\ddot{\mu} = \mu_0 \begin{vmatrix} 1+\chi & j\kappa & 0 \\ -j\kappa & 1+\chi & 0 \\ 0 & 0 & 1+\chi_{\|} \end{vmatrix} = \mu_0 (\ddot{U} + \ddot{\chi}), \tag{7.29}$$

where $\ddot{\boldsymbol{U}}$ is the unit diagonal tensor.

The tensor nature of permeability of ferrites results in electromagnetic *anisotropy*, i.e. EMWs propagating in different directions and with different polarizations have different velocities, and *gyrotropy* due to the presence of off-diagonal components having opposite signs in tensor (7.29), which leads to nonreciprocity in ferrite media and Faraday rotation of the polarization plane. In addition, the effect of gyromagnetic resonance, results in *frequency selectivity*. Therefore, there are "three whales," on which performance of the majority of gyromagnetic devices stands on: anisotropy, gyrotropy, and resonance phenomena.

These effects are used to build various nonreciprocal devices for microwave engineering, e.g. circulators, phase shifters, isolators, directional couplers [11, 13, 15]. Indeed, if there is a plane EMW propagating in a magnetized lossless unbounded ferrite (magneto-dielectric) medium, Maxwell's equations result in two fundamental modes with opposite circular polarizations, left and right.

These two modes have the different phase constants, corresponding to different effective permeability values for the waves with two polarizations. At the fixed travel distance within the ferrite, these two modes will gain different rotations of the polarization planes, which results in Faraday rotation. Also, assume that a ferrite sample placed in the circular polarized RF/microwave field, e.g. along the line of a rectangular waveguide at the distance $x = a/4$ from the narrow walls, where a is the size of the wide wall of the waveguide. Then the magnetized ferrite resonator (FR) at frequencies close to its resonance frequency will interact effectively with one polarization, which coincides with the direction of the magnetization vector precession, and will have much weaker interaction with the opposite polarization. As a result, there will be nonreciprocal interaction of the EMWs propagating in the opposite directions. The electrodynamics of a gyromagnetic medium, resulting in Faraday rotation, and double refraction will be considered in detail in Section 7.4.

7.2.5 Frequency Dispersion of Magnetic Susceptibility of Ferrites

In this section, we will consider the frequency behavior of complex high-frequency magnetic susceptibility if there is a nonzero relaxation term in MME (7.17). The two most widely forms of relaxation terms are the LLG and the Bloch–Bloembergen.

The *MME with the LLG relaxation term* is usually used to describe the behavior of materials with spontaneous magnetization [13–16,18, 19]. The LLG in the SI system is:

$$\frac{d\boldsymbol{M}}{dt} = -\boldsymbol{M}\times\left(\mu_0\gamma\boldsymbol{H} - \frac{\alpha}{M}\cdot\frac{d\boldsymbol{M}}{dt}\right). \tag{7.30}$$

The relaxation field in LLG case is:

$$\boldsymbol{H}_r = \frac{\alpha}{M\gamma}\boldsymbol{H}\times\frac{d\boldsymbol{M}}{dt}. \tag{7.31}$$

The field (7.31) is proportional to the speed of variation of magnetization moment $\boldsymbol{M}$, similar to friction proportional to the speed of motion. The sign of this "friction" is opposite to the variation of speed. This field can be expressed in terms of all the known phenomenological relaxation parameters in MME [14]. The tip of the magnetization moment $\boldsymbol{M}$, in this case moves toward the equilibrium position (along the internal magnetic field, or bias magnetic field in the case of absence of crystallographic anisotropy) drawing a spiral on a spherical surface so that the length of the vector $|\boldsymbol{M}| = M$ is conserved. This is illustrated in Figure 7.4. Note that the vectors $\frac{d\boldsymbol{M}}{dt}$ and $\boldsymbol{M}$ are perpendicular to each other, and the damping happens due the precession torque only.

The components of the RF/microwave susceptibility tensor in this case can be directly obtained from (7.27) by a simple substitution of the real resonance frequency with the complex one:

$$\omega_0 \to \omega_0 + j\alpha\omega, \tag{7.32}$$

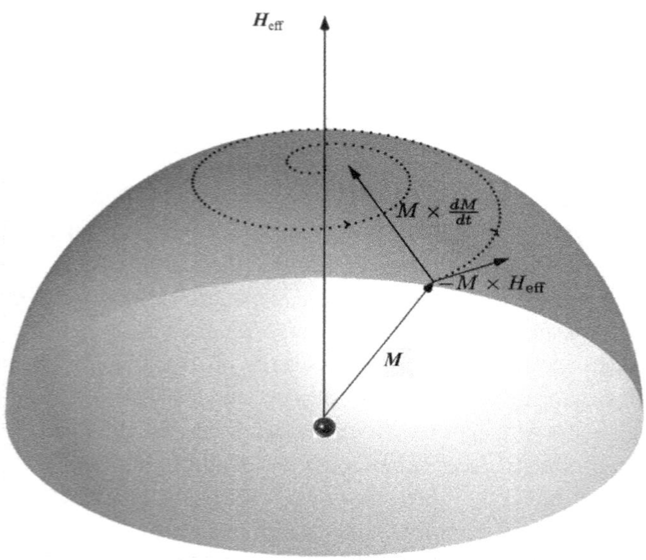

Figure 7.4 Damping in the LLG case results in conservation of the magnetization vector magnitude and motion of its end tip on a surface of a sphere.

and the susceptibility tensor components with Gilbert damping term become:

$$\chi = \frac{\omega_M(\omega_0 + j\alpha\omega)}{\omega_0^2 - \left(1+\alpha^2\right)\omega^2 + 2j\omega\omega_0}; \kappa = \frac{\omega_M\omega}{\omega_0^2 - \left(1+\alpha^2\right)\omega^2 + 2j\omega\omega_0}. \tag{7.33}$$

The resonance frequency corresponding to the LLG dissipation (if loss is such that it is still possible to speak about a resonance) is calculated as:

$$\omega_{res} = \frac{\omega_0}{\sqrt{1+\alpha^2}}, \tag{7.34}$$

which is different from the resonance frequency ω_0 at zero loss. However, at small loss, $\alpha^2 << 1$, one can neglect this difference.

Moreover, with the LLG dissipation, the third diagonal susceptibility tensor component $\chi_{\|}$ in (7.28) becomes nonzero:

$$\chi_{\|} = -\frac{j\omega_M \alpha}{\omega\left(1+\alpha^2\right) - j\omega_0 \alpha}. \tag{7.35}$$

The Gilbert damping parameter α of real materials depends on their crystalline structure and many other physical factors, which will be described further when discussing the resonance line width of ferrites. Herein, it is worth to note that in order to obtain α experimentally, measurements should be done in intense bias field so that the sample under test is uniformly magnetized to saturation. In nearly perfect single crystals, where practically only spin-spin and spin-crystalline relaxation types take place, the typical values of $\alpha \sim 0.01$. The maximum experimentally measured values do not exceed 0.01 [16].

Let us separate real and imaginary parts in the susceptibility components (7.33) as $\chi = \chi' - j\chi''$ and $\kappa = \kappa' - j\kappa''$. The minus sign is chosen before the imaginary part, because then the absorption of energy will be at positive χ'' and κ'' for the time dependence with $e^{j\omega t}$. The imaginary parts are responsible for absorption of EM energy by ferrites:

$$\begin{aligned} \chi' &= \frac{\omega_0 \omega_M}{\Delta}\left(\omega_0^2 - (1-\alpha^2)\omega^2\right); \quad \chi'' = \frac{\omega\omega_M}{\Delta}\left(\omega_0^2 + (1+\alpha^2)\omega^2\right) \\ \kappa' &= \frac{\omega\omega_M}{\Delta}\left(\omega_0^2 - (1+\alpha^2)\omega^2\right); \quad \kappa'' = \frac{2\omega^2 \omega_M \omega_0 \alpha}{\Delta} \end{aligned} \tag{7.36}$$

where $\Delta = (\omega_0^2 - (1+\alpha^2)\omega^2)^2 + 4\alpha^2\omega^2\omega_0^2$.

The corresponding resonance values are $\chi'_{res} = \dfrac{M_0}{2H_0} = \chi_0 / 2$, $\chi''_{res} = \dfrac{M_0}{\Delta H}$, $\kappa''_{res} = \chi''_{res}$, and $\kappa'_{res} = 0$, where ΔH is the width of the resonance line at ½ level from the maximum loss, which is equal to:

$$\Delta H = \frac{\alpha\, \omega_0}{\mu_0 \gamma} = \alpha\, H_0 \tag{7.37}$$

Figure 7.5 shows the calculated magnetic field dependences for the components of the magnetic susceptibility for the following parameters (in Gaussian system of magnetic units): $M_0 = M_S$=1600 G, $f = \omega / (2\pi)$=9.4 GHz, and α = 0.044.

The second type of the relaxation term, the modified Bloch, or *Bloch–Bloembergen (BB)*, is often used as a convenient way of describing resonance behavior of ferrites and loss [20]:

$$\frac{d\boldsymbol{M}}{dt} = -\mu_0 \gamma \boldsymbol{M} \times \boldsymbol{H} - \omega_{rel}(\boldsymbol{M} - \chi_0 \boldsymbol{H}), \tag{7.38}$$

where $\chi_0 = M_0 / H_0$ is static magnetic susceptibility and ω_{rel} is the relaxation frequency.

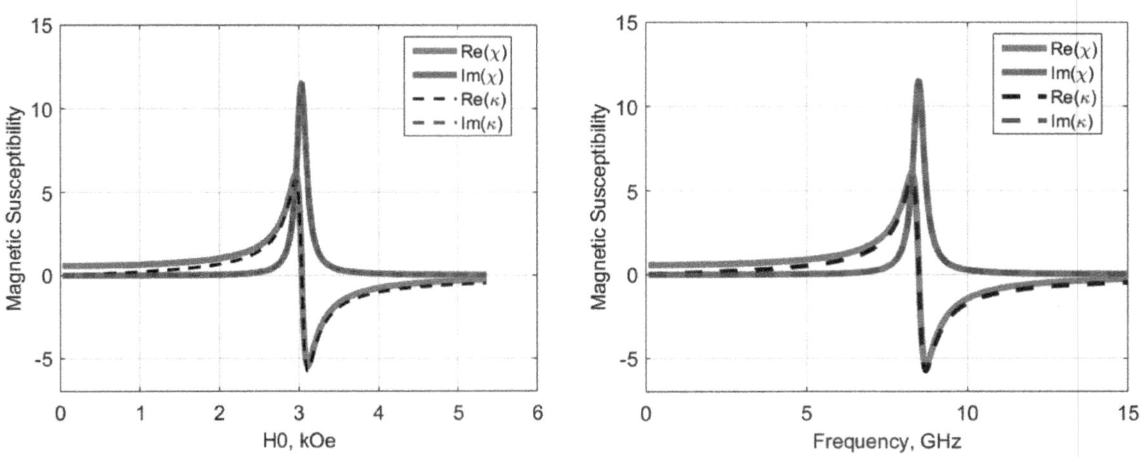

Figure 7.5 Components of magnetic susceptibility tensor near ferrimagnetic resonance as a function of bias magnetic field.

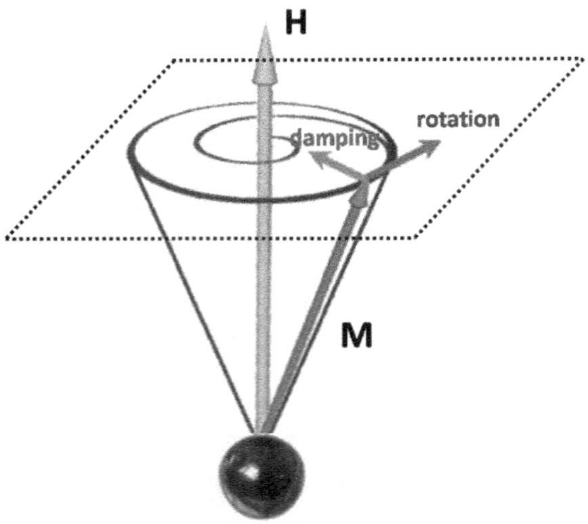

Figure 7.6 Damping in the BB case results in the magnitude variation of magnetization vector so that its end tip moves on a plane surface.

As is mentioned before, the magnitude $|\boldsymbol{M}|$ in the MME with the BB relaxation term (7.38) is not conserved. Indeed, the magnetization vector $\boldsymbol{M}$ decays toward the direction of the bias magnetic field with a tip of $\boldsymbol{M}$ moving in a plane (see Figure 7.6). Therefore, the equilibrium is valid only for a small-signal limit so that the resonance frequency ω_0 is unaffected by the dissipation:

$$\omega_0 = \mu_0 \gamma H_0. \tag{7.39}$$

The susceptibility tensor components obtained from small-signal approximation (7.38) turn out to follow the Lorentzian frequency dispersion:

$$\chi = \frac{\omega_M \omega_0}{\omega_0^2 - \omega^2 + j\omega\omega_{rel}}; \kappa = \frac{\omega_M \omega}{\omega_0^2 - \omega^2 + j\omega\omega_{rel}}. \tag{7.40}$$

Herein, the parameter $\omega_{rel} = 2\Delta\omega$ determines the full width of the resonance line at the level of 0.5 of the resonance values. Shapes of the resonance curves obtained using (7.40) are very close to those shown in Figure 7.5, provided the loss in both cases – LLG and BB – is very $low(\omega_{rel} \ll \omega_0; \alpha \ll 1)$.

The frequency dependences (7.40) follow the Lorentzian form of dispersion. This form is used in number of publications [21], to describe the behavior of complex permeability spectra. Though the BB relaxation term is not applicable to the description of magnetic materials with constant $|\boldsymbol{M}|$, in many practical cases, Lorentzian dependences describe the actual frequency behavior of magnetic materials with sufficient accuracy [22].

The third susceptibility tensor component is nonzero:

$$\chi_{||} = -\frac{j\omega_M \omega_{rel}}{\omega_0 \left(\omega - j\omega_{rel}\right)}. \tag{7.41}$$

One can see that the susceptibility tensor components (7.40) are different from the lossless case (7.27): the latter would result in infinite resonance values. The components (7.40) also formally differ from the LLG case (7.27). However, for small-signal limit and sufficiently low damping parameters ($\alpha \ll 1$), the solutions for the LLG and BB cases become close, and the relation between damping parameters for small-signal limit can be written as:

$$\alpha = \frac{\omega_{rel}}{\omega_0}. \tag{7.42}$$

The resonance frequency in LLG (7.34) is shifted with respect to the BB (Lorentzian) equation similarly to the shift in the lossy case with respect to the lossless one. However, for $\alpha \ll 1$, the difference between resonance frequencies in LLG and BB (Lorentzian) cases could be neglected.

From the Lorentzian dependences (7.40), one can derive that the maximum values of imaginary parts of susceptibility tensor at small dissipation are reached at $\omega = \omega_0$ and are equal to:

$$\chi''_{\text{res}} = k''_{\text{res}} = \frac{\mu_0 \gamma M_0}{2\Delta\omega}, \tag{7.43}$$

where M_0 approximately equals to saturation magnetization M_s. These resonance values increase as the resonance line width decreases. As follows from (7.40), the off-resonance values of χ'' and κ'' decrease with the departure from the resonance. The resonance line width then can be obtained from the known magnetic saturation of the ferrite and by measuring the resonance peak value of its susceptibility:

$$2\Delta\omega = \frac{\mu_0 \gamma M_0}{\chi''_{res}}, \text{ or } 2\Delta\omega = \frac{\mu_0 \gamma M_0}{\kappa''_{res}}. \tag{7.44}$$

In many cases, typically in the ferrite industry, a ferrite's FMR line width is characterized in terms of the magnetic field values, where ΔH is the full resonance line width between the two values of $\chi'' = \frac{\chi''_{res}}{2}$ on the resonance curve, or the two values of $\kappa'' = \frac{\kappa''_{res}}{2}$:

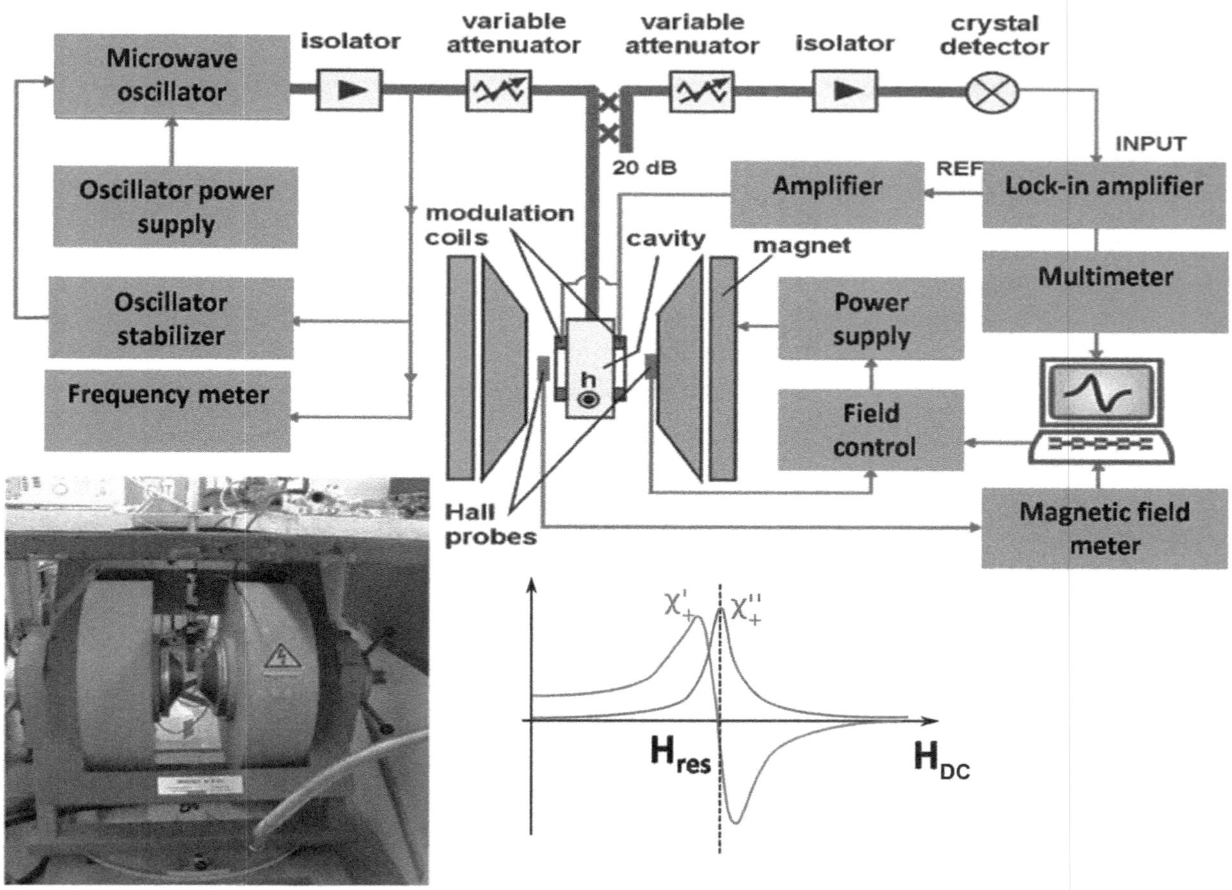

Figure 7.7 Measurement of ferromagnetic resonance characteristics in a ferrite sample.

$$\Delta H = \frac{2\Delta\omega}{\mu_0\gamma}. \tag{7.45}$$

Indeed, the FMR frequency is typically varied within some range by the linear scanning of the bias magnetic field, as $\omega = \mu_0\gamma H$ (if anisotropy fields are omitted). The FMR frequency practically instantaneously follows the variations of the bias field [23]. The installation to measure FMR in a ferrite sample is schematically shown in Figure 7.7.

However, accurate detection of the peak values of magnetic susceptibility is not always technically possible. To do this, very stable bias magnetic fields are needed, but fields of massive electromagnets are usually unstable due to heating, therefore complex thermal stabilization or cooling are required. The alternative way of measuring FMR line width is measuring the derivative of the absorbed signal with respect to the magnetic field. In this method, the FMR line width is determined by measuring an interval of field values $\Delta H = H_+ - H_-$, which correspond to the positions of the maximum H_+ and minimum H_- of the first derivative $\frac{dP}{dH}$ of the absorbed RF/microwave signal $P(H)$. The resonance field value H_{res} always lies between H_- and H_+.

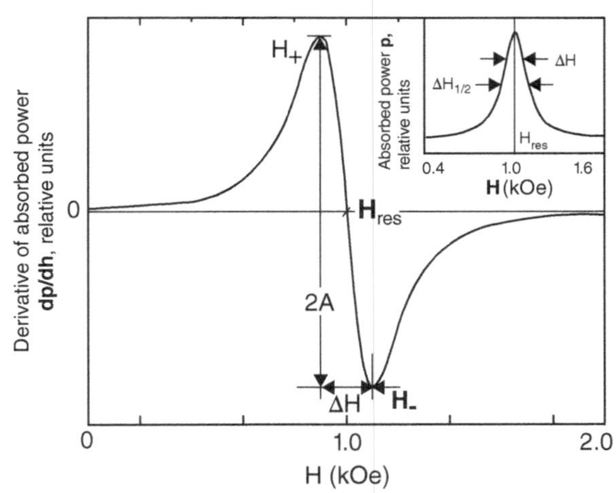

Figure 7.8 Measurement of resonance line width at FMR using derivative of the absorbed power by the bias field.

However, the value of ΔH in this case depends on the angles between the resonance field and the ferrite crystallographic axes. The ferrite sample shape also affects the line width, as there is demagnetization at its surfaces. Just this simple example shows that the measurement of the resonance line width in reality is a very complex problem (see Figure 7.8).

Typical values of FMR resonance line widths of microwave monocrystalline YIG ferrites are $\Delta H = 0.1\ldots10$ Oe at frequencies above 5 GHz; their magnetic saturation is $4\pi M_s = 1750$ G in Gauss magnetic unit system. The minimum resonance frequency is determined by $\omega_M = \mu_0\gamma M_s$, which in the Gauss system corresponds to 2.8×10^{-3} GHz/Oe$\times$1750 G$=$4.9 GHz. For garnet ferrites with the lower magnetic saturation, the minimum resonance frequency can be reduced. For example, the Ca–Bi–V doped garnets can exhibit FMR at frequencies of 300 MHz and even lower. Uniaxial single-crystal doped hexaferrites of BaM type have $\Delta H = 10\ldots100$ Oe, and their minimum $4\pi M_s = 3\ldots4$ kG and crystallographic anisotropy field is at least $H_A = 10$ kOe, so their natural FMR (NFMR) is expected at frequencies above 20 GHz.

The Lorentzian dependences for real and imaginary parts of the magnetic susceptibility components in (7.40) are practically symmetrical with respect to ω_0 for the high-Q resonances with the quality factor $Q = \frac{\omega_0}{\omega_{rel}} \gg 1$. The equivalent linear circuit corresponding to such a resonance is the high-Q *RLC* circuit. Its impulse response is the long-lasting oscillation behaving as $\sim e^{-\alpha t}\sin\beta t$ [24]. As the Q factor of the circuit decreases to $Q \sim 1$, the material frequency response becomes asymmetric. As the Q factor further decreases below 1, the resonance frequency shifts to the lower frequencies ($\omega_0 \to 0$), forming the wideband Lorentzian dispersion, the impulse response of which behaves as $\sim e^{-\alpha t}\sinh\beta t$ and cannot form even a single period of an oscillation. Eventually, at the very low Q, the Lorentzian curve transforms to the *Debye frequency dispersion*, expressed as:

$$\chi = \frac{A}{1 + j\omega\tau}, \tag{7.46}$$

where the coefficient A is associated with the static magnetic susceptibility ($A = \chi_s$) and τ is the relaxation time constant corresponding to the maximum loss for the imaginary part of the Debye curve at the relaxation frequency $\omega_{Drel} = 1/\tau$. At the same frequency ω_{Drel}, the magnitude of the real part drops two times as compared to the static susceptibility. The equivalent linear circuit corresponding to this case is the nonresonance *RC* (or *RL*) circuit, and the impulse response is a decaying exponential function $\sim e^{-t/\tau}$ [24]. The Debye dispersion is used to describe frequency behavior of comparatively low-frequency RF ferrites, mainly spinels, employed in various ferrite cores and common-mode chokes.

For the other relaxation terms $\boldsymbol{R}$, including the LLG form, frequency dispersion of magnetic susceptibility cannot be accurately described by the Lorentzian dispersion, especially with the significant values of the relaxation parameters affecting the magnetization motion precession. This is the case of a comparatively wide peak of magnetic loss, and as is mentioned in [22], the actual frequency dispersion of magnetic susceptibility can be represented as a sum of multiple Lorentzian terms:

$$\chi = \sum_{i=1}^{N} \frac{\chi_{s,i}}{1 + \frac{j\omega}{\omega_{r,i}} - \left(\frac{\omega}{\omega_{0,i}}\right)^2}. \tag{7.47}$$

In (7.47), the three curve-fitting parameters are used for each ith term: partial static susceptibility $\chi_{s,i}$, relaxation frequency $\omega_{r,i}$ associated with the widths of the loss peaks, and "resonance" frequency $\omega_{0,i}$, which determines the position of the ith loss peak.

Note that the relaxation frequency $\omega_{r,i}$ in (7.47) is not the same as ω_{rel} in (7.40). They are related as:

$$\omega_{r,i} = \frac{\omega_0^2}{\omega_{rel}}. \tag{7.48}$$

It is known that a sum of a sufficient number Lorentzian terms can describe an arbitrary frequency dependence of any real material over the limited frequency range with any desirable accuracy [25]. This becomes obvious if considering partial Lorentzian curves with infinite Q factors, where loss peaks turn into δ-functions, from which the real part of permeability is obtained using Kramers–Krönig relations [26].

For this reason, the sum of multiple Lorentzian terms is used to describe magnetic spectra with splitting, broadening, and all kind of distortions. Note that the presence of two fitting parameters in each Lorentzian term of (7.47) (see Figure 7.9), relaxation frequency $\omega_{r,i}$, and resonance frequency $\omega_{0,i}$, may result in ambiguity of curve fitting, and therefore, unphysical resonances may appear. In spite of possible unphysical resonances, the main objective of using such fitting is as accurate as possible description of the permeability dispersion over the frequency range of interest. This may be helpful for numerical EM modeling of complex structures containing ferrite materials, especially for time-domain solvers, as such curve fitting provides causal frequency responses [27].

Resonance losses in a bulk ferrite material depend on many factors. The most important are the morphology of the ferrite material – its purity, homogeneity, presence of pores and undesirable inclusions; its crystallographic structure; direction of the bias field with respect to the crystallographic axes and/or planes; presence of crystallites if the material is polycrystalline; presence of domains if the material is not totally magnetically saturated; ambient temperature; and power level of a continuous or pulse RF/microwave signal. Substantial losses in microwave ferrites may also take place above a certain power level due to spin-wave instability.

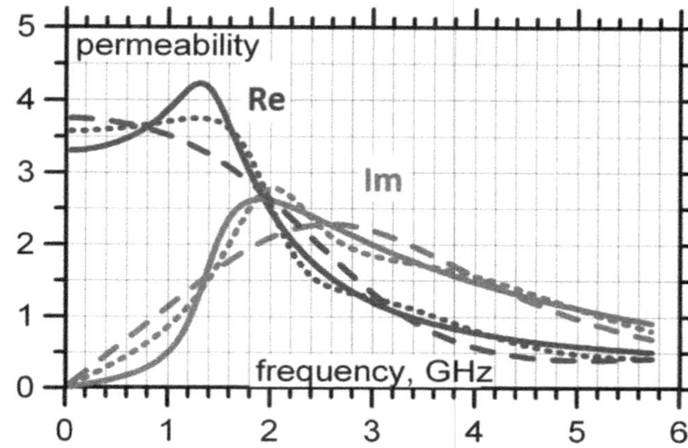

Figure 7.9 *An example of curve fitting of resonance characteristics for a ferrite using (7.47): solid lines – calculated using LLG; dashed lines – curve fitted using one Lorentzian term; and dotted lines – curve fitted using two Lorentzian terms* [22].

Relaxation and various physical factors determining resonance line width of ferrites at the FMR will be discussed in detail in Chapter 6 in Volume 2. Herein, we just mention that the process of magnetic relaxation in ferrites can be understood as the sum of three distinct processes:

i) The direct coupling between the uniform precession of magnetic moment determined by electron spins and the crystal lattice
ii) Dissipation via spin-wave generation
iii) Coupling of spin waves to the crystal lattice.

The classical representation of the energy absorption by a ferrite medium is the following. The MME with the relaxation term (7.17) determines what time it will take for the magnetization vector to return to the thermodynamically equilibrium state, i.e. what the relaxation time is, if the external forces which had excited the system have been removed. If the system remains under the impact of external periodic action, i.e. microwave magnetic fields, magnetization vector precession will continue, and the system will sustain the stationary non-equilibrium status, in which the amplitudes of the forced oscillations will exceed the equilibrium values. This means that the relaxation processes continuously take away the energy from the forced oscillations, and therefore let the system continuously absorb energy of the external sources. The relaxation processes will determine such dissipative properties of the system as the Q factors of the imaginary parts of the magnetic susceptibility tensor, or the width of the corresponding resonance peaks (usually called "resonance line width").

Therefore, the relaxation time $\tau_{rel} = \dfrac{1}{\omega_{rel}}$ determines the losses in the ferrite medium and is related to the resonance line width ($2\Delta\omega$). Strictly speaking, $\omega_{rel} \neq 2\Delta\omega$ because of many factors, including crystallographic anisotropy and shape anisotropy.

Resonance line widths can be extracted from the solutions of the MME with different forms of the relaxation term $\boldsymbol{R}$. However, different relaxations terms will result in different expressions for relaxation times and, hence, line widths. Usually, the parameters of relaxation terms are chosen in such a way that the measured results agree with those obtained from the MME.

7.2.6 FMR in a Small Ellipsoid and Demagnetization Factors

So far, we have only mentioned briefly that the behavior of uniform precession of a magnetization vector is affected by the shape anisotropy and that there is always demagnetization at the interface of the ferrite medium with ambient environment. In reality, shape of a ferrite sample significantly affects its magnetic spectrum – it shifts resonance frequencies as compared to the same unbounded medium, widens the resonance lines, and may result in numerous resonance lines due to harmonics of uniform magnetostatic oscillations excited in a sample. Only perfect spherical FRs have isotropic shape, and their magnetostatic spectrum has only one mode corresponding to the uniform precession of magnetization vector. All the other shapes – disks, cylinders, spheroids of rotation, or general shaped ellipsoids exhibit form anisotropy, and their magnetostatic spectrum may be quite rich over the frequency range of interest resulting in harmonics of the uniform precession. For this reason, measurements of microwave FMR properties of ferrite materials are usually done using spherical FRs with the simplest one-peak resonance spectrum.

Strictly speaking, internal fields and magnetization, both RF/microwave and DC, in the case of a ferrite particle should be found from the corresponding boundary problems for Maxwell and magnetostatic equations. These are typically difficult mathematical problems, and they can be rigorously solved only for a few particular cases. Otherwise, numerical modeling can be used, or simplifications should be done for solving problems analytically.

The simplest case is the boundary problem for uniform oscillations in a small ferrite ellipsoid. Let us assume that a ferrite sample is uniformly magnetized to saturation. Small size of a ferrite sample means that the external bias DC and microwave fields can be defined as the fields in the same point, where the ferrite is placed, but without the ferrite. A ferrite small compared to the wavelength means that it is possible to consider quasistatic (magnetostatic) approximation.

From magnetostatics, it follows that the internal (also called "effective") magnetic field $\boldsymbol{H}$ in the ellipsoid placed in the external field $\boldsymbol{H}_e$, is homogeneous, and comprises two parts: $\boldsymbol{H} = \boldsymbol{H}_e + \boldsymbol{H}_M$, where $\boldsymbol{H}_M = -\vec{\boldsymbol{N}} \cdot \boldsymbol{M}$ is the *demagnetization field*. The tensor $\vec{\boldsymbol{N}}$ is called demagnetization tensor. In the general case, the axes of an ellipsoid (x', y', z') do not coincide with the coordinate axes (x, y, z). Then, the tensor contains $3 \times 3 = 9$ components N_{ij}, which are called *demagnetization factors*, or *form factors*:

$$\vec{\boldsymbol{N}} = \begin{bmatrix} N_{11} & N_{12} & N_{13} \\ N_{12} & N_{22} & N_{23} \\ N_{13} & N_{23} & N_{33} \end{bmatrix}. \tag{7.49}$$

Then, for the DC components of the internal field:

$$\boldsymbol{H}_0 = \boldsymbol{H}_{e0} - \vec{\boldsymbol{N}} \cdot \boldsymbol{M}_0, \tag{7.50}$$

and for microwave components:

$$\boldsymbol{h} = \boldsymbol{h}_e - \vec{\boldsymbol{N}} \cdot \boldsymbol{m}. \tag{7.51}$$

The equilibrium condition is then:

$$\boldsymbol{M}_0 \times (\boldsymbol{H}_{e0} - \vec{\boldsymbol{N}} \cdot \boldsymbol{M}_0) = 0, \tag{7.52}$$

and the linearized LLG MME is:

$$j\omega\boldsymbol{m} + \mu_0\gamma\boldsymbol{m} \times \boldsymbol{H}_0 + \mu_0\gamma\vec{\boldsymbol{N}} \cdot \boldsymbol{m} \times \boldsymbol{M}_0 + \frac{j\omega\alpha}{M_0}\boldsymbol{m} \times \boldsymbol{M}_0 = -\mu_0\gamma\boldsymbol{M}_0 \times \boldsymbol{h}_e. \tag{7.53}$$

First, we will consider *natural, or free uniform oscillations* of magnetization moment in a small ferrite ellipsoidal particle and assume that there is no loss. Therefore, (7.53) becomes:

$$j\omega\boldsymbol{m} + \mu_0\gamma\boldsymbol{m} \times \boldsymbol{H}_0 + \mu_0\gamma\vec{\boldsymbol{N}} \cdot \boldsymbol{m} \times \boldsymbol{M}_0 = 0. \tag{7.54}$$

Let us assume that the sample is magnetized to saturation ($M_0 = M_s$) by the bias magnetic field in the z-direction and that the axis of rotation of the ellipsoidal particle coincides with the z-direction. The RF/microwave field has only the x and y components. Then, from (7.53) and (7.54), we can get a system of two linear equations for microwave magnetization:

$$\begin{aligned} &\left(j\omega + \mu_0\gamma N_{12}M_0\right)m_x + \mu_0\gamma\left(H_{e0z} - N_{33}M_0 + N_{22}M_0\right)m_y = 0; \\ &-\mu_0\gamma\left(H_{e0z} - N_{33}M_0 + N_{11}M_0\right)m_x + \left(j\omega - \mu_0\gamma N_{12}M_0\right)m_y = 0. \end{aligned} \tag{7.55}$$

The resonance frequency is found by equating the determinant of the system to zero:

$$\omega_{res}^2 = (\omega_{01} + N_{11}\omega_M)(\omega_{01} + N_{22}\omega_M) - N_{12}^2\omega_M^2, \tag{7.56}$$

where

$$\omega_{01} = \mu_0\gamma(H_{0ez} - N_{33}M_0) = \omega_0 - N_{33}\omega_M. \tag{7.57}$$

Obviously, absorption of EM energy by a small ellipsoid will take place at frequencies close to the resonance frequency ω_{res}, but not ω_0, which is the characteristic of the material.

If the external magnetic field is directed along one of the axes of ellipsoid, e.g. the z-axis, then the tensor $\tilde{N}$ is diagonal:

$$\tilde{N} = \begin{bmatrix} N_x & 0 & 0 \\ 0 & N_y & 0 \\ 0 & 0 & N_z \end{bmatrix}. \tag{7.58}$$

Therefore, in this case, $N_{11} = N_x;\ N_{22} = N_y; N_{33} = N_z; N_{21} = N_{12} = 0$ Equation (7.49) and Equations (7.55)–(7.57).

The sum of the demagnetization factors for an ellipsoid is always unity [28]:

$$N_x + N_y + N_z = 1. \tag{7.59}$$

Note that the tensor $\tilde{N}$ can be composed for samples of any shape, but only in case of an ellipsoid, the components are numbers, i.e. independent of coordinates. Rigorous formulas to calculate form factors of any ellipsoids are based on the ratio of their main axes and can be found in [28]. Some canonic cases of demagnetization factors are given in Table 7.2 and illustrated in Figure 7.10.

In the case of the sphere, the condition of FMR has the simplest form independent on the DC magnetization. The dissipation, for example, in the LLG form, can be taken into account using the substitution $\omega_0 \to \omega_0 + j\alpha\omega$. Then, the resonance frequency with taking into account dissipation can be represented as:

$$\begin{aligned} \omega_{res\,(\alpha)}^2 &= \omega_0^2 + f_1(\omega_0,\omega_M,N) + j\omega\alpha f_2(N,\omega_0,\omega_M) - (\alpha\,\omega)^2 = \\ &\omega_{res(0)}^2 - (\alpha\,\omega)^2 + j\omega\alpha\, f_2(N,\omega_0,\omega_M). \end{aligned} \tag{7.60}$$

where the resonance frequency in the ellipsoid without loss can be rewritten as

$$\omega_{res(0)}^2 = \omega_0^2 + f_1(\omega_0,\omega_M,N), \tag{7.61}$$

Table 7.2 Demagnetization factors for ellipsoids of rotation.

		Demagnetization factors			
Geometry	**Field direction**	N_z	N_x	N_y	**Resonance frequency**
Infinitely thin plane (disk)	Tangential (a)	0	0	1	$\omega_{res}^2 = (\mu_0\gamma)^2 H_{e0}(H_{e0} + M_0)$
	Normal (b)	1	0	0	$\omega_{res} = \mu_0\gamma(H_{e0} - M_0)$
Infinitely thin cylinder	Tangential (c)	0	1/2	1/2	$\omega_{res} = \mu_0\gamma(H_{e0} + M_0/2)$
	Normal (d)	1/2	1/2	0	$\omega_{res}^2 = (\mu_0\gamma)^2 H_{e0}(H_{e0} - M_0/2)$
Sphere	(e)	1/3	1/3	1/3	$\omega_{res} = \mu_0\gamma H_{e0}$

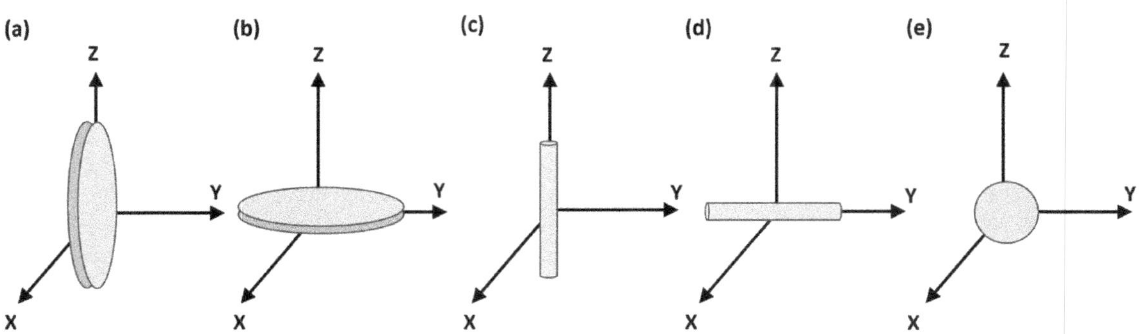

Figure 7.10 Illustration of demagnetization factors for various ellipsoids of rotation. The letters (a)...(e) correspond to the letters denoting field directions in Table 7.2.

and the functions $f_{1,2}$ are

$$f_1(\omega_0,\omega_M,N)=\omega_0\omega_M(N_{11}+N_{22}-2N_{33})+\omega_M^2(N_{11}-N_{33})(N_{22}-N_{33})-\omega_M^2N_{12}^2;$$
$$f_2(\omega_0,\omega_M,N)=\omega_M(N_{11}+N_{22}-2N_{33})+2\omega_0. \tag{7.62}$$

For a small dissipation parameter, the resonance frequency $\omega_{res(\alpha)}$ can be estimated using the representation of the square root of (7.60) as $\sqrt{1+x}\,|_{x<<1}\approx 1+x/2$. The resonance frequency is a complex value:

$$\omega_{res(\alpha)}=\omega'_{res(\alpha)}+j\omega''_{res(\alpha)}. \tag{7.63}$$

The real part of the resonance frequency is the frequency of natural (free) oscillations of the magnetization vector. It can be seen that the presence of dissipation shifts this resonance frequency ω_{res} to the lower frequencies:

$$\omega'_{res(\alpha)}=\omega_{res(0)}-\frac{\alpha^2\omega^2}{2\omega_{res(0)}}. \tag{7.64}$$

The imaginary part is responsible for attenuation of these free oscillations:

$$\omega''_{res(\alpha)}=\frac{\alpha\,\omega(2\omega_0+\omega_M(N_{11}+N_{22}-2N_{33}))}{2\omega_{res(0)}^2}=\frac{\alpha\,\omega\omega_1}{\omega_{res(0)}^2}, \tag{7.65}$$

where

$$\omega_1=(2\omega_0+\omega_M(N_{11}+N_{22}-2N_{33}))/2. \tag{7.66}$$

It is known that the dissipation of oscillations in linear circuits is described by the Q-factor:

$$Q_0=\frac{\omega'_{res(\alpha)}}{2\omega''_{res(\alpha)}}\approx\frac{\omega_{res(0)}}{2\alpha\omega_1}. \tag{7.67}$$

This value is called the *internal* Q-factor of the FR, because it takes into account only loss in the ferrite ellipsoid. If the ellipsoid is placed in a transmission line (a waveguide) or a cavity, the loaded Q-factor, which accounts for the coupling between the transmission line or the cavity and the FR, is considered.

The next step is to consider *forced oscillations*. The right-hand side in MME (7.53) associated with the external microwave field $\boldsymbol{h}_e$ is set as nonzero. Due to linearity of Equation (7.53), its solution is:

$$\boldsymbol{m}=\ddot{\chi}^{\text{ext}}\cdot\boldsymbol{h}_e, \tag{7.68}$$

where $\ddot{\chi}^{ext}$ is the tensor of *external magnetic susceptibility* of the ferrite ellipsoid magnetized in the z-direction. This susceptibility tensor is different from *internal susceptibility* tensor $\ddot{\chi}$ for the material itself. The tensor $\ddot{\chi}^{ext}$ arises from the solution of (7.54) in the magnetostatic approximation and is valid for the homogeneous (uniform) oscillations of magnetization in the sample of the specific form, an ellipsoid. The internal $\ddot{\chi}$ and external $\ddot{\chi}^{ext}$ tensors are correlated through the inverse tensors as:

$$\left[\ddot{\chi}^{ext}\right]^{-1}=\left[\ddot{\chi}\right]^{-1}+\ddot{\boldsymbol{N}}. \tag{7.69}$$

However, finding the external tensor $\ddot{\chi}^{ext}$ may be a cumbersome procedure. Another way, which is simpler, is to solve the linearized MME (e.g. with the LLG dissipation) directly for $\ddot{\chi}^{ext}$ using the external fields, written in terms of demagnetization factors: $\boldsymbol{H}_{e0}=\boldsymbol{H}_0+\ddot{\boldsymbol{N}}\cdot\boldsymbol{M}_0$ and $\boldsymbol{h}_e=\boldsymbol{h}+\ddot{\boldsymbol{N}}\cdot\boldsymbol{m}$.

For the case of the ellipsoid axis coinciding with the direction of the bias field, e.g. along the z-direction, the external susceptibility tensor is:

$$\ddot{\chi}^{ext}=\begin{vmatrix}\chi_{11}^{ext} & j\kappa^{ext} & 0\\ -j\kappa^{ext} & \chi_{22}^{ext} & 0\\ 0 & 0 & 0\end{vmatrix}. \tag{7.70}$$

The components of the tensor with nonzero loss parameter α are:

$$\chi_{11,22}^{\prime\text{ext}} = \frac{\omega_{11,22}\omega_M\left(\omega_{\text{res}}^2 - (1+\alpha^2)\omega^2\right) + 2\omega_M\alpha^2\omega^2\omega_{\text{res}}}{\Delta},$$
$$\chi_{11,22}^{\prime\prime\text{ext}} = \frac{2\omega_{11,22}\omega_M\alpha_G\omega_{\text{res}} - \alpha\,\omega\omega_M\left(\omega_{\text{res}}^2 - (1+\alpha^2)\omega^2\right)}{\Delta} \tag{7.71}$$

$$\kappa^{\prime ext} = \frac{\omega\omega_M}{\Delta}\left(\omega_{res}^2 - (1+\alpha^2)\omega^2 + 2\alpha\,\omega_{res}\omega_{12}\right),$$
$$\kappa^{\prime\prime} = \frac{2\omega^2\omega_M\omega_{res}\alpha - \omega_M\omega_{12}\left(\omega_{res}^2 - (1+\alpha^2)\omega^2\right)}{\Delta} \tag{7.72}$$

where the determinant and the resonance frequency are given by:

$$\Delta = \left(\omega_{res}^2 - (1+\alpha^2)\omega^2\right)^2 + 4\alpha^2\omega^2\omega_{res}^2;$$
$$\omega_{res}^2 = \omega_{11}\omega_{22} - \omega_{12}^2;$$
$$\omega_{11,22} = \omega_0 + \omega_M N_{22,11} \text{ and } \omega_{12} = -\omega_M N_{12} \tag{7.73}$$

If there is a diagonal tensor of demagnetization factors, then indices ij in the form factors N_{ij} can be as in (7.58). Note that although we have obtained the solution for the ellipsoidal ferrite samples taking into account demagnetization for the LLG case, the resultant magnetic susceptibility components can be approximated using (7.47) with multiple Lorentizan terms. The representation (7.47) is a general one for any ferrite with any demagnetization factors [22].

7.2.7 Peculiarities of Ferrimagnetic Resonance

So far, we were using ferromagnetic formalism to describe resonance phenomena and susceptibility spectra in ferrites. This is correct, as FMR indeed takes place in a ferrite.

However, at the very beginning, we have mentioned that ferrites (=ferrimagnetic materials), unlike ferromagnets, contain at least two crystal sublattices, the magnetic moments of which are opposite in sign, and generally, they are not equal (equality takes place in antiferromagnetic materials only). The majority of soft ferrites, used in RF/microwave engineering, return to their unmagnetized state with practically zero net magnetic moment as soon as bias magnetic fields are removed. At temperatures above Néel temperature, which is often referred as Curie temperature, when ferr**i**magnetic materials are considered, the magnetic order of ferrites is destroyed. The three types of magnetic order in materials are illustrated in Figure 7.11.

In this section, we will mainly qualitatively consider some effects which make ferrites different from ferromagnets. The more detailed analysis of ferrimagnetic resonance is given in [13, 15, 29]. Ferrimagnetic materials are described in the same

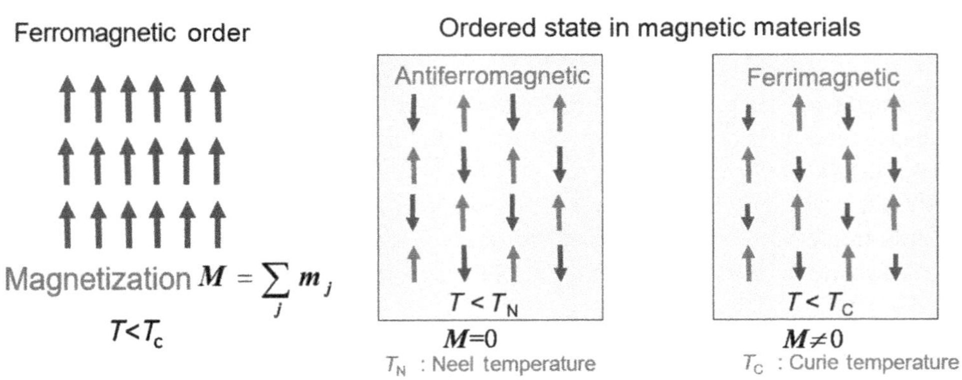

Figure 7.11 Three types of magnetic order – ferromagnetic, antiferromagnetic, and ferrimagnetic.

way using energy density as a function of all vectors of magnetic moments $\boldsymbol{M}_n, n=1,2\ldots$ Landau–Lifshitz MMEs with effective magnetic field and with any form of relaxation terms can be used to describe their behavior:

$$\frac{d\boldsymbol{M}_n}{dt} = -\mu_0\gamma_n \boldsymbol{M}_n \times \boldsymbol{H}_{ef\ n} + \boldsymbol{R}_n. \tag{7.74}$$

The gyromagnetic ratios γ_n can be different for different sublattices (this comes from the different Lande factors g_n for different magnetic ions and hence sublattices [30]). If the LLG form of relaxation is used, then in (7.74) instead of the effective fields, there will be terms with different α_n.

In ferrimagnetics, there are two special conditions – *magnetic compensation*, when sum of static magnetizations of sublattices is zero:

$$\sum_n \boldsymbol{M}_{n0} = 0, \tag{7.75}$$

and *mechanical compensation*, when the following sum is zero:

$$\sum_n \boldsymbol{M}_{n0} / \gamma_n = 0. \tag{7.76}$$

These conditions may occur at certain compositions of ferrites and at some temperatures and do not coincide in the general case.

There are two types of energy in the ferrimagnetic system – *Zeeman energy* and *exchange energy*:

$$E_{\text{mag}} = E_Z + E_{\text{exch}}. \tag{7.77}$$

Zeeman energy E_Z is the magnetic internal energy, which arises due to magnetic dipole–dipole interactions and depends on the total magnetization $\boldsymbol{M}$ and the external bias field $\boldsymbol{H}_{ext}$, i.e.:

$$E_Z = -\boldsymbol{M} \cdot \boldsymbol{H}_{ext}. \tag{7.78}$$

For a ferrimagnetic system, Zeeman energy is the same as for ferromagnetic system. The effective demagnetization fields acting on all sublattices mixed together in a ferrimagnetic sample of some shapes are identical with those in a ferromagnetic sample.

The exchange energy describes the energy of magnetic interactions between sublattices and inside each sublattice and embeds all types of anisotropy, including anisotropy due to crystallographic structure:

$$E_{exch} = -\frac{1}{2}\boldsymbol{M} \times \vec{\boldsymbol{\Lambda}} \times \boldsymbol{M}, \tag{7.79}$$

where $\vec{\boldsymbol{\Lambda}}$ is the tensor of exchange interactions.

Anisotropy energy of ferrimagnetic materials, similar to ferromagnets, depends on the angles between magnetic moments and crystal axes and can be written as power series with respect to the components of $\boldsymbol{M}_n$, $n=1,2\ldots$, or, equivalently, spatial angles [31]. For example, for a magneto-uniaxial ferrite, e.g. trigonal, tetragonal, or hexagonal, the crystal anisotropy energy is expressed as [15, 31]:

$$E_A = K_1 \sin^2\theta + K_2 \sin^4\theta + K_3 \sin^6\theta + \ldots, \tag{7.80}$$

where θ is the angle between the vector $\boldsymbol{M}$ and the principal axis of the crystal (see Figure 7.12 for a uniaxial hexagonal ferrite structure as an example). The magnitude of $|K_1| >> |K_2|, |K_3| \ldots$ When $K_1 > 0$, this is a crystal with easy axis; if $K_2 < 0$, the crystal has an easy plane. This is illustrated in Figure 7.13.

The exchange energy is usually treated as isotropic. But actually it has anisotropic part due to crystallographic structure depending on the angles, and this anisotropic part is typically considered as a separate energy term. Often, for simplicity, energy of anisotropy for sublattices is assumed to be additive, i.e. as the algebraic sum of these energy components, neglecting their mutual product terms.

According to Schlömann [32], the energy density for a ferrimagnetic material with two sublattices comprises:

$$E = -\frac{1}{2}(\Lambda_{11}|\boldsymbol{M}|_1^2 + \Lambda_{22}|\boldsymbol{M}|_2^2 + 2\Lambda_{12}\boldsymbol{M}_1 \cdot \boldsymbol{M}_2) - \boldsymbol{H}_0 \cdot (\boldsymbol{M}_1 + \boldsymbol{M}_2), \tag{7.81}$$

where Λ_{11} and Λ_{22} are constants of exchange interaction inside sublattices and Λ_{12} between the sublattices.

The equilibrium angles θ_1 and θ_2 for vectors $\boldsymbol{M}_1$ and $\boldsymbol{M}_2$ are found from the minimum of energy (81), which results in two solutions:

$$(1)\ \sin\theta_1 = \sin\theta_2;$$
$$(2)\ \cos\theta_{1,2} = \frac{H_0^2 + H^+H^-}{2H_0\Lambda M_{10,20}}, \tag{7.82}$$

where $\Lambda = -\Lambda_{12}$, and the exchange fields are:

$$H^{\pm} = \Lambda(M_{10} \pm M_{20}), \tag{7.83}$$

so that

$$M_{10}\sin\theta_1 = M_{20}\sin\theta_2. \tag{7.84}$$

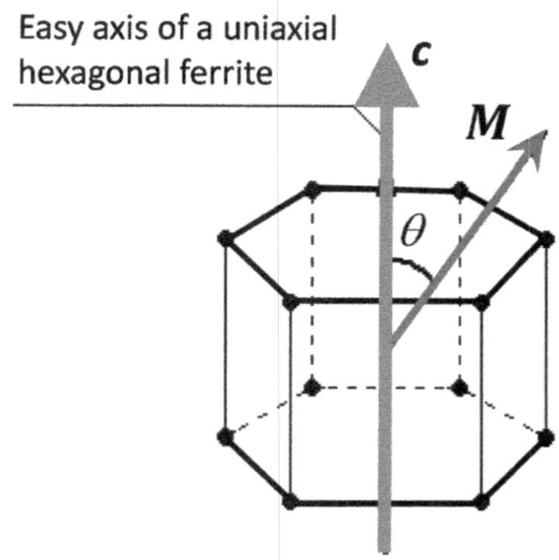

Figure 7.12 Illustration of angle θ in (7.80) as an example for a hexagonal crystalline structure.

For the majority of ferrites, $\Lambda \sim 10^3$; $H^- \sim 10^2$ kOe, and $H^+ >> H^-$.

It can be shown that for a ferrimagnetic material with two sublattices, there could be three states:

i) Antiparallel (when magnetic moments lie on the same axis)
ii) Noncollinear (do not lie on the same axis)
iii) Parallel (closed).

For any state, the total magnetic moment $\boldsymbol{M}_0 = \sum_{i=1}^{2} \boldsymbol{M}_{n0}$ is parallel to the bias field $\boldsymbol{H}_0$.

In antiparallel state, it can be shown that the high-frequency magnetization moments rotate in opposite directions with circular frequencies ω^+ (right-hand precession) and ω^- (left-hand precession) [33], represented through the effective gyromagnetic ratio, which is the ratio of the sum of magnetic moments to the sum of gyromagnetic ratios:

$$\begin{aligned}&\text{Solution 1: } \omega^+ = \mu_0\gamma_{ef}^+ H_0;\\&\text{Solution 2: } \omega^- = \omega_{exch} - \gamma_{ef}^- H_0,\end{aligned} \tag{7.85}$$

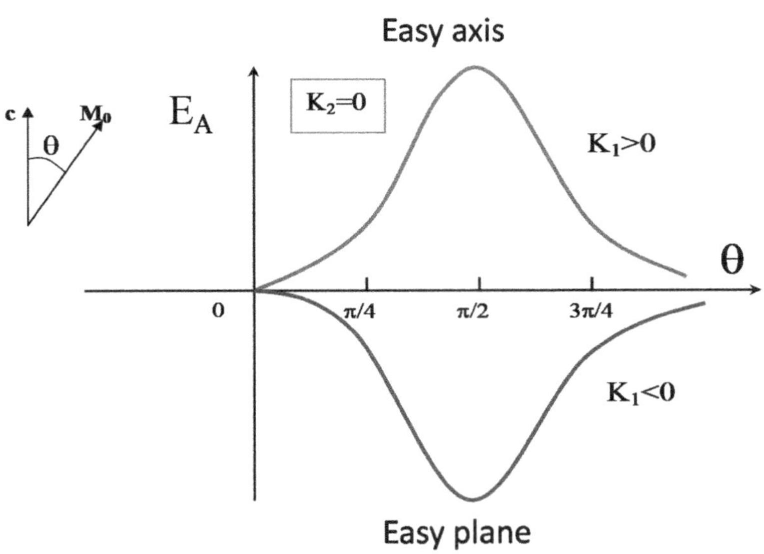

Figure 7.13 Anisotropy energy as a function of angle for crystals with easy plane and easy axis.

where

$$\gamma_{ef}^{+} = \frac{M_{10} - M_{20}}{M_{10}/\gamma_1 - M_{20}/\gamma_2}; \quad (7.86)$$

$$\gamma_{ef}^{-} = \frac{M_{10}\gamma_2/\gamma_1 - M_{20}\gamma_1/\gamma_2}{M_{10}/\gamma_1 - M_{20}/\gamma_2}; \text{ and} \quad (7.87)$$

$$\omega_{exch} = \mu_0\Lambda(\gamma_2 M_{10} - \gamma_1 M_{20}). \quad (7.88)$$

The ratios of the magnitudes of magnetic moments of sublattices for $H_0 << H^-$ are:

$$\begin{aligned} m_1^+/m_2^+ &= -M_{10}/M_{20}; \\ m_1^-/m_2^- &= -\gamma_1/\gamma_2 \end{aligned} \quad (7.89)$$

The precession with ω^+ corresponds to Zeeman energy and is called the *ferromagnetic type of precession*. It results in resonance similar to that in ferromagnets (FMR). The second precession with ω^- mainly depends on the exchange energy and is called *exchange precession*.

It is important that the two frequencies are normally far from each other: at the comparatively low-bias fields H_0, which is usually a case in practice, the exchange frequency ω^- is in the infrared range, whereas ω^+ is in the RF/microwave range. It is interesting that when $\omega^+ \to 0$, the system is close to the magnetic compensation point, whereas when $\omega^+ \to \infty$, it is at the mechanical compensation point. The exchange frequency ω^-, which is typically very high, becomes lower as the system approaches both compensation points. However, in practice, even away from the compensation points, it can be still considered that only ferromagnetic precession exists in ferrites.

In the *noncollinear state*, there also exists the ferromagnetic type of precession. This is the right-hand precession of the total magnetic moment, and it is at exactly the same frequency as ferromagnets have. There is also a zero-frequency precession (infinitely slow), which appears from the cylindrical symmetry of the energy in the system.

The similar situation is observed in the *parallel (closed) state*: one of the solutions is the ferromagnetic type of precession ω^+.

Let us consider FMR in *antiparallel* case. Assume that loss parameters in the MME with LLG relaxation term are low ($\alpha_1, \alpha_2 < 1$), bias field is low and far from both compensation points ($H_0 << H^-$).

Then for the right-hand oscillations the following condition is fulfilled: $\omega^+/(\mu_0\gamma_{1,2}) << H^-$.

In this case:

$$\begin{aligned} \omega_1^+ &= \mu_0\gamma_{ef}H_0(1+\alpha_{ef}); \\ \omega_2^+ &= \alpha_{ef}\mu_0\gamma_{ef}H_0; \end{aligned} \quad (7.90)$$

where γ_{ef}^+ is defined in Equation (7.86) and:

$$\alpha_{ef} = \frac{\gamma_1 M_{10} - \gamma_2 M_{20}}{M_{10}/\gamma_1 - M_{20}/\gamma_2}. \quad (7.91)$$

The effective dissipation α_{ef} contains contributions of both sublattices and tends to infinity as approaching the mechanical compensation point. This is just the trend, because Equation (7.91) becomes invalid near compensation points. Close to the point of mechanical compensation, the effective loss constant becomes very large $\alpha_{ef} \to \infty$; but this contradicts the initial assumption of low loss.

For forced oscillations, introducing circular polarized components of high-frequency field and magnetic moment (expressed in the complex form) and also the right-hand and left-hand susceptibilities, the following expressions can be obtained:

$$m^{\pm} = m_x \pm jm_y; h^{\pm} = h_x \pm jh_y; \text{ and } m^{\pm} = \chi^{\pm}h^{\pm}. \quad (7.92)$$

Then the total right-hand high-frequency magnetic moment is expressed as:

$$m^+ = m_1^+ + m_2^+ = \chi^+ h^+. \quad (7.93)$$

The resultant right-hand susceptibility is:

$$\chi^+ = \mu_0 \gamma_{ef} \frac{M_{10} - M_{20}}{\omega^+ - \omega + j\alpha_{ef}\omega}, \tag{7.94}$$

where $\omega^+ = \mu_0 \gamma_{ef} H_0$ is the frequency of the ferromagnetic type of precession. These results formally coincide with the corresponding results for FMR with circular right-hand polarization, but with the substitution of $\gamma \rightarrow \gamma_{ef}$ and $\alpha \rightarrow \alpha_{ef}$. Then the FMR line width is

$$\Delta H = 2\omega \frac{\alpha_1 M_{10} / \gamma_1 - \alpha_2 M_{20} / \gamma_2}{M_{10} - M_{20}}. \tag{7.95}$$

Formally, ΔH increases as approaching the magnetic compensation. Note that such increase of linewidth ΔH, similar to the increase of α_{ef} and linewidth in terms of frequency is not related to appearance of any new relaxation mechanism. This phenomenon was observed in some ferrogarnets close to the mechanical compensation points, and it is due to possible increase of the effective g-factors at comparatively low ΔH.

Strictly, the theory of magnetic oscillations in ferrimagnetic materials, as well as antiferromagnetics, can be developed only taking anisotropy into account [34]. In the most important FMR in the main antiparallel state with $H_0 << H^-$, for a ferromagnet with any type of anisotropy, the resonance conditions and the components of the magnetic susceptibility tensor are the same as for a regular ferromagnetic. Therefore, ferrimagnetic materials with two sublattices can be treated as the ferromagnets in terms of the FMR formulation. The peculiarity for ferrimagnets is that the effective anisotropy constants should be included. Since anisotropy energy of sublattices are additive, the effective anisotropy constants are just summed up. The effective anisotropy fields are found by dividing these effective anisotropy constants by the static magnetization ($M_{10} - M_{20}$) and increase as approaching compensation points.

Consideration of the shape anisotropy in ferrimagnetic ellipsoids does not change the FMR formalism, as the transition from the internal magnetic susceptibility $\ddot{\chi}$ to the external one $\ddot{\chi}_{ext}$ is a magnetostatic problem valid for any magnetic medium. The conditions for this transition are the same as we have considered before: ferromagnetic type of precession (with ω^+ only); low-bias field ($H_0 << H^-$); and staying away from both compensation points.

Ferrimagnetic materials are the most widely used type of magnetic materials in RF/microwave applications. They can be treated as *ferromagnetic* materials at the aforementioned conditions, which typically are fulfilled in practice, and therefore, all the results of the theory of ferromagnetism can be directly applied to them. Deviations from these conditions (IR frequency range, where exchange precessions exist; strong bias magnetic fields around H^-; and regions near points of compensation) require exchange precession modes to be taken into account.

7.2.8 FMR in a Ferrite with Crystallographic Anisotropy

Ferrites are solids, which typically have crystal structure. Therefore, they have some special directions called *crystallographic axes*. As has been mentioned in Section 7.1, RF and microwave ferrites can have spinel, garnet, magnetoplumbite (hexagonal), and perovskite crystal structures.

Crystals exhibit their anisotropy only if external fields (forces) are applied. This means that anisotropy is dependent on the physical properties of a medium on the direction of the applied forces (or fields). We have already seen that magnetic properties of even an unbound ferrite medium are described by a tensor of magnetic susceptibility $\ddot{\chi}$, or permeability $\ddot{\mu} = \ddot{\boldsymbol{U}} + \ddot{\chi}$, where $\ddot{\boldsymbol{U}}$ is the unit diagonal tensor. In Section 7.2.5, when considering a ferrite sample of finite size, the shape anisotropy was taken into account using demagnetization tensor $\ddot{\boldsymbol{N}}$ for the ferrite sample. The resultant external tensor $\ddot{\chi}^{ext}$ depends not only on the shape of the sample but on orientation of the sample's axes with respect to the bias static magnetic field.

There is another type of anisotropy in ferrites – *crystallographic anisotropy*. In addition to this anisotropy, there may also be the magnetoelastic anisotropy of magnetic properties of ferrites, and it is caused by an external mechanical stress.

Crystallographic anisotropy is determined by the combined action of a few factors. Energy of crystallographic anisotropy comprises:

i) anisotropic interactions of magnetic moments (dipole and quadrupole) in the spin system
ii) exchange spin–orbital interactions of magnetic moments of electrons and partial "freezing" of orbital moments in crystalline field

iii) the sum of energy spectra of spin moments of separate ions. Magnetic anisotropy follows from the symmetry group of the crystal.

Independent of the nature of crystallographic anisotropy, its free energy can be represented as a series of projections of the magnetic moment $\boldsymbol{M}$ on the axes of the crystal or with respect to the trigonometric functions of angles between $\boldsymbol{M}$ and the crystal axes. For example, for magneto-uniaxial ferrites (trigonal, tetragonal, or hexagonal), the anisotropy energy can be expressed as Equation (7.80).

When talking about small magnetic oscillations, we mentioned the equilibrium position of the magnetization vector $\boldsymbol{M}$. If only magnetic crystallographic anisotropy is considered, then the equilibrium position is determined by the minimum of the energy E_A. Such directions are called *easy axes* (of magnetization); the directions where E_A is maximum are called *hard axes*.

For example, in spherical samples, shape anisotropy is absent. If we consider a uniaxial crystal, the orientation of equilibrium $\boldsymbol{M}_0$ is determined by the minimum of energy. The latter is the sum of Zeeman energy (of interaction of the ferrite with external magnetic field) and that of the crystallographic anisotropy – see Equation (7.77) in Section 7.2.6, where we discussed the internal magnetic energy of a ferrimagnetic system. If only K_1 term is taken into account in Equation (7.80), the energy for the crystal with its own Cartesian system ($x'\ y'\ z'$) associated with axes is represented as [15]:

$$E = E_A + E_z = K_1 \sin^2\theta - M_0 H_0 (\sin\theta \sin\theta_H \cos(\varphi - \varphi_H) + \cos\theta \cos\theta_H), \tag{7.96}$$

where φ is the angle between the x' axis and the projection of $\boldsymbol{M}_0$ in the plane $(x'y')$; θ is the angle between the axis z' and $\boldsymbol{M}_0$; and the angles φ_H and θ_H determine the position of the external bias field vector $\vec{H}_0$ with respect to x' and z' axes, as shown in Figure 7.14. The energy minima are determined by the derivatives $\frac{\partial E}{\partial \varphi} = \frac{\partial E}{\partial \theta} = 0$. From these derivatives, one can get two conditions:

$$\text{(a)}\ \varphi_0 = \varphi_H \ \text{and (b)}\ \sin(2\theta_0) = \frac{H_0}{H_{A1}} \sin(\theta_H - \theta_0), \tag{7.97}$$

where anisotropy field $H_{A1} = \frac{K_1}{M_0}$ is determined just by the first constant of anisotropy K_1.

The analogous expressions, but more complex, can be obtained if considering the second anisotropy constant K_2 and higher, with the components of anisotropy field defined as $H_{A2} = \frac{K_2}{M_0}$, etc. [31, 35].

It can be seen that the directions of $\boldsymbol{M}_0$ and bias field $\boldsymbol{H}_0$ coincide for $K_1 > 0$, i.e. along the easy axis, and for $K_1 < 0$, i.e. in the easy plane. If the field $\boldsymbol{H}_0$ is directed along the hard axis or lies in the hard plane, then $\boldsymbol{M}_0$ will be in the same direction as $\boldsymbol{H}_0$ only if $H_0 > |2H_{A1}|$:

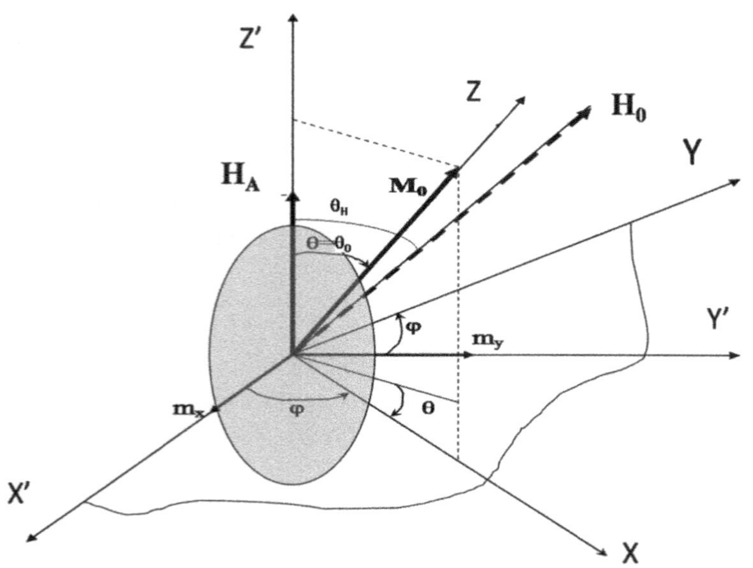

Figure 7.14 Coordinate axes and angles in an ellipsoid of magneto-uniaxial ferrite ($K_1 > 0$), the easy axis is z'.

In a uniformly magnetized (single domain – no domain structure) spherical sample, the direction of the equilibrium magnetization always coincides with the external bias field ($\theta_H = \theta_0$; $\varphi_H = \varphi_0$), if it is along the main easy axis, or lies in the easy plane.

Let us consider the MME with LLG dissipation, taking into account crystallographic anisotropy in a ferrite uniformly magnetized up to saturation. In this case, let us assume that there is an effective internal magnetic field $\boldsymbol{H}_{ef}$, which includes crystal anisotropy. The condition of equilibrium is:

$$\boldsymbol{M} \times \boldsymbol{H}_{ef} = 0, \tag{7.98}$$

i.e. the magnetization vector and the effective magnetic field should be parallel.

The LLG equation is:

$$\frac{d\boldsymbol{M}}{dt} = -\mu_0 \gamma \boldsymbol{M} \times \boldsymbol{H}_{ef} + \frac{\alpha}{M} \boldsymbol{M} \times \frac{d\boldsymbol{M}}{dt} \tag{7.99}$$

The total magnetization moment is:

$$\boldsymbol{M} = \boldsymbol{M}_0 + \boldsymbol{m} e^{j\omega t}, \tag{7.100}$$

and the effective magnetic field is:

$$\boldsymbol{H}_{ef} = \boldsymbol{H}_{ef\,0} + \boldsymbol{h}_{ef} e^{j\omega t}. \tag{7.101}$$

The DC effective magnetic field is:

$$\boldsymbol{H}_{ef\,0} = \boldsymbol{H}_0 - \vec{\boldsymbol{N}}_{ef}(M_0) \cdot \boldsymbol{M}, \tag{7.102}$$

where $\vec{\boldsymbol{N}}_{ef}$ is the effective demagnetization tensor, the components of which depend on the equilibrium magnetization vector $\boldsymbol{M}_0$. The high-frequency effective field is also associated with effective demagnetization as:

$$\boldsymbol{h}_{ef} = \boldsymbol{h} - \vec{\boldsymbol{N}}_{ef}(M_0) \cdot \boldsymbol{m}, \tag{7.103}$$

where $\boldsymbol{h}$ is the external high-frequency field.

Assume that in the Cartesian coordinate system, vectors $\boldsymbol{M}_0$ and $\boldsymbol{H}_{ef\,0}$ are aligned along the z-axis, and $\boldsymbol{h}_{ef}$ and $\boldsymbol{m}$ have only x and y components. The effective DC field expressed through the depolarization factors will have only z omponent (associated with $N_{ef\,33}$):

$$H_{ef\,0} = H_{0z} - N_{ef\,33}(M_0) \cdot M_0. \tag{7.104}$$

The linearized MME written with the effective magnetic field in the left-hand side is:

$$j\omega \boldsymbol{m} + \mu_0 \gamma \boldsymbol{m} \times \boldsymbol{H}_{ef\,0} + \mu_0 \gamma \boldsymbol{M}_0 \times \boldsymbol{h}_{ef} + \frac{j\omega\alpha}{M_0} \boldsymbol{m} \times \boldsymbol{M}_0 = -\mu_0 \gamma \boldsymbol{M}_0 \times \boldsymbol{h}, \tag{7.105}$$

The solution of the similar problem, but with just shape anisotropy, has been obtained in the Section 7.2.5 [see Equations (7.56)–(7.67)]. Equation (7.105) can be rewritten in terms of the external fields, and the external tensor of magnetic susceptibility can be retrieved. This indicates that the components of $\vec{\chi}^{ext}$ for a ferrite sample with crystallographic anisotropy can be easily obtained by substituting [15]

$$\begin{aligned} &N_{ij}(\text{form}) \to N_{ij}(\text{form}) + N_{ef\,ij}(\text{crystal}) \text{ for } i, j = 1,2; \\ &N_{33}(\text{form}) \to N_{33}(\text{form}) + N_{ef\,33}(\text{crystal}). \end{aligned} \tag{7.106}$$

This is because the effective demagnetization tensor is a sum of two tensors – due to crystallographic anisotropy and, if there is a ferrite sample of finite size, due to demagnetization form factors. If there is an unbounded ferrite medium with significant crystallographic anisotropy, then to find the components of its magnetic susceptibility tensor (in this case, this would be the regular susceptibility tensor $\vec{\chi}$, not $\vec{\chi}^{ext}$, which does not include form factors), the following substitution should be done:

$$\begin{aligned} &N_{ij}(\text{form}) \to N_{ef\,ij}(\text{crystal}) \text{ for } i, j = 1,2; \\ &N_{33}(\text{form}) \to N_{ef\,33}(\text{crystal}). \end{aligned} \tag{7.107}$$

For a ferrite uniaxial monocrystal, the anisotropy energy is expressed as:

$$E_A = K_1 \sin^2\theta + K_2 \sin^4\theta, \tag{7.108}$$

and the effective (internal) magnetic field is found from the condition:

$$H_{ef} = -\frac{\partial E_A}{\partial \boldsymbol{M}}, \tag{7.109}$$

then the components of the effective crystal anisotropy demagnetization tensor are obtained by projecting the effective anisotropy field from the system of coordinates $(x'y'z')$, whose z' axis coincides with the easy axis of the crystal to the (xyz) coordinates associated with the direction of equilibrium magnetization $\boldsymbol{M}_0$ along the z-axis, as in [15]:

$$\begin{aligned} N_{ef11} &= N_{ef12} = 0 \\ N_{ef22} &= -\frac{2K_1}{M_0^2}\sin^2\theta_0 + \frac{4K_2}{M_0^2}\left(2\sin^2\theta_0 - 3\sin^4\theta_0\right), \\ N_{ef33} &= -\frac{2K_1}{M_0^2}\cos^2\theta_0 - \frac{4K_2}{M_0^2}\left(\cos^2\theta_0 - \cos^4\theta_0\right) \end{aligned} \tag{7.110}$$

Then the components of the external tensor [see Equations (7.71)–(7.73)] for a uniaxial monocrystalline sample (ellipsoid) are found by substitution of Equation (7.110) as suggested in Equation (7.106).

Using the equations similar to Equations (7.56)–(7.57), but with the new demagnetization factors, the frequency of free oscillations or the resonance frequency can be obtained. In the case of a spherical ferrite sample, form factors can be omitted (they are all equal to 1/3):

$$\left(\frac{\omega}{\mu_0\gamma}\right)^2 = (H_{0z} + (N_{ef11} - N_{ef33})M_0)(H_{0z} + (N_{ef22} - N_{ef33})M_0) - N_{12}^2 M_0^2, \tag{7.111}$$

Then substituting Equation (7.110) in Equation (7.111) results in:

$$\left(\frac{\omega}{\mu_0\gamma}\right)^2 = (H_{0z} + 2H_{A1}\cos^2\theta_0 - H_{A2}\sin^2 2\theta_0)(H_{0z} + 2H_{A1}\cos 2\theta_0 + 4H_{A2}\sin^2\theta_0(1 + 2\cos 2\theta_0)), \tag{7.112}$$

where two anisotropy field components are related to the anisotropy constants:

$$H_{A1} = \frac{K_1}{M_0};\ H_{A2} = \frac{K_2}{M_0}. \tag{7.113}$$

To find the resonance frequency, one needs to know the angle θ_0. But we know only θ_H, which is the angle between the direction of the bias field and the z' axis of the crystal. The needed angle θ_0 is determined by minimizing the internal energy of the crystal [see equation (b) in Equation (7.97)]. This is a transcendental equation, which is easily solved numerically or graphically.

It can be seen that for $K_1 >> 0$ (*easy axis*), if $\theta_H = 0$, then $\theta_0 = 0$.

For small angles $\theta_H << \frac{\pi}{2}$:

$$\theta_0 \approx \frac{\theta_H}{1 + \dfrac{2H_{A1}}{H_0}}. \tag{7.114}$$

For $\theta_H = \frac{\pi}{2}$, there are two cases:

$$\theta_0 = \arcsin\left(\frac{H_0}{2H_{A1}}\right) \text{for } H_0 \le 2H_{A1} \text{ and } \theta_0 = \frac{\pi}{2} \text{for } H_0 > 2H_{A1} \tag{7.115}$$

Note that these relations are valid for uniform magnetization only; if the sample sizes are larger than some critical values, the existence of domain structure may be preferable for some values and orientation of bias field.

Also, if the terms with K_2 are omitted, then Equation (7.112) for the frequency of free oscillations (and the corresponding resonance frequency) is simplified to:

$$\left(\frac{\omega}{\mu_0\gamma}\right)^2 = (H_0\cos(\theta_0-\theta_H)+2H_{A1}\cos^2\theta_0)(H_0\cos(\theta_0-\theta_H)+2H_{A1}\cos 2\theta_0), \tag{7.116}$$

For $K_1 >> 0$ (easy axis), if $\theta_H = \theta_0 = 0$, the well-known Kittel formula is obtained for a ferrite with an anisotropy field [36]:

$$\omega = \mu_0\gamma(H_0 + 2H_{A1}). \tag{7.117}$$

or $\theta_0 = \frac{\pi}{2}$ (field and magnetization moment both lie in the *hard plane*) and $H_0 > 2H_{A1}$, then:

$$\omega = \mu_0\gamma\sqrt{(H_0 - 2H_{A1})H_0}. \tag{7.118}$$

For $\theta_0 \neq 0$; $\theta_H \neq 0$:

$$\left(\frac{\omega}{\mu_0\gamma}\right)^2 = H_0\frac{\sin\theta_H}{\sin\theta_0}(H_0\cos(\theta_0-\theta_H)+2H_{A1}\cos 2\theta_0). \tag{7.119}$$

Ferrite spinels and garnets are used as low-anisotropy gyromagnetic materials. For example, YIG has $|H_{A1}| << \frac{\omega}{\mu_0\gamma}$, and $K_1 < 0$; this is a low-anisotropy case. Hexagonal ferrites exhibit strong crystallographic magnetic anisotropy.

Figure 7.15 shows the Weiss–Pollak curves for determining the resonance frequency in uniformly magnetized uniaxial ($K_1 > 0$) monocrystalline hexagonal ferrites for various orientations $\theta = \theta_H$ of the ferrite axis with respect to the external bias field [37, 38]. It is seen that in the right-hand side of Figure 7.15, where the bias magnetic field H_0 is positive, the curves at angles $\theta > 75°$ are not monotonic. There is a hysteresis in achieving the resonance frequency after some values of bias fields. If there is a small negative bias magnetic field H_0, its magnitude is insufficient to re-magnetize the particle (previously magnetized in the positive H_0 field, and the FMR conditions remain the same as for the positive Ho field (right-hand side of Figure 7.15). This results in the curves in the left-hand side of Figure 7.15, where the dashed lines indicate the magnetization turnover. After these "jumps", the further trend of the curves as $|H_0|$ increases is practically mirror to the trends of the corresponding curves in the right-hand part of the figure. From these curves, it is seen that the resonance

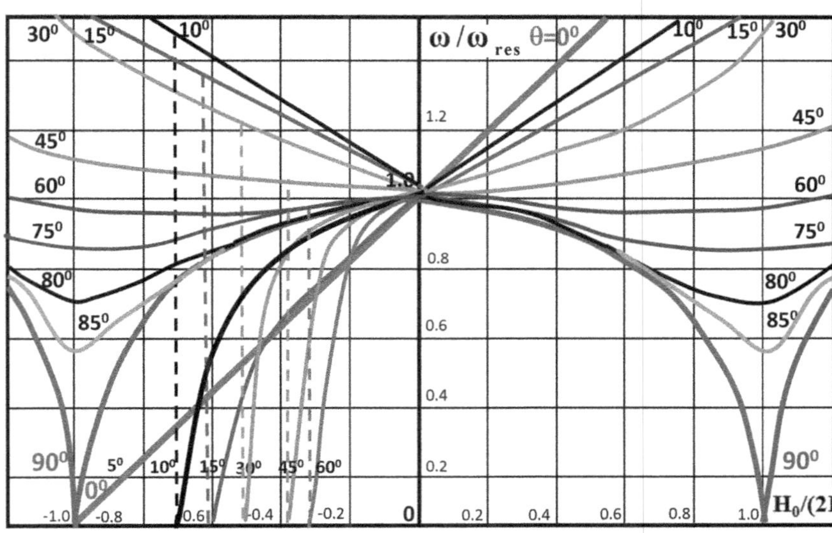

Figure 7.15 Weiss–Pollak curves to determine resonance frequency of uniaxial hexagonal ferrites [37, 38]. FMR resonance conditions in a single-domain particle of a magneto-uniaxial ferrite. Angle θ in these plots is the angle between the easy magnetization axis of the ferrite and the external bias field.

frequency of a hexagonal ferrite may be varied by changing the orientation of its easy axis with respect to the bias magnetic field. Variation of the resonance frequency of a monocrystalline hexagonal ferrite is possible by using a piezoelectric element, which will control the orientation of the ferrite sample.

The components of the nine-component magnetic susceptibility tensor for a magneto-uniaxial monocrystalline hexagonal ferrite are obtained in [35, 39].

7.2.9 FMR in a Polycrystalline Ferrite Medium

Though FMR, or gyromagnetic resonance, is widely used in microwave engineering, polycrystalline ferrites remain the most widely used magnetic materials at RF and microwave frequencies. Polycrystalline ferrite is an agglomerate of tiny crystallites, or grains, of the diameter from about 1 to 10 μm. Their shapes may be arbitrary, and they have pores between them (up to 20%) of the total volume. Crystallites are typically single-domain particles. Their crystallographic axes are usually randomly distributed; therefore, a polycrystalline material is isotropic as a rule, if there is no intended texturization by applying an external magnetic field or due to spontaneous alignment of crystallite axes as it may be possible in densely packed polycrystalline hexagonal ferrites. The *texture axis* is the specific direction along which the majority of crystallites are oriented. The gyromagnetic properties of a texturized ferrite are close to those of a monocrystal; however, the grain structure and pores may significantly affect the FMR in polycrystalline structures as compared to the monocrystals of the same chemical composition. The FMR line widths of polycrystalline ferrites are usually significantly wider than those of the monocrystals.

In the strong magnetic fields, when there is magnetic saturation of the material, the value $\chi_{||} = \chi_{33}$ tends to zero (and this is also related to the comparatively low relaxation term for precession of magnetization vector). Therefore, the only two components of the magnetic susceptibility tensor are sufficient to characterize polycrystalline ferrites, $\chi = \chi_{11} = \chi_{22}$ and $\chi_a = \chi_{12} = -\chi_{21}$:

$$\vec{\chi} = \begin{vmatrix} \chi & j\chi_a & 0 \\ -j\chi_a & \chi & 0 \\ 0 & 0 & \chi_{||} \end{vmatrix}. \tag{7.120}$$

Van Vleck was the first who proposed to treat a polycrystalline ferrite as a set of noninteracting independent grains, in which the FMR is determined by only orientation of their magnetic crystal axes [40]. Shapes of grains, as well as interactions between magnetic oscillations of neighboring grains are neglected in this model. These assumptions are valid for magnetic monocrystal spheres placed sparsely in a nonmagnetic medium, as well as for any grains with comparatively high anisotropy field $|H_A| \gg M_0$, as this takes place in hexaferrites.

To obtain the susceptibility components of a polycrystalline ferrite, one should consider susceptibility tensors for each grain individually and then average them using the distribution of orientations. However, if losses in grains are finite, the problem becomes cumbersome. Therefore, Schlömann assumed zero loss and calculated only $\chi''(\omega, H_0)$ and $\chi_a''(\omega, H_0)$ with infinite peaks at resonances for polycrystals of cubic structure [41] and for hexagonal ferrites [42]; the resultant frequency dependences are significantly asymmetric and have two peaks due to crystalline anisotropy. The finite FMR line widths of grains result in smoothened peaks. The resonance line width of the polycrystalline ferrite just due to the distribution of crystallites orientations is approximately:

$$\Delta H_{pf\,c} \approx \frac{2|K_1|}{M_0}, \tag{7.121}$$

and the resonance peak shift to the lower frequencies as compared to the monocrystal is:

$$\left|H_{pf}^{max} - H_{mf}^{max}\right| \approx \frac{|K_1|}{2M_0}. \tag{7.122}$$

Porosity of a ferrite sample may affect the resonance line width much more than the crystal anisotropy in the assumption of independent grains. Very rough estimation assumes that the line width is on the order of the averaged over the sample demagnetizing field, which is calculated as:

$$\Delta H_{pf\,p} \approx p_{\text{pore}} M_0, \tag{7.123}$$

where $p_{pore} = \frac{V_{pore}}{V_{total}}$ is the porosity factor.

If the polycrystalline ferrite is dense and low anisotropic (like spinels and many garnets with $|H_A| \ll M_0$) so that the grains cannot be considered as independent, there is interaction between the grains due to the demagnetization fields, which causes the "dipole shrinking" of the resonance curve. The resultant resonance width of the polycrystalline ferrite is calculated as:

$$\Delta H_{pf\,c1} \approx k_{ds} \Delta H_{pfc}, \tag{7.124}$$

where the factor $k_{ds} < 1$. This factor can be estimated as:

$$k_{ds} \approx \frac{|H_A|}{M_0}. \tag{7.125}$$

Let us consider polycrystalline ferrite with strong anisotropy, like hexaferrite material. The magnetic susceptibility tensor can be calculated through the statistical distribution:

$$\begin{aligned}&\langle \vec{\chi}(\omega, H_0) \rangle \\ &= \frac{1}{2\pi} \int_{(H_{Ac}=0)}^{\infty} \int_{\theta_c=0}^{\pi/2} \int_{\varphi_c=0}^{2\pi} p_c \vec{\chi}_c(\omega, H_0, H_{Ac}, \theta_c, \varphi_c) P_H(H_{Ac}) P_\theta(cos\theta_c) P_\varphi(cos\varphi_c) sin\theta_c dH_{Ac} d\theta_c d\varphi_c,\end{aligned} \tag{7.126}$$

where $\rho_c = \frac{V_c}{V_{total}}$ is the volume fraction of crystallites (grains) in the total volume of ferrite; in the general case, it is in the integrand but can be pulled out of integral signs if the medium is homogeneous. Assuming that the distributions for H_{Ac}, φ_c, and θ_c are statistically independent, the resultant probability comprises a product of partial probabilities. $P_H(H_{Ac})$ is the distribution function for anisotropy fields of crystallites; $P_\theta(\cos\theta_c)$ and $P_\varphi(\cos\varphi_c)$ are the distribution functions for the angles of separate crystallites with respect to the texture axis position – this way, the cone of magnetic dipole orientations can be taken into account. In Equation (7.126), the tensor is used – this means that this formula should be applied for any nonzero component of the tensor $\vec{\chi}_c$.

This distribution function for angles θ_c can be represented using the normal law as in [43]:

$$\begin{aligned}&P_\theta(\cos\theta_c) = q^+(t)\cdot(1-\nu_t) + q^-(u)\cdot\nu_t;\\ &q^+(t) = \frac{1}{\sqrt{2\pi}\sigma_t^+} \exp\left(-\frac{t-t_0^+}{2(\sigma_t^+)^2}\right);\\ &q^-(t) = \frac{1}{\sqrt{2\pi}\sigma_t^-} \exp\left(-\frac{t-t_0^-}{2(\sigma_t^-)^2}\right);\\ &t = |\cos\theta_c|;\ \ \nu_t = \{0; 0.5; 1\}.\end{aligned} \tag{7.127}$$

For the coefficient ν_t=0, the upward direction is dominating; for $\nu_t = 1$, the downward direction is dominating, and for $\nu_t = 0.5$, the position of the vector in the first and fourth quarters are equal (see Figure 7.16a). Standard deviations $\sigma_t^\pm$ may be identical, but may be different, depending on the symmetry of the distribution.

In [43], a similar distribution was introduced for the angle φ_c with the difference that this angle is distributed from 0 to 2π:

$$\begin{aligned}&P_\varphi(\varphi_c) = q^+(u)\cdot(1-\nu_u) + q^-(u)\cdot\nu_u;\\ &q^+(u) = \frac{1}{\sqrt{2\pi}\sigma_u^+} \exp\left(-\frac{u-u_0^+}{2(\sigma_u^+)^2}\right);\\ &q^-(u) = \frac{1}{\sqrt{2\pi}\sigma_u^-} \exp\left(-\frac{u-u_0^-}{2(\sigma_u^-)^2}\right);\\ &u = \cos\varphi_c;\ \ \nu_u = \{0; 0.5; 1\}.\end{aligned} \tag{7.128}$$

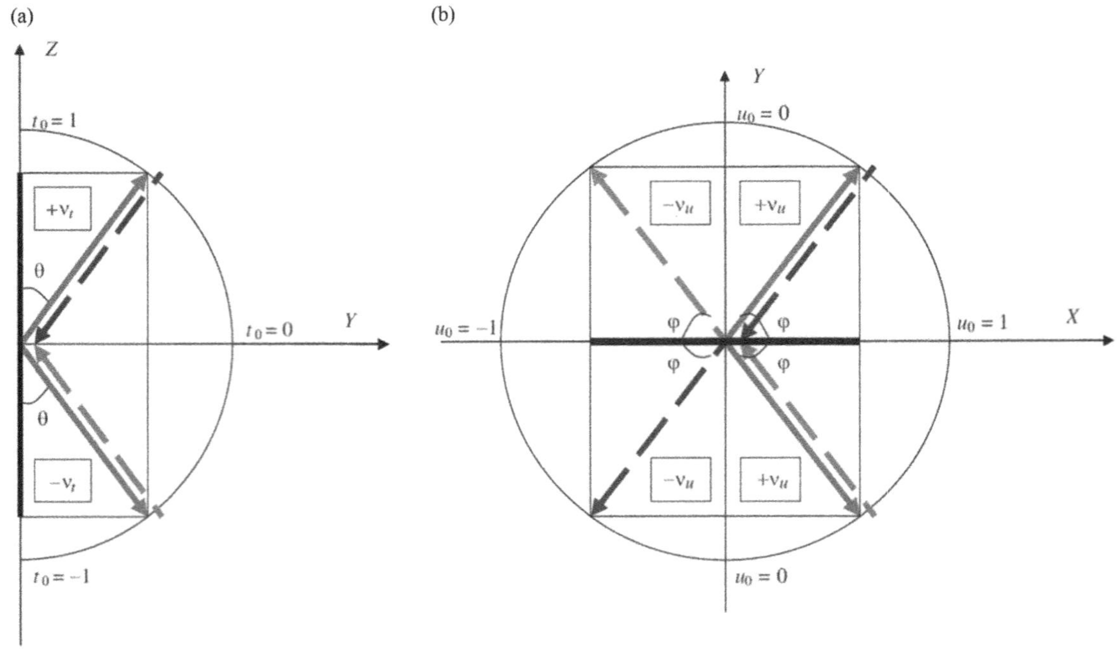

Figure 7.16 Orientation of magnetic dipole vectors (a) with respect to angle θ_c and (b) with respect to angle φ_c.

For the coefficient v_u=0, the direction in the upper half-plane is dominating upward is dominating; for v_u=1, the direction, and for v_u=0.5, the position of the vector in the upper and lower half-planes are equal (see Figure 7.16b). Standard deviations $\sigma_u^{\pm}$ may be identical, but may be different, depending on the symmetry of the distribution.

Obviously, if all the crystallites have identical crystal anisotropy fields H_A, then $F(H_{Ac}) = \delta(H_{Ac} - H_A)$ is the delta-function, and H_A becomes a constant factor in Equation (7.126). Then the only distribution with respect to angles θ_c and φ_c can be considered. If there is a uniform distribution with the same probability for any angle (isotropic case), then $P_\theta(\cos\theta_c)P_\varphi(\cos\varphi_c) = 1$, and only the distribution of anisotropy field magnitudes is included in Equation (7.126).

If all the crystallites are perfectly aligned along some texture axes, then $P(\cos\theta_c) = 2\delta(\cos\theta_c - 1)$, Equation (7.126) becomes:

$$\langle \ddot{\chi} \rangle = \rho_c \ddot{\chi}_c. \tag{7.129}$$

This means that the resonance curves of this ferrite material are of the same as those of a single crystal, just the magnitudes have the factor ϱ. In particular, the resonance field for a single-crystal hexaferrite with $K_1 > 0$ is determined from Equation (7.115) as:

$$H_0 = H_\omega - 2H_{A1}, \tag{7.130}$$

where $H_\omega = \dfrac{\omega}{\mu_0 \gamma}$.

The imaginary susceptibility at the resonance is:

$$\chi''_{res} = \rho_c \chi''_{res\,c} = \frac{M_0}{\Delta H}, \tag{7.131}$$

where $\Delta H = \Delta H_{pf\,c} = 2\alpha H_\omega$ is the width of the resonance line determined by the intrinsic loss in a crystallite ($\alpha \ll 1$ herein is the LLG relaxation factor), and $M_0 = \rho_c M_{sc}$ is the saturation magnetization of the of the entire polycrystalline ferrite comprises the values for individual crystallites.

The resonance line width of a high-quality monocrystal hexagonal ferrite is on the order of 10–20 Oe. The intrinsic resonance line width of crystallites may be slightly higher, but on the same order of magnitude. However, the actual resonance line width of the polycrystalline ferrite is much wider reaching ~ 1 kOe. This means that the intrinsic resonance line of crystallites in reality does not determine the entire line width. This agrees with Schlömann's assumption [42] so that the resonance absorption by each crystallite can be represented through the delta-function:

$$\chi''_{res\,c} = S_c \delta(H_0 - H_{res\,c}), \tag{7.132}$$

where $S_c = \int_{-\infty}^{\infty} \chi_c'' dH_0$ is the area under the curve of loss for any crystallite.

As was mentioned before, the main contribution to the resonance line width is coming from the two factors: the imperfect alignment of crystallites axes and due to the distribution of values of intrinsic crystal anisotropy fields.

The case of the statistically isotropic ferrite comprises uniaxial crystallites was considered in [44], and it was shown that the resonance line width near the frequency of natural FMR (NFMR), that is, without external bias magnetic field, can be calculated as

$$\Delta H = |H_{res}| \sqrt{\frac{H_\omega^2 + H_A^2}{H_\omega^2 - H_A^2}}, \tag{7.133}$$

where $|H_{\text{res}}| = |H_\omega - 2H_{\text{A}}| \ll 2H_{\text{A}}$.

From Equation (7.133), it follows that the width ΔH depends on frequency. Near the NFMR, the texturization should not contribute to the resonance line width. However, experiments show that real isotropic polycrystalline hexaferrites with uniaxial grains do not have significant dependence of their resonance line width on frequency; the width near the NFMR is practically the same as far from it. This means that though there is a difference in the resonance line width for texturized and not texturized ferrites, it is not that huge, which has been shown experimentally [45]. Therefore, the only factor that may significantly affect the resonance line width is the distribution of H_A values of crystallites, i.e. $F(H_{Ac})$.

The following approximate expressions for the components of the tensor are obtained in [46]:

$$\chi = A_1 \frac{\chi^+ + \chi^-}{2}; \ \chi_a = A_2 \frac{\chi^+ - \chi^-}{2}; \ \chi_{||} = A_3 \frac{\chi^+ + \chi^-}{2}; \tag{7.134}$$

$$\chi^\pm = \frac{M_0}{2H_A + kH_0 \mp H_\omega + j\Delta H_c}, \tag{7.135}$$

where $M_0 = \rho_c M_{sc}$; ΔH_c is the intrinsic resonance line width of grains (may be considered identical for all the grains); H_0 is the applied bias field (if this is not NFMR); $H_\omega = \dfrac{\omega}{\mu_0 \gamma}$ is the normalized by $(\mu_0\gamma)$ angular frequency. These components are obtained using the transition to the susceptibility tensor components $\chi^\pm$ corresponding to two circular polarizations instead of linear polarizations; this is done for convenience of the diagonal form of the susceptibility tensor [13, chapter 4]. The coefficients A_1, A_2, and A_3 in Equation (7.134) depend on the texture quality and H_0, and k in Equation (7.135) depends on the rate of alignment in the texture. They are:

$$\begin{aligned} A_1 &= \frac{1+x^2}{2} + \frac{H_0}{2H_A + H_0}(1 - \langle x^2 \rangle); \\ A_2 &= \langle x \rangle + \frac{H_0}{2H_A + H_0}(1 - \langle x^2 \rangle); \\ A_3 &= \left(\frac{2H_A}{2H_A + H_0} \right)^2 (1 - \langle x^2 \rangle); \\ k &= \frac{1}{2 - \langle x \rangle}, \end{aligned} \tag{7.136}$$

with $\langle x \rangle$ and $\langle x^2 \rangle$ being the corresponding moments of probability:

$$\langle x \rangle = 2\int_0^1 xp(x)dx; \ \langle x^2 \rangle = 2\int_0^1 x^2 p(x)dx, \tag{7.137}$$

where $p(x)$ is any distribution function adopted to describe statistical spread of the anisotropy fields of crystallites, and the argument x is proportional to these anisotropy fields.

Experimental studies show that anisotropy field values of crystallites are actually distributed according to the normal law [47, 48]:

$$P_H(H_{Ac}) = p(x) = \frac{1}{\sigma_H\sqrt{2\pi}} exp\left(-\frac{x-1}{2H_A\sigma_H^2}\right), \tag{7.138}$$

where $=\frac{H_{Ac}}{H_A}$; H_{Ac} is the anisotropy field of crystallites, H_A is the average anisotropy field of the sample, and σ_H is the dispersion. The coefficients defined in (7.136) can be calculated numerically based on the assumed distribution law.

In [46], it is proposed to use Cauchy distribution, as it is close to the normal distribution, but simpler for integration:

$$P_H(H_{Ac}) = p(x) = \frac{1}{\pi}\times\frac{\Delta}{(x-1)^2+\Delta^2}, \tag{7.139}$$

where Δ is the half width of the distribution law at the level of 0.5 of maximum; $\Delta \ll 1$ if the range of possible H_{Ac} values is narrow.

Numerical modeling using statistical analysis analogous to the presented earlier [49, 50] shows that the major factors responsible for the comparatively wide resonance lines of polycrystalline ferrites with uniaxial magnetic anisotropy are the spread of intrinsic anisotropy fields of crystallites H_{Ac} and texture quality (see Figure 7.17). To reduce the resonance line

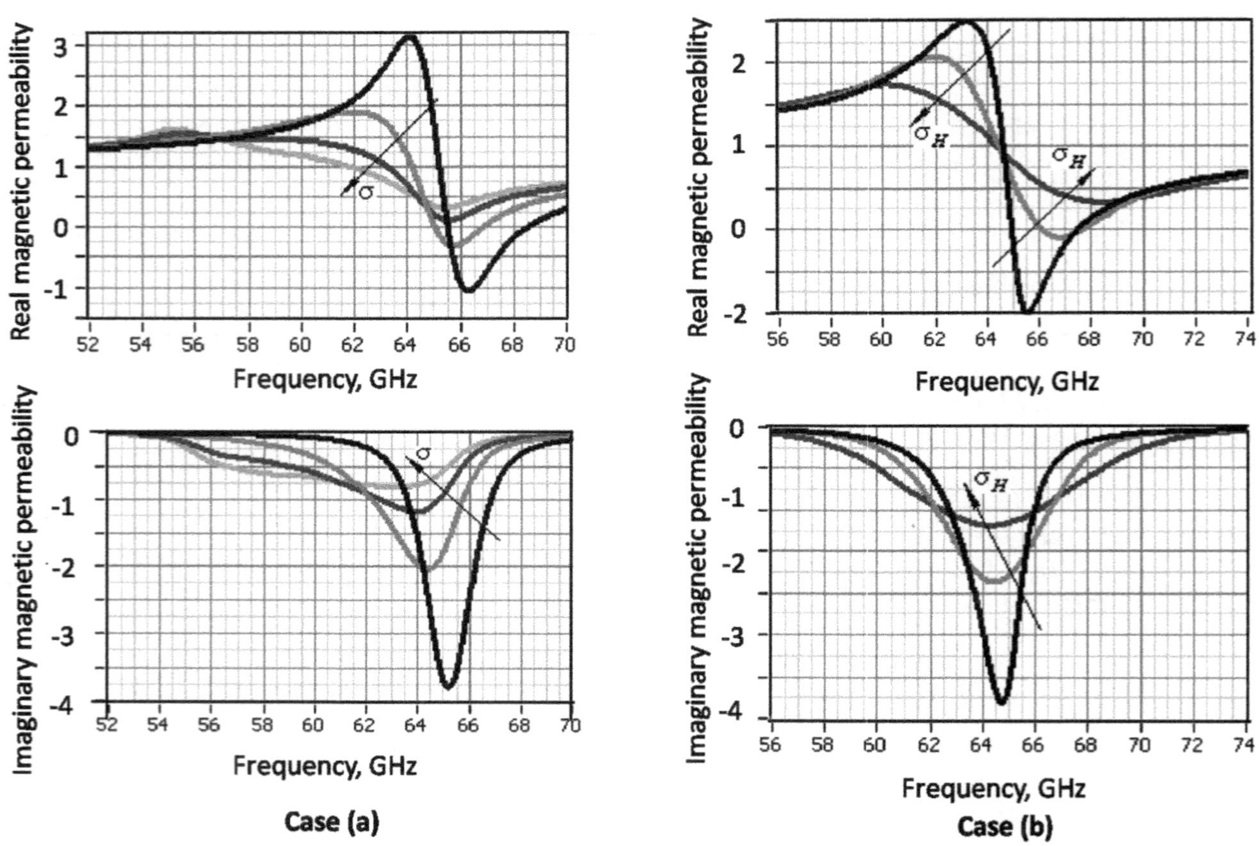

Figure 7.17 Real and imaginary parts of complex permeability in the form $\mu_r = \mu_r' + j\mu_r''$ for polycrystalline hexaferrite with statistical distribution of texture angles, case (a), and with scatter of crystallographic anisotropy fields, case (b), NFMR frequency f_e = 56 GHz; bias field H_0 = 6.27 kOe in both cases; dispersion for the angle is $\sigma = 0.01, 0.48, 1.0, 70.0$; dispersion for the fields of anisotropy $\sigma_H = 0.01, 0.5, 1\text{kOe}$.

width of polycrystalline ferrites, and hence the spread of H_{Ac}, the homogeneity of the ferrite grain structure is needed. Usually, it is easier to achieve more homogeneous grain structure with planar hexagonal ferrites rather than with uniaxial. This explains why the resonance line width of polycrystalline ferrites with planar anisotropy is lower by an order of magnitude than the one of the uniaxial ferrites.

7.2.10 Effect of Domains on FMR in Ferrites

Majority of gyromagnetic devices, which use the phenomenon of FMR in ferrites, employ uniform magnetostatic oscillations in FRs. These are typically frequency-selective devices, and for their operation, the minimum possible resonance line width is needed. This takes place in the uniformly magnetized ferrites, and the domain structure in ferrites is undesirable. When a homogeneous bias magnetic field sufficient for uniform magnetization to saturation is applied, an FR becomes a single domain, and uniform precession of its magnetic moment is possible. When the frequency of precession coincides with the frequency of the RF/microwave forced oscillations, FMR will take place.

If the bias field is insufficient and significantly inhomogeneous, and if a ferrite sample is comparatively large, then there could be domains in the ferrite sample. Magnetic moments in different parts (domains) of the sample will have different orientations with respect to the crystallographic axes and the bias field (if the latter is applied), and therefore, the magnetic resonance conditions will be different in different parts of the sample. This will result in a wider resonance line than needed for a device operation.

For this reason, FRs are always employed at frequencies higher than their NFMR determined by Polder–Smit formula [51]:

$$\omega > \omega_{\text{NFMR max}} = \mu_0\gamma(2H_{A1} + M_s). \tag{7.140}$$

To reach this, some bias field should be applied:

$$H_0 > \frac{\omega - \omega_{\text{NFMR max}}}{\mu_0\gamma}. \tag{7.141}$$

Therefore, to extend the frequency range of high-Q FMR application to the lower frequencies, one needs ferrites with comparatively low values M_s and crystallographic anisotropy fields. This can be done by doping ferrites with such ions that the values of M_s and $2H_A$ decrease. For example, monocrystal YIG ferrites used for spectrum analysis have NFMR at about 5 GHz. To be able to use ferrite garnets at the lower frequencies, low-saturation Ca–Bi–V ferrites have been synthesized; their NFMR is at about 100–200 MHz. BaM hexaferrites have $2H_A \sim 17$ kOe, and $4\pi M_s$=3 kG, which allows to use their monocrystals in frequency-selective gyromagnetic applications based on FMR at frequencies higher than 56 GHz. For applications at the lower frequencies, the ferrites with the lower anisotropy field are needed, and Sc- and TiCo-doped uniaxial BaM hexaferrites have much lower anisotropy field ($2H_A$), which shifts NFMR to frequencies as low as a few gigahertzs [12, 45].

In the previous sections, when relaxation in ferrites at the FMR was considered, it was assumed that the ferrites are magnetized to saturation. Herein, we will consider (mainly qualitatively) how the domain structure of unsaturated (or incompletely saturated) ferrites affects the FMR resonance frequency and relaxation.

The total energy in a ferrite monocrystal not affected by a bias field consists of a few parts, including exchange energy, crystallographic anisotropy energy, elastic and magnetoelastic energy, and magnetic energy related to demagnetization fields. There is no Zeeman term, because the bias field is absent. Equilibrium of magnetic moment (magnetization) in a sample corresponds to the minimum of all these energy contributions. If a sample substantially large, the magnetization distribution is not necessarily homogeneous; if there are domains with antiparallel magnetic moments, this will be the more energy saving configuration. As indicated in [15], for a ferrite plate the magnetic energy per surface unit is on the order of:

$$E_m \sim M_0^2 d, \tag{7.142}$$

where d is the domain thickness, and M_0 is the domain magnetization. As d increases, domains merge together and E_m reduces. However, other parts of energy may increase. As a result, an equilibrium domain structure is due to the minimum of the total energy.

There are walls between the separate domains. The domain wall (DW) region is a transition layer where magnetization $\boldsymbol{M}_i$ of the ith domain transforms to $\boldsymbol{M}_{i+1}$ of the neighboring domain. Anisotropy energy would be minimal if there is a jump from $\boldsymbol{M}_i$ to $\boldsymbol{M}_{i+1}$, but exchange energy would increase. Therefore, this transition is gradual. If the transition is in the plane of the boundary, no additional demagnetization fields appear due to the normal to the boundary component of magnetization; this case is called *Bloch domain wall*. In thin films, this may be not the optimum transition from energy point of view, as magnetic vectors may leave the film plane. Then another type of a wall appears – this is *Néel domain wall*, in which rotation of magnetization is in the plane normal to the boundary. Landau and Lifshitz were the first to describe the behavior of the angle of magnetic moment rotation θ at the DW dynamics [8]. The similar problem was solved for a magneto-uniaxial ferrite with the easy axis [28]. The law of the magnetic wall rotation is:

$$\cos\theta = -\tanh\left(x \times \sqrt{K_1/(M_0^2 q)}\right) = -\tanh(\pi x / w), \tag{7.143}$$

where $K_1 = H_{A1} M_s$ is the first anisotropy constant of the crystal, and q is the lattice constant, or a constant of nonuniform exchange (its units are [length]^2). Assuming that the boundary between two domains is in plane (yz); x is the coordinate along which rotation of $\boldsymbol{M}$ takes place; and the angle θ is counted from the axis z. The value:

$$w = \pi M_0 \sqrt{q/K_1}, \tag{7.144}$$

is the DW thickness (units are [length]). If Equation (7.143) is satisfied, the exchange and anisotropy energies are equal and the DW energy per unit surface area is:

$$E_w = 2\mu_0 w K_1. \tag{7.145}$$

If there is a ferrite plate, according to Landau and Lifshitz [8], domains are formed inside the plate as layers parallel to the easy axes but have antiparallel magnetic moments. At the surfaces of the sample, where it borders with surrounding environment (e.g. air), closing prismatic domains appear due to demagnetization. If there are inhomogeneities inside a sample (e.g. bubbles and pores), domains also appear on the interfaces with them, and the domains structure may be very complex, e.g. serpentine or labyrinth. The domain size d depends on the thickness of the ferrite plate D as $d \sim \sqrt{D}$, and if the sample is small enough with $D < D_{min} \approx E_w / M_0^2$ in all three dimensions, the sample will be single domain. Indeed, for a YIG, the $D_{\min} = 0.1\,\mu\text{m}$ [52].

The processes of magnetization are considered in many textbooks [53, 54]. As soon as the bias field $\boldsymbol{H}_0$ is applied, a Zeeman term appears in the magnetic energy of the sample. Then the average equilibrium magnetization will be directed close to the direction of the bias field. There are two mechanisms for an individual magnetic moment $\boldsymbol{M}_i$ of a domain to get in this new equilibrium state. The first is shift, or *motion of magnetic walls*, which results in the growth of the domains whose magnetization direction is close to $\boldsymbol{H}_0$ (or at least forms the sharp angle). The second is *rotational dynamics*, when the vectors $\boldsymbol{M}_i$ turn inside a domain, as is illustrated in Figure 7.18. It is important that the magnetic wall motion dominates at the lower bias fields and finishes when the main domain structure disappears. The rotation processes also start at the low field but continue after the domain structure disappears due to wall motion. Rotation stops when the overall magnetization moment $\boldsymbol{M}$ is parallel to the bias field $\boldsymbol{H}_0$.

The magnetization process may be complicated, e.g. in the films with high uniaxial crystallographic anisotropy with the easy axis perpendicular to the film plane, if the bias field is applied in the same direction, isolated cylindrical domains may arise in place of the planar layers in some intervals of the applied bias field [55].

When AC magnetic field is applied to a ferrite sample with domain structure, then first, at the low frequencies, remagnetization of the sample is quasistatic, i.e. magnetization at each moment depends on the magnitude of the AC field, as well as the sample's magnetic previous state. This is characterized at the low field magnitudes by the static (or initial) magnetic susceptibility and at the high field intensity by the static hysteresis loop. Magnetic loss, in this case, is determined by the area of this loop and frequency.

As the frequency of AC field increases, the wall motion will be the first to experience "friction," and then the rotational dynamics cannot follow fast alternations of magnetic field because of the ferromagnetic (gyromagnetic) resonance effect. The three types of frequency dispersion: resonance type due to the DW vibration, relaxation type due to the DW rotation, and resonance type due to gyromagnetic (ferromagnetic) resonance are indicated in Figure 7.19.

The wall motion dynamics is based on the rotation of the magnetic moment; therefore, its theoretical description is based on the MME with dissipation, e.g. using LLG relaxation term. It turns out that in the region of the interface, there is a

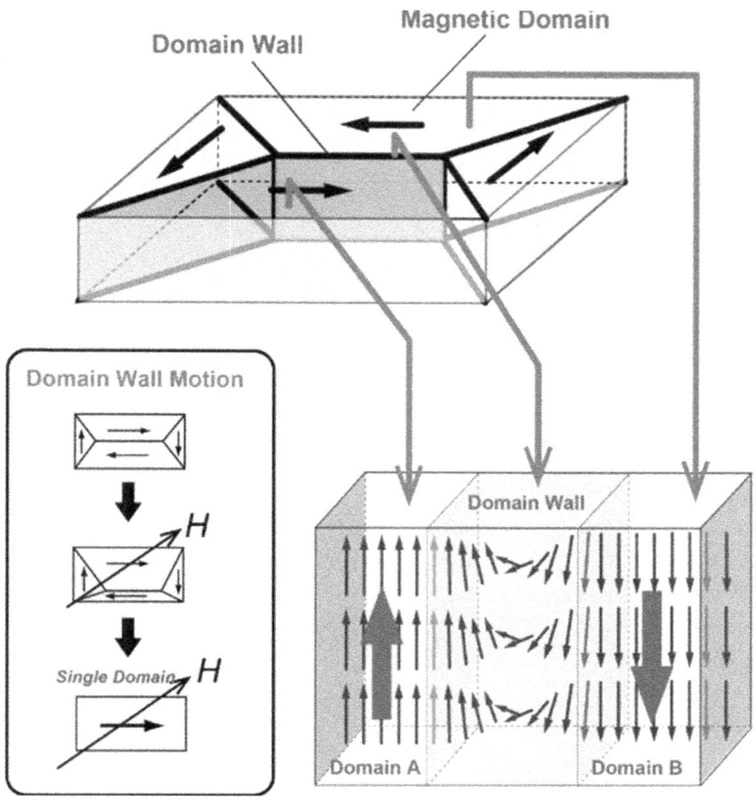

Figure 7.18 Illustration of DW dynamics: rotation of magnetization moment in the transitional DW region.

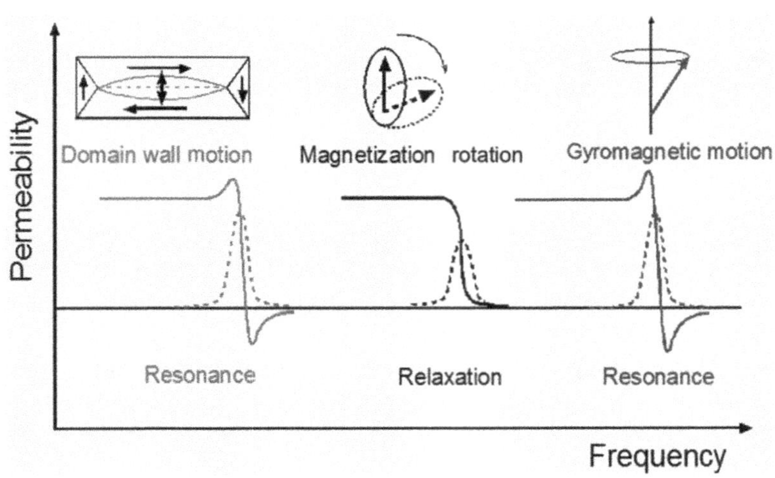

Figure 7.19 Illustration of frequency behavior of permeability in a ferrite as frequency increases.

solution, of the DW dynamics which depends on the coordinate perpendicular to the interface plane and on time as $(x - vt)$, where v is the speed of the DW shift. It was shown by Becker [56] that the equation for the wall motion coincides with the equation of the mechanical harmonic oscillator [15, 57]:

$$m_w \frac{d^2x}{dt^2} + \beta_w \frac{dx}{dt} + \xi_w x = \mu_0 M_0 H \tag{7.146}$$

where m_w is the DW mass normalized to the wall surface area (w is the DW thickness) [57]:

$$m_w = \frac{1}{A\gamma^2}\sqrt{|K_1|} \approx \frac{1}{w\gamma^2}, \tag{7.147}$$

where $A = \frac{E_{\text{exch}}}{a_0}$ is the spin exchange energy per lattice constant; the coefficient of the second term is:

$$\beta_w = \frac{m_w}{\tau_w} \tag{7.148}$$

where τ_w is the wall damping time constant; and the coefficient of the third term is:

$$\xi_w = \frac{2M_0^2}{\chi_s d}, \tag{7.149}$$

where d is the domain thickness and χ_s is static magnetic susceptibility. In Equation (7.146), the first term in the left-hand side is the term associated with the kinetic energy of the wall, i.e. the inertial term; the second corresponds to friction at the motion; and the third is elastic force, where ξ_w serves as a coefficient of elasticity. The right-hand side is the force causing the wall motion. Evaluations show [15] that the DW thickness and mass of the DW normalized to the surface area for YIG are $w = 5 \times 10^{-10}$ cm and $m_w = 5 \times 10^{-10}$ g/cm^2, respectively.

As follows from Equation (7.146), the shift along the coordinate x in the static case is:

$$x = M_0 H_0 / \xi_w. \tag{7.150}$$

The average magnetic moment in the small magnetic field H_0 is expressed as:

$$\hat{M} = \frac{2xM_0}{d}, \tag{7.151}$$

and appears as a result of the DW shifting in opposite sides at the distance x.

If the force in the right-hand side of Equation (7.146) is associated with the AC magnetic field $H = he^{j\omega t}$, these will be forced oscillation of the DW back and forth along the coordinate x. The corresponding average AC magnetic moment is $\hat{M} = me^{j\omega t}$.

The oscillations of the DWs will contribute to all the diagonal components of the magnetic susceptibility tensor so that the new tensor will be a superposition:

$$\ddot{\chi} = \ddot{\chi}_{\text{satur}} + \chi_{\parallel} \times \ddot{U}, \tag{7.152}$$

where $\ddot{U}$ is the unit tensor and $\ddot{\chi}_{\text{satur}}$ is the tensor for the saturated ferrites. As is seen from Equation (7.26), the $\ddot{\chi}_{\text{satur}}$ at small loss can be considered as having only four nonzero components; the new tensor $\ddot{\chi}$ in has the component $\chi_{\parallel}$ appearing, as is seen from Equation (7.28). Indeed, this is a nine-component tensor in the general case, and each component would be affected by all the intrinsic resonances of the sample divided by domains.

Then expressing x through $\hat{M}$ from Equation (7.151) and using the expression for the elasticity coefficient ξ_w fron Equation (7.149), one can get the high-frequency component of the magnetic susceptibility tensor due to oscillations of magnetic walls as:

$$\chi_{\parallel} = \frac{\widehat{M}}{h} = \chi_s \left(1 - \frac{\omega^2}{\omega_{0w}^2} + j\frac{1}{Q_w} \times \frac{\omega}{\omega_{0w}}\right)^{-1}. \tag{7.153}$$

The intrinsic resonance frequency for the wall domain resonance is:

$$\omega_{0w} = \sqrt{\xi_w / m_w}, \tag{7.154}$$

and the Q factor is:

$$Q_w = \frac{\sqrt{\xi_w m_w}}{\beta_w}. \tag{7.155}$$

For $Q_w > 1$, $\chi_{||}(\omega)$ has the resonance (Lorentzian) behavior, and for $Q_w < 1$, there is relaxation (Debye) type of $\chi_{||}(\omega)$ characteristics. The line width of the resonance for $Q_w > 1$ is:

$$\Delta\omega_w = \mu_0\gamma\pi\alpha_{LL} = \mu_0\gamma\pi\frac{\alpha}{1+\alpha}. \tag{7.156}$$

Computations for polycrystalline YIG with the grain size of d ~ 1,...,10 μm, and static magnetic susceptibility of χ_s ~10 result in the DW resonance frequency $f_{0w} = \omega_{0w}/(2\pi)$ in the range from 25 to 45 MHz. For polycrystal Co ferrite with χ_s ~ 0.3, the DW resonance frequency f_{0w} is from 200 to 350 MHz. The DW resonance line widths for YIG ferrite $\Delta f_w = \Delta\omega_w/(2\pi)$ ~10 MHz and that of the Co ferrite is $\Delta f_w = \Delta\omega_w/(2\pi)$~25 MHz. This means that $Q_w > 1$ in both cases.

It should be mentioned that rigorous derivation of uniform magnetic oscillations in unsaturated samples with the domain structure is possible for only a limited number of cases with simplifying assumptions on their structure, e.g. simplest regular parallel-plate domain structure independent of bias field. DWs were considered static and infinitesimally thin. Nagamiya [58] calculated these for a disk of a tetragonal crystal. Smit and Beljers [59] calculated magnetic spectrum for magneto-uniaxial crystal, and Artman considered a sphere of a cubic crystal [60]. The analysis of an ellipsoid of a uniaxial ferromagnetic material with easy axis and parallel-plate domain structure is considered in [59]. It is shown that there are two types of oscillations of high-frequency magnetic moment; one type with intrinsic frequency ω_n is excited by the components of magnetic high-frequency field normal to the bias field, and the second type, with ω_p, is excited by the components parallel to the bias field. The bias field in these cases was normal to the crystalline anisotropy axis and domain boundaries. The two frequencies at the NFMR (when bias field is zero) are always finite and distinct ($\omega_p \neq \omega_n$). The oscillations excited by normal field never result in the zero ω_n; there are two distinct regions of the resonance frequency behavior ω_n as a function of the applied bias field: first nonlinearly decreasing, and at some value of bias field, it starts linearly increasing. The ω_p frequency nonlinearly decreases from some maximum value at zero bias field to zero frequency; this is the break point where ω_n reaches minimum and starts increasing.

The analogous results with two types of oscillations, ω_n and ω_p, were obtained for the cubic crystal [60, 61] for crystals with both $K_1 < 0$ and $K_1 > 0$.

The spectral model of gyromagnetic interaction indicating regions of transmission and absorption over the wide frequency range starting from VHF-UHF to sub-millimeter range is given in [57, figure 5]. Also, [57, table I] summarizes the frequency ranges for the main mechanisms determining frequency behavior of ferrites: domain rotation, DW movement (vibration), and gyromagnetic resonances, for cubic (low-anisotropy) and hexagonal (high anisotropy ferrites, both axial and planar).

In particular, as is given in [57], the domain rotation takes place from DC to $\omega_\tau = 1/\tau$, where τ is the effective spin-lattice relaxation time. The DW movement takes place from DC to $\omega_{max} = \sqrt{-\frac{1}{2\tau_w^2} + \omega_w^2}$, where ω_w is the natural undamped wall oscillation frequency and τ_w is the wall damping time constant. The gyromagnetic resonance for cubic crystals is possible in the frequency range from $\omega_{\min} = \mu_0\gamma H_A$ to $\omega_{\max} = \mu_0\gamma(H_A + M_s)$ in multidomain case, and $\omega_{\max} = \mu_0\gamma\sqrt{H_A(H_A + M_s)}$ in the single-domain, planar case. For uniaxial hexagonal ferrites, gyromagnetic resonance exists in the range determined by the same formulas as for cubic crystals, from $\omega_{\min} = \mu_0\gamma H_{A\theta}$ to $\omega_{\max} = \mu_0\gamma(H_{A\theta} + M_s)$; and for the planar hexaferrites, the range is from $\omega_{\min} = \mu_0\gamma\sqrt{H_{A\theta}H_{A\varphi}}$ to $\omega_{\max} = \mu_0\gamma\sqrt{(H_{A\theta} + M_s)(H_{A\varphi} + M_s)}$. In all these limits, H_A represents the effective maximum anisotropy field for the particular physical scenario and is $H_A \approx K_1/M_s$.

Overall, when designing various devices of gyromagnetic electronics operating over microwaves at frequencies close to the FMR, it is usually desirable to magnetize ferrites up to saturation by the bias field with the magnitude around M_S (or $4\pi M_S$ in theGauss unit system). This results in reduction of losses at FMR. Alternatively, ferrites with high intrinsic magnetic crystallographic anisotropy H_A should be used for self-biasing of the ferrite. An example of this is using a hexaferrite for designing circulator for a 4-mm wave band [62].

7.3 Experimental Study of Microwave Spectra of Ferrite Materials

In this section, some methods of the complex permeability and permittivity measurements over RF/microwave range will be presented, and EM properties of the ferrite materials (spinel, garnet, and hexagonal ferrites) will be discussed based on the experimental results of complex permeability and permittivity. The contribution of the DW vibration and magnetization rotation will be considered in the analysis of permeability spectra of ferrites, as well as the gyromagnetic effect. Complex permittivity spectra of ferrite materials will also be presented.

7.3.1 Measurements of Complex Permeability and Permittivity Spectra of Ferrite Materials

The magnetic and electric response of ferrites in the EM field can be explained by the complex permeability and permittivity:

$$\mu = \mu_0(\mu_r' - j\mu_r'') \text{ and } \varepsilon = \varepsilon_0(\varepsilon_r' - j\varepsilon_r'') \tag{7.157}$$

where ε_r and μ_r are the relative permittivity and permeability of the material, while ε_0 and μ_0 are the permittivity and permeability of vacuum, or free space. When the excitation fields (magnetic field $\boldsymbol{H}$ and electric field $\boldsymbol{E}$) have harmonic time variation:

$$\boldsymbol{H} = H_0 e^{j\omega t}\boldsymbol{u}_H,\ \boldsymbol{E} = E_0 e^{j\omega t}\boldsymbol{u}_E, \tag{7.158}$$

where the $\boldsymbol{u}_H$ and $\boldsymbol{u}_E$ are the unit vectors along each field direction, the flux densities $\boldsymbol{B}$ and $\boldsymbol{D}$ can be described by:

$$\boldsymbol{B} = B_0 e^{j(\omega t - \theta_B)}\boldsymbol{u}_H,\ \boldsymbol{D} = D_0 e^{j(\omega t - \theta_D)}\boldsymbol{u}_E. \tag{7.159}$$

Hence, μ_r and ε_r are represented by:

$$\mu_r = \frac{1}{\mu_0}\frac{B_0 e^{j(\omega t - \theta_B)}}{H_0 e^{j\omega t}} = \frac{B_0}{H_0}e^{-j\theta_B} = \mu_{r0}e^{-j\theta_B} \tag{7.160}$$

and

$$\varepsilon_r = \frac{1}{\varepsilon_0}\frac{D_0 e^{j(\omega t - \theta_D)}}{E_0 e^{j\omega t}} = \frac{D_0}{E_0}e^{-j\theta_D} = \varepsilon_{r0}e^{-j\theta_D} \tag{7.161}$$

as the results of the magnetic and electric responses of materials. The values μ_{r0} and ε_{r0} are the amplitude of the relative permeability and permittivity; θ_B and θ_D are the phase difference between magnetic or electric fields and their flux densities. These phase differences are related to the energy dissipation of EM field in the materials by several mechanisms; generally, the EM energy will be converted to the heat energy.

In the permeability and permittivity measurements, a coaxial line is often used. The coaxial line is composed of the inner and outer conductors as shown in Figure 7.20a; EM waves propagate in the TEM mode as shown in Figure 7.20b in the frequency range below the cutoff frequency f_c. The propagation constant γ and the impedance Z of the TEM waves in the coaxial line are given as follows by the use of permeability and permittivity [63]:

$$\gamma = j\omega\sqrt{\varepsilon\mu} = \gamma_0\sqrt{\varepsilon_r\mu_r};\gamma_0 = j\omega\sqrt{\varepsilon_0\mu_0} = j\frac{2\pi}{\lambda_0} \tag{7.162}$$

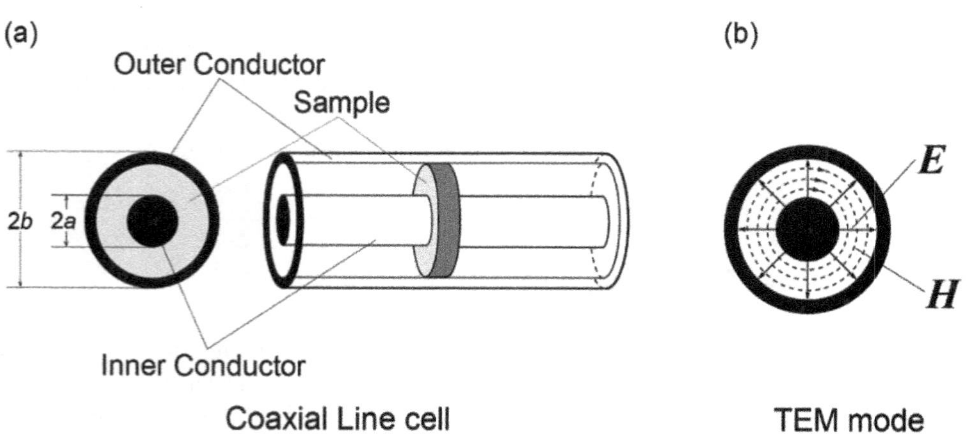

Figure 7.20 (a) Coaxial line cell and (b) the electromagnetic field (TEM mode) in the coaxial line.

$$Z = \sqrt{\frac{\mu}{\varepsilon}} \ln\frac{b}{a} = Z_0\sqrt{\frac{\mu_r}{\varepsilon_r}};\; Z_0 = \sqrt{\frac{\mu_0}{\varepsilon_0}} \ln\frac{b}{a}, \tag{7.163}$$

where γ_0, λ_0, and Z_0 are the propagation constant, wavelength, and impedance of free space, respectively; a and b are the inner and outer radii of the coaxial line.

For the RF/microwave measurements of the complex permeability and permittivity spectra of ferrites, transmission lines such as coaxial lines or waveguides are used; several measurement methods have been developed so far [63]. In this section, "*S*-parameter method" and "open- and short-circuited method" using coaxial lines will be presented.

7.3.1.1 *S*-parameter Method

In the *S*-parameter method, the scattering matrix parameters (*S*-parameters) are measured to determine the complex permeability and permittivity by the calculation of reflection and transmission coefficients of the EM waves through the sample in the coaxial line. A schematic diagram of the measurement setup for the *S*-parameter method using a vector network analyzer (VNA) and a coaxial line is shown in Figure 7.21 [64]. As a device under test (DUT), a coaxial line cell (see Figure 7.20) can be used; transmission and reflection signals from the DUT are measured by the two *S*-parameters, S_{11} and S_{21} [63–65]. As shown in Figure 7.20a, a toroidal sample with the thickness d is loaded in the coaxial line. The reflection $\Gamma = \Gamma' + j\Gamma''$ and transmission coefficients $T = T' + jT''$ of the sample can be obtained from the complex *S*-parameters by:

$$\Gamma = K \pm \sqrt{K^2 - 1} \tag{7.164}$$

and

$$T = e^{-\gamma d} = \frac{S_{11} + S_{21} - \Gamma}{1 - (S_{11} + S_{21})\Gamma}, \tag{7.165}$$

where

$$K = \frac{S_{11}^2 - S_{21}^2 + 1}{2S_{11}}.$$

Then complex relative permittivity ε_r and permeability μ_r of the toroidal sample can be calculated. To do this, let us first introduce a parameter A defined as:

$$A = \varepsilon_r \mu_r = \left(\frac{\gamma}{\gamma_0}\right)^2. \tag{7.166}$$

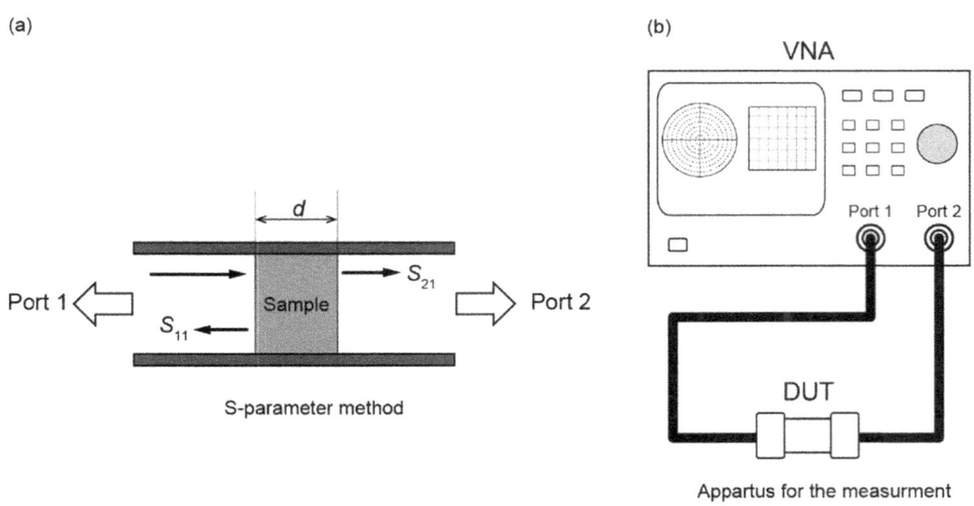

Figure 7.21 *S*-parameter method.

The ratio (γ/γ_0) is also derived from transmission coefficient as:

$$\left(\frac{\gamma}{\gamma_0}\right)^2 = -\left\{\frac{\lambda_0}{2\pi d}\ln\left(\frac{1}{T}\right)\right\}^2. \tag{7.167}$$

As the reflection coefficient Γ is given by the input impedance Z as:

$$\Gamma = \frac{Z/Z_0 - 1}{Z/Z_0 + 1}, \ \frac{Z}{Z_0} = \frac{1+\Gamma}{1-\Gamma} = \sqrt{\frac{\mu_r}{\varepsilon_r}}, \tag{7.168}$$

the parameter B is defined as:

$$B = \frac{\mu_r}{\varepsilon_r} = \left(\frac{1+\Gamma}{1-\Gamma}\right)^2. \tag{7.169}$$

Hence, the values of relative permittivity ε_r and permeability μ_r can be calculated by simultaneously solving the two equations:

$$\varepsilon_r = \sqrt{\frac{A}{B}}, \ \mu_r = \sqrt{AB}. \tag{7.170}$$

7.3.1.2 Open- and Short-Circuited Method

In the open and short-circuited method, the complex permittivity ε_r and permeability μ_r are determined from the input impedance Z_{in} of the samples which are loaded by open- or short-ended coaxial line cells [66, 67]. The sample thickness is assumed to be very small compared to the wavelength in the coaxial line cell; the electric or magnetic field at the sample position can be regarded as uniform. The sample setup in the coaxial line is schematically shown in Figure 7.22. As the open-ended transmission line cannot be made due to the effect of opening edge, the sample is loaded at the position at which the length from the short end is quarter wave length $\lambda/4$ in the transmission line, as shown in Figure 7.22a. From the following relations,

$$Z_{in} = Z\tanh\gamma d = Z\tanh j\frac{2\pi}{\lambda}\frac{\lambda}{4} = jZ\tan\frac{\pi}{2} = \infty, \tag{7.171}$$

the input impedance at the $\lambda/4$ distance from the short end is regarded as infinity in the transmission line. It should be noted that the value λ is the wavelength of the EMW in the transmission line, not in the free space. Meanwhile, in the permeability measurement, the sample is located at the short end of the transmission line as shown in Figure 7.22b. The input impedance Z_{open} and Z_{short} can be measured by the impedance analyzer or network analyzer as a function of frequency; these are given by:

$$Z_{open} = \frac{Z}{\tanh\gamma d} \text{ and } Z_{short} = Z\tanh\gamma d, \tag{7.172}$$

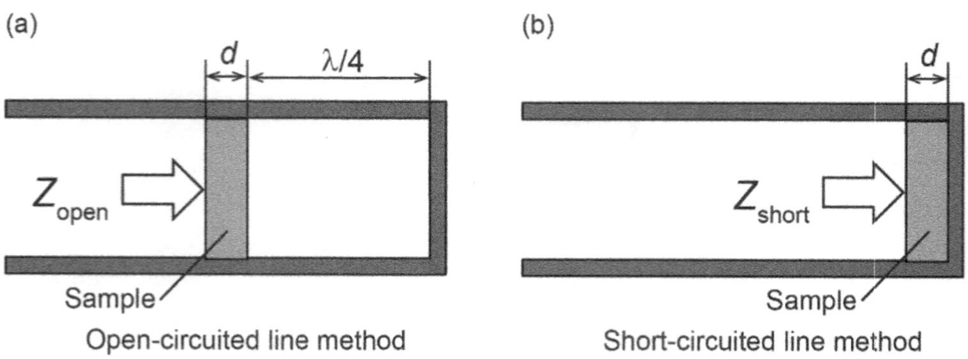

Figure 7.22 Open- and short-circuited line method.

where the characteristic impedance Z and the propagation constant γ in the sample loaded coaxial line are derived as:

$$Z = \sqrt{Z_{\text{short}} Z_{\text{open}}} = Z_0 \sqrt{\frac{\mu_r}{\varepsilon_r}}; \tag{7.173}$$

$$\gamma = \frac{1}{d} \tanh^{-1} \sqrt{\frac{Z_{\text{short}}}{Z_{\text{open}}}} = \gamma_0 \sqrt{\varepsilon_r \mu_r}. \tag{7.174}$$

Hence, the relative permittivity and permeability of the sample are:

$$\varepsilon_r = \frac{Z_0}{Z} \frac{\gamma}{\gamma_0}, \ \mu_r = \frac{Z}{Z_0} \frac{\gamma}{\gamma_0}. \tag{7.175}$$

As discussed earlier, in the open- and short-circuited methods, the sample thickness should be thin enough to the wavelength. Hence, this method cannot be used at the very high frequencies; the measurement frequency is normally restricted below 1 GHz. Further, the sample thickness should be small to avoid the dimensional resonance effect in the coaxial line even in the low-frequency range.

7.3.2 Permeability Spectra of Ferrites

Magnetic materials can be classified into the two categories. One is the hard magnetic materials, and another is the soft magnetic ones. This refers to their low or high magnetic coercivity. For example, M-type hexagonal ferrites with magnetoplumbite structure $MFe_{12}O_{19}$ (M = Ba, Sr, etc.) are often used for the permanent magnet or magnetic recording media [68–70]. Generally, the M-type hexaferrites are hard; the Y-, Z-, and U-type ferrites are soft; and the W- and X-type ferrites can vary between these two extremes, but all have large intrinsic magnetic crystallographic anisotropy fields and magnetization vector magnitudes [71]. In the RF device applications, spinel ferrites represented by the chemical formula MFe_2O_4 with the cubic crystal structure have been used in the bulk or composite forms [72, 73]. The spinel ferrites can be used in the microwave applications such as the EMW absorbers, too, in the composite forms (see Chapter 6 in Volume 2 of this book series). The garnet ferrites, e.g. the YIG with the chemical formula $Y_3Fe_5O_{12}$, are used for the microwave devices as the circulator or isolator. The YIG is in the garnet structure with the space group Ia_3d in which the atoms locate on the four different crystallographic sites [74].

7.3.2.1 Spinel Ferrite

A typical permeability spectra of spinel ferrites are shown in Figure 7.23a for Mn–Zn ferrite and in Figure 7.23b for Ni–Zn ferrite [21]. These ferrites show a soft magnetic property in the magnetization curves. In the Mn–Zn ferrite, the real part of relative permeability μ_r' is almost constant, and the imaginary one μ_r'' is zero in the low frequency range. The value of μ_r'

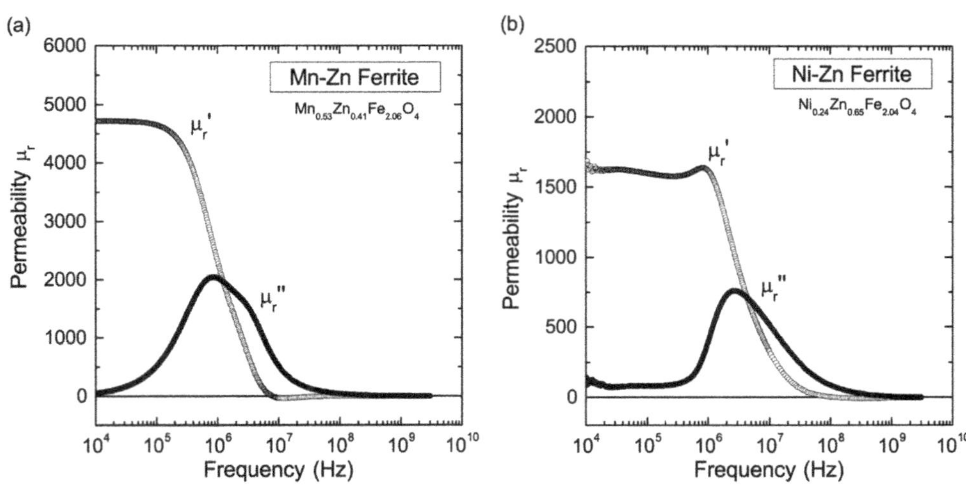

Figure 7.23 Permeability spectra of spinel ferrites (a) Ni–Zn ferrite and (b) Mn–Zn ferrite [21].

starts to decrease from about 200 kHz; a significant drop takes place in the several megahertz range. In this frequency range, the μ_r'' increases with frequency and has a broad maximum. Meanwhile, the Ni–Zn ferrite indicates smaller permeability at the lower frequency as compared to that of Mn–Zn ferrite, as is shown in Figure 7.23b; the permeability dispersion occurs at about 1 MHz, which is higher than that of Mn–Zn ferrite. The magnetic permeability dispersion in the spinel ferrites has been widely investigated and described by the magnetic resonance due to the DW vibration, and the magnetic relaxation caused by the uniform magnetization rotation in the external EM field.

As is discussed in Chapter 6 in Volume 2 of this book series, the permeability spectrum of spinel ferrites can be attributed to the model with the resonance-type DW vibration and the relaxation type magnetization rotation [75–77]. Further, the gyromagnetic spin rotation can also be taken into account in the permeability spectra of spinel ferrite and their composite materials by the Lorentz type and Landau–Lifshitz–Gilbert-type resonance formulas [21]. The contribution of the DW vibration and the gyromagnetic spin motion to the permeability spectrum has been investigated by the numerical analysis in the spinel ferrites and their composite materials. The dispersion parameters, such as the resonance frequencies, DC (typically called static) magnetic susceptibilities, and damping factors are evaluated in [21].

7.3.2.2 Garnet Ferrite

Typical complex relative permeability spectra (μ_r' and μ_r'') of YIG are shown in Figure 7.24a [78]. The YIG has a relatively high low-frequency permeability (μ_r' is about 60) indicating a soft magnetic property. In the absence of the external bias DC magnetic field, two distinct peaks at about 10 and 200 MHz can be observed in the imaginary part of its permeability μ_r''. At these frequencies, the real part μ_r' shows a rapid decrease. The double peaks in the μ_r''' spectrum indicate the existence of the different permeability dispersion mechanisms in YIG. These frequency dispersions can be attributed to the DW vibration and the gyromagnetic spin rotation; the high-frequency permeability dispersion accompanies the negative μ_r' around 1 GHz.

As is discussed later, the YIG shows significant resonance frequency dispersion in the permeability spectrum under the DC bias magnetic fields below 1 kOe due to gyromagnetic resonance [78].

7.3.2.3 Hexagonal Ferrite

Hexagonal ferrites generally have a large magnetocrystalline anisotropy and often show a hard magnetic property. Magnetization curve tends to have a large hysteresis with relatively high coercivity; the initial permeability μ_i is small due to the large anisotropy field. Hence, the gyromagnetic effect in hexagonal ferrites takes place at the higher frequencies than that of spinel or garnet ferrites; the gyromagnetic spin resonance frequency f_s of hexaferrites is typically on the order of a few tens gigahertz [68]. However, depending on the contents, the frequency range of hexaferrite gyromagnetic applications can range from as low as 2.5 GHz to as high as 300 GHz [12].

The permeability spectra of a hexagonal ferrite $BaFe_{12}O_{19}$ are shown in Figure 7.24b in the frequency range from 100 kHz to 20 GHz [79]. The permeability value is very small as compared with that of YIG or spinel ferrites, indicating the hard

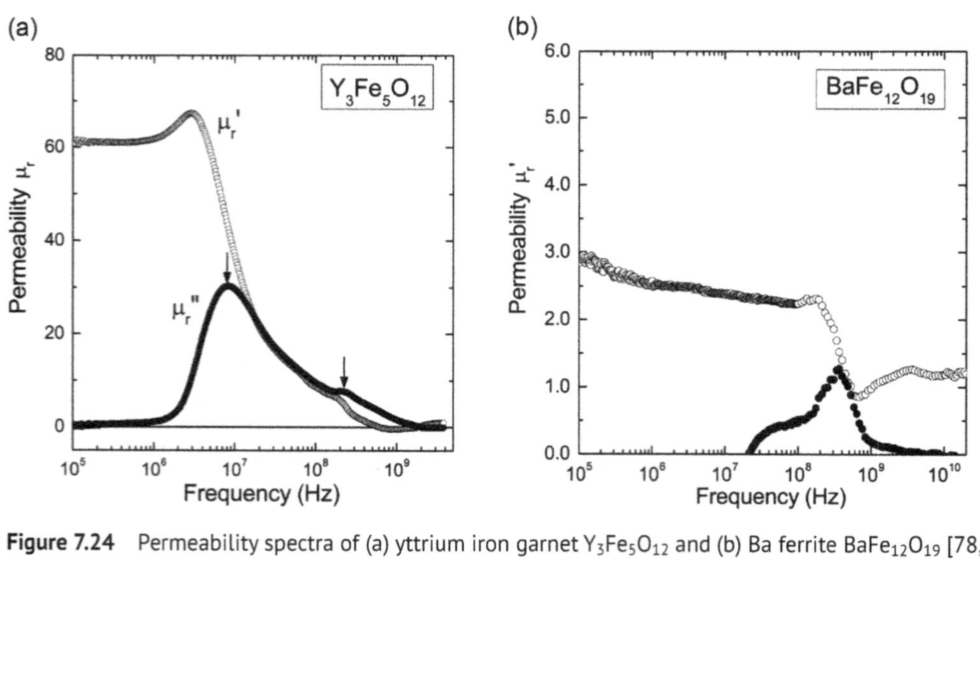

Figure 7.24 Permeability spectra of (a) yttrium iron garnet $Y_3Fe_5O_{12}$ and (b) Ba ferrite $BaFe_{12}O_{19}$ [78, 79].

magnetic property of the BaM hexaferrite. The value of μ_r' becomes almost unity at frequencies above 10 GHz. Interestingly, the permeability spectra, $\mu_r = \mu_r' - j\mu_r''$, indicate a distinct frequency dispersion from about 200 MHz to 1 GHz; and the low temperature permeability has a relatively large value about $\mu_r' = 6$. This type of permeability dispersion has been observed in the several Fe substituted Ba ferrites ($BaFe_{1-x}M_xO_{19}$) [80–82]. As the gyromagnetic natural resonance frequency has been reported to locate at about 50 GHz in $BaFe_{12}O_{19}$ [68], it is hard to associate this permeability dispersion is caused by the natural resonance caused by the anisotropy field H_A. Thus, the contribution of the DW vibration or magnetization rotation process can be considered. From the permeability measurements of $BaFe_{12}O_{19}$ powders, the μ_r' value was reported to be unity from 8 to 26 GHz [83]. As the ferrite powders contain very few magnetic domains, the DW vibration component in the permeability spectrum can be ignored. Further, the permeability spectrum of the composite materials containing Ba ferrite powders also indicate the small permeability about unity from 2 to 10 GHz [84]. From aforementioned discussion, the DW vibration can play an important role for the permeability dispersion of bulk $BaFe_{12}O_{19}$ in the RF range.

7.3.2.4 Snoek's Law

The relative permeability μ_r' of spinel ferrites becomes unity in the microwave frequency range as shown Figure 7.26. This indicates that the gyromagnetic spin rotation as well as the DW vibration of spinel ferrites cannot induce the magnetic flux along the alternating magnetic field direction in the microwave range. As is discussed earlier regarding the permeability spectra of spinel ferrites, a ferrite which has the higher static magnetic permeability, e.g. shows the more rapid decrease of μ_r' in the low-frequency range as compared to the ferrite with a lower static permeability. This phenomenon was first pointed out by J. L. Snoek [85]; there exists a frequency limitation of high permeability in the permeability spectrum. The following relation has been proposed as an empirical law (written in the Gauss system):

$$f_0(\mu_{ri} - 1) = \frac{4}{3}\gamma_g M_s, \tag{7.176}$$

where the values $\gamma_g = 2.8\,\text{GHz/kOe}$ and M_s are the gyromagnetic ratio and the saturation magnetization of ferrite $4\pi M_s/4\pi$ in Gauss respectively, and μ_{ri} is relative initial (static) permeability [72]. The permeability spectra of spinel ferrites which have different static permeability (Mn–Zn and Ni–Zn ferrites) are summarized in Figure 7.25a, as well as the permeability spectra of the three Ni–Zn ferrites, which have different static permeability in Figure 7.25b [86]. It can be confirmed that the Snoek's law is maintained in the different spinel ferrites, as well as within the same ferrite system. The Snoek's law and the limitation of the high permeability in ferrite materials will be discussed in Chapter 6 in Volume 2 in more detail.

7.3.2.5 Analysis of the Complex Permeability of Ferrites

As described in Sections 7.2.5 and 7.2.9, the permeability spectrum of ferrites at the higher frequencies can be fitted by the Lorentzian or the Landau–Lifshitz–Gilbert type resonance formulas describing frequency dispersion of magnetic

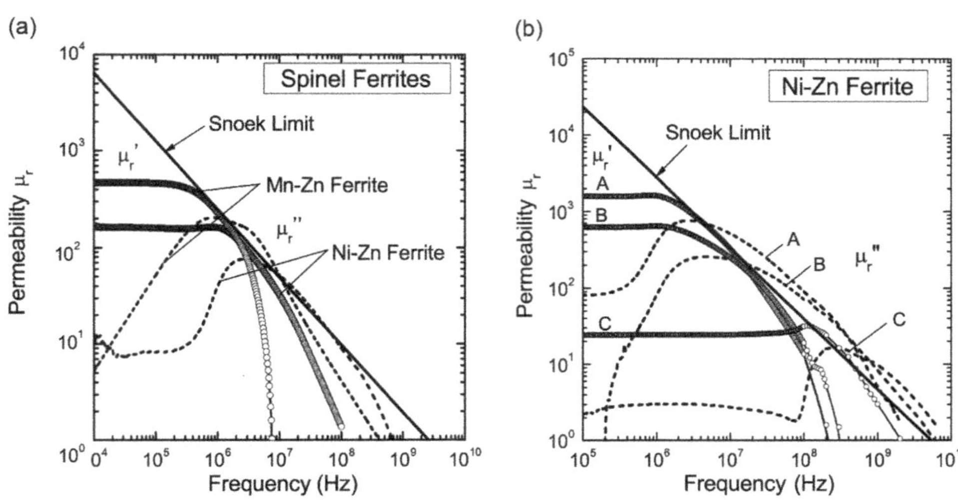

Figure 7.25 Permeability spectra of spinel ferrites. Solid line indicates the Snoek's limit [86].

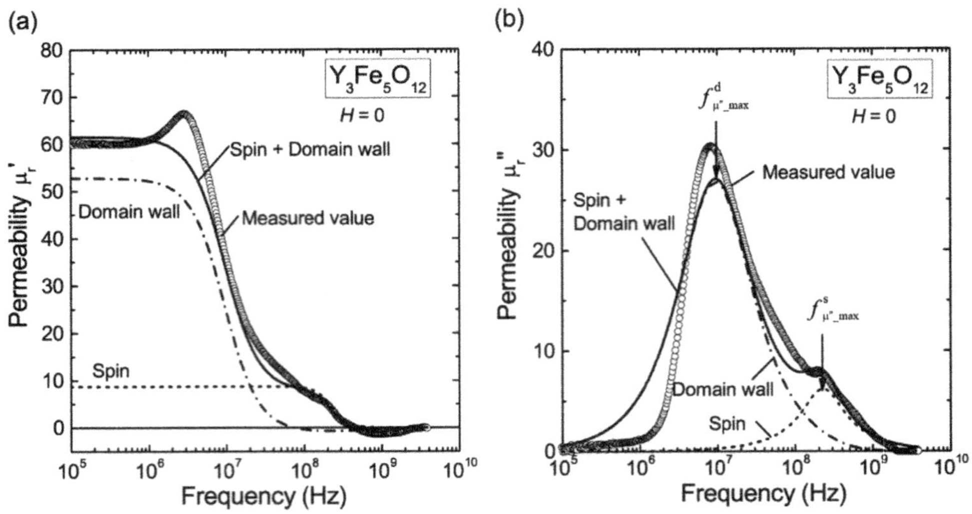

Figure 7.26 Permeability spectra of YIG under zero external field [78].

susceptibility. For the interpretation of the measured permeability spectrum over the RF/microwave range, the permeability dispersion $\mu_r(\omega)$, where $\omega = 2\pi f$ is the angular frequency of applied EM field, has been analyzed using the two terms associated with DW and spin susceptibility, $\chi_d(\omega)$ and $\chi_s(\omega)$, in the resonance form [21, 78]:

$$\mu_r(\omega) = 1 + \chi_d(\omega) + \chi_s(\omega) = 1 + \frac{\chi_{d0}\omega_d^2}{\omega_d^2 - \omega^2 + j\beta\omega} + \frac{\chi_{s0}\omega_s(\omega_s + j\alpha\omega)}{(\omega_s + j\alpha\omega)^2 - \omega^2}, \quad (7.177)$$

where χ_{d0} and χ_{s0} are the static susceptibilities, ω_d and ω_s are the resonance frequencies for DW and spin components; β and α are the damping constants of DW and spin resonance, respectively. The further numerical analysis of the permeability dispersion was carried out using the nonlinear least mean square fitting.

7.3.2.5.1 YIG under External Field

The results of the numerical fitting of the permeability spectra of YIG, which is indicated in Figure 7.24a, are shown in Figure 7.26. From the analysis results, the low-frequency dispersion below 10 MHz is mainly originated by the DW resonance; the high-frequency dispersion from 100 MHz to 1 GHz is considered to be a result of the gyromagnetic resonance. The large μ_r'' maximum around 10 MHz suggests that the relatively large permeability of YIG is mainly attributed to the DW vibration under zero external field. The frequencies $f^d_{\mu''\_\max}$ and $f^s_{\mu''\_\max}$, at which the calculated μ_r'' takes a maximum, correspond to the corresponding characteristic frequencies for the DW and gyromagnetic spin resonance calculated by $d\mu_r''/d\omega = 0$.

The following formulas are derived for the DW component [78]:

$$f^d_{\mu''\_\max} = \frac{f_d}{6}\sqrt{12 - 6\left(\frac{\beta}{f_d}\right)^2 + 6\sqrt{16 - 4\left(\frac{\beta}{f_d}\right)^2 + \left(\frac{\beta}{f_d}\right)^4}}, \quad (7.178)$$

and the gyromagnetic spin component:

$$f^s_{\mu''\_\max} = \frac{f_s}{\sqrt{\alpha^2 + 1}}. \quad (7.179)$$

These formulas suggest that the frequencies $f^d_{\mu''\_\max}$ and $f^s_{\mu''\_\max}$ are determined by the damping of the DW vibration and gyromagnetic spin rotation.

On the other hand, when the bias magnetic field is applied perpendicular to the high-frequency alternating field, a distinct resonance characteristic is observed in the permeability spectra of YIG, as shown in Figure 7.27. The μ_r'

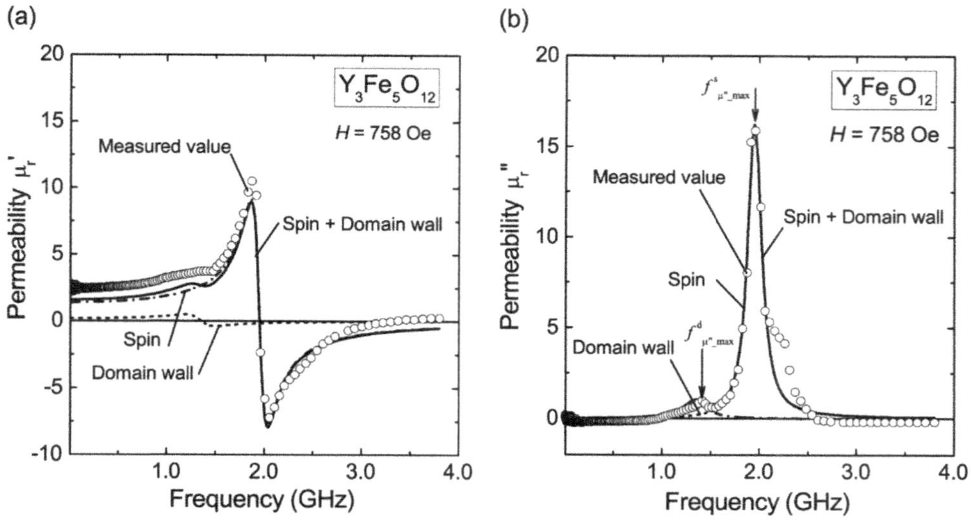

Figure 7.27 Permeability spectra of YIG under 758 Oe external field [78].

spectrum of YIG demonstrates a large frequency dispersion around 2.0 GHz; simultaneously, the μ_r'' spectrum indicates a sharp peak at 1.9 GHz under 756 Oe bias field. These peculiarities of the permeability frequency dispersion can be described by the magnetic resonance. In the spectrum of μ_r'', a small peak is observed at about 1.4 GHz; a small anomaly in the value of μ_r' is also recognized near the same frequency. This anomaly suggests that the DW vibration contribution remains, even when the bias field reaches 756 Oe. As discussed earlier, the two susceptibility components $\chi_d(\omega)$ and $\chi_s(\omega)$ can be separated from the permeability spectrum; the different frequency dispersion mechanism in the magnetic permeability can be discussed by the numerical analysis. As the bias field increases, the frequencies $f^d_{\mu''\_max}$ and $f^s_{\mu''\_max}$ become higher, and the permeability dispersion by the DW motion disappears under several kOe by the achievement of almost single-domain state [78].

The increase of $f^d_{\mu''\_max}$ with the increasing bias field might be attributed to the increase of the DW stiffness caused by the DC magnetic field. This phenomenon can be explained as follows. Assume that an external bias field $\boldsymbol{H}$ is applied along the direction indicated in Figure 7.18. This bias field generates the movement of DWs and the magnetic domain structure in the ferrite changes. As this DW motion can be disturbed by the lattice defects or grain boundaries in the crystal, the DWs can be pinned at the positions of these under the external static or alternating fields [87, 88]. When the alternating magnetic field is applied to this domain structure without a DC bias field, a kind of mechanical vibration can be induced; the pinned DWs can vibrate as the membranes. This DW vibration has its own intrinsic oscillation frequency, which is determined by the elasticity (stiffness) and the effective mass of DWs, as discussed in Section 7.2.9. When the DC bias field is applied to the vibrating DWs, the magnetostatic force will be generated, and this force tries to displace the DW. As the pinning force generated by the pinning of DW at the defect acts against this force, a kind of tension of DW will be enhanced by the external DC bias field, and the stiffness of DW will increase. As a result, the DW resonance frequency can be increased by the external DC bias field.

Meanwhile, the shift of $f^s_{\mu''\_max}$ can originate due to the change of gyromagnetic resonance frequency.

The permeability dispersion parameters of YIG under 0 and 758 Oe bias field, which have been obtained by the nonlinear least mean square fitting, are summarized in Table 7.3. The estimated gyromagnetic resonance frequency f_s is 320 MHz under zero bias field and shifts to 1.95 GHz under 758 Oe. As is discussed in Section 7.2, the gyromagnetic spin resonance angular frequency is determined by the Larmor resonance frequency given by Equation (7.3), $\omega_s = \mu_0 \gamma H_A$, where H_A is the internal magnetic field created by the magnetocrystalline anisotropy. When the bias field H_{DC} is applied along the internal anisotropy field H_A, the resonance frequency increases with H_{DC} by the following equation $\omega_s = \mu_0 \gamma_g (H_A + H_{DC})$. Under the 758 Oe, damping factor of gyromagnetic component α becomes 0.04 from that of 0.99 in the zero external field. Hence, from Equation (7.180), the resonance frequency f_s is almost the same as the peak frequency of μ_r'', i.e. $f^s_{\mu''\_max}$.

Table 7.3 Permeability dispersion parameters of YIG under 0 and 758 Oe external field.

$H = 0$	**Domain wall component χ_d**	**Gyromagnetic component χ_s**
Resonance frequency (GHz)	72.0	320
DC susceptibility	52.75	8.72
Damping sonstants (s^{-1})	6.52×10^8	0.99
$H = 758$ Oe	Domain wall component χ_d	Gyromagnetic component χ_s
Resonance frequency (GHz)	1.35	1.95
DC susceptibility	0.19	1.38
Damping constants (s^{-1})	2.97×10^8	0.04

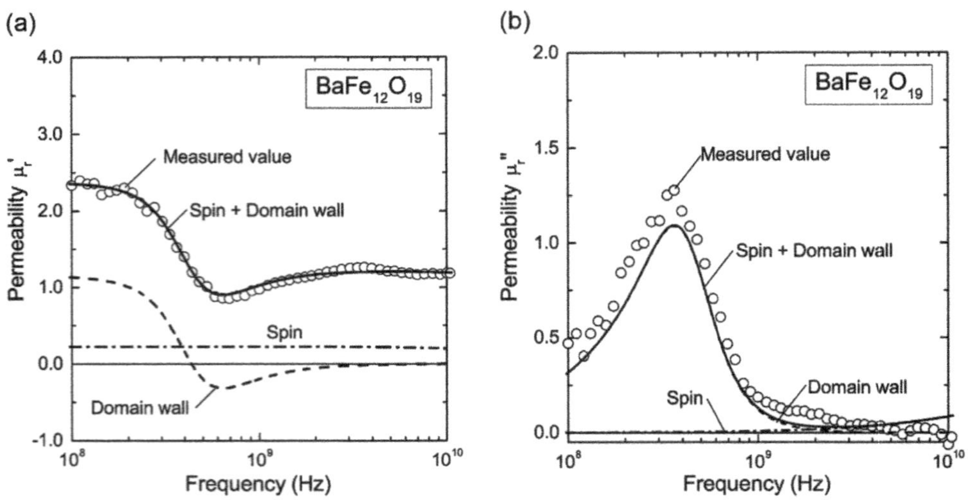

Figure 7.28 Permeability spectra of $BaFe_{12}O_{19}$ [79].

7.3.2.5.2 Barium Ferrite

As discussed in Section 7.2, in $BaFe_{12}O_{19}$, a relatively large permeability in RF range is attributed to the DW vibration. The numerical analysis was carried out assuming that the gyromagnetic resonance takes place around 50 GHz. The obtained results are shown in Figure 7.28 [79]. The real part of permeability spectra can be explained by the DW resonance and the constant gyromagnetic spin components. Estimated permeability dispersion parameters for the DW components are $\chi_{d0} = 1.15$, $f_d = 435$ MHz, and $\beta = 501 \times 10^6$, respectively.

Meanwhile the gyromagnetic spin components are $\chi_{s0} = 0.22$, $f_d = 50$ GHz, and $\alpha = 2.00$, respectively.

Several studies on the substitution of Fe atoms in the M-type Ba ferrite by the other elements have been carried out in the viewpoint of the development of EM absorbers in the microwave range. The Fe atom substitution by the other tetravalent elements results in the decrease of magnetic anisotropy; gyromagnetic resonance frequency shifted down to the K-band (18–27 GHz) or X-band (8–12 GHz) from 50 GHz [82, 89]. In the substituted Ba ferrite, $BaFe_{12-x}(Ti_{0.5}Co_{0.5})_xO_{19}$, the low-frequency permeability increases with x, i.e. (Ti–Co) content, takes a maximum at around $x = 3$ and then decreases [79]. The relative permeability μ_r' value at 100 MHz reaches about 16 at $x = 6$. From the numerical analysis mentioned earlier, it can be concluded that the large relative permeability below 1 GHz in the substituted Ba ferrite can be mainly originated by the DW vibration; gyromagnetic susceptibility component in this frequency range remains about unity at $x = 6$. The contribution of the spin susceptibility to the relatively high permeability is not so large even in the K-band up to 20 GHz in spite of that the magnetic anisotropy was suppressed by the substitution [79].

As discussed earlier, the M-type hexagonal ferrites have been investigated not only for the hard ferrite applications such as the magnetic recording or permanent magnet, but also for the soft ferrites, such as the EM wave absorber materials, or for extending frequency range of various gyromagnetic devices to mm waves [12].

7.3.3 Complex Permittivity of Ferrites

As ferrites are composed of iron(III) oxide combined chemically with one or more additional metallic elements, they have a large electrical resistance and shows dielectric property. However, the dielectric constant, i.e. the electric permittivity of ferrites is not so large compared to the ferroelectric oxides such as $BaTiO_6$. Under alternating electric field $\boldsymbol{E}$, the dielectric response of ferrites can be described by the relative complex permittivity $\varepsilon_r = \varepsilon_r' - j\varepsilon_r''$, as discussed in Section 7.3.1.

Complex permittivity spectra of spinel ferrites are shown in Figure 7.29: the case (a) corresponds to the Mn–Zn ferrite, and the case (b) to Ni–Zn ferrite. In the Mn–Zn ferrite, the low-frequency permittivity ε_r' is very large compared to that of the Ni–Zn ferrite, as shown in the figure; and ε_r'' decreases rapidly with increasing frequency. This property is attributed to the conductive loss. As a matter of fact, it is reported that the Mn–Zn ferrite with relatively low Fe^{2+} ions shows a semiconductive property in the variation of electrical conductivity σ with temperature [90]. The inset in Figure 7.29 shows an electrical conductivity σ of the Mn–Zn ferrite as a function of frequency. Though the conductivity σ is about 20 S/cm at 1 kHz, indicating a relatively large value, it decreases with frequency starting from several kilohertzs. Corresponding to this property, the high-frequency permittivity is small and almost constant. However, in the several gigahertz range, a permittivity dispersion can be observed from about 4 GHz. This indicates a dielectric property; the Debye-type permittivity dispersion may be originated by the dielectric relaxation of dipoles, which are created by the space charges on the grain boundary [91]. On the other hand, the permittivity spectra of Ni–Zn ferrite have no frequency dispersion from 10 kHz to 10 GHz; dielectric loss is almost zero. The dielectric constant ε_r' is about 12. This characteristic is generally seen in the Ni–Zn ferrites [92].

The permeability spectra of YIG ($Y_3Fe_5O_{12}$) and Ba ferrite, ($BaFe_{12}O_{19}$), are shown in Figure 7.30a and b, respectively. The YIG has almost the same property as Ni–Zn ferrite; dielectric constant is about 10 up to 6 GHz, indicating a dielectric

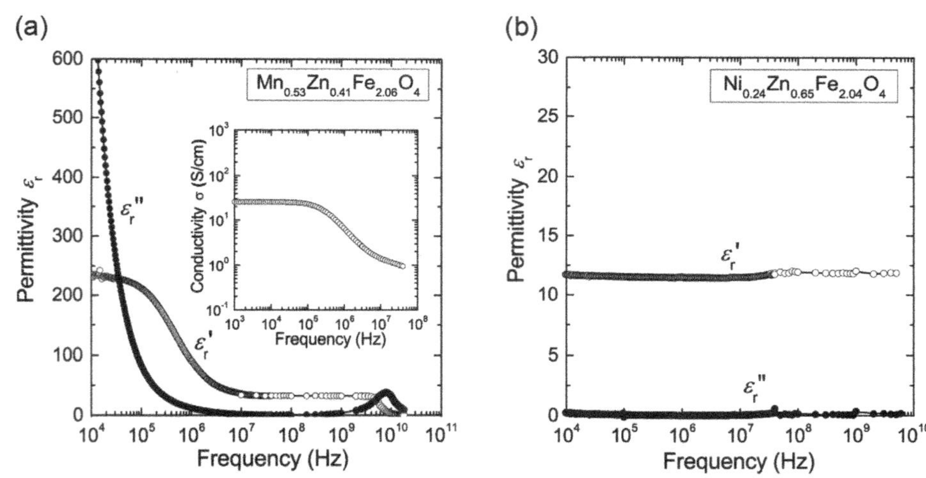

Figure 7.29 Permittivity spectra of spinel ferrites (a) Mn–Zn ferrite and (b) Ni–Zn ferrite [86].

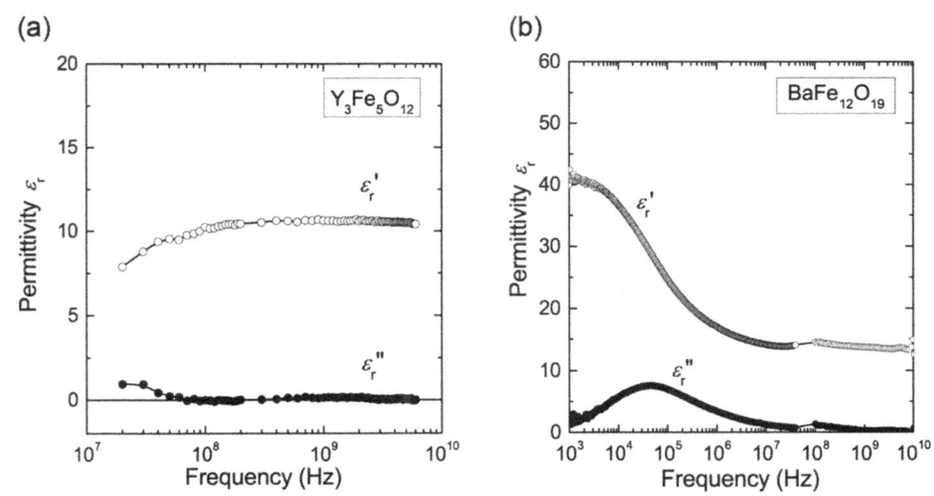

Figure 7.30 Permittivity spectra of (a) YIG and (b) Ba ferrite [78, 79].

insulator. Meanwhile, $BaFe_{12}O_{19}$ shows relatively large permittivity about 40 at 1 kHz; a Debye-type frequency dispersion is observed in the frequency range from 1 kHz to about 10 MHz. In the microwave range above 1 GHz, the permeability spectra of $BaFe_{12}O_{19}$ have no frequency dispersion as reported before [83]. The low-frequency permittivity dispersion in this Ba ferrite can be attributed to the relaxation process of the orientational polarization (dielectric moment induced by the space charges at the grain boundary), as discussed earlier, too [79].

7.4 Gyromagnetic Electrodynamics

In this chapter, we will consider peculiarities of electrodynamics and gyromagnetic effects in ferrites useful for designing devices of gyromagnetic electronics. Ferrite media possess unique properties from EM s point of view: anisotropic behavior of their magnetic susceptibility (or permeability), resulting in gyrotropy and nonreciprocity; frequency selectivity due to gyromagnetic resonance due to spins of electrons in magnetic atoms of a ferrite; possibility of frequency sweeping by an external electromagnet; as well as stable nonlinear effects. All these phenomena are widely used to build various devices operating in microwave and millimeter-wave devices. Two groups of gyromagnetic devices can be considered: one group employs extended ferrite media, and the other uses "lumped" ferrite resonators (FRs). Ferrite media or tiny resonators are typically placed in/on some waveguiding or cavity structures based on some transmission lines segments.

The *first group* of the devices uses *extended (continuous) ferrite media* for their operation, and the EMW propagation effects should be considered. Ferrite samples completely or partially filling a waveguide/transmission line-cross section and having some extended length along the direction of waves propagation are used in the various nonreciprocal and reciprocal devices, for Faraday rotation of wave polarization plane, for waveguide filters of harmonics, for matched loads, etc. EMW propagation in such media is a topic of many publications [13, 15, 93–102].

The *second group* employs miniature "lumped" FRs. Their characteristic dimensions are usually much smaller than the shortest wavelength in the ferrite medium. Therefore, static approximation of internal field in the resonator, resulting in uniform magnetostatic oscillations can be considered in the FR. Due to small sizes of ferrite samples (resonators), when considering their interaction with EM fields of a waveguide, transmission line, or external resonator where they are placed in, the approaches based on small perturbation theory are usually applicable. Such kinds of structures are described in publications [13, 15, 103–107]. To study interaction of magnetic spheres with microwaves in cavities beyond the magnetostatic and rotating wave approximations, Mie scattering theory can be applied [108]. Numerical models have proven to be powerful in considering interaction of FRs used in filters [109].

Tiny FRs are used as the main element of frequency-selective tunable filters, for example, in spectrum analyzers and power meters of microwave signals. The resonators must have high Q factor (narrow resonance line width) to increase frequency resolution. They must have such saturation magnetization and crystal anisotropy field that the resonance exists in the required frequency range of the device operation.

For gyromagnetic devices, specifically employing FMR in ferrites, the lower frequency is usually determined by the saturation magnetization, $\omega_M = \mu_0\gamma M_s$, and for the ferrites with high values of the field of magnetic crystallographic anisotropy, like hexagonal ferrites, is determined by the anisotropy field $\omega_A = \mu_0\gamma H_A$ [11], where $H_A \approx 2H_{A1}$. The latter is called the frequency of NFMR, which occurs without any bias magnetic field. The upper frequency limit is determined by the achievable resonance frequency, the value of which is mainly determined by the dimensions and weight of the needed bias magnetic field system, as well as frequency ranges of transmission lines on which the ferrite devices are built.

In this section, we will consider a few basic problems of EMWs interaction with both groups of ferrites – continuous media and small resonators. Special attention will be paid to the perturbation technique which has proven to be comparatively simple yet powerful method in solving many complex electrodynamic problems where ferrite media are involved.

7.4.1 EMWs in Unbounded Gyrotropic Media

In any electromagnetically anisotropic medium, the relationship between the pair of high-frequency vectors: $\boldsymbol{d}$ (displacement vector) and $\boldsymbol{e}$ (electric field vector); or $\boldsymbol{b}$ (magnetic induction, or magnetic flux density vector) and $\boldsymbol{h}$ (magnetic field vector) are more complex depending on which parameter – permittivity or permeability exhibits anisotropy. Ferrites, as we have seen before, are anisotropic media, which have a relationship between $\boldsymbol{b}$ and $\boldsymbol{h}$ through the permeability tensor $\vec{\mu}$:

$$\boldsymbol{b} = \vec{\mu} \times \boldsymbol{h}. \tag{7.180}$$

Let us consider an unbounded ferrite medium magnetized by the bias magnetic field $\boldsymbol{H}_0$ so that the equilibrium magnetic moment $\boldsymbol{M} = M_0\boldsymbol{z}^0$ is aligned with the axis z so that the permeability tensor is written as:

$$\vec{\mu} = \begin{vmatrix} \mu & j\kappa & 0 \\ -j\kappa & \mu & 0 \\ 0 & 0 & \mu_{\|} \end{vmatrix}, \tag{7.181}$$

or

$$\begin{aligned} b_x &= \mu h_x + j\kappa h_y; \\ b_y &= -j\kappa h_x + \mu h_y; \\ b_z &= \mu_{\|} h_z. \end{aligned} \tag{7.182}$$

Let us substitute these relations into the second of Maxwell's equations in the complex amplitude form (omitting the time-dependent $e^{j\omega t}$ term):

$$\nabla \times \boldsymbol{h} = j\omega \boldsymbol{d}, \tag{7.183}$$

$$\nabla \times \boldsymbol{e} = -j\omega \boldsymbol{b}. \tag{7.184}$$

In Equations (7.183) and (7.184), both electric and magnetic impressed currents $\boldsymbol{j}_{e,m}$ are omitted. Then the second Maxwell's equation will be represented as a set of scalar equations:

$$\begin{aligned} \frac{\partial e_z}{\partial y} - \frac{\partial e_y}{\partial z} &= -j\omega\mu h_x + \omega\kappa h_y; \\ \frac{\partial e_x}{\partial z} - \frac{\partial e_z}{\partial x} &= -\omega\kappa h_x - j\omega\mu h_y; \\ \frac{\partial e_y}{\partial x} - \frac{\partial e_x}{\partial y} &= -j\omega\mu_{\|} h_z. \end{aligned} \tag{7.185}$$

If there is no bias magnetic field, the ferrite medium usually becomes isotropic with $\kappa = 0$ and $\mu = \mu_{\|}$; therefore, the system (7.185) turns into Maxwell's equations for isotropic medium.

There is typically no anisotropy of the dielectric properties of ferrites so that the first Maxwell's equation is simply written as:

$$\begin{aligned} \frac{\partial h_z}{\partial y} - \frac{\partial h_y}{\partial z} &= j\omega\varepsilon e_x; \\ \frac{\partial h_x}{\partial z} - \frac{\partial h_z}{\partial x} &= j\omega\varepsilon e_y; \\ \frac{\partial h_y}{\partial x} - \frac{\partial h_x}{\partial y} &= j\omega\varepsilon e_z. \end{aligned} \tag{7.186}$$

7.4.1.1 Longitudinal Magnetization of Ferrite Medium and Faraday Effect

Let us consider the propagation of a plane EMW, with the Poynting vector aligned along the z-axis. In the case of the *longitudinal magnetization* of a ferrite medium, the bias field in the z-direction, too. Assume that this is homogeneous unbounded ferrite medium. Let us, for simplicity, neglect the dielectric and magnetic loss, which will not affect the main physics results we are going herein to obtain.

The plane transverse EMW propagating in the z-direction has the following components:

$$\begin{aligned} & e_x, e_y, h_x, h_y \neq 0; \\ & e_z = h_z = 0; \\ & \frac{\partial}{\partial z} \neq 0, \frac{\partial}{\partial x} = \frac{\partial}{\partial y} = 0. \end{aligned} \tag{7.187}$$

Then the following set of equations describes the plane EMW propagation in the longitudinally magnetized medium:

$$\frac{dh_y}{dz} = j\omega\varepsilon e_x;$$
$$\frac{dh_x}{dz} = j\omega\varepsilon e_y;$$
$$\frac{de_y}{dz} = j\omega\mu h_x - \omega\kappa h_y; \tag{7.188}$$
$$\frac{de_x}{dz} = -\omega\kappa h_x - j\omega\mu h_y.$$

Let us represent the solutions of (7.188) as the waves propagating in the positive z-direction with the real (no loss) phase constant β:

$$h_x = H_{x0}e^{-j\beta z};$$
$$h_y = H_{y0}e^{-j\beta z}; \tag{7.189}$$
$$e_x = E_{x0}e^{-j\beta z};$$
$$e_y = E_{y0}e^{-j\beta z}.$$

Substituting solutions (87.189) into (7.188), one can get:

$$\beta H_{y0} = \omega\varepsilon E_{x0}; \tag{7.190}$$

$$\beta H_{x0} = -\omega\varepsilon E_{y0}; \tag{7.191}$$

$$-\beta E_{y0} = \omega(\mu H_{x0} + j\kappa H_{y0}); \tag{7.192}$$

$$-\beta E_{x0} = \omega(-j\kappa H_{x0} + \mu H_{y0}). \tag{7.193}$$

Let us substitute E_{x0} from Equation (7.190) to (7.193), and E_{y0} from Equation (7.191) to (7.192). Then the two equations are obtained:

$$\beta H_{y0} = \frac{\omega^2\varepsilon}{\beta}(-j\kappa H_{x0} + \mu H_{y0}), \tag{7.194}$$

and

$$\beta H_{x0} = \frac{\omega^2\varepsilon}{\beta}(\mu H_{x0} + j\kappa H_{y0}). \tag{7.195}$$

From Equation (7.195), one can get:

$$H_{x0} = \frac{j\kappa\omega^2\varepsilon}{\beta^2 - \omega^2\mu\varepsilon}H_{y0}. \tag{7.196}$$

Let us multiply Equation (7.194) by β and substitute Equation (7.196) into Equation (7.194), then the equation for the phase constant of a plane wave in a longitudinally magnetized ferrite can be obtained:

$$\beta^4 - (2\omega^2\varepsilon\mu)\beta^2 - \omega^4\varepsilon^2\mu^2 + \omega^4\varepsilon^2\kappa^2 = 0. \tag{7.197}$$

This equation is easily solved as:

$$\beta = \pm\omega\sqrt{\varepsilon(\mu \pm \kappa)}. \tag{7.198}$$

The first "$\pm$" sign (before the square root sign) corresponds to the waves propagating in positive (+) and in negative direction (–). As the medium is unbounded, there are no reflections, and the solutions with the "–" sign can be omitted. Therefore, there are two phase constants for a plane wave in a ferrite medium:

$$\beta = \omega\sqrt{\varepsilon(\mu \pm \kappa)}, \tag{7.199}$$

each of them propagates with its own phase constant,

$$\begin{aligned} v^+ &= \omega / \beta^+; \\ v^- &= \omega / \beta^-. \end{aligned} \tag{7.200}$$

If β from Equation (7.199) is substituted in Equation (7.196), it can be shown that:

$$H_{x0} = \pm jH_{y0}. \tag{7.201}$$

The existence of two unequal phase velocities corresponds to the rotation of the polarization plane. The total magnetic field components in the time domain can be expressed as:

$$h_x(t) = 2H_{x0}\cos(\omega t - \beta_1 z)\cos(\beta_2 z). \tag{7.202}$$

and

$$h_y(t) = 2H_{x0}\cos(\omega t - \beta_1 z)\sin(\beta_2 z), \tag{7.203}$$

where

$$\begin{aligned} \beta_1 &= (\beta^+ + \beta^-)/2; \\ \beta_2 &= (\beta^+ - \beta^-)/2. \end{aligned} \tag{7.204}$$

The angle ψ of the total magnetic vector with respect to the x-axis is expressed as:

$$\tan\psi(t) = h_y(t)/h_x(t) = \tan\beta_2 z. \tag{7.205}$$

The angle ψ increases as the plane wave propagates in the positive direction. The total vector of the high-frequency field $\boldsymbol{h}_\Sigma(t) = h_x(t)\boldsymbol{x}^0 + h_y(t)\boldsymbol{y}^0$ rotates as the wave propagates. The rotation is clockwise if looking in the positive direction of the z-axis. The rotation of field vectors in a gyrotropic medium is called *Faraday effect*, and the value β_2, which determines the rotation of the magnetic field vector per unit length in a gyrotropic medium is called *Faraday constant*.

So far, we have considered the plane EMW propagating in the positive z-direction. If the wave propagates in the negative z-direction, this case corresponds to the opposite signs before β^+ and β^-. Therefore, the rotation angle is negative:

$$\tan\psi(t) = -\tan\beta_2 z = \tan(-\beta_2 z). \tag{7.206}$$

The angle increases as the wave propagates in the negative direction. The rotation of the total field $\boldsymbol{h}_\Sigma(t)$ is in the opposite direction as compared to the case when the wave propagates in the positive direction. Therefore, an effect of a ferrite medium on the EMW propagation depends on the direction of this EMW. This is a property of *nonreciprocal media*.

Faraday effect and nonreciprocity are widely used to build different microwave devices, including phase shifters, circulators, and isolators.

7.4.1.2 Transverse Magnetization of Ferrite Medium and Cotton–Mouton Effect

Another interesting case is when the Poynting vector of the plane EMW propagating through the unbounded ferrite medium is perpendicular to the magnetic moment vector of the medium (due to the corresponding direction of the applied bias magnetic field). This is the case of *transverse magnetization* of the ferrite medium. In this case, let us assume that $\boldsymbol{M} = M_0\boldsymbol{z}^0$, and the wave propagates along the positive x-axis. In this case, there are the following field components:

$$\begin{aligned} & e_y, e_z, h_y, h_z \neq 0; \\ & e_x = h_x = 0; \\ & \frac{\partial}{\partial x} \neq 0, \frac{\partial}{\partial y} = \frac{\partial}{\partial z} = 0. \end{aligned} \tag{7.207}$$

For a ferrite medium magnetized along the z-direction, there will be the following scalar equations following from Maxwell's equations:

$$-\frac{dh_z}{dx} = j\omega\varepsilon e_y; \tag{7.208}$$

$$\frac{dh_y}{dx} = j\omega\varepsilon e_z; \tag{7.209}$$

$$0 = -j\omega\mu h_x + \omega\kappa h_y; \tag{7.210}$$

$$-\frac{de_z}{dx} = -\omega\kappa h_x - j\omega\mu h_y; \tag{7.211}$$

$$\frac{de_y}{dx} = -j\omega\mu_{\|} h_z. \tag{7.212}$$

As there is $h_y \neq 0$ in the incident wave, it cannot be zero inside the ferrite medium. Due to anisotropy ($\kappa \neq 0$), in the ferrite medium a new component $h_x \neq 0$ appears – there was no such a component in the incident wave. This component is shifted by the angle $-\pi/2$ with respect to h_y:

$$h_x = -j\frac{\kappa}{\mu} h_y. \tag{7.213}$$

Such phase shift results in the rotation of polarization plane.

Equations (7.208) and (7.212) are independent of (7.209), (7.210), and (7.211) and so can be solved separately. When Equation (7.212) is differentiated by x, then using Equation (7.208), one can get the following differential equation:

$$\frac{d^2 e_y}{dx^2} + \omega^2 \mu_{\|} \varepsilon e_y = 0. \tag{7.214}$$

The standard way of solving this equation is to represent it as a sum of two waves propagating in the opposite directions:

$$e_y = A_1 e^{-j\beta_o x} + A_2 e^{j\beta_o x}, \tag{7.215}$$

where

$$\beta_o = \omega\sqrt{\mu_{\|}\varepsilon}. \tag{7.216}$$

The ferrite medium is uniform and unbounded; therefore, the reflected wave is absent so that $A_2 = 0$, and the so-called ordinary wave is expressed as:

$$e_y = A_1 e^{-j\beta_o x}. \tag{7.217}$$

The second system of equations is formed of Equations (7.209) and (7.211). The component h_x can be excluded using Equation (7.210). Then one can get:

$$\frac{de_z}{dx} = -j\omega\left(\mu - \frac{\kappa^2}{\mu}\right) h_y. \tag{7.218}$$

Differentiating Equation (7.218) by x and using Equation (7.209), one can get:

$$\frac{d^2 e_z}{dx^2} + \omega^2 \varepsilon \left(\frac{\kappa^2}{\mu} - \mu\right) e_z = 0. \tag{7.219}$$

This differential equation has the solution as two waves propagating in the opposite directions:

$$e_z = B_1 e^{-j\beta_e x} + B_2 e^{j\beta_e x}, \tag{7.220}$$

where

$$\beta_e = \omega\sqrt{\varepsilon\left(\frac{\kappa^2}{\mu} - \mu\right)}. \tag{7.221}$$

In the case of the uniform unbounded ferrite medium, there are no reflections; therefore:

$$e_z = B_1 e^{-j\beta_e x}. \tag{7.222}$$

This solution corresponds to the *extraordinary wave.*

As the solution for e_y and e_z are obtained, the solution for h_z is found analogously to e_y; and the solution for h_y is analogous to e_z. Therefore, the magnetic field in the case of the extraordinary wave has two components. Moreover, the solution for the extraordinary wave has the component h_x due to Equation (7.210), which is phase shifted in time with respect to h_y by $\pi/2$. Therefore, in the extraordinary wave, there is a rotation of the field vectors as the wave progresses along the x-axis. The appearance of the extraordinary wave at the transverse magnetized ferrite medium with respect to the propagation axis is called *Cotton–Mouton effect,* well known in physical optics and related to *birefringence* in a liquid medium due to bias transverse magnetic field (the analogous Voigt effect exists in gases).

7.4.2 Perturbation of Cavities and Waveguides by Ferrite Samples

The presence of anisotropic, and at the same time frequency dispersive media such as ferrites, usually results in an increased complexity of electrodynamic problems. When considering such media, it is reasonable to find approximate solutions rather than trying to solve complex boundary problems rigorously. The *perturbation techniques* have proven to be very useful and comparatively general in solving problems containing ferrites. Perturbation techniques are based on the variational analysis. The parameters, which are the functionals of EM fields, e.g. eigenfrequencies of resonators and eigenmodes in waveguides, can be derived using perturbation techniques. Assume that a solution of a problem for an unperturbed (not containing any ferrite) electrodynamic system is known. Then, small variation, or perturbation, can be introduced to model the system of interest which contains some amount of ferrite. Perturbation could be done in two ways. The first is the slight variation of the material parameters of the medium, e.g. an appearance of small components responsible for the gyrotropic effects in magnetic susceptibility tensor. The second way allows for significant variations of material parameters, but only locally in a small volume.

It is known that the presence of any amount of ferrite in a waveguide affects the EM field in the waveguide causing phase shift and attenuation. However, it is difficult to quantify which part of the effect is caused by dielectric and which part is caused by magnetic properties of the ferrite sample. This problem is usually solved by placing a small ferrite sample into a hollow cavity with the intrinsic resonance frequency ω_0 and the unperturbed quality factor Q_0. The sample will change characteristics of the cavity: will shift the resonance frequency and change the Q factor. To measure magnetic properties of the sample, it is usually placed in such location where the electric field is minimum, while magnetic field is maximum. To measure dielectric properties, it is done vice versa – the electric field should be maximum, and magnetic field should be minimum in the points where the sample is inserted. The variation of Q factor determines the dissipative part of either permeability or permittivity, and the resonance frequency shift usually is an indication about the dispersive part of one of the components (usually, the diagonal, if the material parameter is characterized by a tensor).

For *a cavity with a small ferrite sample,* the relative frequency shift is equal to the relative variation of the energy in the cavity:

$$\frac{\Delta\omega}{\omega_0} = -\frac{\Delta W}{W_0}, \tag{7.223}$$

where W_0 is the energy of the unperturbed cavity. The energy W_0 stored in an empty (air-filled) cavity of the volume V is calculated in a standard way as:

$$W_0 = \frac{1}{2}\int_V \left(\varepsilon_0 \boldsymbol{e}_0 \cdot \boldsymbol{e}_0^* + \mu_0 \boldsymbol{h}_0 \cdot \boldsymbol{h}_0^*\right) dV = \frac{1}{2}\int_V \left(\varepsilon_0 |e_0|^2 + \mu_0 |h_0|^2\right) dV \tag{7.224}$$

where (*) sign denotes complex conjugation.

If there is a small ferrite sample of the volume $\Delta V \ll V$ in the point of the cavity, where the electric field is zero, but magnetic field is maximum, then the variation of energy will be determined by the magnetic induction $\boldsymbol{b}$ and magnetic field $\boldsymbol{h}$ inside the ferrite sample only:

$$\Delta W = \int_{\Delta V} (\boldsymbol{b}\cdot\boldsymbol{h}^*)dV. \tag{7.225}$$

Maxwell's equations in complex amplitudes (with omitted $e^{j\omega_0 t}$ term) in an empty cavity are:

$$\begin{aligned}\nabla\times\boldsymbol{h}_0 &= j\omega_0\varepsilon_0\boldsymbol{e}_0;\\ \nabla\times\boldsymbol{e}_0 &= -j\omega_0\mu_0\boldsymbol{h}_0\end{aligned} \tag{7.226}$$

If a small ferrite sample is placed inside this cavity, the field components will become $\boldsymbol{h}$ and $\boldsymbol{e}$; the resonance frequency will change (ω vs. ω_0), and Maxwell's equations will be:

$$\begin{aligned}\nabla\times\boldsymbol{h} &= j\omega\varepsilon_0\boldsymbol{e}+\boldsymbol{j}_e;\\ \nabla\times\boldsymbol{e} &= -j\omega\mu_0\boldsymbol{h}+\boldsymbol{j}_m,\end{aligned} \tag{7.227}$$

where $\boldsymbol{j}_{e,m}$ are impressed electric and magnetic current densities; they are nonzero only within the region ΔV, where the small ferrite sample is placed. In the perturbing media, these current densities are determined as:

$$\begin{aligned}\boldsymbol{j}_e &= j\omega\varepsilon_0\vec{\chi}_e\boldsymbol{e} = j\omega\left(\vec{\varepsilon}-\varepsilon_0\vec{\boldsymbol{U}}\right)\boldsymbol{e};\\ \boldsymbol{j}_m &= -j\omega\mu_0\vec{\chi}_m\boldsymbol{h} = -j\omega\left(\vec{\mu}-\mu_0\vec{\boldsymbol{U}}\right)\boldsymbol{h},\end{aligned} \tag{7.228}$$

where $\vec{\chi}_{e,m}$ are the dielectric and magnetic susceptibility tensors and $\vec{\boldsymbol{U}}$ is the unit tensor.

Let us scalar multiply the first line in Equation (7.227) by $\boldsymbol{e}_0^*$ and the second by $\boldsymbol{h}_0^*$. Then let us scalar multiply the first line of Equation (7.226) by $\boldsymbol{e}^*$ and the second by $\boldsymbol{h}^*$. Then by adding together the first two of the resultant equations and subtracting the other two equations, one can get:

$$\begin{aligned}&\int_V \left(\boldsymbol{e}_0^*\cdot\nabla\times\boldsymbol{h}+\boldsymbol{e}\cdot\nabla\times\boldsymbol{h}_0^*-\boldsymbol{h}_0^*\cdot\nabla\times\boldsymbol{e}-\vec{h}\cdot\nabla\times\boldsymbol{e}^*\right)dV =\\ &j(\omega-\omega_0)\int_V\left(\varepsilon_0\boldsymbol{e}_0^*\cdot\boldsymbol{e}+\mu_0\boldsymbol{h}_0^*\cdot\boldsymbol{h}\right)dV+\int_{\Delta V}\left(\boldsymbol{j}_e\cdot\boldsymbol{e}-\boldsymbol{j}_m\boldsymbol{h}_0^*\right)dV.\end{aligned} \tag{7.229}$$

The left-hand side of Equation (7.229) can be simplified using the known equalities of vector algebra, and the volume integral can be reduced to the surface one using Stokes theorem from vector calculus. Considering the walls of the cavity as perfect electric conductors, the tangential components of electric field on their surfaces can be set as zero, and after some simplifications, the following expression for the cavity resonance frequency shift can be derived:

$$\omega-\omega_0 = j\frac{\int_{\Delta V}\left(\boldsymbol{j}_e\cdot\boldsymbol{e}_0^*-\boldsymbol{j}_m\cdot\boldsymbol{h}_0^*\right)dV}{\int_V\left(\varepsilon_0\boldsymbol{e}_0^*\cdot\boldsymbol{e}+\mu_0\boldsymbol{h}_0^*\cdot\boldsymbol{h}\right)dV}. \tag{7.230}$$

If distribution of fields in the cavity is known, this expression can be used to calculate frequency shift. But the distributions for $\boldsymbol{e}$ and $\boldsymbol{h}$ inside the ferrite are unknown. However, if the ferrite sample is extremely small as compared to the size of the cavity, as well as compared to the wavelengths in the cavity and in the ferrite material, the fields anywhere inside the cavity can be considered as quasistatic so that $\boldsymbol{e}\approx\boldsymbol{e}_0$ and $\boldsymbol{h}\approx\boldsymbol{h}_0$. Note that the field inside the ferrite sample is related to the external field through demagnetization factors. Also, the average in time electric and magnetic energy parts in a cavity are practically equal, so either the magnetic or the electric field part in the denominator of Equation (7.230) can be doubled.

Then, when measuring magnetic susceptibility in the points, where $\boldsymbol{e}\approx\boldsymbol{e}_0=0$, the observed frequency shift is:

$$\omega - \omega_0 \approx -\frac{\omega_0 \int\limits_{\Delta V} \left(\ddot{\chi}_m \cdot \boldsymbol{h}_0\right) \cdot \boldsymbol{h}_0^* dV}{2\int\limits_V \left(\boldsymbol{h}_0^* \cdot \boldsymbol{h}_0\right) dV}. \tag{7.231}$$

Similarly, when measuring the dielectric susceptibility in the points, where $\boldsymbol{h} \approx \boldsymbol{h}_0 = 0$, the frequency shift is calculated as:

$$\omega - \omega_0 \approx -\frac{\omega_0 \int\limits_{\Delta V} \left(\ddot{\chi}_e \cdot \boldsymbol{e}_0\right) \cdot \boldsymbol{e}_0^* dV}{2\int\limits_V \left(\boldsymbol{e}_0^* \cdot \boldsymbol{e}_0\right) dV}. \tag{7.232}$$

Equations (7.231) and (7.232) can be also represented as relative frequency shifts:

$$\frac{\omega - \omega_0}{\omega_0} \approx -\frac{\int\limits_{\Delta V} \left(\Delta\ddot{\mu} \cdot \boldsymbol{h}_0\right) \cdot \boldsymbol{h}_0^* dV}{2\mu_0 \int\limits_V \left|h_0\right|^2 dV} \tag{7.233}$$

and

$$\frac{\omega - \omega_0}{\omega_0} \approx -\frac{\int\limits_{\Delta V} \left(\Delta\ddot{\varepsilon} \cdot \boldsymbol{e}_0\right) \cdot \boldsymbol{e}_0^* dV}{2\varepsilon_0 \int\limits_V \left|e_0\right|^2 dV}, \tag{7.234}$$

where the increments in permittivity and permeability are $\Delta\ddot{\varepsilon} = \ddot{\varepsilon} - \varepsilon_0 \ddot{\boldsymbol{U}}$ and $\Delta\ddot{\mu} = \ddot{\mu} - \mu_0 \ddot{\boldsymbol{U}}$, respectively. Note that these are tensors.

Equations (7.233) and (7.234) agree with Equation (7.223), as numerators are the variations of the stored energy in the cavity due to the presence of the ferrite sample and denominators are the stored energy of unperturbed cavity.

Now, let us consider the effect of dissipative parts of complex magnetic and dielectric susceptibilities. Let us first consider an *isotropic ferrite material*. When a small sample of this material is placed in the points where either magnetic, or electric fields are maximum, then Equations (7.231) and (7.232) transform to:

$$\omega - \omega_0 \approx -\omega_0 \frac{\chi_m h_{\max}^2 \Delta V}{2\int\limits_V \left|h_0\right|^2 dV} \tag{7.235}$$

and

$$\omega - \omega_0 \approx -\omega_0 \frac{\chi_e e_{\max}^2 \Delta V}{2\int\limits_V \left|e_0\right|^2 dV}, \tag{7.236}$$

where $\chi_{m,e}$ are the diagonal components of magnetic and dielectric susceptibility tensors, respectively. As susceptibilities are complex, $\chi_{m,e} = \chi'_{m,e} - j\chi''_{m,e}$, frequency ω can also be represented as complex through some variation around resonance frequency of the empty cavity as:

$$\omega = \omega_0 + \Delta\omega_0 + j(2\delta), \tag{7.237}$$

where $\Delta\omega_0$ is the real part of resonance frequency variation, and (2δ) is its imaginary part. Obviously, $\delta = \delta_{m,e}$ is related to either magnetic or dielectric loss in the ferrite medium.

Splitting the real and imaginary parts in Equations (7.235) and (7.236) after setting the frequency as in Equation (7.237), one can get the corresponding pairs of expressions:

$$\frac{\Delta\omega_0}{\omega_0} \approx -\frac{1}{2}\frac{\chi'_m h_{\max}^2 \Delta V}{\int\limits_V h_0^2 dV} \Leftrightarrow 2\delta_m \approx \frac{\chi''_m h_{\max}^2 \Delta V}{\int\limits_V h_0^2 dV} \tag{7.238}$$

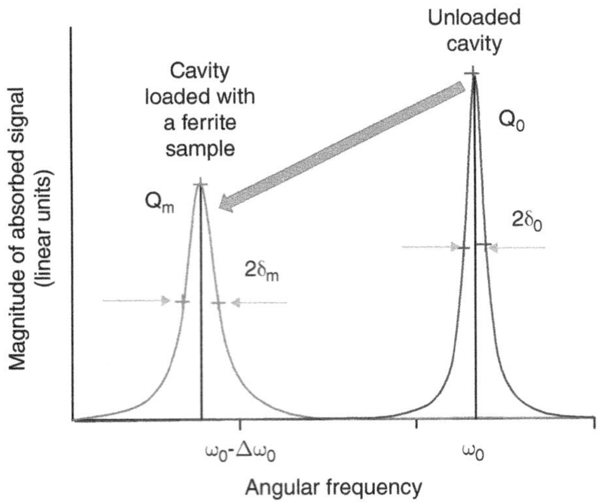

Figure 7.31 Shift of resonance and reduction of Q factor in the loaded by a ferrite sample cavity as compared to the unloaded one.

and

$$\frac{\Delta\omega_0}{\omega_0} \approx -\frac{1}{2}\frac{\chi_e' e_{\max}^2 \Delta V}{\int\limits_V e_0^2 dV} \Leftrightarrow 2\delta_e \approx \frac{\chi_e'' e_{\max}^2 \Delta V}{\int\limits_V e_0^2 dV} \tag{7.239}$$

The parameters $(2\delta_{m,e})$ are associated with the loss in the loaded cavity due to the ferrite sample and hence to the variations of the corresponding Q factors:

$$2\delta_{m,e} = \Delta\left(\frac{1}{Q_{m,e}}\right) = \frac{1}{Q_{m,e}} - \frac{1}{Q_0}, \tag{7.240}$$

where Q_0 is the quality factor of the unloaded cavity and $Q_{m,e}$ are the loaded Q factors due to the magnetic and dielectric loss of the ferrite sample, respectively. Measuring the variations in the Q factor of the cavity, it is possible to measure dissipative parts of magnetic and/or dielectric losses of a ferrite sample. This is illustrated in Figure 7.31.

Also, directly using Equations (7.238) and (7.239), one can calculate the resonance frequency shift and the variation of quality factor, if the EM field distribution in the cavity is known; the parameters of the ferrite sample, including its volume, are known; and the coordinates of the points where the sample is placed (with either zero magnetic field, or zero electric field) are also known.

For example, consider a small ferrite sample placed in a point $P_1(x = a/2; y = b/2; z = 0)$ in the center of a wall of a rectangular metal cavity with the dimensions $(x = a; y = b; z = L)$ supporting the fundamental mode TE_{10n} [110]. This is illustrated in Figure 7.32.

The EM field components are:

$$\begin{aligned} &e_x = e_z = h_y = 0;\\ &e_y = \sin\left(\frac{\pi x}{a}\right)\sin\left(\frac{\pi z}{L}\right);\\ &h_x = \frac{a}{a^2+L^2}\sin\left(\frac{\pi x}{a}\right)\cos\left(\frac{\pi z}{L}\right);,\\ &h_z = \frac{L}{a^2+L^2}\cos\left(\frac{\pi x}{a}\right)\sin\left(\frac{\pi z}{L}\right). \end{aligned} \tag{7.241}$$

Assume that the bias magnetic field is normal (along the y-axis) to the high-frequency magnetic field (h_x, h_z). The only nonzero field component at $P_1(a/2, b/2, 0)$ is:

$$h_x = \frac{a}{a^2+L^2}. \tag{7.242}$$

This gives the following term in Equation (7.237):

$$\frac{h_{\max}^2 \Delta V}{2\int\limits_V h_0^2 dV} = \frac{2a^2 \Delta V}{\left(a^2+L^2\right)V}. \tag{7.243}$$

Then Equations (7.238) and (7.239) result in:

$$\begin{aligned} &\frac{\Delta\omega_0}{\omega_0} \approx -\frac{\chi_m' \Delta V}{V}\cdot\frac{2a^2}{a^2+L^2};\\ &\Delta\left(\frac{1}{2Q_m}\right) = \frac{\chi_m'' \Delta V}{V}\cdot\frac{2a^2}{a^2+L^2}, \end{aligned} \tag{7.244}$$

where $\chi_m = \chi_{xx} = \chi_{zz}$ is the diagonal component of the magnetic susceptibility tensor (note that $\chi_{yy} = 0$ in this geometry).

To measure dielectric susceptibility, one must place the sample in the point with maximum electric field. In this case, Equation (7.239) becomes:

$$\frac{e_{y\max}^{2}\Delta V}{2\int_{V} e_{y}^{2}dV}=\frac{2\Delta V}{V}, \tag{7.245}$$

and then Equations (7.238) and (7.239) result in:

$$\begin{aligned}\frac{\Delta\omega_0}{\omega_0} &\approx -\chi_e'\cdot\frac{2\Delta V}{V};\\ \Delta\left(\frac{1}{2Q_m}\right) &= \chi_e''\cdot\frac{4\Delta V}{V}.\end{aligned} \tag{7.246}$$

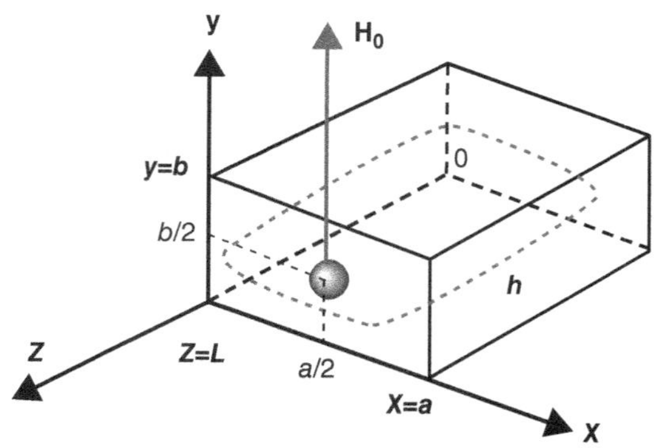

Figure 7.32 Rectangular cavity with a small ferrite sample in it.

So far, the small perturbation derivations were done for scalar values of magnetic susceptibility. However, if the *components of the susceptibility tensor* are needed to be retrieved, a cavity with two modes should be used [101]. The components of the fields in these modes should be perpendicular to each other and to the bias field, but resonance frequencies of these modes should be the same. The examples of such cavity is a resonator with a square cross-section (with resonance frequencies of the modes TE_{110} and TE_{101} being degenerate), or cylindrical resonator with the mode TE_{11n}, where the field in the cavity can be represented as a linear combination of the two modes with equal frequencies [110]. As the sample causes only a small perturbation, in the first-order approximation, it does not excite any other modes.

The perturbation technique is applied in [13, 15] to a cylindrical two-mode resonator with opposite directions of circular polarization of the magnetic field and two practically equal resonance frequencies. This structure is perturbed by a small ferrite sample magnetized to saturation by a bias field, perpendicular to the field components. For simplicity, it is assumed that the sample is placed on the axis of the cylinder. It is shown that if the resonances are degenerate in the empty resonator, when a ferrite sample is placed there, resonance frequencies of the modes split. This is due to the appearance of small off-diagonal components of the magnetic susceptibility tensor of the magnetized ferrite. Note that the problems like that can be solved for isotropic susceptibility first, and then the off-diagonal components can be introduced as small perturbation, too, substituting the scalar susceptibility by a tensor one.

Depolarization and demagnetization factors should also be taken into account. Equation (7.230) contains current densities $\boldsymbol{j}_{e,m}$. To calculate them using Equation (7.228), one needs to know the perturbed fields $\vec{e}$ and $\vec{h}$ inside the ferrite sample. However, in Equations (7.231) and (7.232), the internal susceptibilities and fields correspond to an unbounded ferrite medium. Therefore, it is important to use the internal fields inside the ferrite. One approach is to express the internal fields through the external fields using depolarization and demagnetization form factors of the sample. The other way is to represent current densities through the external fields and external, or effective, susceptibility tensor and then express this tensor through the internal tensor, but using form factors. The results of these two approaches should be the same, so these two approaches are equivalent.

The effective magnetic susceptibility diagonal and off-diagonal components are calculated as in Section 7.2 using the corresponding demagnetization factors, and the resonance frequency of the ferrite sample is also affected by these demagnetization factors. It is interesting that effective dielectric susceptibility of a ferrite sphere decreases due to depolarization by a factor of about 5 as compared to the dielectric susceptibility of the unbounded ferrite medium [13], as it can be calculated for a ferrite sample through the dielectric susceptibility of the ferrite medium χ_e as

$$\chi_e^{eff}=\frac{\chi_e}{1+N\chi_e}, \tag{7.247}$$

where N is the depolarization factor (for a sphere $N=1/3$, therefore, $\chi_e^{eff}=\frac{3\chi_e}{3+\chi_e}$).

Perturbation theory is useful for solving problems when *small ferrite samples are placed in waveguides.* Comparatively simple results can be obtained for a waveguide with a thin plate or a rod extended along the direction of propagation (see Figure 7.33). The necessary condition of using perturbation theory is quasistatic approximation of the internal field in a ferrite sample, i.e. at least one of the dimensions should be small as compared to the wavelength of EMW in this sample.

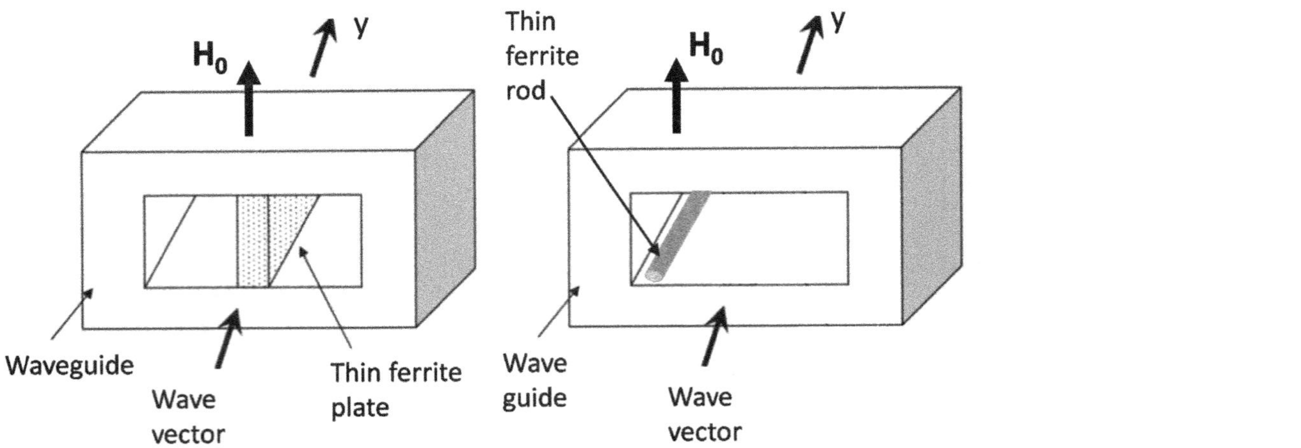

Figure 7.33 Thin ferrite samples (a plate or a rod) along the waveguide. Bias magnetization can be transverse as shown, or longitudinal.

As the transverse diagonal components of the tensors $\ddot{\mu}$ and $\ddot{\mu}$ are typically greater than the other components, the condition for quasistatic applicability can be written as [15]:

$$\beta_0 d\sqrt{|\varepsilon_{\max}|\cdot|\mu_{\max}|} < \frac{\pi}{6}, \tag{7.248}$$

where d is the characteristic dimension (diameter of a sphere, diameter of a rod, or thickness of a plate). In [15, chapter 4], a summary of formulas describing relations between unperturbed ($\boldsymbol{h}_0$) and perturbed ($\boldsymbol{h}$) magnetic fields in quasistatic approximation for a plate and a rod (for both tangential and normal bias fields), as well as a sphere, taking into account demagnetization form factors is given.

The problem of a small (at least in one dimension) ferrite sample has practical importance not only because it allows for measuring properties of these ferrite samples but also because ferrite samples in waveguides are used to design many microwave devices, e.g. isolators, phase shifters, and filters. This problem was considered in 1950s–1960s in many works related to Faraday rotation and nonreciprocal phenomena in waveguides with ferrite samples given in [13] and [15] and references therein. Many problems were treated using perturbation technique for rectangular and circular waveguides [103–105].

Briefly, Maxwell's equations are written for an empty waveguide in terms of the waves propagating in both directions along the waveguide axis. Then the same waveguide is considered with a cross-section totally or partially filled with a ferrite. The problem is translationally invariant along the propagation direction, i.e. cross-section of the waveguide is the same, and ferrite fills each cross-section identically. Using some vector algebra and vector calculus operations with Maxwell's equations in the cases of the empty waveguide and the one perturbed by a sample, the equation for the complex propagation constants in the empty waveguide γ_0 and with ferrite γ_f is obtained in the form [111]:

$$\gamma_f + \gamma_0^* = \frac{\int_{\Delta S}\left(\boldsymbol{j}_e\cdot\boldsymbol{e}_0^* + \boldsymbol{j}_m\cdot\boldsymbol{h}_0^*\right)dS}{\int_S \boldsymbol{z}^0\cdot\left(\boldsymbol{e}\times\boldsymbol{h}_0^* + \boldsymbol{e}_0^*\times\boldsymbol{h}\right)dS} = \frac{j\omega\int_{\Delta S}\left(\Delta\ddot{\varepsilon}\boldsymbol{e}\cdot\boldsymbol{e}_0^* + \Delta\ddot{\mu}\boldsymbol{h}\cdot\boldsymbol{h}_0^*\right)dS}{\int_S \boldsymbol{z}^0\cdot\left(\boldsymbol{e}\times\boldsymbol{h}_0^* + \boldsymbol{e}_0^*\times\boldsymbol{h}\right)dS}, \tag{7.249}$$

where S and ΔS are the cross-sectional areas of the waveguide and of the ferrite rod, respectively.

The derivation of Equation (7.249) is analogous to the case of a cavity with a small ferrite sample.

The variation of phase constant $(\beta - \beta_0)$ in the lossless case (neither ferrite, nor waveguide have losses), when $\gamma_0 = \alpha_0 + j\beta_0$ and $\gamma_f = \alpha_f + j\beta_f$, where $\alpha_0 = \alpha_f = 0$, can be expressed as:

$$\gamma_f - \gamma_0 = \beta_f - \beta_0 = \frac{\omega\int_{\Delta S}\left(\varepsilon_0\left(\ddot{\chi}_e\times\boldsymbol{e}\right)\cdot\boldsymbol{e}_0^* + \mu_0\left(\ddot{\chi}_m\times\boldsymbol{h}\right)\cdot\boldsymbol{h}_0^*\right)dS}{\int_S \boldsymbol{z}^0\cdot\left(\boldsymbol{e}\times\boldsymbol{h}_0^* + \boldsymbol{e}_0^*\times\boldsymbol{h}\right)dS}. \tag{7.250}$$

Using this approach and Equation (7.249), for example, for a case of a square waveguide (having the degenerate modes TE_{10} and TE_{01}) with a thin longitudinally magnetized ferrite rod passing through the center of the cross-section, the expressions for phase constant variation have been obtained [13]. The propagation of the wave is along the z-axis; the rod is magnetized in the same direction, and x and y are the coordinates in the cross-section of the waveguide. Losses in the waveguide and in the ferrite rod were neglected.

The difference in the phase velocities in this case is [13]:

$$\beta_f^{\pm} - \beta_0 = \frac{\Delta S}{S\beta_0}\left(\omega^2 \varepsilon_0 \mu_0 \chi_e^{eff} + \beta_0^2 (\chi_{xx}^{eff} \pm j\chi_{xy}^{eff})\right), \tag{7.251}$$

where ($\pm$) signs correspond to the positively and negatively circular polarized components of the wave propagating along the z-axis. The effective components of the magnetic susceptibility tensor and effective dielectric susceptibility are used to take into account demagnetization and depolarization. The effective dielectric susceptibility and the effective magnetic susceptibility of a long rod are:

$$\chi_e^{eff} = \frac{2\chi_e}{2+\chi_e} \text{ and } \chi_{\pm}^{eff} = \frac{2\chi_m^{\pm}}{2+\chi_m^{\pm}}, \tag{7.252}$$

where $\chi_m^{\pm} = \dfrac{\omega_M}{\omega_0 \mp \omega}$ for right (positive) and left (negative) circular polarizations of the incident wave, as discussed in Section 7.2. Note that loss in the ferrite rod is neglected. Therefore, Faraday rotation can be obtained due to the different phase velocities for two circular polarized waves with opposite polarizations. The rotation angle per unit length for a ferrite rod is:

$$\theta = \frac{\beta_f^- - \beta_f^+}{2} = -jC\beta_0 \chi_{xy}^{eff} \frac{\Delta S}{S}, \tag{7.253}$$

where $C$is the constant which depends on the waveguide cross-sectional geometry (C =1 for a square waveguide).

The similar effect can be obtained using a circular cylindrical waveguide with a ferrite rod in the center, as there are also two degenerate modes with the same frequencies [101].

Faraday rotation is usually well described using perturbation theory if the ferrite cross-section is much smaller than the cross-section of the waveguide (filling factor should be less than 3%...4%). As filling factor increases, experimental rotation exceeds theoretical predictions due to undesirable nonreciprocal field shift [112]. This effect is undesirable in the devices using Faraday rotation, as may cause instability of their operation; however, there are many devices which employ this field shift as a useful effect.

However, low filling factor is not a limitation for perturbation technique. For example, a circular or square waveguides completely filled with a ferrite medium can be easily treated using perturbation. To do this, ferrite is first considered as just a dielectric medium, and the propagation constant in the waveguide filled with this dielectric is calculated. Then, the magnetic susceptibility tensor is added as a perturbation. All the integrals are calculated similar to those for a thin ferrite rod. Depolarization and demagnetization factors do not need to be accounted for in this case. However, as ferrite has comparatively high relative permittivity and permeability, the product ($\mu\varepsilon$) may be significantly higher than ($\mu_0\varepsilon_0$), and the higher modes may be excited in the waveguide.

A nonreciprocal phase shifter can be designed using a thin transversally magnetized ferrite rod passing through the rectangular waveguide (fundamental mode TE_{10}) of any cross-section, placed anywhere in the waveguide. Complex phase constants for the waves propagating in opposite directions are different – both phase velocities and attenuations. An example of this is given in [111] for a lossless waveguide (with the mode TE_{10}) having the propagation y-axis and cross-sectional coordinates x (horizontal, the waveguide dimension a) and z (vertical, the waveguide dimension b), with a ferrite rod magnetized transversally (along the z-axis). The difference in phase constants is:

$$\beta_f^{\pm} - \beta_0 = \frac{\Delta S}{S\beta_0}\begin{pmatrix} \omega^2 \varepsilon_0 \mu_0 \cdot \chi_e' \sin^2(k_x x) + \chi_{xx}' \beta_0^2 \sin^2(k_x x) + \\ \chi_{yy}' k_x^2 \cos^2(k_x x) \pm \chi_{xy}' k_x \beta_0 \sin(2k_x x) \end{pmatrix}, \tag{7.254}$$

and the loss constant, assuming that there are only magnetic losses in the ferrite, is:

$$\alpha_f^{\pm} = \frac{\Delta S}{S\beta_0}\begin{pmatrix} \chi_{xx}'' \beta_0^2 \sin^2(k_x x) + \\ \chi_{yy}'' k_x^2 \cos^2(k_x x) \mp \chi_{xy}'' k_x \beta_0 \sin(2k_x x) \end{pmatrix}, \tag{7.255}$$

where $k_x = \frac{\pi}{a}$. The difference of the phase constants for the right and left polarizations is:

$$\beta_f^+ - \beta_f^- = 2\chi'_{xy} k_x \frac{\Delta S}{S} \sin(2k_x x). \tag{7.256}$$

The maximum phase shift will be if the ferrite rod is placed at $x = a/4$ or $x = 3a/4$, which are the points of the circular polarization of the magnetic field of the TE_{10} mode. This is valid for $\frac{\Delta S}{S} < 3\%$.

There is also a position of a rod in the waveguide, where the isolation ratio, or the ratio of the return loss to the direct loss, $R = \frac{\alpha_f^-}{\alpha_f^+}$, is maximum. To find this position, it is necessary to calculate the derivative $\frac{dR}{dx} = 0$ to determine the maximum of the function. Then the optimal position of the rod is determined from the transcendental equation

$$\cos(2k_x x) = \frac{\chi''_{xx}\beta^2 - \chi''_{yy}k^2}{\chi''_{xx}\beta^2 + \chi''_{yy}k^2}. \tag{7.257}$$

For a resonance isolator, the dissipative parts of susceptibility tensor are determined at the FMR, and therefore, the optimal position of the rod in the rectangular waveguide is determined by the ferrite type, resonance frequency, and the waveguide dimensions. Maximum $\boldsymbol{R}$ is related to the Q factor of the resonance isolator, or the ferrite resonance line width:

$$R_{\max} = \frac{\sqrt{\chi''_{xx}\chi''_{yy}} + \chi''_{xy}}{\sqrt{\chi''_{xx}\chi''_{yy}} - \chi''_{xy}} \approx \frac{4\omega_{\text{res}}}{\mu_0 \gamma \Delta H}. \tag{7.258}$$

To achieve as high $R_{\max}$ as possible, ferrites with narrow FMR line width are required; also, increase of ω_{res} results in the increase of the isolation ratio. Therefore, resonance isolators are very useful at mm and sub-mm wavelengths. Magneto-uniaxial polycrystalline hexagonal ferrites with oriented magnetic moments in crystallites are especially important for that, as they allow for obtaining high phase shifts and high resonance isolation ratios; moreover, they do not need high bias fields due to significant intrinsic crystallographic anisotropy fields [12]. Some hexaferrite resonance isolators on different transmission lines for mm wave band are shown in Figure 7.34. They are operating over frequency ranges from 25 to 150 GHz on metal and dielectric waveguides and from 40 to 150 GHz on metal waveguiding structures. The bias magnetic field ranges from 0 to 3 kOe. The insertion loss for these isolators is from 0.3 to 1 or 2 dB (does not exceed 3 dB), the return loss is from 20 to 30 dB depending on the isolator, and the voltage standing wave ratio is less than 1.3 for any designed isolator. Characteristics are preserved at average power up to 200 mW, over the temperature range from −40 to +60 °C [12].

Application of perturbation technique is also useful for a design of devices, which employ Cotton–Mouton effect, or birefringence [113]. This effect is used, e.g. in 180° phase shifters, which are used as modulators of single-sided frequency bands [114, 115]. In this case, a ferrite rod is placed in a square or circular waveguide with mode degeneration, either in the center of the waveguide or near one of the walls. Bias field is transverse to the ferrite rod. For one of the two degenerate modes, the high-frequency magnetic field is normal to the bias field and for the other is parallel, and phase constants for the modes are different. This results in the double refraction. Also, the field polarization rotation about the waveguide axis is accompanied by the frequency shift. When a ferrite rod is placed near one of the walls of the square waveguide,

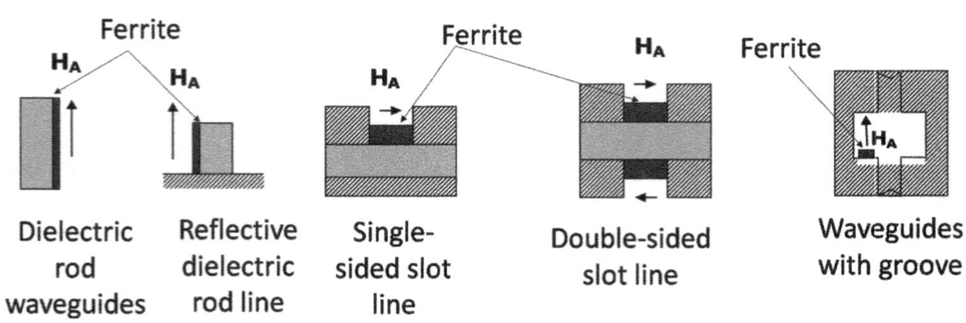

Figure 7.34 Some types of resonance isolators for millimeter waves based on polycrystalline hexagonal ferrites [12].

the difference of phase constants of the two modes can be 90^0. If there is a linearly polarized wave incident at the angle of 45° with respect to the horizontal plane at the input, it will form a circular polarized wave at the output. The similar effect can be obtained with a ferrite rod in the center of the waveguide.

Reciprocal phase shifters can also be designed using small perturbation theory. In this case, the longitudinally magnetized ferrite plate is placed along the waveguide axis. Perturbation theory can be used even in the case of comparatively high filling factor for a ferrite in the waveguide (~25%). In this case, the ferrite plate is substituted by a dielectric one, and the rigorous electrodynamic boundary problem is solved. Then components of magnetic susceptibility tensor are introduced as a perturbation. The problem with scalar permeability is solved rigorously; then off-diagonal tensor components are taken into account as perturbations. Reciprocal phase shifters are considered [116–119].

Some of the modern ferrite microwave devices based on physical peculiarities of propagation of EMWs in gyromagnetic media are overviewed in papers [120–123].

7.4.3 Small Ferrite Ellipsoid in a Waveguide

In this section, we will consider a small FR of an ellipsoid shape placed in a regular (with translational invariance in its cross-section) waveguide; this may also be a transmission line with TEM or quasi-TEM mode. As the FR dimensions are much less than the wavelength in the ferrite medium, the small perturbation technique with magnetostatic approximation of the internal field inside the ferrite sample can be applied. We are interested in finding the EM field outside this sample, which is related to the reflection (R), transmission (T), and absorption (A) coefficients in the cross-section, where this FR is placed.

To solve this problem, we will employ the *self-matched field approach* proposed by A.G. Gurevich [15,105,106,124]. This means that the magnetization vector of the FR is determined through the total magnetic field which is the sum of the known incident field $\boldsymbol{h}_0$ and the field radiated by the sample into the transmission line in both directions of propagation $\boldsymbol{h}_r$. This is illustrated in Figure 7.35.

As the ellipsoid is small, its magnetization is considered to be uniform and is determined through the external tensor of magnetic susceptibility:

$$\boldsymbol{m} = \ddot{\chi}^{\text{ext}}(\boldsymbol{h}_0 + \boldsymbol{h}_r). \tag{7.259}$$

In order to find the coefficients R, T, and A, it is sufficient to find the relation between $\boldsymbol{h}_0$ and $\boldsymbol{h}_r$, excluding vector $\boldsymbol{m}$. The radiated field can be written as:

$$\boldsymbol{h}_r = -j\ddot{\boldsymbol{w}} \times \boldsymbol{m}. \tag{7.260}$$

where $\ddot{w}$ is a tensor, which depends on frequency, geometries of a waveguide and FR, and the point in the waveguide where the FR is placed. This field can be represented as a sum of eigenmodes in the waveguide:

$$\boldsymbol{h}_r^{\pm} = \sum_{n=1}^{N} C_n^{\pm} \boldsymbol{h}_n^{\pm}, \tag{7.261}$$

where the coefficients of this series are determined through the integral of the scalar product:

$$C_n^{\pm} = -\frac{1}{N_n} \int_{V_0} (\boldsymbol{j}^m \cdot \boldsymbol{h}_n^{\mp}) dV, \tag{7.262}$$

where N_n are the norms of the eigenmodes determined through the condition of orthogonality of the full set of modes and is proportional to the complex power of the mode passing through the waveguide cross-section S:

$$N_n = -4P = -2 \int_S \boldsymbol{e}_n \times \boldsymbol{h}_n^* ds, \tag{7.263}$$

where $\boldsymbol{e}_n$ and $\boldsymbol{h}_n^*$ are the microwave electric and conjugated magnetic field of the nth mode, respectively.

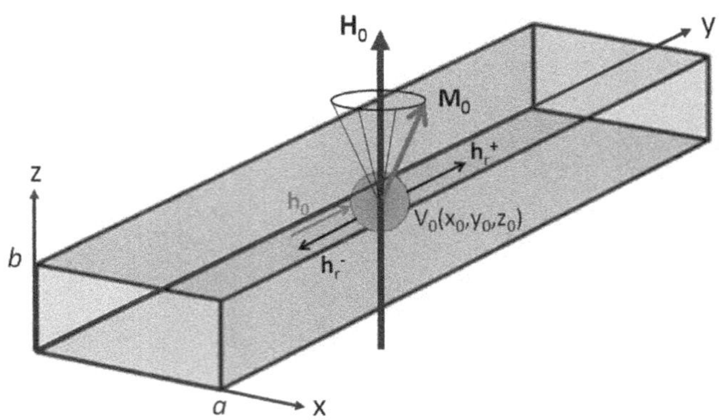

Figure 7.35 Geometry describing small ferrite resonator in transmission line for the self-matched field approach.

Herein, all the RF/microwave fields are written in terms of complex amplitudes, but the exponential terms are omitted. The signs of $C_n^{\mp}$ and magnetic field components determine the direction of waves (modes) propagation – positive ($e^{-j\beta_y y}$) and negative ($e^{+j\beta_y y}$), respectively.

Assume that the FR of the volume V_f is so small can be considered as an elementary magnetic dipole with current density $\boldsymbol{j}^m$ placed in the point $V_0(x_0, y_0, z_0)$ in the waveguide described as:

$$\boldsymbol{j}^m = j\omega\mu_0 V_f \boldsymbol{m} \times \delta(V - V_0). \tag{7.264}$$

Then, the coefficients (7.262) can be calculated as:

$$C_n^{\pm} = -\frac{1}{N_n}\int_{V_0} V_f j\omega\mu_0 \boldsymbol{m} \times \delta(V - V_0)\boldsymbol{h}_n^{\mp} dV = -\frac{j\omega\mu_0 V_f}{N_n}(m_x h_{xn}^{\mp} + m_y h_{yn}^{\mp} + m_z h_{zn}^{\mp})_{V_0} \tag{7.265}$$

The radiated fields in the positive and negative directions of a regular transmission line are expressed as:

$$\boldsymbol{h}_r^{\pm} = -j\omega\mu_0 V_f \times \sum_{n=1}^{N} \frac{\boldsymbol{h}_n^{\pm}}{N_n}(m_x h_{xn}^{\mp} + m_y h_{yn}^{\mp} + m_z h_{zn}^{\mp})_{V_0} \tag{7.266}$$

According to the self-matched field approach, the ferrite ellipsoid, which is situated in the point $V_0(x_0, y_0, z_0)$ is affected by its own total radiated field taken as an average of fields emitted in both directions:

$$\boldsymbol{h}_r(V_0) = \frac{\boldsymbol{h}_r^{+}(V_0) + \boldsymbol{h}_r^{-}(V_0)}{2}. \tag{7.267}$$

Hence, from Equations (7.266) and (7.267), one can get the expression for the radiated field:

$$\boldsymbol{h}_r(V_0) = -\frac{j\omega\mu_0 V_f}{2} \times \sum_{n=1}^{N} \frac{1}{N_n}\begin{vmatrix} \boldsymbol{h}_n^{+}(V_0)\left(m_x h_{xn}^{-} + m_y h_{yn}^{-} + m_z h_{zn}^{-}\right)_{V_0} + \\ \boldsymbol{h}_n^{-}(V_0)\left(m_x h_{xn}^{+} + m_y h_{yn}^{+} + m_z h_{zn}^{+}\right)_{V_0} \end{vmatrix}. \tag{7.268}$$

Further, we will omit notations that the fields are taken in the point V_0. The microwave magnetic moment of the ellipsoid in the point V_0 is then obtained by substituting Equation (7.268) in Equation (7.259):

$$\boldsymbol{m} = \vec{\chi}_m^{\text{ext}} \times (\boldsymbol{h}_0 + \boldsymbol{h}_r) = \vec{\chi}_m^{\text{ext}} \times \left(\boldsymbol{h}_0 + \frac{\boldsymbol{h}_r^{+} + \boldsymbol{h}_r^{-}}{2}\right). \tag{7.269}$$

However, it would be convenient for the further analysis to represent this magnetic moment in the form

$$\boldsymbol{m} = \vec{\chi}^{\text{eq}} \times \boldsymbol{h}_0, \tag{7.270}$$

where $\vec{\chi}^{\text{eq}}$ is the equivalent tensor accounting for demagnetization factors in the FR and the geometry of the waveguide.

For simplicity, let us consider the transmission line with just a single mode of the order n; then the entire solution can be combined as a superposition of these modes. We can find the geometry-related tensor $\vec{w}_n$ included in Equation (7.260) using Equations (7.268) and (7.269) as:

$$\vec{\boldsymbol{w}}_n = \frac{\omega\mu_0 V_f}{2N_n} \times \begin{bmatrix} h_{xn}^{+}h_{xn}^{-} & h_{yn}^{+}h_{xn}^{-} & h_{zn}^{+}h_{xn}^{-} \\ h_{xn}^{+}h_{yn}^{-} & h_{yn}^{+}h_{yn}^{-} & h_{zn}^{+}h_{yn}^{-} \\ h_{xn}^{+}h_{zn}^{-} & h_{yn}^{+}h_{zn}^{-} & h_{zn}^{+}h_{zn}^{-} \end{bmatrix}. \tag{7.271}$$

Then the final expression for the geometry-related tensor is:

$$\vec{\boldsymbol{w}}_n = \begin{bmatrix} w_{11} & w_{12} & w_{13} \\ w_{12} & w_{22} & w_{23} \\ w_{13} & w_{23} & w_{33} \end{bmatrix} = \frac{\omega\mu_0 V_f}{2} \times \begin{bmatrix} \sum_{n=1}^{N} \frac{h_{xn}^{+}h_{xn}^{-}}{N_n} & \sum_{n=1}^{N} \frac{h_{yn}^{+}h_{xn}^{-}}{N_n} & \sum_{n=1}^{N} \frac{h_{zn}^{+}h_{xn}^{-}}{N_n} \\ \sum_{n=1}^{N} \frac{h_{xn}^{+}h_{yn}^{-}}{N_n} & \sum_{n=1}^{N} \frac{h_{yn}^{+}h_{yn}^{-}}{N_n} & \sum_{n=1}^{N} \frac{h_{zn}^{+}h_{yn}^{-}}{N_n} \\ \sum_{n=1}^{N} \frac{h_{xn}^{+}h_{zn}^{-}}{N_n} & \sum_{n=1}^{N} \frac{h_{yn}^{+}h_{zn}^{-}}{N_n} & \sum_{n=1}^{N} \frac{h_{zn}^{+}h_{zn}^{-}}{N_n} \end{bmatrix}. \tag{7.272}$$

This tensor turns out to be symmetrical: the propagation terms containing $e^{\mp j\gamma x}$ for the field propagating in the opposite directions along the x-axis cancel out. Then the expressions for the equivalent tensor $\ddot{\chi}^{\text{eq}}$ can be obtained by solving Equation (7.259) for $\boldsymbol{m}$ as:

$$\boldsymbol{m} = \left[\ddot{\boldsymbol{U}} + j\ddot{\chi}_m^{\text{ext}} \times \ddot{\boldsymbol{w}}\right]^{-1} \times \ddot{\chi}_m^{\text{ext}} \times \boldsymbol{h}_0, \tag{7.273}$$

where $\ddot{\boldsymbol{U}}$ is the unit tensor. This means that the equivalent tensor of magnetic susceptibility of an FR in a waveguide is

$$\ddot{\chi}^{\text{eq}} = \left[\ddot{\boldsymbol{U}} + j\ddot{\chi}_m^{\text{ext}} \times \ddot{\boldsymbol{w}}\right]^{-1} \times \ddot{\chi}^{\text{ext}} \tag{7.274}$$

Assume that the propagation direction is y, the ferrite ellipsoid is magnetized to saturation in the z-direction, and the magnetic field of the EMW propagating in the waveguide has only the components $h_{x,y}$. Then the equivalent magnetic susceptibility tensor corresponds to a 2×2 matrix:

$$\ddot{\chi}^{\text{eq}} = \left(\begin{bmatrix} 1 & 0 \\ 0 & 1 \end{bmatrix} + j\begin{bmatrix} \chi_{11}^{\text{ext}} & j\kappa^{\text{ext}} \\ -j\kappa^{\text{ext}} & \chi_{22}^{\text{ext}} \end{bmatrix} \times \begin{bmatrix} w_{11} & w_{12} \\ w_{12} & w_{22} \end{bmatrix}\right)^{-1} \times \begin{bmatrix} \chi_{11}^{\text{ext}} & j\kappa^{\text{ext}} \\ -j\kappa^{\text{ext}} & \chi_{22}^{\text{ext}} \end{bmatrix}, \tag{7.275}$$

which is simplified to

$$\ddot{\chi}^{\text{eq}} = \begin{bmatrix} \dfrac{1 + w_{12}\kappa^{\text{ext}} + jw_{22}\chi_{11}^{\text{ext}}}{\Delta} & \dfrac{w_{22}\kappa^{\text{ext}} - jw_{12}\chi_{11}^{\text{ext}}}{\Delta} \\ \dfrac{-w_{11}\kappa^{\text{ext}} - jw_{12}\chi_{22}^{\text{ext}}}{\Delta} & \dfrac{1 - w_{12}\kappa^{\text{ext}} + jw_{11}\chi_{11}^{\text{ext}}}{\Delta} \end{bmatrix} \times \begin{bmatrix} \chi_{11}^{\text{ext}} & j\kappa^{\text{ext}} \\ -j\kappa^{\text{ext}} & \chi_{22}^{\text{ext}} \end{bmatrix}, \tag{7.276}$$

where the determinant is

$$\Delta = 1 + (\chi_{11}^{\text{ext}}\chi_{22}^{\text{ext}} - \kappa^2)(w_{12}^2 - w_{11}w_{22}) + j(\chi_{11}^{\text{ext}}w_{11} + \chi_{22}^{\text{ext}}w_{22}). \tag{7.277}$$

Then the equivalent susceptibility tensor relating RF/microwave magnetic moment with incident magnetic field in Equation (7.270) is expressed as:

$$\ddot{\chi}^{\text{eq}} = \begin{bmatrix} \dfrac{\chi_{11}^{\text{ext}} + jw_{22}(\chi_{11}^{\text{ext}}\chi_{22}^{\text{ext}} - (\kappa^{\text{ext}})^2)}{\Delta} & \dfrac{j\kappa^{\text{ext}} - jw_{12}(\chi_{11}^{\text{ext}}\chi_{22}^{\text{ext}} - (\kappa^{\text{ext}})^2)}{\Delta} \\ \dfrac{-j\kappa^{\text{ext}} - jw_{12}(\chi_{11}^{\text{ext}}\chi_{22}^{\text{ext}} - (\kappa^{\text{ext}})^2)}{\Delta} & \dfrac{\chi_{11}^{\text{ext}} + jw_{11}(\chi_{11}^{\text{ext}}\chi_{22}^{\text{ext}} - (\kappa^{\text{ext}})^2)}{\Delta} \end{bmatrix}. \tag{7.278}$$

As a result, for the microwave magnetic moment vector (7.270) written in the matrix form, we have:

$$\begin{bmatrix} m_x \\ m_y \end{bmatrix} = \begin{bmatrix} \chi_{11}^{\text{eq}} & \chi_{12}^{\text{eq}} \\ \chi_{21}^{\text{eq}} & \chi_{22}^{\text{eq}} \end{bmatrix} \times \begin{bmatrix} h_{0x} \\ h_{0y} \end{bmatrix} = \begin{bmatrix} \chi_{11}^{\text{eq}}h_{0x} + \chi_{12}^{\text{eq}}h_{0y} \\ \chi_{21}^{\text{eq}}h_{0x} + \chi_{22}^{\text{eq}}h_{0y} \end{bmatrix}. \tag{7.279}$$

Note that for small imaginary terms $\left|jw_{ij}(\chi_{11}^{\text{ext}}\chi_{22}^{\text{ext}} - (\kappa^{\text{ext}})^2)\right| << \left|\chi_{11,22}^{\text{ext}}\right|; \left|j\kappa^{\text{ext}}\right|$, any component of the equivalent tensor can be represented as:

$$\chi_{ij}^{\text{eq}} = \frac{\chi_{ij}^{\text{ext}}}{1 + \eta_c}, \tag{7.280}$$

where η_c can be considered as a complex coupling coefficient:

$$\eta_c = j(\chi_{11}^{\text{ext}}w_{11} + \chi_{22}^{\text{ext}}w_{22}) = \Delta - 1 \tag{7.281}$$

Using expressions for the components w_{11} and w_{22} from Equation (7.272), the coupling coefficient can be written through the magnetic field components and properties of the FR:

$$\xi_c = j\frac{\omega\mu_0 V_f}{2} \times \left(\chi_{11}^{\text{ext}} \times \sum_{n=1}^{N} \frac{h_{xn}^2}{N_n} + \chi_{22}^{\text{ext}} \times \sum_{n=1}^{N} \frac{h_{yn}^2}{N_n}\right). \tag{7.282}$$

The physical meaning of the coupling coefficient is seen from the loaded Q factor of the resonator:

$$Q_{\text{load}} = \frac{Q_0}{1+|\eta_c|}. \tag{7.283}$$

For a transmission line with a single mode $(n=1)$ propagating in the y-direction with only the $h_{x,y}$ components of the microwave magnetic field, it is possible to obtain the following coupling coefficient:

$$\xi_c = j\frac{\omega\mu_0 V_f}{2N_1}(\chi_{11}^{\text{ext}} h_x^- h_x^+ + \chi_{22}^{\text{ext}} h_y^- h_y^+ + j\kappa^{\text{ext}}(h_x^- h_y^+ - h_y^- h_x^+)) = j\frac{\omega\mu_0 V_f}{2N_1}\left(\chi_{11}^{\text{ext}} h_x^2 + \chi_{22}^{\text{ext}} h_y^2\right). \tag{7.284}$$

Now, the expressions for the complex transmission and reflection coefficients can be obtained. The transmission coefficient in the cross-section where the FR is placed is derived from the incident field and the total field propagating in the positive direction:

$$T \times \boldsymbol{h}_0 = \boldsymbol{h}_0 + \boldsymbol{h}_r^+ \tag{7.285}$$

so that the magnitude of the transmission coefficient is

$$|T| = \left|1 + \frac{\boldsymbol{h}_r^+}{\boldsymbol{h}_0}\right|. \tag{7.286}$$

The reflection coefficient is determined in the same cross sections from the incident field and the field propagating in the negative direction:

$$R \times \boldsymbol{h}_0 = \boldsymbol{h}_r^-, \tag{7.287}$$

so that the magnitude of the reflection coefficient is:

$$|R| = \left|\frac{\boldsymbol{h}_r^-}{\boldsymbol{h}_0}\right|. \tag{7.288}$$

Let us write Equation (7.285) in the matrix form. Substituting the components of the radiated field as in Equation (7.268), we get:

$$T \times \begin{bmatrix} h_{0x} \\ h_{0y} \\ h_{0z} \end{bmatrix} = \begin{bmatrix} h_{0x} + h_{rx}^+ \\ h_{0y} + h_{ry}^+ \\ h_{0z} + h_{rz}^+ \end{bmatrix} = \begin{bmatrix} h_{0x} - j\omega\mu_0 V_f \cdot \sum_{n=1}^{N} \frac{h_{xn}^+}{N_n}\left(m_x h_{xn}^- + m_y h_{yn}^- + m_z h_{zn}^-\right)_{V_0} \\ h_{0y} - j\omega\mu_0 V_f \cdot \sum_{n=1}^{N} \frac{h_{yn}^+}{N_n}\left(m_x h_{xn}^- + m_y h_{yn}^- + m_z h_{zn}^-\right)_{V_0} \\ h_{0z} - j\omega\mu_0 V_f \cdot \sum_{n=1}^{N} \frac{h_{zn}^+}{N_n}\left(m_x h_{xn}^- + m_y h_{yn}^- + m_z h_{zn}^-\right)_{V_0} \end{bmatrix}. \tag{7.289}$$

Substituting the components of the microwave magnetic moment (7.279) in (2.789) for the components x and y and introducing the polarization coefficient for the incident magnetic field:

$$\lambda_p = \frac{h_{0y}}{h_{0x}}, \tag{7.290}$$

we get:

$$\begin{aligned} T &= 1 - j\omega\mu_0 V_f \left[(\chi_{11}^{\text{eq}} + \lambda_p \chi_{12}^{\text{eq}})\sum_{n=1}^{N}\frac{h_{xn}^2}{N_n} + (\chi_{21}^{\text{eq}} + \lambda_p \chi_{22}^{\text{eq}})\sum_{n=1}^{N}\frac{h_{yn}h_{xn}}{N_n}\right] = \\ &1 - \frac{j\omega\mu_0 V_f}{1+\xi_c}\left[(\chi_{11}^{\text{ext}} + \lambda_p \chi_{12}^{\text{ext}})\sum_{n=1}^{N}\frac{h_{xn}^2}{N_n} + (\chi_{21}^{\text{ext}} + \lambda_p \chi_{22}^{\text{ext}})\sum_{n=1}^{N}\frac{h_{yn}h_{xn}}{N_n}\right] \end{aligned} \tag{7.291}$$

Analogously, for the reflection coefficient:

$$R \times \begin{bmatrix} h_{0x} \\ h_{0y} \\ h_{0z} \end{bmatrix} = \begin{bmatrix} h_{rx}^- \\ h_{ry}^- \\ h_{rz}^- \end{bmatrix} = \begin{bmatrix} -j\omega\mu_0 V_f \sum_{n=1}^{N} \frac{h_{xn}^-}{N_n} \left(m_x h_{xn}^+ + m_y h_{yn}^+ + m_z h_{zn}^+ \right)_{V_0} \\ -j\omega\mu_0 V_f \sum_{n=1}^{N} \frac{h_{yn}^-}{N_n} \left(m_x h_{xn}^+ + m_y h_{yn}^+ + m_z h_{zn}^+ \right)_{V_0} \\ -j\omega\mu_0 V_f \sum_{n=1}^{N} \frac{h_{zn}^-}{N_n} \left(m_x h_{xn}^+ + m_y h_{yn}^+ + m_z h_{zn}^+ \right)_{V_0} \end{bmatrix}, \tag{7.292}$$

one can get the following expression for the reflection coefficient considering just two components x and y of the microwave magnetic moment:

$$\begin{aligned} R = & -j\omega\mu_0 V_f \left[(\chi_{11}^{\text{eq}} + \lambda_p \chi_{12}^{\text{eq}}) \sum_{n=1}^{N} \frac{h_{xn}^2}{N_n} + (\chi_{21}^{\text{eq}} + \lambda_p \chi_{22}^{\text{eq}}) \sum_{n=1}^{N} \frac{h_{yn} h_{xn}}{N_n} \right] = \\ & -\frac{j\omega\mu_0 V_f}{1+\xi_c} \left[(\chi_{11}^{\text{ext}} + \lambda_p \chi_{12}^{\text{ext}}) \sum_{n=1}^{N} \frac{h_{xn}^2}{N_n} + (\chi_{21}^{\text{ext}} + \lambda_p \chi_{22}^{\text{ext}}) \sum_{n=1}^{N} \frac{h_{yn} h_{xn}}{N_n} \right]. \end{aligned} \tag{7.293}$$

It can be seen from Equations (7.291) and (7.293) that complex transmission and reflection coefficients for fields are related as:

$$T = 1 + R. \tag{7.294}$$

But ferrites are known to absorb EM power. Equation (7.294) has absorption embedded so that the power balance is always satisfied:

$$|T|^2 + |R|^2 + \kappa_{\text{abs}} = 1, \tag{7.295}$$

where the absorption coefficient in the general case is calculated as:

$$\kappa_{\text{abs}} = -2\,\text{Re}(R) - 2|R|^2. \tag{7.296}$$

In terms of power in the waveguide or a transmission line:

$$|R|^2 = \frac{P_{\text{refl}}}{P_{\text{inc}}}; \tag{7.297}$$

the transmission coefficient in terms of power is:

$$|T|^2 = \frac{P_{\text{tr}}}{P_{\text{inc}}}; \tag{7.298}$$

and the absorption coefficient in terms of power is:

$$\kappa_{\text{abs}} = \frac{P_{\text{abs}}}{P_{\text{inc}}}. \tag{7.299}$$

The power balance (7.295) is always fulfilled for any transmission line and any polarization (circular, linear, or elliptical) of the magnetic field, and power absorption by an FR in the general case is determined by κ_{abs} calculated as Equation (7.299).

As an example, let us consider the case when a ferrite is placed in a point with linearly polarized magnetic field, e.g. in a standard rectangular metal waveguide with the fundamental mode TE_{10}, which has the only component h_{0x} propagating in the y-direction. The expressions for the transmission, reflection, and absorption coefficients can be reduced to:

$$T = \frac{1}{1+\xi_c}; R = \frac{-\xi_c}{1+\xi_c}; \text{ and } \kappa_{\text{abs}} = \frac{2\,\text{Re}(\xi_c)}{|1+\xi_c|^2}. \tag{7.300}$$

It can be easily seen that in this case, the power balance (7.295) is satisfied.

It is important that there exists an optimum coupling coefficient at which the absorption is maximum. As $\kappa_{abs}>0$, then $\mathrm{Re}(\xi_c)>0$. Taking the derivative $\frac{\partial \kappa_{abs}}{\partial \xi_c}=0$, one can easily find from Equation (7.300) that the optimum coupling coefficient between the FR and the field in the waveguide is for $|\xi_c|=1$. In this case, half of radiated power is absorbed, the rest is equally transmitted and reflected. Note that this is a point with linear polarization of magnetic field, which is on the waveguide axis. However, maximum absorption will be achieved in the case of the circular polarization, which coincides with the direction (positive) of the magnetic moment rotation of the ferrite. Placing an FR with high Q factor (e.g. for YIG with ΔH~0.3 Oe, $Q \sim 10^4$ at 10 GHz) in the same waveguide at $x = a/4$, where a is the size of the wide wall of the waveguide, will result in practically 100% absorption at the FMR with reflection and transmission practically absent.

The absorbed power is determined as:

$$\kappa_{abs}, dB = 10\log_{10}\left(\frac{\chi''_{res}\mu_0\omega_{res}V_f}{Z_w S}\right), \tag{7.301}$$

where $\chi''_{res} = \frac{M_S}{\Delta H}$, S is the cross-section area of the waveguide, and Z_w is the wave impedance of the waveguide. For the standard metal rectangular waveguide with the only TE_{10} mode propagating, the wave impedance is $Z_{TE10} = \frac{\eta_0}{\sqrt{1-(\lambda/(2a))^2}}$, where $\eta_0 = \sqrt{\mu_0/\varepsilon_0} = 377\ \Omega$.

For a monocrystalline YIG spherical resonator with saturation magnetization $4\pi M_S = 1750$ G and $\Delta H = 0.3$ Oe, the resonance absorption in a point of circular polarization may reach as much as 60 dB.

Note that such a structure is nonreciprocal. For the waves incident from the second port of the waveguide, the FR position will correspond to the opposite circular polarization, and there will be no absorption in the FR.

This type of analysis is very important for designing rejection filters, bandpass filters, and filter circulators based on the effect of FMR in tiny FRs. FRs are typically spheres (with isotropic shape) of small diameter, on the order of 0.5–1.0 mm. Bandpass filters are often designed on the crossed waveguides with a hole for an FR, or on the other types transmission lines, e.g. microstrips, striplines, or coaxial lines with inductively coupled loops [15, 125–129] (see also Figure 7.36). Ferrite spherical resonators are also used to design devices which employ stable nonlinear phenomena at the FMR – magnetic detectors (MDs), cross-multipliers, and various power and frequency converters used in power meters of spectral power

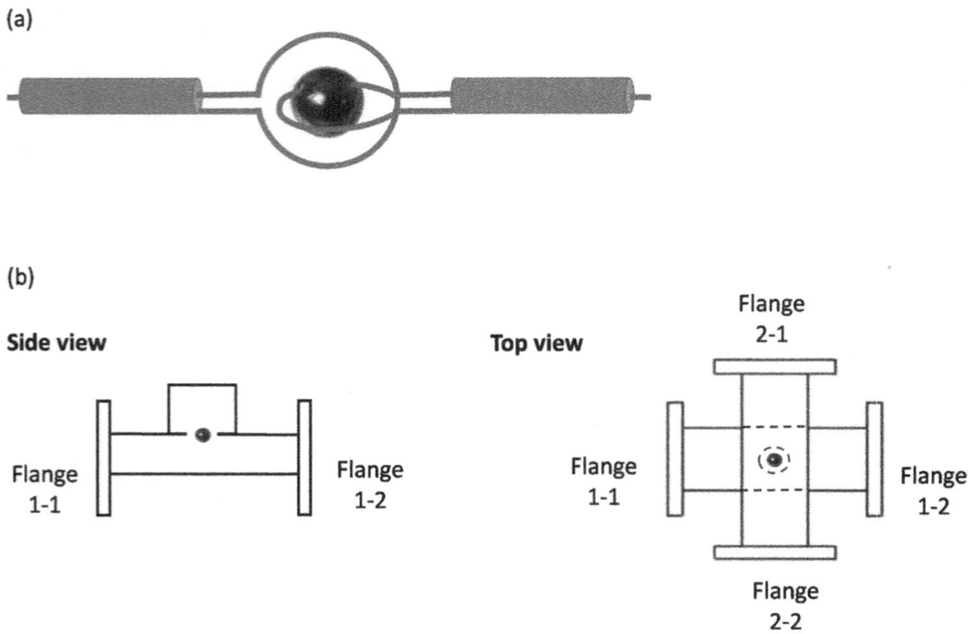

Figure 7.36 Tunable passband ferrite filters on (a) coupled loops of coaxial lines and (b) two crossed waveguides coupled through the common hole with a ferrite resonator in it.

density, peak power, and average power of signals. They operate at high and medium intensity levels (from tens microwatts to kilowatts in continuous wave regime and even megawatts in pulse regime) and over wide frequency range – from ~ 100 MHz to 40 GHz [130–135]. These devices will be considered in more detail in Section 7.6.

The presented herein self-matched field approach was extended for calculating reflection, transmission, and absorption coefficients from a half-space occupied by a periodic structure comprises small ferrite ellipsoids, including those of uniaxial hexaferrites [136]. This structure works as a metamaterial. Mutual EM coupling between separate ellipsoids treated as a combination of elementary electric and magnetic dipoles is taken into account.

Note that if a spherical ferrite sample is substituted by a thin disk of the same material and approximately the diameter of a few millimeters so that magnetostatic conditions are still fulfilled, the main resonance may have Q factor higher than that of sphere. However, the spectrum of magnetostatic oscillations (Walker spectrum) will be richer than that of a sphere with more harmonics of the lowest uniform oscillation [137–142]. Planar ferrite filters can be built using these disk-type FRs [143–146].

7.5 Gyrovector Model of Instantaneous Energy Absorption

In this chapter, we will describe an interesting and unconventional *pure geometrical model of gyromagnetism* proposed by a Russian scientist L. K. Mikhailovsky in 1980s [23]. He was developing his *gyromodel* from 1970s till the end of his life in 2014. The gyromodel explains how quanta of EM energy interact with and get absorbed by a gyromagnetic medium. The gyromodel is based on a number of preceding fundamental works, which have existed in classical and quantum physics for many decades, some even for more than 150 years.

The fundamental theoretical concepts of gyromodel herein will be discussed mainly qualitatively, as its actual mathematical description is comparatively complex. Our goal is to let a reader have a general idea about this original and useful theory. It is important to stress that this model laid the basis to the development of the new field of technology, *Currentless Spin Electronics*. In this area of electronics, the material carrier of energy quanta is not an electron (or a charge in the general case), but a spin of a material point mass. There are many papers on this topic by the founder of this theory, L. K. Mikhailovsky. However, most of the papers were published in Russian in 1970–2014, and just a few have been translated into English. Special attention should be paid to works [23, 147–153], which contain the most comprehensive overview of the mathematical and physical fundamentals of the gyromodel.

7.5.1 Basis for Gyromodel

Any physical phenomenon may be described by a number of different mathematical models, depending on the mathematical formalism used. The gyromodel is a physical model, which describes absorption of EM energy by a gyromagnetic currentless medium, a particular case of which is a microwave ferrite. This physical model uses *gyrovector formalism* as a mathematical language.

The gyromodel is based on a number of preceding fundamental works of classics of EM theory and quantum theory.

The first key idea, which lays the basis for the gyromodel, is that of *currentless magnetism* in a uniform geometry. It was J. C. Maxwell who proposed this model in 1855–1856; it is based on a concept of a unit cell in Faraday's force tubes and scalar pressure of incompressible fluid [154]. Maxwell's model included a concept of single-point magnetism of a microscopic particle without any current, just in terms of the potential energy. This Maxwell's model was later proven by P.N. Lebedev's experiments (1899) with pressure of light. Acceptance or loss of mass–energy portions by a medium through some boundaries is similar to the flow of incompressible fluid between unit cells in Faraday's force tube, where a unit volume of fluid passes from the surface with pressure p to the surface with pressure $(p-1)$ in a unit of time. Integer portions, or quanta of mass–energy, correspond to Maxwell's unit cells in Faraday's tube, and mass–energy absorption is due to pure mechanical currentless "resistance" of a medium to a work of a single-point pressure. Unfortunately, later on Maxwell abandoned his initial idea of the currentless nature of magnetism and accepted Ampere's explanation of magnetic phenomena using molecular electric currents and magnetic energy as kinetic energy.

The second important for the development of the gyromodel idea is the concept of a *single-point quantum of potential energy*, or more specifically, mass–energy [155]. This idea was not fully developed, but bequeathed by Einstein in 1953 [156]. In his gyromodel, Mikhailovsky combined the representation of EM quanta as Einstein's "potential mass–energy" quanta and Maxwell's unit cell model. A work of EM forces, which requires at least two points in space–time, in the gyromodel was replaced by a force of pressure exerted by the "potential mass–energy" quanta upon a volume of a single point, which Einstein called "*a singular point*." Therefore, in the gyromodel, the concept of *energy* is taken as the first principle, and such vector force characteristic as a field is secondary.

The third key idea that laid the basis of the gyromodel is *Hubbard's model* for quantitative description of a one-dimensional (1D) degree of freedom for rotational (spin) motion of a material point [157]. This spinning produces the potential mass–energy, which is associated with Einstein's single-point potential energy quantum. Gyromodel has its own concept of "a material point" as some volume, in which an EM quantum is absorbed, or from which it is radiated. It is thought as a minimum acceptable volume surrounded by a spherical Fermi surface [158]. This "material point," according to Mikhailovsky, is considered as a purely gyromagnetic, i.e. associated with spin magnetic moment, but not with any translational motion of electrons or other charges. A microwave monocrystalline ferrite is an example of such a medium: conductivity loss in it is known to be negligibly small, therefore, this medium can be considered as "currentless."

In Mikhailovsky's gyromodel, there are concepts of *macrospace* and *microspace*. The unified mass–energy produced by an intrinsic rotation (spinning) of a material point with some mass concentrated in its own volume is associated with Einstein's quantum of potential energy. However, there is a difference. Einstein's quantum occupies the volume of *macrospace* of the singular point. Indeed, a quantum of Einstein's potential mass–energy of a material point is represented by a product $h\nu$, but the harmonic oscillation at the frequency ν is both infinite in time and space. To avoid this contradiction, Mikhailovsky introduced *microspace* of a material point with the intrinsic spinning of its mass. Moreover, in quantum mechanics, $h\nu$ is associated with the kinetic energy of a quantum traveling from one point in space–time to another. In the gyromodel, mass–energy is associated with potential energy only.

The traditional classical EM models of absorption and radiation of EM energy are built using such physical concepts as "field strength" and "electromagnetic parameters of a medium." These models assume multipoint (or at least two-point) forces acting in a medium, and therefore, the mechanisms of EM energy absorption or radiation are of kinetic type, and eventually, they are associated with currents. Conductivity currents result either in resistive loss and conversion to heat, or radiate as miniature "antennas."

Phenomenological mathematical models, which are used to describe interaction between EM field and a gyromagnetic medium, employ the magnetization vector motion equation with different relaxation terms (Landau–Lifshitz–Gilbert, Bloch, Bloch–Bloembergen, Callen, etc.) [13–15]. All of them use multiport vector formalism and, therefore, assume force interactions.

In quantum mechanics, absorption of an energy quantum is described by a single-point model, where electron's spin moment turns over, i.e. its direction changes by 180° [2, 158]. This is associated with "carrying" the electron through the Fermi surface. This single-point quantum mechanics model still cannot explain some experimentally observed phenomena in ferrite media, e.g. practically instantaneous "phaseless" energy absorption at the gyromagnetic resonance in an almost perfect monocrystalline ferrite (YIG), which does not have free electrons.

To the contrast of classical electrodynamics, phenomenological and quantum mechanics models, the energy-based single-point currentless mechanism of absorption and radiation does not deal with any conductivity electrons, currents, vector field strengths, or EM parameters of media. A motionless, or "frozen" in space–time, gyromodel proposed by Mikhailovsky, directly employs Maxwell's idea of 1856 for energy quanta absorption and radiation. The single-point gyrovector formalism, or *gyromodel*, is a representation of Einstein's mass–energy quantum.

7.5.2 Difference between Vectors and Gyrovectors

Vector calculus is known to be widely used in physics – not only in the classical theory of EM field, but also in quantum mechanics. "Classical" quantum mechanics, based on the application of statistical Ψ-function for Louis de Broglie's wave particle and Schrödinger's equation, still uses variation in time of quantum objects in the vector force fields [159, 160]. Traditional vectors are defined between two points in space and thus allow for translational motion between these points, which occurs in time. In vector formalism time varies continuously, while sufficiently small spatial volumes are squeezed into the discrete points of infinitesimally small volumes. The main instruments of vector calculus are differential and integral operations, which in the majority of cases rely on continuous harmonic functions. Postulates and axioms, including the principles of linear combination, or superposition; notion of distance between two points; and rules of vector algebra are applied to a linear Euclidean space. Riemannian, including non-Euclidean (curved) geometry has its own postulates, axioms, and rules, but also deals with vectors, angles between them, and concept of distance, or metric [161].

Gyrovector formalism, or *gyromodel*, also comprises "vectors." However, unlike any vector-based model, gyromodel deals with relatively immobile geometrical elements – *single-point gyrovectors*. Each gyrovector denoted as $\boldsymbol{G} = \boldsymbol{R}\mathrm{L}^{V}\boldsymbol{R}$ contains two orthogonal "vectors": vector of spin $^{V}\boldsymbol{R}$ and rotating vector of state $\boldsymbol{R}$, the end tip of which draws a circle on the plane normal to the vector of spin. This is illustrated in Figure 7.37. A gyrovector can be associated with the precessing spin magnetic moment of a gyromagnetic particle described by the phenomenological equilibrium magnetization motion vector equation.

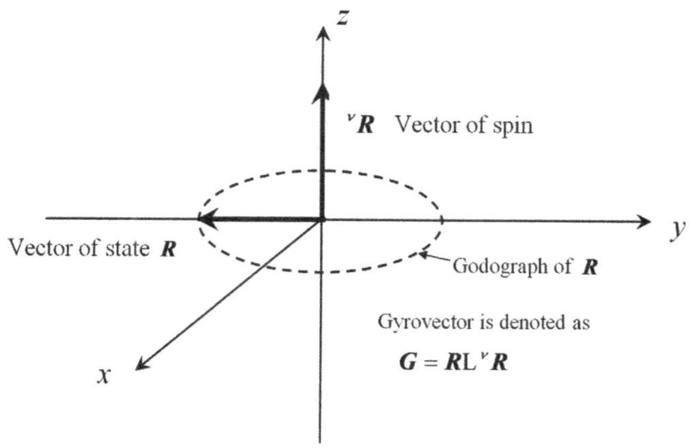

Figure 7.37 Gyrovector comprises two orthogonal "vectors" – vector of spin and vector of state [23].

The peculiarity of the gyromodel is that its vectors of spin and of state do not follow the superposition principle, i.e. they may not be added according to vector algebra rules. The minimum length of the vector of state is the "length quantum" h (this was first introduced by Maxwell to denote the minimum length in the Faraday's force tube). The lengths of the vector of spin and the vector of state determine the intrinsic volume assigned to the material point for a center of absorption and radiation (CAR). This volume is immobile in space–time; but it gives degrees of freedom for the possible own phaseless rotational motion of a material point mass – either clockwise or counterclockwise. Phaselessness means that the pure spinning is always considered exactly "in phase" with a period (360°) of internal revolution of the material point mass. It is also assumed that the volume of the material point does not allow for any incremental translational motion of the material point mass; however, this volume is sufficient for representing intrinsic spinning. The degree of freedom for a spin is associated with the "excitation of Fermi's nature having the spin of ½, or spinon" [162, 163]. Therefore, the gyromodel assumes intrinsic phaseless magnetism of masses of material points, the one which is described with Lande factor $g = 2$. Maxwell's model of 1856 agrees with such intrinsic magnetism, while the model based on Ampere's currents contradicts the very existence of such magnetism.

7.5.3 Concept of Material Point in Gyromodel

The gyromodel deals with its own concept of a *material point*. The intrinsic volume of a material point in a gyromodel is enclosed in the *Einstein's singular point*, where the energy quantum is localized. Note that Einstein defined a quantum as "a special point surrounded by a strong vector field" [155]. At the same time, the gyromodel operates with the term "*a field of a spin of a material point rest mass*," which means that the material point has a degree of freedom to spin, as in Hubbard's model [157]. Note that Hubbard's model is the simplest model to describe interacting particles in a crystal lattice. The Hamiltonian in Hubbard's model has only two terms – for kinetic energy and potential energy. The kinetic term deals with electron tunneling, or hopping, between atoms. The potential term is associated with Coulomb's repulsion. As interatomic distance tends to infinity, the 1D chain of atoms simply converges to a set of isolated magnetic moments. The 1D degree of freedom results in the intrinsic rotation of a material point mass about its axis of symmetry (*spinning*) and, therefore, is associated with spins $\pm 1/2$ and with the corresponding Fermi energy difference.

A material point in a gyromodel is a relatively immobile finite volume in a structural (without motion) geometry. This material point is assigned a *volume*, and the volume is assigned a *mass*, or *rest mass*. This mass rotates, or spins, about its axis of symmetry. This spinning determines the *intrinsic time* within this volume. The time instants are *discrete*, and they correspond to the period of rotation, such that at these instants the material point remains exactly in the same state.

The *material point mass* in the single-point geometry of the gyromodel is the minimum possible mass of a material body within its own volume, which has the only 1 degree of freedom for its motion. This degree of freedom allows for only intrinsic, pure spinning of the mass, while any translational motions, even incremental, are prohibited. This is the fundamental difference between the material point in the gyromodel and in Hubbard's model [157].

The volume of Einstein's singular point in the gyromodel is the volume of macrospace, which encloses microspace. *Microspace* resembles the Fermi sphere [2, 158] filled with two components: the first is the *rest mass* of the material point, and the second is *mass–energy*. The mass–energy is phaseless due to its intrinsic spinning, and corresponds to the potential energy in Hubbard's model [157]. The remainder of the Einstein's singular point macrospace is associated with kinetic energy in Hubbard's model, i.e. the energy due to the orbital rotation as a particular case of translational motion, and has nothing to do with spinning.

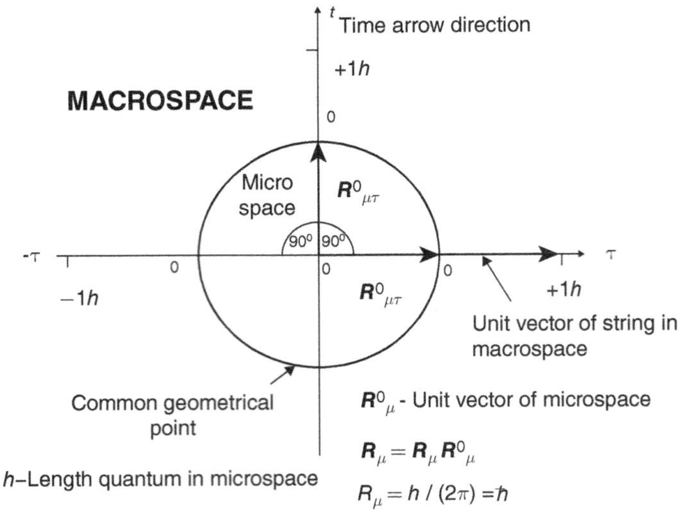

Figure 7.38 A material point in a gyromodel: microspace inside macrospace volume [23].

(Note that the *Fermi surface* is an abstract boundary derived in the general case from the periodicity and symmetry of the crystalline lattice and from the occupation of electronic energy bands. The existence of a Fermi surface is a direct consequence of Pauli's exclusion principle, which allows for having only one electron per quantum state. The Fermi sphere is the shape in the k-space [momentum–space] in the case of the quadratic dispersion law. The Fermi sphere encloses the wave vectors of the lowest energy electrons in the crystal. Any electron in a state outside the Fermi sphere can lower its energy by adopting a wave vector at the surface of the sphere [158]).

In the 1D Hubbard's model, the degrees of freedom of *spinons* (quasiparticles carrying spins of electrons) *and holons* (spinless noninteracting fermions associated with charges) can be assigned some volumes inside a common volume of a point. The volume of spin's degree of freedom of a material point with some "rest mass" is set as unity. Exactly this volume is used in the gyromodel as a material point and is called *microspace*. Hubbard's volume for charge's degree of freedom of finite magnitude is to some extent is analogous to the *macrospace* of the gyromodel. In the gyromodel, this macrospace is associated with the gyrovector's *unobservable motion* of the material point rest mass along the circular line (see Figure 7.37). The combination of macro- and microspaces of the gyromodel are illustrated in Figure 7.38. The distance h is the "length quantum" in microspace, i.e. the minimum possible circumferential length common for both microspace and macrospace of the gyromodel. Thus, a material point in the gyromodel includes both microspace and macrospace.

7.5.4 Geometrical Peculiarities of Gyromodel

It is well known in quantum physics that $h\nu$ is the energy of Planck's quantum of frequency ν. This is a "wave" quantum described in terms of the Ψ-function. Formally, Plank's quantum should be unlimited in space–time. For this reason, D. Blokhintsev in "Fundamentals of Quantum Mechanics" [159] underlined that a quantum may not be represented as a "fishing bob floating on waves," but it should be understood as a wave infinite both in space and in time. He reasonably stated that "...a quantum of light described by $E = h\nu$ and $p = \hbar k$, according to its definition is associated with a monochromatic plane wave. Such wave is a pure periodic process, infinite both in space and in time. An assumption that the quantum is located somewhere contradicts the perfect periodicity of the wave: a sinusoidal wave, if somehow deformed, is not a sinusoidal wave anymore, but a superposition of many sinusoidal waves." However, along with this mathematically rigorous statement, in the models with space quantization, the time required to absorb and radiate quanta of EM energy is always finite. In the case of gyromagnetic media, there is the condition of "instantaneous" stepwise change (flip over) of the spin direction at both absorption and radiation of an energy quantum. Therefore, Planck's quantum which is not localized in space–time cannot be used for adequate description of absorption and radiation in gyromagnetic media.

The gyromodel also deals with quantum formally denoted similar to Planck's quantum as $h\nu$. According to Mikhailovsky's definition, $h\nu$ in the gyromodel is a polarized phaseless quantum of potential mass–energy (Einstein's mass–energy quantum) surrounded by the Fermi energy surface in the microspace of the gyromodel, as is illustrated in Figure 7.38. Unlike Planck's quantum, the potential mass–energy quantum is *always localized in space–time*. Instead of the continuous, multipoint, statistical, and, therefore, ambiguous (according to Einstein) Ψ-function used in quantum mechanics, the gyromodel operates with the so-called *closed harmonic function* in its gyrovector representation. This purely geometrical function is determined in a single point of the unified Euclid–Riemann–Lobachevsky three-geometry (3G) [164]. There is no motion in the 3G, time is not included; therefore, Figure 7.39 illustrates the static 3G picture.

In the gyromodel, the microspace is limited by the energy surface of the Fermi sphere. The macrospace corresponds to the Riemann geometry (+) and the Lobachevsky geometry (–), as shown in Figure 7.39. In Figure 7.38, the unit vector of string (in macrospace) is introduced as a continuation of the horizontal unit vector of space (in microspace). This string in macrospace is associated with the *closed harmonic function* shown in Figure 7.40, which starts and ends on the Fermi

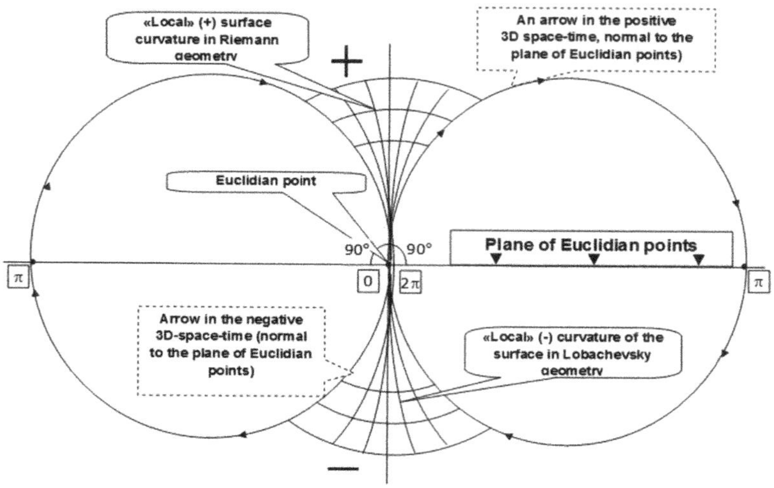

Figure 7.39 Three-geometry inside the material point in the gyromodel [23].

energy surface. The length $h\nu$ in the 3G of macrospace is assigned to the string. As a cycle of the closed harmonic function along the string, the length $h\nu$ disappears, when a quantum is radiated, i.e. leaves the volume limited by the Fermi surface. Analogously, the length $h\nu$ appears, when a quantum is absorbed, i.e. becomes contained within the Fermi surface. As is known, hopping from one energy state of a quantum system to another is always associated with radiation or absorption of energy quanta $h\nu$.

The gyromodel value h indicated in Figure 7.40 is the length of the vector of state. The value h is the "microquantum" for counting out discrete time intervals in their discrete sequence. This is the *system unity* for counting out the intrinsic time for a material point. Note that the intrinsic time is individual for each material point and for different material points may be different. The physical meaning of the value h is the time of one revolution (spin) of the material point mass within the Fermi surface. Therefore, it determines the minimum length of the circumference, common for the microspace and macrospace of the gyromodel. The positive integer number ν is the number of such revolutions (spins). From the macrospace point of view, h corresponds to the distance, which is the Maxwellian minimal length in the Faraday tube, or the "length quantum." The variation of the energy state of the system by $h\nu$ corresponds to an absorption or radiation of an energy quantum.

The the time arrow $\boldsymbol{t}$ and 3G space arrow $\boldsymbol{\tau}$ are orthogonal in the volume of m**i**crospace of Einstein's singular point, but they are parallel in its macrospace and in the Minkowski space, i.e. between the volumes of singular points in *Einstein's Closed Universe* (this is Einstein's term [165]). Therefore, the volume of microspace limited by the Fermi surface inside the volume of the Einstein's singular point, is the minimum possible volume in the Einstein's Closed Universe.

In the gyromodel, a quantum of spinning polarized potential mass–energy of a material point in the microspace is represented by single-point discrete quantum formalism, which does not deal with the continuous Ψ- function.

As for *polarization of the potential energy quantum*, it can be understood as follows. Any potential energy quantum is created by the degree of freedom for the intrinsic Hubbard's spinning of the material point mass. The axis of rotation determines the direction of the time axis. The direction of the spinning is the same as the direction of the time axis within the Fermi sphere, or in the

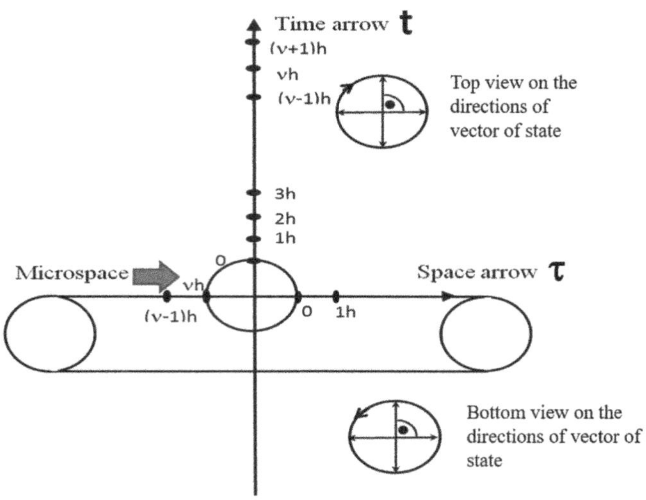

Figure 7.40 Intrinsic coordinate system of the volume of the point in a gyromodel: h is the Maxwellian minimal length (length quantum), and ν is a positive integer determining a scale of the system unit for counting intrinsic time of a polarized in the macrospace of the gyromodel point volume [23].

microspace of the gyromodel. This direction is determined in the coordinate system of Einstein's singular point in the macrospace. It also determines the polarization of the potential energy quantum within the Fermi sphere. In the other words, the potential energy quantum polarization in the macrospace is the same as the direction of the quantization axis, or, according to Vonsovsky [2], the spin vector projection axis.

7.5.5 Geometrical Description of Energy Absorption and Radiation in a Gyromodel

Let us have a medium containing nano- or microsize gyromagnetic particles at conditionally infinite distance between the volumes of material points, where only electron spin rotations take place (there is no translational motion of charges, no conductivity currents, and no Ohmic losses). In a gyromodel, a material point, where EM energy is absorbed or emitted, is called a *center of absorption and radiation (CAR).* In Faraday's force tube, this point is like a *sink* or a *source.* EM energy absorption by a medium is understood as a result of energy pressure on the volumes of these material points. This is phaseless single-point work by Maxwell's pressure force, completely different from the traditional two-point work of the Lorentz force. In the latter, electric and magnetic vectors are always associated with phase relations in space–time, and the work of Lorentz force always acts on moving charges.

There are four consecutive-in-time quantized states for a polarized quantum of potential energy as are shown in Figure 7.41. The corresponding directions of the spin vector of a gyrovector indicate these states. Let us consider these states, giving them some conventional numeration.

The first is the *absorbed state* of the potential mass–energy quantum at the material point. This state is relatively stationary, either both in space and time, or only in time. In this state, the vector of spin of the intrinsic gyrovector determines the *positive direction of the unit vector of time* ($\bar{t}_+$) in the microspace volume.

The second is the *preradiated state.* This state is bound with microspace. The potential energy quantum is leaving the microspace, but is still within the macrospace of Einstein's singular point. The vector of spin determines the positive

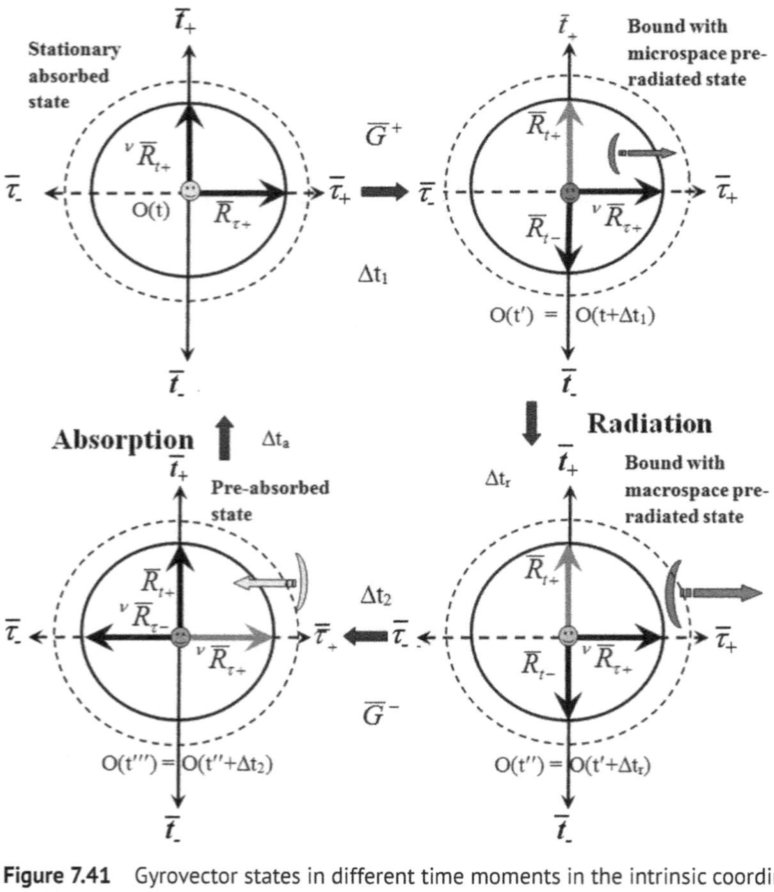

Figure 7.41 Gyrovector states in different time moments in the intrinsic coordinate system of a material point in the microspace (solid line) and in the macrospace (dashed line) [23].

direction of the unit vector of space–time in this volume ($\overline{\tau}_+$) and the direction of the future incremental translational (orbital) motion of the potential mass–energy quantum created by the spinning of the material point mass.

The third is the *radiated state*, which is bound with the macrospace of Einstein's singular point. The quantum of potential energy is leaving the macrospace for the external Einstein's Closed Universe. The latter comprises numerous single-point volumes of Einstein's singular points, i.e. by a finite number of "infinitely distant" from each other CARs. Then, the direction of the spin vector of the radiated quantum of the potential mass–energy determines the positive direction of the space–time unit vector in the closed Einstein's Closed Universe.

The fourth is the *preabsorbed state*. It is also bound with macrospace. The potential energy quantum enters the volume of the macrospace from the Einstein's Closed Universe to be absorbed (get in the absorbed state as described at the first step). In this fourth state, the spin vector of the intrinsic gyrovector of the transferred quantum of mass–energy comprises an arbitrary angle with respect to the unit vector of the intrinsic time of the CAR, which absorbs this quantum. The value of this angle determines the probability of this absorption.

It is important that if there is no space–time variation of the system state, i.e. we deal with the "frozen" gyromodel, then there is no translational (including orbital) motion. If there is such variation, then the direction of the unit vector of space–time in the macrospace of the gyromodel determines the direction of the energy quantum motion. If variation is in time only, but not in space, then the orientation of the spin vector (either parallel, or antiparallel with the unit vector of time in the microspace) determines the polarization of the potential mass–energy quantum in the macrospace.

Positive gyrovectors $\bar{\boldsymbol{G}}^+$ are used to describe quanta of "free energy" both inside the single-point *sources*, and outside, when the quanta are transmitted between the gyromodel points. Negative gyrovectors $\bar{\boldsymbol{G}}^-$ are used to describe single-point *sinks* in nonconducting gyromagnetic media, where spin degrees of freedom become "excited" due to absorbed energy quanta. One can imagine sinks in the gyromodel as "tornado vortices," while sources as "gushing springs."

Note that the *energy balance* in the Closed Universe (without observer's participation) is sustained by the transfer of energy quanta between intrinsic volumes of points in the Euclidian plane (see Figure 7.39). This is the consequence of the requirement of the First Principle of Thermodynamics to be fulfilled in Einstein's Closed Universe.

An important peculiarity of the gyromodel is the simultaneity of observations and invariance in time of the distances between the intrinsic volumes of all Euclidean points and Einstein's singular points in the 3G. This is a consequence of not allowing for incremental translational motion inside each gyromodel material point.

7.5.6 Experimental Validation and Practical Applications of Gyromodel

7.5.6.1 Existence of Single-Point Currentless Absorption in Gyromagnetic Media

The existence of currentless and phaseless, i.e. without any work of forces associated with vectors of the EM field, energy absorption by the *CAR* in gyromagnetic media, for the first time was proven experimentally using the Mössbauer effect [166]. Microwave ferrite was exposed to microwave radiation at FMR, and the kinetic energy of a quantum of EM field was transformed to the potential energy of an electron. The revolution of the spin of the d-electron varied the resultant magnetic field at the iron (Fe) nuclei, which in turn caused variation of Mössbauer spectrum shape [167].

Energy absorption by each point of a gyromagnetic medium individually was also obtained by creating gyromagnetic aerosols, i.e. clouds of noninteracting with each other particles. The total absorbed energy was equal to the sum of individually absorbed energy by each ferrite particle [152].

Another evidence of single-point absorption of energy by a gyromagnetic medium can be found in [168]. The energy of the continuous microwave oscillation in the 8-mm wave band was absorbed at the FMR by a high-Q hexagonal ferrite monocrystalline resonator with the FMR line width of 10 Oe (30 MHz). The diameter of the resonator was only 0.5 mm. The resonator heated up to 100 °C practically in a single point.

7.5.6.2 Experimental Comparison of Energy Absorption Time with Relaxation Time for Magnetization Moment

Frequency-selective absorption of energy of microwave pulses by high-Q gyromagnetic resonators (GRs) has been studied experimentally [169, 170]. The experimental setup and waveforms of incident, absorbed, and transmitted power are shown in Figure 7.42. The objective of this study was to determine the time of relatively "instantaneous" absorption of quanta of microwave energy in a GR and compare it with the relaxation time of the magnetization moment. In these experiments, the Einstein's singular points in the GR were activated as usual by the external bias magnetic field. There were two GRs, GR-1 and GR-2, which were the tiny spheres of the YIG monocrystals (0.3 mm in diameter). Each of them has a full FMR line width of around 0.4 Oe. GR-1 was placed in the bias magnetic field H_{res}, and it was absorbing energy of the EMW at the FMR. This

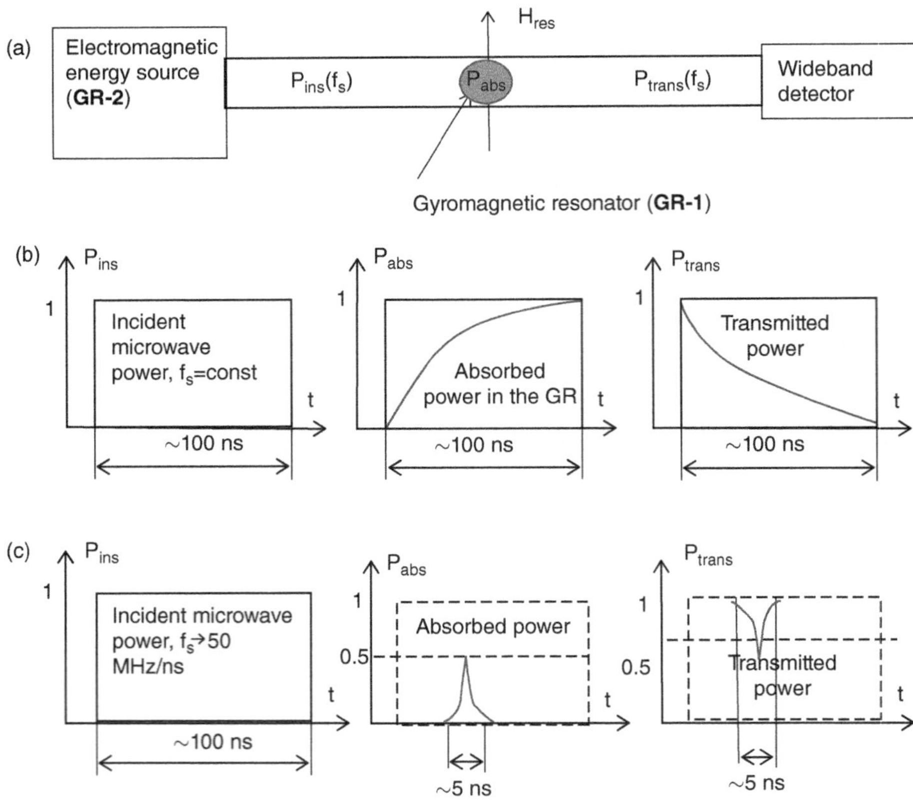

Figure 7.42 Resonance energy absorption from electromagnetic pulses: (a) schematic of the experimental setup; (b) relatively slow absorption observed without preliminary excitation of a gyromagnetic resonator ($f_s = const$); (c) almost instantaneous frequency-selective energy absorption observed with preliminary excitation of the gyromagnetic resonator ($f_s \rightarrow$ 50 MHz/ns) [23].

EM energy was produced by another resonator, GR-2, which changed the carrier frequency of the microwave signal in time. The rate of the carrier frequency variation was about 50 MHz/ns. The EM energy was radiated due to the freely damping precession of the magnetization vector of GR-2 during the time interval over 100 ns. This freely damping precession of GR-2 was produced by an "auxiliary" bias magnetic field, which formed a sharp angle with the main bias magnetic field of GR-2.

The observed practically instantaneous (1–5 ns) energy absorption in GR-1 can be explained from the standpoint of the gyromodel. The time of the energy absorption in GR-1 is significantly less than the relaxation time for the magnetization vector (5 ns ≪ 100 ns) of GR-1. This time of absorption of individual energy quanta (not the total energy) is associated with the re-orientation by 180° of the individual projections of the spin vectors of gyrovectors on the direction H_{res}.

This phenomenon allows for measuring the carrier of single (nonperiodic) ultra-short pulses using FRs, while it is impossible to do using traditional narrowband microwave filters. The reason is that the duration of the pulse under test is so short that it cannot fully excite (100%) the measuring resonance structure. Moreover, varying statistically the carrier of aperiodic series of pulses using GRs, as described earlier, one can code the transmitted information and protect it from detection.

7.5.6.3 Cross-multiplication and Frequency Downconversion without Combinational Frequencies

Another validation and practical application of the gyromodel is the *gyromagnetic cross-parametric converter*, also called *gyromagnetic cross-multiplier*, which allows for frequency downconversion of the microwave carrier of the EM signals to RF frequency range without producing combinational frequencies [171, 172]. The peculiarity of such frequency conversion is linearity of the input impedances of the cross-multiplier at both ports, for a microwave signal to be converted and the RF signal of the heterodyne oscillator [173]. Linearity of input impedances means that the spectra of the corresponding reflected signals from the ports are not "enriched" by any new spectral components due to combinational frequencies. The spectrum of the converted signal contains only harmonics of the RF heterodyne signal, and the amplitudes of these harmonics are determined by the microwave energy absorbed by the FR of the gyromagnetic converter (cross-multiplier) at its FMR frequency.

The "heart" of a gyromagnetic converter is a *magnetic detector* (MD), which is a tiny high-Q monocrystalline resonator (typically garnet) surrounded in its equatorial plane by a flat spiral microcoil [130, 132]. If there are variations of the longitudinal component of magnetization vector due to short EM (microwave) pulses or wideband random signals (noise), there will be a low-frequency signal induced in the microcoil. The magnitude of this signal is proportional to the spectral power density of the microwave signal at the frequency of the FMR. Varying the bias magnetic field, it is possible to scan the frequency range of interest and get information about power spectral density over a wide frequency range. This way of measuring power spectral density is called *magnetic detection*; it is applicable to short-pulse and random signals, but not suitable for continuous harmonic signals. To measure power density of harmonic signals, a stable RF modulation signal is launched into the microcoil. Then, when the microwave signal frequency corresponds to the FMR frequency in the FR, there will always be nutational oscillations due to the RF (a few MHz) modulation of the magnetization vector accompanying the precessional motion. Due to the gyromagnetic properties of the FR, the nutational oscillations will comprise only harmonics of the RF modulation signal; there are no combinational frequencies in principle. Therefore, there will be only the harmonics of the RF modulation signal in the spectrum of the output signal in the microcoil. The magnitudes of the nutational oscillations, and hence, of the harmonics of the microcoil signal, are proportional to the magnitude of the microwave signal. Usually, the second harmonic of RF modulation is used to measure the power of the microwave signal. This is done to distinct the input modulation signal and the resultant output oscillation due to the interaction between the microwave signal and the FR with varying in time bias field. This process is called *cross-multiplication* (CM). As has been mentioned earlier, a cross-multiplier has linear input impedances for microwave signal and for RF modulation. Still, *CM* resulting in the appearance of only harmonics of the modulation signal is possible. No combinational frequencies appear in this process. Note that the frequency of the RF modulation signal can be extremely stable, e.g. using quartz crystal oscillator, and so do the harmonics of the output signal. There are no frequency instability issues due to the traditional heterodyne frequency downconversion.

A new class of microwave devices exploiting this effect has been developed. This includes devices for measuring peak power of short-pulse microwave signals, spectral power density meters of microwave noise signals, and devices for detection and measurement of small deterministic signals on the background of intense noise [130–132]. Also, power meters of the fundamental and higher order modes in microwave waveguides and planar transmission lines at medium and high power levels have been developed on the basis of an MD with an external crystalline detector [174]. These devices based on gyromagnetic converters and MDs operate within a few standard subranges covering a wide frequency range, 100 MHz to 40 GHz. They can operate even at higher frequencies in millimeter wave-band, if applying high-*Q* monocrystalline hexagonal FRs, which have high internal fields of crystallographic magnetic anisotropy and do not need intense bias magnetic fields [39, 40, 131]. The devices with gyromagnetic converters and ferrite transducers are intended for operation at medium and high microwave power levels. Though their sensitivity is typically ranges from 10^{-7} W to 10^{-1} W, which is not very high compared to the high-sensitivity receivers of the modern spectrum analyzers, they are able to frequency-selectively measure up to a few kilowatts of continuous microwave signal power, and in some cases, it may be up to 20 MW in the pulse regime [175]. An important advantage of gyromagnetic converters/transducers operating in the *CM* regime in the devices for frequency-selective power measurements is the absence of parasitic channels of reception, which always arise in conventional heterodyne-type nonlinear frequency downconverters of spectrum analyzers due to the combinational frequencies. A comprehensive review of the devices built on the basis of gyromagnetic converters based on MDs is given in [176].

7.5.6.4 Phase Shift of 180° between Incident Microwave Signal and Signal Radiated by FR

An interesting effect has been observed in experiments with a ferrite bandpass filter. A bandpass filter is typically designed using two shorted crossed perpendicular waveguides with the common opening. A spherical high-*Q* monocrystalline FR is placed in this opening symmetrically with respect to both waveguides (identical distances to the shorted ends). Another design is based on two coaxial cables with perpendicular coupling loops at the open ends of the center conductor, and an FR is surrounded by both loops. The microwave power transmission from one port of the filter to another is possible only when the FR absorbs (at the FMR) the EM energy from the first line and re-radiates it into the second line [15, pp. 141–142]. It has been demonstrated that there is a phase shift by the half of the period of the EMW between the phases of the incident microwave oscillation and the wave re-radiated by the FR [177]. It is important that the size of the ferrite sphere is significantly less than the wavelength in the ferrite material. This effect confirms that in the "single-point" FR there is a stepwise overturn of the spin magnetic moment to the opposite direction, which causes the by 180° phase jump in the re-radiated wave as compared to the incident.

7.5.6.5 Currentless Absorbing Materials

The most important practical result that was achieved due to gyromodel is the development of the currentless absorbing materials (CAMs). An exhaustive review of physics of CAMs and coatings on their basis and also a description of their distinction from the traditional "resistive" absorbing materials and coatings are presented in [152] and references therein. The CAMs (in the case of comparatively low concentration of gyromagnetic particles, like in aerosols) may have frequency characteristics omnidirectionally matched with the impedance of free space up to the level of −100 dB [178]. Another advantage is that there is no necessity of matching phase diagrams when applying CAMs on arbitrary surfaces. Moreover, to form frequency characteristics and calculate bandwidths of CAMs, for example, containing several types of magnetically uniaxial hexaferrites with different anisotropy fields, it is sufficient just to use simple algebraic methods for adding up frequency characteristics, due to the summation of energy absorbed in all hexaferrite particles individually [177, 178]. Frequency characteristics of power absorption by some types of engineered CAMs are presented in [23, Figure 9] and also discussed in Chapter 6 in Volume 2 in this book series. The experimental resonance curves of absorption are asymmetric in the general case. The experimental frequency dependencies agree well with the theoretical "resonance curves" obtained from the pure geometrical gyromodel. Note that the absorption resonance curves have probabilistic nature. They show probability absorption of polarized quanta of energy by polarized centers of absorption and radiation.

7.6 Stable Nonlinear Phenomena at FMR and Devices on Its Basis

Stable nonlinear phenomena at FMR in ferrites take place at the microwave power less than the spin-wave instability threshold.

As is mentioned in the previous section, frequency-selective microwave devices for measuring power parameters of both deterministic (narrowband or pulse) and random (wideband noise – white noise or with frequency-limited Gaussian spectra) have been developed based on these stable nonlinear phenomena [130–133, 174–176]. These devices operate over the wide frequency range from 200 MHz to 40 GHz using ferrite garnets. The frequency range of these devices can be extended to the higher frequencies into millimeter-wave band using monocrystalline uniaxial BaM hexagonal ferrites with relatively narrow resonance line width [12, 39, 131].

7.6.1 Frequency Conversion – Resonance Detection and CM

Let us consider the frequency conversion due to stable nonlinear effects in more detail. Though the phenomena are nonlinear, but there is linear dependence between the components of ferrite's microwave magnetic moment $\boldsymbol{m}$ and microwave magnetic field $\boldsymbol{h}$ acting on the ferrite, i.e. FR operates at the small-signal regime. The nonlinearity is associated with the square dependence between the transverse and longitudinal components of microwave magnetic moment. Due to this nonlinearity, frequency conversion of microwave oscillations to the lower frequencies is possible.

The CM as a nonheterodyne technique of frequency conversion is based on this principle [171–173]. It is realized on the basis of an MD [179]. An MD is typically a tiny spherical FR surrounded by a spiral coil in its equatorial plane and placed on a dielectric substrate (typically glass textolite, or glass cloth laminate) into microwave field (usually in an air-filled waveguide or a stripline) parallel to the magnetic field force lines, as shown in Figure 7.43. An MD can operate without any modulation in the spiral microcoil (this regime is called *resonance detection [RD]*), or with modulation (this is *CM* regime). Modulation is usually accomplished by an RF (a few megahertz) signal from a local highly stable oscillator, typically with quartz crystal. An MD and its principle of operation are illustrated in Figure 7.44. The geometry of the microcoil (internal and external diameters, type of microwire, and number of turns) are chosen depending on the size of the FR and based on the requirement of maximum useful output voltage.

In the regime of *RD*, when there is no modulation signal in the microcoil, a microwave short-pulse or wideband random signals near the frequency of the FMR induce voltage in the microcoil, which is proportional to the time variations of the longitudinal component of the ferrite's magnetic moment:

$$V = Z_v \frac{dm_z}{dt}, \tag{7.302}$$

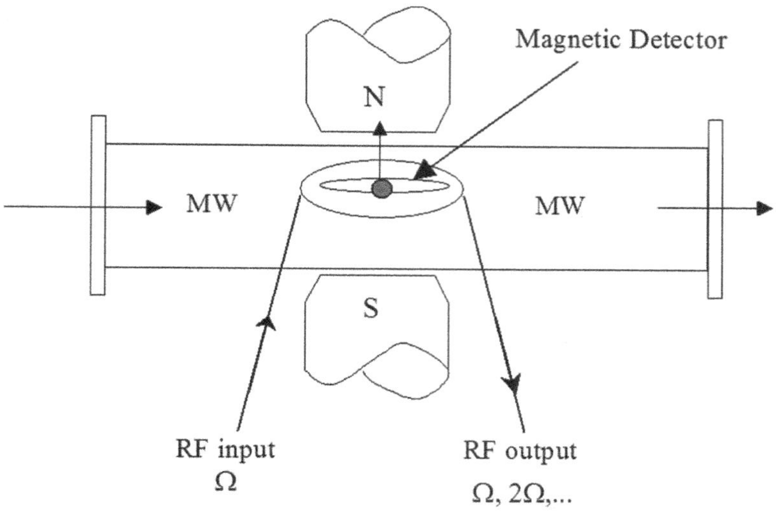

Figure 7.43 A magnetic detector in a waveguide affected by microwave (MW) field and RF modulation in a microcoil.

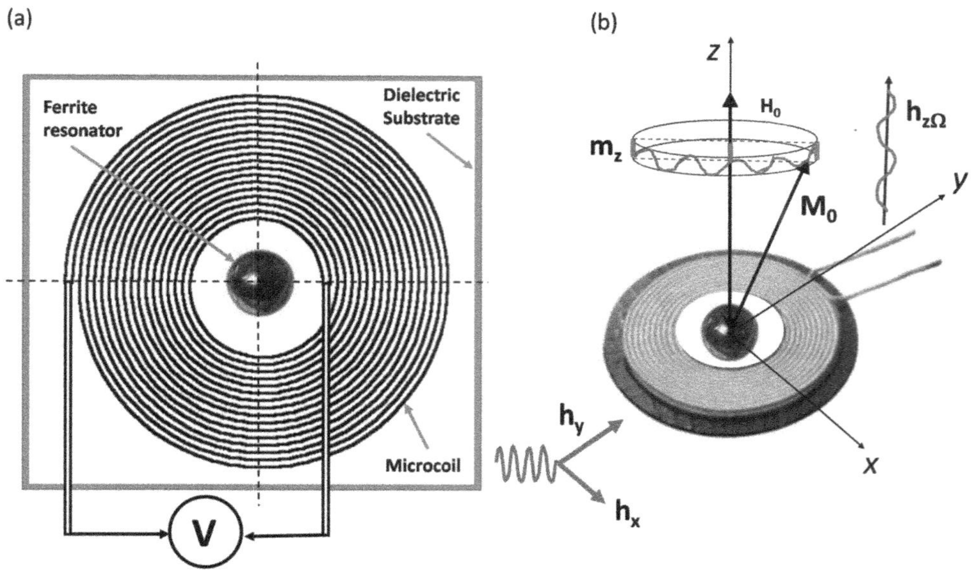

Figure 7.44 (a) Magnetic detector and (b) physics of its operation.

where Z_v is the coefficient which depends on the microcoil geometry and the position of the FR with respect to it. Assume that the ferrite is magnetized to saturation in z direction, and the microwave field has components $h_{x,y}$. The microwave magnetic moment should also have components $m_{x,y}$, if the microwave signal were harmonic. However, a short pulse (or a random signal) causes a deviation of the equilibrium magnetic moment from pure circular precession and, hence, an appearance of the longitudinal component m_z in the vicinity of the FRM. The time-varying magnetic flux associated with this variation induces the voltage (Equation 7.302). The magnitude of this voltage is proportional to the power absorbed by the ferrite at the FMR. Changing the FMR by linearly varying in time bias magnetic field, one can do spectral analysis in the frequency range of interest and measure spectral power density of the microwave signal.

In the regime of *CM*, there is an RF modulating signal with the angular frequency Ω in the microcoil. This signal produces magnetic field parallel with the bias magnetic field H_{0z}, i.e. the modulating field $h_{\Omega z}$ has only z-component, too. In this case, there are nutational (in addition to precessional) oscillations of the equilibrium magnetic moment of the ferrite even if the incident microwave signal is a continuous monochromatic (harmonic) oscillation at angular frequency ω.

The harmonics of the frequency Ω in the spectrum of m_z, as well as in the spectra of envelopes of the waves re-radiated by the ferrite into the microwave transmission line (waveguide), contain information about power parameters of signals at FMR in the ferrite. Therefore, the voltage V induced in the microcoil, will be proportional to the power of the microwave signal, if $\omega = \omega_{res}$. But this voltage will contain the harmonics of the modulating signal $n\Omega$, where $n = 1,2,3...$ Therefore, there is a frequency conversion from the microwave frequency ω to the lower (intermediate) RF frequency. There is a modulator instead of the heterodyne. Usually, for CM, the second or higher harmonics of Ω are selected. The first harmonic at Ω is not used at the microcoil output, as it is important to separate the converted signal from the initial modulating one.

If an envelope of the re-radiated field by the ferrite with the modulated resonance frequency is captured by a microwave detector, i.e. a crystal detector, then the harmonics of Ω can be selected from this detected signal, and they also contain information about the microwave signal at the frequency $\omega = \omega_{res}$. This approach is usually more sensitive than using the signal from the microcoil; but the dynamic range of such conversion is limited by the linearity of the crystal detector characteristic.

For a wideband noise signal, modulation at the frequency Ω in the coil does not change the spectrum of the m_z component, as it remains random. Therefore, for a wideband the regimes of CM and RD produce practically the same output voltage.

7.6.2 Spectrum of Magnetic Moment at Harmonic Microwave Signal and Modulated FMR

Now let us consider the spectrum of the ferrite's magnetic moment. The monocrystalline FR is magnetically saturated and small as compared to the microwave wavelength in ferrite; it can be considered as a single-domain particle. This particle has a magnetic moment determined by its equilibrium magnetization $\boldsymbol{M}$. The precessional dynamics of this described by MME with a relaxation term, for example, in modified Bloch (or BB) form, as described earlier in Section 7.2,

$$\frac{d\boldsymbol{M}}{dt} = -\mu_0\gamma \boldsymbol{M}\times\boldsymbol{H} + \omega_{rel}(\chi_0\boldsymbol{H} - \boldsymbol{M}). \tag{7.303}$$

For small angles of precession, there is square relation between the longitudinal and transverse components of the vector $\boldsymbol{M}$ [15, 105]:

$$m_z(t) = M_0 - \frac{m_x^2(t) + m_y^2(t)}{2M_0}, \tag{7.304}$$

where $M_0 = M_s$ is the magnitude of the equilibrium magnetic moment equal to the saturation magnetization.

The standard limitations when solving the equation (7.303) are the following: the ferrite is isotropic; precession of the vector $\boldsymbol{M}$ is uniform; exchange interactions are neglected; there are no propagation of EMWs inside the ferrite sample; the sample is spherical; and AC components of magnetic field and magnetization are much smaller than the DC components. Also assume that the microwave magnetic field is linearly polarized, with the only component $h_y(t) = h_{ym}\cos\omega t$, e.g. the FR is placed somewhere in the middle of the wide side of the standard rectangular waveguide. The direction of propagation is y. Then the following equations can be obtained:

$$\begin{aligned} &\frac{d^2m_x}{dt^2} + 2\delta\frac{dm_x}{dt} + \omega_\perp^2 m_x = \omega_M\frac{dh_y}{dt};\\ &\frac{d^2m_y}{dt^2} + 2\delta\frac{dm_y}{dt} + \omega_\perp^2 m_y = Ah_y + B\frac{dh_y}{dt}, \end{aligned} \tag{7.305}$$

where $\delta = \omega_{rel}\left(1 + \frac{\chi_0}{3}\right)$ is the relaxation parameter related with the demagnetization form factors of the sphere, $\omega_\perp^2 = \delta^2 + \omega_{res}^2$, $\omega_M = \mu_0\gamma M_0$, $\omega_0 = \mu_0\gamma H_0$, $A = \chi_0\omega_{rel}\delta + \omega_M\omega_0$, and $B = \chi_0\omega_0$.

Then solving Equations (7.305) for m_x and m_y and substituting in Equation (7.304), one can get the expression for the longitudinal component of the ferrite's magnetic moment varying in time:

$$m_z(t) = M_0 - \frac{B^2 + \omega_M^2}{2M_0}e^{-2\delta t}(I_{1y}^2 + I_{2y}^2), \tag{7.306}$$

where $I_{1y}(t)$ and $I_{2y}(t)$ are the following integrals:

$$I_{1y}(t)=\int_0^t e^{\delta\tau}h_y(\tau)\sin\alpha(\tau)d\tau;$$
$$I_{2y}(t)=\int_0^t e^{\delta\tau}h_y(\tau)\cos\alpha(\tau)d\tau; \tag{7.307}$$

where the derivative of the phase angle α in time is:

$$\frac{d\alpha}{dt}=\omega_0+\mu_0\gamma H_z(t), \tag{7.308}$$

and the total bias magnetic field, including RF modulation, is:

$$H_z(t)=H_{0z}+h_{z\Omega}\cos\Omega t. \tag{7.309}$$

Then the total phase angle is:

$$\alpha(t)=\omega_0 t+q\sin\Omega t, \tag{7.310}$$

where

$$q=\mu_0\gamma h_{z\Omega}/\Omega, \tag{7.311}$$

is the normalized modulation "amplitude" for the bias magnetic field.

Similarly, the normalized detuning value can be introduced:

$$a=(\omega_0-\omega)/\Omega, \tag{7.312}$$

and the relative (with respect to the ferrite resonance half line width) modulation frequency is:

$$p=\Omega/\delta. \tag{7.313}$$

Then, assuming that the microwave frequency is close to the FMR of the ferrite ($\omega\approx\omega_{res}=\omega_0$) with $\omega_0>>\delta$ for the high-Q FR, the longitudinal magnetic moment component is expressed as follows:

$$m_z(t)=M_0-\frac{B^2+\omega_M^2}{2M_0}\left(\frac{h_{ym}}{2\delta}\right)^2\left(\Psi_0(a,p,q)+\sum_{n=1}^{\infty}\Psi_n(a,p,q)\cos\left(n\Omega t-\varphi_n(a,p,q)\right)\right), \tag{7.314}$$

where the functions $\Psi_n(a,p,q)$ are the amplitudes of the harmonics of the modulation frequency and $\varphi_n(a,p,q)$ are the corresponding phases of these harmonics; their expressions are given next.

If there is elliptically polarized microwave magnetic field with two components – $h_x(t)$ and $h_y(t)$, the longitudinal magnetic moment is described by:

$$m_z(t)=M_0-\frac{B^2+\omega_M^2}{2M_0}e^{-2\delta t}\left(\left(I_{1x}-I_{1y}\right)^2+\left(I_{2x}-I_{2y}\right)^2\right), \tag{7.315}$$

and the integrals $I_{1x}(t)$ and $I_{2x}(t)$, in addition to those for y-components (7.307), are:

$$I_{1x}(t)=\int_0^t e^{\delta\tau}h_x(\tau)\sin\alpha(\tau)d\tau;$$
$$I_{2x}(t)=\int_0^t e^{\delta\tau}h_x(\tau)\cos\alpha(\tau)d\tau. \tag{7.316}$$

Then Equation (7.314) for $m_z(t)$ will be formally the same, but will include the sum of the squares of magnitudes of the magnetic field components, $h_{ym}^2\rightarrow h_{xm}^2+h_{ym}^2$.

The modulation spectrum components $\Psi_n(a,p,q)$ and $\varphi_n(a,p,q)$ can be calculated as follows:

$$\Psi_n(a,p,q)=\sqrt{C_n^2+S_n^2};\qquad \tan\varphi_n(a,p,q)=S_n/C_n, \tag{7.317}$$

where

$$\begin{aligned}
C_n&=2\left[\sum_{k=0}^{\infty}J_k(q)J_{k+n}(q)\left(R_{kn}^{+}(a,p)+(-1)^n R_{kn}^{-}(a,p)\right)+\sum_{i=1}^{n-1}(-1)^{n-i}J_i(q)J_{k-i}(q)Q_{ni}(a,p)\right],\\
R_{kn}^{+}(a,p)&=\frac{1+(kp+a)((n+k)p+a)}{(1+(kp+a)^2)(1+((k+n)p+a)^2)};\\
R_{kn}^{-}(a,p)&=\frac{1+(kp-a)((n+k)p-a)}{(1+(kp-a)^2)(1+((k+n)p-a)^2)};\\
Q_{ni}(a,p)&=\frac{1-((n-i)p-a)(ip-a)}{(1+(ip+a)^2)(1+((n-i)p-a)^2)}
\end{aligned} \tag{7.318}$$

and

$$\begin{aligned}
S_n&=2np\left[\sum_{k=0}^{\infty}J_k(q)J_{k+n}(q)\left(T_{kn}^{+}(a,p)+(-1)^n T_{kn}^{-}(a,p)\right)+\sum_{i=1}^{n-1}(-1)^{n-i}J_i(q)J_{n-i}(q)P_{ni}(a,p)\right],\\
T_{kn}^{+}(a,p)&=\frac{1}{(1+(kp+a)^2)(1+((k+n)p+a)^2)};\\
T_{kn}^{-}(a,p)&=\frac{1}{(1+(kp-a)^2)(1+((k+n)p-a)^2)};\\
P_{ni}(a,p)&=\frac{1}{(1+(ip+a)^2)(1+((n-i)p-a)^2)}.
\end{aligned} \tag{7.319}$$

$J_k(q)$ in these formulas are Bessel functions of the first kind. In the formulas above, i and k are the summation indices.

7.6.3 RD of Microwave Wideband Random Signal

Now assume that the microwave field $h_{x,y}(t)$ is the normal stationary random process distributed in the frequency range much wider than the FMR line width, and its spectral power density W is uniform. Let us find the spectrum of the longitudinal magnetic vector component m_z. The energy spectrum of m_z, $S_{mz}(\omega)$ is associated with the corresponding correlation function of the random process $K_z(\tau)$ as:

$$S_{mz}(\omega)=\int_{-\infty}^{+\infty}K_z(\tau)e^{-j\omega\tau}d\tau, \tag{7.320}$$

according to Wiener–Khinchin theorem [180, 181].

The correlation function is found as:

$$K_z(\tau)=\int_{-\infty}^{+\infty}m_z(t)m_z(t-\tau)dt-\langle(m_z(t))^2\rangle. \tag{7.321}$$

As there is a nonlinear (square) dependence between longitudinal and transverse components of the magnetic moment (7.306), the correlation function is expressed as [181, 182]:

$$K_z(\tau)=\frac{K_x(\tau)+K_y(\tau)+K_{xy}(\tau)+K_{yx}(\tau)}{4M_0^2}. \tag{7.322}$$

The microwave field is normally distributed, and so do the components $m_x(t)$ and $m_y(t)$, and their mutual distribution is also normal. Then, according to the theory of random processes, using spectral approach, the partial correlation functions and then $K_z(\tau)$ in Equation (7.230) are found. Omitting the cumbersome derivations, we will give the resultant correlation function for m_z as [182]:

$$K_z(\tau)=\frac{\left(\sigma_x^2+\sigma_y^2\right)^2}{4M_0^2}+\frac{\sigma_x^4R_x^2(\tau)}{2M_0^2}+\frac{\sigma_y^4R_y^2(\tau)}{2M_0^2}+\frac{\sigma_x^2\sigma_y^2R_{xy}^2(\tau)}{2M_0^2}+\frac{\sigma_x^2\sigma_y^2R_{yx}^2(\tau)}{2M_0^2}, \tag{7.323}$$

where:

$$\sigma_x^2=\frac{W\omega_M^2}{8\delta};\ \sigma_y^2=\frac{W}{8\delta\omega_\perp^2};\ R_x(\tau)=e^{-\delta|\tau|}\left(\cos\omega_0\tau-\frac{\delta}{\omega_0}\sin\omega_0|\tau|\right);$$
$$R_y(\tau)=e^{-\delta|\tau|}\left(\cos\omega_0\tau-G\frac{\delta}{\omega_0}\sin\omega_0|\tau|\right);\ G=\frac{A^2-\omega_\perp^2B^2}{A^2+\omega_\perp^2B^2}. \tag{7.324}$$

The mutual correlation functions $K_{xy}(\tau), K_{yx}(\tau)$ can be found through the two-dimensional probability density for $m_{x,y}(t)$; for normal statistical distribution, with Hermitian polynomials as set of orthogonal functions to expand this 2D probability density in a series [181]. However, herein, these correlation functions are not of interest. Indeed, if Equation (7.323) is substituted in Equation (7.320), there will be three spectral regions:

$$S(\omega)=S_0+S_{LF}(\varpi)+S_{HF}(\omega), \tag{7.325}$$

The first term in Equation (7.325) is the DC component S_0 arising from the first term of Equation (7.323); the term $S_{LF}(\varpi)$ is a comparatively low-frequency continuous spectrum $\varpi=2\pi f_{LF}$ note that these "low frequencies" are typically in the HF frequency band, and the third term $S_{HF}(\omega)$ is the high-frequency continuous spectrum near the frequency $2\omega_\perp$. For the analysis of MD interacting with microwave noise, we are interested in the term $S_{LF}(\varpi)$ only, as only this part determines the output voltage in the microcoil. This part of spectrum is comparatively low-frequency and is expressed as [182]:

$$S_{LF}(\varpi)=\frac{\left(\sigma_x^2+\sigma_y^2\right)^2}{M_0^2}\times\frac{2\delta}{(2\delta)^2+\varpi^2}. \tag{7.326}$$

The energy spectrum of voltage induced in the microcoil of MD from Equation (7.3032) is:

$$S_{EV}(\varpi)=(Z_V\varpi)^2S_{LF}(\varpi). \tag{7.327}$$

Then, from Equation (7.326):

$$S_{EV}(\varpi)=\frac{Z_e^2\left(\sigma_x^2+\sigma_y^2\right)^2}{M_0^2}\times\frac{2\delta\varpi^2}{(2\delta)^2+\varpi^2}. \tag{7.328}$$

The spectral part (7.324) in a different form is represented as:

$$S_{LF}(\varpi)=\frac{\left(\omega_M^2+B^2\right)^2}{4M_0^2}\times\left(\frac{W}{2\delta}\right)^2\times\frac{\Lambda(\nu)}{2\delta}, \tag{7.329}$$

where $\Lambda(\nu)=1/\left(1+(\nu/2)^2\right)$ and $\nu=\varpi/\delta$ is the normalized frequency of the signal after frequency conversion.

7.6.4 RD of Additive Random and Deterministic Microwave Signals

Assume that there is an additive sum of a microwave deterministic (harmonic) signal and the random signal as considered in the previous section. Then it can be shown that the correlation function for such statistical process is [183]:

$$K(\tau)=\frac{\left(\omega_M^2+B^2\right)^2}{4M_0^2}\times\left\{\begin{matrix}\left(\frac{W}{2\delta}\right)^2+\frac{\left(h_{xm}^2+h_{ym}^2\right)^2}{4\delta^2(1+a^2)^2}+\frac{W\left(h_{xm}^2+h_{ym}^2\right)}{8\delta^2(1+a^2)}\\+\frac{W\left(h_{xm}^2+h_{ym}^2\right)}{8\delta^3(1+a^2)}e^{-\delta|\tau|}\times\cos\left(\omega-\omega_0\right)\tau+\left(\frac{W}{4\delta}\right)^2e^{-2\delta|\tau|}\end{matrix}\right\}. \tag{7.330}$$

The corresponding useful low-frequency spectrum of m_z contains two components:

$$S_{LF}(\varpi)=S_R(\varpi)+S_{DR}(\varpi). \tag{7.331}$$

The first term in (7.331) associated with random signal only; the second term is associated with the mixture of deterministic signal and random.

The random component is determined by Equation (7.329), and the mixed deterministic–random spectral component is calculated as:

$$S_{DR}(\varpi)=\frac{\left(\omega_M^2+B^2\right)^2}{4M_0^2}\times\left(\frac{W}{2\delta}\right)^2\times\frac{\left(h_{xm}^2+h_{ym}^2\right)}{2\delta^3(1+a^2)}\left(\frac{1}{1+(\varpi-a)^2}+\frac{1}{1+(\varpi+a)^2}\right) \tag{7.332}$$

The spectrum of the voltage induced in the microcoil of MD is calculated from this spectrum using Equation (7.327).

Note that in the RD regime, the low-frequency spectrum of m_z does not have any separate components due to only deterministic signal. The DC components, as well as the high-frequency (microwave) components, are not needed to be considered, because we are interested in only converted comparatively low-frequency signal.

7.6.5 CM Spectra at Additive Random and Deterministic Microwave Signals

The correlation function for m_z in the case of the modulated by AC bias magnetic field depends on time, $K(t,\tau)$. Therefore, to calculate energy spectrum of random process using Wiener–Khinchin theorem, it is needed to average it in time [181]:

$$K_{av}(\tau)=\lim_{T\to\infty}\int_{-T/2}^{T/2}K(t,\tau)dt. \tag{7.333}$$

If there is an additive sum of microwave random and deterministic signals acting on the MD with modulated resonance frequency, the correlation function $K_{av}(\tau)$ contains components associated with random signal, with deterministic signal, and with their interaction due to nonlinear dependence between longitudinal and transverse components of ferrite's magnetic moment.

The low-frequency energy spectrum contains the following components:

$$S_{LF}(\varpi)=S_D(\Omega)+S_R(\varpi)+S_{DR}(\varpi), \tag{7.334}$$

where the first term is associated with discrete components at the harmonics of the modulation frequency Ω; the second term is associated with random signal, and the this is with the mixture of deterministic and random spectra.

As before, the DC and high-frequency components are not considered, as only low-frequency spectrum will induce the voltage of interest in the microcoil.

The expressions for these components are the following. The discrete spectrum for the deterministic harmonic signal is expressed as:

$$S_D(\Omega)=\pi\left(\frac{h_{xm}^2+h_{ym}^2}{(2\delta)^2}\right)^2\times\frac{\left(\omega_M^2+B^2\right)^2}{4M_0^2}\times\sum_{n=1}^{\infty}\Psi_n(a,p,q)(\delta(\varpi-n\Omega)+\delta(\varpi+n\Omega)), \tag{7.335}$$

where $\delta(\varpi\pm n\Omega)$ are delta functions in the points $\varpi=\mp n\Omega$.

The continuous spectrum due to random signal is the same as in Equation (7.329), and the continuous spectrum due to deterministic-random mixture is:

$$S_{DR}=\frac{\left(\omega_M^2+B^2\right)^2}{4M_0^2}\times\frac{W}{8\delta^3}\times\left(\frac{h_{xm}^2+h_{ym}^2}{2\delta}\right)^2\times\Phi(a,p,q,\nu), \tag{7.336}$$

where:

$$\Phi(a,p,q,\nu)=\sum_{n=1}^{\infty}\frac{2J_n^{*2}(q)}{1+(np+a)^2}\times\left(\frac{1}{1+(\nu-np-a)^2}+\frac{1}{1+(\nu+np+a)^2}\right). \quad (7.337)$$

Bessel functions in Equation (7.335) are $J_n^*(q)=\begin{cases}J_n(q)/\sqrt{2},\ n=0\\ J_n(q),\ n>0\end{cases}$.

The spectrum of voltage in the microcoil of MD is found using Equation (7.326), and in the case of CM, it contains two continuous components, similar to the RD, and discrete components at frequencies of modulation.

Figure 7.45 schematically shows the components of the low-frequency energy spectrum of m_z for a random signal, a deterministic signal, and their mixture. In the RD regime, there are only two continuous spectra components, and in the regime of CM discrete spectra at harmonics of the modulation frequency appear.

This difference in the spectra in two regimes is important for detection of deterministic components in random wideband noise, even though noise is significantly more intense than discrete signal harmonics. Switching modulation on and off in the MD allows for detecting these deterministic signal components, selecting them from the intense noise background, and measuring their power. This is illustrated in Figure 7.46.

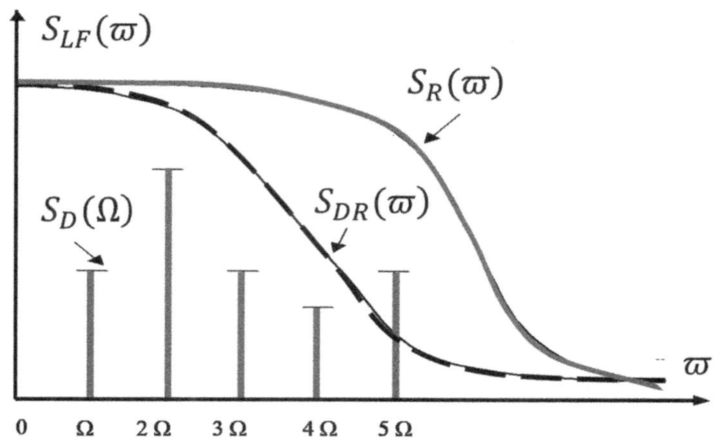

Figure 7.45 Low-frequency part of energy spectrum of m_z when additive narrowband deterministic and wideband random microwave signals interact with an MD in the regime of cross-modulation.

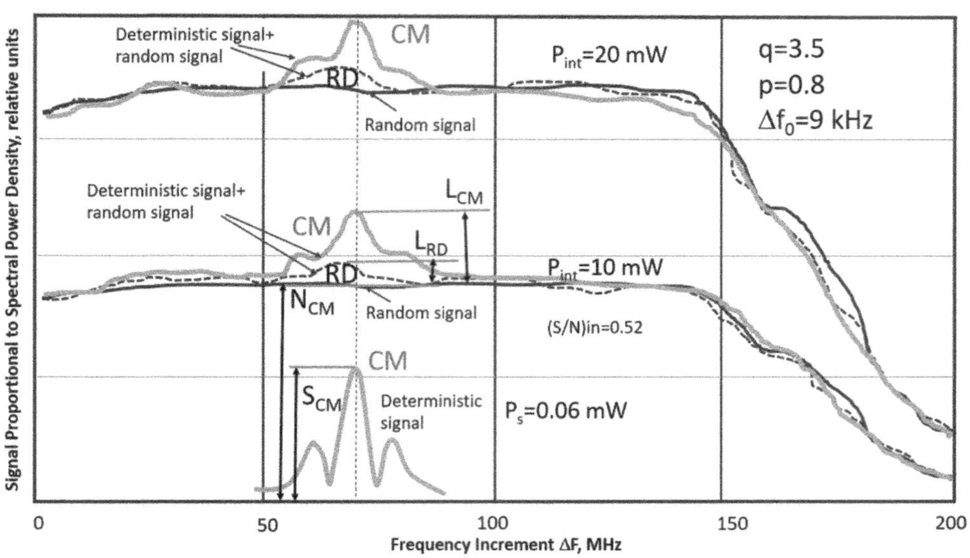

Figure 7.46 Detection of low deterministic signal in the wideband random signal by switching between two regimes of magnetic detection – resonance detection (RD) and cross-multiplication (CM).

Figure 7.46 shows the captured envelopes of wideband random signal (spectrum spread is 200 MHz) and deterministic signal falling within the 2.5 MHz resonance line of Ca–Bi–V garnet FR in RD (no modulation in MD) and CM (with modulation in MD). The power of the deterministic microwave signal is $P_S = 0.06$ mW, and the two wideband random signals have integral power of $P_{int} = 10$ and 20 mW, respectively. The induced voltage in the microcoil of the MD, in this case, is processed by a narrowband amplifier with the bandwidth of 9 kHz tuned to the most intense second harmonic of the modulation frequency (2Ω). The modulation regime in this example is optimized. As is found from the theoretical formulas and numerical optimization described earlier that the maximum increase of "signal-to-noise" ratio corresponds to the maximum possible magnitude of the second harmonic of modulation frequency $\Psi_2(a, p, q)$ (7.316). This is achieved when the normalized modulation frequency is $p = 0.8$ and the normalized amplitude of modulation $q = 3.5$. With this regime of modulation, $\Psi_2 \approx 0.4$ at the FMR (detuning $a = 0$).

The ratios of the separate (not mixed together) measured at the FRM frequency initial amplitudes of signals after CM for the deterministic signal (S_{CM}) and for the random signal, or "noise," N_{CM}, are $S_{CM} / N_{CM} = 0.52$ and 0.23, respectively. With modulation turned off (RD regime), the ratios of amplitudes are L_{RD} / N_{RD}=0.103 and 0.037, respectively. With modulation turned on (CM regime), the ratios of amplitudes are $L_{CM} / N_{CM} = 0.31$ and 0.12, respectively, for two "noise" signals. It is seen that the "signal-to-noise" ratio significantly increases when there is a modulation in MD microcoil. Moreover, the captured envelope of the power spectrum has characteristic shape with a peak and two side lobes, which is due to the shape of the second harmonic of modulation frequency $\Psi_2(a)$ as a function of detuning from resonance. The appearance of such shape in the power spectrum envelope when modulation is on is an indication of the presence of the deterministic signal on the background of the wideband random "noise." This is especially important when the inhomogeneities in the spectrum envelope are comparable to the magnitude of the deterministic signal so that without modulation it would have been difficult detect such a signal. Only the RD regime does not allow for such detection.

A reduction of the bandwidth of the narrowband amplifier can increase sensitivity to low-intensity deterministic signals in the wideband random signals. The resonance amplifiers with bandwidth as low as 0.2 kHz were tested when using very stable modulation frequency. The highest achieved experimentally sensitivity for detecting a "signal" of $P_S = 0.027$ mW on the background of wideband "noise" of $P_{int} = 15$ mW spread within 100 MHz.

This technique and device on its basis are described in [131–133]. A "panoramic" spectral power density measurer with a channel for low-intensity deterministic signal detection has been designed for the frequency range from 0.3 to 40 GHz; the dynamic range of the measured spectral power density is from 3 mW/MHz to 30 W/MHz.

7.7 Conclusion

Richard Feynman once stated that ferrites were one of the most difficult areas for theoretical description, but the most interesting for studies and practical applications [184]. Ferrites mainly gained popularity after World War II, when there happened the breakthrough in materials science and their sintering, and ferrites started being synthesized on mass scale. This could be explained by then rapid development of military radar and space technology, nuclear engineering, telecommunications, and other areas of RF and microwave engineering. Novel designs of functional electronics required applications of resonance, nonlinear, nonreciprocal, and polarization-sensitive phenomena in solid state. Ferrites due to their unique physical properties turned out to be very favorable materials for the development of new functional electronics. Based on their gyromagnetic properties (employing FMR), a new branch of electronics appeared – currentless spin electronics.

Various areas of applications of ferrites, including technological and societal areas, are summarized in Figure 7.47.

In the 1950s–1960s, there was a boom of microwave waveguide engineering. In the 1970s–1980s, there was a transition to planar transmission lines, and integration of ferrites in planar technology was considered as very perspective. However, there was an important limitation in employing ferrites for frequency-selective applications. Their best narrowband performance could be achieved only at their magnetization to saturation. In the case of ferrites with low crystallographic anisotropy – spinels or garnets – to operate at microwaves, one needs a massive magnetic system to provide bias fields. The higher the operating frequency, the more intense bias fields are needed. As frequencies increase higher than about 40 GHz, employment of spinel and garnet ferrites becomes unreasonable. However, development of technology of sintering and crystal growth of hexagonal ferrites, which possess high intrinsic magnetic crystallographic fields and can be self-biased, allowed for extending frequency ranges of microwave devices with ferrites to millimeter and even sub-millimeter ranges.

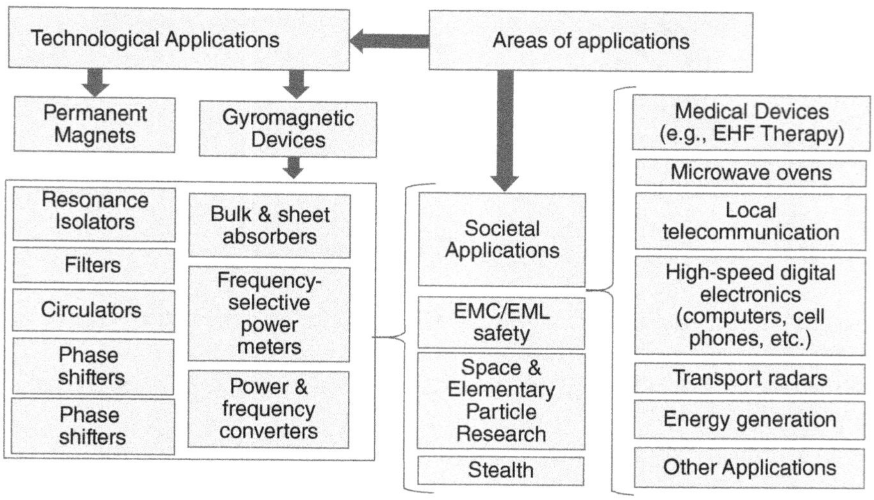

Figure 7.47 *Areas of applications of ferrites* [12].

As is seen from Figure 7.47 and discussions in this chapter and in the entire book, ferrites are still important materials for present-day technology. Ferrites have huge impact on our everyday life, security, safety, and scientific research, which opens new horizons to mankind.

References

1 Néel, L. (1948). Properiétés magnétiques des ferrites: Ferrimagnétisme et antiferromagnétisme (Magnetic properties of ferrites: Ferrimagnetism and antiferromagnetism). *Annals of Physics Paris* 3: 137–198.

2 Vonsovskii, S.V. (1966). *Ferromagnetic Resonance: The Phenomenon of Resonant Absorption of a High-Frequency Magnetic Field in Ferromagnetic Substances*. London: Pergamon.

3 Weiss, P. and Foëx, G. (1926). *Le Magnétisme*. Paris: Armand Colin.

4 Arkadiev, V.K. (1936). *Electromagnetic Processes in Metals, Part II – Electromagnetic Field*. Moscow-Leningrad: NKTP-ONTI SSSR. (in Russian).

5 Dorfman, I.G. (1923). Einige Bemerkungen zur Kenntnis des Mechanismus magnetischer Erscheinungen. *Zeitschrift für Physik* (Journal of Physics) 17 (2): 98–111 (in German).

6 Landau, L.D. and Lifshitz, E.M. (1935). On the theory of the dispersion of magnetic permeability in ferromagnetic bodies. *Physikalische Zeitschrift der Sowjetunion* 8: 153–169. In Landau, L.D. (1967). *Collected Papers* (ed. D. ter Harra), 101–117. New York: Gordon and Breach.

7 Zavoisky, E.K. (1946). Spin magnetic resonance in the decimeter-wave region. *Journal of Physics USSR* 10: 197–198.

8 Griffiths, J.H.K. (1946). Anomalous high frequency resistance of ferromagnetic material. *Nature* 158 (4019): 670–671.

9 Okamoto, A. (2009). The invention of ferrites and their contribution to miniaturization of radios. In: *Proc. IEEE* Global Communications Conference (*GLOBECOM) Workshops*, 30 November–4 December 2009, Honolulu, Hawaii.

10 Smit, J. and Wijn, H.P.J. (1959). *Ferrites. Physical Properties of Ferrimagnetic Oxides in Relation to Their Technical Applications*. New York: John Wiley and Sons.

11 Pardavi-Horvath, M. (2000). Microwave applications of soft ferrites. *Journal of Magnetism and Magnetic Materials* 215–216: 171–183.

12 Koledintseva, M.Y., Khanamirov, A.E., and Kitaitsev, A.A. (2011). Advances in engineering and applications of hexagonal ferrites in Russia. In: *Ceramic Materials* (ed. C. Sikalidis), 61–86. Rijeka: InTech.

13 Lax, B. and Button, K.J. (1962). *Microwave Ferrites and Ferrimagnetics*. New York: McGraw-Hill.

14 Patton, C.E. (1975). Microwave resonance and relaxation. In: *Magnetic Oxides* (ed. D.J. Craik), 575–648. London: Wiley and Sons.

15 Gurevich, A.G. and Melkov, G.A. (2000). *Magnetic Oscillations and Waves*. Boca Raton: CRC Press.

16 Gilbert, T.L. (2004). A phenomenological theory of damping in ferromagnetic materials. *IEEE Transactions on Magnetics* 40 (6): 3443–3449.

17 Callen, H.B. (1958). A ferromagnetic dynamical equation. *Journal of Physics and Chemistry of Solids* 4: 256–270.

18 Neo, C.P., Yang, Y., and Ding, J. (2010). Calculation of complex permeability of magnetic composite materials using ferromagnetic resonance model. *Journal of Applied Physics* 107: art. 083906.

19 Wallace, J.L. (1993). Broadband magnetic microwave absorbers: Fundamental limitations. *IEEE Transactions on Magnetics* 29: 4209–4214.

20 Acher, O. and Adenot, A.L. (2000). Bounds on the dynamic properties of magnetic materials. *Physical review B* 62 (17): 11324–11327.

21 Tsutaoka, T. (2003). Frequency dispersion of complex permeability in Mn–Zn and Ni–Zn spinel ferrites and their composite materials. *Journal of Applied Physics* 93 (5): 2789–2796.

22 Rozanov, K.N. and Koledintseva, M.Y. (2016). Application of generalized Snoek's law over a finite frequency range: A case study. *Journal of Applied Physics* 119 (7): 073901.

23 Mikhailovsky, L.K. and Koledintseva, M.Y. (2012). A quantum of potential mass-energy in single-point geometrical representation based on currentless magnetism and gyrovector formalism. In: *Proc. 20th Int. Conf. Electromagnetic Fields and Materials*, 16-18 November 2012, Moscow, 98–141.

24 Koledintseva, M.Y., Drewniak, J.L., Pommerenke, D.J. et al. (2005). Wideband Lorentzian media in FDTD algorithm. *IEEE Transactions on Electromagnetic Compatibility* 47 (2): 392–39.

25 Koledintseva, M.Y., Rozanov, K.N., and Drewniak, J.L. (2011). Engineering, modeling and testing of composite absorbing materials for EMC applications. In: *Advances in Composite Materials - Ecodesign and Analysis* (ed. B. Attaf), 291–316. InTech.

26 Milton, G.W., Eyre, D.J., and Mantese, J.V. (1997). Finite frequency range Kramers-Kronig relations: Bounds and dispersion. *Physical Review Letters* 97 (16): 3062–3065.

27 Koledintseva, M., Drewniak, J., Zhang, Y. et al. (2009). Modeling of ferrite-based materials for shielding enclosures. *Journal of Magnetism and Magnetic Materials* 321: 730–733.

28 Landau, L.D., Lifshitz, E.M., and Pitaevskii, L.P. (1984). Electrodynamics of continuous media. In: *Course of Theoretical Physics*, vol. 8, 2e. Oxford-New York: Pergamon Press.

29 Srinivasan, G. and Slavin, A.N. (eds.) (1995). *High-frequency Processes in Magnetic Materials.* Singapore: World Scientific.

30 Landé, A. (1921). Uber den anomalen Zeemaneffekt (On the anomalous Zeeman effect). *Zeitschrift für Medizinische Physik* (Journal of Medical Physics) 5 (4): 231–241. (*in German*).

31 Akulov, N.S. (1939). *Ferromagnetism.* Moscow-Leningrad: ONTI SSSR. (in Russian).

32 Schlömann, E. (1960). Ferrimagnetic resonance in the compensated region. In: *In Solid State Physics in Electronics and Telecommmunications: Magnetic and Optical Properties*, vol. 3 (eds. M. Desirant and J. L. Michiels), 322. New York: Academic Press.

33 Wangsness, R.K. (1953). Sublattice effects in magnetic resonance. *Physical Review* 91 (5): 1085–1091.

34 Geschwind, S. and Walker, L.R. (1959). Exchange resonances in Gadolinium Iron Garnet near magnetic compensation temperature. *Journal of Applied Physics* 30 (4), Suppl: 163s–170s.

35 Pollak, B.P. and Khanamirov, A.E. (1969). Peculiarities of ferromagnetic resonance in monocrystals of hexagonal ferrites. In: *Rep. Sc.-Research Conf. (Radio Engineering, Ferrite Microwave Radio Physics) in 1968-1969*, Moscow: Moscow Power Engineering Institute, 90–100. (in Russian).

36 Kittel, C. (2005). *Introduction to Solid State Physics*, 8e. Hoboken: John Wiley and Sons. 361–392.

37 Mikhailovsky, L.K., Pollak, B.P., and Sokolov, O.A. (1966). On the problem of ferromagnetic resonance in uniaxial single-domain particle of a ferromagnetic. *Physics of Metals and Physical Metallurgy* 21 (4): 524–528. (in Russian).

38 Mikhailovsky, L.K., Pollak, B.P., Balakov, V.F. et al. (1965). Properties and application of magneto-uniaxial ferrites at millimeter waves (Review). *Radiotekhnika i Elektronika (Radio Engineering and Electronics)* 10: 1739–1752. (in Russian).

39 Kitaytsev, A.A. and Koledintseva, M.Y. (1999). Physical and technical bases of using ferromagnetic resonance in hexagonal ferrites for electromagnetic compatibility problems. *IEEE Transactions on Electromagnetic Compatibility* 41 (1): 15–21.

40 Van Vleck, J.H. (1951). Ferromagnetic resonance. *Physica* 17 (3): 234–252.

41 Schloemann, E. (1958). Ferromagnetic resonance in polycrystalline ferrites with large anisotropy – I. General theory and application to cubic materials with a negative anisotropy constant. *Journal of Physics and Chemistry of Solids* 6: 257–266.

42 Schloemann, E. and Jones, R.V. (1959). Ferromagnetic resonance in polycrystalline ferrites with hexagonal crystal structure. *Journal of Applied Physics Supplement* 30 (4): 177S–178S.

43 Koledintseva, M., DuBroff, R.E., and Schwartz, R.W. (2009). Maxwell Garnett rule for dielectric mixtures with statistically distributed orientations of inclusions. *Progress in Electromagnetics Research* 99: 131–148.

44 Polivanov, K.M. and Pollak, B.P. (1964). Resonance characteristics of magneto-uniaxial polycrystalline ferrite in microwave field. *Izvestiya AN SSSR (News of the USSR Academy of Sciences), Ser. Physics* 28 (3): 470–480. (in Russian).

45 Nedkov, I., Cheparin, W., and Hanamirov, A. (1988). Ferromagnetic resonance of polycrystalline AL-substituted M-type hexagonal ferrite. *Journal de physique Colloque C8, Supplement*, 49 (12): C8-945–C8-946.

46 Pollak, B.P. (1977). Analysis of peculiarities of magnetic susceptibility tensor of a polycrystalline hexaferrite. In: *Trans. of Moscow Power Engineering Institute, Gyromagnetic Electronics and Electrodynamics*, 320, 45–53. (in Russian).

47 Hoselitz, K. and Nolan, R.D. (1969). Anisotropy-field distributions in barium ferrite micropowders. *British Journal of Applied Physics* 2 (2): 115–123.

48 Golubenko, Z.V., Kamzin, A.S., and Ol'khovik, L.P. (1998). Magnetic anisotropy of a system of nanocrystalline particles $BaO{\cdot}6Fe_2O_3$. *Fizika Tverdogo Tela (Solid State Physics)* 40 (10): 1894–1897. (in Russian).

49 Kolodin, P.S., Pollak, B.P., and Rakov, A.V. (2014). Imitation computer model of gyromagnetic material on the basis of hexaferrite particles. *Journal of Radio Electronics: Electronic Journal*, 3:1-17. http://jre.cplire.ru/jre/mar14/6/text.pdf. (in Russian).

50 Kolodin, P.S. and Pollak, B.P. (2014). Experimentally correctable computer model of a gyromagnetic resonator based on polycrystalline hexaferrite material. In: *Proc. 22nd Int. Conf. Electromagnetic field and Materials (Fundamental Research in Physics)*, Moscow, 499–510. (in Russian).

51 Polder, D. and Smit, J. (1953). Resonance phenomena in ferrites. *Reviews of Modern Physics* 25 (1): 89–90.

52 Brown, W.F., Jr. (1963). *Micromagnetics*. New York: Interscience Publishers.

53 Cullity, B.D. and Graham, C.D. (2009). *Introduction to Magnetic Materials*, 2e. Hoboken: John Wiley and Sons, IEEE Press.

54 Morrish, A.H. (2001). *The Physical Principles of Magnetism*. New York: An IEEE Press Classic Reissue.

55 Thiele, A.A. (1969). The theory of cylindrical magnetic domains. *Bell System Technical Journal* 48: 3287–3335.

56 Becker, R.A. (1954). *Introduction to Theoretical Mechanics*. New York: Mc-Graw Hill. 144.

57 Dione, G. (2003). Magnetic relaxation and anisotropy effects on high-frequency permeability. *IEEE Transactions on Magnetics* 39 (5): 3121–3126.

58 Nagamiya, T. (1953). A tentative interpretation of Bickford's observation of the resonance absorption in magnetite below its transition point. *Progress of Theoretical and Experimental Physics Oxford Journals* 10 (1): 72–82.

59 Smit, J. and Beljers, H.G. (1955). Ferromagnetic resonance absorption in $BaFe_{12}O_{19}$, A highly anisotropic crystal. *Philips Research Reports* 10 (2): 113–130.

60 Artman, J.O. (1957). Microwave resonance relations in anisotropic single-crystal ferrites. *Physical Review* 105 (1): 62–73.

61 Artman, J.O. (1956) Microwave resonance relations in anisotropic single crystal ferrites. *Proceedings of the IRE* 44 (10): 1284–1293.

62 Akaiwa, Y. and Okazaki, T. (1974). An application of a hexagonal ferrite to millimeter-wave Y circulator. *IEEE Transactions on Magnetics* 10: 374–378.

63 Baker-Jarvis, J., Janezic, M.D., Grosvenor, J.J.H. et al. (1993). Transmission/reflection and short-circuit line methods for measuring permittivity and permeability. In: *NIST Technical Note 1355-R*. Boulder: National Institute of Standards and Technology. U.S. Department of Commerce.

64 Pozar, D.M. (2012). *Microwave Engineering*, 4e. Hoboken: John Wiley and Sons.

65 Baker-Jarvis, J., Vanzura, E.J., and Kissick, W.A. (1990). Improved technique for determining complex permittivity with the transmission/reflection method. *IEEE Transactions on Microwave Theory and Techniques* 38: 1096–1103.

66 Takeda, S., Hotchi, T., Motomura, S. et al. (2015). Theoretical consideration on short- & open-circuited transmission lines for permeability & permittivity measurement. *Journal of the Magnetics Society of Japan* 39: 116–120.

67 Radonic, V., Blaz, N., and Zivanov, L. (2010). Measurement of complex permeability using short coaxial line reflection method. *Acta Physica Polonica A* 117: 820–824.

68 Kojima, H. (1982). Fundamental properties of hexagonal ferrites with magnetoplumbite structure. In: *Ferromagnetic Materials – A Handbook on the Properties of Magnetically Ordered Substances* (ed. E.P. Wohlfarth), 305–391. New York: North-Holland Publishing Company.

69 Bate, G. (1980). Recording materials. In: *Ferromagnetic Materials – A Handbook on the Properties of Magnetically Ordered Substances* (ed. E.P. Wohlfarth), 381–507. Amsterdam: Elsevier Science Publishers B.V.

70 Stablein, H. (1982). Hard ferrites and plastomerrites. In: *Ferromagnetic Materials – A Handbook on the Properties of Magnetically Ordered Substances* (ed. E.P. Wohlfarth), 441–602. New York: North-Holland Publishing Company.

71 Pullar, R.C. (2016). Hexagonal ferrite fibres and nanofibres. In: *Solid State Phenomena, Ferrites and Ceramic Composites II* (ed. R.B. Jotania and H.S. Virk), vol. 241, 1–68.

72 Slick, P.I. (1980). Ferrites for non-microwave applications. In: *Ferromagnetic Materials – A Handbook on the Properties of Magnetically Ordered Substances* (ed. E.P. Wohlfarth), 189–241. Amsterdam: Elsevier Science Publishers B. V.

73 Krupicka, S. and Novak, P. (1982). Oxide spinels. In: *Ferromagnetic Materials – A Handbook on the Properties of Magnetically Ordered Substances* (ed. E.P. Wohlfarth), 189–304. New York: North-Holland Publishing Company.

74 Gilleo, M.A. (1980). Ferromagnetic insulators: Garnets. In: *Ferromagnetic Materials – A Handbook on the Properties of Magnetically Ordered Substances* (ed. E.P. Wohlfarth), 1–53. Amsterdam: Elsevier Science Publishers B. V.

75 Nakamura, T., Tsutaoka, T., and Hatakeyama, K. (1994). Frequency dispersion of permeability in ferrite composite materials. *Journal of Magnetism and Magnetic Materials* 138: 319–328.

76 Tsutaoka, T., Ueshima, M., Tokunaga, T., Nakamura, T., and Hatakeyama, K. (1995). Frequency dispersion and temperature variation of complex permeability of Ni–Zn ferrite composite materials. *Journal of Applied Physics* 78 (6): 3983–3991.

77 Nakamura, T. (1997). Low-temperature sintering of Ni-Zn-Cu ferrite and its permeability spectra. *Journal of Magnetism and Magnetic Materials* 168: 285–291.

78 Tsutaoka, T., Kasagi, T., and Hatakeyama, K. (2011). Permeability spectra of yttrium iron garnet and its granular composite materials under DC magnetic field. *Journal of Applied Physics* 110: 053909.

79 Tsutaoka, T., Tsurunaga, A., and Koga, N. (2016). Permeability and permittivity spectra of substituted barium Ferrites BaFe12−x($Ti_{0.5}Co_{0.5}$)xO_{19}(x=0 to 5). *Journal of Magnetism and Magnetic Materials* 399: 64–71.

80 Sugimoto, S., Haga, K., Kagotani, T. et al. (2005). Microwave absorption properties of Ba M-type ferrite prepared by a modified coprecipitation method. *Journal of Magnetism and Magnetic Materials* 290–291: 1188–1191.

81 Haijun, Z., Xi, Y., and Liangying, Z. (2002). The preparation and microwave properties of $Ba_2Zn_zCo_2$-z$Fe_{12}O_{22}$ hexaferrites. *Journal of the European Ceramic Society* 22: 835–840.

82 Kagotani, T., Fujiwara, D., Sugimoto, S. et al. (2004). Enhancement of GHz electromagnetic wave absorption characteristics in aligned M-type barium ferrite $Ba_{1-x}\ La_xZn_xFe_{12-x-y}(Me_{0.5}Mn_{0.5})_yO_{19}$ (x=0.0–0.5; y=1.0–3.0, Me: Zr, Sn) by metal substitution. *Journal of Magnetism and Magnetic Materials* 272-276: E1813–5.

83 Bahadoor, A., Wang, Y., and Afsar, M.N. (2005). Complex permittivity and permeability of barium and strontium ferrite powders in X, KU, and K-band frequency ranges. *Journal of Applied Physics* 97: 10F105.

84 Singh, P., Babbar, V.K., Razdan, A. et al. (2004). Magnetic, dielectric and microwave absorption studies of Ba-Co Ti hexaferrite - Epoxy composites. *Indian Journal of Pure and Applied Physics* 42: 221–228.

85 Snoek, J.L. (1948). Dispersion and absorption in magnetic ferrites at frequencies above one Mc/s. *Physica* 14: 207–217.

86 Tsutaoka, T., Hatakeyama, K., and Nakamura, T. (2003). Complex permeability spectra of spinel ferrite composite materials. In: *Recent Research Developments in Magnetism & Magnetic Materials* (ed. S.G. Pandalai), 123–145. Trivandrum: Transworld Research Network.

87 Globus, A. (1977). Some physical considerations about the domain wall size theory of magnetization mechanisms. *Journal de Physique: Colloques, Magnetization Mechanisms* 38 (C1): 1–15.

88 Rado, G.T. (1953). Magnetic spectra of ferrites. *Reviews of Modern Physics* 25: 81–89.

89 Sugimoto, S., Okayama, K., Kondo, S. et al. (1998). Barium M-type ferrite as an electromagnetic microwave absorber in the GHz range. *Materials Transactions JIM* 39: 1080–1083.

90 Ravinder, D. and Latha, K. (1994). Electrical conductivity of Mn–Zn ferrites. *Journal of Applied Physics* 75: 6118–6120.

91 Maleknejad, Z., Gheisari, K., and Raouf, A.H. (2016). Structure, microstructure, magnetic, electromagnetic, and dielectric properties of nanostructured Mn–Zn ferrite synthesized by microwave-induced Urea–Nitrate process. *Journal of Superconductivity and Novel Magnetism* 29 (10): 2523–2534.

92 Krishna, K.R., Ravinder, D., Kumar, K.V. et al. (2012). Dielectric properties of Ni–Zn ferrites synthesized by citrate gel method. *World journal of condensed matter physics* 2: 57–60.

93 Baden Fuller, A.J. (1987). *Ferrites at Microwave Frequencies*. IEE Electromagnetic Wave Series, **23**. London: Peregrinus.

94 Tyras, G. (1969). *Radiation and Propagation of Electromagnetic Waves*. New York: Academic Press.

95 Potekhin, A.I. (1973). *Radiation and Propagation of Electromagnetic Waves in Anisotropic Media*. Washington: NASA. 1–61.

96 Hogan, C.L. (1952). The ferromagnetic Faraday effect at microwave frequencies and its applications: The microwave gyrator. *Bell Labs Technical Journal* 31 (1): 1–31.

97 Suhl, H. and Walker, L.R. (1952). Faraday rotation of guided waves. *Physical Review* 86: 122–123.

98 Gamo, H. (1953). The Faraday rotation of waves in a circular waveguide. *Journal of the Physical Society of Japan* 8: 176–181.

99 Suhl, H. and Walker, L.R. (1954). Topics in guided-wave propagation through gyromagnetic media: Part I – The completely filled cylindrical guide. *Bell System Technical Journal* 33 (3): 579–659.

100 Suhl, H. and Walker, L.R. (1954). Topics in guided-wave propagation through gyromagnetic media: Part II –Transverse magnetization and non-reciprocal helix. *Bell System Technical Journal* 33 (4): 939–986.

101 Suhl, H. and Walker, L.R. (1954). Topics in guided-wave propagation through gyromagnetic media: Part III –Perturbation theory and miscellaneous results. *Bell System Technical Journal* 33 (5): 1133–1194.

102 Epstein, P.S. (1956). Theory of wave propagation in a gyromagnetic medium. *Reviews of Modern Physics* 28 (1): 3–17.

103 Nikolskiy, V.V. (1957). Gyrotropic perturbation of a waveguide. *Radiotekhnika i Electronika (Radio Engineering and Electronics)* 2 (2): 157–171. (in Russian).

104 Nikolskiy, V.V. (1957). Calculation of phase shifts of gyrotropic inhomogeneities in a waveguide using perturbation method. *Radiotekhnika i Electronika (Radio Engineering and Electronics)* 2 (7): 833. (in Russian).

105 Gurevich, A.G. (1963). *Ferrites at Microwave Frequencies*. New York: Consultants Bureau Enterprises Inc.

106 Gurevich, A.G. (1952). *Hollow Resonators and Waveguides*. Moscow: Sov. Radio. (in Russian).

107 Gurevich, A.G. (1973). *Magnetic Resonance in Ferrites and Antiferrimagnetics*. Moscow: Nauka. (in Russian).

108 Rameshti, B.Z., Cao, Y., and Bauer, G.E.W. (2015). Magnetic spheres in microwave cavities. *Physical Review B* 91: 214430.

109 Krupka, J., Salski, B., Kopyt, P. et al. (2016). Electrodynamic study of YIG filters and resonators. Scientific Reports 6: 1-9 http://www.nature.com/srep34739.

110 Artman, J.O. and Tannenwald, P.E. (1955). Measurement of susceptibility tensor in ferrites. *Journal of Applied Physics* 26 (9): 1124–1132.

111 Lax, B. (1956). Frequency and loss characteristics of microwave ferrite devices. *Proceedings of the IRE* 44 (10): 1368–1386.

112 Rowen, J.H. (1953). Ferrites in microwave applications. *Bell Systems Technical Journal* 32 (6): 1333–1369.

113 Weiss, M.T. and Fox, A.G. (1952). Magnetic double refraction at microwave frequencies. *Physical Review* 88: 146–147.

114 Cacheris, J. (1954). Microwave single-sideband modulator using ferrites. *Proceedings of the IRE* 42 (8): 1242–1247.

115 Cacheris, J. and Dropkin, H.A. (1956). Compact microwave single-sided modulator. *IEEE Transactions on Microwave Theory and Techniques* 4 (3): 152–155.

116 Reggia, F. and Spenser, E.G. (1957). A new technique in ferrite phase shifting for beam scanning of microwave antennas. *Proceedings of the IRE* 45 (11): 1510–1517.

117 Boyd, C.R. (2005). Duplexing ferrite reciprocal phase shifters. In: *IEEE MTT–S International Microwave Symposium Digest*, 347–350.

118 Saadoun, M.M.I. and Engheta, N. (1992). A reciprocal phase shifter using novel pseudochiral or omega medium. *Microwave and Optical Technology Letters* 5 (4): 184–188.

119 Vendik, O.G. and Yegorov, Y.V. (2000). The first phased-array antennas in Russia: 1955–1960. *IEEE Antennas and Propagation Magazine*, 42 (4), 46–52.

120 Lebed, B.M. and Voronkov, V.D. (1996). State of the art millimeter and sub-millimeter wave ferrite components and devices. In: *26th Europ. Microwave Conf., EuMC*, 2, 816–822.

121 Dionne, G.F. (2005). Ferrite properties and technology issues for improved microwave systems. *Materials Research Society Symposium Proceedings* 833: 153–164.

122 Schloemann, E. (2000). Advances in ferrite microwave materials and devices. *Journal of Magnetism and Magnetic Materials* 209 (1–3): 15–20.

123 Harris, V.G., Geiler, A., Chen, Y. et al. (2009). Recent advances in processing and applications of microwave ferrites (Invited Review). *Journal of Magnetism and Magnetic Materials* 321: 2035–2047.

124 Gurevich, A.G. (1963). Ferrite ellipsoid in a waveguide. *Radiotekhnika I Elektronika (Radio Engineering and Electronics)* 8: 780–790. (in Russian).

125 Tokheim, R.E., Greene, C.K., Hoover, J.C. et al. (1967). Nonreciprocal YIG filters. *IEEE Transactions on Magnetics* 3 (3): 383–392.

126 Röschmann, P. (1971). YIG filters. *Philips Technical Review* 32: 322–327.

127 Il'chenko, M.E., Melkov, G.A., and Mirskikh, G.A. (1977). *Solid-state Microwave Filters*. Kiev: Tekhnika. (in Russian).

128 Clarke, J. and Dunsmore, M.R.B. (1985). High-power tunable YIG filters. *Proceedings of the IEE* 132-H (4): 251–254.

129 Aigle, M., Hechtfischer, G., Hohenester, W. et al. (2007). A systematic way to YIG-filter-design. In: *Proc. 37 Europ. Microw. Conf.*, Munich, Germany, 668–671.

130 Balakov, V.F., Kartsev, V.A., Kitaitsev, A.A. et al. (1986). Application of gyromagnetic effects in ferrite monocrystals for electromagnetic signals parameters measurement. In: *Proc. 5th Int. Conf. Microwave Ferrites*, 3, Moscow: MPEI, 86–99. (in Russian).

131 Koledintseva, M.Y., Mikhailovsky, L.K., and Kitaytsev, A.A. (2000). Advances of gyromagnetic electronics for EMC problems. In: *Proc. IEEE Int. Symp. Electromag. Compat.*, 2, Washington, DC, 773–778.

132 Koledintseva, M., Kitaitsev, A., Konkin, V., and Radchenko, V. (2004). Spectrum visualization and measurement of power parameters of wide-band noise. *IEEE Transactions on Instrumentation and Measurement* 53 (4): 1119–1124.

133 Koledintseva, M.Y., Kitaytsev, A.A., and Konkin, V.A. (2001). High-power microwave wideband random signal measurement and narrowband signal detection against the noise background. In: *Proc. Int. IEEE Symp. Electromag. Compat*, 2, Montreal, 1027–1029.

134 Bil'ko, M.I. and Tomashevsky, A.K. (1986). *Power Measurement at Microwaves*, 2e. Moscow: Radio i Svyaz (Radio and Communication). (in Russian).

135 Bogdanov, G.B. (1973). *Frequency-selective Systems with Ferrites and Their Application in Microwave Engineering*. Moscow: Sov. Radio. (in Russian).

136 Koledintseva, M.Y. (2006). Electromagnetic modelling of 3D periodic structure containing magnetized or polarized ellipsoids. *Opto-Electronics Review* 14 (3): 253–262.

137 Walker, L.R. (1957). Magnetostatic modes in ferromagnetic resonance. *Physical Review* 105 (2): 390–399.

138 Damon, R.W. and Eschbach, J.R. (1961). Magnetostatic modes of ferromagnetic slab. *Journal of Chemical Physics of Solids* 19 (3–4): 308–320.

139 Damon, R.W. and van der Vaart, H. (1965). Propagation of magnetostatic spin waves at microwave frequencies in normally magnetized disk. *Journal of Applied Physics* 36 (11): 3453–3459.

140 Bornmann, S., Schönecker, A., and Haubenreisser, W. (1972). Magnetostatic mode pattern and mode spectrum analysis in finite ferromagnetic circular disks. *Physica Status Solidi (A)* 11 (1): 207–217.

141 Schönecker, A., Bornmann, S., and Haubenreisser, W. (1974). Magnetostatic modes in finite ferromagnetic circular disks in perpendicular resonance. *Physica Status Solidi (A)* 22 (1): 53–64.

142 Kurawa, Y. and Abe, K. (1977). Resonant modes of magnetostatic waves in a normally magnetized disk. *Journal of Applied Physics* 48 (7): 3001–3007.

143 Yakovlev, Y.M. and Salyganov, V.I. (1987). FMR in planar structures and problems of planar ferrite filters design. *Reviews on Electronics Engineering, Ser. I, Microwave Electronics* 17 (1309): 1–64. (in Russian).

144 Celinski, Z., Harward, I.R., Anderson, N.R. et al. (2016). Planar magnetic devices for signal processing in the microwave and millimeter wave frequency range. *Handbook of Surface Science* 5 (10): 421–457.

145 Adam, J.D., Davis, L.E., Dionne, G.F. et al. (2002). Ferrite devices and materials. *IEEE Transactions on Microwave Theory and Techniques* 50 (3): 721–737.

146 Adam, J.D. (2014). Mitigate the interference: nonlinear frequency selective ferrite devices. *IEEE Microwave Magazine*, 15(6): 45–56

147 Mikhailovsky, L.K. (1994). Gyromodel (multitime gyrovector formalism) in the theory of electromagnetic field interaction with gyromagnetic media (Review). In: *Proc. 12th Int. Conf. on Microwave Ferrites*, Gyulechitsa, Bulgaria, 38–57.

148 Mikhailovsky, L.K. (1996). Geometrical representation of the formulae for energy quantum, action quantum, momentum and mass of the electromagnetic field energy in gyromodel (in "geometrostatics"). In: *Proc. 13th Int. Conf. Microw. Ferrites*, Busteni, Romania, 41–48.

149 Mikhailovsky, L.K. (2002). Gyromagnetic electronics (gyromodel with microparticle "gyron"). *Journal of Electrical Engineering (Elektrotechnický časopis) Special Issue "Electromagnetic Fields and Materials"*, Slovak Centre of IEE, 53 (9/s): 37–46.

150 Mikhailovsky, L.K. (2002). Gyrovectors for the Planck quanta and the Einstein quanta in the absorption and radiation centres of spin medium. *Journal Electrical Engineering (Elektrotechnický Časopis) Special Issue "Electromagnetic Fields and Materials"*, Slovak Centre of IEE, 53 (9/s): 47–52.

151 Mikhailovsky, L.K. (2009). Logical and mathematical bases for phaseless (quantum gyrovector) electrodynamics. In: *Proc. 17th Int. Conf. Magnetism, Long and Short-range Spin-spin Interaction*, Moscow: MPEU-TU, Appendix, 1–127. (in Russian).

152 Mikhailovsky, L.K. (2000). Radioabsorbing spin media, materials and coatings (Review of electromagnetic properties and practical applications). *Uspekhi Sovremennoi Radioelektroniki (Progress of Modern Radio Electronics)* 9: 21–30. (in Russian).

153 Mikhailovsky, L.K. (2010). Fundamentals of the single-point physics, or a gyromodel. In: *Proc. 18th Int. Conf. Electromagnetic Fields and Materials*, Moscow: MPEI (TU), 33–46.

154 Maxwell, J.C. (1890). On Faraday's lines of force. In: *Scientific Papers of James Clerk Maxwell, Part 1* (ed. W.D. Niven), 155–229. New York: Dover Publications.

155 Einstein, A. (1909). Über die Entwicklung unserer Anschauungen über das Wesen und die Konstitution der Strahlung (On the development of our views on the composition and essence of radiation). *Physikalische Zeitschrift* 10: 817–825.

156 Einstein, A. (1953). Einleitende Bemerkungen über Grundbegriffe" (in German), "Remarques préliminaires sur les concepts foundamentaux (in French), Introductory comments on fundamental concepts. In: *Louis de Broglie, Physician Et Penseur* (ed. A. George), 4–14. Paris: Albin Michel.

157 Hubbard, J. (1963). Electron correlations in narrow energy bands. *Proceedings of the Royal Society of London* 276 (1365): 238–257.

158 Ashcroft, N. and Mermin, N.D. (1976). *Solid-State Physics*. New York: Rinehart and Winston Holt.

159 Blokhintsev, D.I. (1968). *Fundamentals of Quantum Mechanics*. Moscow: Nauka. 16–17. (in Russian).

160 Shankar, R. (1994). *Principles of Quantum Mechanics*, 2e. New York: Kluwer Academic/Plenum Press.

161 Gray, J. (1989). *Ideas of Space: Euclidean, Non-Euclidean and Relativistic*, 2e. Oxford: Clarendon Press.

162 Kim, C., Matsuura, A.Y., Shen, Z.-X. et al. (1996). Observation of spin-charge separation in one-dimensional $SrCuO_2$. *Physical Review Letters* 77: 4054–4057.

163 Skryabin, Y.N. (1966). Exactly solved models. In *Physics Encyclopedia Dictionary* (ed. B.A. Vvedenskiy), vol. 5, 153. (in Russian).

164 Wheeler, J.A. (1963). *Geometrodynamics*. New York: Academic Press.

165 Einstein, A. (1922). Uber die gegenwartigeb Krise der theoretische Physik ("On the present crisis of theoretical physics"). *Kaizo (Reconstruction)* Kaizo-sha, Kyoto, 4: 1–8.

166 Mikhailovsky, L.K. (1993). Search of the possibility of validating the local absorption of microwave energy quanta in gyromagnetic media using Mössbauer effect. In: *Proc. XI Int. Conf. Gyromagn. Electronics and Electrodynamics*, Moscow: MPEI, 5, 7–8. (in Russian).

167 Astakhov, M.V., Muratov, V.A., and Mikhailovsky, L.K. (1995). Mössbauer studies of ferrites on microwave field. In: *Proc. Int. Conf. Currentless Spin-Electronics (ICCSE)*, Moscow: MPEI (TU), 241–245. (in Russian).

168 Koledintseva, M.Y. and Kitaitsev, A.A. (2007). Analysis of interaction between a crystallographically uniaxial ferrite resonator and a Hall-effect transducer. *Progress In Electromagnetic Research (PIER)* 74: 1–19.

169 Mikhailovsky, L.K., Alekseichik, L.V., and Kazantsev, Y.A. (1972). Interaction of microwave electromagnetic oscillations with independently excited precession of the ferrite magnetization. In: *Proc. of Conf., Electrotechnical Institute of Slovak Academy of Sciences*, ČSSR, 1, 75–86. (in Russian).

170 Mikhailovsky, L.K. and Alekseichik, L.V. (2008). On the question of microwave energy absorption by preliminary excited gyromagnetic resonator. In: *Proc. 16th Int. Conf. "Radar Engineering and Radio Communication*, Moscow: MPEI(TU), 524–533 (in Russian).

171 Mikhailovsky, L.K. (1976). Method of converting the electromagnetic spectrum carrier frequency and electromagnetic energy receiver for same. U.S. Patent 4059803 A, Apr. 20, 1976, Intern. Class H03D 7/00, National Class 325/445.

172 Mikhailovsky, L.K. (1988). Cross-parametric converters using gyromagnetic media. In: *Proc. 11th Int. Conf. on Gyromagnetic Electronics and Electrodynamics, Int. Conf. Microw. Ferrites ICMF-88*, Estergom, Hungary, 332–343. (in Russian).

173 Mikhailovsky, L.K. (1967). Cross-modulation frequency conversion on ferrite. In: *Physical Properties of Ferrites* (ed. N.N. Sirota), 235–241. Minsk: Nauka i Technika. (in Russian).

174 Gabeyeva, I.K. and Puchkov, V.S. (2002). Ferrite transducers for microwave power conversion in the regime of cross-multiplication. In: *Proc. 11th Int. Conf. Spin-Electronics and Gyrovector Electrodynamics*, Moscow: MPEI(TU), 590–601. (in Russian).

175 Mikhailovsky, L.K., Kitaitsev, A.A., Savchenko, N.I. et al. (1999). Stability of the devices of spin (currentless) electronics to the impact of intense and super-intense electromagnetic fields. In: *Proc. 8th Int. Conf. Spin-Electronics*, Moscow: MPEI (TU), 68–70. (in Russian).

176 Mikhailovsky, L.K. and Kitaitsev, A.A. (2009). Ferrite systems and devices, developed in OPLF MPEI. In: *Proc. 17th Int. Conf. Magnetism, Long and Short-Range Spin-Spin Interaction*, Moscow: MPEI (TU), 31–38. (in Russian).

177 Mikhailovsky, L.K. (2009). Results of experiments and peculiarities of parameters of currentless radioabsorbing materials created in OPLF MPEI(TU) and the designed microwave measurement equipment on the basis of isotropic and anisotropic monocrystal and polycrystal ferrites. *Proc. 17th Int. Conf. Magnetism, Long and Short-Range Spin-spin Interaction*, Moscow: MPEI (TU), 25–26. (in Russian).

178 Abarenkova, S.G., Ivanova, V.I., Karpov, V.N. et al. (1980). Resonance properties of a conglomerate of independent gyromagnetic particles. In: *Proc. 5th Int. Conf. on Microwave Ferrites*, 3, Moscow: MPEI, 170–174. (in Russian).

179 Mikhailovsky, L.K. (1988). Magnetic detector, certificate of authorship for invention SU 337061 of Sept. 27, 1967. Intern. Class H01L31/00. In *Byulleten' Izobreteniy (Bulletin of Inventions)*, 11 (in Russian). http://patents.su/3-337061-magnitnyjj-detektor.html Available online as of 07/31/2022 at the USSR Patent Database.

180 Lathi, B.P. (1998). *Modern Digital and Analog Communication Systems*, 3e. New York, Oxford: Oxford University Press.

181 Levin, B.R. (1966). *Theoretical Bases of Statistical Radio Engineering*, vol. 1. Moscow: Sovetskoye Radio (Soviet Radio) (in Russian).

182 Kitaitsev, A.A. (1969). Conversion of microwave noise by a magnetic detector. In: *Proc. Scientific and Engineering Conference on Research in 1968-1969*, Radio Engineering, Microwave Ferrite Radio Physics, Moscow: MPEI, 30–35. (in Russian).

183 Kitaitsev, A.A. (1969). Oscillation of magnetization at impact of microwave signal and noise on magnetic detector. In: *Proc. Scientific and Engineering Conference on Research in 1968-1969*, Radio Engineering, Microwave Ferrite Radio Physics, Moscow: MPEI, 36–39. (in Russian).

184 Feynman, R.P., Leighton, R.B., and Sands, M. (2006). *The Feynman Lectures on Physics*, vol. 2, 2e. Boston: Addison–Wesley.

8

Nonlinear Excitations in Ferrites

Pavel Kabos, John D. Adam, and Boris A. Kalinikos

8.1 Introduction

The basic theoretical principles of the ferrite response at high power levels that leads to nonlinear dynamics in ferrites are introduced in this chapter. Excitation by either a homogeneous or an inhomogeneous external pumping (driving) field will be discussed, but only modes excited when the applied microwave field amplitude $\boldsymbol{h}_p$ is at, or slightly above, the onset threshold of parametrically excited spin waves will be investigated. The presented material should introduce the understanding of ferrite-based devices when operating in the nonlinear regime, which is important for the design and applications of novel nonlinear ferrite devices operating at microwave and millimeter-wave frequencies.

The problem treated in this chapter is essentially quantum mechanical, but most of the practical device applications can be understood based on macroscopic spin-wave theory where the spin waves represent the perturbation of the continuous magnetization at a given position. References to the quantum mechanical approach will be introduced, if necessary.

The classical approach to the investigation of excitations in ferromagnetic materials follows from the equation of motion for the magnetization in the form [1]:

$$\frac{\mathrm{d}\boldsymbol{M}}{\mathrm{d}t} = -\mu_0 |\gamma| (\boldsymbol{M} \times \boldsymbol{H}_{\mathrm{eff}}) + \boldsymbol{R} \tag{8.1}$$

where $\boldsymbol{H}_{\mathrm{eff}}$ is the effective field accounting for the sample shape, anisotropy and nearest-neighbor interactions. $|\gamma| = \mu_0 |-e| g / 2\mathrm{m} = \mu_0 g \mu_B / \hbar = 1.76085963023 \times 10^{11}\,\mathrm{rad.T^{-1}.s^{-1}}$ is the gyromagnetic ratio, g is the Landé spectroscopic splitting factor, e is the electron charge, μ_B is the Bohr magneton, and $\mu_0 = 4\pi \cdot 10^{-7} \dfrac{\mathrm{Vs}}{\mathrm{Am}}$ is the permeability of vacuum. $\boldsymbol{R}$ is the damping term that will be discussed later in the text.

The effective field is expressed in terms of total energy E_{tot} of the system as [58]:

$$\boldsymbol{H}_{\mathrm{eff}} = -\frac{1}{\mu_0} \nabla_{\mathrm{M}} E_{\mathrm{tot}} = \boldsymbol{H}_{\mathrm{ext}} + \boldsymbol{h} + \boldsymbol{H}_{\mathrm{dip}} + \boldsymbol{h}_{\mathrm{ex}} + \boldsymbol{H}_{\mathrm{aniz}} + \boldsymbol{H}_{\mathrm{me}} + \boldsymbol{H}_{\mathrm{mel}} + \boldsymbol{H}_{\mathrm{th}} + \ldots \tag{8.2}$$

with $\boldsymbol{H}_{\mathrm{ext}}$, $\boldsymbol{H}_{\mathrm{aniz}}$, $\boldsymbol{H}_{\mathrm{dip}}$, $\boldsymbol{h}_{\mathrm{ex}}$, $\boldsymbol{H}_{\mathrm{me}}$, $\boldsymbol{H}_{\mathrm{mel}}$, and $\boldsymbol{H}_{\mathrm{th}}$ corresponding to the external, anisotropy, dipolar, exchange, magnetoelastic, magnetoelectric, and thermal fields, respectively, and $\boldsymbol{h}$ is the external RF excitation field. The essential fields required to understand the nonlinear effects in ferrites are the dipolar, exchange, and anisotropy fields. Magnetoelastic, magnetoelectric, and thermal fields, which are listed in Equation (8.2), will be neglected.

The relevant fields in the laboratory coordinate system, as shown in Figure 8.1 that coincides with the crystallographic anisotropy axes are as follows.

The dipolar field $\boldsymbol{H}_{\mathrm{dip}}$ must fulfill the Maxwell's equations:

$$\nabla \cdot (\boldsymbol{H}_{\mathrm{dip}}(\boldsymbol{r},t) + \boldsymbol{M}(\boldsymbol{r},t)) = 0 \tag{8.3a}$$

and

$$\nabla \times \boldsymbol{H}_{\mathrm{dip}}(r,t) = 0 \tag{8.3b}$$

Modern Ferrites: Basic Principles, Processing and Properties, Volume 1, First Edition. Edited by Vincent G. Harris.

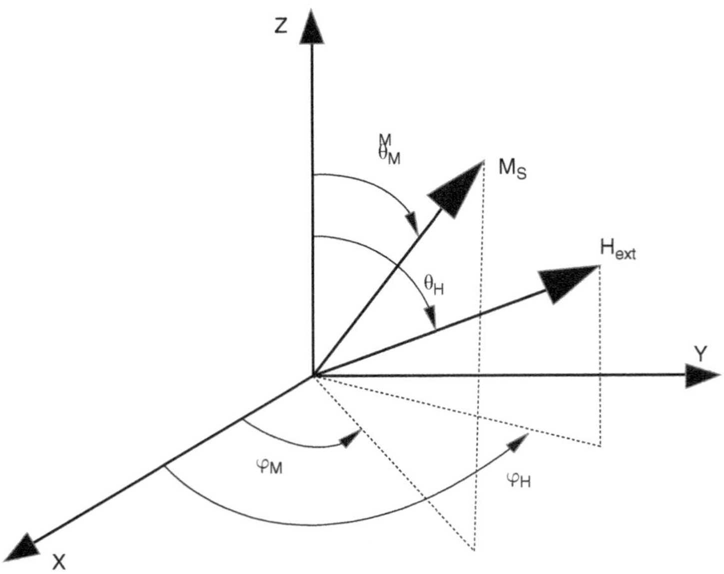

Figure 8.1 The laboratory (*X*, *Y*, *Z*) reference frame for sample with anisotropy. The external static field H_{ext} is applied at a polar angle θ_H and an azimuthal angle φ_H. The static magnetization $\mathbf{M}_s$ lies at a polar θ_M and an azimuthal angle φ_M [15].

The solution of the static problem is the uniform field inside the ellipsoid:

$$\boldsymbol{H}_i = \boldsymbol{H}_{\text{ext}} - [\boldsymbol{N}] \cdot \boldsymbol{M}_s \tag{8.4}$$

where [**N**] is the demagnetizing tensor, $\boldsymbol{H}_i$ is the internal field, and $|\boldsymbol{M}_s| = M_s$ is the saturation magnetization in SI units corresponding to the saturation induction, $4\pi M_s$ in CGS units. The solution for the uniform dynamic term, or uniform precession, is:

$$\boldsymbol{h}_{0dip}(t) = -[\boldsymbol{N}] \cdot \boldsymbol{m}_0(t). \tag{8.5}$$

The nonuniform dynamic fields are the solutions of Equations (8.3a) and (8.3b), and their form will be presented later when discussing spin-wave excitations leading to nonlinear effects.

The static part of the anisotropy field can be expressed as [1]:

$$H_{ai} = -\frac{1}{\mu_0}\frac{\partial F_a}{\partial M_i} = -\frac{1}{\mu_0 M_s}\frac{\partial F_a}{\partial \alpha_i}\Big|_{\alpha_{eq}}$$

where α_i represent the directional cosines of $\boldsymbol{M}$ in the laboratory frame.

The dynamic anisotropy field contribution is [2]:

$$\boldsymbol{h}_A(\boldsymbol{r},t) = \frac{1}{\mu_0 M_s^2}\sum_{i=x}^{z}\frac{\partial}{\partial \alpha_i}\left\{-\frac{\partial F_a}{\partial \boldsymbol{\alpha}}\right\}\Big|_{\alpha_{eq}} \cdot \boldsymbol{m}_i(\boldsymbol{r},t) = [\boldsymbol{A}]\cdot\{\boldsymbol{m}_0(t) + \boldsymbol{m}(\boldsymbol{r},t)\} = \boldsymbol{h}_{A0}(t) + \boldsymbol{h}_A(\boldsymbol{r},t) \tag{8.6}$$

F_a is the free energy density, $[\boldsymbol{A}]$ is a 3 × 3 anisotropy tensor with $A_{ij} = A_{ji}$, and

$$A_{ij} = -\frac{1}{\mu_0 M_s^2}\frac{\partial^2}{\partial\alpha_i\partial\alpha_j}F_a\Big|_{\alpha_{eq}} \tag{8.7}$$

$\boldsymbol{h}_{A0}$ represents the dynamic anisotropy contribution induced by the uniform precession and $\boldsymbol{h}_A(\boldsymbol{r},t)$ represents the dynamics anisotropy contribution due to spin waves. Here, a unit vector in the magnetization direction $\boldsymbol{\alpha} = \dfrac{\boldsymbol{M}}{M_s}$ is introduced. Phenomenologically, the free energy is expressed as a power series of directional cosines of the magnetization vector, leaving only the terms that are invariant to symmetry operations of the crystalline structure in the expansion. The specific form of the anisotropy matrix depends on the type of crystallographic anisotropy of the sample. Here, only a few special cases that frequently appear in the application of ferrites are presented. For a cubic crystal, such as yttrium iron garnet (YIG) or a spinel ferrite, the free energy can be expressed as [1, 2]:

$$F_a^{acb} = K_1(\alpha_X^2\alpha_Y^2 + \alpha_X^2\alpha_Z^2 + \alpha_Y^2\alpha_Z^2) + K_2\alpha_X^2\alpha_Y^2\alpha_Z^2 + \ldots \tag{8.8}$$

where K_1 is the first and K_2 the second cubic anisotropy constant. For a crystal with a uniaxial anisotropy the free energy can be expressed as [1, 2]:

$$F_a^{au} = K_U(1-\alpha_Z^2) \tag{8.9}$$

where K_U is the uniaxial anisotropy constant and for simplicity it is assumed that the uniaxial axis (*c*-axis) is Z in the laboratory frame. Depending on the sign of the anisotropy constant K_U, this expression represents either a uniaxial or an easy-plane anisotropy energy.

For hexagonal symmetry anisotropy about the main axis, the free energy is usually expressed in polar a coordinate system as [1, 3]:

$$F_a^{\text{hex}} = K_1 \sin^2\theta + K_2 \sin^4\theta + K_3 \sin^6\theta + \ldots \tag{8.10}$$

where θ is the angle between the magnetization and the main axis.

The exchange field can be expressed as [1, 4]:

$$\boldsymbol{h}_{ex}(\boldsymbol{r},t) = \frac{2A}{\mu_0 M_s^2}\nabla^2\boldsymbol{M}(\boldsymbol{r},t) = \frac{D}{\boldsymbol{M}_s}\nabla^2 M(\boldsymbol{r},t) \tag{8.11}$$

A = 10 pJ/m is the exchange stiffness constant and $D = \dfrac{2A}{\mu_0 M_s}$ is the exchange parameter.

Note that the static magnetization is along the effective field and the equilibrium position is given by $\boldsymbol{M} \times \boldsymbol{H}_{\text{eff}}^S = 0$, where $H_{\text{eff}}^S = \boldsymbol{H}_{\text{ext}} - [\boldsymbol{N}] \cdot \boldsymbol{M}_s - \boldsymbol{H}_a^S$ represents the static part of the effective field. The applied static external field and the equilibrium position of the magnetization are not, in general, collinear. Note also that the dipolar and anisotropy fields have both static and dynamic components, while the exchange field is dynamic, only.

$\boldsymbol{R}$ in Equation (8.1) is the damping term that can have several different forms. The form:

$$\boldsymbol{R} = -\frac{\lambda}{M_s^2}(\boldsymbol{M} \times (\boldsymbol{M} \times \boldsymbol{H}_{\boldsymbol{eff}})) \tag{8.12}$$

was first introduced in 1935 [5] and then extended [6]. The corresponding equation is called the Landau–Lifschitz equation of motion where λ is the damping parameter. An alternative damping term that is proportional to the change of total magnetization was proposed as [7]:

$$\boldsymbol{R} = \frac{\alpha}{M_s}\left(\boldsymbol{M} \times \frac{d\boldsymbol{M}}{dt}\right) \tag{8.13}$$

where α, a phenomenological constant, is often called the Gilbert damping parameter. Note, we will use the scalar α, without subscript, to distinguish it from the direction cosines, as a damping parameter. Equation (8.1) with damping parameters from Equations (8.12) and (8.13) represents the situation when the coupling between the neighboring spins is strong and the total magnetization magnitude is preserved. However, if the coupling between the spins is weak and the magnetization amplitude is not conserved, the equation of motion is modified to:

$$\frac{dM_z}{dt} = -\mu_0|\gamma|(\boldsymbol{M} \times \boldsymbol{H}_{\mathbf{eff}})_z - \frac{M_z - M_s}{T_1} \tag{8.14a}$$

$$\frac{dM_{x,y}}{dt} = -\mu_0|\gamma|(\boldsymbol{M} \times \boldsymbol{H}_{\mathbf{eff}})_{x,y} - \frac{M_{x,y}}{T_2} \tag{8.14b}$$

Equation (8.14) introduced two relaxation times, one for the relaxation of the z-component T_1 and the second for the relaxation of the transverse component T_2 of magnetization. They were initially derived in [8] and modified for ferromagnetic resonance in [9]. T_1 is the spin–lattice relaxation time, and T_2 is the spin–spin relaxation time. These equations must be complemented by Equations (8.3a) and (8.3b) and the appropriate boundary conditions.

The linearized equation of motion, obtained in the small signal limit, is extensively used for investigation of material properties at low power levels because its solution gives the relation between the driving field and the dynamic magnetization in the form of the magnetic susceptibility, which is independent of the amplitude of the applied driving field. The fundamental solution in the linear regime is called the uniform precession and the corresponding response is called ferromagnetic resonance (FMR) that is valid only to first order if the dipolar and other interactions are included. The corresponding derivation of susceptibility and its applications in linear microwave devices can be found in (Chapter 13).

8.2 General Instability Theory

Experimental investigation of FMR at higher power levels [11] showed a surprising line broadening at resonance (see Figure 8.2). The imaginary part of the susceptibility remains at its small signal value up to certain power level, but after exceeding this power level, the imaginary part of the susceptibility starts to decline. This decline in the imaginary part of the susceptibility is called saturation of the main resonance.

To extend the solution of the equation of motion to higher power levels, it was necessary to retain the higher order terms in $\boldsymbol{h}$. As the magnetization at these power levels did not seem to be conserved, the predictions were derived from the Bloch–Bloembergen equations, Equations (8.14a) and (8.14b). The following prediction for the ratio of the z-component of the magnetization to the saturation magnetization and of the susceptibility at resonance to the initial susceptibility is given by [10]:

$$\frac{\chi''_{res}}{\chi''_0} = \frac{M_z}{M_s} = \left(1 + \frac{1}{4}\gamma^2 |h_0|^2 T_1 T_2\right)^{-1} \tag{8.15}$$

However, early experiments [11, 12] showed that the onset of FMR saturation starts at power levels orders of magnitude smaller than the decline of the z-component of the magnetization as predicted by the extension of the small signal theory

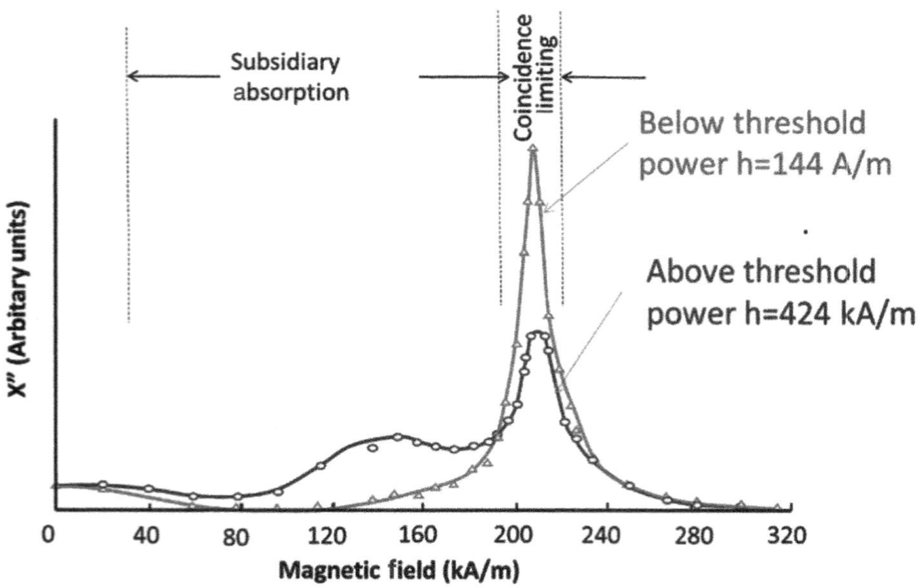

Figure 8.2 Variation of the imaginary part of the susceptibility (power absorption) with magnetic bias field, low power (triangles), and high power (circles), for a single-crystal nickel ferrite sphere at 9.4 GHz [11].

represented by Equation (8.15). An additional broad absorption also appeared at field levels several hundreds of A/m (Oe) below the FMR in Figure 8.2. As already mentioned, the uniform precession is a natural mode of the system only under special conditions but with dipolar and exchange included the higher order modes, spin waves, are coupled to the uniform precession and provide an explanation for these unexpected results [13].

The basic results of the nonlinear excitations in ferrites will be addressed, without loss of generality, by following the classical approach of Equation (8.1), with the damping term in the form of Equation (8.13). Initially, it will be assumed that the sample has the form of an ellipsoid magnetized to saturation in the z-direction by an external static magnetic field $\boldsymbol{H}_0$. The information that is necessary for an understanding of the nonlinear processes in ferrites will be introduced, and the formulas that can be used for immediate application discussed. The references should enable interested readers to get a deeper insight into nonlinear excitations in ferrites. The problem will be treated in two steps. First, it will be assumed that the wavelength of the spin waves is much shorter than the dimensions of the sample. In this case, the effects of the sample dimensions and surface effects can be neglected. Next, the case of nonlinear excitations when the spin-wave wavelength is comparable with the sample dimensions will be considered. It will be seen that this can significantly influence the dispersion and the power levels for the onset of nonlinear effects.

A detailed derivation of the formulas will not be presented, but only the most important components that are relevant to the derivation will be introduced. Details of the derivation can be found in the listed references.

In the following, a reference precession frame with the z-axis in the direction of the equilibrium position of the magnetization $\boldsymbol{M}$ is introduced. The precession frame is obtained from the laboratory frame using the rotational matrix with angles indicated in Figure 8.1. As the transformation is unitary, the final demagnetizing and anisotropy tensors are still symmetric and details of the transformation can be found in [4, 15]. Under the stated assumptions, it is possible to write the magnetization as:l

$$\boldsymbol{M}(\boldsymbol{r},t) = M_s\boldsymbol{u}_z + \boldsymbol{m}(\boldsymbol{r},t) \tag{8.16}$$

$\boldsymbol{u}_z$ is a unit vector in the z-axis and $\boldsymbol{m}(\boldsymbol{r},t)$ represents the dynamic magnetization having a small amplitude fulfilling the relation $|\boldsymbol{m}(\boldsymbol{r},t)| << M_s$. The dynamic magnetization can be expanded in spatial Fourier series:

$$\boldsymbol{m}(\boldsymbol{r},t) = \boldsymbol{m}_0(t) + \sum_{k=0} \boldsymbol{m}_k(t)e^{-j\boldsymbol{k}\cdot\boldsymbol{r}} \tag{8.17}$$

where $\boldsymbol{m}_0(t)$ represents the uniform mode and $\boldsymbol{m}_k(t)$ define the nonuniform, in general propagating, modes of dynamic magnetization response, or spin waves. Note that $\boldsymbol{m}_k^*(t) = \boldsymbol{m}_{-k}(t)$. In the following, unless it is mentioned otherwise, harmonic time dependence in the form $e^{j\omega t}$ is assumed that will be omitted in the formulas. In the Cartesian coordinate system, the wave vector $\boldsymbol{k}$ is described by the magnitude k, and angles ϕ_k between the x-axis and the projection of the $\boldsymbol{k}$ into the x-, y-plane and θ_k the angle between the $\boldsymbol{k}$ and the z-axis. Initially, the influence of the first four terms from the effective field in Equation (8.2) will be investigated. Using the expansion in Equation (8.17) and inserting it into the effective fields represented by Equation (8.2) the contributions of nonuniform excitations to the dynamic fields can be expressed as [4]:

$$\boldsymbol{h}_{ex}(r,t) = \frac{D}{M_s}\nabla^2\boldsymbol{m}(r,t) = -\frac{D}{M_s}\sum_{k\neq 0} k^2\boldsymbol{m}_k(t)e^{-j\boldsymbol{k}\cdot\boldsymbol{r}} \tag{8.18}$$

The solution for the nonuniform dynamic term of the dipolar field is [4]:

$$\boldsymbol{h}_{kdip}(t) = -\frac{\boldsymbol{k}}{k^2}\left[\boldsymbol{k}\cdot\boldsymbol{m}_k(t)\right] = -\left[\boldsymbol{N}_k\right]\cdot\boldsymbol{m}_k(t); \tag{8.19}$$

where the dynamic dipolar demagnetizing tensor is defined as [4]:

$$\left[\boldsymbol{N}_k\right] = \frac{1}{k^2}\begin{pmatrix} k_x^2 & k_xk_y & k_xk_z \\ k_yk_x & k_y^2 & k_yk_z \\ k_zk_x & k_zk_y & k_z^2 \end{pmatrix} \tag{8.20}$$

The tensor components are related to the wave vector $\boldsymbol{k}$ coordinates defined in previous section.

The effective field, including only external, exchange, dipolar, and anisotropy fields, can now be expressed as:

$$\begin{aligned}\boldsymbol{H}_{\mathit{eff}}(\boldsymbol{r},t)=\boldsymbol{H}_{\mathit{eff}}^{S}+\boldsymbol{h}(\boldsymbol{t})-[\boldsymbol{N}]\cdot\boldsymbol{m}_0(\boldsymbol{t})-\sum_{k\neq0}[\boldsymbol{N}_K]\cdot\boldsymbol{m}_K(\boldsymbol{t})\boldsymbol{e}^{-\boldsymbol{jk}\cdot\boldsymbol{r}}-\\ \frac{D}{M_S}\sum_{k\neq0}k^2\boldsymbol{m}_k(t)e^{-\boldsymbol{jk}\cdot\boldsymbol{r}}+[\boldsymbol{A}]\cdot\boldsymbol{m}_0(t)+\sum_{k\neq0}[\boldsymbol{A}]\cdot\boldsymbol{m}_k(t)e^{-\boldsymbol{jk}\cdot\boldsymbol{r}}=\boldsymbol{H}_{\mathit{eff}}^{s}+\boldsymbol{h}_{\mathit{eff}}(\boldsymbol{r},t)\end{aligned} \tag{8.21}$$

This equation inserted together with Equations (8.16) and (8.17) into the equation of motion (8.1) represents the general equation of motion for the uniform precession and spin waves. Note that due to the tensor characters of $[\boldsymbol{N}]$ and $[\boldsymbol{A}]$ in the component form, the scalar products will lead to double summations over k and the coordinate variables x, y, z. The solution of the linearized equation of motion is investigated in Chapter 13, but in this chapter, the initial focus will be on the nonlinear solutions of the equation of motion for short-wavelength spin waves that will allow the definition of the threshold field amplitude h_{crit}, for the onset of nonlinear effects. Later in the chapter, long wavelength spin waves will be considered where it will be necessary to include the boundary conditions. First, the formulas will be introduced for h_{crit} as a function of the spin-wave damping represented by the spin-wave linewidth ΔH_k, the spin-wave wave vector $\boldsymbol{k}$, material parameters, external static field value, and pumping microwave field configuration.

As follows from the experimental results of [11, 12], magnetization is preserved to much higher power levels than those necessary for the nonlinear decline of the susceptibility. Therefore, it is possible to describe the onset of the nonlinear effects from Equation (8.1), which preserves the magnitude of magnetization $\boldsymbol{M}$. Thus, the z-component of magnetization can be expressed through the other two components as:

$$\begin{aligned}M_z=M_s\sqrt{1-\frac{1}{M_s^2}(m_x^2+m_y^2)}\approx M_s\left[1-\frac{(m_x^2+m_y^2)}{2M_s^2}\right]\ \text{or}\\ m_z(\boldsymbol{r},t)\approx-\frac{1}{2M_s}\left(m_x^2(\boldsymbol{r},t)+m_y^2(\boldsymbol{r},t)\right)\end{aligned} \tag{8.22}$$

if the assumption $|\boldsymbol{m}(\boldsymbol{r},t)|<<M_s$ holds. For simplicity, the usual approach is to obtain the equation of motion for $k\neq0$ from the torque equation and add the damping term later. Due to the complexity of the equations, a significantly shortened approach following [15] will be presented because it includes an arbitrary pumping field orientation and anisotropy. Details for the isotropic case can be found in [13, 14, 16, 17]. Equation (8.1) for the x- and y-components becomes:

$$\frac{dm_x}{dt}=-\mu_0|\gamma|(m_y(H_{\mathit{eff}}^S+h_{\mathit{effz}})-(M_s+m_z)H_{\mathit{effy}}) \tag{8.23a}$$

$$\frac{dm_y}{dt}=-\mu_0|\gamma|(m_x(H_{\text{eff}}^S+h_{\mathit{effz}})-(M_s+m_z)H_{\mathit{effx}}) \tag{8.23b}$$

Multiplying Equation (8.23b) by the imaginary unit j and adding (subtracting) it to (from) Equation (8.23a), we get

$$\frac{dm_\pm}{dt}=\mp j\mu_0|\gamma|(m_\pm(H_{\text{eff}}^S+h_{\mathit{effz}})\pm(M_s+m_z)h_{\text{eff}\pm}) \tag{8.24}$$

where $m_\pm=m_x\pm jm_y$; $h_{\text{eff}\pm}=h_{\mathit{effx}}\pm jh_{\mathit{effy}}$. Then m_z, from Equation (8.22) can be expressed as $m_z=-\frac{1}{2M_s}m_+m_-$. The next step is to introduce new variables in the Fourier spatial series as:

$$m_+=M_s\sum_k a_k e^{-\boldsymbol{jk}\cdot\boldsymbol{r}}$$

$$m_-=M_s\sum_k a_{-k}^* e^{-\boldsymbol{jk}\cdot\boldsymbol{r}}$$

These summations include the uniform precession $\mathbf{k} = 0$ and spin-wave terms with $\mathbf{k} \neq 0$. After tedious algebra, the general equation for spin-wave amplitude is given in the form [15]:

$$-j\frac{da_k}{dt} = (A_k + C_k)a_k + (B_k + D_k)a^*_{-k} + \mu_0|\gamma|h_z a_k \tag{8.25}$$

with coefficients defined in [14] as:

$$A_k = \mu_0|\gamma|\left[H^S_{eff} + Dk^2 - M_s(A_{xx} + A_{yy})/2\right] + \omega_M \sin^2\theta_k / 2 \tag{8.26a}$$

$$B_k = \omega_M \sin^2\theta_k \exp(2j\varphi_k)/2 - \mu_0|\gamma|\omega_M(A_{xx} - A_{yy} + 2jA_{xy})/2 \tag{8.26b}$$

$$\begin{aligned} C_k = {} & -\omega_M \sin(2\theta_k)\left[a_0 \exp(-j\varphi_k) + a_0^* \exp(j\varphi_k)\right]/4 \\ & +\left[\omega_M\left(A_{zx} - jA_{zy}\right) - \omega_M(\tilde{N}_{zx} - j\tilde{N}_{zy}\right]a_0/2 + \\ & \left[\omega_M\left(A_{zx} + jA_{zy}\right) - \omega_M\left(\tilde{N}_{zx} + j\tilde{N}_{zy}\right)\right]a_0^* \\ & +\mu_0|\gamma|h_\perp a_0^*/2 + +\omega_M\left(A_{xx} + A_{yy} - 2A_{zz}\right)a_0 a_0^*/4 \\ & +\omega_M\left(2\tilde{N}_{zz} - \tilde{N}_{xx} - \tilde{N}_{yy} + 3\cos^2\theta_k - 1\right) \\ & a_0 a_0^*/4 - -\omega_M\left(\tilde{N}_{xx} + 2j\tilde{N}_{xy} - \tilde{N}_{yy}\right)a_0 a_0^*/4 \end{aligned} \tag{8. 26c}$$

$$\begin{aligned} D_K = {} & \left[-\omega_M \sin(2\theta_k)\exp(j\varphi_k)/2 + \omega_M\left(A_{zx} + jA_{zy}\right)\right]a_0 \\ & +\mu_0|\gamma|h_\perp a_0 - \omega_M \sin^2\theta_k \exp(j2\varphi_k)a_0 a_0^*/4 - -\omega_M\left(\tilde{N}_{xx} + 2j\tilde{N}_{xy} - \tilde{N}_{yy}\right) \\ & a_0 a_0^*/4 + \omega_M\left(A_{xx} + 2jA_{xy} - A_{yy}\right)a_0 a_0^*/4 + \omega_M\left(2\cos^2\theta_k - \tilde{N}_{xx} - \tilde{N}_{yy}\right) \\ & a_0 a_0^*/4 + \mu_0|\gamma|\left(Dk^2 - M_S A_{zz}\right)a_0 a_0/2 \end{aligned} \tag{8.26d}$$

In the aforementioned equations, θ_k and $\varphi_k = \arctan(k_y/k_x)$ are the polar angles of the spin-wave vector $\mathbf{k}$ in the x, y, z precession frame. $h_\perp = h_x + jh_y$ and, $\omega_M = \mu_0|\gamma|M_s.\tilde{N}_{ij} = N_{ij} - A_{ij}$ which is different to that in [15] due to the use of SI units in this chapter. Simplified equations for an arbitrary pumping orientation and uniaxial anisotropy are discussed in [18]. The coefficients A_k and B_k do not depend on the amplitude of the driving field or the dynamic magnetization and therefore represent the linear response of the system. C_k and D_k contain the driving field and magnetization amplitudes and are the source of the nonlinear interactions. In general, the linearized part of Equation (8.25), obtained by leaving out the nonlinear coefficients and the driving field term, represents elliptically polarized spin waves. This part can be diagonalized using the Holstein–Primakoff transformation [19], leading to a description of spin waves with circular polarization. The transformation equations to this new set of spin-wave variables are:

$$a_k = \lambda_k b_k - \mu_k b^*_{-k}$$

$$a^*_{-k} = \lambda_k b^*_{-k} - \mu^*_k b_k$$

with

$$\lambda_k = \sqrt{\frac{A_k + \omega_k}{2\omega_k}} \tag{8.27}$$

$$\mu_k = \sqrt{\frac{A_k - \omega_k}{2\omega_k}} \frac{B_k}{|B_k|} \tag{8.28}$$

and the spin-wave frequency is:

$$\omega_k = \sqrt{A_k^2 - B_k^2}$$

For a sphere, this equation is:

$$\omega_k = \mu_0|\gamma|\sqrt{(H_i + Dk^2)(H_i + Dk^2 + M_s \sin^2\theta_k)}. \tag{8.29}$$

The transformation into this new spin-wave amplitude represents the critical step in the establishment of the nonlinear response of the system. The equation of motion after the transformation into the new variables is

$$-j\frac{db_k}{dt} = \omega_k b_k + F_k b_k + G_k b^*_{-k}$$

$$j\frac{db^*_{-k}}{dt} = \omega_k b^*_{-k} + F_k b^*_{-k} + G_{-k}{}^* b_k \tag{8.30}$$

with coefficients

$$F_k = \lambda_k^2\left(\mu_0|\gamma|h_z + C_k\right) + |\mu_k|^2\left(\mu_0|\gamma|h_z{}^* + C_k{}^*\right) - \lambda_k\left(D_k\mu_k{}^* + \mu_k D_k{}^*\right) \tag{8.31a}$$

$$G_k = \lambda_k^2 D_k + \mu_k^2 D_k^* - \lambda_k\mu_k\left(\mu_0|\gamma|h_z + \mu_0|\gamma|h_z{}^* + C_k + C_k{}^*\right) \tag{8.31b}$$

Note that C_k and D_k and, hence, F_k and G_k include the uniform precession amplitude a_0 and its complex conjugate a_0^*. Therefore, it is necessary to address the uniform precession solution to obtain the critical threshold fields. For this, it is necessary to keep only the $\boldsymbol{m}_0(t)$ term from the expansion in Equation (8.16). The effective field from Equation (8.21) is now:

$$\boldsymbol{H}_{eff}(\boldsymbol{r},t) = \boldsymbol{H}_{eff}^S + \boldsymbol{h}(t) - [\tilde{\boldsymbol{N}}]\cdot\boldsymbol{m}_0(t) \tag{8.32}$$

where the demagnetizing and anisotropy terms are combined into $[\tilde{\boldsymbol{N}}]$. The equation of motion for the uniform precession amplitude can now be written in the form [15]:

$$-j\dot{a}_0 = \mu_0|\gamma|H_{eff}^S - |\mu_0\gamma|h_\perp + \omega_M\left[\left(\tilde{N}_{xx} + \tilde{N}_{yy}\right)a_0 + \left(\tilde{N}_{xx} + 2j\tilde{N}_{xy} - \tilde{N}_{yy}\right)a_0^*\right]/2 \tag{8.33}$$

This equation must be solved together with its complex conjugate, and the complex uniform mode amplitude a_0 is then obtained as:

$$a_0(t) = \mu_0|\gamma|h_p\left[q_{L0}e^{j\omega t} + q_{A0}e^{-j\omega t}\right]/2 \tag{8.34}$$

The q_{L0} and q_{A0} coefficients represent the amplitudes of Larmor and anti-Larmor parts of the uniform precession as driven by the applied external pumping field and are given by [14]:

$$q_{L0} = \frac{\left(\omega_y + \omega_p - j\omega_{x,y}\right)\alpha_x e^{j\delta_x} + j\left(\omega_x + \omega_p + j\omega_{x,y}\right)\alpha_y e^{j\delta_y}}{\omega_0^2 - \omega_p^2} \tag{8.35a}$$

and

$$q_{A0} = \frac{\left(\omega_y - \omega_p - j\omega_{x,y}\right)\alpha_x e^{-j\delta_x} + j\left(\omega_x - \omega_p + j\omega_{x,y}\right)\alpha_y e^{-j\delta_y}}{\omega_0^2 - \omega_p^2} \tag{8.35b}$$

with

$$\omega_x = \mu_0|\gamma|H_{\text{eff}}^S + \omega_M\tilde{N}_{xx} \tag{8.36a}$$

$$\omega_y = \mu_0|\gamma|H_{\text{eff}}^S + \omega_M\tilde{N} \tag{8.36b}$$

$$\omega_{xy} = \omega_M\tilde{N}_{xy}$$

and

$$\omega_0 = \sqrt{\omega_x \omega_y - \omega_{xy}{}^2} \tag{8.37}$$

is the FMR frequency. The selection of α_i's and δ_i's defines the polarization of the pumping external RF field. These equations do not include the loss that can be introduced as before through $\omega_0 \to \omega_0 + j\eta_0$, with η_0 being the uniform mode relaxation rate. These equations will be discussed further when considering the critical threshold in hexagonal ferrites.

To understand the influence of the terms in Equation (8.30), following [10] let's first omit the G_k terms in Equation (8.30). In this case, the coefficients at b_k and $b_{-k}{}^*$ can be combined, and therefore, the F_k term in these equations only shifts and modulates the spin-wave frequency but does not contribute to an exponential increase in the spin-wave amplitude and thus does not have to be considered in the onset of the nonlinear process. The G_k term, on the other hand, provides the coupling between the modes and the external driving field, leading to nonlinear effects. Note also that the terms in coefficients C_k and D_k are time dependent due to higher orders of spin-wave amplitudes. The time dependence can be expressed as $e^{jn\omega t}$, with n having values between −2 and 2, including 0 for the expansion used. Because the critical coefficient for the nonlinear effects is G_k, the explicit time dependence can be expressed as:

$$G_k(t) = \sum_{-2}^{2} G_k^{(n)} e^{jn\omega t}$$

with

$$G_k^{(n)} = G_{-k}^{(n)}$$

Analysis shows that the $G_k^{(0)}$ term does not lead to an exponential increase of the spin-wave amplitude, and therefore, it will not be considered further. As will be seen later, only the $n = 1$ and $n = 2$ terms lead to a nonlinear increase in the spin-wave amplitude. At this point, it is also necessary to include the damping that was up to now omitted. This can be done following [13] by adding the imaginary part to the frequency $\omega_k \to \omega_k + j\eta_k$ in Equation (8.30), where η_k is the spin-wave relaxation. Expressing the time dependence of the spin-wave amplitude as $b_k = b_{k0}{}^n(t) e^{j\omega_k t}$, the equation of motion for the spin-wave amplitude has the form:

$$\frac{db_{k0}{}^n(t)}{dt} = -\eta_k b_{k0}{}^n(t) + jb_{-k0}^{n}{}^* \sum_{2}^{-2} G_k^{(n)} e^{j(n\omega - 2\omega_k)t} \tag{8.38}$$

The relevant terms are those that are changing slowly with the time. They are the only the terms that will have the difference between the driving and spin-wave frequencies close to zero, reducing the sum in Equation (8.38) to $n = 1$ and $n = 2$ terms. The negative n lead to small fast time modulation of the slow-in-time varying $b_{k0}(t)$. Therefore, only the $n = 1$ and $n = 2$ terms will be considered. Taking the complex conjugate of Equation (8.38), differentiating it, and using the complex conjugate, Equation (8.38) leads to:

$$\left[\frac{d^2}{dt^2} + 2\eta_k \frac{d}{dt} - \left|G_k^{(n)}\right|^2 + (n\omega - 2\omega_k)^2 + \eta_k{}^2\right] b_{k0}^n = 0 \tag{8.39}$$

with $n = 1, 2$. The solution of this equation is in the form $b_{k0}^n \propto e^{\kappa t}$ with:

$$\kappa = -\eta_k + \sqrt{\left|G_k^{(n)}\right|^2 - (n\omega - 2\omega_k)^2} \tag{8.40}$$

For $\kappa > 0$, the amplitude of the spin waves increases exponentially. $\left|G_k^n\right|$ contains the field amplitude, and therefore, this condition also defines the amplitude and orientation of the RF field at which there is the onset of the nonlinear effect. This amplitude is called the critical field, h_{crit}. There are two frequency conditions that fulfill this requirement. For $n = 1$, the condition is $\omega_k = \dfrac{\omega}{2}$ and corresponds to the first-order instability process. The second condition $n = 2$ is $\omega_k = \omega$, and the process is called the second-order instability. To obtain the critical field, it is necessary to include the specific forms of the coupling coefficient, which are defined by Equations (8.31b) together with Equations (8.26a–8.26d). Details of this procedure will not be provided, and only the results for specific cases will be listed. The interested reader can either derive them or find them in [10, 13, 15]. To reiterate the important parameters entering the critical field condition Equation (8.40) are frequency, static magnetic bias field, anisotropy fields, spin-wave relaxation time or spin-wave linewidth $\Delta H_k = \frac{2\eta_k}{\mu_0|\gamma|}$, and the sample shape. In general, the critical threshold field in the plane-wave approximation can be expressed as:

$$h_{\text{crit}} = \min_k \left\{ \frac{\left|\eta_k + j(n\omega - 2\omega_k)\right|}{\left|G_k^n\right|} \right\}_{2\omega_k = n\omega_p} \tag{8.41}$$

In the following sections, this condition will be specified for the first- and second-order instabilities and different pumping configurations. In the next step, $\left|G_k^n\right|$ is expressed in terms more suitable for the derivation of the critical threshold fields. It is easier to understand this representation for the special cases of short spin-wave excitations, or in other words, for cases when the boundary conditions do not have to be considered.

8.2.1 First-Order Instability Processes in Spheres

First, spin-wave subsidiary absorption (first-order instability) at half of the driving frequency will be addressed. Equation (8.39) now becomes:

$$\left[\frac{d^2}{dt^2} + 2\eta_k \frac{d}{dt} - \left|G_k^{(1)}\right|^2 + (\omega - 2\omega_k)^2 + \eta_k{}^2\right] b_{k0} = 0 \tag{8.42}$$

where only terms of the first order in amplitude a_0 are retained in Equation (8.31b) for $G_k^{(1)}$. Note that these terms contain the pumping field amplitude h_p. For short-wavelength spin waves, the fulfillment of the critical field condition in Equation (8.40) will depend on the position of the spin-wave manifold (Chapter 13) with respect to the driving frequency. This position is controlled by the static bias field H_0. In subsidiary absorption, the driven uniform mode decays by a three-magnon process into spin waves at half of the pumping frequency. The frequency and internal magnetic field regions in which ω to $\omega_k = \omega/2$ spin-wave coupling is possible are shown in Figure 8.3 for different sample configurations and the RF field applied perpendicularly to the biasing static magnetic field. This configuration is usually referred to as *perpendicular or transverse pumping*. Half-frequency spin waves exist above the line $\omega = 2\mu_0|\gamma|H$ (labeled $f/2$ "spin waves" in Figure 8.3). Lines corresponding to perpendicular resonance (labeled "Perp. Res." in Figure 8.3) in a thin film or slab $\omega = \mu_0\gamma H$, resonance in a sphere (labeled "Sphere" in Figure 8.3) $\omega = \gamma(H + M_s/3)$ and parallel resonance, (labeled "Par. Res." in Figure 8.3) in a thin film or slab $\omega = \mu_0\gamma(H(H + M_s))$ are also shown. With an internal magnetic field of 16 kA/m (200 Oe), subsidiary absorption occurs for frequencies higher than 1120 MHz (A). In the low internal field region, where the coincidence limiting condition that $\omega_k = \omega/2$ and $\omega_{res} = \omega_p$ are satisfied simultaneously, the critical field can reach extremely low values. The upper limit for fulfilling the condition that $\omega_k = \omega_p/2$ will be above resonance is [13]:

$$N_T\omega_M > \frac{\omega_p}{2} \tag{8.43}$$

with N_T being the transverse demagnetizing factor. From the lower end, the limit is $H_i \geq 0$: the threshold field is on the order of $\Delta H \Delta H_k / M_s$ and the coincidence threshold occurs at 1753 MHz (B) for a thin film or slab magnetized in its plane and at 2203 MHz (C) for a sphere. If the internal magnetic field is increased to 64 kA/m (800 Oe), the coincidence regime is not possible and for frequencies above 4480 MHz (D) only the first-order instability or subsidiary absorption can be observed. These limits depend on the saturation magnetization and the numbers listed above apply to YIG with saturation magnetization M_s = 140 kA/m (1750 Oe). The theory will now be applied to the special case, widely used in the original experiments, of a sphere for having different RF field orientations with respect to the static field.

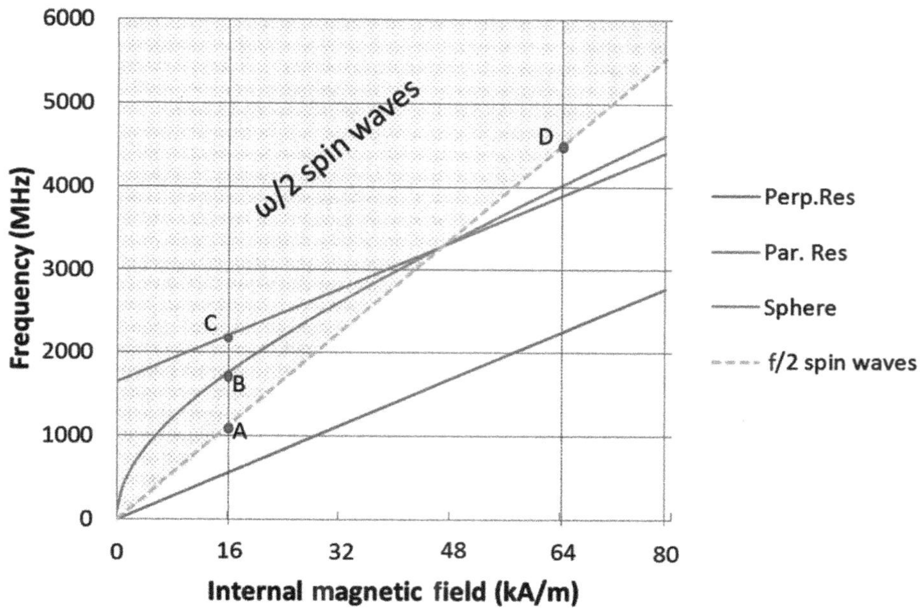

Figure 8.3 Variation with the internal magnetic bias field (H_{int}) of ferromagnetic resonance (FMR) frequencies for sphere ($\omega_0 = \mu_0\gamma(H_{int} + M_s/3)$), and parallel ($\omega_0 = \mu_0\gamma(H_{int}(H_{int} + M_s))^{1/2}$) and perpendicular ($\omega_0 = \mu_0\gamma H_{int}$) resonance in ferrite thin films. $\omega/2$ spin waves are available at frequencies higher than $\omega = \mu_0\gamma H_{int}/2$.

8.2.1.1 Transverse Pumping

For this case, the static and dynamic anisotropy fields in the coefficients $G_k^{(1)}$ are zero, the laboratory and precession frames are the same and the internal field is in the direction of the applied external field. $G_k^{(1)}$ from Equations (8.26c) and (8.26d) is then reduced to:

$$G_k^{(1)} = -(\frac{\omega_M}{2\omega_k})\frac{\mu_0|\gamma|h_p}{2}e^{2j\varphi_k}W_k(\omega_k,k,\ \theta_k,\varphi_k) \tag{8.44}$$

where the reduced coupling parameter [4]:

$$W_k(\omega_k,k,\theta_k,\varphi_k) = \omega_k \sin\theta_k \cos\theta_k \left[\frac{e_k(\theta_k)+1}{\omega_0-\omega_p}e^{-j\varphi_k} + \frac{e_k(\theta_k)-1}{\omega_0+\omega_p}e^{j\varphi_k}\right] \tag{8.45}$$

The coupling parameter $W_k(\omega_k,k,\theta_k,\varphi_k)$ for an arbitrary pump configuration can be expressed as [17]:

$$W_k(k,\theta_k,\varphi_k) = \sin\theta_k \cos\theta_k \left\{q_{L0}\left[f(\theta_k)+\omega_p/2\right]e^{-j\varphi_k} + q_{A0}{}^*\left[f(\theta_k)-\omega_p/2\right]e^{j\varphi_k}\right\} + \alpha_z \sin^2\theta_k \tag{8.46}$$

where

$$f(\theta_k) = \frac{1}{2}\left[(\omega_M^2 \sin^4\theta_k + \omega_p^2)^{1/2} - \omega_M \sin^2\theta_k\right]$$

The linearly polarized RF field is perpendicular to the external field and can be set, without loss in generality into the x-direction. This leads to the derivation of the threshold field h_{crit}, in the case of transverse pumping for a sphere in a form [4, 20]:

$$h_{crit} = \min_k \left(\frac{\omega_p}{\omega_M}\frac{\Delta H_k}{|W_k(\theta_k,\varphi_k)|}\right); \tag{8.47}$$

where

$$|W_k(\theta_k,\varphi_k)| = \frac{\left[1+(\omega_0/\omega_p)e_k(\theta_k)\right]}{1-(\omega_0/\omega_p)^2}\sin\theta_k\cos\theta_k \tag{8.48a}$$

and

$$e_k(\theta_k) = \sqrt{1+\left(\frac{\omega_M}{2\omega_p}\sin^2\theta_k\right)^2} - \frac{\omega_M}{\omega_p}\sin^2\theta_k \tag{8.48b}$$

$\omega_M = \mu_0|\gamma|I$, $\omega_0 = \mu_0|\gamma|$. is the FMR frequency, ω_p is the pump frequency, and θ_k is the angle of propagation of the $\omega/2$ spin waves relative to the internal magnetic field direction. Realizing that to maximize the denominator requires $\varphi_k = 0° or 180°$ in Equation (8.46), the solution to Equation (8.47) is found by minimizing h_{crit} by varying θ_k only. The spin-wave manifold for YIG, plotted as the wave number variation with external bias field, is shown in Figure 8.4 for $\omega_k = \omega_p/2 = 4580\text{MHz}$ with the [16] three regions – I, IIa, and IIb –indicated. In region I, the full range of θ_k, $0° \le \theta_k \le 90°$, is available but because of the $\sin\theta_k\cos\theta_k$ factor on the denominator of Equation (8.45), the minimum threshold occurs for $\theta_k^{\text{crit}} = \theta_k^{\text{min}} \approx 45°$ as indicated by the solid circles. Region II is divided into two subranges, IIa and IIb. In region IIa, where $H_i > (((M_s)^2 + 4(\omega_k/\mu_0|\gamma|)^2)^{1/2} - M_s)/2$ and $H_i = H_0 - \frac{M_s}{3}$ for a sphere, $\theta_k = 90°$ is not available, and $\theta_k^{crit} = \theta_k^{\text{min}} \approx 45°$ with increasing H_0 until $\theta_k = 45°$ is no longer available. $\theta_k = 45°$ is not available in region IIb and here:

$$\theta_k = \sin^{-1}\left(\frac{(\omega_k/\mu_0|\gamma|)^2 - H_i^2}{M_s H_i}\right)^{1/2} \tag{8.49}$$

until $H_i = \omega_k/\mu_0|\gamma|$ is reached.

The calculated variation of h_{crit} with external bias field and $\omega_p = 9.16\text{GHz}$ is shown by a full line in Figure 8.5. A constant spin-wave linewidth, $\Delta H_k = 23.2\frac{\text{A}}{\text{m}}$ (0.29 Oe) was assumed to match measured data given by the filled circles in Figure 8.5 and a correction of –4 kA/m (–50 Oe) was also applied to H_i to account for the anisotropy field of the YIG sphere used in the measurement. In Figure 8.5, the calculated h_{crit} decreases slowly with increasing H_i until the wave number $k_{\text{crit}} = 0$ corresponding to regions I and IIa. At higher H_i in region IIb, h_{crit} increases more rapidly until it becomes very

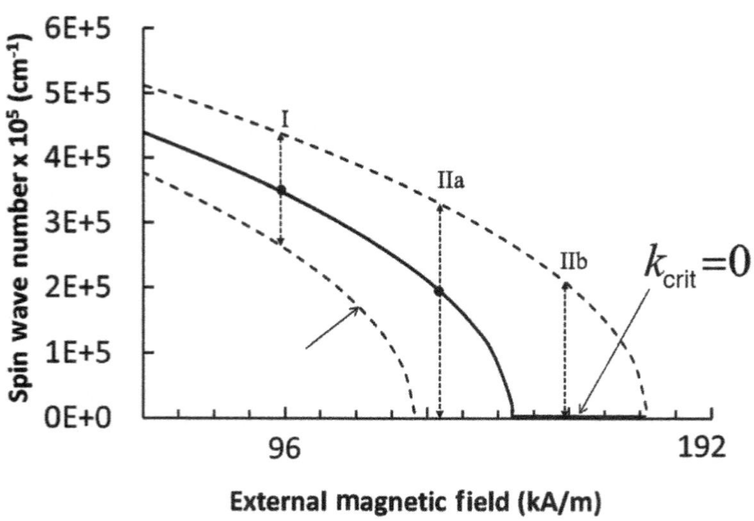

Figure 8.4 Spin-wave manifold at half the pump frequency, $\omega_p/2 = 4580MHz$, showing transverse pumping wave number and polar angle variations for (I) low and (IIa and IIb) intermediate static external field.

large at $H_i \approx 172$ kA/m (2150 Oe). However, the agreement between the measured and calculated variations of h_{crit} in Figure 8.5 is not good. The major discrepancies are a flat region with some fine structure around the h_{crit} minimum, and a rapid increase in h_{crit} below the flat region followed by a sharp "kink." However, the slope of the measured data for H_i below the kink is close to that calculated. These discrepancies could be resolved, except for the fine structure, by invoking a wave number–dependent spin-wave linewidth function, $\Delta H_k(k)$. The fine structure was recognized to be due to magnetostatic mode interactions [20] at $\omega_k = \omega_p/2$.

Based on the physical reasoning and spin-wave relaxation theory [21], a suitable function for $\Delta H_k(k)$ was obtained [20] containing several adjustable constants, A_0, A_1, B_1, and k_c in the form

$$\Delta H_k^{\text{sa}} = A_0 + A_1 \sin^2 2\theta_k + B_1 L(k/k_c) \tag{8.50}$$

The "sa" superscript is used to distinguish the wave number-dependent spin-wave linewidth for subsidiary absorption from that for parallel pumping that will be discussed later. The "kink" in the $\Delta H_k(k)$ function is provided by:

$$L(x) = \frac{0.72}{x} \ln\left[1 + 8x^2 \frac{1+2x^2}{1+8x^2}\right]$$

Substituting Equation (8.50) for $\Delta H_k(k)$ in Equation (8.47) and minimizing h_{crit} with respect to θ_k yields a variation in h_{crit} that mimics the measured data in Figure 8.5. Best fit to the data, for that particular case, was obtained with $A_0 = 13.6\frac{\text{A}}{\text{m}}$ (0.17 Oe), $A_1 = 13.6\frac{\text{A}}{\text{m}}$ (0.17 Oe), $B_1 = 12.8\frac{\text{A}}{\text{m}}$ (0.16 Oe) and $k_c = 10^5 cm^{-1}$. The results with these parameters are shown in Figure 8.6. The agreement is much improved, except for the flat region between 128 and 152 kA/m (1600 and 1900 Oe). The corresponding variation in spin-wave number k_{crit} and propagation angle θ_{crit}, shown in Figure 8.7, are, however, significantly different from those obtained with a constant ΔH_k in Figure 8.4. As H_0 is increased from 64 kA/m (800 Oe) to approximately 121.6 kA/m (1520 Oe), θ_k decreases slowly from approximately 38° to 35°, then with further increase in H_0, θ_k increases abruptly to approximately 72° as k_{crit} drops to zero. θ_{crit} then follows the $k = 0$ spin-wave dispersion relation until $\theta_k = 0$. The infinite media approach to the spin-wave theory is not valid in the region where $k \approx 0$ due to the neglect of the boundary conditions that gives rise to magnetostatic modes. Several authors have discussed theories for magnetostatic mode nonlinearity in spheres [22–25]. The influence of the boundary conditions will be discussed later in the context of thin film geometry samples.

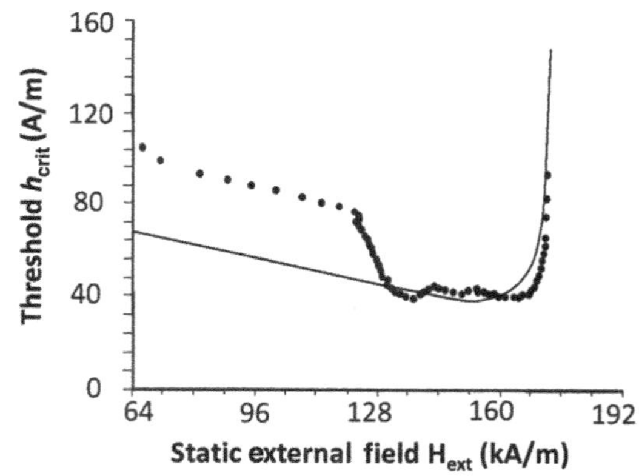

Figure 8.5 Calculated threshold assuming ΔH_k is constant, and measured data for subsidiary absorption in a YIG sphere.

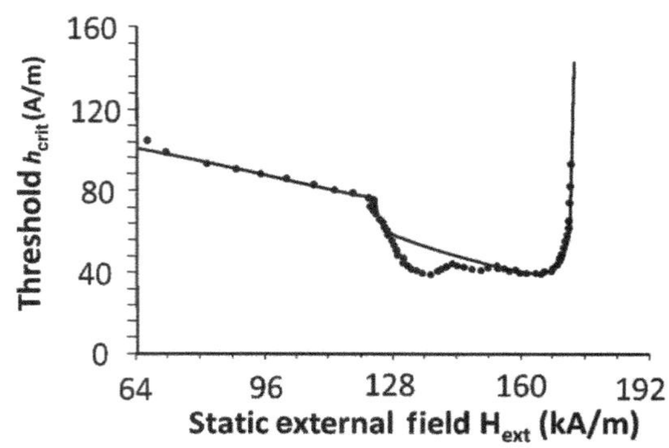

Figure 8.6 Calculated threshold with best fit $\Delta H_k(k)$ coefficients, and measured data for subsidiary absorption in a YIG sphere.

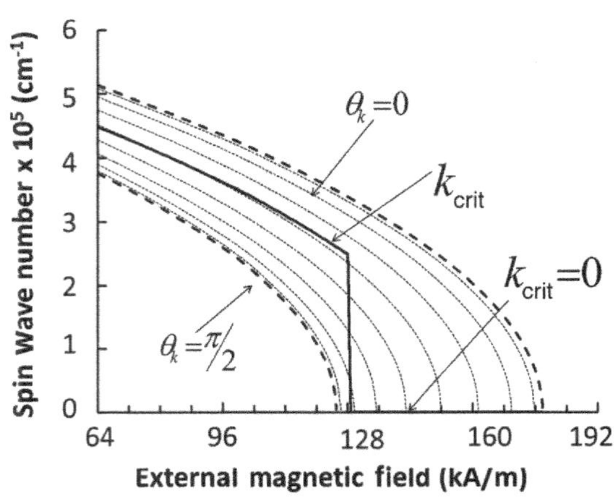

Figure 8.7 Spin-wave manifold at half the pump frequency, $\omega_p/2 = 4580$MHz, showing transverse pumping wave number and polar angle variations resulting from the best-fit $\Delta H_k(k)$ coefficients.

8.2.1.2 Parallel Pumping

The next special case is when a linearly polarized RF driving field is parallel to the DC bias field, which is referred to as parallel pumping. Theoretical predictions [26 and 10] and experimental studies [10] showed that parallel pumping

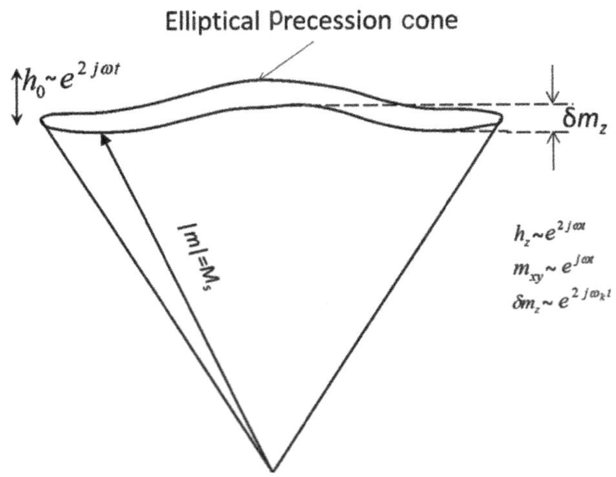

Figure 8.8 Parallel pump at ω_p couples to elliptical spin precession at $\omega_p / 2$.

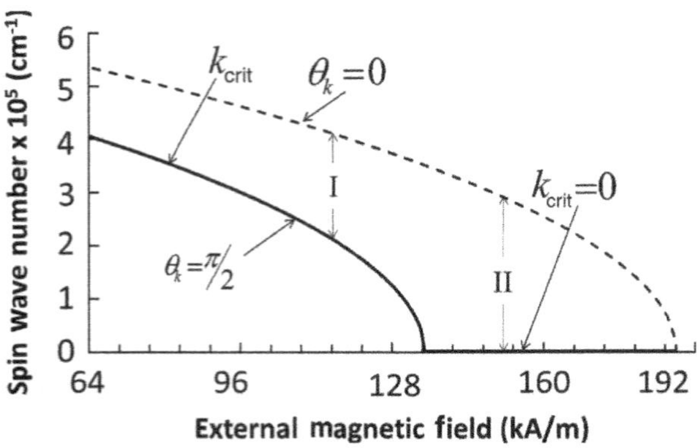

Figure 8.9 Spin-wave manifold at half the pump frequency, $\omega_p / 2 = 4580 MHz$, showing parallel pumping wave number and polar angle variations for (I) low and (II) intermediate static external field. Starting at low fields with increasing field, k_{crit} follows the $\theta_k = \pi / 2$ limit (region I) until $k_{crit} = 0$, which it maintains (region II) until the $\theta_k = 0$ limit is reached.

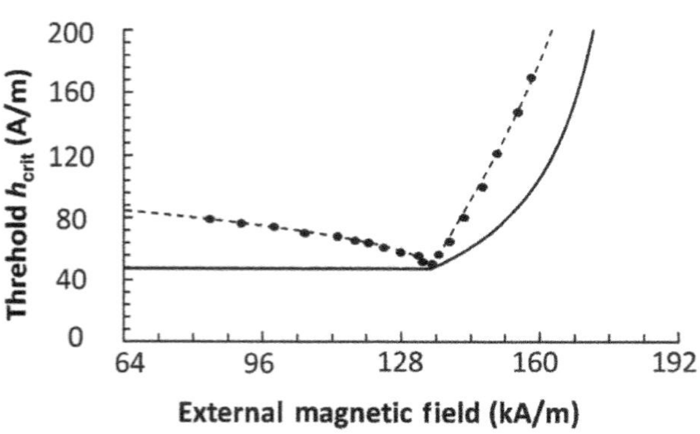

Figure 8.10 Calculated threshold assuming constant ΔH_k, full line, and with best-fit $\Delta H_k(k)$ coefficients, broken line, and experimental data for parallel pumping of a YIG sphere.

resulted in a similar frequency-selective nonlinear response at half of the pumping frequency. This result appears counterintuitive, with a strictly circular precession but a coupling mechanism between the pump magnetic field and the $\omega_k = \omega_p / 2$ spin-wave becomes apparent when elliptical precession is considered, as shown in Figure 8.8 [16]. Here, δm_z goes through two cycles for each rotation of $|m|$. Thus, the coupling between the pump field $h_z e^{j\omega_p t}$ and the spin-wave precession, with $\omega_k = \omega_p / 2$, increases with the ellipticity of the precession. Then parallel pump instability occurs when the energy coupled from the pump to the $\omega_k = \omega_p / 2$ precession exceeds the loss of energy due to spin-wave relaxation. This result can be obtained from the previously introduced Equations (8.45) and (8.46) for the pump field in the direction of the DC bias field and if the anisotropy is neglected. Values of h_{crit} for parallel pumping are found to be [4]:

$$h_{crit} = \min_k \left(\frac{\omega_p}{\omega_M} \frac{\Delta H_k}{\sin^2\theta_k} \right) \tag{8.51}$$

where h_{crit} is minimized relative to all available k values at $\omega_k = \omega_p / 2$. For this case, $|W(\theta_k, \varphi_k)| = 2\sin^2\theta_k$. The spin-wave manifold at a frequency of $\omega_k = \omega_p / 2 = 4710 MHz$ is given in Figure 8.9 and shows two regions I and II [16]. In range I, all values of $0 \le \theta_k \le 90^\circ$ are available and h_{crit} is minimal for $\theta_k = 90^\circ$ at some minimum value of wave number. In range II, $\theta_k = 90^\circ$ is not available and h_{crit} is minimized from Equation (8.47), becoming very large as $H_i = \omega_k / \mu_0 |\gamma|$ is approached. Calculated h_{crit} assuming constant ΔH_k is shown as a full line in Figure 8.10. As expected from Equation (8.51), h_{crit} is constant with increasing magnetic field until $H_i \ge (((M_s)^2 + 4(\omega_k / \mu_0 |\gamma|)^2)^{1/2} - M_s)/2$; when $\theta_k = 90^\circ$, spin waves are no longer available and increase until $H_i = \omega_k / \mu_0 |\gamma|$ is reached. Measured data are also shown in Figure 8.10 by the filled circles and show a significant difference to the calculated h_{crit}, with ΔH_k constant. As in the case of transvers pumping, a wave number–dependent wave number, $\Delta H_k(k)$, is required.

A simple expression for $\Delta H_k(k)$ that provides good agreement with measured parallel pump data is [4]:

$$\Delta H_k^{pp}(k) = A_0 + A_1 \sin^2 2\theta_k + A_2 k \tag{8.52}$$

The constant term A_0 determines the minimum value of $\Delta H_k(k)$, which occurs at the cusp in the measured data where $\theta_k = 90^\circ$ and $k = 0$. Above the cusp, where $k = 0$, the $\Delta H_k(k)$ variation is determined by A_1, and below the cusp where $\theta_k = 90^\circ$, the variation is determined by A_2. Substituting Equation (8.52) into Equation (8.51) for ΔH_k yields the broken line curve in Figure 8.10 when $A_0 = 24.8\ \mathrm{A/m}(0.31\mathrm{Oe})$; $A_1 = 17.6\ \mathrm{A/m}(0.22\mathrm{Oe})$ and $A_2 = 4\times10^{-7}\mathrm{A\cdot m}(5\times10^{-7}\mathrm{Oe\cdot cm})$ for a YIG sphere.

8.2.1.3 Oblique Pumping

The final sphere example addresses the arbitrary orientation of the pumping field with respect to the DC bias field. This configuration has been investigated by several authors [14, 17, 27, 28, 29]. Again, the presented result can be obtained from the previously introduced general equations for the pumping field with all three coordinate components in Equations (8.26c) and (8.26d). The detailed calculation for a general case of arbitrary polarization of the pumping field is quite involved even for an ellipsoid of revolution and the reader can find the details in [17, 29]. For the linearly polarized oblique pumping field applied to a spherical sample, the situation simplifies and the conclusion from the theoretical studies is that the parallel and perpendicular components of the oblique linearly polarized pumping field are acting together and the resulting threshold field can be expressed as:

$$h_{\mathrm{crit}}^{\mathrm{obl}} = \min\left(\frac{\cos\psi}{h_{\mathrm{crit}}^{\parallel}} + \frac{\sin\psi}{h_{\mathrm{crit}}^{\perp}}\right)^{-1} \tag{8.53}$$

where $h_{\mathrm{crit}}^{\parallel}$ and $h_{\mathrm{crit}}^{\perp}$ are expressions in the parenthesis of Equations (8.47) and (8.51). ψ is the angle between the linearly polarized pumping field and the direction of the static external field. However, experimental results from a high-quality, single-crystal YIG sphere [29] showed that the critical fields for the perpendicular and parallel instabilities seemed to act independently and the experimental results could be best explained by:

$$h_{\mathrm{crit}}^{\mathrm{obl}} = \min\left(\frac{h_{\mathrm{crit}}^{\parallel}}{\cos\psi}, \frac{h_{\mathrm{crit}}^{\perp}}{\sin\psi}\right) \tag{8.53a}$$

In fitting the experimental data, the wave vector dependence of the spin-wave linewidth in the form of Equation (8.50) was used for the $h_c^{//}$ and $h_c^{\perp}$ expressions, but it was necessary to adjust the B_1 parameter for each angle. The variation of the instability threshold with the applied bias field at a 40° angle is shown in Figure 8.11. Filled circles indicate measured data with the full lines giving the results of the calculations. The calculated results closely match the measured data except over the 128 kA/m (1600 Oe) to 160 kA/m (2000 Oe) range where the discrepancy is attributed to magnetoelastic effects in the parallel pumping region and magnetostatic modes in the transverse pumping region. Even today, there are some open questions that require additional theoretical and experimental work to explain why the independent action of $h_c^{//}$ and $h_c^{\perp}$ gives better agreement with measurements than the collaborative action, which is predicted from the theory [17]. The need for a variation in the B_1 parameter with angle is also not resolved.

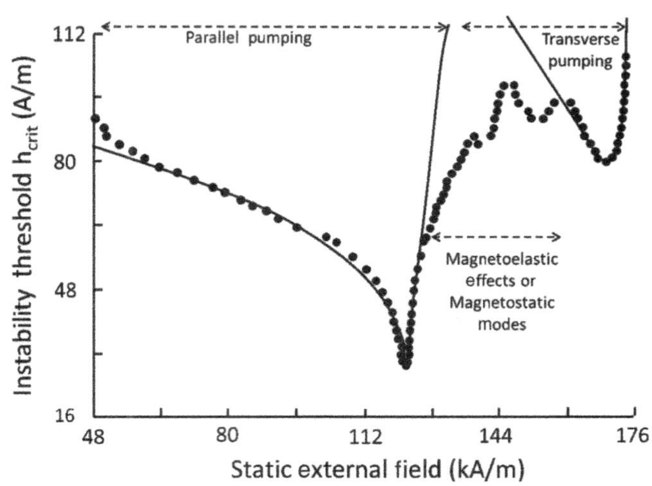

Figure 8.11 Experimental (filled circles) and theoretical threshold field variation for oblique pumping at $\psi = 40°$.

8.2.1.4 Magnetostatic Modes

A detailed experimental study of nonlinear effects in spheres [4, 20] revealed, in the $k \approx 0$ region, a fine structure in the critical field dependence with external bias field. The theoretical investigation proved that in the short-wavelength spin-wave limit, the corresponding expansion into plane waves does not describe the fine structure observed in these experiments. Recall (Chapter 13) that for the $k \approx 0$ region, the wavelength of the spin-wave is of the order of the sample dimensions, and therefore, it is necessary to consider the conditions at the sample boundary. A useful discussion about the relation of the spin-wave wavelength to sample dimensions is given in [30]. As observed experimentally [31], the threshold fields for subsidiary absorption as well as for parallel pumping show closely spaced resonance features as the wave vectors of the excited spin waves are quantized due to the finite dimensions of the sample. Therefore, the plane-wave expansion in Equation (8.17) must be replaced by an expansion into the normal (eigen-) modes of the system that consider the sample boundary conditions [22]. To avoid any confusion, we include the complex conjugates of the variables to keep the expression for magnetization real:

$$\boldsymbol{M}(\boldsymbol{r},t) = M_s\boldsymbol{u}_z + \left[m_0(\boldsymbol{r})e^{j\omega_p t} + c.c\right] + \sum_{l\neq 0}[\boldsymbol{m}_l(\boldsymbol{r})e^{j\omega_l t} + c.c.] \tag{8.54}$$

Inserting this form of magnetization and the corresponding static and dynamic dipolar fields, including boundary conditions, into the equation of motion gives a set of coupled equations for the magnetization eigenmodes. The solutions for the RF dipolar field under these conditions are called magnetostatic or Walker modes and for the sphere were first derived in [32, 33]. These solutions will not be repeated in detail, but the information will be included that is necessary to understand the impact of boundary conditions on the nonlinear effects. More detailed information and derivations can be found in [32, 33, 34]. The eigen modes that include the boundary conditions are usually solved in the magnetostatic approximation of Maxwell's equations with the electromagnetic propagation effects neglected. The RF dipolar field then can be expressed through potential equation:

$$\boldsymbol{h}_{dipk} = -\mu_0 \nabla \grave{E}_k \tag{8.55}$$

and the corresponding equation for dynamic field $\mathbf{h}_{0dip}$ (Equation 8.5) that couples the magnetization with the field. The magnetization in this approximation can be expressed through the permeability tensor discussed in Chapter 13. The solution of this equation for a ferromagnetic sphere is [23, 24, 25, 32, 34]:

$$\psi_{n,m,\theta_r} = e^{jm\varphi} \cdot \begin{cases} r^{-(n+1)} P_n^m(\cos\theta_r) \text{ outside of the sphere} \\ P_n^m(\xi) P_n^m(\eta) \text{ inside of the sphere} \end{cases} \tag{8.56}$$

Note the change in notation of the coordinates in Equation (8.56). They result from the use of a spherical coordinate system, with variables (r,θ,φ) outside the sphere, while inside of the sphere an elliptical coordinate system with variables (η,ξ,φ) is used that is related to the cylindrical coordinate system (ϱ,z,φ) through relations [25]:

$$\varrho^2 = \frac{1}{\sin^2\theta_r}(1-\xi^2)(1-\eta^2); z = \frac{1}{\cos\theta_r}\xi\eta$$

The excited magnetostatic modes are identified by indices $n = 1, 2, 3,...,$ and $m = -n, -n + 1,...n$, and by the angle θ_r given by the characteristic equation [32]:

$$\frac{P_n^{m'}(\cos\theta_r)}{P_n^m(\cos\theta_r)} = -\frac{m\nu(\theta_r)+n+1}{\cos\theta_r}; \nu(\theta_r) = -\frac{\sqrt{H_i(H_i+M_s\sin^2\theta_r)}}{H_i} \tag{8.57}$$

P_n^m are associated Legendre spherical functions. The index m can be thought of as an angular momentum number of the spherical mode. The frequency of the magnetostatic mode is given by:

$$\omega_{\theta_r}^2 = \mu_0|\gamma|^2 H_i(H_i + M_s\sin^2 6$$

The exchange field can be introduced into these equations in the limit of $H_{ext} << H_i$, which is satisfied in most of our cases of interest through substitution $H_i \rightarrow H_i + \langle H_{ext}\rangle$, where $\langle H_{ext}\rangle$ represents the average exchange field of the magnetostatic mode and can be expressed as:

$$\langle H_{ex}\rangle = Dk_{eff}^2;\ k_{eff} = n\frac{\pi}{d}$$

where d is the diameter of the sphere. This leads to the dispersion relation for magnetostatic modes in the form:

$$\omega_{n,\theta_r}^2 = \mu_0 |\gamma|^2 (H_i + Dk_{\text{eff}}^2)(H_i + Dk_{\text{eff}}^2 + M_s \sin^2\theta_i \tag{8.58}$$

Note that the solution of Equation (8.57) with the appropriate boundary conditions, necessary for obtaining the resonant frequencies from Equation (8.58) and thus the critical fields, is usually quite involved. The corresponding magnetostatic modes obey the orthogonality relations that can be found in [32].

The main points of the theory and its implication are introduced by restricting the discussion to isotropic samples. Extension of the theory to include anisotropy should be quite straightforward following the discussion introduced earlier in this chapter. Magnetostatic modes, as the eigenmodes of the system, were introduced into the nonlinear theory for cylindrical samples in [35] and for spheres in [24, 35]. Detailed analysis of the critical fields is given in [23, 25, 35]. In the plane-wave analysis the pump field couples spin waves with the equal and opposite wave numbers only. However, the basic idea for the nonlinear excitations in finite size samples is that the uniform precession couples to a pair(s) of magnetostatic modes. As follows from the referenced analysis, in the case of magnetostatic modes the pump field couples any given mode m to many m' mode pairs characterized by the corresponding coupling coefficients. As it turns out, these coefficients are nonzero only if certain selection rules are satisfied. In principle, more than one pair of modes can contribute to the instability, but the present analysis considers only two modes. If more modes are involved this usually lowers the instability threshold [35]. The selection rules depend on the shape of the sample, and for a sphere, from symmetry considerations, the rules for modes n, n' and m, m' are $n = n'$ and $m + m' = 1$ for the first-order Suhl instability and $m + m' = 0$ for parallel pumping. At the instability threshold, the frequency condition is now $\omega_{n,\theta_r} + \omega_{n,\theta_r'} = \omega_p$. Because of the limiting boundary conditions, the magnetostatic mode pairs are constrained to the low wave number region. For high wave number spherical modes, the critical threshold fields follow the plane-wave results. These conditions limit the frequency and field ranges where the development of the spin-wave instability is possible. This region was labeled in [22] as a "magic diamond." The detailed derivations for h_{crit} are straightforward but quite tedious.

The condition for the critical field for parallel pumping is now [35]:

$$h_{crit}^{ms} = min_{\theta_r,\theta_r'} \left(\frac{\Delta H_{\theta_r} \Delta H_{\theta_r'}}{W_{\theta_r,\theta_r'} W_{\theta_r',\theta_r}^*} \right)$$

ΔH_{θ_r} and $\Delta H_{\theta_r'}$ denote the corresponding magnetostatic mode linewidths and W represent the corresponding coupling coefficients. The lowest threshold must satisfy the aforementioned selection rules. The coupling coefficients depend on the shape of the sample and are calculated for the given modes using the orthogonality relations. As mentioned, these calculations are involved but straightforward. The readers can find those for the shapes addressed here in [35]. As shown in [35], for the m = 0 mode, because it is not a pair but a single mode, the critical field happens at half of the pumping frequency and is the same as in the case of the plane-wave approximation. For $m \neq 0$, the instability is prohibited at exactly half of the pumping frequency, but the frequencies of the critical modes can be close to it. Obtaining the critical fields is complicated and can be obtained from numerical calculations [23, 25]. Application of YIG spheres in nonlinear devices is limited at present, and therefore, we will focus on the practically more important case of thin films.

8.2.2 First-Order Instability Processes in Thin Films

In many practical ferrite devices (discussed in Chapter 13), transmission lines such as microstrip, stripline, or coplanar waveguides with small dimensions are used to maximize the RF magnetic field and, thus, in the case of nonlinear devices, to minimize the threshold power. The ferrite material is represented not by a sphere but as a thin ferrite (in most cases, YIG) film. Thus, it is necessary to consider the boundary conditions when investigating nonlinear effects in ferromagnetic films. As is the case for spheres, it is important first to understand the dispersion relations of spin excitations in thin film structures. These are discussed in (Chapter 13), and therefore, the focus here is on nonlinear effects referring, if necessary, to the results given therein. The thin film configuration supports the propagation of spin waves that have different dispersion depending on the orientation of the external static magnetic field with respect to film geometry, film thickness, and the orientation of the spin-wave propagation wave vector $\boldsymbol{k}$. This is important for the nonlinear effects because the dispersion allows control of the excitation conditions through changing the direction and the value of the bias DC magnetic field, film thickness, state (boundary conditions) of the spins on the surface, and at the interface of the films with the substrate.

Due to the multimode nature of spin-wave excitations in thin films, a pair or multiple pairs of modes [40], at their corresponding frequencies, are excited at the nonlinear threshold, as in the case of spheres [23, 25]. The pumping field excites a certain mode at frequency $\omega_1(\boldsymbol{k}_1)$, which at the threshold couples to a pair of critical modes that due to the difference in their wave numbers will have frequencies slightly different from the half of pumping field frequency. The corresponding conditions are $\omega_2(\boldsymbol{k}_2)+\omega_3(\boldsymbol{k}_3)=\omega_1(\boldsymbol{k}_1)$ together with the condition for wave vectors $\boldsymbol{k}_2+\boldsymbol{k}_3=\boldsymbol{k}_1$. Depending on the geometry of the problem, the pump field may be periodic, $k_1 \neq 0$, when associated with a propagating or standing wave or a periodic transducer or, in the case of uniform pumping, such as in resonance cavities, $\omega_1=\omega_p$ and $k_1=0$. The dispersion relations represent the discrete spin-wave spectrum of the film. Here, we restricted ourselves only to two modes without loss of generality. Experimental data show [36, 37] that for thin films excited by a uniform RF field, the threshold fields can be derived with a reasonable precision for both the subsidiary and second-order instabilities from the bulk spin-wave theories described in previous sections. The situation is different for laterally small ferrite samples driven by uniform RF fields and/or for the nonlinear response in thin films excited with localized nonuniform fields driven by finite size antennas.

8.2.2.1 Transverse Pumping

The nonlinear behavior in films excited by a uniform RF field in microwave cavities, will be addressed first. Theory for the threshold magnetic field h_{crit} for subsidiary absorption, or transverse pumping, in ferrite films with mm-cm size lateral dimensions was based on earlier sphere work [38, 39]. The coordinate system and the configuration of the bias field and RF magnetic field relative to the YIG film for three cases of interest are shown in Figure 8.12. In Figure 8.12a, the static bias field, H_0, is applied along the z-axis and fully saturates the magnetization in that direction. The spin-wave wave vector $\boldsymbol{k}$ is specified, as before, by the wave number k, polar angle θ_k, and azimuthal angle φ_k. Figure 8.12b–d shows the orientations of the YIG film and the RF magnetic pump field h_p relative to the coordinate system for the DC bias field in the plane of the YIG film and h_p normal to the YIG film (IP), DC bias field H_0 normal to the YIG film and h_p in plane (PI), DC bias field and h_p in the plane of the film (II), respectively. The elliptical spin precession cones are also shown in Figure 8.12b–d for the uniform mode at the pump frequency ω_p. In short, I and P designate an in-plane and a normal field, respectively, and the first letter refers to the bias field and the second to the RF pump field. The ellipticity factors, ϵ_x and ϵ_y, show the relative lengths of the principal axes and strongly influence the coupling between the uniform precession and spin-wave magnetizations, and thus, the critical threshold field h_{crit}. Their influence is discussed in detail in [38]. The threshold field depends on the coupling coefficients in the x- and y-directions. If the coupling factor in the x-direction is larger than in the y-direction, the critical modes will have $\varphi_k=0$, and in the opposite case, the critical mode azimuthal angle will flip to $\varphi_k=\pi/2$.

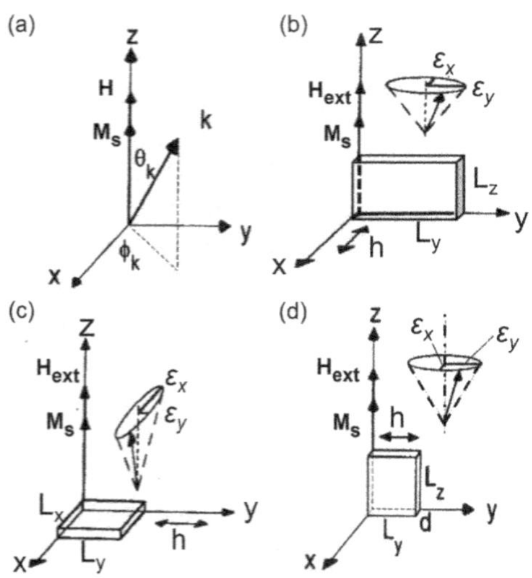

Figure 8.12 Coordinate system and directions for external bias field H_{ext}, magnetization M_s, spin-wave wave number k and propagation directions θ_k, ϕ_k, for different sample orientation and pumping configurations: (a) coordinate system, (b) IP configuration, (c) PI configuration, and (d) II configuration. ε_x and ε_y are the relative amplitudes of the uniform precession magnetization [38, 39].

Measured data on the threshold field in thick YIG films obtained in microwave cavities with uniform h_p and magnetic bias field are not available. It has been found that modification to the bulk theory is required to explain experimental data on thin films of thickness ≈ 13 μm [38, 39], and thus, the rough demarcation between the two ranges can be defined as $\delta\omega_k >> \eta_k$ for thin films and $\delta\omega_k << \eta_k$ for thick films. The analysis in [40] shows that the condition $\delta\omega_k << \eta_k$ can be used as a general criterion and that the film can be regarded as a bulk medium. Here, $\delta\omega_k$ is the frequency spacing of the spin-wave modes and η_k is the spin-wave relaxation frequency. In thin films, where the condition $\delta\omega_k >> \eta_k$ applies, the critical spin waves are expected to always propagate in the thin film plane. As $\delta\omega_k \propto d^{-2}$ [41], where d is the film thickness, it is reasonable to assume that in practical ferrite devices the bulk theory is applicable to ≈ 100 μm films used in transmission-line-based frequency-selective limiters (FSLs), whereas thin film theory is required for MSW FSLs and NEs where films < 30 μm are typically used.

In plane-wave (bulk medium) limit for an infinite, homogeneous, film of thickness d, magnetized in the film plane by external field, H_0, the dispersion relation is [42]:

$$\omega_k = \mu_0|\gamma|\left\{\left[H_0 + Dk^2 + M_s(1-N_k)\sin^2\theta_k\right]\left[H_0 + Dk^2 + M_s N_k\right]\right\}^{1/2} \tag{8.59}$$

N_k is the effective dynamic demagnetizing factor $N_k = (1-e^{-kd})/kd$. Then, for the first-order critical field, it is possible to write:

$$h_{\text{crit}} = \frac{\omega_p}{\omega_M}\min\left[\frac{\Delta H_k(k,\theta_k)}{\left|W_k^{(1)}(k,\theta_k)\right|}\right]$$

where the minimization procedure is over k and θ_k for $\omega_k = \omega_p/2$ and:

$$W_k^{(1)} = (1-N_k)sin\theta_k\cos\theta_k\left\{(A_k - B_k + \omega_p/2)q_{L0} + (A_k - B_k - \omega_p/2)q_{A0}^*\right\} \tag{8.60}$$

In this case:

$$A_k = \omega_H + \mu_0|\gamma|Dk^2 + \frac{1}{2}\omega_M\left[(1-N_k)\sin^2\theta_k + N_{,}\right. \tag{8.61a}$$

$$B_k = \frac{1}{2}\omega_M\left[(1-N_k)\sin^2\theta_k - N_k\right] \tag{8.61b}$$

the general equations for q_{L0} and q_{A0} simplify to

$$q_{L0} = \frac{\omega_H + \omega_M + \omega_p\left[1 + j2\eta_0/(2\omega_H + \omega_M)\right]}{\omega_0^2 - \omega_p^2 + j2\omega_p\eta_0} \tag{8.62a}$$

$$q_{A0}^* = \frac{\omega_H + \omega_M - \omega_p\left[1 + j2\eta_0/(2\omega_H + \omega_M)\right]}{\omega_0^2 - \omega_p^2 + j2\omega_p\eta_0} \tag{8.62b}$$

with $\omega_0 = \sqrt{\omega_H(\omega_H + \omega_M)}$ and uniform precession relaxation rate $\eta_0 = \frac{\mu_0|\gamma|^2}{4}\frac{2H_{\text{res}} + M_s}{\omega_0}\Delta H_0$. ΔH_0 is the uniform precession linewidth.

Details of the tedious but more or less straightforward derivations are not given here, but the resulting expressions for the bulk threshold field h_{crit} for the three YIG thick film configurations shown in Figure 8.13 are presented. For IP and PI [38]:

IP

$$h_{\text{crit}}^{\text{IP-bulk}} = min_{\theta_k}\left[\frac{\omega_p\Delta H_k(k)\left(\omega_p^2 - \omega_0^2\right)}{\omega_M\sin\theta_k\cos\theta_k\left(\omega_H\left(\left(\left(\omega_M\sin^2\theta_k\right)^2 + \omega_p^2\right)^{1/2} - \omega_M\sin^2\theta_k\right) + \omega_p^2\right)}\right] \tag{8.63}$$

PI

$$h_{\text{crit}}^{\text{PI-bulk}} = min_{\theta_k}\left[\frac{\omega_p\Delta H_k(k)\left(\omega_p^2 - \omega_0^2\right)}{\omega_M\sin\theta_k\cos\theta_k\left(\omega_0\left(\left(\left(\omega_M\sin^2\theta_k\right)^2 + \omega_p^2\right)^{1/2} - \omega_M\sin^2\theta_k\right) + \omega_p^2\right)}\right] \tag{8.64}$$

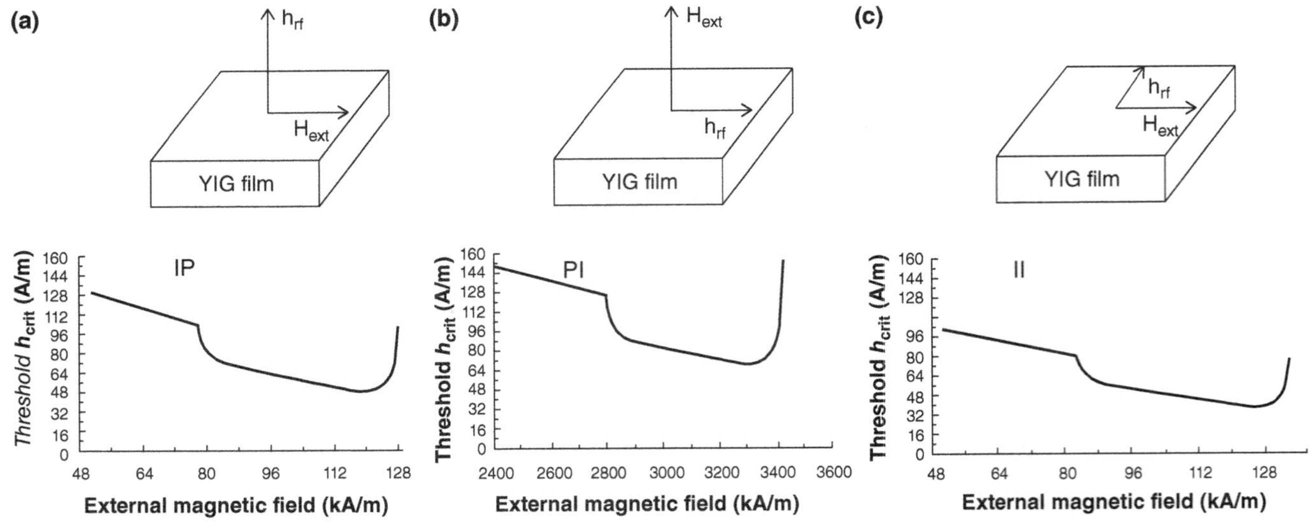

Figure 8.13 Variation of threshold field h_{crit} with external magnetic field H_{ext} for subsidiary absorption in thick YIG films at 9.4 GHz: (a) IP configuration, (b) PI configuration, and (c) I configuration.

And for I configuration [39]
I

$$h_{crit}^{II-bulk} = \min_{\theta_k}\left[\frac{\omega_p \Delta H_k(k)\left(\omega_p^2-\omega_0^2\right)}{\omega_M \sin\theta_k \cos\theta_k\left(\left(\frac{\omega_0^2}{\omega_H}\right)\left(\left(\left(\omega_M \sin^2\theta_k\right)^2+\omega_p^2\right)^{1/2}-\omega_M \sin^2\theta_k\right)+\omega_p^2\right)}\right] \tag{8.65}$$

The variation of h_{crit} with external field, or butterfly curves, for the three configurations are given in Figure 8.13. ΔH_k is obtained from Equation (8.50) for subsidiary absorption in spheres as before. The same fitting parameters, $A_0 = B_1 = 15.2\frac{\mathrm{A}}{\mathrm{m}}(0.19Oe)$, $A_1 = 17.2\frac{\mathrm{A}}{\mathrm{m}}(0.215)$, and $k_c = 10^6 cm^{-1}$ were used in all cases and were obtained from thin film measurements. The fitting parameters are close to those used earlier with YIG spheres, except for k_c, which differs by a factor of 10. The three cases, IP, PI and II, show similar characteristics except for the difference in h_{crit}, which is due to the difference in the coupling between the uniform precession and the $\omega_k = \omega_p/2$ spin waves. The variation of k_{crit} and θ_{crit} within the spin-wave manifold for the three thick film cases are like those of a sphere shown in Figure 8.7.

Next, the difference in the behavior of out-of-plane and in-plane magnetized thin films for the IP and PI configurations in Figure 8.14a and b will be discussed. Experimental studies of subsidiary absorption in thin YIG films showed that the variation of h_{crit} was not satisfactorily explained by the thick film or plane-wave theory [38, 39]. The experimental data were obtained on a 12.8-µm-thick YIG films at 9.4 GHz. The difference in these two cases between the observed threshold field dependence on the applied DC bias field, called the butterfly curve, and theory is due to excitation of different critical modes, depending on the film and pumping field geometry. The theoretical approach follows the procedure developed for the spherical samples in [40] and extended to think films [39, 40]. The geometry of the problem is shown in Figure 8.12, where the film is magnetized to saturation in the z-direction. From the theory developed here, it follows that the critical parameters are the θ_k and φ_k angles that determine the propagation direction and strongest coupling of the modes with the lowest threshold value. The basic result is that the coupling is strongest for the φ_k angle where the ellipticity of the excited mode matches the orientation of the uniform mode elliptical precession cone. For the PI configuration, in the bulk theory the orientation of the critical mode wave vector k is out-of-plane. However, for thin films, for a given θ_k near $\pi/4$, no φ_k rotation can reduce $k_\perp$, that is approximately $k/\sqrt{2}$, to 0. Therefore, out-of-plane k vectors, limited to discrete values $k_\perp = n\pi/d$, where n is a positive integer and d is the film thickness, must be included in the critical mode consideration. This leads to the observed absence of the kink in the butterfly curve for the in-plane bias (PI case; see Figure 8.14b).

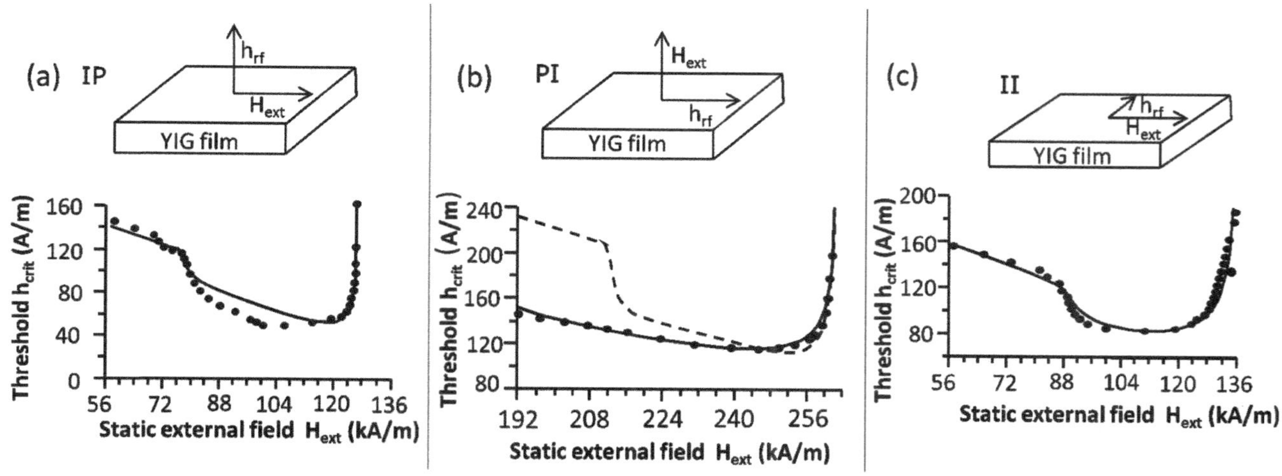

Figure 8.14 Variation of threshold field h_{crit} with external magnetic field H_{ext} for subsidiary absorption in thin YIG films at 9.4 GHz: (a) IP configuration, (b) PI configuration, and (c) II configuration [38, 39].

For the thin film IP configuration, the critical mode k vector is in plane as in the thick film case and still exhibits the "kink" (see Figure 8.14a).

The variation of h_{crit} with the bias field for thin film IP is given by [38]:

$$h_{\text{crit}}^{IP-thin\ film} = min_{\theta_k}\left[\frac{\Delta H_k(k)\left(\omega_p^2-\omega_0^2\right)}{\omega_M \sin\theta_k \cos\theta_k\left(\left(\left(\omega_M \sin^2\theta_k\right)^2+\omega_p^2\right)^{1/2}-\omega_M \sin^2\theta_k\right)+\omega_H}\right] \tag{8.66}$$

The $\Delta H_k(k)$ expression and fitting parameters were the same as used in the thick-film (bulk) cases. Although the agreement between the measured and calculated data in Figure 8.14a is generally good, the measured h_{crit} is lower than calculated when $80 < H_0 < 112\ kA/m(1000 < H_0 < 1400Oe)$, which may be due to the effects of excited magnetostatic wave mode pairs as in the case of a sphere [22, 25].

The variation of h_{crit} with a bias field for the thin film PI case is given by [38]:

$$h_{\text{crit}}^{PI-thin\ film} = \frac{\Delta H_k(k)\omega_p\left(\omega_p^2-\omega_0^2\right)}{\omega_M \sin\theta_k \cos\theta_k\left(2\omega_0\left(\omega_H+Dk^2\right)+\omega_p^2\right)} \tag{8.67}$$

In the PI case, it is not necessary to minimize the expression for h_{crit} with respect to θ_k as the condition $\partial\omega_k/\partial k_\perp = 0$ can be satisfied and thus differentiating the spin-wave dispersion relation, Equation (8.29), which is also applicable to thin films, gives $(1+\varepsilon_k^2)Dk^2 = M_s sin^2\theta_k$, where $\varepsilon_k = \omega_k/(\omega_M + Dk^2)$. This expression, together with the spin-wave dispersion, defines the lowest threshold spin-wave modes. The resulting variation in h_{crit} for a thin film in the PI configuration is compared with measured data in Figure 8.14b. Note the excellent agreement between the calculated, full line, and measured, filled circles, data and the absence of the "kink." Also shown, a broken line, is the calculated variation in h_{crit} if the $\partial\omega_k/\partial k_\perp = 0$ condition is not applied and minimization of the h_{crit} expression is required. The expression for h_{crit} in a thin film in the PI configuration is then the same as that of the bulk or thick film case, given by Equation (8.64), but with a $\pi/2$ multiplier. The $\pi/2$ multiplier is a result of the $k_\perp = n\pi/d$ mode quantization when the $\partial\omega_k/\partial k_\perp = 0$ condition is not applied. Both expressions for h_{crit} in the PI thin film configuration are valid, but the one resulting in the lowest threshold provides the best fit to measured data. The $\Delta H_k(k)$ expression and fitting parameters were the same as used in the thick-film (bulk) cases.

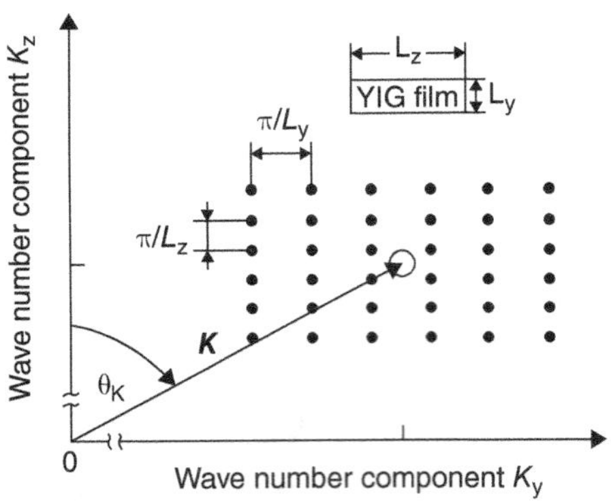

Figure 8.15 Schematic reciprocal lattice wave vector diagram showing 6 × 6 array of lateral standing modes for a YIG film. Solid dots indicate the discrete standing wave number components, k_y and k_z for the array centered at a wave number K and polar propagation angle θ_K. The wave number separations in the y- and z-directions are π / L_y and π / L_z [39].

The previous discussion considered only the configurations of the pumping and static external fields shown in Figure 8.14a and b. However, it turns out that when the sample has a small rectangular shape, there is a difference in the butterfly curve for the in-plane II configuration when the static magnetic field is applied along the short and long axes of the rectangle as shown in Figure 8.14c. The extension of the theory that includes the quantization of the spin-wave numbers to these II configurations was done in [39] and enabled an excellent agreement between calculated and measured h_{crit} variation with bias field for a 7-μm-thick rectangular YIG film with sides of unequal length (1.1 mm × 2.0 mm; see Figure 8.14c). Measured results showed a difference in h_{crit} depending on the orientation of the sides of the sample relative to the bias field, which was accurately reflected in the theory. The coordinate configurations for the discussed case are shown in Figure 8.12a and d. In the previous PI case, the standing modes across the film thickness played an important role. However, here the standing modes across the film thickness are not important because the wavevector k is in the plane of the film but because of the limited lateral dimensions, the extension of the theory requires that the lateral standing modes are considered. The lateral standing modes have y and z wave vector components (see Figure 8.15) in the form $k_y = n_y\pi / L_y; k_z = n_z\pi / L_z$, where n_y and n_z are positive integers and L_y and L_z are the length of the y- and z-sides of the rectangular ferrite film. Details of the derivation are quite complicated, and an iterative procedure is required with consideration of a certain number of coupled modes close to the minimum threshold wave number K, at around half of the pumping frequency. The alternative approach is to use a limited $N \times N$ cluster of lateral modes centered at some $\omega_K = \omega_K(\omega_p / 2, K, \theta_K)$, as shown in Figure 8.15. Now, at a specific value of the external applied field H_0, the threshold is obtained by scanning the available ω_K states and finding the minimum critical field value. The procedure is still involved but offers more insight into the solution. An interested reader can find the details in [36]. Using this approach, only a simple addition to the thick film or bulk theory, Equation (8.65), is necessary for the investigated II thin film case. The h_{crit} expression for the thin film II configurations is:

$$h_{\text{crit}}^{\text{II-thin film}} = \min_{\theta_k} \left(h_{\text{crit}}^{\text{II-bulk}} G(\Delta\omega_K / \eta_K)\right) \tag{8.68}$$

where $h_y^{\text{II-bulk}}$ is the complete expression within the brackets in Equation (8.65) before minimization, the G factor is:

$$G(\Delta\omega_K / \eta_K) = \left(\frac{\pi}{2}\right)^2 - \frac{(\pi/2)^2 - 1}{cosh\left(\frac{3}{4}\Delta\omega_K / \eta_K\right)} \tag{8.69}$$

and

$$\Delta\omega_K = \frac{\omega_M\left(\omega_H + |\gamma| Dk^2\right)}{\omega_p} \frac{sin(2\theta_K)}{K} \sqrt{(\pi / L_z)^2 \sin^2\theta_K + (\pi / L_y)^2 \cos^2\theta_K)} \tag{8.70}$$

The uppercase subscript K is the center wave vector number indicating an operating point in the center of an array of coupled spin-wave modes within $\Delta\omega_K$ and not as previously a lowercase k indicating an actual single spin-wave mode in the array. L_y and L_z are the lengths of the sides of the YIG film parallel and perpendicular to the bias field direction, respectively. In this case, $\Delta H_k(k) = A_0 + B_1 L(k / k_c)$. This is simpler than the expression for $\Delta H_k(k)$ used for spheres or bulk films due to the absence of the $sin(2\theta_k)$ term as it is already incorporated in the $\Delta\omega_K$ expression. The parameters for the theoretical curves in Figure 8.14c are $A_0 = 28$ A/m(0.35 Oe), $B_1 = 46.4$ A/m(0.58 Oe) and $k_c = 8.5 \times 10^5\, 1/cm$. These values are similar those used earlier in the bulk-film case, and k_c is the same as a cited [20] theoretical value.

8.2.2.2 Parallel Pumping

Parallel pumping results for thin ferrite films will now be discussed. Studies of parallel pumping in YIG films [43] showed that, at least for films of thickness >10 μm, the gross variation of h_{crit} with the magnetic bias field could be explained by the bulk theory developed for YIG spheres. The results of measurements on a 15.9-μm-thick YIG film at a pump frequency of 9.4 GHz and magnetized in-plane are shown by the filled circles in Figure 8.16. Also, shown by the continuous line, are the results of calculation using Equations (8.51) and (8.52) with the constants $A_0 = 16\frac{A}{m}(0.2\ Oe)$, $A_1 = 9.6\frac{A}{m}(0.12\ Oe)$, and $A_2 = 7.6\times10^{-7}A(9.5\times10^{-7}Oe\cdot cm)$. Closer examination of the measured data showed the presence of a fine periodic structure, which was attributed to standing spin-wave modes in the YIG film thickness [43]. This fine structure is observed below the cusp in the butterfly curve, as indicated in Figure 8.16 [43]. The fine structure in thinner 0.5 and 3.8 μm films had been reported earlier and similarly attributed to spin-wave resonance in the film thickness [45]. The fine structure implies that φ_k is rotated from 0° (in-plane) to 90° (normal to plane), which due to the combination of volume dipole–dipole and thin film shape effects [43] results in maximum ellipticity and thus minimum threshold. Interpretation of the fine structure also requires the quantization of the wave vector perpendicularly to the film thickness similar to that discussed earlier for the spheres. Fine structure is not expected in transversely magnetized thick films when $\delta\omega_k >> \eta_k$ or in normally magnetized films where the value of φ_k is unimportant and θ_k is always in-plane.

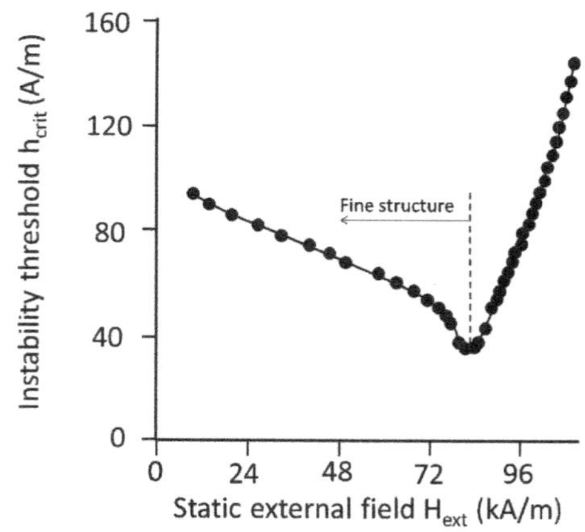

Figure 8.16 Variation of threshold field h_{crit} with an external magnetic field H_{ext} for parallel pumping at 9.4 GHz in a 15.9-μm-thick YIG film [43].

8.2.2.3 Dipole–Exchange Interactions

A more rigorous theory for the parallel pumping threshold in thin films has been presented [46], which considers exchange interactions in addition to the dipolar interactions already discussed. Dipolar interactions are weakly dispersive, but exchange interactions in thin films can, depending on the excitation frequency and corresponding wave number, demonstrate both weakly and highly dispersive properties. These spin-wave excitations are called dipole–exchange spin waves. For these waves, it is necessary to consider also the exchange boundary conditions for the dynamic magnetization at the surfaces of the film in addition to electromagnetic boundary conditions. These exchange boundary conditions are also called surface spin pinning conditions. The considered geometry is given in Figure 8.17, where a ferromagnetic film of thickness d in the ξ direction is shown. The film is infinite in the η and ζ directions and the ζ-axis is parallel to the direction of the spin-wave propagation. The film is magnetized to saturation by an external field H_0 applied in an arbitrary direction. Another coordinate system (x,y,z) is introduced in which the z-axis is parallel to the direction of magnetization M_0. Transition from the (ξ,η,ζ) coordinate system to the (x,y,z) system is obtained by orthogonal transformation through the angles φ and θ, which are determined by the equilibrium orientation of M_0. The boundary conditions can be expressed in general in the form [47]:

$$\frac{\partial \boldsymbol{m}}{\partial \xi} + p_{1,2}\boldsymbol{m}|_{\xi=\pm d/2} \tag{8.71}$$

where ξ is the coordinate perpendicular to the film surface, $p_{1,2}$ are the surface spin pinning parameters, and d is the film thickness. There are two often used extreme cases: one is when the spins at the surface are pinned, then:

$$\boldsymbol{m}|_{\xi=\pm d/2} = 0 \tag{8.72}$$

and when the spins at the surface are free to move – unpinned surface spins condition:

$$\frac{\partial \boldsymbol{m}}{\partial \xi}|_{\xi=\pm d/2} = 0 \tag{8.73}$$

The detailed discussion of the dipole and exchange interactions and anisotropy effects on the critical field

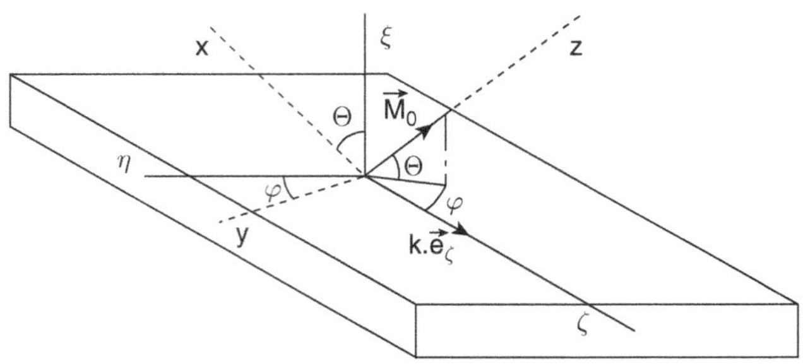

Figure 8.17 Definition of the coordinate system for the dipole–exchange mode calculation [46].

conditions is too extensive to be included here and the reader can find the details in [45, 46, 47]. Only, the essential equations for subsidiary absorption and parallel pumping that include dipole–exchange modes are presented.

The critical fields for the subsidiary absorption considering the dipole–exchange interaction were derived in [48, 49, 50]:

$$h_{\text{thn},n'}^{\text{pinned}} = \frac{2n'\pi\omega_n \left\{ \left[\left(\Omega_{n'0} + \omega_M \Omega_{n'0} \right) - \omega_p^2 \right]^2 + 4\omega_p^2 \omega_{rn'0}^2 \right\}^{1/2} \left[\left(\frac{\omega_p}{2} - \omega_n \right)^2 + \omega_{rnk}^2 \right]^{1/2}}{\mu_0 \gamma \omega_M \left[(\Omega_{n'0} + \omega_M)(\Omega_{nk} + \omega_M - \omega_M P_{nk}) + \omega_p \omega_n \right] Q_{nn'} \sin 2\theta} \quad (8.74)$$

for pinned surface spins and:

$$h_{\text{thn},n'}^{\text{free}} = \frac{4\sqrt{2}\omega_n \left\{ \left[\omega_H (\omega_H + \omega_M) - \omega_p^2 \right]^2 + 4\omega_p^2 \omega_{r00}^2 \right\}^{1/2} \left[\left(\frac{\omega_p}{2} - \omega_n \right)^2 + \omega_{rnk}^2 \right]^{1/2}}{\mu_0 |\gamma| \omega_M \left[(\omega_H + \omega_M)(\Omega_{nk} + \omega_M - \omega_M P_{nk}) + \omega_p \omega_n \right] Q_{nn'} \sin 2\theta} \quad (8.75)$$

for free surface spins. Film thickness d is along the x-axis and with the film magnetized in-plane. The spin-wave propagation is along ζ, as shown in Figure 8.17. The angle θ is between the direction of propagation and the orientation of the static field H_0. $n,\ n' = 0,1,2,\ldots$ represent the excited spin-wave eigenmodes of the tangentially magnetized film. $\Omega_{nk} = \omega_H + Dk_n^2$ and $k_n^2 = \kappa_n^2 + k^2, \kappa_n = n\pi / d$, and $\omega_{rnk} = \mu_0 |\gamma| \Delta H$ is the relaxation frequency of a spin-wave mode with number n. Here, we assume that k is in the z-propagation direction. The corresponding functions in the aforementioned equations are:

$$P_{n,k} = \frac{k^2}{k_n^2} + \frac{k^2 \kappa_n^2}{k_n^4} \frac{2}{kd} \left[1 - (-1)^n e^{-kd} \right] \quad (8.76)$$

$$Q_{n,n'} = \frac{4k^2 \kappa_n^3}{\left[k^2 + (\kappa_n - \kappa_{n'})^2 \right] \left[k^2 + (\kappa_n + \kappa_{n'})^2 \right]} \frac{\kappa_{n'}}{dk_n^2} \left[1 + (-1)^n e^{-kd} \right] \sqrt{\frac{\pi}{2}} \quad (8.77)$$

For spins pinned at the surface and:

$$P_{n,k} = \frac{k^2}{k_n^2} - \frac{k^4}{k_n^4} \frac{2}{kd} \left[1 - (-1)^n e^{-kd} \right] \frac{1}{1 + \delta_{0n}} \quad (8.78)$$

$$Q_{n,n'} = \left\{ \frac{k^2}{k_n^2} + \frac{2}{k_n^4} \frac{k^3}{d} \left[1 - (-1)^n e^{-kd} \right] \right\} \frac{\sqrt{\pi}}{(1 + \delta_{0n})2} \quad (8.79)$$

for free spins at the surface.

It follows from the $\sin(2\theta)$ term in the denominator of Equations (8.74) and (8.75) that the lowest parametric threshold (critical field) is for spin waves propagating at an angle $\theta = 45^\circ$ with respect to static field.

For parallel pumping configuration, the threshold field can be expressed as:

$$h_{th}^{pp} = \frac{4\omega_n \left[\left(\frac{\omega_p}{2} - \omega_n \right)^2 + \omega_{rnk}^2 \right]}{\mu_0 \gamma \omega_M \left| P_{nk} (1 + \sin^2 \theta) - \right.} \quad (8.80)$$

For parallel pumping, the lowest threshold happens for the $k = 0$ spin waves. This threshold is the same for standing waves with different numbers n with $\Delta H_{n,k} = \text{const}(n)$. If the static field has the value such that $\omega_n(k) = \omega / 2$ is not satisfied, the generation of $k = 0$ waves is difficult, and then a pair of short-wavelength parallel or transverse to static magnetization spin waves is generated. The waves with strongest ellipticity are generated and this increase of ellipticity appears only if the exchange interactions are considered. Note that the ellipticity of nonexchange spin waves decreases monotonously with increasing k. The closer the ellipticity of these waves is to circular polarization, the higher is the threshold field. The detailed description of the dipole–exchange mode thresholds within and outside regions of dipole–dipole mode hybridization is complex and investigated in [46].

8.2.3 First-Order Instability – Hexagonal Ferrites

The first-order instability theory for an easy-plane hexagonal ferrite as a function of the DC bias field angle with respect to the plane of the disk was presented in [51] and for arbitrary pumping configuration and general anisotropy in [15]. The critical fields can be derived from the general theory discussed earlier in this chapter using the Equations (8.26a–8.26d) and the corresponding equations of motion Equation (8.30). Detailed calculations are given in [15].

An example of the calculation of the first-order thresholds for an easy-plane Zn–Y ferrite thin disk is presented here. Because of the planar anisotropy, the anisotropy energy for this case is given by Equation (8.9) with a negative anisotropy constant K_U. A similar approach would also apply for the uniaxial anisotropy case. The relevant geometry is shown in Figure 8.18 where the bias field and the magnetization are in the Y–Z plane. For the given configuration, the anisotropy field has only one nonzero component $H_{AZ} = H_A \cos\theta_M = \dfrac{2K_U}{M_s}\cos\theta_M$. The angle θ_M is obtained from the static equilibrium condition $\boldsymbol{M} \times \boldsymbol{H}_{\text{eff}} = 0$ that for this case is

$$H_0 \sin(\theta_H - \theta_M) = (H_A - M_s)\sin(2\theta_M)/2. \tag{8.81}$$

In the short spin-wave wavelength limit, the condition of critical mode frequency for the first-order instability of $\omega_k = \omega_p/2$ is fulfilled. The condition for the threshold field comes from Equation (8.26). It is to note that coefficients in Equation (8.26) contain the uniform precession amplitude a_0 and its complex conjugate $a_0{}^*$. As mentioned before, the derivation is quite involved and we will present only the results following [15].

The threshold for the first-order instability can be expressed as:

$$h_{\text{crit}}^{(1)} = \min_k \left(\frac{\Delta H_k}{\left| Y_k^1 \right|} \right) \tag{8.82}$$

where:

$$Y_k^{(1)} = \left[\lambda_k^2 q_{L0} d_k + \mu_k^2 q_{A0}^* d_k^* - \lambda_k \mu_k (2\alpha_z + q_{L0} c_k + q_{A0}^* c_k^* \right] \tag{8.83}$$

The coefficients d_k and c_k are obtained from Equations (8.26c) and (8.26d), including only the first order of a_0 and its complex conjugate. For special pumping configurations, Y_k^1 can be expressed in terms of W, we used before. As an example, we present the results of a calculation of the first-order instability threshold for an easy-plane Zn–Y hexagonal ferrite disk shown in Figure 8.18. For these calculations, the saturation magnetization M_s was set to 224 kA/m (2800 G) and the anisotropy field H_a was set to −720 kA/m (−9.0 kOe). The calculations were done for 9 GHz driving frequency. Results of the numerical calculation are detailed in Figure 8.19 [15] for θ_H = 5°, 10°, and 90°. Chart (a) gives the critical field as a function of the external bias field, and (b) and (c) present the dependence of the critical mode spin-wave number k and the polar angle θ_k as a function of the external field. As follows from Figure 8.19a, the h_{crit} curve holds a constant value for $\theta_H = 90°$ up to about 20 kA/m (250 Oe), while the critical mode k-value is nonzero. For other angles, h_{crit} starts at the same value as for H_{ext} = 0 and increases slightly as the external field is increased. A constant spin-wave linewidth $\Delta H_k = 80$ A/m(1 Oe) was considered in calculations shown in Figure 8.19. Note that $\theta_H = 90°$ (leftmost curves in Figure 8.19) corresponds to the in-plane parallel pumping configuration and

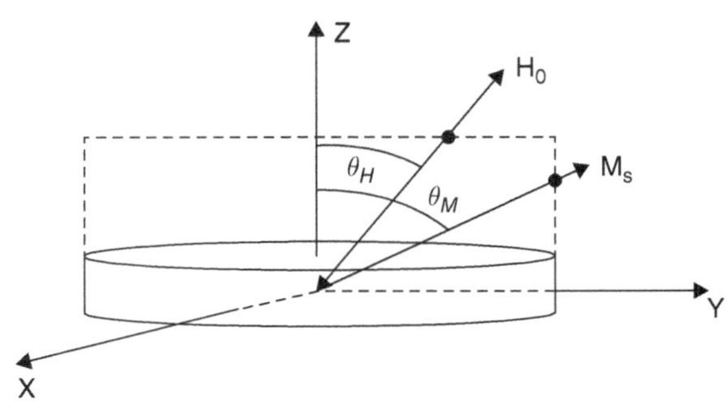

Figure 8.18 Geometry for the easy-plane hexagonal ferrite disk threshold analysis. The disk is in the X–Y plane of the laboratory frame. The external static field $\boldsymbol{H}_0$ is in the Z–Y plane at an angle θ_H relative to Z. The static magnetization M_s is in the Y–Z plane at an angle θ_M relative to Z [15].

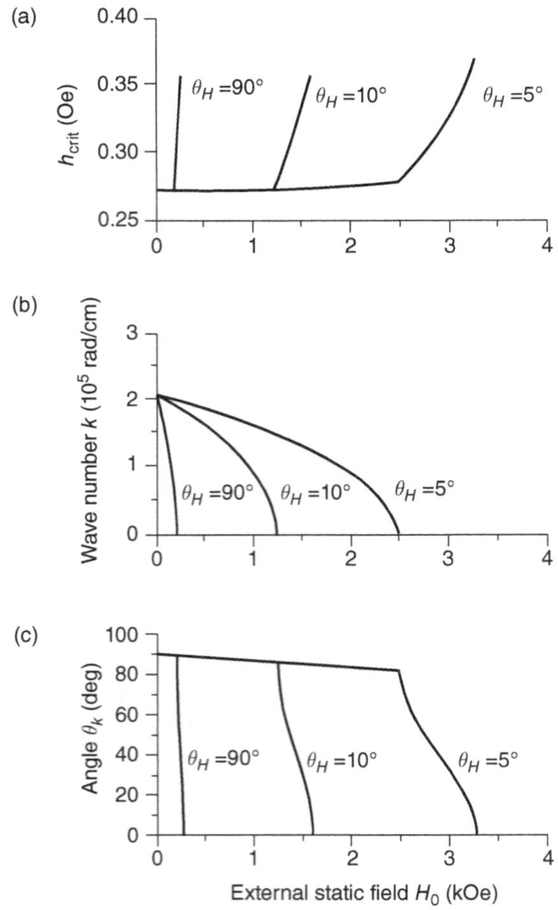

Figure 8.19 Threshold and critical mode parameters as a function of the external field $\boldsymbol{H}_{ext}$ for an easy-plane hexagonal ferrite disk with 9 GHz microwave excitation. The graphs show (a) first-order spin-wave instability threshold h_{crit} vs. H_{ext}, (b) critical mode spin-wave wave number k vs. H_{ext}, and (c) polar spin-wave propagation angle θ_k vs. H_{ext}. The parameter is θ_H, as indicated [15].

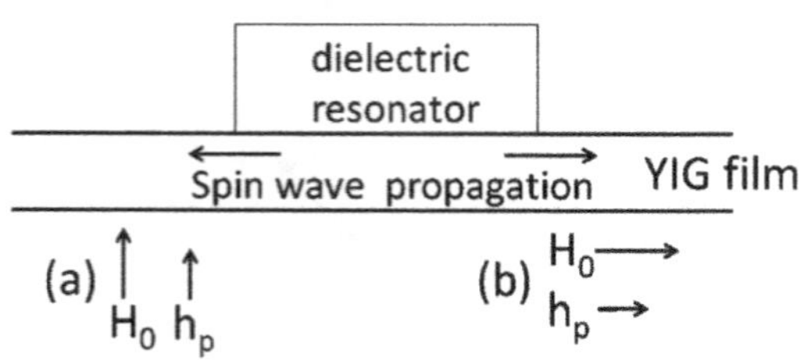

Figure 8.20 Experimental configuration for homogeneous local parallel pumping with a dielectric resonator bonded to the surface of the YIG film. (a) Pumping perpendicular to the YIG film. (b) Pumping in-plane of the YIG film. [55].

that the results are similar to those for parallel pumping in YIG (see Figure 8.10) without taking the wavevector dependence of the spin-wave linewidth into account.

8.2.4 Nonlinear Excitations in Thin Films under Local Pumping

The general first- and second-order nonlinear threshold excitations that have been addressed so far have historical relevance and are important for understanding the nonlinear effects in ferrite materials that were excited by uniform or quasi-uniform pumping fields such as those in resonant cavities. Measuring small spherical or thin film samples in the uniform RF magnetic field of a microwave cavity biased by a uniform static DC magnetic bias field provides an ideal approach to establishing the basic theory for the instability threshold fields in ferrites. The critical phenomena above and well beyond the threshold have not been addressed, but the derivations discussed earlier in this chapter can be extended to aforementioned threshold and are important for understanding the onset of chaotic behavior that can be triggered in these systems. The topic of chaos is out of the scope of this chapter, but the interested reader can find a review of this problem in [52]. The nonlinear devices of current interest, e.g. FSLs, Chapter 13, use planar transmission lines such as microstrip, stripline, or coplanar waveguides with narrow strips as excitation antennas to maximize the RF magnetic field and hence minimize the threshold power. Another approach to local excitation of propagating spin waves was presented in [53–55] where a dielectric resonator was used to locally excite the propagating spin waves. In these devices, the spin-wave propagation out of the region of the excitation antenna has a significant influence on the device performance. The narrow transmission lines and dielectric resonators represent a local excitation of these propagating spin waves because the source of the excitation is concentrated in a spatially limited area of the sample.

In the case of linear excitation by a narrow antenna, the excited spin waves propagate perpendicular to the antenna aperture and details of the calculation of the radiation resistance and the corresponding group velocities, for different orientations of the applied static DC bias field with respect to the direction of propagation of the excited waves are given in Chapter 13. Here, the results related to nonlinear excitations are addressed that are critical for understanding the performance of locally excited thin film ferrite devices. First, the effect of the dimensions of a local antenna on the nonlinear response of the system are investigated.

8.2.4.1 Direct Excitation by an Antenna

Dielectric resonators operating at 9.37 GHz and designed to provide either an in-plane or normal RF magnetic field were attached to the surface of the YIG film as shown in Figure 8.20 [55]. The internal bias, H_i, and the RF magnetic field, h_p, were parallel for longitudinal pumping. It was assumed that the RF magnetic field was uniform over the area of the dielectric resonator in contact with the YIG film and zero elsewhere, i.e. locally uniform pumping. The dimensions of the dielectric resonator area in contact with the YIG were $a = b = 3.5$ mm and $a = b = 1.5$ mm for parallel pumping normal to the YIG film and $a = b = 3.5$ mm and $a = 1.5$ mm and $b = 1$ mm for parallel pumping in-plane.

Measurements of the threshold power in thin YIG films with localized pumping provided by small dielectric resonators or narrow wire antennas showed that localized pumping resulted in increased and additional variation in threshold power [53–57]. This is not surprising as the excited waves under the antenna can propagate out of the antenna region. In the first approximation, the spin waves propagating out of the pumping area can be considered as contributing to the spin-wave loss term represented by the spin-wave linewidth in the earlier threshold equation (Equation 8.51). This is a reasonable assumption because the energy/power necessary for the excitation of propagating waves has to be supplied by the external pumping RF field.

Experimental results with pumping normal to a YIG film are shown in Figure 8.21 for three configurations: 1, cavity resonator; 2, $a = b = 3.5$ mm dielectric resonator; and 3, $a = b = 1.5$ mm dielectric resonator. The YIG film dimensions were 5.5 × 7 mm with thickness 16 μm and spin-wave linewidth, $\Delta H_k = 16\frac{A}{m}(0.2\ \text{Oe})$. Data for the three different resonators were normalized for $H_i = H_r - 4kA/m(50\ \text{Oe})$, where. In this case, the threshold magnetic fields for $H_i \le H_r$ in all three resonator configurations were identical within the experimental error. The lower curve, 1, in Figure 8.21, for $H_i > H_r$ shows the lowest thresholds and was obtained with uniform pumping in a cavity. Curves 2 and 3, obtained with dielectric resonators, show threshold fields that increase inversely with the pumping area. The increased threshold field for curves 2 and 3 is, as mentioned before, ascribed to the propagating spin waves carrying energy out of the pumping area and increasing the effective damping of the system. This effect is not significant when $H_i \le H_r$ as exchange dominated, half-frequency, spin waves with high wave number, and low group velocity, $v_g = {}^{\partial\omega}\!/\!_{\partial k}$, are generated in this region. However, when $H_i > H_r$, low wave number spin waves, or magnetostatic waves, with a larger group velocity are excited and carry significant energy away from the pumping region. Equation (8.51) can then be modified to give the parallel pumping threshold for locally homogeneous pumping in the form [55]:

$$h_{\text{crit}} = \min\left(\Delta H_k \frac{\omega_p}{\omega_M \sin^2\theta_k}\sqrt{1+({}^{\beta l}\!/\!_{w})^2}\right) \tag{8.84}$$

where β is a numerical coefficient between π, when $\frac{l}{w} \ll 1$ and $\pi/2$ in the opposite limit, w is the width of the pumping region, and $l = {}^{2v_g}\!/\!_{\gamma\Delta H_k}$ is the characteristic distance traveled by the spin-wave.

The long wavelength $\omega_k = \omega_p/2$ spin waves, excited when $H_i > H_r$, are magnetostatic forward volume waves (MSFVWs), which propagate isotropically in the plane of the normally magnetized YIG film. The lowest order MSFVW mode, $n = 1$, has an approximately half-wavelength field variation through the YIG film thickness, with higher order modes having multiple half wavelengths corresponding to their mode number n. h_{crit} is determined by the MSFVW mode with the lowest threshold. If the wave number normal to the film, i.e. parallel to H_i and h_p, is $k_x = n\pi/d$ and the propagating wave number in the film is k_T, then $k = \sqrt{k_x^2 + k_T^2}$ and $\sin\theta_k = {}^{k_T}\!/\!_{k}$. Thus, minimization of Equation (8.96) involves the mode number. The lowest order mode does not necessarily have the lowest threshold.

Further explanation follows [54], presented in connection with an investigation of kinetic instability of spin waves. As far as the authors are aware, a quantitative comparison of results similar to those presented in [55] with theory has not been published. While the data in Figure 8.21 are presented as a function of magnetic bias field, as cavity and dielectric resonators have high Q and operate over a narrow frequency range, the following explanation is provided in the frequency domain. However, the characteristic frequencies and magnetic fields are easily translated from bias field to frequency domains.

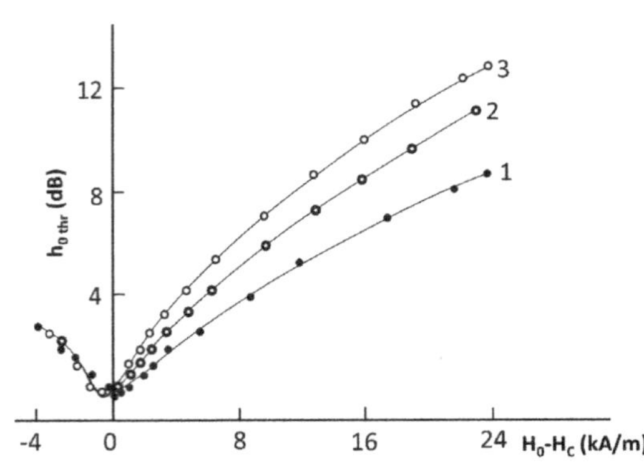

Figure 8.21 Variation of the threshold field for parallel pumping with the magnetic bias field applied normal to the YIG film. Pump frequency is 9.37 GHz. (1) Cavity resonator. (2) Dielectric resonator with $a = b = 3.5$ mm. (3) Dielectric resonator with $a = b = 1.5$ mm [55].

The frequency dispersion of the $\omega_k = \omega_p / 2$ MSFVW–spin waves as a function of a normalized wave number *kd* is shown in Figure 8.22a, $d = 20\mu$m and external field $H_0 = 24$kA / m(3000 Oe). Mode numbers $n = 1$, $n = 100$, $n = 190$, and $n = 240$ are shown. $f_{\min} = \mu_0|\gamma|(H_0 - M_s) = \mu_0|\gamma|H_i = 3.5$, with $M_s = 140$kA / m(1750 G) and $f_r = \mu_0|\gamma|(H_i(H_i + M_s))^{1/2} = 5.422$GI. f_r in this figure translates to H_r in Figure 8.21. The bold line was obtained by calculating the *n* and *kd* values that minimize Equation (8.96) at different frequencies. We can see from Figure 8.22a that for $f_k > f_r$ (corresponding to $H_i < H_r$ for constant f_p), the parallel pump instability occurs for relatively large wave number, $k > 10^5$cm^{-1} with $n = 1$, but when $f_{\min} < f_k < f_r$ ($H_1 > H_i > H_r$ for constant f_p, where $H_1 = f_k / \mu_0|\gamma|$), Equation (8.84) is minimized at larger *n*, where the group velocity is smaller. For example, when $f_k = f_p / 2 = 4.75$GHz, the $n = 70$ mode has the lowest threshold.

The calculated path length l_x, for $\omega_k = \omega_p / 2$ waves normal to the plane of the film, as a function of frequency, is shown in Figure 8.22b. Waves with $f_k > f_r$ ($H_i < H_r$) have a very short path length and are not influenced by the YIG film surfaces. They thus behave as uniform spin waves in an infinite medium. In contrast, for $f_{\min} < f_k < f_r$, l_x is greater than the film thickness *d* and the threshold is determined by the mode number. The path length in-plane is also much longer than for $f_k > f_r$, which results in increased sensitivity to the size of the pumping region, as seen in Figure 8.21 for $H_i > H_r$.

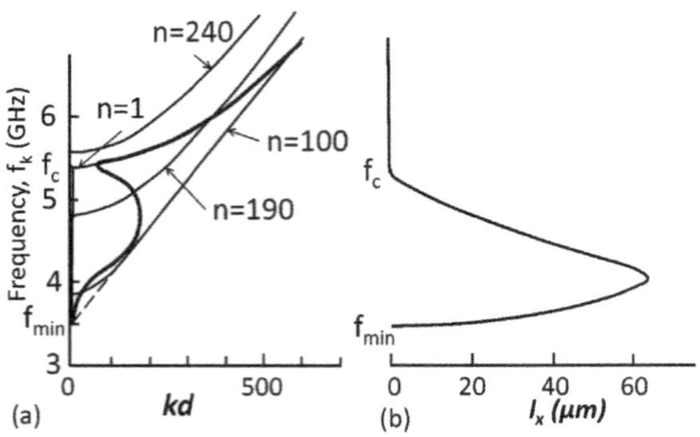

Figure 8.22 (a) Dispersion of magnetostatic volume waves in a normally magnetized YIG film for modes *n* = 1, 100, 190, and 240. The YIG film thickness is 20 μm, and the external bias field is 240 kA/m (3000 Oe). Bold line corresponds to modes with a minimum threshold for local parallel pumping. (b) Path length of spin waves directed normal to the YIG film surface for waves indicated by the bold line in (a) [54].

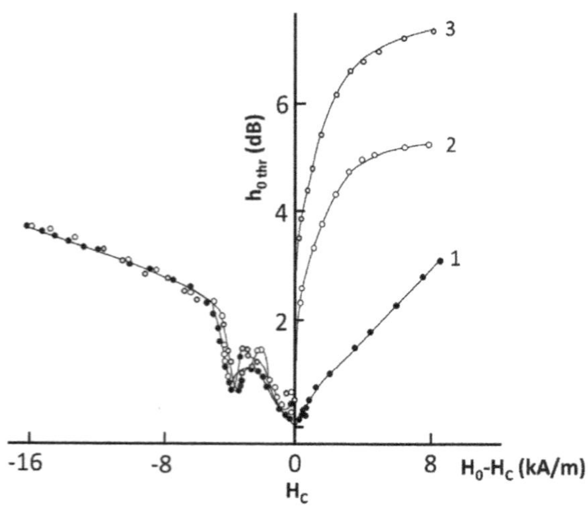

Figure 8.23 Variation of threshold field for parallel pumping with magnetic bias field applied in the plane of the YIG film. Pump frequency is 9.37 GHz. 1, cavity resonator; 2, dielectric resonator with *a* = *b* = 3.5 mm; and 3, dielectric resonator with *a* = 1.5 mm and *b* = 1 mm [55].

The variation of the threshold field, h_{0_thr}, for locally uniform parallel pumping of a YIG film magnetized in-plane is shown in Figure 8.23 [55]. Curve 1 was obtained with a cavity and the other curves with dielectric resonators of dimensions, curve 2 $a = b = 3.5$mm, and curve 3 $a = 1.5$mm and $b = 1$mm. The measurements from the cavity and dielectric resonators were normalized at $H_i = \left(H_r - 16\frac{\mathrm{kA}}{\mathrm{m}}\right)$ (200 Oe). In this transversely magnetized case, there is a stronger dependence of the threshold field on the size of the dielectric resonator for curves 2 and 3 with $H_i > H_r$. The variation of h_{0_thr} for $-4.8\frac{kA}{m} < (H_r - H_i) < 0$ is more complicated than in the previous normally magnetized case due to the interaction with magnetostatic surface waves (MSSWs) at $f_k = f_p / 2 = 4.685$GHz, which are not present in a perpendicularly magnetized film. These show a small variation in threshold with the resonator size. The waves excited above $H_i = H_r$ are magnetostatic backward volume waves (MSBVWs), which also exhibit multiple thickness dependent modes. Again, the determination of the threshold fields requires minimization of Equation (8.84). However, MSBVWs exhibit anisotropic propagation over a range of angles relative to the bias field direction in the plane of the film, which must be considered in the threshold field calculation.

8.2.4.2 Excitation by Magnetostatic Surface Waves

The previous section considered only the direct parallel pump excitation of $\omega_p / 2$ spin-wave by the RF magnetic field of the resonator. Here, nonlinear excitation of $\omega_p / 2$ spin-wave by transverse pumping through an intermediate stage of MSSW is discussed. This excitation is relevant to the theory of MSSW FSLs and signal to noise enhancers. If the excitation is represented by a strip antenna of width *w* and aperture (length) *l* (see Figure 8.24a), it can linearly excite spin waves, with the wave number perpendicular to the antennas aperture in the interval $0 < k < 2\pi / w$. The wave number components parallel to the antenna aperture are defined by the antenna length l. Here, the derivation of the quasi-surface spin-wave threshold in thin ferromagnetic films excited by localized pumping is summarized [50]. The threshold fields for other magnetostatic wave excitations, although computationally involved, can be introduced in a similar way. The sample is a thin film of thickness d in the x-direction (see Figure 8.24a and b) with unpinned surface spins. The shaded area represents the antenna on the surface of the film. The film was considered infinite in the Y–Z plane and magnetized to saturation in the z-direction. The excitation antenna was along the z-direction with width w. The analysis considers a two-stage process. First, the quasi-surface wave at frequency ω_1 and in-plane wave vector $\boldsymbol{k}_1$ is excited, and in the second stage, when the amplitude of this wave exceeds a certain critical level this wave decays into two waves with frequencies ω_2 and ω_3 and in-plane wave vectors $\boldsymbol{k}_2'$ and $\boldsymbol{k}_3'$ (see Figure 8.24a). The details of the calculation can be found in [50], and the threshold field for the process $1 \rightarrow 2,3$ can be expressed as:

$$h_{th1,23} = \frac{\left\{\left[\left(\omega_1 - \omega_p\right)^2 + \left(\mu_0|\gamma|\Delta H_1\right)^2\right]\left[\left(\omega_2 + \omega_3 - \omega_1\right)^2 + \mu_0\left(|\gamma|\Delta H\right)^2\right]\right\}}{|U_1 V_{1,23}|} \tag{8.85}$$

where in the case of degenerate decay, $\omega_2 = \omega_3 = \omega_1/2$, $n_2 = n_3 = n$:

$$|U_1 V_{1,23}| = \omega_M R_{1,22} P_{nn} sin2\theta;\ R_{1,22} = \left[\left(A_{n_1k_1} + |B_{n_1k_1}|\right)\left(A_{n_2k_2} + |B_{n_2k_2}|\right) + \omega_1\omega_2\right] / \left(\omega_1\omega_2\right) \tag{8.86}$$

and for the case of non-degenerate decay, $\omega_2 \neq \omega_3 \neq \omega_1/2$, $n_2 = n$, $n_3 = m$:

$$|U_1 V_{1,23}| = \omega_M R_{1,23} Q_{nm} (\cos\phi_2 + \cos\phi_3). \tag{8.87}$$

coefficient $R_{1,23}$ is on the order of unity.

Here:

$$2A_{nk} = 2\Omega_{nk} + \sin^2\theta + \omega_M P_{nn}\left[1 - \sin^2\theta(1 + \cos^2\phi)\right] \tag{8.88a}$$

$$B_{nk} = A_{nk} - \Omega_{nk} - \omega_M P_{nn}\sin^2\phi + j(\omega_M/2)P_{nn}cos\theta sin2\phi \tag{8.88b}$$

$$\Omega_{nk} = \omega_H + D(k^2 + \kappa_n^2) \tag{8.88c}$$

This determines the general dispersion relation $\omega_{nk}(k, \kappa_n) = \sqrt{A_{nk}{}^2 - |B_{nk}|^2}$ [58]. The dipole–exchange spectrum calculated from dispersion relation illustrating the process of parametric decay of a quasi-surface spin-wave with $n = 0$, $\phi = \pi/2$ into a pair of short $k \cong k'$ or long $k \cong k''$ volume spin waves is shown in Figure 18.24b. The calculation is done for $d = 1.6 \cdot 10^{-3}$cm, $H_0 = 29.6\frac{\text{kA}}{m}$ (370 Oe), $M_s = 140\frac{\text{kA}}{m}$ (1750G) for f = 3.3 GHz and the linewidth of $\Delta H = 40\frac{A}{m}$ (0.5Oe). As before, Equation (8.85) has to be minimized to get the critical threshold field. The influence of local pumping is introduced into these equations through a simple, but effective, model that is based on the fact that the minimum threshold field is achieved when the conservation laws for frequency and wave vector are fulfilled. For localized pumping, this leads to the conditions [50]:

$$\omega_1 - \omega_p \cong v_{g1}\Delta k_p \ and\ \omega_2 + \omega_3 - \omega_1 \cong v_{g2}\Delta k_p + v_{g3}\Delta k_p \tag{8.89}$$

where $v_{gi} = \partial\omega_i/\partial k_i$ are the group velocities and $\Delta k_p = 2\pi\varsigma/w$, where $\varsigma < 1$ is the "localization" parameter. Note that this leads to a similar expression to that for parallel pumping discussed earlier (see Equation 8.84). The three-way parametric process can happen either in the short-wave ($k >> 1/d$) or long-wave ($k < 1/d$) regions of the spin-wave spectrum. In the first case, the film boundaries do not play an important role, and the parametric instability process is essentially the same as in the bulk ferromagnet. A detailed analysis shows that the lowest threshold corresponds to degenerate decay of quasi-surface spin-wave, with $n = 0$ and $\phi = \pi/2$ into short-wavelength spin waves at $\omega_2 = \omega_3 = \omega_1/2$ propagating at $\theta = \pi/4$ to

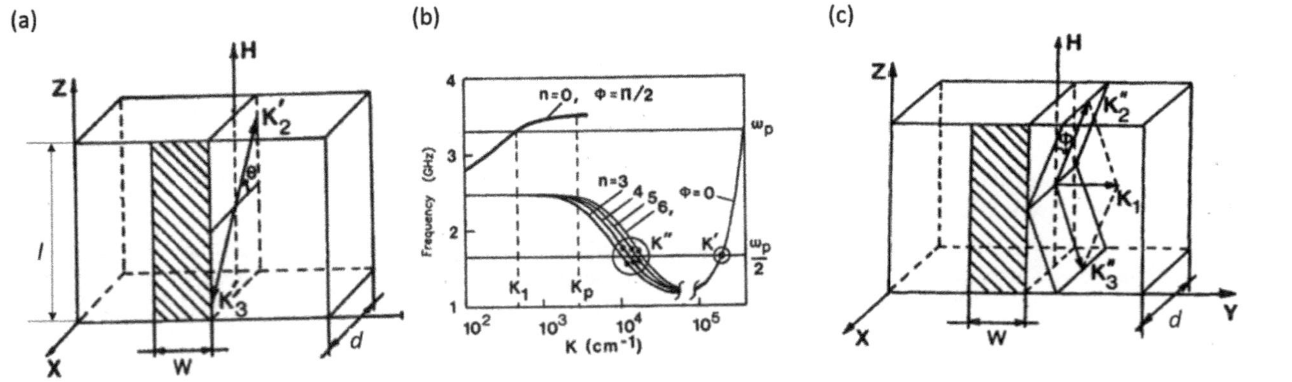

Figure 8.24 (a) and (c) The ferromagnetic film showing the coordinate axes, the direction of bias magnetic field H, and the zone of localization of microwave pumping field. $k2'$ and $k3'$ are the wave vectors of the spin waves parametrically excited in the short-wave region of the SW spectrum. The wave vector of quasi-surface SW k_1, which is small compared to $k2'$, $k3'$, is not shown. (b) The dipole–exchange spectrum, calculated from dispersion relation illustrating the process of parametric decay of a quasi-surface spin-wave with $n = 0, \phi = \pi/2$ into a pair of short $k \cong k'$ or long $k \cong k''$ volume spin waves, is shown in Figure 18.24b. The calculation is done for $d = 1.6 \cdot 10^{-3}$cm, $H_0 = 29.6\frac{kA}{m}$ (370 Oe), $M_s = 140\frac{kA}{m}$ (1750G) for f = 3.3 GHz, and the linewidth of $\Delta H = 40\frac{A}{m}$ (0.5Oe) [47]

the film normal. In the long wavelength region, the lowest threshold corresponds to a nondegenerative decay into pair of modes, with $m = n + 1$ propagating parallel to the film surface at the angles $\phi_2 \cong \pi - \phi_3 \cong \Delta k_p / k''$, as shown in Figure 8.24c, where $|k_2''| = |k_3''| = k''$. Because these spin waves carry energy away from the pumping region, the threshold is significantly increased as in the former case.

Although the topic of localized pumping is complex, these solutions are required for accurate modeling of practical devices such as FSLs that are of current interest.

8.2.5 Second-Order Instability Processes

Up to now, the first-order instability (or three magnon) processes that happen at half, or close to half, of the pumping frequency has been investigated. Most of the nonlinear device applications discussed in Chapter 13 utilize the first-order instability processes. Therefore, we will only briefly address the nonlinear effects that happen at the pumping frequency and that are known as the second-order Suhl instability (or four magnon) processes. This means that two uniform mode magnons, each at driving frequency ω_p, transform into two oppositely propagating magnons each at frequency $\omega_k = \omega_p$ (if the boundary conditions are not considered). As in the case of the first-order nonlinear processes, the energy and momentum conservation conditions must be fulfilled. However, in contrast to the first-order processes, which may not be allowed for certain frequencies, the second-order processes are always allowed for fields at or near the FMR frequency. The theory of second-order instability was developed by Suhl [14]. For the second-order instability, Equation (8.39) becomes:

$$\left[\frac{d^2}{dt^2} + 2\eta_k \frac{d}{dt} - \left|G_k^{(2)}\right|^2 + (\omega - \omega_k)^2 + {\eta_k}^2\right] b_{k0}^2 = 0 \tag{8.90}$$

The relevant task is to obtain the expression for G_k^2. Only terms of the second order in a_0 are retained in the coefficients from Equations (8.26a–d), and only spin waves for which $\omega_k \sim \omega_p$ are of interest. The process of obtaining the relations for the critical fields is straightforward but tedious as was the case of the first-order instability. Without going into details of this derivation, the equations for the second-order instability are listed in the following sections.

8.2.5.1 Second-Order Instability – Isotropic Sphere

The critical threshold field for the second-order instability in spheres excited by linearly polarized RF field at FMR frequency can be expressed as [10, 13]:

$$h_{\text{crit}} = \Delta H_{\text{FMR}} \sqrt{\frac{\Delta H_k}{M_s}} \tag{8.91}$$

where ΔH_{FMR} is the FMR linewidth.

8.2.5.2 Second-Order Instability – Isotropic Thin Film

The working equations for the second-order instability in thin films at FMR were discussed in [49] for three configurations: in-plane static field and out-of-plane microwave field, IP; both fields in-plane and perpendicular to each other, II; and out-of-plane static field and in-plane microwave field, PI. The corresponding threshold field expressions obtained at FMR and using the bulk theory, which neglects the boundary conditions at the film surfaces are [59]:

$$h_{\text{crit}}^{(\text{IP})} = \Delta H \frac{\left(2\mu_0|\gamma|H_{\text{res}} + \omega_M\right)}{\left(\omega_p + \mu_0|\gamma|H_{\text{res}}\right)} \left(\frac{\mu_0|\gamma|\Delta H_k}{\omega_M - \mu_0|\gamma|H_{\text{res}} + \omega_p}\right)^{1/2} \tag{8.92}$$

$$h_{\text{crit}}^{(\text{II})} = \Delta H \frac{\left(2\mu_0|\gamma|H_{\text{res}} + \omega_M\right)}{\left(\omega_p + \omega_M + \mu_0|\gamma|H_{\text{res}}\right)} \left(\frac{\mu_0|\gamma|\Delta H_k}{\omega_M - \mu_0|\gamma|H_{\text{res}} + \omega_p}\right)^{1/2} \tag{8.93}$$

$$h_{\text{crit}}^{(PI)} = \Delta H \left(\frac{\Delta H_k}{M_s}\right)^{1/2} \tag{8.94}$$

Here, H_{res} and the driving frequency ω_p are connected through the corresponding dispersion relations. Note that the bulk theory has significant problems as was shown by Brillouin light scattering (BLS) measurement of the second-order instability in a thin YIG film [60]. The main problem is that the bulk theory predicts $\theta_{k-\text{min}} = 0°$ and k_{min} values in the range of $3-4\times10^5$ rad/cm, while the BLS measurements reveal a well-defined propagation angle of $\theta_k = 80° - 90°$ and $k_{\text{min}} = 10^2 - 4\times10^4$ rad/cm, a much lower value that expected from the bulk theory.

A modified bulk theory that gave a better agreement with the observed experimental data in thin films was presented in [61], where the thin film dispersion relation in form of Equation (8.29) was considered. The second-order critical field can then be expressed as:

$$h_{\text{crit}}^{(2)} = 2\cdot\min\left[\sqrt{\frac{\omega_p \Delta H_k}{\mu_0|\gamma|\left|W_k^{(2)}(H,\theta_k)\right|}}\right]_{\omega_k=\omega_p} \tag{8.95}$$

where the available k and $\theta_k$are obtained from the dispersion relation (Equation 8.29). Because of this constraint, the critical field can be obtained as a minimum of only one variable. Then $W_k^{(2)}(H,\theta_k)$ can be expressed as [61]:

$$\begin{aligned} W_k^{(2)}(H,\theta_k) &= q_{L0}^2\left[r\left(A_k+\omega_p\right) - B_k v\right] + q_{A0}^{*2}\left[r\left(A_k - \omega_p\right) - B_k v\right] \\ &+ q_{A0}^* q_{L0}\left[s\left(\left(A_k+\omega_p\right)\right) + s\left(A_k - \omega_p\right) - B_k u\right] + \frac{1}{2} q_{L0}\left(A_k + \omega_p - B_k\right) \\ &+ \frac{1}{2} q_{A0}^*\left(A_k - \omega_p - B_k\right) \end{aligned} \tag{8.96}$$

Here, A_k and B_k are given by Equations (8.26a) and (8.26b) and the r, v, s, and u functions are defined as:

$$r = \frac{1}{4}\omega_M\left[2(1-N_k)\cos^2\theta_k - 1\right] + \frac{1}{2}Dk^2 \tag{8.97a}$$

$$v = \frac{1}{4}\omega_M \tag{8.97b}$$

$$s = -\frac{1}{4}\omega_M\left[(1-N_k)\sin^2\theta_k - (1+N_k)\right] \tag{8.97c}$$

and

$$u = \frac{1}{2}\omega_M\left[(1-N_k)(3\cos^2\theta_k - 1) - (1+N_k)\right] \tag{8.97d}$$

8.2.5.3 Second-Order Instability – Hexagonal Ferrites

Discussion of the basic expressions for the first- and second-order instabilities excited by a uniform RF field is completed with the expressions for the second-order instability in hexagonal ferrites. Again, the approach obtained in [15] is followed. The threshold for the second-order instability in the bulk approximation can be written in a similar form to the first-order instability earlier in this chapter as:

$$h_{\text{crit}}^{(2)} = 2\cdot\min\left[\sqrt{\frac{M_s\Delta H_k}{\left|Y_k^{(2)}\right|}}\right]_{\omega_k=\omega_p} \tag{8.98}$$

with similar constrains as discussed earlier. The coupling factor $Y_k^{(2)}$ is defined as:

$$\begin{aligned} Y_k^{(2)} = \omega_M\Big\{ & \lambda_k^2\left[q_{L0}^2 f_k + q_{L0}\overset{*}{q}_{A0} g_k + q_{L0}\left(\alpha_x e^{j\delta_x} + j\alpha_y e^{j\delta_y}\right)\right] \\ & + \mu_k^2\left[\overset{*}{q}_{A0}{}^2 f_k + q_{L0}\overset{*}{q}_{A0}\overset{*}{g}_k + \overset{*}{q}_{A0}\left(\alpha_x e^{j\delta_x} - j\alpha_y e^{j\delta_y}\right)\right] \\ & - \lambda_k\mu_k\left[2r_k q_{L0}\overset{*}{q}_{A0} + s q_{L0}^2 + s^* \overset{*}{q}{}_{A0}^2 + \overset{*}{q}_{A0}\left(\alpha_x e^{j\delta_x} + j\alpha_y e^{j\delta_y}\right) + q_{L0}\left(\alpha_x e^{j\delta_x} - j\alpha_y e^{j\delta_y}\right)\right]\Big\} \end{aligned} \tag{8.99}$$

The coefficients are given by:

$$f_k = \mu_0|\gamma| Dk^2 + \omega_M\left[\cos^2\theta_k - A_{zz} - \frac{\left(\tilde{N}_{xx} + \tilde{N}_{yy}\right)}{2}\right] \tag{8.100}$$

$$g_k = \frac{\omega_M}{2}\left[(\tilde{N}_{xx} + 2j\tilde{N}_{xy} - \tilde{N}_{yy} + \sin^2\theta_k e^{2j\varphi_k}) + (A_{xx} + 2jA_{xy} - A_{yy})\right] \tag{8.101a}$$

$$r_k = \frac{\omega_M}{2}\left[2\cos^2\theta_k + 2\tilde{N}_{zz} + (A_{xx} + A_{yy}) - 2A_{zz} - (\tilde{N}_{xx} + \tilde{N}_{yy} + \sin^2\theta_k)\right] \tag{8.101b}$$

and

$$s = -\frac{\omega_M}{2}(\tilde{N}_{xx} - 2j\tilde{N}_{xy} - \tilde{N}_{yy}) \tag{8.101c}$$

The other parameters entering these equations were defined previously. For isotropic ellipsoids of revolution and pure Larmor pumping $|Y_k^{(2)}|$ reduces to the expression obtained by Suhl [13] and the second-order instability threshold field is then

$$h_{\text{crit}}^{(2)} = \frac{\left[(\omega_p - \omega_k)^2 + \eta_k{}^2\right]^{1/4}\left[(\omega_p - \omega_{res})^2 + \eta^2\right]^{1/2}}{\mu_0|\gamma|\left[\frac{\lambda_k{}^2}{2}\left(Dk^2 - \omega_M N_T + \omega_M\frac{k_z{}^2}{k^2}\right)\right]^{1/2}} \tag{8.98a}$$

In the case that $\omega_{res} = \omega_p = \omega_k$, the threshold is:

$$h_{crit}^{(2)} = \frac{\eta\sqrt{\eta_k}}{\mu_0|\gamma|\left[\frac{\lambda_k^{\ 2}}{2}\left(Dk^2 - \omega_M N_T + \omega_M \frac{k_z^{\ 2}}{k^2}\right)\right]^{1/2}} \tag{8.98b}$$

The application of this theory to hexagonal ferrites with in-plane anisotropy can be found in [13, 16].

8.2.5.4 Power Absorption – First-Order Instability

To be able to make a meaningful comparison of the theory with experiment, it is necessary to introduce some relation between the measured absorbed power (peak powers in the case of propagating pulses) and the corresponding spin-wave amplitudes. The power relevant relations for power absorption just above the threshold will be considered here. Nonlinear behavior significantly above the threshold will not be considered here as it is complicated by many processes.

The power absorbed above the threshold level is an important parameter in the operation of FSLs and can be expressed in terms of the imaginary part of the susceptibility, χ'' as:

$$P_{abs} = \omega_p h_p^2 V \chi'' \tag{8.102}$$

where ω_p is the pump frequency in radians/s, V is the ferrite volume, and h_p is the driving RF magnetic field. Discussion so far has been concerned with the calculation of the threshold field strength for instability with nonlinear signal propagation and nonessential nonlinear terms in the equations eliminated. Failure to include these nonlinear terms results in power absorption that would increase indefinitely, which is physically unrealistic. Higher order terms, or in other words interactions between nonlinearly excited spin waves and their feedback on the excited modes, must now be included to investigate the above-threshold state. At least three processes have been identified, which characterize the power absorption and place limits on the power absorbed [1]: (i) reaction of the parametric spin waves on the uniform precession, (ii) phase mechanism, and (iii) nonlinear damping. An excellent discussion of this topic is given in [62].

The reaction of the parametric spin-wave mechanism [1, 11, 13] applies only to perpendicular pumping as parallel pumping does not involve the uniform precession. The amplitude of the uniform precession is a balance between the RF driving field and spin-wave losses at both ω_p and $\omega_p/2$. With subthreshold pumping, spin-wave losses at ω_p (FMR) dominate and losses due to ω_k spin waves are negligible. However, at threshold, the energy transfer to ω_k spin waves become very large, which prevents further growth in the precession amplitude, i.e. the precession angle is "stuck" or "frozen" at its threshold value. The above-threshold energy transfer to spin waves, which can be looked at as a power loss, can be represented by an additional damping parameter, which when inserted into the expression for the uniform precession allows calculation of the above-threshold susceptibility [1, 11, 13]:

$$\chi''_{P>P_{th}} = \chi''_0 \frac{P_{th}}{P}\left[\left(\frac{P}{P_{th}} - 1\right)\left(\frac{\omega_0 - \omega_p}{\eta_0}\right)^2 + \frac{P}{P_{th}}\right]^{1/2} \tag{8.103}$$

where χ''_0 is the imaginary part of the small-signal RF susceptibility, ω_0 is the resonance frequency, and η_0 is the relaxation time defined previously. For subsidiary absorption, far from FMR, Equation (8.103) becomes:

$$\chi''_{P>P_{th}} \approx \chi''_0 \frac{[\omega_0 - \omega_p]}{\eta_0}\frac{P_{th}}{P}\sqrt{\frac{P}{P_{th}} - 1} \tag{8.104}$$

and when FMR and subsidiary absorption coincide, Equation (8.103) becomes:

$$\chi''_{P>P_{th}} = \chi''_0 \frac{P_{th}}{P} \tag{8.105}$$

The measured and calculated variations of $\chi''_{P>P_{th}}/\chi''_0$, with $\frac{P}{P_{th}}$, are shown in Figure 8.25a for an Mn ferrite sphere at 9.3 GHz with subsidiary absorption far from resonance (see Equation 8.104) and in Figure 8.25b for an Mn ferrite disc

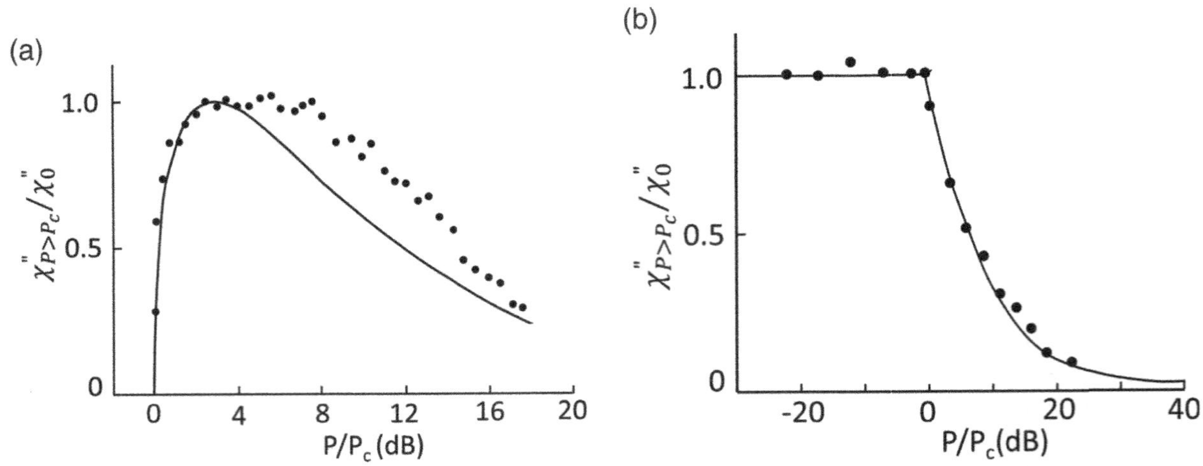

Figure 8.25 Imaginary parts of the susceptibility- above- threshold vs pump power above-threshold for first-order parametric spin-wave excitation under transverse pumping: (a) off-resonance at 9.3 GHz with a Mn ferrite sphere and (b) on resonance at 4.04 GHz with a Mn ferrite disk magnetized in-plane. Full circles represent experimental data [13].

magnetized in-plane at 4.04 GHz where subsidiary absorption coincides with the main resonance (see Equation 8.105). These results mimic the behavior observed in Figure 8.2. The discrepancy between measured and calculated data in Figure 8.25a is due to the additional contribution of the phase mechanism which is discussed next. Substituting for χ''_0 in Equation (8.104) and inserting the result into Equation (8.102) yield an expression for the power absorbed due to the reaction of the spin waves on the uniform precession for subsidiary absorption:

$$P_{\mathrm{abs}}^{\mathrm{react}} = \mu_0 \omega_{\mathrm{p}} h_{\mathrm{c}}^2 V \frac{\omega_{\mathrm{M}}}{(\omega_{\mathrm{p}} - \omega_0)} \sqrt{\left(h / h_{\mathrm{c}} \right)^2 - 1} \tag{8.106}$$

The phase mechanism of limiting the growth of parametric spin waves above-threshold [1] consists of the change in phase shift between the pump and the parametric spin waves, which leads to a decrease in the coupling between the pump and the spin waves as the pump amplitude is increased. This mechanism plays the principal role in parallel pumping, which will be the focus of the discussion here. Expressions have been derived for the real $\chi_c{}'$ and imaginary $\chi_c{}'$ parts of the susceptibility [1, 63] for parallel pumping due to the phase mechanism:

$$\chi_c{}' = \frac{\omega_M}{\left(\sqrt{\left(\omega_p^2 - \omega_M^2 \right)} - \omega_M + N_z \omega_M \right)} \frac{\left(P/P_{th} - 1 \right)}{P/P_{th}} \tag{8.107}$$

$$\chi_c{}'' = \frac{\omega_M}{\left(\sqrt{\left(\omega_p^2 - \omega_M^2 \right)} - \omega_M + N_z \omega_M \right)} \frac{\sqrt{\left(P/P_{th} - 1 \right)}}{P/P_{th}} \tag{8.108}$$

Here, N_z is the demagnetizing factor, which is 1/3 for a sphere, unity for a normally magnetized film, and 0 for a film magnetized in-plane.

The variation of measured and calculated (broken lines) $\chi_c{}'$ and $\chi_c{}''$ as a function of P/P_{th} is shown in Figure 8.26 for a YIG sphere at 9.4 GHz, magnetized along <1 1 1>. The internal field was $H_i = H_r - 8\frac{kA}{m}(100Oe)$ where $H_r = \left(\left(M_s^2 + \left(\frac{\omega}{\mu_0 \gamma} \right)^2 \right)^{1/2} - M_s \right) / 2$. The phase shift mechanism also operates with transverse pumping and is responsible for the difference between measured and calculated data when $P / P_c \geq 5\text{dB}$ in Figure 8.25a. The power absorbed in parallel pumping due to the phase mechanism is given by:

$$P_{\text{abs}}^{\text{phase-mech}} = \mu_0 \omega_{\text{p}} h_{\text{c}}^2 V \frac{\omega_{\text{M}}}{\left(\sqrt{\left(\omega_{\text{p}}^2 + \omega_{\text{M}}^2\right)} - \omega_{\text{M}} + N_z \omega_{\text{M}}\right)} \sqrt{\left(h / h_{\text{c}}\right)^2 - 1} \quad (8.109)$$

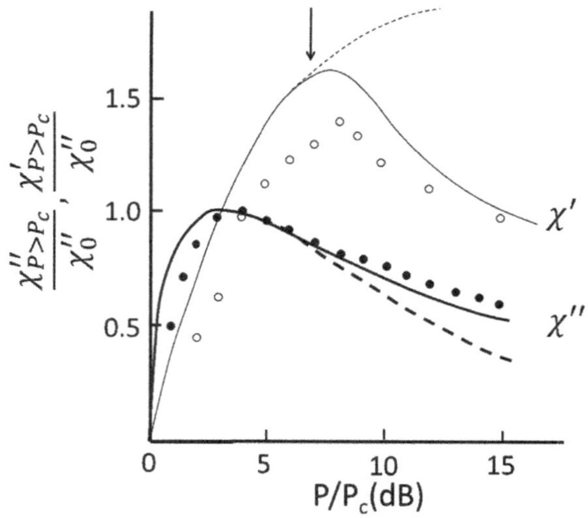

Figure 8.26 Variation of the real and imaginary parts of susceptibility above-threshold vs pump power above-threshold, for parallel pumping. Solid curves include both the phase mechanism and excitation of the second group of spin waves. The broken lines are for the phase mechanism alone. Full circles show experimental data for a YIG sphere at 9.4 GHz [63].

Nonlinear damping is due to the increase in the effective temperature of the magnon (or phonon) modes that interact with the unstable spin waves [64] and the resulting increase in damping of these modes. Studies of the different interactions, three, four, five magnons, etc., showed that different interactions were possible only over specific magnetic field (or frequency) ranges [1, 64, 65]. Here, only the three-magnon interaction will be discussed, which is one of coalescence in which a spin-wave with wavevector k combines with a spin-wave, with wave vector k' to give a spin-wave with $\omega_{k+k'} = \omega_k + \omega_{k'}$ [66]. The full lines in Figure 8.26 show the calculated dependence of the real χ' and imaginary χ'' parts of susceptibility on the pump power for a YIG sphere with the combined action of the phase mechanism and nonlinear damping. Also, shown by the broken line is the dependence of χ' and χ'' on pump power for one magnon pair (phase mechanism above). The threshold of the second magnon pair, $h_2 / h_1 = 2.3 = 7.2dB$, is shown by the arrow with a propagation angle $40^\circ < \theta_{k,2} < 50^\circ$. h_1 and h_2 are the threshold fields for the phase mechanism and the second magnon pair, respectively. The agreement with measured data is good and in the case of χ' changes from a continuous growth in the one magnon pair case to a maximum value followed by a decrease in χ' with further increase in pump power [63]. Depending on the magnetic bias field and frequency, different orders of interaction are possible leading to different dependence on pump power.

8.3 Soliton Propagation in Thin Ferromagnetic Films

Due to the excellent low-loss properties of YIG films, spin-wave excitations can propagate through significant distances within these films. This led to an investigation of the propagation of spin-wave pulse packages and demonstration of microwave devices based on spin-wave propagation. Linear devices utilizing the spin-wave propagation are discussed in Chapter 13. Here, the propagation of nonlinear wave packets with potential applications in a broad range of microwave devices are briefly discussed. The initial focus on nonlinear phenomena with propagating spin-wave packages at microwave frequencies was on the propagation of solitons in thin magnetic films. The interest in this propagating nonlinear excitation was due to their unique property of shape preservation during propagation and collision. This soliton property is well known

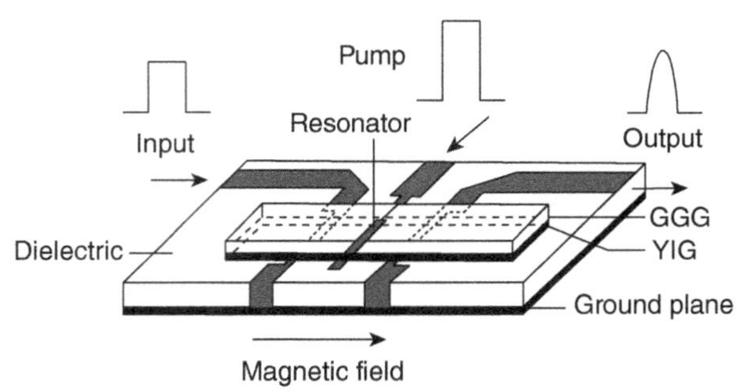

Figure 8.27 Diagram of YIG film and transducer setup for soliton transmission and pumping experiments. The structure consists of two 30 μm wide microstrip transducers and a few tens of micrometers wide microstrip pumping resonator on a single ground plane dielectric substrate. The YIG film on a gadolinium gallium garnet (GGG) substrate is placed across the transducers, with input, output, and parametric pumping signals as indicated. The static magnetic field was parallel to the propagation direction for backward magnetostatic wave excitation. [85]

and observed in shallow waters, nonlinear transmission lines and as nonlinear excitations explored for optical communication. Observation of bright temporal solitons in magnetic films with pinned surface magnetization was first reported in [67] and extended to dipole–exchange modes in [68]. The experiment was performed with a low-loss YIG film, and the spin-wave propagation was in a known delay line arrangement that included 30-μm-wide and 4-mm-long microstrip antennas for excitation and reception (see Figure 8.27). The pump electrode in the middle of the propagation path was not included in initial experiments. The results of several investigations on the propagation and properties of these excitations have been published [69–78]. The solitons are space-localized eigen solutions of the nonlinear dispersive dynamic equation and in the magnetic case they are stable nonlinear excitations of magnetization. One-dimensional propagation is considered initially with the propagation direction in the y-axis, and as discussed previously, the dispersion relation is a function of the static field orientation and the microwave amplitude that is of the first order in the form $\Psi^2 = m^2/2M_s^2$. Following the established procedure [69, 70, 71], it is possible to derive the constitutive nonlinear dynamic equation through an expansion of the dispersion relation into a Taylor series around the working point ($\omega_k = \omega_0, k = k_o, |\Psi|^2 = 0$) in the form:

$$\omega(k,|\Psi|^2) = \omega_0(k_{0y}) + \omega'\Delta k_y + \frac{1}{2}\omega''(\Delta k_y)^2 + \frac{\partial\omega}{\partial|\Psi|^2}|\Psi|^2 + \dots \tag{8.110}$$

Through application of the method of envelopes [72] and with $\Delta\omega \to j\frac{\partial}{\partial t}$ and $\Delta k \to -j\frac{\partial}{\partial y}$ and $\frac{\partial}{\partial t} \to -v_g\frac{\partial}{\partial y}$, Equation (8.110) represents an operator that when applied to a new slow-varying envelope variable, $U(y,t) = \Psi(y,t)\exp(-j(\omega_0 t - k_{0y}y))$ leads to a nonlinear Schrödinger equation in the form [69]:

$$j\left(\frac{\partial U}{\partial t} + v_g\frac{\partial U}{\partial y} + \eta U\right) + \frac{D}{2}\frac{\partial^2 U}{\partial y^2} - N|U|^2 U = 0 \tag{8.111}$$

Here, $v_g = \omega' = \partial\omega/\partial k_y|_{k_{0y}}$ is the group velocity, $\omega'' = D = \partial^2\omega/\partial k_y^2|_{k_{0y}}$ (note that here D is the dispersion and not the exchange constant) the dispersion coefficient, $N = \partial\omega/\partial|U|^2|_{k_{0y}}$ is the so-called nonlinear coefficient and $\eta = \mu_0|\gamma|\Delta H$, is the relaxation rate, which, as before, represents the dissipation through the half linewidth of the FMR. In general Equation (8.111) represents the propagation of a pulse in a one-dimensional dispersive and nonlinear medium. It can be extended to more dimensions and an arbitrary direction of propagation. If the information carried by the propagating pulse is in its positive amplitude, then the signals are called bright. If the pulse propagation is in the form of a dip the signal is labeled as dark. For the present purpose, one-dimensional propagation will provide the necessary insight into for the spin-wave soliton propagation in thin ferromagnetic films. All the parameters introduced depend on the excited spin-wave spectra, which are discussed as a function of the bias static magnetic field orientation with respect to spin-wave propagation and the YIG film surface in Chapter 13.

The nonlinear coefficient N can be expressed for an in-plane magnetized film (surface and backward volume waves) as [73]:

$$N = -\frac{\omega_H\omega_M}{4\omega_B}\left[1 + \frac{\omega_H^2}{\omega_B^2}\right] \tag{8.112}$$

with $\omega_B = \mu_0|\gamma|\sqrt{H_i(H_i + M}$. For MSBVW, $v_g = -\frac{\omega_H\omega_M d}{4\omega_B}$, $D = \frac{5}{48}\frac{\omega_H\omega_M}{4\omega_B}(1 + \frac{3}{5}\frac{\omega_H^2}{\omega_B^2})$. Here, only one-dimensional propagation parallel to **M** along the y-axis is assumed. The condition $DN < 0$ is fulfilled for the selected operating point on the dispersion curve for MSBVW solitons and therefore this configuration supports the propagation of bright solitons in thin ferromagnetic films. MSBVW solitons in thin YIG films were experimentally investigated in [73].

For surface magnetostatic waves in the same approximation, it follows from [71] that $v_g = \frac{\omega_M{}^2 d}{4\omega_B}$, $D = -\frac{\omega_M^4 d^2}{16\omega_B^3}\left(1+8\frac{\omega_B^2}{\omega_M^2}\right)$. In this case, D and N, for a given operation point, have the same sign with the product $D \cdot N > 0$, and therefore, the surface waves do not support the propagation of bright solitons. On the other hand, this configuration supports the propagation of dark solitons (propagation of a dip or depression) that like the bright solitons preserve for a certain distance their shape without distortion.

The amplitude of the dark soliton can be expressed as [74, 75]:

$$U(y,t) = \frac{(D/N)^{1/2}}{v_g T_0}\left[\frac{(1-A^2)^{1/2}}{A} + j\tanh\left(\frac{t-t_0-y/v_g}{T_0}\right)\right] e^{-j(\tilde{\kappa}y - \tilde{\Omega}t)} \tag{8.113}$$

where the soliton group velocity and the other parameters in equation are:

$$v_g = \omega_k' + \kappa D, \quad \tilde{\kappa} = \kappa - \frac{(1-A^2)^{1/2}}{v_g T_0 A} \quad \text{and} \quad \tilde{\Omega} = \kappa\omega_k' + \frac{1}{2}\kappa^2 D + \frac{D(3-A^2)}{2v_g^2 T_0^2 A^2} - \frac{(1-A^2)^{1/2}}{T_0 A} \tag{8.114}$$

κ represents the soliton wave number shift, t_0 is the soliton center position at $y = 0$ and A denotes the dark soliton contrast. A is in within the range $-1 \le A \le 1$.

For forward volume waves (MSFVW), when the film is magnetized perpendicularly to the film surface all directions of propagation within the film are equivalent. In this case, the nonlinear Schrödinger equation has the form:

$$j\left(\frac{\partial U}{\partial t} + v_g\frac{\partial U}{\partial y} + \eta U\right) + \frac{D}{2}\frac{\partial^2 U}{\partial y^2} + S\frac{\partial^2 U}{\partial x^2} - N|U|^2 U = 0 \tag{8.115}$$

where S is the diffraction coefficient. For MSFVW [71], $S = {}^{v_g}\!/_{2k}$, $v_g = {}^{\omega_M d}\!/_{4}$, $D = -\frac{\omega_M{}^2 d^2}{16\omega_H}\left(1+\frac{8}{3}\frac{\omega_H}{\omega_M}\right)$, and $N = \omega_M$. The amplitude of the forward volume wave soliton with unpinned surface spins and small damping can be expressed as [76]:

$$|U(y,t)| = U_0 e^{-2\eta t}\operatorname{sech}\left[U_0 e^{-2\eta t}\sqrt{\left|\frac{N}{D}\right|}(y - v_g t)\right] \tag{8.116}$$

under the assumption that the linear and soliton group velocities are the same. From Equation (8.116), it follows that the initial amplitude at $t = 0$, U_0, and initial pulse width are related as:

$$T_0^2|U_0|^2 = \frac{1}{v_g^2}\left|\frac{D}{N}\right| \tag{8.117}$$

demonstrating the nonlinear characteristic of the soliton. As follows from relations for the dispersion and nonlinear coefficients introduced below Equation (8.115), the product $D \cdot N < 0$ is negative, and therefore, the MSFVW configuration supports the propagation of bright soliton waves. Remember that all the relevant parameters discussed here have to be calculated at the given operating point of interest.

Based on Equations (8.115) and (8.116), it is possible to introduce some important times related to the spin-wave propagation. The propagation time between the excitation and receiving antenna separated by a distance l is:

$$T_p = \frac{l}{v_g} \tag{8.118}$$

The dissipation is represented by the decay time defined as:

$$T_r = \frac{1}{\eta} \tag{8.119}$$

The increase in the width of the propagating wave packet from its initial value T_0 due to dissipation can be, in a good approximation, expressed as a dissipation time [72, 73]:

$$T_d = \frac{{v_g}^2 T_0^2}{2\pi |D|} \tag{8.120}$$

and the nonlinear response time defined as the time for the nonlinear frequency shift to yield 1 radian of phase change between the original low-power carrier signal and the high-power level of interest is [77, 78]:

$$T_{nl} = \frac{\pi}{|N||U_0|^2} \tag{8.121}$$

It is interesting to note that the condition $T_{nl} = T_d$ leads to the condition of Equation (8.117) presented earlier in the chapter. Any pulse (signal) propagating in a dispersive medium will decay and change it shape with distance. To be able to excite stable propagation that conserves the shape of the propagating pulse the dispersion must be compensated by the nonlinearity of the medium. Note that this does not mean that the amplitude will stay constant. The criteria for bright soliton propagation are the Lighthill criterion [79, 68] $D \cdot N < 0$, and the dissipation criteria $|N||U|^2 \gg \mu_0|\gamma|\Delta$ and $Dq^2 >> \mu_0|\gamma|\Delta$, where $q = 2\pi / v_g T_0$ is the wave number of the envelope of the input pulse. (For two-dimensional propagation, an additional condition for the so-called diffraction coefficient S defined as D_d but with the derivative with respect to wave number in the perpendicular to propagation direction must be considered.) The power threshold for the temporal soliton formation is given by [80]:

$$|U|_{th}{}^2 = \frac{\mu_0|\gamma|\Delta H}{|N|} \tag{8.122}$$

The threshold amplitude for the formation of a soliton of the nth eigenvalue from a square input pulse of width T_0 can be expressed as [77, 81] in the limit of the zero damping as:

$$|U_{0,n}|_{th} = \frac{1}{2}\frac{(2n-1)\pi}{|v_g|T_0}\sqrt{\left|\frac{D}{N}\right|} \tag{8.123}$$

And the corresponding velocity of the soliton of order n can be expressed as:

$$v_s^{(n)} = v_g + \frac{(2n-1)\pi D}{v_g T_0} \tag{8.124}$$

Detailed numerical analysis of the soliton phase behavior [82] also shows that when the soliton is formed the phase is constant for all the envelope solitons of n-th order (number) across the soliton power profile. Figure 8.28c–f shows the evolution of the soliton power profiles and the corresponding phase behavior as a function of the input power. Note, as follows from the presented analysis that the evolution of solitons with higher numbers are dependent on the initial width of the excitation pulse T_0 and the initial amplitude of the pulse.

From the soliton amplitude dependence discussed earlier, it follows that in the limit of small damping, the soliton amplitude decays at twice the rate of the linear dissipation coefficient, although it keeps its shape during the propagation in the nonlinear dispersive medium. This limits the potential device applications of solitons in low-loss YIG films to single soliton propagation lifetime on the order of a few hundred nanoseconds. Therefore, there is a question whether it is possible to compensate for the soliton loss during its propagation along the thin film and thus increase the propagation length or the propagation time. It was demonstrated that this could be achieved through a parametric pumping configuration or active feedback loop. Parallel pumping compensation of the soliton propagation loss was proven experimentally in [71, 83], and the experimental configuration is shown in Figure 8.27, with a pump electrode in the middle of the soliton propagation pathway. The amplification of a spectrally narrow backward volume wave packet by parallel pumping was experimentally investigated in [84, 85] and the full theory presented in [80]. Soliton amplification by parallel parametric pumping is shown in Figure 8.28a and b. In this case, the input pulse with a carrier frequency of 4.01 GHz was 22 ns wide and the 8.05 GHz pump pulse was applied midway through the pulse propagation with 5 W of input power. The dashed pulses at $t = 400$ ns represent the pulse profiles with and without pumping. The experiment also demonstrated that it is possible to generate a soliton pulse from a linear propagating pulse with this approach. The detailed theory presented in [84] gives for the gain coefficient G of the pulse propagating through the pumping zone of width ΔW:

$$G = \frac{Kv_g}{\eta \sinh(K\Delta W) + Kv_g \cosh(K\Delta W)} \tag{8.125}$$

with

$$K = \frac{\sqrt{\eta^2 - h^2 V^2}}{v_g}. \tag{8.125a}$$

Here, h is the amplitude of the microwave pumping field and V is the parallel pumping coupling parameter that for the backward volume wave can be expressed as:

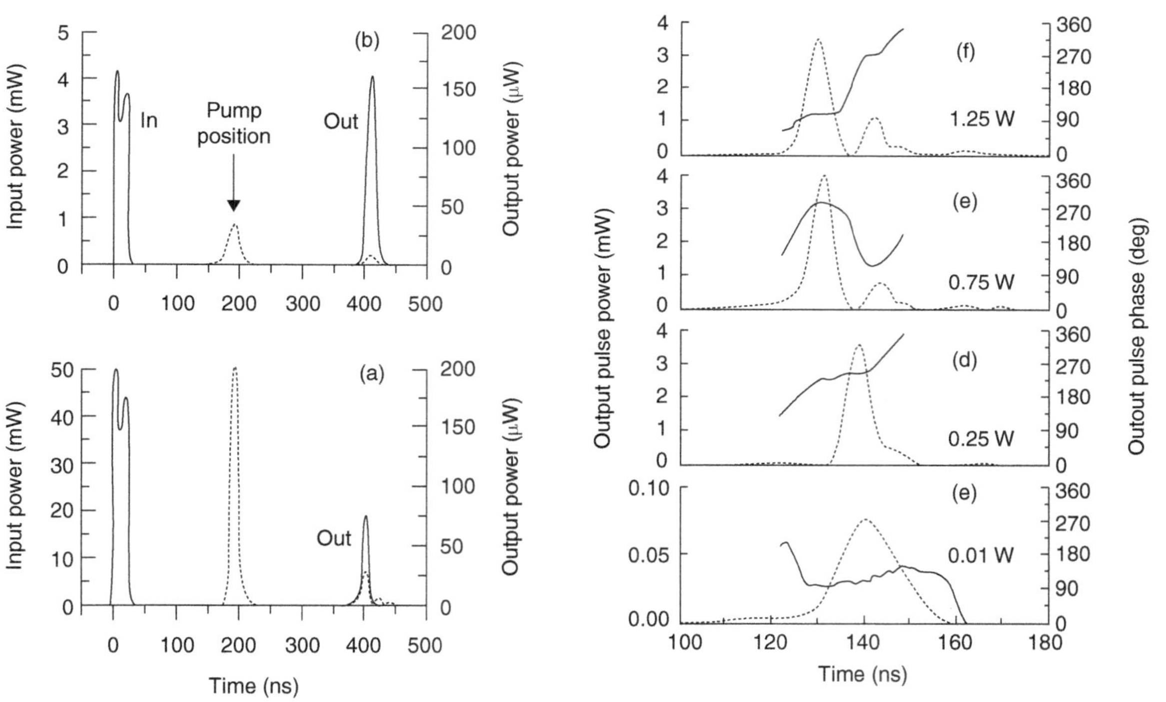

Figure 8.28 Representative soliton pulse data with and without parametric pumping. The 4.01 GHz input pulse was 22 ns wide. The resonator pulse was at 8.05 GHz and 5 W. The dashed lines at 200 ns show the MME pulse profile at the resonator position. The dashed pulses at 400 ns show the output signals without pumping. The solid line pulses at 400 ns show the output signal with pumping. For (a), the input pulse power was 50 mW and the resonator pulse width was 15 ns. For (b), the input pulse power was 4 mW and the resonator pulse width was 23 ns [85]. (c)–(f) Experimental phase and power profiles for an MSBVW signal detected at the output transducer. The YIG film thickness was 8.2 μm. The solid curves show phase, and the dashed curves show power. The input pulse width was 26 ns, the carrier frequency was 5 GHz, and the static field was 87.04 kA/m (1.088 kOe) [82].

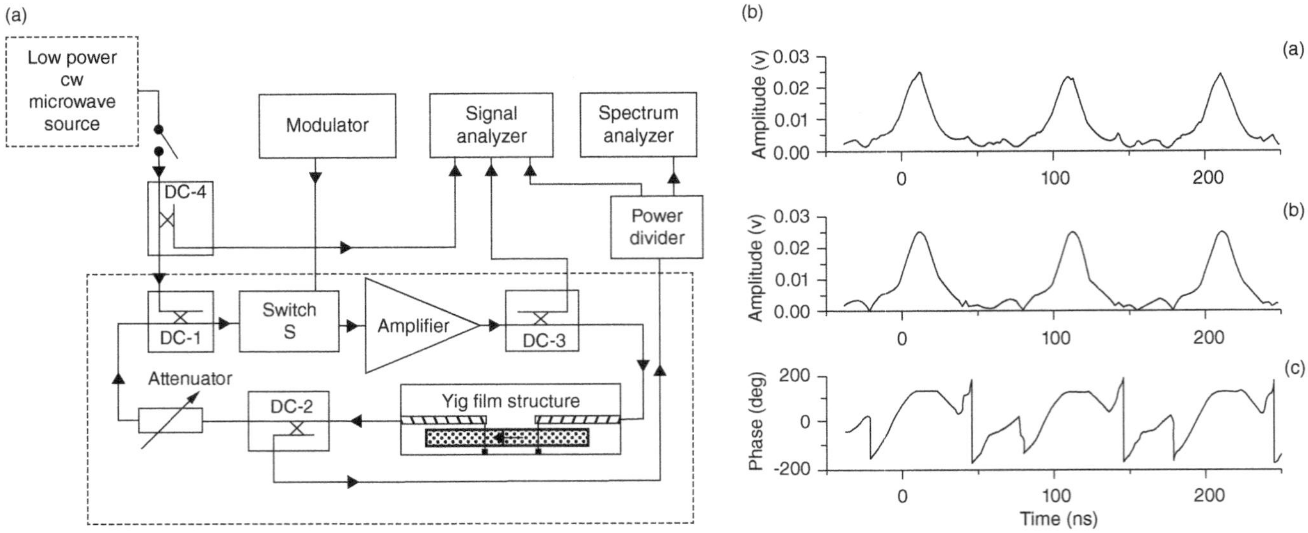

Figure 8.29 (a) Diagram of the YIG film system for ring amplification and self-generation of microwave magnetic envelope solitons and (b) charts. Charts (a) and (b) shows detected microwave voltage vs. time for the signal in the MSBVW ring structure for a 22 ns pulse modulation at a 10 MHz rate with a loop gain of 14 dB. Chart (c) shows the phase profile corresponding to (b). For charts (b) and (c), a 0.2 mW CW microwave synchronization signal at 5.1555 GHz was applied to the ring [86].

$$V = \frac{\omega_M}{8\omega}\frac{1}{kd}\left(1 - e^{-kd}\right) \tag{8.126}$$

The effective pumping zone ΔW, as obtained from Equation (8.125), was calculated to be on the order of 2.3 mm. This is a reasonable value because it is expected that the stray field of a microstrip pumping antenna, a few tens of micrometers wide, can extend well beyond its physical parameters.

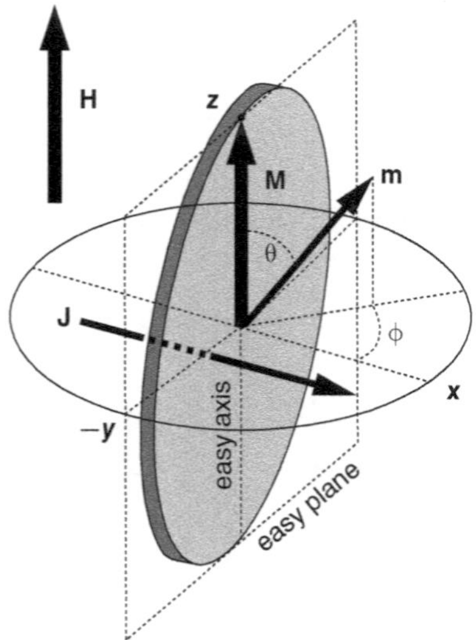

Figure 8.30 The ellipsoidal cross section of the free layer (shaded) lies in the *Y–Z* plane. We represent its magnetization by a macrospin ***m*** that can point in any direction. The fixed layer (not shown) is represented by a fixed macrospin ***MPz*** [90].

Another option for compensation of the soliton decay is to use an active feedback loop [86]. The experimental setup is shown schematically in Figure 8.29, with the amplifier feeding the signal from the output antenna of the YIG structure back to the input antenna. The experiments were done in an MSBVW configuration that supported soliton formation. A stable sequence of soliton pulses with constant amplitude was generated by regular interruption of the feedback circuit in time intervals corresponding to the delay time of the MSBVW signal around the ring (100 ns in the referenced case). The generated sequence is shown in Figure 8.29b. As already mentioned, self-generation of the solitons was obtained when the modulation period was given by:

$$t_m = L / v_g + t_e$$

where L is the separation of the antennas, v_g is the group velocity, and t_e is the electronic delay for the rest of the ring.

8.4 Spin-Torque Nonlinear Excitations in Ferrites

Recent development in nonlinear magnetic excitations in ferrites due to current-induced spin-torque will be addressed in this section. Studies of spin-torque excitations in ferromagnets started with seminal papers [87], [88], and [89], which showed that when spin-polarized current passes through a thin ferromagnetic film, it can introduce magnetization precession or switching of the films. This effect is due to transfer of the spin-torque from the free spin-carrying electrons to the spins of electrons in the lattice of the ferromagnetic metals. This spin-torque transfer can be phenomenologically expressed by a simple generalization of the Landau–Lifschitz equation with Gilbert damping (LLG) that includes a spin-transfer torque term. A thorough investigation of the generalized Landau–Lifschitz equation for the magnetization dynamics in a macrospin model approximation [90] explained many but not all the results obtained in the experiments. It is sufficient for the purpose of this chapter to follow that approach. The configuration for the double-layered structure is shown in Figure 8.30, where the fixed layer is not shown but is represented by a fixed macrospin $\mathbf{M}$ along the z-direction. For the selected configuration, the generalized LLG equation can be expressed as:

$$\frac{d\boldsymbol{u_m}}{dt} = -\mu_0|\gamma|\boldsymbol{u_m}\times[\boldsymbol{H}_{eff}+\boldsymbol{H}_T]+\alpha\boldsymbol{u_m}\times\frac{d\boldsymbol{u_m}}{dt}+\frac{\mu_0\gamma}{\mu_0 M_s}\mathbf{T} \tag{8.127}$$

where $\boldsymbol{u_m} = \dfrac{\boldsymbol{m}}{M_s}$ is the unit vector in the direction of the dynamic magnetization, $\boldsymbol{H}_{eff}$ is the effective field introduced previously, and $\boldsymbol{H}_T$ represents a stochastic effective vector that simulates the influence of the finite temperature. The first two terms in the equation represent the classical LLG terms (if $\boldsymbol{H}_T = 0$). The quantity $\mathbf{T}_{st}$ in Equation (8.127) represents the spin-transfer torque density through the spin valve system considered in Figure 8.30 and represents the concept that a spin-polarized current flowing through the free layer will have a component of the angular momentum that is transverse to the magnetization of the free layer. The transfer component is largely absorbed by the ferromagnet [91, 92]. This absorbed angular momentum produces a torque, $\mathbf{T}_{st}$, in Equation (8.127). This torque can be expressed as [90, 91, 93–95]:

$$\boldsymbol{T}_{st} = \eta(\theta)\frac{\hbar}{2e}\frac{J}{d}\boldsymbol{u_m}\times[\boldsymbol{u_m}\times\boldsymbol{u_M}] \tag{8.128}$$

where $\boldsymbol{u_M} = \dfrac{\boldsymbol{M}}{M_s}$ and $cos\theta = \boldsymbol{u_m}\cdot\boldsymbol{u_M}$. J is the current density, and d is the free layer thickness. $\eta(\theta)$ covers different choices that can be found in [90]:

$$\eta(\theta) = \frac{q}{A+Bcos\theta} \tag{8.129}$$

where q, A, and B are the material and geometric parameters. This equation applies for a symmetric case when the ferromagnetic layer thicknesses are the same. In a general spin valve case [90]:

$$\eta(\theta) = \frac{q_+}{A+\mathrm{B}\cos\theta}+\frac{q_-}{A-\mathrm{B}\cos\theta} \tag{8.130}$$

Equations (8.129) and (8.130) represent the minimal model for the description of the transfer torque. In practical applications, the spin-transfer is not 100%, and in that case, it is necessary to modify η by an additional factor. The additional terms for the extended models are discussed in detail in [90] and the references presented there. Note that this process is nonlinear and that there is a threshold current density (current) that has to be applied to observe the discussed spin-torque effects. This limits the practical applications to magnetic nanostructures so that critical currents are in a reasonable range. Most research in this area has been done on metallic multilayers [96], but the effect was also observed in magnetic tunnel junctions with nonmagnetic spacers [97–100].

8.5 Spin-Hall Effects

Other relevant nonlinear effects in ferromagnetic insulators will be only briefly addressed in this chapter. These processes are based on the so-called spin-Hall or inverse spin-Hall effects. The spin-Hall effect describes the conversion of a longitudinal charge current density J_c into a transverse spin current density $J_S\mathrm{h}/2e$ originating from spin–orbit scattering leading to deflection of conduction electrons with opposite spin orientation into opposite directions. We are dealing with ferromagnetic insulators, and it was shown in [101] that spin waves can be generated and detected in a magnetic insulator through the spin-Hall effect in hybrid magnetic insulator/normal metal structure. Note that no electric current is flowing through the magnetic insulator. The spin-wave generation occurs through the conversion of angular momentum from the electric current, $\boldsymbol{J}_c$, in a normal metal with large spin–orbit coupling such as Pt or Pd, into a pure spin current $\boldsymbol{J}_s$ that can be pumped into the magnetic insulator or vice versa. If the spin current is injected from the magnetic insulator into the normal metal it is converted there into electric voltage through the inverse-spin-Hall effect. Therefore, these hybrid structures may have an important role in potential linear and nonlinear electronic devices discussed in Chapter 13.

Modeling of spin pumping into a magnetic insulator from the attached metal, or vice versa, is also based on the generalized LLG equation discussed earlier. The only difference is that in this case, the last term in Equation (7.127) has to be replaced by the term that represents the spin current injected into the system, which can be expressed by the term:

$$\boldsymbol{T}_{st} = \frac{\hbar}{2ed}\boldsymbol{J}_s \tag{8.131}$$

where $\boldsymbol{J}_s$ is a spin current that has field-like, as well as anti-damping-like, torque terms [102]. As in the case of the spin-torque, there are several different representations of the spin current $\boldsymbol{J}_s$. Models of spin transport introduce the spin mixing conductance $G^{\uparrow\downarrow}$ that describes the magnitude and phase of the interface spin current [103]. The corresponding spin current can be expressed as:

$$\boldsymbol{J}_s = \frac{Re\left(G^{\uparrow\downarrow}\right)}{e}\boldsymbol{M}\times(\boldsymbol{M}\times\boldsymbol{\mu}_s) + \frac{Im\left(G^{\uparrow\downarrow}\right)}{e}(\boldsymbol{M}\times\boldsymbol{\mu}_s) \tag{8.132}$$

$\boldsymbol{\mu}_s$ is the spin accumulation at the magnetic insulator metal interface. A detailed calculation is given in [104]. If it is assumed that the direct spin pumping by the magnetization precession into the neighboring magnetic material excites FMR, the spin current can be expressed as [105]:

$$\boldsymbol{T}_{st} = \frac{C\boldsymbol{J}_s}{d}\left(\boldsymbol{u_m}\times\left[\boldsymbol{\sigma}\times\boldsymbol{u_m}\right]\right) \tag{8.133}$$

$\boldsymbol{\sigma}$ is the direction of the injected spin moment, C is a phenomenological coefficient that takes into account the properties at the normal metal magnetic isolator interface such as the mixing conductance, and $J_s = \eta\theta_{SH}A_{F/N}(\mathrm{h}/e)(J_c/A_N)$ is the density of spin current related to spin-Hall angle, with $A_{F/N}$ the area of interface, A_N the cross section of the normal metal, η the injection efficiency, and J_c the current density in the normal metal. The injection efficiency can be expressed as [106]:

$$\eta = \left[\frac{2\sin\mathrm{h}^2(d_N/2\lambda_N)}{\cosh(d_N/\lambda_N)}\right] / \left[1 + (\lambda_F/\lambda_N)(\sigma_c^N/\sigma_c^F)\frac{\tanh(d_N/\lambda_N)}{\tanh((d_F/\lambda_F)}\right] \tag{8.134}$$

with thicknesses $d_N(d_F)$, the electrical conductivity $\sigma_c^N(\sigma_c^F)$, and the spin diffusion length $\lambda_N(\lambda_F)$ of the normal (N) and magnetic (F) layer. A more detailed calculation of $\boldsymbol{J}_s$ for this case is given in [104]. In an investigation of the voltage generated in a Pt layer by the inverse-spin-Hall effect, the spin current at FMR generated by the spin pumping effect [103], which results in the excitation of FMR and related linear and nonlinear effects was defined as [107]:

$$\boldsymbol{J}_s = \left(\frac{\mathrm{h}}{2e}\right)^2 G^{\uparrow\downarrow}\left[\boldsymbol{u_M}\times\frac{\partial\boldsymbol{u_M}}{\partial t}\right] \tag{8.135}$$

For the interested reader, another detailed calculation of spin–orbit-induced magnetization dynamics in multilayered nanostructures is given in [108]. For practical applications, the important parameter is the critical spin current required to excite the magnetization oscillations. Due to the spin–orbit interaction, the charge current flowing in the plane of a Pt film produces spin current flowing out of its plane. Through this process, a spin angular momentum is transferred through the interface into an insulating oxide such as YIG, observed for the first time in [101]. The authors of [101] also reported that a spin current produced by FMR can be converted into a charge current through the inverse-spin-Hall effect. The reciprocal effect of a spin current produced in Pt and transferred to YIG though the direct spin Hall effect resulting in damping compensation was also predicted but could not be confirmed, at that time. Experimental results on metallic structures [101] showed that the critical current densities necessary for the onset of spin-torque-induced oscillations were about an order of magnitude larger than could be injected into Pt. One of the reasons was the relative large dimensions of the investigated devices. This issue was addressed in [109] where measurement and calculation of the threshold spin current for the onset of auto-oscillations of magnetization were investigated on a 5 μm in diameter patterned disk. The predicted critical spin current is $J_{sc} = \frac{1}{\tau}\frac{dM_s}{\gamma}$, where τ represents the natural relaxation time of the material. The authors demonstrated that the proper device configuration and prediction of the critical spin current can control the spin-wave damping but additional improvements can be expected by further decreasing the sample dimensions. Nevertheless, the utilization of the spin–orbit interactions at the metal/oxide interface opens the possibility of an improved performance of dielectric oxide devices.

In [110], it was shown that this effect also applies for nonlinear excitations. Excited spin waves in subsidiary absorption process were detected by the inverse-spin-Hall effect. The results in the short-wavelength exchange spin-wave region indicated a strong coupling between exchange spin waves and spin currents at the interface due to efficient spin-transfer. The inverse-spin-Hall measurements showed hysteresis effects like those observed in thin films and discussed in [111] for nonlinear device applications.

References

1 Gurevich, A.G. and Melkov, G.A. (1996). *Magnetization Oscillations and Waves.* CRC Press.

2 Kabos, P. (1995). High frequency excitations in thin magnetic films. In: *High Frequency Processes in Magnetic Materials* (ed. G. Srinivasan and A.N. Slavin). World Scientific.

3 Krupicka, S. (1969). Physics of ferrites and magnetic oxides. (in Czech), Academia, Prague., Krupička, S. and Novák, P. (1982). Chapter 4 Oxide Spinels. In: *Handbook of Ferromagnetic Materials 3* pp.189–304 (ed. E.P. Wohlfarth). Elsevier.

4 Chen, M. and Patton, C. (1994). Spin wave instability processes in ferrites. In: *Nonlinear Phenomena and Chaos in Magnetic Materials* (ed. P.E. Wigen), 33–813. Singapore: World Scientific Publishing Company, Ltd.

5 Landau, L.D. and Lifshitz, E.M. (1935). On the theory of the dispersion of magnetic permeability in ferromagnetic bodies. *Physikalische Zeitschrift der Sowjetunion* 8: 153–169.

6 Kittel, C. (1947). Interpretation of the anomalous larmor frequencies in ferromagnetic resonance experiment. *Physical Review* 71: 270; Kittel, C. (1948). On the theory of ferromagnetic resonance absorption. *Physical Review* 73: 155.

7 Gilbert, T.L. (2004). A phenomenological theory of damping in ferromagnetic materials. *IEEE Transactions on Magnetics* 40: 3443–3449.

8 Bloch, F. (1996). Nuclear induction. *Physical Review* 70: 460–474.

9 Bloembergen, N. (1950). On the ferromagnetic resonance in nickel and supermalloy. *Physical Review* 78: 572–580.

10 Schlömann, E. (1959). Raytheon Technical Report No. R-48, Research Division, Raytheon Company; Schlömann, E., Green, J.J., and Milano, U. (1960). Recent developments in ferromagnetic resonance at high power levels. *Journal of Applied Physics* 31: 386S–395S.

11 Damon, R.W. (1953). Relaxation effects in ferromagnetic resonance. *Reviews of Modern Physics* 25: 239; Damon, R. W. (1963). Ferromagnetic resonance at high power. *Magnetism* 1, (ed. G.T. Rado and H. Suhl). New York: Academic Press.

12 Bloembergen, N. and Wang, S. (1954). Relaxation effects in para-and ferromagnetic resonance. *Physical Review* 93: 72–83.

13 Suhl, H. (1956). The nonlinear behavior of ferrites at high microwave signal levels. *Proceedings of the IRE* 44(10): 1270–1284

14 Suhl, H. (1957). The theory of ferromagnetic resonance at high signal powers. *Journal of Physics and Chemistry of Solids* 1(4): 209–227.

15 Nazarov, A.V., Patton, C.E., Cox, R.G. et al. (2002). General spin wave instability theory for anisotropic ferromagnetic insulators at high microwave power levels. *Journal of Magnetism and Magnetic Materials* 248: 164–180.

16 Patton, C.E. (1975). Chapter 10, Magnetic Oxides, D. J. Craik, editor (John Wiley & Sons, London,-New York, 575–648.

17 Patton, C.E. (1979). Spin-wave instability theory in cubic single crystal magnetic insulators I. General theory. *Physica Status Solidi B* 92: 211–221; Patton, C.E. (1969). Theory for the first-order-spin-wave instability threshold in ferromagnetic insulators of ellipsoidal shape with arbitrary pumping configuration. *Journal of Applied Physics* 40: 2837–2843.

18 Kogekar, M.V., Kabos, P., and Patton, C.E. (1984). Spin-wave instability theory for hexagonal ferrites. *Journal of Applied Physics* 55: 2524–2526.

19 Holstein, T. and Primakoff, H. (1940). Field dependence of the intrinsic domain magnetization of a ferromagnet. *Physical Review* 58: 1098–1113.

20 Patton, C.E. and Jantz, W. (1979). Anomalous subsidiary absorption in single crystal YIG and evaluation of spin-wave linewidth. *Journal of Applied Physics* 50: 7082–7088.

21 Sparks, M., Loudon, R., and Kittel, C. (1961). Ferromagnetic relaxation. I. Theory of the relaxation of the uniform precession and the degenerate spectrum in insulators at low temperatures. *Physical Review* 122: 791–802.

22 Chen, M. and Patton, C.E. (1991). First-order instability theory for magnetostatic modes in ferromagnetic spheres. *Journal of Applied Physics* 69: 5724.

23 Bryant, P.H. (1989). Parametric excitation of walker modes in ferromagnetic spheres. *Physical Review B* 39: 4363.

24 Wiese, G. and Benner, H. (1990). Multistability and chaos by parametric excitation of longwave modes in a YIG sphere. *Z. Phys. B* 79: 119–131.

25 Wiese, G. (1991). Parametric excitation of spherical modes in ferromagnetic spheres by perpendicular and parallel pumping. *Zeitschrift für Physik B Condensed Matter* 82: 453–463.

26 Morgenthaler, F.R. (1960). Survey of ferromagnetic resonance in small ferrimagnetic ellipsoids . *Journal of Applied Physics* 31: 95S–97S.

27 Green, J.J., Patton, C.E., and Stern, E. (1969). Threshold microwave-field amplitude for the unstable growth of spin waves under oblique pumping. *Journal of Applied Physics* 40: 172–177.

28 Patton, C.E. (1970). Influence of spin-wave linewidth on the instability threshold for oblique pumping. *Journal of Applied Physics* 41: 431–433.

29 Liu, Y.H. and Patton, C.E. (1982). Spin-wave instability threshold in single-crystal yttrium iron garnet for oblique pumping. *Journal of Applied Physics* 53: 5116–5121.

30 Cherepanov, V.B. and Slavin, A.N. (1995). Size effects in parametric excitation of spin waves in ferrites. In: *High Frequency Processes in Magnetic Materials* (ed. G. Srinivasan and A.N. Slavin), 250. World Scientific.

31 Jantz, W. and Schneider, J. (1971). High resolution spin wave spectroscopy in YIG. *Solid State Communications* 9: 69; Jantz, W., Schneider, J., and Andlauer B. (1972). Fine structure of the parallel pump instability in YIG. *Solid State Commun.* 10: 937; Jantz, W. and Schneider, J. (1975). Fine structure of the subsidiary absorption in YIG. *Physica Status Solidi a* 31: 595.

32 Walker, L.R. (1957). Magnetostatic modes in ferromagnetic resonance. *Physical Review*105: 390–399.

33 Fletcher, P. and Bell, R.O. (1959). Ferrimagnetic resonance modes in spheres. *Journal of Applied Physics* 30: 687–698.

34 Röschmann, P. and Dötsch, H. (1977). Properties of magnetostatic modes in ferrimagnetic spheroids. *Physica Status Solidi B* 82: 11–57.

35 Schlömann, E. and Joseph, R.I. (1961). Instability of spin waves and magnetostatic modes in a microwave magnetic field applied parallel to the dc field. *Journal of Applied Physics* 32: 1006–1014.

36 Wilber, W.D., Booth, J.G., Patton, C.E. et al. (1988). Light scattering observation of anomalous parametric spin wave character in subsidiary absorption. *Journal of Applied Physics* 64: 5477–5479.

37 Kabos, P., Wiese, G., and Patton, C.E. (1994). Measurement of spin wave instability magnon distribution for subsidiary absorption in yttrium iron garnet films by Brillouin light scattering. *Physical Review Letters* 72: 2093–2096.

38 Wiese, G., Kabos, P., and Patton, C.E. (1993). Subsidiary absorption spin wave instability processes in yttrium iron garnet thin films, critical modes, and the "kink" effect. *Journal of Applied Physics* 74: 1218–1228.

39 Wiese, G., Kabos, P., and Patton, C.E. (1995). Subsidiary-absorption spin-wave-instability processes in yttrium iron garnet thin films: Coupled standing modes, critical modes, and the kink effect. *Physical Review B* 51 (15): 85–102.

40 Wiese, G. (1968). Theory for the first order spin wave instability threshold in ferromagnetic insulating thin films. *Z. Phys. B* 91: 57–64.

41 Weber, R. (1968). Spin –wave resonance. *IEEE Transactions on Magnetics* 4: 28–31.

42 An, S.-Y., Krivosik, P., Kraemer, M.A. et al. (2004). High power ferromagnetic resonance and spin wave instability processes in permalloy thin films. *Journal of Applied Physics* 96: 1572–1579.

43 Wiese, G., Buxman, L., Kabos, P. et al. (1994). Parallel pumping fine structure at 9.4 GHz for in-plane magnetized yttrium iron garnet thin films. *Journal of Applied Physics* 75: 1041–1046.

44 Seshadri, S.R. (1970). Surface magnetostatic modes of a ferrite slab. *Proceedings of the IEEE* 58: 506–507.

45 Kalinikos, B.A., Kovshikov, N.G., and Kozhus, N.V. (1984). Determination of the instability threshold of standing spin waves in yttrium iron garnet films subjected to longitudinal lumping. *Soviet Physics: Solid State* 26: 1735–1736.

46 Kostylev, M.P., Kalinikos, B.A., and Dötsch, H. (1995). Parallel pump spin wave instability threshold in thin ferromagnetic films. *Journal of Magnetism and Magnetic Materials* 145: 93–110.

47 Kalinikos, B.A. and Ustinov, A.B. (2013). Nonlinear spin waves in magnetic films and structures: Physics and devices. Chapter 7. In: *Recent Advances in Magnetic Insulators-From Spintronics to Microwave Applications* (ed. M. Wu, A. Hoffmann, R.E. Camley, and R.L. Stamps), 194. Academic Press Elsevier.

48 Vendik, O.G., Kalinikos, B.A., and Chartorizhshkii, D.N. (1977). Instability of spin waves in tangentially magnetized ferromagnetic films. *Soviet Physics: Solid State* 19: 222–227.

49 Kalinikos, B.A. (1980). Excitation of propagating spin waves in ferromagnetic films. *IEE Proceedings-H* 127: 4–10.

50 Kalinikos, B.A. and Slavin, A.N. (1991). First order parametric instability of quasi-surface spin waves in ferromagnetic films. *IEEE Transactions on Magnetics* 27: 5444–5446.

51 Nazarov, A.V., Cox, R.G., and Patton, C.E. (2001). General spin wave instability theory. *IEEE Transactions on Magnetics* 37: 2380–2382.

52 Wigen, P.E. (1994). *Nonlinear Phenomena and Chaos in Magnetic Materials*. World Singapore: Scientific Publishing Company, Ltd.

53 Melkov, G.A. (1976). Size effects in parametric excitation of spin waves in ferrites. *Soviet Physics: JETP* 43: 680–693.

54 Melkov, G.A. and Sholom, S.V. (1991). Kinetic instability of spin waves in thin ferrite films. *Soviet Physics: JETP* 72: 341–346.

55 Melkov, G.A. and Sholom, S.V. (1987). Parametric excitation of spin waves by local pumping. *Soviet Physics Solid State* 29: 1870–1872; Cottam, M.G. and Slavin, A.N. (1994). Chapter 1. In: *Linear and Nonlinear Spin Waves in Magnetic Films*, 64. Singapore: World Scientific.

56 Zil'berman, P.E., Golubev, N.S., and Temiryazev, A.G. (1990). Parametric excitation of spin waves by spatially localized pumping of tangentially magnetized yttrium iron Garnet Films. *Soviet Physics: JETP Letters* 70: 353–358.

57 Gulyaev, Y.V., Zil'berman, P.E., and Lugovsko, A.V. (1997). Parametric excitation of spin waves in ferromagnets by longitudinal pumping localized in space. *Soviet Physics: JETP Letters* 84: 109–119.

58 Kalinikos, B.A. and Slavin, A.N. (1986). Theory of dipole–exchange spin wave spectrum for ferromagnetic films with mixed exchange boundary conditions. *Journal of Physics C* 19: 7013–7033.

59 Zhang, Y.T., Patton, C.E., and Srinivasan, G. (1988). The second-order spin wave instability threshold in single-crystal yttrium iron garnet films under perpendicular pumping. *Journal of Applied Physics* 63: 5433–5438.

60 Kabos, P., Patton, C.E., Wiese, G. et al. (1996). Butterfly curves and critical modes for second-order spin-wave instability processes in yttrium iron garnet films. *Journal of Applied Physics* 80: 3962–3971.

61 Olson, H.M., Krivosik, P., Srinivasan, K. et al. (2007). Ferromagnetic resonance saturation and second order Suhl spin wave instability processes in thin Permalloy films. *Journal of Applied Physics* 102: 023904.

62 Lvov, V.S. (1994). *Wave Turbulence under Parametric Excitation: Application to Magnetics*. Springer Verlag.

63 Zautkin, V.V., Zakharov, V.E., L'vov, V.S. et al. (1962). Parallel spin-wave pumping in yttrium garnet single crystals. *Soviet Physics: JETP* 35: 926–933.

64 Schlömann, E. (1962). Longitudinal susceptibility of ferromagnets in strong rf fields. *Journal of Applied Physics* 33: 527–534.

65 Gottlieb, P. and Suhl, H. (1962). Saturation of ferrimagnetic resonance with parallel pumping. *Journal of Applied Physics* 33: 1508–1514.

66 Melkov, A. and Sholom, S.V. (1990) *Amplification of surface magnetostatic waves by a parametric pump. Soviet Physics-Technical Physics* 35: 943–946.

67 Kalinikos, B.A., Kovshikov, N.G., and Slavin, A.N. (1983). Observation of spin wave solitons in ferromagnetic films . *Soviet Physics: JETP Letters* 38: 413–417.

68 Kalinikos, B.A., Kovshikov, N.G., and Slavin, A.N. (1988). Envelope solitons and modulation instability of dipole exchange magnetization waves in yttrium iron garnets. *JEPT Letters* 69: 303–312; Kalinikos, B.A., Kovshikov, N.G., and Slavin, A.N. (1991). Envelope solitons of highly dispersive and low dispersive spin waves in magnetic films. *Journal of Applied Physics* 69: 5712–5717.

69 Karpman, V.I. and Krushkal, E.M. (1969). Modulated waves in nonlinear dispersive media. *Soviet Physics: JETP Letters* 28: 277–281.

70 Lukomskii, V.P. (1983). Contribution of the theory of diffraction of wave beams in anisotropic media. *Soviet Physics: JETP Letters* 57: 297–303.

71 Zvezdin, A.K. and Popkov, A.F. (1983). Contribution to the nonlinear theory of magnetostatic spin waves. *Soviet Physics: JETP Letters* 57: 350–355.

72 Whitham, G.B. (1974). *Linear and Nonlinear Waves, 367*. John Wiley&Sons Inc.

73 Chen, M., Tsankov, M.A., Nash, J.M. et al. (1994). Backward-volume-wave microwave-envelope solitons in yttrium iron garnet films. *Physical Review B* 49: 12773–12790.

74 Chen, M., Tsankov, M.A., Nash, J.M. et al. (1993). Microwave magnetic-envelope dark solitons in yttrium iron garnet thin films. *Physical Review Letters* 70: 1707–1710.

75 Hasegawa, A. and Tappert, F. (1973). Transmission of stationary nonlinear optical pulses in dispersive dielectric fibers. II. Normal dispersion. *Applied Physics Letters* 23: 171–172.

76 Tsankov, M.A., Chen, M.A., and Patton, C.E. (1994). Forward volume wave microwave envelope solitons in yttrium iron garnet films: Propagation, decay and collision. *Journal of Applied Physics* 76: 4274–4289.

77 Kovshikov, N.G., Kalinikos, B.A., Patton, C.E. et al. (1996). Formation, propagation, reflection, and collision of microwave envelope solitons in yttrium iron garnet films. *Physical Review B* 54: 15210–15223.

78 Staudinger, R.A., Kabos, P., Xia, H. et al. (1998). Calculation of the formation time for microwave magnetic envelope solitons. *IEEE Transactions on Magnetics* 34: 2334–2338.

79 Lighthill, M.J. (1965). Contributions to the theory of waves in non-dispersive systems. *Journal of the Institute of Mathematics and Its Applications* 1: 269–306.

80 Demokritov, S.O., Hillebrands, B., and Slavin, A.N. (2001). Brillouin light scattering studies of confined spin waves: Linear nonlinear confinement. *Physics Reports* 348: 441–489.

81 Xia, H., Kabos, P., Staudinger, R.A. et al. (1998). Velocity characteristics of microwave-magnetic-envelope solitons. *Physical Review B* 58: 2708–2715.

82 Nash, J.M., Kabos, P., Staudinger, R. et al. (1998). Phase profiles of microwave magnetic envelope solitons. *Journal of Applied Physics* 83: 2689–2699.

83 Kalinikos, B.A., Kovaleva, M.K., and Kovshikov, N.G. (1979). Avtorskoe Svidetel'stvo, USSR Patent No. 778606.

84 Kalinikos, B.A. and Kostylev, M.P. (1997). Parametric amplification of spin wave envelope solitons in ferromagnetic films by parallel pumping. *IEEE Transactions on Magnetics* 33: 3445–3447.

85 Kolodin, P.A., Kabos, P., and Patton, C.E. (1998). Amplification of microwave magnetic envelope solitons in thin yttrium iron garnet films by parallel pumping. *Physical Review Letters* 80: 1976–1979.

86 Kalinikos, B.A., Kovshikov, N.G., and Patton, C.E. (1998). Self-generation of microwave magnetic envelope soliton trains in yttrium iron garnet thin films. *Physical Review Letters* 80: 4301–4304.

87 Slonczewski, J.C. (1996). Current-driven excitation of magnetic multilayers. *Journal of Magnetism and Magnetic Materials* 159: L1–L7.

88 Slonczewski, J.C. (1999). Excitation of spin waves by an electric current. *Journal of Magnetism and Magnetic Materials* 195: L261–L268.

89 Berger, L. (1996). Emission of spin waves by a magnetic multilayer traversed by a current. *Physical Review B, Condensed Matter* 54: 9353–9358.

90 Xiao, J., Zangwill, A., and Stiles, M.D. (2005). Macrospin models of spin transfer dynamics. *Physical Review B* 72: 014446.

91 Stiles, M.D. and Zangwill, A. (2002). Anatomy of spin-transfer torque. *Physical Review B* 66: 014407-1–014407-14.

92 Xia, K., Kelly, P.J., Bauer, P.E.W. et al. (2002). Spin torques in ferromagnetic/normal-metal structures. *Physical Review B* 65: 220401-1–220401-4(R).

93 Slonczewski, J. (2002). Currents and torques in metallic magnetic multilayers. *Journal of Magnetism and Magnetic Materials* 247: 324–338.

94 Kovalev, A.A., Brataas, A., and Bauer, G.E.W. (2002). Spin transfer in diffusive ferromagnet–normal metal systems with spin-flip scattering. *Physical Review B* 66: 224424-1–224424-7.

95 Xiao, J., Zangwill, A., and Stiles, M.D. (2004). Boltzmann test of Slonczewski's theory of spin-transfer torque. *Physical Review B* 70: 172405-1–172405-4.

96 Rippard, W.H., Pufall, M.R., Kaka, S. et al. (2004). Direct-current induced dynamics in Co90Fe10/Ni80Fe20 point contacts. *Physical Review Letters* 92: 027201-1-4.

97 Slonczewski, J.C. (1989). Conductance and exchange coupling of two ferromagnets separated by a tunneling barrier. *Physical Review B* 39: 6595–7002.

98 Heiliger, C. and Stiles, M.D. (2008). Ab initio studies of the spin-transfer torque in magnetic tunnel junctions. *Physical Review Letters* 100: 186805-1-4.

99 Deac, A.M., Fukushima, A., Kubota, H. et al. (2008). Bias-driven high-power microwave emission from MgO-based tunnel magnetoresistance devices. *Nature Physics* 4: 803–809.

100 Zeng, Z.M., Khalili Amiri, P., Rowlands, G. et al. (2011). Effect of resistance-area product on spin-transfer switching in MgO-based magnetic tunnel junction memory cells. *Applied Physics Letters* 98: 072512-1-3.

101 Kajiwara, Y., Harii, K., Takahashi, S. et al. (2010). Transmission of electrical signals by spin-wave interconversion in a magnetic insulator. *Nature* 464: 262–266.

102 Jungfleisch, M.B., Ding, J., Zhang, W. et al. (2017). Insulating nanomagnets driven by spin torque. *Nano Letters* 17: 8–14.

103 Tserkovnyak, Y., Brataas, A., and Bauer, G.E.W. (2002). Spin pumping and magnetization dynamics in metallic multilayers. *Physical Review B* 66: 224403-1–10.

104 Chiba, T., Bauer, G.E.W., and Takahashi, S. (2014). Current-induced spin-torque resonance of magnetic insulators. *Physical Review Applied* 2: 034003-1–6.

105 Wang, Z., Sun, Y., Wu, M. et al. (2011). Control of spinwaves in a thin film ferromagnetic insulator through interfacial spin scattering. *Physical Review Letters* 107: 146602-1-5.

106 Ando, K., Takahashi, S., Harii, K. et al. (2008). Electric manipulation of spin relaxation using the spin Hall effect. *Physical Review Letters* 101: 036601-1-4.

107 Hahn, C., de Loubens, G., Viret, M. et al. (2013). Detection of microwave spin pumping using the inverse spin Hall effect. *Physical Review Letters* 111: 217204-1-5.

108 Korostil, A. and Krupa, M. (2016). Spin-orbit induced dynamics in multilayer nanostructures. *American Journal of Nanosciences* 2: 8–20.

109 Hamadeh, A., d'Allivy Kelly, O., Hahn, C. et al. (2014). Full control of the spin-wave damping in a magnetic insulator using spin-orbit torque. *Physical Review Letters* 113: 197203-1-5.

110 Ando, K. and Saitoh, E. (2012). Spin pumping driven by bistable exchange spin waves. *Physical Review Letters* 109: 026602-1-5.

111 Fetisov, Y.K., Patton, C.E., and Synogach, V.T. (1999). Nonlinear ferromagnetic resonance and foldover in yttrium iron garnet thin films—inadequacy of the classical model. *IEEE Transactions on Magnetism* 35: 4511–4521.

9

Chemical Processing and Magnetic Properties of Ferrite Nanoparticles

Sarah E. Smith, Melissa Tsui, Brent Williams, and Everett E. Carpenter

Chemistry Department, Virginia Commonwealth University, Richmond, VA 23284, USA

9.1 Introduction

In recent years, the scientific literature has been replete with reports of magnetic nanoparticle synthesis, properties, and novel applications. This has largely been driven by hope and, in some instances, realization of surface-enhanced properties. By producing materials on nanoscale, there exists the possibility that surface-based properties, for example, catalysis, magnetism, optical, and other electronic properties may be enhanced. In the case of magnetic nanoparticles, the disruption of the crystal lattice at the surface greatly influences magnetic properties. The truncation of the lattice at the surface disrupts magnetic coupling between cations and neighboring particles. In the case of magnetic oxide systems, such as ferrites, magnetic superexchange relies upon cation–anion–cation bond angles that are disrupted by surface truncation. The truncation and bond bending commonly experienced at surfaces can produce profound changes in magnetic properties of nanoparticles and solids consisting of nanograins. Notwithstanding, the hope for surface-enhanced properties fuels much of nanoscience research carried out today.

Magnetite is the prototypical spinel ferrite and comprises iron and oxygen in the formula of $FeO \cdot Fe_2O_3$ or more commonly Fe_3O_4. It has a fairly high saturation magnetization with semiconducting properties owing to Verwey's insulator–metal (semiconductor) transition occurring near 124 K (for bulk). It is very stable and, therefore, lends itself well to nanoscale applications. It is noteworthy that the work of Bohra et al. [1]. revealed that the Verway transition depends sensitively upon the sample strain state and size (surface state), as demonstrated in thin films and nanoparticles, and can be shifted many degrees.

Figure 9.1 is a photograph of a unique specimen of natural Fe_3O_4 discovered in Bolivia, revealing large crystallites of symmetry mimicking that of the atomic symmetry. Such behavior derives from the preferential diffusivity of principle atomic planes leading to pronounced facets.

Additionally, magnetization and resistivity, which can be tailored by replacing some of the iron cations with other divalent cations, manganese, nickel, and/or zinc, are popular choices [1]. The science of doping magnetite with these cations is well understood in bulk compounds; however, this is not the case at the nanoscale. Although a variety of methods have been successfully employed to produce nanosized ferrites, the inclusion of substitutional cations has produced inconsistent results in reported properties. Some of these discrepancies are due to variations in shape and morphology, while others are the result of differences in cation valency and distribution within the crystalline lattice.

This chapter focuses on the design of ferrite nanoparticles and in particular reports details of their processing methodologies. A general review of the chemical methods that have been employed for ferrite synthesis is provided. The analysis of these materials is presented with a discussion regarding the influence of reaction variables on the final products and their functional properties.

9.2 Ferrites

Ferrites are a class of metallic oxides containing at least 50% iron oxide by mass [3]. These materials have various crystal structures, including spinel, hexagonal (magnetoplumbites and ferroxplana), and garnets [4]. The discussion presented here will be confined to the most commonly used ferrites that are based on the structure of the naturally occurring mineral magnetite. Magnetite has the spinel crystal structure, named for the crystal structure of the mineral spinel, $MgAl_2O_4$, and has the chemical formula $FeO \cdot Fe_2O_3$, more commonly expressed as Fe_3O_4. In the spinel crystal structure, oxygen ions form a face-centered

Modern Ferrites: Basic Principles, Processing and Properties, Volume 1, First Edition. Edited by Vincent G. Harris.

Figure 9.1 Cerro Huañaquino, Potosí Department, Bolivia. A collection of magnetite specimens from Bolivia with exceptional size and morphology. Crystals measure to >1 cm on edge and >1.5 cm from tip to tip [2].

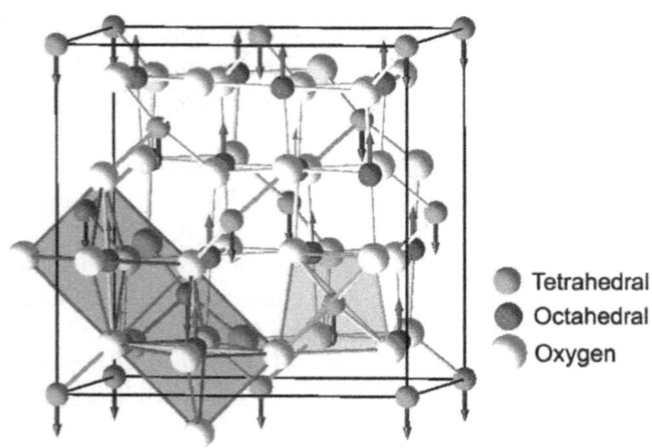

Figure 9.2 The complete unit cell for the spinel crystal structure with tetrahedral and octahedral cations shown in representative polyhedra (after West [3]).

cubic (f.c.c.)*** lattice with transition metal cations residing on the fourfold (i.e. tetrahedra) or sixfold coordinated (i.e. octahedra) interstices, as shown in Figure 9.2.

A complete unit cell of the spinel lattice contains eight AB_2O_4 formula units. The packing efficiency of an f.c.c. array is 74%, leaving available space to accommodate interstitial ions. There are two types of interstitial sites between the oxygen atoms in this type lattice: tetrahedral, which is fourfold coordinated sites, referred to as **A** sites, and octahedral, which is sixfold coordinated sites, known as **B** sites [5]. In the spinel structure, there are 64 **A** sites and 32 **B** sites. However, only one-eighth of the **A** sites and one-half of the **B** sites are occupied by cations. The cations are larger in diameter than the interstitial sites, so their presence slightly distorts the surrounding oxygen ions from their ideal lattice positions.

The type, valence, and distribution of cations strongly influence the material's magnetic and electronic properties. Therefore, by doping magnetite with transition metal cations, it is possible to substantially change intrinsic properties. The generic chemical formula of a doped ferrite is $MeFe_2O_4$, in which Me is normally a divalent first row transition metal ion, although Li, Mg, Cu, and Cd have also found utility.

Spinels can be classified as *normal*, *inverse*, or *mixed*. The *normal* spinel has only divalent cations residing on the eight **A** sites, with trivalent ions filling the 16 available **B** sites. An example of a *normal* ferrite is the Zn spinel ferrite ($[Zn^{II}]\{Fe^{III}\}_2O_4$), where divalent Zn fill the **A** sites and trivalent Fe fill the **B** sites. Zinc preferentially occupies **A** sites. Because divalent zinc has no free valence electrons (i.e. d^{10}), its inclusion increases the imbalance of moments between **A** and **B** sublattices, providing an overall increase to the overall magnetization. However, magnetic coupling of the **A** sites with the **B** sites occurs via superexchange (J_{AB}) through neighboring oxygen anions. As such, there is a limit to the enhanced magnetization by the continued doping with zinc. Studies have shown that the magnetic moment begins to decline once the inclusion of nonmagnetic zinc degrades J_{AB} at values greater than ~45% [4].

Alternatively, when divalent ions fill 8 of the 16 **B** sites with trivalent ions occupying the remaining **A** and **B** sites, an *inverse* spinel result. An example of an inverse spinel is Ni-ferrite ($[Fe^{III}]\{Ni_{0.5}{}^{II}Fe_{0.5}{}^{III}\}_2O_4$). Other popular inverse spinels are ($[Fe^{III}]\{Fe_{0.5}{}^{II}Fe_{0.5}{}^{III}\}_2O_4$) or Fe_3O_4, ($[Fe^{III}]\{Cd_{0.5}{}^{II}Fe_{0.5}{}^{III}\}_2O_4$), ($[Fe^{III}]\{Cu_{0.5}{}^{II}Fe_{0.5}{}^{III}\}_2O_4$), and ($[Fe^{III}]\{Mg_{0.5}{}^{II}Fe_{0.5}{}^{III}\}_2O_4$). As will be shown, these designations are not strictly adhered to as the often nonequilibrium nature of ferrite processing lead to a degree of cation disorder in which cations redistribute from their equilibrium sites.

Finally, *mixed* spinels have different cations of mixed valence occupying both **A** and **B** sublattices. Because each species of ion has thermodynamically preferred lattice distributions based predominantly on ionic radii, electrostatic energy, and electronic configuration, this class of ferrite commonly results from nonequilibrium processing such as those that involve physical vapor quenching (e.g. pulsed laser deposition and sputter deposition) and high kinetic energy transfer (e.g. ball milling), among other techniques. Cation species that exist naturally in a multiplicity of valence states are more susceptible to this type of structure. For example, Mn ions with common valences of 2, 3, 4, 6, and 7, and are of comparable size to Fe ions, allow for such a distribution as $[Mn^{II,III}_{1-\delta}Fe^{II,III}_{\delta}]\{Mn^{II,III}_{\delta}Fe^{II,III}_{2-\delta}\}[6]O_4$ to exist (where δ represents the cation inversion parameter). The tendency to stabilize defects in such structures adds yet another degree of complexity. As one might imagine, determining such a structure with convincing measurement statistics represents a significant challenge to the experimentalist [5].

Exchange interaction between **A–B** sites is strong and negative ($J_{AB}\sim -20$ K) whereas the $J_{BB}\sim +/-\ 3$ K **B–B** couple can be parallel (double exchange) or antiparallel (superexchange) in nature [4]. Because of the indirect exchange interaction, the magnetization of ferrites is greatly reduced compared with 3d ferromagnetic metal systems (i.e. Fe, Co, and Ni) that experience positive direct exchange. For this reason, the magnetic properties in ferrite systems are controlled by cation valency and occupancy on the two lattices [7]. For example, in the magnetite system, the trivalent iron has five unpaired electrons, while the divalent iron has four. As magnetite is an inverse spinel, the overall magnetic moment is decreased when compared to a theoretical normal spinel of the same material. In order to increase the overall magnetic moment, ferrites such as MFe_2O_4 (M = Mn, Ni) are often doped with zinc to produce $(Mn_xZn_{1-x})Fe_2O_4$ and $(Ni_xZn_{1-x})Fe_2O_4$, respectively [5].

As mentioned previously, there are several factors that influence site occupancy of cations, including ionic radii, electrostatic energy, and electronegativity. These competing factors can explain the cation distribution observed in spinel systems. However, the influence of these factors is often of secondary importance when compared to the intrinsic energy of the synthesis route [7–9]. High-temperature ceramic methods promote a high degree of cation migration during processing from thermal agitation, resulting in the rearrangement of cations into near-equilibrium distributions. Alternatively, rapid quenching such as is the case of vapor quenching common to sputter, pulsed laser deposition, or mechanochemical synthesis produces ferrites with unusual cation distributions. Nonequilibrium cation distribution in ferrite unit cells can lead to enhanced magnetic, electronic, optical, biological, and chemical properties.

9.3 Synthesis Techniques

Many disciplines, including biology, physics, chemistry, and engineering, pursue a wide variety of methodologies for producing nanomaterials. Various techniques for producing nanoparticles are generally classified as either physical or chemical in nature [10]. Physical methods are commercially attractive because they are generally easier and more cost effective to produce large volumes of product, but they tend to impart physical stresses and introduce contaminants to the material that can degrade the final product. Chemical techniques are generally better for the synthesis of nanoparticles because they provide superior control over particle stoichiometry, shape, size distribution, crystallinity, and phase purity. However, scale-up may be challenging if the processing requires unconventional conditions such as the introduction of heat, light, ultrasound, etc. [11].

Numerous chemical processing methods have been developed that have proven viable for the synthesis of ferrite nanoparticles [12–15]. In contrast to the previously described physical processing methods, which often require a top-down approach, for example, through mechanical attrition, chemical processing methods are bottom-up techniques that assemble particles or clusters from reactant atoms. In general, these synthesis routes utilize some form of solution chemistry, thereby providing pathways to kinetically stable phases. As a result, chemical methods are often used to produce a precursor material that is then annealed to yield a final product.

As magnetism in ferrites derives from superexchange interactions, it depends sensitively upon bond angles and lengths between cation–anion–cation arrangements. Therefore, particles must be highly crystalline to experience high values of magnetism. By its very nature, the surface and grain boundary of any nanomaterial represents a dramatic disruption to the lattice in chemistry and structure thus negatively impacts superexchange and magnetic properties. A successful synthesis technique should produce nanoparticles that are single crystallites.

9.3.1 Precipitation

The most common and simplest synthesis route for the processing of Fe_3O_4 is by coprecipitation that was first published by Massart in 1981 [15]. In this method, a base is added to an aqueous solution containing a 2:1 stoichiometric ratio of Fe^{3+}:Fe^{2+} under an inert atmosphere resulting in a precipitate of Fe_3O_4. The reaction is written in the form:

$$2Fe^{2+} + Fe^{3+} 8OH^{-} \rightarrow Fe_3O_4 + 2H_2O \quad (9.1)$$

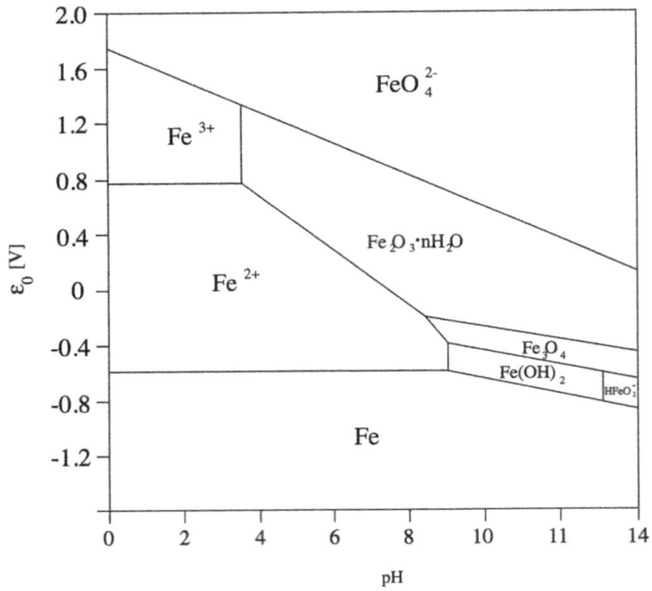

Figure 9.3 Pourbaix diagram of iron in water at various pH values.

Based on Equation (9.1), a precipitate of Fe_3O_4 requires an aqueous basic environment.

Pourbaix experimentally determined that a pH ranging between 8 and 14, with a chemical potential ranging between 0 and −0.4 V is required to synthesize pure phase Fe_3O_4; this is shown in Figure 9.3 [16].

Massart synthesized Fe_3O_4 nanoparticles having a crystallite size of ~12 nm from a solution of iron chloride salts precipitated with ammonium hydroxide. The particle size was approximated based on X-ray diffraction (XRD) measurements and application of the Scherer equation [3]. Research investigating the various effects of bases [17], pH [18–20], temperature [19, 21], and iron precursor concentrations [22] on the properties of Fe_3O_4 nanoparticles has been reported. The nucleation and growth kinetics of Fe_3O_4 was studied using low-temperature transmission electron microscopy (TEM) suggesting that crystallites of Fe_3O_4 form through the accumulation of smaller primary particles [23]. Gribanov et al. revealed that when strong bases such as NaOH, KOH, or LiOH are used, the pH increases rapidly and the chemical potential drops to −0.6 V, forming $Fe(OH)_2$ or $HFeO_2^-$ [17]. NH_4OH changes the pH to a value of 8.5–10 with a chemical potential of −0.3 V, forming an environment conducive to the formation of Fe_3O_4 [17]. Gnanaprakash et al. studied the effect of the initial pH of iron salt solutions on the resulting nanoparticles after the addition of the base solution. Selected results are presented in Table 9.1 [19]. These authors found that if the initial pH of the iron salt solution was below a pH of 5, Fe_3O_4 was synthesized; above a pH of 5, α-FeO(OH) resulted. As the pH was decreased, the diameter of the Fe_3O_4 nanoparticles reduced. The concentration of iron precursors and the molar ratio between Fe^{2+} and Fe^{3+} also affect the resulting nanoparticles. By increasing the iron precursor concentration, the average particle size increases [22]. As synthesis temperature increases, the nanoparticle diameter decreases [21, 24].

The role of mixing rate of the two solutions has also been studied [25–27]. Strom et al. performed an *in situ* study of the magnetic susceptibility during several coprecipitation reactions under different mixing conditions, bases, and salt and base precursor concentrations (see Figure 9.4). They determined that with rapid mixing rates, the precipitation is completed within 1 minute, resulting in a reduced nanoparticle size with a reduced standard deviation [26, 27]. It was speculated that the rapid mixing decreased side reactions such as the condensation of Fe^{3+} [26].

While coprecipitation methods are advantageous for the synthesis of Fe_3O_4 due to the ease of reaction and the ability to produce at large scales compared with other methods, there are disadvantages. Due to the *burst* nucleation stage followed by a slow growth stage, the resulting Fe_3O_4 nanoparticles are polydispersed, with control over particle size challenging [17]. In order to gain greater control of size and shape of the final product, surfactants are often employed.

Table 9.1 Estimated percentage and particle size of nanoparticles synthesized at different initial pH values. Adapted from the literature [19].

pH of iron solution	Temperature (°C)	α-FeO(OH) (wt. %)	Fe_3O_4 (wt. %)	Fe_3O_4 particle size (nm)
0.7	30	0	100	6.47
0.7	60	0	100	8.2
1.5	30	0	100	7.68
3.0	30	0	100	9.94
4.7	30	0	100	9.94
5.7	30	35	65	—

A diverse group of materials can be used as surfactants, including proteins, starches, nonionic species, and ionic species [24, 28–35]. With polyvinyl alcohol as a surfactant, Fe_3O_4 nanoparticles with a diameter of 4–7 nm can be synthesized at a pH of 13.8 [33]. When sodium dodecyl sulfate is used as a surfactant, in a basic solution of iron (II) sulfate and iron (III) sulfate and in the presence of visible light (i.e. $\lambda = 400–750$ nm), the morphology of Fe_3O_4 nanoparticles transition from spheres to ellipsoids (of high aspect ratio) to cubes as shown in Figure 9.5. With increased irradiation time, the resulting nanoparticle diameters increase.

The applications of Fe_3O_4 nanoparticles greatly depend on the choice of surfactants. When coated with phosphatidylcholine, Fe_3O_4 is used to make magnetoliposomes that can be used in magnetic hypothermia treatments [28]. With chitosan as a coating, Fe_3O_4 is used to treat hypothermia [34].

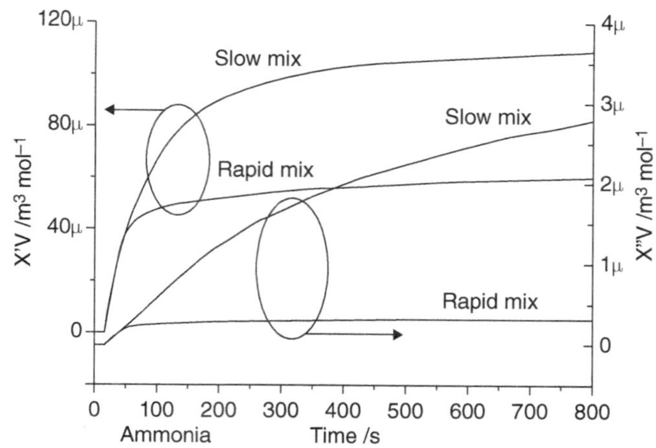

Figure 9.4 In-phase susceptibility (left axis) and out-of-phase susceptibility (right axis) of iron solutions reacted with ammonia under different mixing conditions (adapted from the literature [26]).

The point during the reaction upon which the surfactant is added affects the size of the resulting nanoparticles. When citric acid is added immediately after the introduction of the base, the resulting Fe_3O_4 nanoparticles have a crystallite size of 6–7 nm compared to 9 nm when citric acid is employed in a later step [36]. The presence of the surfactant during the growth period limits the growth, resulting in smaller diameters.

Most coprecipitation reactions have basic solutions added dropwise to the metal solution during stirring. In a reverse coprecipitation method, the metal solution is added dropwise to the base solution [37, 38]. Aono *et al.* determined that a reverse coprecipitation method yield particles of smaller diameters at various process temperatures compared to those synthesized with a traditional coprecipitation protocol as presented in Table 9.2 [37].

All the reactions described above involve the use of both Fe^{2+} and Fe^{3+} as precursors. It is possible to synthesize Fe_3O_4 from just a single Fe^{2+}precursor, but this can take up to 24 hours for the reaction to complete, which is often impractical [39–41]. In these instances, the reaction is:

$$FeCl_2 + NaNO_2 \rightarrow Fe(H_2O)_6^{\ 2+} + Fe(H_2O)_6^{\ 3+} + NaOH \rightarrow Fe(OH)_2 + Fe(OH)_3 \rightarrow Fe_3O_4 \tag{9.2}$$

Nedkov et al. synthesized Fe_3O_4 from a solution of iron (II) chloride at a pH of 13 [40]. By varying the ratio of Fe^{2+} to sodium nitrite and reaction time, the shape and size of the nanoparticles can be made spherical, cubic, or ellipsoidal [40]. Using a similar method, Aphesteguy et al. synthesized 100-nm cubic shape Fe_3O_4 particles from iron (II) sulfate with potassium nitrate [39].

Synthesis of doped ferrites with formula MFe_2O_4 by coprecipitation is very similar to synthesis of Fe_3O_4. In most cases, the Fe^{2+} salt is exchanged for a cation salt of the same oxidation state, for example, ferrites doped with cobalt and manganese [42, 43] have been processed in this fashion.

The coprecipitation method can also be used to synthesize core/shell nanoparticles. Fe_3O_4/ZnO nanoparticles were synthesized in a three-step process that included (i) coprecipitation synthesis of Fe_3O_4 particles, (ii) coating of Fe_3O_4 with citric acid, and (iii) a precipitation of ZnO onto the citric acid–coated Fe_3O_4 [44]. A similar method was used to synthesize Fe_3O_4@SiO_2 [45]. Fe_3O_4@SiO_2 was also synthesized in a two-step process using an ultrasonic horn [46]. Evans et al. synthesized CeTe@CdS-decorated Fe_3O_4 by coprecipitation using 3-meraptoporoanoic acid as a linker. The particles, shown in Figure 9.6, were superparamagnetic with size-tunable luminescence [47].

9.3.2 Microemulsion

Microemulsions are thermodynamically stable systems composed of an aqueous and oil phase that are miscible due to the addition of a surfactant. The surfactant forms a monolayer between the oil and aqueous layers that lowers the interfacial tension by an order of 10^{-2} to 10^{-3} mN/m [48]. In 1954, Winsor predicted four types of emulsion equilibria [49]. Among these types, oil in water (o/w) emulsions are the most studied. o/w is made up of nanodroplets of water that provide

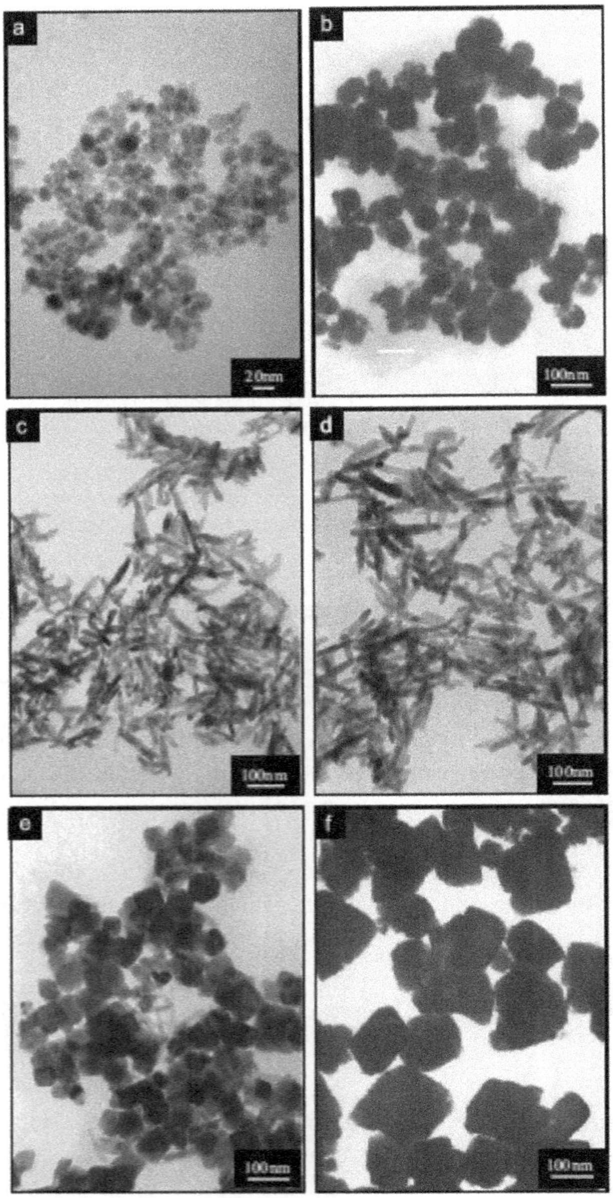

Figure 9.5 TEM images of nanoparticles synthesized at different irradiation times of visible light: (a, c, e) 30 minutes and (b, d, f) 60 minutes with different amounts of SDS of (a and b) 0 g, (c) 0.4 g, (d) 0.3 g, (e) 0.8 g, and (f) 1.0 g. Adapted from the literature [33].

confinement effects that limit particle nucleation and growth [50]. Nanoparticles are formed by mixing two microemulsions, one consists of the metal precursor, and the other a precipitation agent that coalesces to form nanometric-sized particles. Similar to coprecipitation, this system allows for synthesis at near ambient temperatures and pressures. However, unlike coprecipitation this two-phase system allows for a simple route for controlling particle size that can be used to fine-tune magnetic properties of ferrites.

The structure of microemulsions vary depending on the composition of the surfactant. Ionic surfactants generally need a cosurfactant such as aliphatic alcohols or electrolytes due to their high hydrophilic nature. Equilibria phase diagrams help to determine where a stable microemulsion is formed as a function of water/oil/surfactant ratio. Examples of phase diagrams are shown in Figure 9.7, where L_2 represents the range in which the water/oil/surfactant are miscible [50].

Sodium 2-methylsulfosuccinate (AOT) is a well-characterized microemulsion system as well as the first microemulsion system used to synthesize iron oxide in the 1980s [52]. Figure 9.8 shows AOT systems have large microemulsion regions in comparison to other systems. Studies have shown the microemulsion size increases linearly with water and AOT can increase the droplet size from 4 to 8 nm with 0.1 AOT [53]. This general trend has been used to determine magnetic saturation and coercivity values for Fe_3O_4 nanoparticles with 4–16 nm radius [54, 55]. AOT was also used to synthesize superparamagnetic iron oxide for MRI or drug delivery applications [56]. These particles 10–15 nm in diameter were synthesized with aerosol-OT (AOT)/*n*-hexane with a 2:1 1 M solution of ferric and ferrous salt under nitrogen flow before being coated with polyethylene glycol (PEG) to protect its surface.

Cationic microemulsions, mainly CTAB, have grown in popularity during the last decade due in part to their reduced metal interactions. Recently, different synthesis routes were studied for producing iron oxide for protein purification. Three synthesis routes were prepared: route 1 involved mixing two w/o microemulsions, one containing the iron precursor the other consisting of ammonium hydroxide as the precipitating agent; route 2 was a one-pot synthesis of iron

Table 9.2 Crystallite size determined by XRD with particle size estimated from BET. Adapted from [37].

	Normal coprecipitation method				Reverse coprecipitation method			
Synthesized temperature (°C)	35	50	80	100	35	50	80	100
Crystal Diameter (nm)	8.5	9.5	14.3	15.6	7.0	8.6	9.6	10.9
Particle Diameter (nm)	30.4	24.6	20.4	22.4	23.2	23.6	18.1	13.1

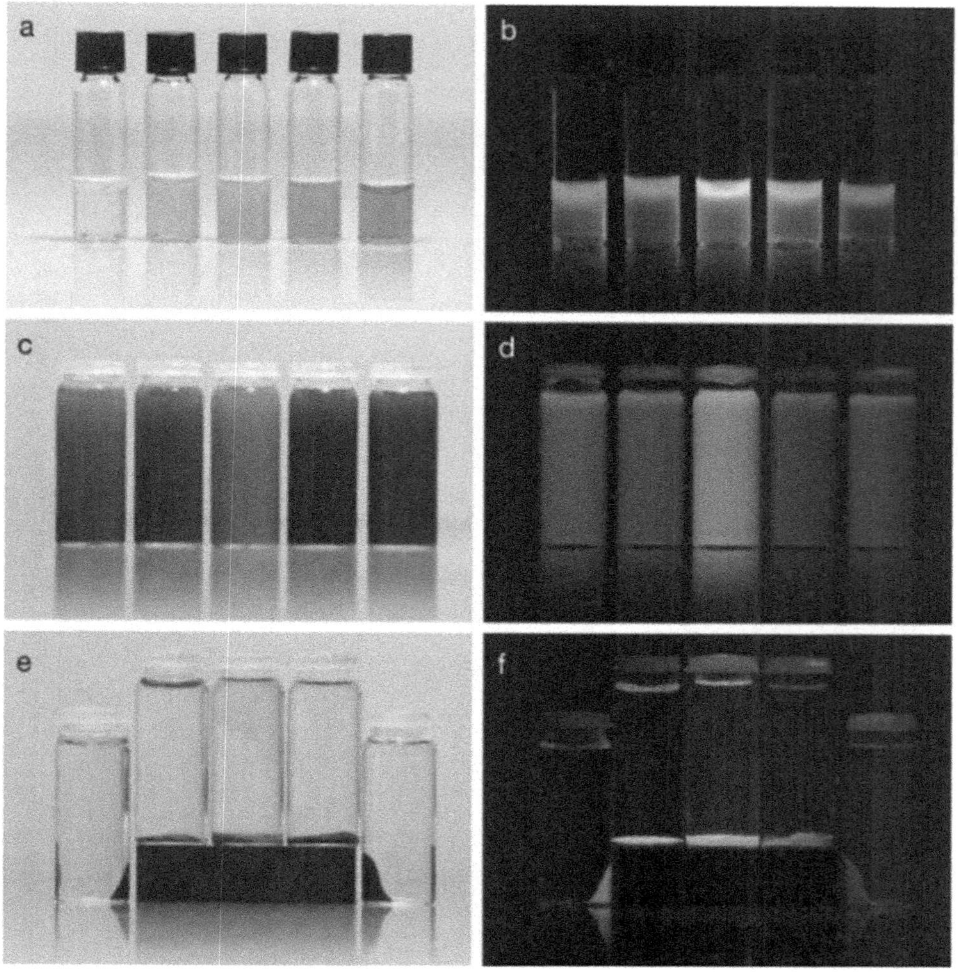

Figure 9.6 CeTe@CdS-decorated Fe_3O_4 under white light and 380 nm light (a and b) colloidal CeTe@CdS, (c and d) CeTe@CdS-decorated Fe_3O_4 in water, and (e and f) CeTe@CdS-decorated Fe_3O_4 in the presence of a magnetic field. Reproduced from [47].

precursor in a microemulsion followed by the addition of ammonium hydroxide solution dropwise; and route 3 was a traditional coprecipitation of iron oxide to compare to the two microemulsion routes. Results shown in Figure 9.9 depict, despite the difference in synthesis methods, both microemulsions showed dimensions with partially agglomerated 7–10 nm particles while the coprecipitation route formed 15 nm nanoparticles. The protein elution showed better elution for the particles synthesized with the microemulsion potentially due to the greater surface area [57].

Another study using cationic surfactant didodecyldimethylammonium bromide used a bicontinuous microemulsion. Unlike normal reverse microemulsions that have water droplets suspended within a continuous oil phase, bicontinuous microemulsions have interconnected aqueous channels that accept a content 50–100% higher than w/o microemulsions. The synthesis was performed at 80°C under argon with 33 wt-% surfactants, methyl methacrylate (MMA), the iron precursor was introduced to the aqueous phase under sonication. The solution was not agitated once ammonium hydroxide was added in order to not interfere with the aqueous channels. TEM images, shown in Figure 9.10, reveal 8 nm nanoparticles with a yield up to 1.16 g per 100 g solution, which has a greater yield than reverse micelle systems (0.1–0.4 g of product) [58, 59].

Recently, a study synthesizing Mn–Zn ferrite nanoparticles was performed to determine the effect cosurfactant 1-hexanol has on the structure of o/w microemulsions and reaction rate with synperonic and isoctooctane. Results showed that the addition of a cosurfactant lowered the microemulsion region on the equilibria phase diagram in Figure 9.11; however, it also increased the reaction rate due to the decrease in the rigidity at interfacial film. TEM showed despite variations in the synthesis, nanoparticles 4 nm in diameter were formed. The exception was the bicontinuous microemulsion provided an almost doubling in particle size [60].

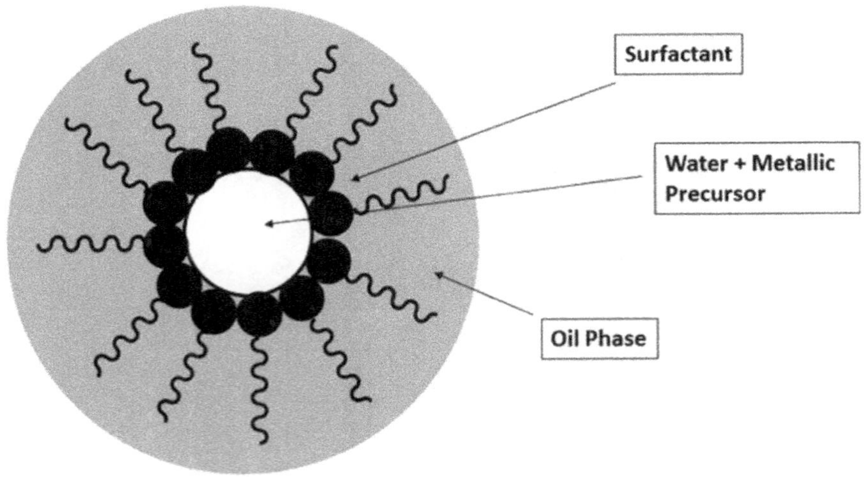

Figure 9.7 A schematic of the w/o microemulsion.

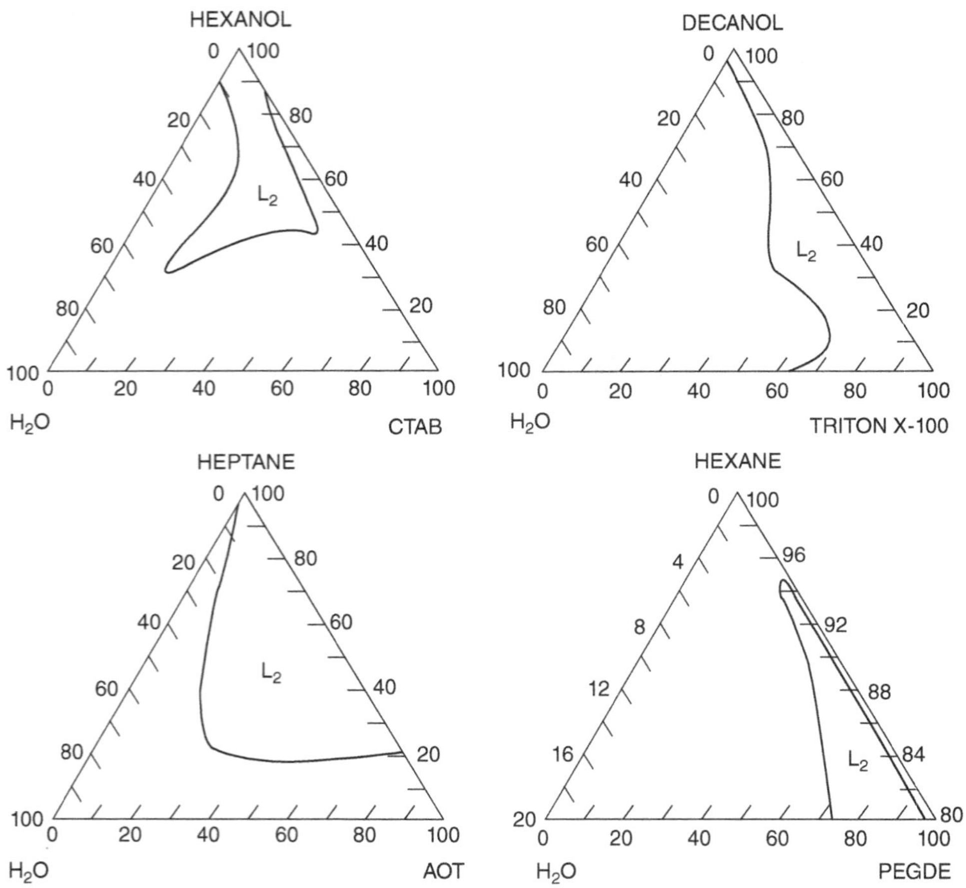

Figure 9.8 Pseudo binary phase diagram as a function of temperature. Reproduced with permission [51].

In addition to previous studies, an oil in water microemulsion was used to synthesize ferrite nanoparticles. The synthesis included iron (III) 2-ethylhexanoate in hexane as the iron precursor in oil phase dispersed in water, and synperonic 10/6 as the nonionic surfactant. A w/o microemulsion with CTAB/butanol as surfactant/cosurfactant were prepared for comparison (see Figure 9.12). Results showed the w/o contained rod-like nanoparticles 5–10 nm in width with 20 to 50 nm in

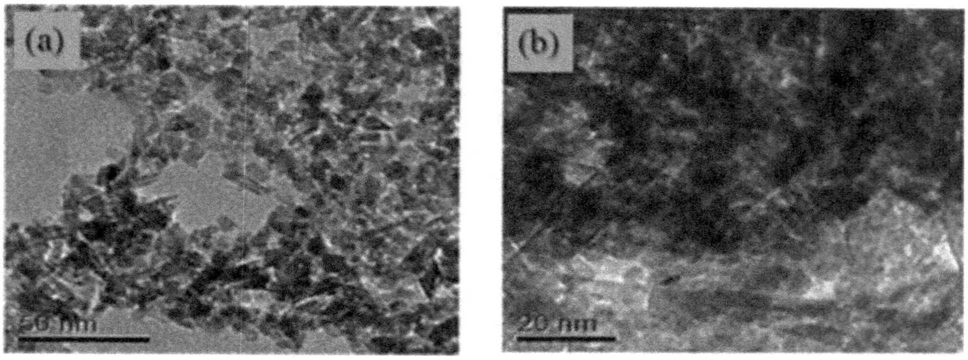

Figure 9.9 Ferrite nanoparticles synthesized with CTAB by (a) mixing two microemulsion solutions containing the precursor and precipitating agent and (b) adding precipitation agent to a microemulsion that contains the iron precursor. Reproduced with permission [57].

length, while the o/w had 3 nm spheroidal nanoparticles. These results match the decrease in magnetic saturation from 35 and 10 emu/g[1] for the o/w and w/o systems, respectively [61].

9.3.3 Sol–Gel

The sol–gel process is a wet chemical process that involves the transformation of a colloidal suspension of inorganic precursor or polymer solution (sol) through hydrolysis and condensation reactions to obtain a continuous network structure (gel). This process began in the 19th century by Ebelman and Graham for the development of glasslike ceramic materials through a series of hydrolysis and condensation reactions of tetraethylorthosilicate (tetraethoxysilane, or TEOS) [62]. The sol–gel process can be summarized in a number of steps: formation of the sol, gelation, aging, drying, and densification. The sol is obtained by a series of hydrolysis and condensation reactions where its kinetics are pH dependent. The hydrolysis reaction is a nucleophilic substitution (Schemas 1 and 2). During this process, additional alkoxy group can be substituted by altering the concentration of metal to water ratio [63]. Similar to hydrolysis, condensation can be acid or base catalyzed where hydroxyl bridges are formed under acidic conditions and oxo bridges are formed under basic conditions. Typically, acid-catalyzed hydrolysis and condensation often result in a cross-linked polymer-like gel as the hydrolysis happens rapidly prior to condensation. On the other hand, base-catalyzed hydrolysis typically results in a colloidal gel as the hydrolysis step is slower than the condensation [64]. As the gel is allowed to age, the sols become more interconnected where the solvent is removed from the pores through the condensation process [64]. At this stage, the pore size is dependent upon this process where the longer it is allowed to age, and more sols will become connected, resulting in a denser monolith [65]. The final drying of the gel involves the removal of solvents from the gel. The final density and the surface area are dependent upon this step, as shown in Figure 9.13. Supercritical CO_2 drying results in a highly porous structure with high surface area and low density; this is known as aerogels [66–68]. On the other hand, drying under atmospheric conditions results in denser gel structures known as xerogels [69]. An additional densification step can be performed if the density of the gel requires to be higher for targeted applications. This step typically involves the calcination of gel at elevated temperatures to further shrink the pores to densify the structure.

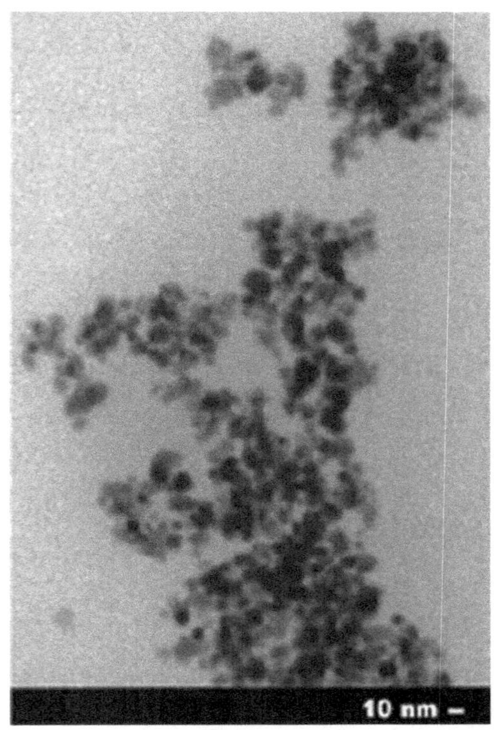

Figure 9.10 Nanoparticles synthesized using bicontinuous microemulsions. Copied with permission [59].

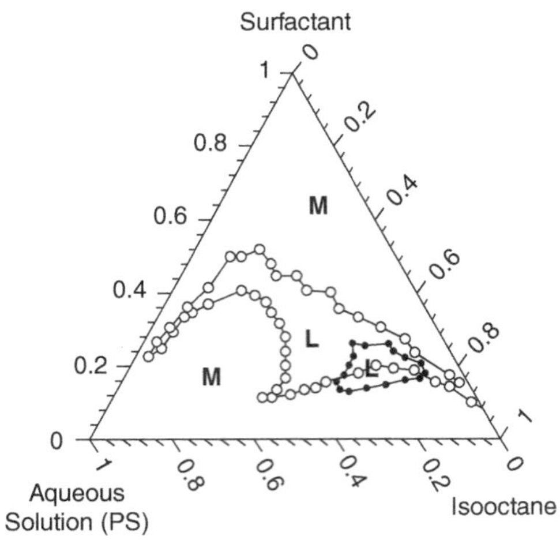

Figure 9.11 Equilibria phase diagram where open symbols define microemulsion region without cosurfactant, while the solid symbols represent the region with 1-hexanol. Copied with permission [60].

The sol–gel method used in the production of ferrite nanoparticles is of interest in recent years as the structure of the product particles can be influenced by experimental conditions (surfactants, metal to water ratio, etc.) [70]. The hydrolysis of Fe (III) ions under acidic conditions (pH < 1) have been well developed and demonstrated to follow Equations (9.3) and (9.4) [71]:

$$\left[Fe\left(H_2O\right)_6\right]^{3+} \rightarrow [Fe(OH)(H_2O)_5]^{2+} + H_3O^+ \tag{9.3}$$

$$[Fe(OH)(H_2O)_5]^{2+} \leftrightarrow \left[(H_2O)_5\,FeOFe(H_2O)_5\right]^{4+} + H_2O \tag{9.4}$$

Various iron precursors had been studied in these systems, and it was determined that the iron oxide phase and surface area are dependent upon the precursor salt and solvent used [72, 73]. However, the ferrite particles synthesized by the sol–gel methods have similar issues encountered by coprecipitation where particles often agglomerate as a high-temperature annealing step is typically required [74, 75]. Silica-based core–shell ferrite particles are among the most popular pursuits

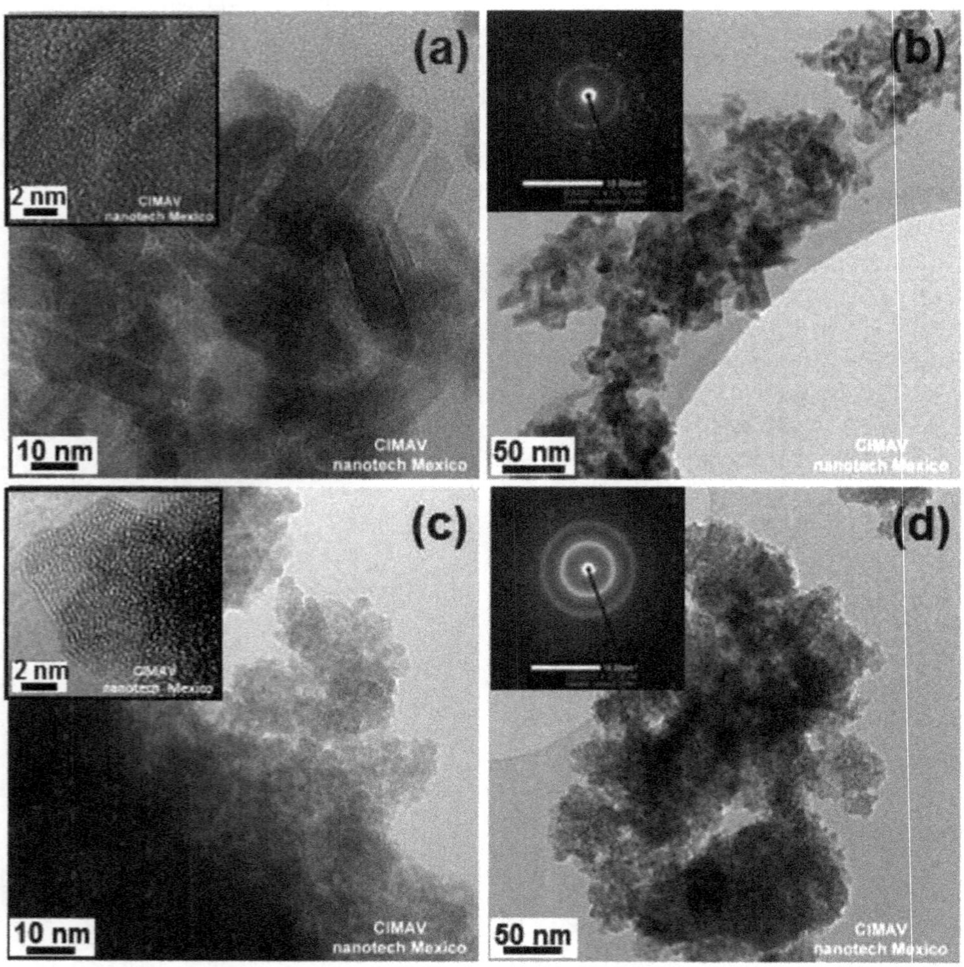

Figure 9.12 (a and b) w/o microemulsion synthesized with synperonic. (c and d) o/w microemulsion synthesized with CTAB [61].

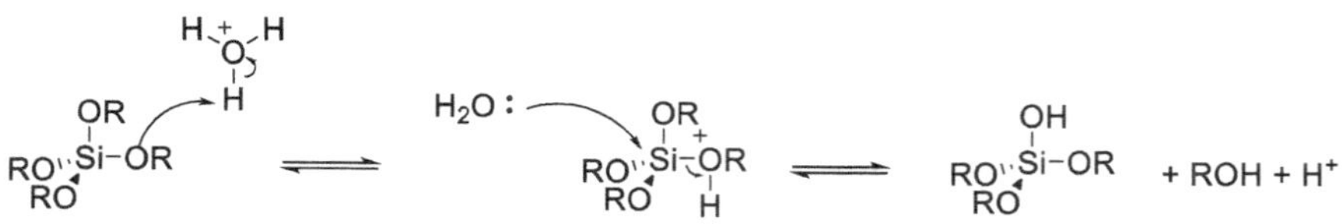

Scheme 9.1 Hydrolysis of silicon alkoxide in acidic medium.

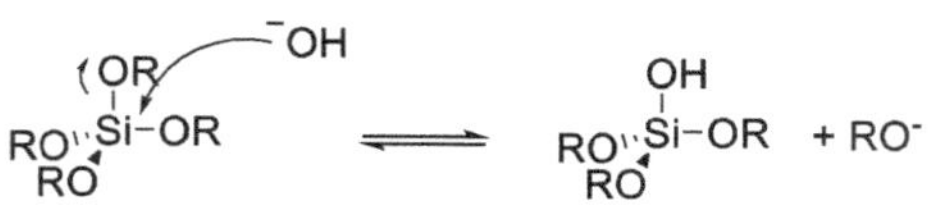

Scheme 9.2 Hydrolysis of silicon alkoxide in basic medium.

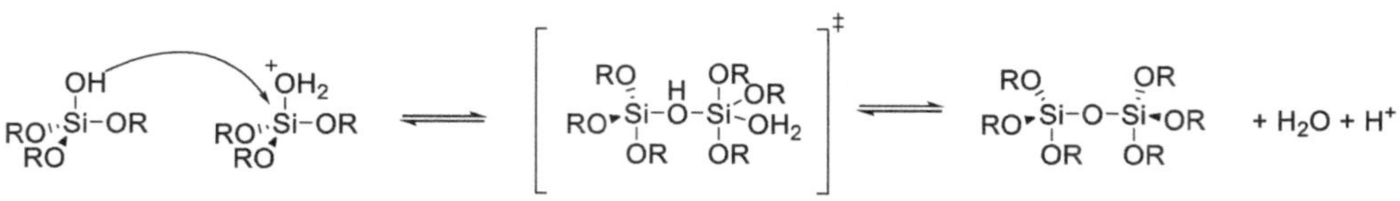

Scheme 9.3 Condensation of silicon alkoxide in acidic medium.

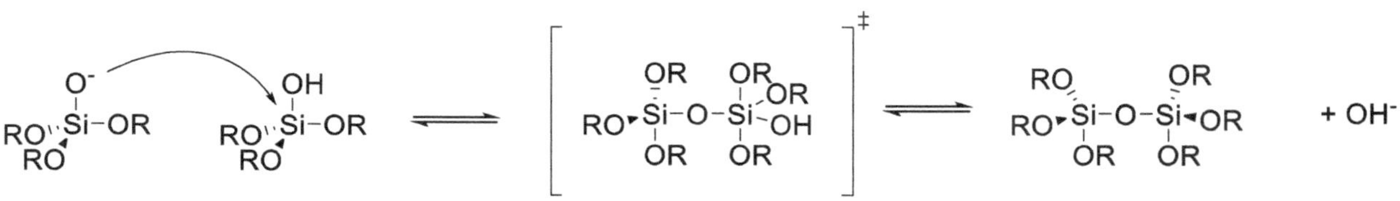

Scheme 9.4 Condensation of silicon alkoxide in basic medium.

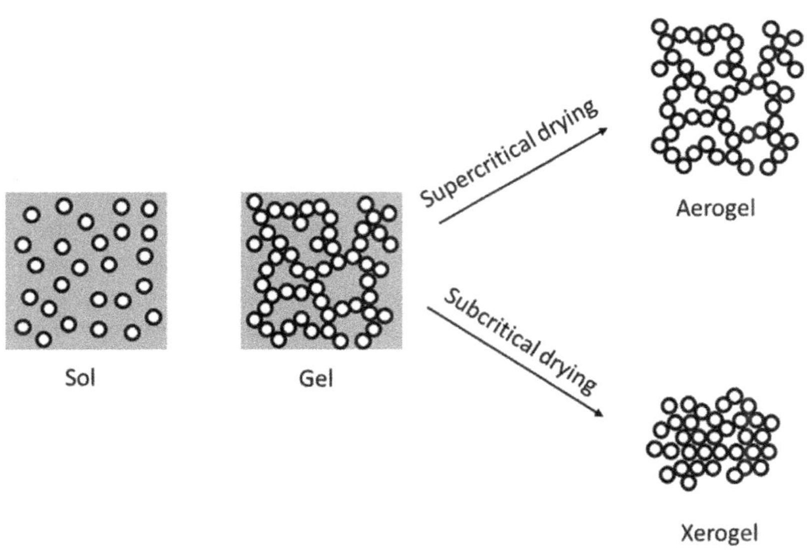

Figure 9.13 Various drying methods of gel.

using the sol–gel methods, as they have been shown to have a higher reactivity than iron oxide particles embedded in a gel matrix. This can be achieved through variation of the Stöber process where silica and iron oxide are formed during sol–gel synthesis [76]. The silica on the surface of the ferrite particles can be adjusted by controlling the silicon precursor concentrations [77, 78]. The resulting particles often possess high surface area, which can be beneficial for various applications, including catalysis [79, 80].

9.3.4 Thermal Decomposition

Previously, we discussed ferrite synthesis using coprecipitation and microemulsion methods. Even though these methods are simple for synthesizing nanoparticles under near ambient conditions, they normally form nanoparticles with a broad distribution of sizes and a low quality of crystallinity. The thermal decomposition method is a simple and quick method for synthesizing nanoparticles that utilizes thermal decomposition of a metal precursor with a surfactant as a capping agent. This method was first used for synthesis of ferrite nanoparticles by Alivisatos and coworkers who used a nonhydrolytic system with iron Cupferron complexes to synthesize 4–10 nm γ-Fe_2O_3 with nonhydroxylated surfaces [81]. Since then, thermal decomposition has become the most reliable system for making monodispersed particles with nanoparticle sizes ranging from 2 to 30 nm. The increase in temperature needed for thermal decomposition in comparison to coprecipitation and microemulsions allows for the clear separation of nucleation and growth stages for monodispersed nanoparticles, as well as a higher quality of crystallinity, required by most applications [33, 82].

Iron pentacarbonyl, iron oleate, and iron oxyhydroxide are the most common precursors for thermal decomposition. Many sources claim that iron oleate is a normal intermediate phase in forming ferrite nanoparticles [83–85]. Mixtures of octadecene, oleylamine, and oleic acid are normally used as stabilizing agents that can be used to control nucleation and growth rates. Recently, work by Hufshid and coworkers evaluated synthesis of iron oxide using iron pentacarbonyl, iron oleate, and iron oxyhydroxide in the presence of oleic acid [86]. Results showed iron pentacarbonyl and iron oleate were sufficient for making smaller nanoparticles (<10 nm) while iron oxyhydroxide forms nanoparticles in the range of 10–25 nm.

As formation of iron oleate appears to play a crucial role in forming monodispersed nanoparticles, a thermal stability study was performed using thermal gravimetric analysis (TGA) and differential scanning calorimetry (DSC), as shown in Figure 9.14. Results showed three distinct points of mass loss on TGA scans at 193.6, 260.4, and 330.5 °C, as well as two thermal deflections on DSC scans at 165 and 240 °C. Temperatures corresponding to mass loss for TGA represent the dissociation of the three oleate ligands, while the DSC temperature deflection points represent the endothermic release of CO_2 from the symmetrically bound oleate ligands and asymmetric ligands. This change in temperature at which iron oleate dissociates is critical for the synthesis of monodispersed iron oxide nanoparticles. Previous work by Park et al. shows by not separating nucleation and growth stages and aging nanoparticles at 240 °C polydispersed ferrite nanoparticles result [87].

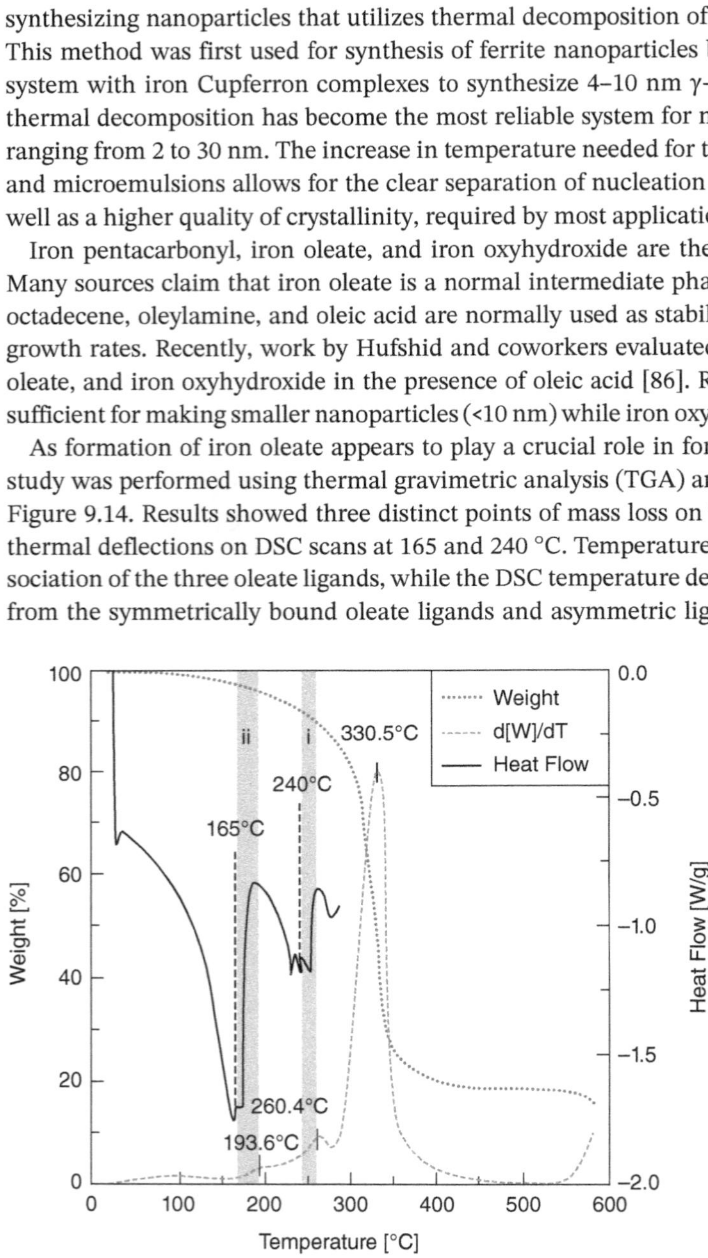

Figure 9.14 TGA (weight) and derivative of TGA (d[*W*]/d*T*, and DSC (heat flow)) versus temperature for iron oleate complexes. Copied with permission [86].

The work of Park et al. also covered many important aspects of nanoparticle processing including scale-up synthesis and particle size control [88]. The synthesis used 36 g of iron oleate and 5.7 g oleic acid dissolved in 200 g water to synthesize 40 g of iron oxide. Park et al. used various solvents with different boiling points to tailor particle size. Increasing the boiling point of the solvent from lowest (hexadecane bp~ 274°C) to highest (trioctylamine bp~ 365 °C) showed an increase in particle size ranging from 5 to 22 nm, with a <4.1% size variation. The reaction time was also shown to increase particle size while decreasing the size distribution, which agreed with digestive ripening where smaller nuclei dissolve to reform as larger particles [89]. Other researchers have manipulated the particle size by controlling nucleation and growth rates through oleic acid concentration where higher concentrations slowed the nucleation process, leading to a greater concentration of nuclei and larger particles. Using oleic acid and a seed-mediated growth synthesis, the control of particle growth within a 1 nm precision was possible and demonstrated for products ranging from 6 to 13 nm (see Figure 9.15) [84, 88]. Mixtures of 1:1, 1:2, and 1:3 ratios of $Fe(CO)_5$ and oleic acid were used to make 4, 8, and 11 nm nanoparticles and for the seed-mediated growth of larger particles.

Synthesis of ferrite nanoparticles can be controlled by an additional oxidation step. Hyeon and coworkers used iron pentacarbonyl and oleic acid in the presence of trimethylamine oxide, a mild oxidant [90]. Two methods were employed: the first method produced iron nanoparticles first followed by their oxidation with trimethylamine. The second method was a one-pot synthesis with the precursor and oxidant. Results showed similar results with 11 nm and 13 nm nanoparticles of γ-Fe_2O_3 produced for the two- and one-step synthesis methods, respectively. Another work covered two oxidation methods, one similar to the previous method with trimethylamine, the other annealed the previously synthesized iron oxide in air at 100 °C. Both synthesis methods show they are capable of converting Würsite phases into Fe_3O_4 or γ-Fe_2O_3 [86].

The ability to control the shape of ferrite nanoparticles has attracted much attention due to applications in magnetic sensing and as catalysts [91]. The changes in particle shape exposes crystal facets of nanoparticles affecting properties [92]. Change in shape also affects surface anisotropy, K, which can cause shifts in the blocking temperature (T_B) [93, 94]. One way of controlling particle shape is by changing the precursor. Sayed and Polshettiwar explored reactions with iron sulfate heptahydrate, iron oxalate dehydrate, iron chloride hexahydrate, iron nitrate nonahydrate, gluconate dehydrate, and iron pentacarbonyl [95]. The resulting morphologies included nanorods, nano-husks, cubes, nanocubes, porous spheres, and self-arranged flowers. Despite the interesting array of shapes, XRD showed most of the morphologies were a mixed phase of α-Fe_2O_3 and γ-Fe_2O_3 that drastically diluted the samples' magnetism. Ho and coworkers performed a modified hot injection method adding 0.2 M solution of iron acetylacetonate [Fe(acac)] in benzyl ether at rates of 10 and 20 mL/h at 290 °C followed by a 2-hour dwell at this temperature [96]. Results showed the 10 mL/h introduction rate formed monodispersed nanocubes terminated at {100} [97] plane while the 20 mL/h injection resulted in rhombicuboctahedral particles terminated at the {1 0 0} and {1 1 0} plane. The change in shape from cubic to rhombicuboctahedral is said to result from increasing the monomer concentration. Shape determination mechanism was studied using TEM that showed monomer and rapid growth occurs in phase II of the LaMer model while the final shape was stabilized in phase III.

Surfactants are another way of controlling particle shape. Bao and coworkers have used trioctylphosphine oxide to control growth of iron oxide nanoparticles [98]. In the work reported on iron oxide, "nanoworms" were synthesized from thermal decomposition of a mixture of iron oleate in 13 mL 1-octadecene and 1 mL oleic acid, followed by the hot addition

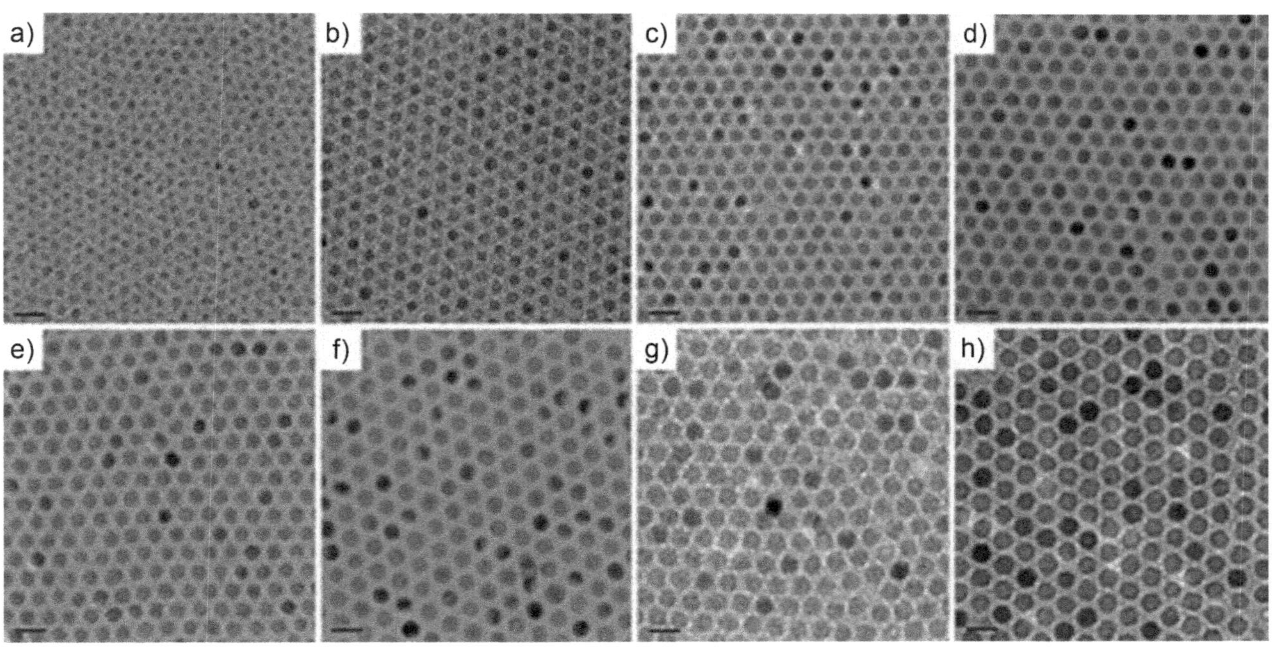

Figure 9.15 TEM of (a) 6 nm, (b) 7 nm, (c) 8 nm, (d) 9 nm, (e) 10 nm, (f) 11 nm, (g) 12 nm, and (h) 13 nm iron oxide nanoparticles. Copied with permission [88].

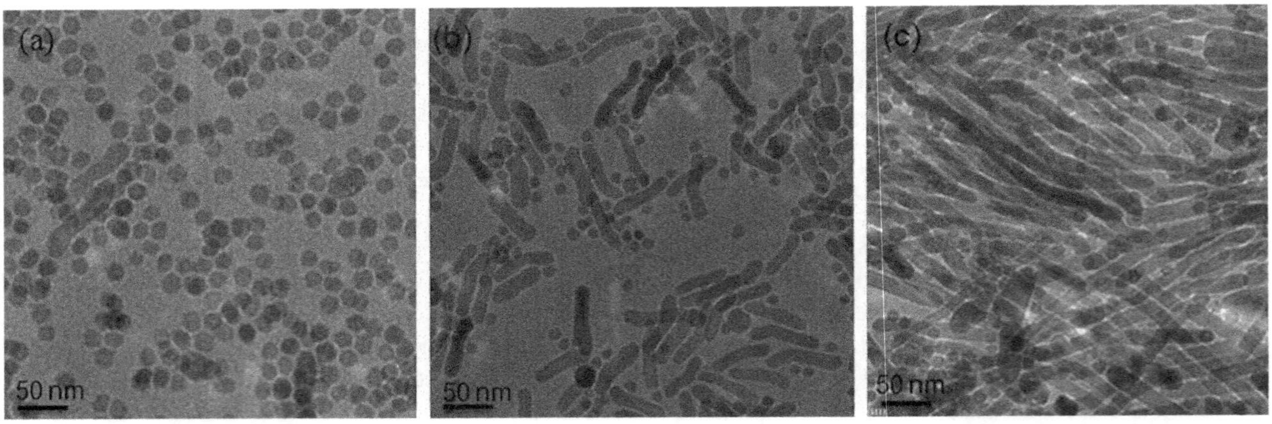

Figure 9.16 TEM of iron oxide nanoworms at different temperatures: (a) 1 hour, (b) 2.5 hours, and (c) 5 hours. Copied with permission [99].

of trioctylphosphine oxide (TOPO) at 290 °C (see Figure 9.16) [99]. A time study showed the nanoparticles grew from 12 nm spheroids to lengths exceeding 200 nm. The proposed mechanism is that the weakly bound surfactant TOPO concentration mediates aggregation of nanoparticles. In a more recent study, various ratios of TOPO to oleylamine from 1.65 to 8.25 were found to alter particle morphology from 3 nm nanoplates to 20 nm nanoflowers [100]. A mechanistic study of both systems showed that the nanoplates are formed due to excess $CH_3CH_2O^-$ acting as a third surfactant while the nanoflowers formed as a result of an excess of monomers. Therefore, TOPO is said to influence the nucleation by forming more monomers.

Oleic acid is a common surfactant for controlling dimensions of a nanoparticle. In recent work, pseudospherical and faceted Fe_3O_4 and γ-Fe_2O_3 particles ranging from 4 to 20 nm were synthesized with oleic acid while larger cubic nanoparticles (~45 nm) were synthesized using decanoic acid [101]. This result may be due to the differences in intermediate complexes. Other works studied how different surfactants affect particle growth. For example, oleic acid was shown to form spherical particles, while sodium oleate formed cubes, and tetraoctylammonium bromide formed octahedrons (see Figure 9.17) [102]. The paper concludes that oleate anions that form once iron oleate dissociates play a crucial role in the formation of cubes. X-ray photoelectron spectroscopy (XPS) suggests sodium may also play a role in the formation of cubes. Cationic surfactants have many potential roles they may play in reactions, but it has become obvious that sodium oleate and tetraoctylammonium bromide adsorb to different crystal planes and affect nanoparticle morphology in very different ways.

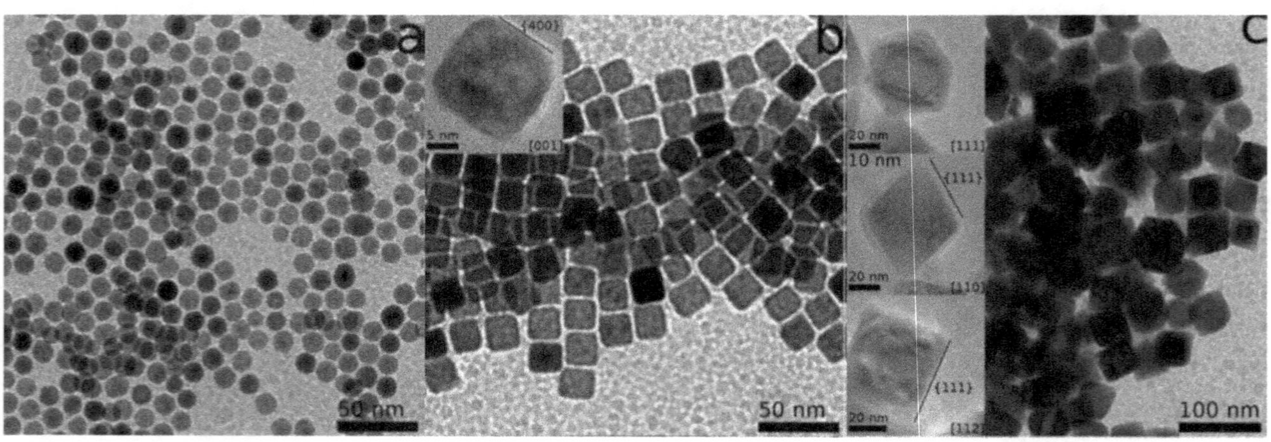

Figure 9.17 TEM image of iron oxide nanoparticles synthesized with (a) oleic acid, (b) sodium oleate, and (c) tetraoctylammonium bromide. Copied with permission [102].

Ge et al. performed an in-depth study on how cationic surfactant cetylpyridinium chloride (CPC) with $Fe(acac)_3$ as the precursor in benzyl ether affected the shape of the nanoparticle product (see Figure 9.18) [103]. Similar to previous work, faceted Fe_3O_4 were formed. However, with changes in CPC concentration from 0.028 to 2 mmol, five different shapes, consisting of cubic, truncated cubic, cuboctahedral, truncated, octahedral, and octahedral nanostructures, were formed. Comparing CPC to other quaternary cationic amines from cetylpyridinium bromide (CPB), cetylammonium chloride (CTAC), cetylammonium bromide (CTAB), NH_4Cl, and pyridinium chloride (PC), the authors came to three distinct influences by which CPC assisted in tuning the particle shape. These three factors include: (i) selective adsorption of Cl^- on {1 1 1} planes, (ii) the synergistic effect of pyridinium, and (iii) the dense packing of long carbon chains for on the {1 1 1} planes.

Figure 9.18 TEM (left) and SEM (right) of Fe_3O_4 synthesized using CPC of concentrations: (a and b) 0, (c and d) 0.028 mmol, (e and f) 0.056 mmol, (g and h) 0.11 mmol, and (i and j) 0.19 mmol. Copied with permission [102].

9.3.5 Solvothermal

Nanoparticles synthesized using solvothermal methods employ various wet synthesis techniques at high pressure ranging from 0.3 to 4 MPa in a stainless-steel autoclave. These methods are suitable for synthesizing crystalline materials above the product melting point. Solvothermal methods pertaining to ferrites normally include synthesis in polyol solvents such as ethylene and diethylene glycol, long aliphatic organic solvents such as oleylamine and octylamine, and aqueous (i.e. hydrothermal) solutions.

Synthesizing nanoparticles using hydrothermal methods normally require a reducing agent in order to partially reduce the Fe^{3+} to Fe^{2+} ions needed to form Fe_3O_4 rather than α-Fe_2O_3. Previous work used sucrose as a method to partially reduce the iron (III) chloride dissolved in an aqueous solution with a pH of 11, followed by heating for 48 hours at 180 °C [104]. Results show sucrose was needed to form pure phased magnetite nanoparticles coated with sucrose ranging from 4 to 16 nm while the absence of sucrose led to formation of larger impure nanoparticles consisting of α-Fe_2O_3. Another study utilizing the hydrothermal approach in the presence of graphene oxide acted to reduce agglomeration [105].

Solvothermal synthesis with polyol solvents has been comprehensively reported in the literature. In most cases, a single iron precursor was dissolved in either ethylene glycol or diethylene glycol and heated to ~200 °C. The morphology consisted of secondary micron-sized iron oxide composed of primary smaller ferrite particles approximately 10 nm in diameter, as shown in Figure 9.19 [106–109]. Solvothermal methods employing polyols have been applied to synthesize hollow Fe_3O_4 nanoparticles [110, 111]. In a normal synthesis, an iron precursor is dissolved in ethylene glycol with directing agents such as sodium acetate, ammonium acetate, and urea. A gas-bubble-assisted Ostwald ripening growth mechanism has been proposed to account for the formation of hollow microspheres [109, 112]. It was proffered that $NH_3{\cdot}H_2O$ evaporates and forms bubbles that act as heterogeneous nucleation centers for newly formed crystallites [113].

Despite advances in solvothermal synthesis of iron oxides, most product results in micron-sized particles. Long aliphatic mixtures such as oleylamine, *n*-octylamine, with surfactants oleic acid and octanol have been used to synthesize monodispersed magnetite with diameters ranging from 4 to 12 nm. Tian and coworkers were able to tune the ferrite nanoparticles in 1 nm increments from 4 to 6 nm (see TEM in Figure 9.19b) by changing the ratio between *n*-octylamine to octanol [114, 115]. Figure 9.20a shows the re-dispersion of the ferrite nanoparticles within cyclohexane and an aqueous CTAB solution (see Figure 9.20b).

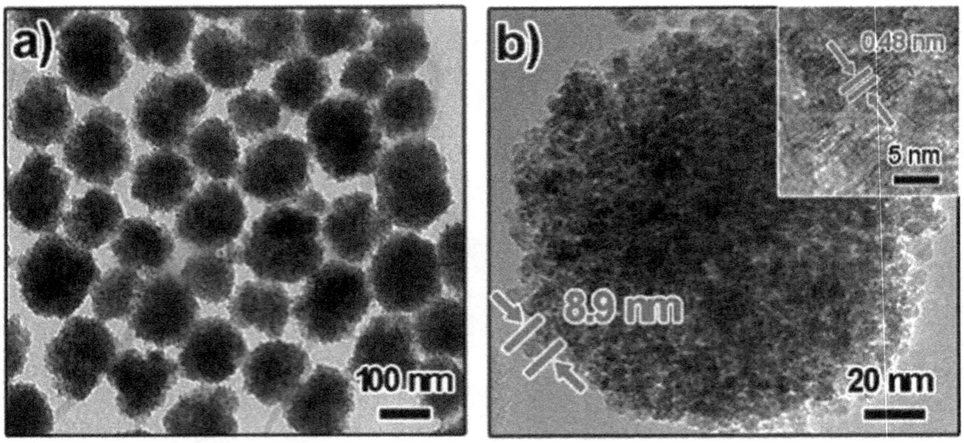

Figure 9.19 (a) 138 nm Fe_3O_4 microstructures and (b) enlarged particle with inset showing primary 8.9 nm nanoparticles [109]. Copied with permission [114].

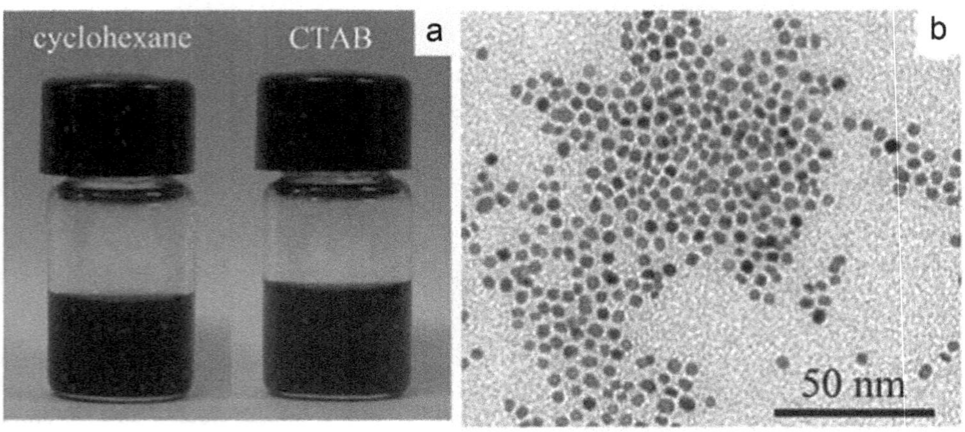

Figure 9.20 (a) Re-dispersion of Fe_3O_4 nanoparticles within cyclohexane and CTAB. (b) TEM of nanoparticles dispersed in CTAB aqueous solution. Copied with permission [114].

9.3.6 Polyol Methods

The polyol methods are a variation of solvothermal syntheses in the production of metal and metal oxides nanoparticles. Polyol typically refers to a polyhydric alcohol such as 1,2-hexadecadiol, ethylene glycol, triethylene glycol, tetraethylene glycol, diethylene glycol, propylene glycol, etc. This class of synthesis methods was first reported in 1989 for the synthesis of elemental metal (Cu, Ni, and Co) and its alloys [116–119]. However, it was soon extended to the synthesis of submicron-size precious metals and metal oxides [120–122]. In a typical polyol synthesis, metal precursor salts (nitrates, hydroxides, sulfates, carboxylates, etc.) are dissolved or partially dissolved in the selected polyol. The solution is then heated to the boiling temperature of the alcohol solution, typically at temperatures ranging from 120 to 300 °C [12, 116, 117, 120, 123]. This condition is held for an extended period of time depending on the desired size and shape of the product particles. The polyhydric alcohol acts as a solvent, reducing agent, and stabilizing agent to control nucleation and growth of particles along with preventing their aggregation. The reduction of metal occurs upon dissociation of the metal precursor salts in the polyol solution at the boiling point of the polyol, followed by the reduction of metal precursors through a redox reaction in which the nucleation and growth of particles take place [117] (see Figure 9.21).

The reduction potential of the polyol is dependent upon the choice of metal precursor salt, the reflux temperature, the concentration of base, and the type of polyol in solution. Commonly used polyols, with their associated boiling points, are

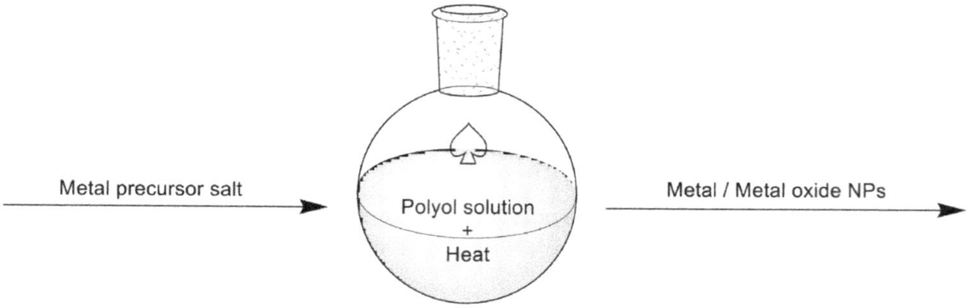

Figure 9.21 Schematic of polyol synthesis.

listed in Table 9.3. Typically, with the absence of a base, easily reducible metals, such as Ag, Au, Cu, Pt, etc., can be reduced using low boiling point polyols such as ethylene glycol [116, 117, 120–122, 124]. Higher boiling point polyols are often required to reduce metals with higher reduction potentials [125]. However, with the introduction of bases, such as NaOH, KOH, etc., to the reaction vessel, the reduction potential of the polyol can be adjusted for reducing metals at lower temperatures. In the polyol process, a seed-mediated (i.e. heterogeneous) process can be employed to reduce the energy requirement for nucleation [126, 127]. However, metal oxides are often obtained without seeding with foreign nuclei as there are a number of modifications that can be made to modify the size, morphology, and crystal quality of the particles.

The synthesis of metal oxide particles in polyol solutions was first reported by Collins et al., where they synthesized approximately 250 nm zinc oxide particles using zinc nitrate and zinc acetate in ethylene glycol under reflux conditions [123]. Jézéquel et al. later elaborated on the effect of heating rate, concentration of water, and precursor concentrations using di(ethylene) on the effect on size and shape of the ZnO particles [123]. Typically, ferrite nanoparticles are obtained via coprecipitation methods due to the simplicity of the synthesis. However, the size and shape of the nanoparticles can be difficult to control, leading to polydispersity. Ferrite nanoparticles synthesized using the polyol method provide a number of advantages over coprecipitation methods due to the ability to control the nucleation and growth of particles.

In the synthesis of ferrites nanoparticles, Cai et al. had first reported a synthetic route for monodispersed, single-phase crystalline, Fe_3O_4 (magnetite) using $Fe(acac)_3$ and triethylene glycol at reflux for 4 hours without the use of other surfactants or reducing agents [128]. The water-soluble nature of these magnetite particles was due to the electrostatic interaction between the triethylene glycol and the surface of the particles [129].

The ferrite particles that are mostly used in medical applications are known as superparamagnetic iron oxides (SPIONs), which include Fe_3O_4 (magnetite) and γ-Fe_2O_3 (maghemite). More specifically, Basti et al. had reported a phase-controlled synthesis of Fe_3O_4 and γ-Fe_2O_3 by controlling the water per metal ion ratio in diethylene glycol using iron (III) acetate as precursor [130]. However, these as-synthesized particles were not stable dispersion in water and would require additional surface modification steps [130]. The polyol method is significant in the synthesis of SPIONs as it yields particles of high-quality crystallinity and high magnetic moments with narrow size distribution in comparison to coprecipitation methods. Additionally, the as-synthesized particles produced by polyol processing are biodegradable, nontoxic, and biocompatible in physiological environments without further modifications [129].

Table 9.3 Properties of commonly used polyol. Boiling point values obtained from PubChem.

Full name	Abbreviation	Structure	Molecular weight (g/mol)	Boiling point (°C)
Ethylene glycol	EG	$OH(CH_2)_2OH$	62.07	198
Diethylene glycol	DEG	$(CH_2CH_2OH)_2O$	106.1	245.8
Triethylene glycol	TEG, 3-EG, TriEG, TREG	$HOCH_2(CH_2CH_2O)_2CH_2OH$	150.2	285
Tetraethylene glycol	TEG, 4-EG, TetraEG	$HOCH_2(CH_2CH_2O)3CH_2OH$	194.2	327.3
Glycerol	-	$HOCH_2CH(OH)CH_2OH$	92.1	290
1,2-dodecanediol	DDE	$C_{12}H_{26}O_2$	202.3	323.5
1,2-hexadecanediol	HDD, HDIOL	$C_{16}H_{34}O_2$	258.4	356.1

Polyol synthesis have also been employed in the synthesis of core–shell ferrite structures for various applications. For example, Fe_3O_4@CoO core–shell structures had been synthesized via polyol synthesis for magnetic exchange bias applications. Core–shell nanostructures can be obtained by three main methods: (i) seed-mediated growth [131–134], (ii) sequential addition [135–137], and (iii) one-pot synthesis method [138]. In seed-mediated growth, the core particles are first synthesized and used as seeds (i.e. nuclei) to disperse into a fresh solution of polyol along with a metal precursor as the source of the shell material. The seed-mediated synthesis can take place in various solvents, for instance, the seed nanoparticles that serve as the core material can be synthesized in a polyol and dispersed in other organic solvent [139]. The sequential method is a variation to the seed-mediated method, where the second metal precursor (or precursor solution) is added to the same solution instead of using a fresh solution. In the one-pot synthesis of core–shell nanoparticles, this method takes advantage of the difference in the reduction potential of the metal precursors where the core metal is reduced first at a certain temperature followed by the reduction of the second precursor creating a core–shell structure [126].

9.3.7 Continuous Flow

Research into large-scale synthesis using continuous flow reactors represents a new area of study. Continuous flow reactors can reduce batch-to-batch variability seen with the use of large reaction vessels. Continuous flow reactors are also advantageous because they require smaller volumes, which decreases the volume of expensive or hazardous chemicals needed as precursors or as byproducts [140]. Heat and mass transfer is more efficient due to the large surface area to volume ratio of the microreactors [140]. Continuous flow reactors also increase reagent mixing, leading to a more homogenous solution and resulting products. Such reactors have been used in the synthesis of metals, semiconductors, zeolites, organic compounds, and semiconductors [140–142]. Multistep reactions can also be completed using microreactors and have been demonstrated in organic synthesis reactions [143].

Many continuous flow reactions synthesizing Fe_3O_4 are based on coprecipitation synthesis methods. Typically, the microreactor uses two feed solutions, one containing the metal precursors and other containing a base solution. These are pumped separately to a mixing reactor where the reaction takes place. The resulting nanoparticles are then collected using a reservoir. A basic schematic is shown in Figure 9.22.

9.3.8 Coprecipitation Methods

Many continuous flow reactions utilize a coprecipitation method. Typically, two feed solutions are pumped separately to a mixing point where the precipitate is formed. Several different types of continuous flow reactors can be used. A droplet type reactor, or plug flow reactor, have both shown promise in synthesizing Fe_3O_4 [144]. Frenz et al. synthesized ~4 nm Fe_3O_4 using hydrodynamically coupled nozzles to deliver precursor solutions to make aqueous microdroplets in an oil containing either an iron precursor or base brought together by electrocoalescence so that reaction can take place in about 2 ms, as depicted in Figure 9.23.

Salazar-Alvarez et al. used a similar method to synthesize Fe_3O_4 but employed a multichannel peristaltic pump with a T-shaped injector for mixing solutions. Running as a continuous flow system, the Fe_3O_4 particles grew to ~3 nm with a polydispersity of 0.26. In aqueous 0.2 M NaCl, the Fe_3O_4 particles of ~5 nm resulted with a polydispersity of 0.17.

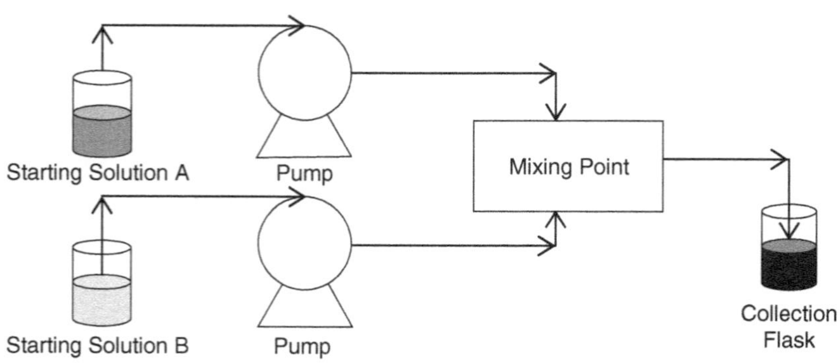

Figure 9.22 Schematic of a typical continuous flow reactor utilizing two feed solutions.

The larger particle size is attributed to the aqueous nature of lowering the base concentration around the nuclei thus forming larger particles. With kerosene as the carrier solvent, particles closely resembled those synthesized from continuous flow. The particles were ~2.6 nm with a polydispersity of 0.3. In a similar system, Simmons et al. was able to synthesize Zn-doped Fe_3O_4 with different amounts of Zn [145].

A continuous stirred-tank reactor using a coprecipitation method has also been used to synthesize Fe_3O_4 [145]. In this type reactor, shown in Figure 9.24 [146], all precursor solutions are pumped into a reaction vessel and aged while stirring. During the synthesis, two feed streams collide and enter the rotating packed bed where the two solutions are mixed on the micron scale and Fe_3O_4 is precipitated [147]. Simeonidis et al. was able to synthesize Fe_3O_4 nanoparticles ranging in diameters of 8–35 nm by adjusting pH, the iron precursors, reaction time, temperature, and surfactant. They also demonstrated the synthesis of $CoFe_2O_4$ and $MnFe_2O_4$ in this manner.

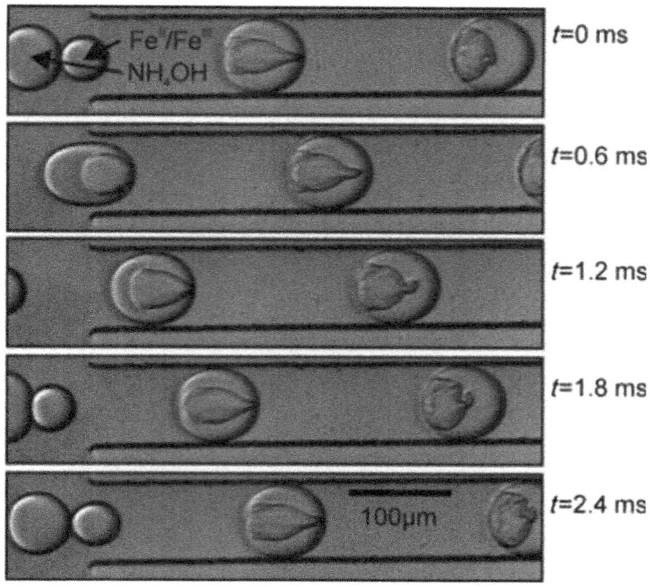

Figure 9.23 Formation of iron oxide precipitates after coalescence of reactive droplet pairs. Copied with permission from [144].

Instead of using an aqueous basic solution, gas can also be used as the precipitating agent [148]. Chin et al. used NH_3 gas as their base source and a continuous flow spinning disc, and they studied the effect of rotation speed, type of disc, precursor concentration, and surfactant use [148]. They found that at higher rotation speeds, the nanoparticles were 3–5 nm, but not uniform in shape. At lower rotation speeds, the nanoparticles were larger at 10 nm and spherical. When a grooved disc was used, a narrower size distribution was observed compared to a smooth disc due to improved mixing of the iron solution within the gas [148].

Flow reactors have successfully synthesized Fe_3O_4 by using heating modules. These reactors are very similar to the setups described above and shown in Figure 9.20, but require only a single pump and the mixing point is heated using either a molten salt or oil bath. If high pressure is required, a back-pressure regulator is added [149, 150]. Glasgow et al. synthesized Fe_3O_4 using iron oleate, oleic acid, and trioctylamine with a resonance time of 86 minutes [149]. By adjusting the flow rate, the reaction temperature, and the iron oleate to oleic acid ratio, the size of nanoparticles was adjusted.

Jiao et al. combined one solution of $Fe(acac)_3$, HOOC-PEG-COOH, and anisole with a solution of olelyamine and anisole using two pumps into a T-mixer operating at 250 °C and 3.3×10^6 Pa with various resonance times ranging from 2 to 20 minutes. They found that at longer resonance times, the yield of Fe_3O_4 increased from 73.2% ($t_R = 2$ minutes) to 85.6% ($t_R = 20$ minutes). As the resonance time increased, they also saw an increase in particle size and polydispersion.

Liang et al. synthesized Fe_3O_4 in isopropanol at 250 °C and 20 MPa (see Figure 9.25) [151]. As the temperature increased, the crystallinity of the Fe_3O_4 also improved.

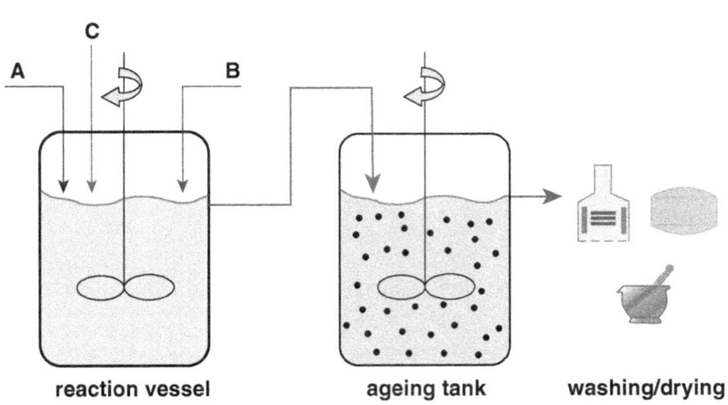

Figure 9.24 Continuous stirred-tank reactor. Two streams are added to a continuously stirred tank. The mixed product is constantly being withdrawn into a separate stirred tank where the nanoparticles are allowed to ripen to the final desired size before being drawn off and dried [146].

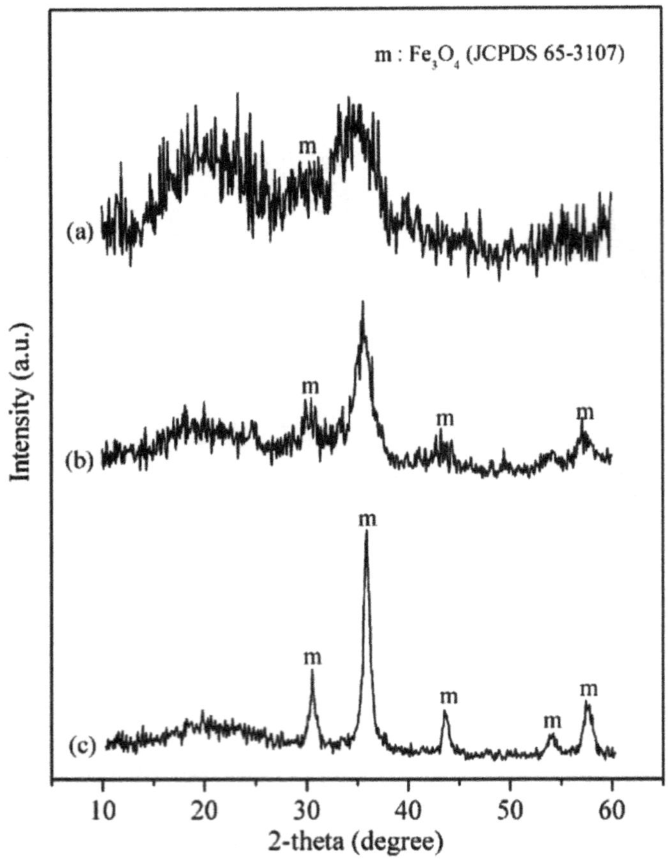

Figure 9.25 XRD of Fe_3O_4 synthesized at (a) 250 °C and 20 MPa, (b) 300 °C and 30 MPa, and (c) 350 °C and 20 MPa. Adapted from [151].

Supercritical fluids can also be accommodated as the solvent and carrier liquids in continuous flow reactors. Maurizi et al. synthesized Fe_3O_4 in supercritical water at 673 K and 25 MPa from Fe (II) sulfate and citrate [152]. Cabanas et al. synthesized Fe_3O_4 from Fe(acac)$_2$ in water at temperatures above 200 °C, resulting in particle sizes ranging from 40 to 92 nm [6]. They also reported the synthesis of $CoFe_2O_4$, $NiFe_2O_4$, and $ZnFe_2O_4$ [6].

References

1 Bohra, M., Agarwal, N., and Singh, V. (2019). A short review on Verwey transition in nanostructured Fe3O4 materials. *Journal of Nanomaterials* 2019: 18.
2 (2020). Available from: http://www.mindat.org/photo-118736.html.
3 West, J.K. (1984). *Solid State Chemistry*.Wiley, Chichester.
4 Smit, J. and Wijn, H. (1959). *Ferrites*. Eindhoven, The Netherlands: Philips Technical Library.
5 Harris, V.G. (2012). Modern microwave ferrites. *IEEE Transactions on Magnetics* 48 (3): 1075–1104.
6 Cabanas, A. and Poliakoff, M. (2001). The continuous hydrothermal synthesis of nano-particulate ferrites in near critical and supercritical water. *Journal of Materials Chemistry* 11 (5): 1408–1416.
7 Rath. (2002).
8 Glavee, G.N. et al. (1995). Chemistry of borohydride reduction of iron(II) and iron(III) ions in aqueous and nonaqueous media. Formation of nanoscale Fe, FeB, and Fe2B powders. *Inorganic Chemistry* 34 (1): 28–35.
9 Harris, V.G. et al. (1995). Direct measurement of octahedral and tetrahedral site environments in NiZn-ferrites. *IEEE Transactions on Magnetics* 31 (6): 3473–3475.

10 Willard, M.A. et al. (2004). Chemically prepared magnetic nanoparticles. *International Materials Reviews* 49 (3–4): 125–170.

11 Cushing, B.L., Kolesnichenko, V.L., and O'Connor, C.J. (2004). Recent advances in the liquid-phase syntheses of inorganic nanoparticles. *Chemical Reviews* 104 (9): 3893–3946.

12 Feldmann, C. and Jungk, H.-O. (2001). Polyol-mediated preparation of nanoscale oxide particles. *Angewandte Chemie International Edition* 40 (2): 359–362.

13 Hochepied, J.F. and Pileni, M.P. (2000). Magnetic properties of mixed cobalt–zinc ferrite nanoparticles. *Journal of Applied Physics* 87 (5): 2472–2478.

14 Huang, C. (1996). Preparation and characterization of ultrafine iron-rich nickel ferrites. *Solid State Ionics* 84 (3–4): 249–258.

15 Massart, R. (1981). Preparation of aqueous magnetic liquids in alkaline and acidic media. *IEEE Transactions on Magnetics* 17 (2): 1247–1248.

16 Delahay, P., Pourbaix, M., and Van Rysselberghe, P. (1950). Potential-pH diagrams. *Journal of Chemical Education* 27 (12).

17 Gribanov, N.M. et al. (1990). Physico-chemical regularities of obtaining highly dispersed magnetite by the method of chemical condensation. *Journal of Magnetism and Magnetic Materials* 85 (1): 7–10.

18 Faiyas, A.P.A. et al. (2010). Dependence of pH and surfactant effect in the synthesis of magnetite (Fe3O4) nanoparticles and its properties. *Journal of Magnetism and Magnetic Materials* 322 (4): 400–404.

19 Gnanaprakash, G. et al. (2007). Effect of initial pH and temperature of iron salt solutions on formation of magnetite nanoparticles. *Materials Chemistry and Physics* 103 (1): 168–175.

20 Mascolo, C.M., Pei, Y., and Ring, A.T. (2013). Room temperature co-precipitation synthesis of magnetite nanoparticles in a large pH window with different bases. *Materials* 6 (12).

21 Mei, F. et al. (2012). Particle size and magnetic properties dependence on growth temperature for rapid mixed co-precipitated magnetite nanoparticles. *Nanotechnology* 23 (14): 145601.

22 Martínez-Mera, I. et al. (2007). Synthesis of magnetite (Fe3O4) nanoparticles without surfactants at room temperature. *Materials Letters* 61 (23–24): 4447–4451.

23 Baumgartner, J. et al. (2013). Nucleation and growth of magnetite from solution. *Nature Materials* 12 (4): 310–314.

24 Petcharoen, K. and Sirivat, A. (2012). Synthesis and characterization of magnetite nanoparticles via the chemical co-precipitation method. *Materials Science and Engineering: B* 177 (5): 421–427.

25 Fang, M. et al. (2011). Rapid mixing: A route to synthesize magnetite nanoparticles with high moment. *Applied Physics Letters* 99 (22): 222501.

26 Strom, V., Olsson, R.T., and Rao, K.V. (2010). Real-time monitoring of the evolution of magnetism during precipitation of superparamagnetic nanoparticles for bioscience applications. *Journal of Materials Chemistry* 20 (20): 4168–4175.

27 Valenzuela, R. et al. (2009). Influence of stirring velocity on the synthesis of magnetite nanoparticles (Fe3O4) by the co-precipitation method. *Journal of Alloys and Compounds* 488 (1): 227–231.

28 Giri, J. et al. (2005). Preparation and characterization of phospholipid stabilized uniform sized magnetite nanoparticles. *Journal of Magnetism and Magnetic Materials* 293 (1): 62–68.

29 Kim, D.K. et al. (2003). Starch-coated superparamagnetic nanoparticles as MR contrast agents. *Chemistry of Materials* 15 (23): 4343–4351.

30 Lee, J., Isobe, T., and Senna, M. (1996). Preparation of ultrafine Fe3O4 particles by precipitation in the presence of PVA at high pH. *Journal of Colloid and Interface Science* 177 (2): 490–494.

31 Mamani, J.B. et al. (2013). Synthesis and characterization of magnetite nanoparticles coated with lauric acid. *Materials Characterization* 81: 28–36.

32 Massart, R. et al. (1995). Preparation and properties of monodisperse magnetic fluids. *Journal of Magnetism and Magnetic Materials* 149 (1): 1–5.

33 Shen, L. et al. (2014). Facile co-precipitation synthesis of shape-controlled magnetite nanoparticles. *Ceramics International* 40 (1, Part B): 1519–1524.

34 Shete, P.B. et al. (2014). Magnetic chitosan nanocomposite for hyperthermia therapy application: Preparation, characterization and in vitro experiments. *Applied Surface Science* 288: 149–157.

35 Wei, Y. et al. (2012). Synthesis of Fe3O4 nanoparticles and their magnetic properties. *Procedia Engineering* 27: 632–637.

36 Frimpong, R.A. et al. (2010). Enhancing remote controlled heating characteristics in hydrophilic magnetite nanoparticles via facile co-precipitation. *Journal of Magnetism and Magnetic Materials* 322 (3): 326–331.

37 Aono, H. et al. (2005). Synthesis of fine magnetite powder using reverse coprecipitation method and its heating properties by applying AC magnetic field. *Materials Research Bulletin* 40 (7): 1126–1135.

38 Mahmed, N. et al. (2011). Room temperature synthesis of magnetite (Fe 3−δ O 4) nanoparticles by a simple reverse co-precipitation method. *IOP Conference Series: Materials Science and Engineering* 18 (3): 032020.

39 Aphesteguy, J.C. et al. (2015). Magnetite nanoparticles prepared by co-precipitation method in different conditions. *Materials Chemistry and Physics* 161: 243–249.

40 Nedkov, I. et al. (2006). Surface oxidation, size and shape of nano-sized magnetite obtained by co-precipitation. *Journal of Magnetism and Magnetic Materials* 300 (2): 358–367.

41 Šutka, A. et al. (2015). Study of the structural phase transformation of iron oxide nanoparticles from an Fe2+ ion source by precipitation under various synthesis parameters and temperatures. *Materials Chemistry and Physics* 149–150: 473–479.

42 Pereira, C. et al. (2012). Superparamagnetic MFe2O4 (M = Fe, Co, Mn) nanoparticles: Tuning the particle size and magnetic properties through a novel one-step coprecipitation route. *Chemistry of Materials* 24 (8): 1496–1504.

43 Sun, S. et al. (2004). Monodisperse MFe2O4 (M = Fe, Co, Mn) nanoparticles. *Journal of the American Chemical Society* 126 (1): 273–279.

44 Hong, R.Y. et al. (2008). Preparation, characterization and application of Fe3O4/ZnO core/shell magnetic nanoparticles. *Materials Research Bulletin* 43 (8–9): 2457–2468.

45 Hong, R.-Y. et al. (2009). Preparation and characterization of silica-coated Fe3O4 nanoparticles used as precursor of ferrofluids. *Applied Surface Science* 255 (6): 3485–3492.

46 Morel, A.-L. et al. (2008). Sonochemical approach to the synthesis of Fe3O4@SiO2 core–shell nanoparticles with tunable properties. *ACS Nano* 2 (5): 847–856.

47 Evans, C.W., Raston, C.L., and Iyer, K.S. (2010). Nanosized luminescent superparamagnetic hybrids. *Green Chemistry* 12 (7): 1175–1179.

48 Cross, J. and Aveyard, R. (1987). Nonionic surfactants: Chemical analysis (surfactant science series, vol. 19): M. Dekker, New York, 1987 (ISBN 0-8247-7626-7). x+ 417 pp. Price $99.25 ($82.75 in the USA and Canada). *Analytica Chimica Acta* 199: 271–272.

49 Winsor, P.A. (1954). *Solvent Properties of Amphiphilic Compounds*. Butterworths Scientific Publications.

50 Pileni, M.P. (1993). Reverse micelles as microreactors. *The Journal of Physical Chemistry* 97 (27): 6961–6973.

51 Destrée, C. and Nagy, J.B. (2006). Mechanism of formation of inorganic and organic nanoparticles from microemulsions. *Advances in Colloid and Interface Science* 123–126: 353–367.

52 Gobe, M. et al. (1983). Preparation and characterization of monodisperse magnetite sols in WO microemulsion. *Journal of Colloid and Interface Science* 93 (1): 293–295.

53 Pileni, M. (1998). Fabrication and properties of nanosized material made by using colloidal assemblies as templates. *Crystal Research and Technology* 33 (7–8): 1155–1186.

54 Liz, L. et al. (1994). Preparation of colloidal Fe3O4 ultrafine particles in microemulsions. *Journal of Materials Science* 29 (14): 3797–3801.

55 López-Quintela, M.A. and Rivas, J. (1993). Chemical reactions in microemulsions: A powerful method to obtain ultrafine particles. *Journal of Colloid and Interface Science* 158 (2): 446–451.

56 Gupta, A.K. and Wells, S. (2004). Surface-modified superparamagnetic nanoparticles for drug delivery: Preparation, characterization, and cytotoxicity studies. *IEEE Transactions on Nanobioscience* 3 (1): 66–73.

57 Okoli, C. et al. (2011). Application of magnetic iron oxide nanoparticles prepared from microemulsions for protein purification. *Journal of Chemical Technology and Biotechnology* 86 (11): 1386–1393.

58 Esquivel, J. et al. (2007). A novel method to prepare magnetic nanoparticles: Precipitation in bicontinuous microemulsions. *Journal of Materials Science* 42 (21): 9015–9020.

59 Loo, A.L. et al. (2008). Synthesis of magnetic nanoparticles in bicontinuous microemulsions. Effect of surfactant concentration. *Journal of Materials Science* 43 (10): 3649–3654.

60 Aubery, C. et al. (2013). Microemulsions as reaction media for the synthesis of mixed oxide nanoparticles: Relationships between microemulsion structure, reactivity, and nanoparticle characteristics. *Langmuir* 29 (6): 1779–1789.

61 Okoli, C. et al. (2012). Comparison and functionalization study of microemulsion-prepared magnetic iron oxide nanoparticles. *Langmuir* 28 (22): 8479–8485.

62 Ebelmen, J. J.. (1846). *Recherches sur les combinaisons des acides borique et silicique avec les éthers*. Paris: s.n.

63 Brinker, C.J. et al. (1986). Sol-gel transition in simple silicates. *Journal of Non-Crystalline Solids* 82 (1): 117–126.

64 Fricke, J. and Emmerling, A. (1991). Aerogels. *Advanced Materials* 3 (10): 504–506.

65 Hench, L.L. and West, J.K. (1990). The sol-gel process. *Chemical Reviews* 90 (1): 33–72.

66 Fricke, J. (1988). Aerogels — Highly tenuous solids with fascinating properties. *Journal of Non-Crystalline Solids* 100 (1): 169–173.

67 Poelz, G. and Riethmüller, R. (1982). Preparation of silica aerogel for Cherenkov counters. *Nuclear Instruments and Methods in Physics Research* 195 (3): 491–503.

68 Tewari, P.H., Hunt, A.J., and Lofftus, K.D. (1985). Ambient-temperature supercritical drying of transparent silica aerogels. *Materials Letters* 3 (9): 363–367.

69 Glaves, C.L. et al. (1989). In situ pore structure studies of xerogel drying. *Chemistry of Materials* 1 (1): 34–40.

70 Woo, K. et al. (2003). Sol-gel mediated synthesis of Fe2O3 nanorods. *Advanced Materials* 15 (20): 1761–1764.

71 Cotton, A.F. and Wilkinson, G. (1988). *Advanced Inorganic Chemistry*, 5e. New York: John Wiley and Sons.

72 del Monte, F. et al. (1997). Formation of γ-Fe2O3 isolated nanoparticles in a silica matrix. *Langmuir* 13 (14): 3627–3634.

73 Gash, A.E. et al. (2001). Use of epoxides in the sol–gel synthesis of porous iron(III) oxide monoliths from Fe(III) salts. *Chemistry of Materials* 13 (3): 999–1007.

74 Szczygieł, I. et al. (2014). The effect of the sol-gel autocombustion synthesis conditions on the Mn–Zn ferrite magnetic properties. *Journal of Alloys and Compounds* 604: 1–7.

75 Winiarska, K., Szczygieł, I., and Klimkiewicz, R. (2013). Manganese–zinc ferrite synthesis by the sol–gel autocombustion method. Effect of the precursor on the ferrite's catalytic properties. *Industrial & Engineering Chemistry Research* 52 (1): 353–361.

76 Stöber, W., Fink, A., and Bohn, E. (1968). Controlled growth of monodisperse silica spheres in the micron size range. *Journal of Colloid and Interface Science* 26 (1): 62–69.

77 Deng, Y. et al. (2008). Superparamagnetic high-magnetization microspheres with an Fe3O4@SiO2 core and perpendicularly aligned mesoporous SiO2 shell for removal of microcystins. *Journal of the American Chemical Society* 130 (1): 28–29.

78 Lu, Y. et al. (2002). Modifying the surface properties of superparamagnetic iron oxide nanoparticles through a sol–gel approach. *Nano Letters* 2 (3): 183–186.

79 Casula, M.F., Corrias, A., and Paschina, G. (2001). Iron oxide–silica aerogel and xerogel nanocomposite materials. *Journal of Non-Crystalline Solids* 293–295: 25–31.

80 Prakash, A., McCormick, A.V., and Zachariah, M.R. (2004). Aero-sol–gel synthesis of nanoporous iron-oxide particles: A potential oxidizer for nanoenergetic materials. *Chemistry of Materials* 16 (8): 1466–1471.

81 Rockenberger, J., Scher, E.C., and Alivisatos, A.P. (1999). A new nonhydrolytic single-precursor approach to surfactant-capped nanocrystals of transition metal oxides. *Journal of the American Chemical Society* 121 (49): 11595–11596.

82 Lamer, V.K. and Dinegar, R.H. (1950). Theory, production and mechanism of formation of monodispersed hydrosols. *Journal of the American Chemical Society* 72 (11): 4847–4854.

83 Vallina, B. et al. (2014). Enhanced magnetic coercivity of α-Fe2O3 obtained from carbonated 2-line ferrihydrite. *Journal of Nanoparticle Research* 16 (3): 1–13.

84 William, W.Y. et al. (2004). Synthesis of monodisperse iron oxide nanocrystals by thermal decomposition of iron carboxylate salts. *Chemical Communications* (20): 2306–2307.

85 Zhao, S. et al. (2013). A novel solvothermal method for the preparation of magnetic monodisperse Fe 3 O 4 nanoparticles II: High-surface-activity ferrihydrite used as precursor. *Materials Research Bulletin* 48 (10): 4385–4389.

86 Hufschmid, R. et al. (2015). Synthesis of phase-pure and monodisperse iron oxide nanoparticles by thermal decomposition. *Nanoscale* 7 (25): 11142–11154.

87 Park, J. et al. (2004). Ultra-large-scale syntheses of monodisperse nanocrystals. *Nature Materials* 3 (12): 891–895.

88 Park, J. et al. (2005). One-nanometer-scale size-controlled synthesis of monodisperse magnetic Iron oxide nanoparticles. *Angewandte Chemie* 117 (19): 2932–2937.

89 Belaïd, S. et al. (2013). A new approach to follow the formation of iron oxide nanoparticles synthesized by thermal decomposition. *Nanotechnology* 24 (5): 055705.

90 Hyeon, T. et al. (2001). Synthesis of highly crystalline and monodisperse maghemite nanocrystallites without a size-selection process. *Journal of the American Chemical Society* 123 (51): 12798–12801.

91 Jun, Y.W., Choi, J.S., and Cheon, J. (2006). Shape control of semiconductor and metal oxide nanocrystals through nonhydrolytic colloidal routes. *Angewandte Chemie International Edition* 45 (21): 3414–3439.

92 Hutchings, G., Polshettiwar, V., and Asefa, T. (2013). *Nanocatalysis: Synthesis and Applications*. John Wiley & Sons.

93 Lu, A.H., Salabas, E.L., and Schüth, F. (2007). Magnetic nanoparticles: Synthesis, protection, functionalization, and application. *Angewandte Chemie International Edition* 46 (8): 1222–1244.

94 Salazar-Alvarez, G. et al. (2008). Cubic versus spherical magnetic nanoparticles: The role of surface anisotropy. *Journal of the American Chemical Society* 130 (40): 13234–13239.

95 Sayed, F.N. and Polshettiwar, V. (2015). Facile and sustainable synthesis of shaped iron oxide nanoparticles: Effect of iron precursor salts on the shapes of iron oxides. *Scientific Reports* 5.

96 Ho, C.-H. et al. (2011). Shape-controlled growth and shape-dependent cation site occupancy of monodisperse Fe3O4 nanoparticles. *Chemistry of Materials* 23 (7): 1753–1760.

97 Babes, L. et al. (1999). Synthesis of iron oxide nanoparticles used as MRI contrast agents: A parametric study. *Journal of Colloid and Interface Science* 212 (2): 474–482.

98 Palchoudhury, S. et al. (2010). Platinum attachments on iron oxide nanoparticle surfaces. *Journal of Applied Physics* 107 (9): 09B311.

99 Palchoudhury, S. et al. (2011). Synthesis of iron oxide nanoworms. *Journal of Applied Physics* 109 (7): 07E314.

100 Palchoudhury, S. et al. (2012). Controlled synthesis of iron oxide nanoplates and nanoflowers. *Chemical Communications* 48 (85): 10499–10501.

101 Guardia, P., Labarta, A., and Batlle, X. (2010). Tuning the size, the shape, and the magnetic properties of iron oxide nanoparticles. *The Journal of Physical Chemistry C* 115 (2): 390–396.

102 Shavel, A. and Liz-Marzán, L.M. (2009). Shape control of iron oxide nanoparticles. *Physical Chemistry Chemical Physics* 11 (19): 3762–3766.

103 Ge, W. et al. (2015). Simple surfactant concentration-dependent shape control of polyhedral Fe3O4 nanoparticles and their magnetic properties. *ChemPhysChem* 16 (15): 3200–3205.

104 Sun, X. et al. (2009). Size-controlled synthesis of magnetite (Fe3O4) nanoparticles coated with glucose and gluconic acid from a single Fe (III) precursor by a sucrose bifunctional hydrothermal method. *The Journal of Physical Chemistry C* 113 (36): 16002–16008.

105 Liu, Y.-W. et al. (2012). Facile and straightforward synthesis of superparamagnetic reduced graphene oxide? Fe3O4 hybrid composite by a solvothermal reaction. *Nanotechnology* 24 (2): 025604.

106 Zheng, J. et al. (2012). One-step solvothermal synthesis of Fe3O4@ C core–shell nanoparticles with tunable sizes. *Nanotechnology* 23 (16): 165601.

107 Gao, J. et al. (2013). One-step solvothermal synthesis of highly water-soluble, negatively charged superparamagnetic Fe 3 O 4 colloidal nanocrystal clusters. *Nanoscale* 5 (15): 7026–7033.

108 Ji, R. et al. (2014). Solvothermal synthesis of Co x Fe 3− x O 4 spheres and their microwave absorption properties. *Journal of Materials Chemistry C* 2 (29): 5944–5953.

109 Yang, P. et al. (2016). Gram-scale synthesis of superparamagnetic Fe 3 O 4 nanocrystal clusters with long-term charge stability for highly stable magnetically responsive photonic crystals. *Nanoscale*. 8: 19036–19042.

110 Hu, P. et al. (2008). Fabrication of monodisperse magnetite hollow spheres. *The Journal of Physical Chemistry C* 113 (3): 900–906.

111 You, L.-J. et al. (2012). Ultrafast hydrothermal synthesis of high quality magnetic core phenol–formaldehyde shell composite microspheres using the microwave method. *Langmuir* 28 (28): 10565–10572.

112 Yu, D. et al. (2006). Oriented assembly of Fe3O4 nanoparticles into monodisperse hollow single-crystal microspheres. *The Journal of Physical Chemistry B* 110 (43): 21667–21671.

113 Lin, X. et al. (2012). Formation mechanism and magnetic properties of hollow Fe 3 O 4 nanospheres synthesized without any surfactant. *CrystEngComm* 14 (24): 8658–8663.

114 Tian, Y. et al. (2011). Facile solvothermal synthesis of monodisperse Fe 3 O 4 nanocrystals with precise size control of one nanometre as potential MRI contrast agents. *Journal of Materials Chemistry* 21 (8): 2476–2481.

115 Gao, G. et al. (2010). Solvothermal synthesis and characterization of size-controlled monodisperse Fe3O4 nanoparticles. *Journal of Materials Science* 45 (13): 3483–3489.

116 Fievet, F. et al. (1993). Controlled nucleation and growth of micrometre-size copper particles prepared by the polyol process. *Journal of Materials Chemistry* 3 (6): 627–632.

117 Fievet, F. et al. (1989). Homogeneous and heterogeneous nucleations in the polyol process for the preparation of micron and submicron size metal particles. *Solid State Ionics* 32: 198–205.

118 Kurihara, L., Chow, G., and Schoen, P. (1995). Nanocrystalline metallic powders and films produced by the polyol method. *Nanostructured Materials* 5 (6): 607–613.

119 Toneguzzo, P. et al. (1998). Monodisperse ferromagnetic particles for microwave applications. *Advanced Materials* 10 (13): 1032–1035.

120 Ducamp-Sanguesa, C., Herrera-Urbina, R., and Figlarz, M. (1992). Synthesis and characterization of fine and monodisperse silver particles of uniform shape. *Journal of Solid State Chemistry* 100 (2): 272–280.

121 Ducamp-Sanguesa, C., Herrera-Urbina, R., and Figlarz, M. (1993). Fine palladium powders of uniform particle size and shape produced in ethylene glycol. *Solid State Ionics* 63: 25–30.

122 Silvert, P.Y. and Tekaia-Elhsissen, K. (1995). Synthesis of monodisperse submicronic gold particles by the polyol process. *Solid State Ionics* 82 (1): 53–60.

123 Jézéquel, D. et al. (1995). Submicrometer zinc oxide particles: Elaboration in polyol medium and morphological characteristics. *Journal of Materials Research* 10 (1): 77–83.

124 Silvert, P.-Y. et al. (1996). Preparation of colloidal silver dispersions by the polyol process. Part 1-Synthesis and characterization. *Journal of Materials Chemistry* 6 (4): 573–577.

125 Ammar, S. et al. (2001). Magnetic properties of ultrafine cobalt ferrite particles synthesized by hydrolysis in a polyol medium. *Journal of Materials Chemistry* 11 (1): 186–192.

126 Carroll, K.J. et al. (2010). Selective nucleation and growth of Cu and Ni core/shell nanoparticles. *Chemistry of Materials* 22 (7): 2175–2177.

127 Yu, H. et al. (2001). Heterogeneous seeded growth: A potentially general synthesis of monodisperse metallic nanoparticles. *Journal of the American Chemical Society* 123 (37): 9198–9199.

128 Cai, W. and Wan, J. (2007). Facile synthesis of superparamagnetic magnetite nanoparticles in liquid polyols. *Journal of Colloid and Interface Science* 305 (2): 366–370.

129 Maity, D. et al. (2009). Studies of magnetite nanoparticles synthesized by thermal decomposition of iron (III) acetylacetonate in tri(ethylene glycol). *Journal of Magnetism and Magnetic Materials* 321 (19): 3093–3098.

130 Basti, H. et al. (2010). Catechol derivatives-coated Fe3O4 and γ-Fe2O3 nanoparticles as potential MRI contrast agents. *Journal of Colloid and Interface Science* 341 (2): 248–254.

131 Gaudisson, T. et al. (2014). Exchange-biased oxide-based core-shell nanoparticles produced by seed-mediated growth in polyol. *Journal of Nanoparticle Research* 16 (4): 8.

132 Gaudisson, T. et al. (2016). On the exact crystal structure of exchange-biased Fe3O4-CoO nanoaggregates produced by seed-mediated growth in polyol. *CrystEngComm* 18 (21): 3799–3807.

133 Ourry, L. et al. (2016). A tandem polyol process and ATRP used to design new processable hybrid exchange-biased CoxFe3-xO4@CoO@PMMA nanoparticles. *RSC Advances* 6 (55): 49973–49979.

134 Souza, C.G.S., Beck, W., and Varanda, L.C. (2013). Multifunctional luminomagnetic FePt@Fe3O4/SiO2/Rhodamine B/SiO2 nanoparticles with high magnetic emanation for biomedical applications. *Journal of Nanoparticle Research* 15 (4): 1545.

135 Teng, X. et al. (2003). Platinum-maghemite core–shell nanoparticles using a sequential synthesis. *Nano Letters* 3 (2): 261–264.

136 Teng, X. and Yang, H. (2003). Synthesis of face-centered tetragonal FePt nanoparticles and granular films from Pt@Fe2O3 core–shell nanoparticles. *Journal of the American Chemical Society* 125 (47): 14559–14563.

137 Wan, J., Li, H., and Chen, K. (2009). Synthesis and characterization of Fe3O4@ZnO core–shell structured nanoparticles. *Materials Chemistry and Physics* 114 (1): 30–32.

138 Lai, C.-W. et al. (2008). One-pot solvothermal synthesis of FePt/Fe3O4 core-shell nanoparticles. *Chemical Communications* (42): 5342–5344.

139 Xia, J. et al. (2011). Preparation and characterization of bifunctional, Fe3O4/ZnO nanocomposites and their use as photocatalysts. *Applied Surface Science* 257 (23): 9724–9732.

140 Zhao, C.-X. et al. (2011). Nanoparticle synthesis in microreactors. *Chemical Engineering Science* 66 (7): 1463–1479.

141 Marre, S. and Jensen, K.F. (2010). Synthesis of micro and nanostructures in microfluidic systems. *Chemical Society Reviews* 39 (3): 1183–1202.

142 Wagner, J. and Köhler, J.M. (2005). Continuous synthesis of gold nanoparticles in a microreactor. *Nano Letters* 5 (4): 685–691.

143 Webb, D. and Jamison, T.F. (2010). Continuous flow multi-step organic synthesis. *Chemical Science* 1 (6): 675–680.

144 Frenz, L. et al. (2008). Droplet-based microreactors for the synthesis of magnetic iron oxide nanoparticles. *Angewandte Chemie International Edition* 47 (36): 6817–6820.

145 Simmons, M.D. et al. (2015). Doping of inorganic materials in microreactors – preparation of Zn doped Fe3O4 nanoparticles. *Lab on a Chip* 15 (15): 3154–3162.

146 Simeonidis, K. et al. (2016). A versatile large-scale and green process for synthesizing magnetic nanoparticles with tunable magnetic hyperthermia features. *RSC Advances* 6 (58): 53107–53117.

147 Jiao, W., Liu, Y., and Qi, G. (2010). A new impinging stream–rotating packed bed reactor for improvement of micromixing iodide and iodate. *Chemical Engineering Journal* 157 (1): 168–173.

148 Chin, S.F. et al. (2008). Size selective synthesis of superparamagnetic nanoparticles in thin fluids under continuous flow conditions. *Advanced Functional Materials* 18 (6): 922–927.

149 Glasgow, W. et al. (2016). Continuous synthesis of iron oxide (Fe3O4) nanoparticles via thermal decomposition. *Particuology* 26: 47–53.

150 Jiao, M. et al. (2015). Flow synthesis of biocompatible Fe3O4 nanoparticles: Insight into the effects of residence time, fluid velocity, and tube reactor dimension on particle size distribution. *Chemistry of Materials* 27 (4): 1299–1305.

151 Liang, M.-T. et al. (2010). Iron oxide synthesis using a continuous hydrothermal and solvothermal system. *Ceramics International* 36 (3): 1131–1135.

152 Maurizi, L. et al. (2014). Fast and continuous synthesis of nanostructured iron spinel in supercritical water: Influence of cations and citrates. *RSC Advances* 4 (86): 45673–45678.

10

Ferrite Films: Deposition Methods and Properties in View of Applications

Pieter J. van der Zaag[1], *O. Fitchorova*[2,3], *A. Sokolov*[4], *and Vincent G. Harris*[3]

[1] *Zernike Institute for Advanced Materials, University of Groningen, Nijenborgh 4, 9747 AG Groningen, The Netherlands.*
[2] *KRI at Northeastern University, LLC, 141 South Bedford Street, Burlington, MA 01803, USA.*
[3] *Electrical and Computer Engineering Department, Northeastern University, 360 Huntington Ave, Boston, MA 02115, USA.*
[4] *Metamagnetics, Inc., 115 Flanders Rd Suite 135, Westborough, MA 01581, USA.*

10.1 Introduction

Due to their magnetic character, in combination with their high resistivity, ferrites have played an important role in the electronic industry as fundamental electronic components, such as inductors and transformer cores. Their commercialization was first promoted by Philips in 1948. An important reason for their near-immediate success was their high resistivity that enabled applications at frequencies in excess of 10 kHz, where metallic magnetic components are no longer effective due to eddy current losses [1, 2]. Moreover, the fact that these ferrites, most notably MnZn- and NiZn ferrites, could be made by conventional ceramic processing added significant cost advantage. The unique properties of ferrites, which include high permeability, high resistivity, high permittivity, and nonreciprocal electromagnetic behavior, along with low magnetic and dielectric losses at high frequencies, make them also ideally suited for a wide range of high-frequency microwave applications as key materials in isolators/circulators, limiters, phase shifters, and filters. For these applications, both single-crystal and highly crystalline phase pure ferrites are regularly employed, the selection depending upon performance demands.

The past half century has seen rapid, and continuous advances in the performance and functionality of microelectronics and IC/chips due to the persistent drive for increased integration of more electronic solid-state components on a chip following the renowned Moore's law. These innovations were facilitated by the unabated demand for smaller and faster electronic components principally through miniaturization and integration made possible by the development of novel materials and process technologies. These developments have made IC/chips prevalent in many new applications affecting everyday lives as personal computers with ever-increasing computational power, GPS positioning technology and smartphones, sensor technology in cars, video gaming and entertainment platforms, and artificial intelligence.

Similar to the remarkable evolution witnessed circa 1980–2010 with respect to silicon-based commercial electronics, new and burgeoning markets in high-frequency, high-power wireless, and satellite communications, such as GPS, Bluetooth, wireless local area network (WLAN), commercial radar, etc., are witnessing unprecedented growth. This ongoing revolution is being advanced by component integration leading to technologies including: new magnetic and electrostatic (i.e. flash) nonvolatile data storage; spintronics; and magnetic-wide-bandgap (WBG) microwave integrated circuits (MICs), e.g. transmit and receive module a system-on-wafer (SoW) technology. SoW will redefine sensing, tracking, and communication technologies for the next several decades across many commercial and defense sectors.

For these applications to be realized, ferrite layers must be integrated for the first time, with semiconductor heterostructures requiring new materials and processing protocols. In contrast to Si- and GaAs-based devices, ferrite-based microwave and millimeter-wave devices, such as isolators, circulators, phase shifters, filters, limiters, etc., have remained largely unchanged from their 1970s three-dimensional (3D) topologies and form factors. With the sustained demands from commercial markets for enhanced integration and thus miniaturization, the inevitable transition of ferrite radio frequency (RF) components from 3D to two-dimensional (2D) topologies necessitates their deposition as high-quality films.

Figure 10.1 provides an overview relating various applications and technologies, which have acted as key drivers in the development of ferrite materials and film processing over time. Initially, magnetic recording was the main driver as ferrites were developed as magnetic layers in disk- and tape-based recording media. It was the discovery of the giant magnetoresistance (GMR) effect in 1988 [3] that triggered a wave of new research in the physical properties of magnetic layers. This led to the notion that the spin of the electron could be used in new electronic logic applications, leading to a completely new field of research coined spintronics. At the same time, significant expertise in the deposition of oxides had been established

Modern Ferrites: Basic Principles, Processing and Properties, Volume 1, First Edition. Edited by Vincent G. Harris.

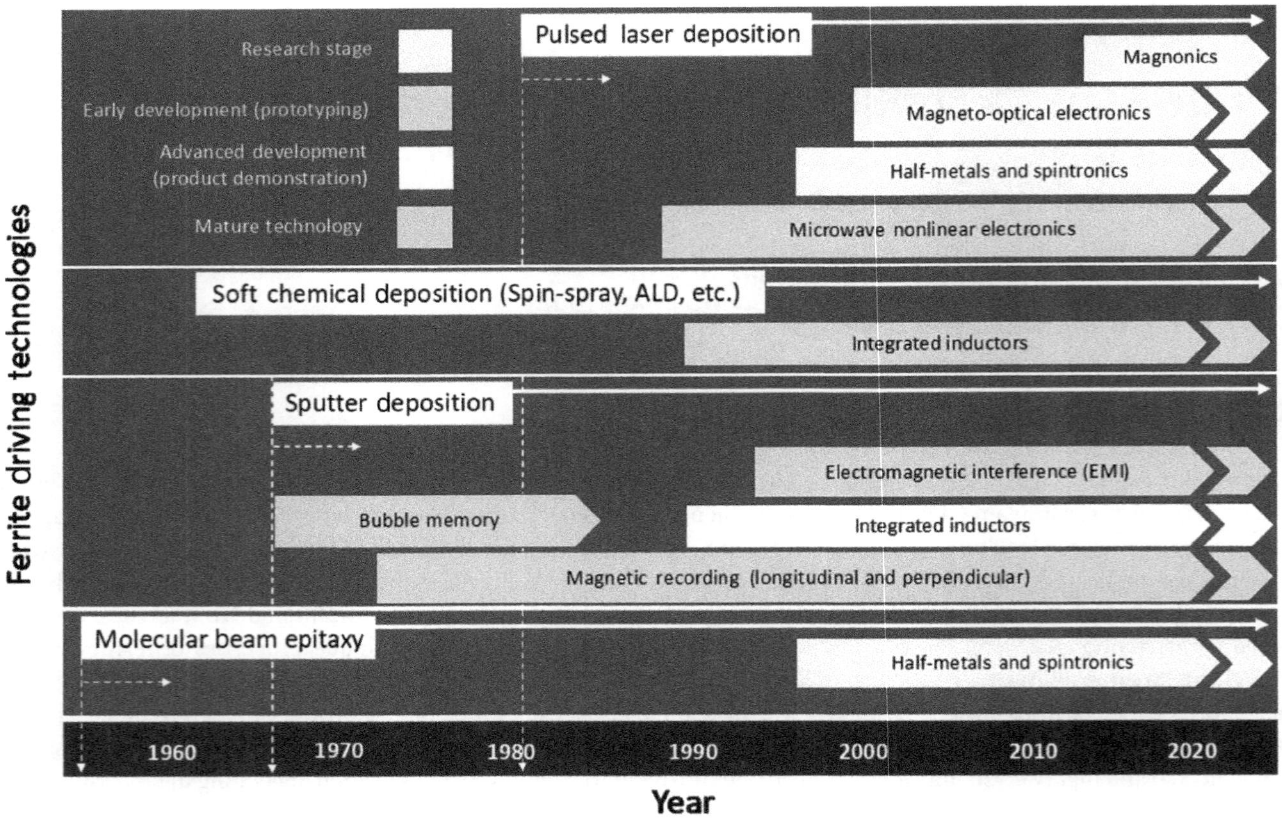

Figure 10.1 Overview of the various applications that have driven the research into ferrite layers and their deposition. Listed for the various ferrite deposition technologies considered (molecular beam epitaxy, MBE; sputter deposition; pulsed-laser deposition, PLD; and atomic layer deposition, ALD) are the key application investigated. The color code indicates the current status of this endeavor (see inset).

as a results of the discovery of high-T_c cuprate-based superconductors in 1986 [4]. This enabled research on magnetic oxide films, and the exploration of their application potential as integrated inductors in support of the telecom industry and the boom in mobile telephony of the mid-1990s. The need for possible integration of passive components that had not yet undergone the miniaturization that active components had became a limiting factor to system miniaturization. In addition, due to the rapid development in spintronics, the half-metallic properties predicted of ferrites became of intense interest circa 2000. Finally, novel magnetic memory concepts, e.g. magnetic RAM, from late 1990s onward. During the same period also new research sensor technology started in part for biomedical application, in which chemically inert layers such ferrites are attractive.

In this chapter, we explore the deposition of ferrite films using a myriad of physical, chemical, and mechanical deposition technologies, including MBE, PLD, chemical vapor deposition (CVD), sputtering, ALD, screen printing, spin-spray, and others. Challenges such as process compatibility, substrate lattice matching, crystal quality, and process yields are explored in the context of realizing the potential of new technologies that embrace the powerful multifunctionality of magneto dielectric ferrites.

10.1.1 General Introduction Ferrites and Ferrite Films

10.1.1.1 Structure and Chemistry of Ferrites

Ferrites are magnetic oxides that exist in spinel, garnet, and hexaferrite structures. Here, we present a brief discussion of each material phase in terms of their crystal structure, chemistry, and magnetic properties. This discussion is intended to provide the reader a broad overview of the fundamental physical and chemical principles required to better understand the functional properties and utility of ferrites. Other chapters of this text provide a more detailed discussion of the chemistry, structure, and magnetism of ferrites. The ferrite structure, be it spinel, garnet, or hexaferrite, has as its structural backbone a close-packed arrangement of oxygen anions. Metallic cations, magnetic and nonmagnetic, and typically divalent and trivalent, reside on the interstices of the close-packed oxygen lattice, in some cases filling all available sites, while in others preferentially filling select sites.

10.1.1.1.1 Spinels

The spinel structure, space group *Fd*3*m*, is typically expressed in the form (A)[B]$_2$O$_4$ where ( ) indicates divalent cations occupying tetrahedral lattice sites with fourfold symmetry and [ ] indicates trivalent cations occupying octahedral lattice sites with sixfold symmetry. (A)[B]$_2$O$_4$ is a single formula unit that represents one-eighth of the spinel unit cell (see depiction in Figure 10.2a). Cations have either fourfold or sixfold coordination forming tetrahedral and octahedral sublattices that are arranged in a close-packed arrangement with respect to each other. *A* cations reside on 8 of the 64 available tetrahedral sites, whereas *B* cations reside on 16 of the 32 available octahedral sites. Oxygen anions form a close-packed structure and contribute 32 anions that electrically charge balance the unit cell. The most popular and abundant of the magnetic spinels is magnetite, Fe_3O_4, or $(Fe^{III})[Fe^{II}Fe^{III}]_2O_4$, which has eight trivalent Fe ions on the *A* sublattice and a mix of divalent (eight) and trivalent (eight) ions on the *B* sublattice.

Depending on their cation distribution, spinel ferrites can be classified as *normal*, *inverse*, or *mixed* variants of the spinel structure. The *normal* spinel has divalent cations residing on the eight *A* sites with trivalent ions filling the 16 *B*-sites. Alternatively, when divalent ions fill 8 of the 16 *B*-sites with trivalent ions occupying the remaining sites the structure is said to be an *inverse* spinel. As we will see, these designations are not strictly adhered to, as the propensity of ferrite processing is often toward a nonequilibrium nature and leads to cation disorder, in which cations redistribute from their thermodynamically preferred sites. This is especially common for thin film deposition processes where arriving adatoms experience high energies leading to increased mobilities and subsequent cation disorder during film growth.

Finally, the *mixed* spinel has different ions of mixed valence occupying both *A* and *B* sublattices. Because each species of ion has thermodynamically preferred lattice distributions based predominantly on cation ionic radius, electrostatic energy, and electronic configuration, this class of ferrite commonly results from nonequilibrium processing. Cation species that exist naturally in a multiple of valence states are more susceptible to the mixed ferrite structure. For example, Mn ions with common valences of 2, 3, 4, 6, and 7, and of comparable ionic radii to Fe ions, allow for distributions such as to $\left(Mn_{1-\delta}^{2+,3+}\ Fe_{\delta}^{2+,3+}\right)\left\{Mn_{\delta}^{2+,3+}\ Fe_{2-\delta}^{2+,3+}\right\}O_4$ to exist (where δ represents the cation inversion parameter or the cation disorder parameter). The tendency to stabilize defects in such structures adds yet another degree of complexity. As one might imagine, determining such a structure with robust measurement statistics represents a significant challenge.

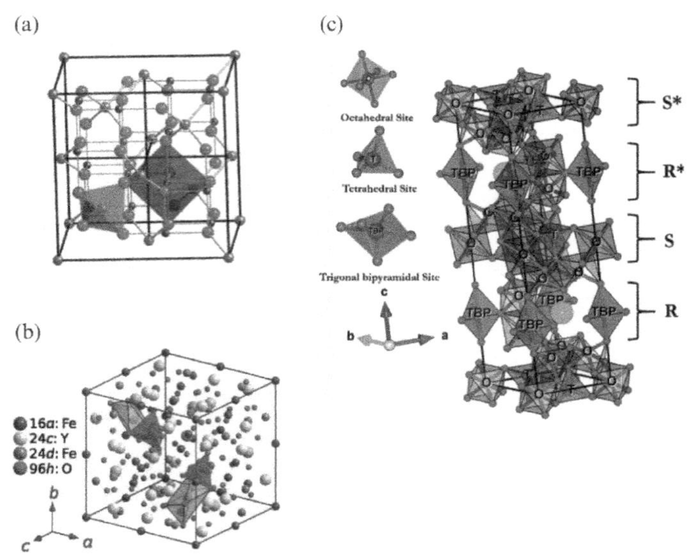

Figure 10.2 The lattice structures for three key ferrites classes: (a) the spinel lattice for conventional spinel ferrites, (b) garnet structure, and (c) magnetoplumbite structure in which hexaferrites occur (figures taken from [5–7], respectively).

10.1.1.1.2 Garnets

When considering cubic ferrites, garnets hold an important position as superior insulators, and as such, are the materials of choice for applications in which the minimization of conduction losses is an important consideration, i.e. RF and optical sensing, tracking, and communication technologies. In 1951, Yoder and Keith [8] were the first to show that substitutions of Y^{III} and Al^{III} for Mn^{II} and Si^{IV} ($Mn_3Al_2Si_3O_{12}$) allow for the stabilization of a Si-free structure, $Y_3Al_5O_{12}$. This led Bertaut and Forrat to report the structure and magnetic properties of $Y_3Fe_5O_{12}$, yttrium–iron–garnet (YIG), of the space group *Ia3d* in 1956 [9].

The garnet structure has five trivalent Fe cations that occupy two octahedral and three tetrahedral sites, with trivalent Y cations coordinated by eight oxygen ions (see Figure 10.2b). The Fe ions in the two distinct coordination sites exhibit different spin, resulting in net magnetization [10]. By substituting specific sites with rare earth elements, for example, interesting magnetic properties can be obtained [11]. In addition, YIG has a high Verdet constant that provides a large magneto-optical response, high *Q* factor at microwave frequencies, low absorption in the infrared (up to 600 nm), and low ferromagnetic resonance (FMR) linewidth [12]. YIG is used in RF, microwave, optical, and magneto-optical applications, for example, as microwave filters. It also finds utility in solid-state lasers, Faraday rotators, and in nonlinear optical applications. In the YIG structure, trivalent Fe ions occupy two octahedral and three tetrahedral sites. The yttrium ions are too large to occupy the tetrahedral and/or octahedral sites and too small to substitute for the oxygen anion. They instead occupy sites that are irregularly surrounded by eight oxygen ions in a distorted cube. In the garnet unit cell, all possible cation positions are filled, leading to uncommon chemical and structural stability as well as high insulating properties.

The Fe ions on the tetrahedral and octahedral sites are coupled ferrimagnetically by superexchange as in the spinel systems. The net moment derives from the excess on tetrahedral sites. The Curie temperature also derives from superexchange and is nearly the same for all garnets 560 K [1]. However, the saturation magnetization and the magnetic anisotropy field are rather low and are, respectively, compared with other ferrite systems resulting in a low FMR frequency. For these reasons, YIG is usually biased by permanent magnets for microwave devices applications. Notwithstanding the need for bias magnets, YIG is a choice material for many applications at frequencies up to and including X-band (i.e. 8.0–12 GHz), above which the bias magnet size, weight, and cost become prohibitive. At higher frequencies, ferrites with substantially higher magnetization and magnetic anisotropy are required. For this, we turn to hexaferrites.

10.1.1.1.3 Hexaferrites

In contrast to the spinel and garnet ferrites, the hexaferrites have hexagonal crystal symmetry leading to high magnetocrystalline anisotropy energies and magnetic anisotropy fields and subsequently high FMR frequencies. Among the most popular of the microwave hexaferrites are those derived from the barium M-type (BaM) hexaferrites (see Figure 10.2c). BaM has the magnetoplumbite structure and space group *Pb*3/*mmc* and consists of 10 closely packed oxygen layers in which 4 successive layers, each containing four oxygen anions, are followed by a fifth layer containing three anions and a Ba cation. The structure is constructed of spinel (S) blocks containing two oxygen layers separated by an R-block containing the Ba ion. Every other S and R blocks are rotated 180° with respect to the other and is denoted as S* and R*, respectively. Trivalent Fe ions occupy tetrahedral and octahedral sites (derived from spinel blocks) as well as a trigonal bipyramidal (TBP) site. The TBP site resides on the Ba-containing layer amidst the triagonal arrangement of oxygen anions [1]. There exist other hexaferrite structural variants designated W, Y, W, X, Z, and U, each having different structures and oxygen stoichiometries of markedly different magnetic properties. The reader is directed to the review of hexaferrites by Pullar [13] for details.

10.1.1.1.4 Ferrite Magnetism

Circa 1948, Néel put forward an expansion of the Curie–Weiss molecular field theory [14] introducing for the first time the use of multiple molecular fields to describe coupled aligned and anti-aligned spin lattices to explain complex thermomagnetic behavior in magnetic oxide systems. Thus, Néel introduced to the magnetics community the concept of ferrimagnetic and antiferromagnetic materials.

In magnetic oxide systems such as ferrites, due to the closely packed oxygen backbone structure, the distance between metal ions is too large to support direct exchange, such as that experienced in ferromagnetic metals of the 3d transition elements Fe, Co, and Ni. The exchange in ferrites, therefore, derives from the indirect exchange between adjacent cations within the same sublattice or neighboring sublattices mediated through a bridging oxygen anion and is thus short range in nature and extremely sensitive to local bonding arrangements. The first postulation of this indirect exchange in oxide systems was put forth by Kramers [15] and expanded upon by Anderson [2, 16]. Today, this theory is referred to as Kramer–Anderson superexchange (KA-SE). Phenomenological rules that relate the rigorous physics and mathematics of KA-SE to electron filling, ligand interaction, and bonding geometries in molecular structures were developed by Goodenough and Kanamori [3–6,

17–20]. Today, these rules go by the Goodenough–Kanamori rules for superexchange [21]. The GK rules relate bridging angles and orbital filling to exchange strength. Three principal factors are shown to affect the strength of Me–O–Me exchange, these include the distance, direction, and angle of the cations with respect to the bridging anion. While the direction and angle are defined by the relationship between Me–O–Me, the critical distance is that between Me–O and not Me–Me, the latter having little impact. All three factors determine the degree of orbital overlap between the 2p orbitals of oxygen and the 3d orbitals of the cations and hence the magnitude of the superexchange interaction represented by J. By applying the GK rules to p- and d-orbital alignments, one can readily conclude that larger Me–O–Me angles produce strong and antiferromagnetic superexchange interactions, while angles that approach 90° produce super exchange interaction that are weak and ferromagnetic. In spinel ferrites, this theory explains that the interaction between magnetic ions on the A- and B-sites, which make an angle of around 125° (see Figure 10.2a), is much stronger than between those on the B-site, which make an angle of around 90° [1]. The magnitude of J plays a direct and dominant role in determining magnetization and Curie temperature among other important intrinsic magnetic properties [22]. An important consequence is that in (spinel) ferrites the dominant interaction is the AB interaction, which is typically 20 times stronger than the BB interaction. This causes the magnetic moments on the A- and B-sites to couple in an antiparallel orientation, whereas those on the B-sites to align parallel to one another [23, 24]. A direct result of this ferrimagnetic alignment is that the magnetization values in ferrites are smaller than those in metallic magnetic materials. Typically, spinel ferrites have saturation magnetization values, M_S, between 143 and 150 kA/m at 0 K and 120 and 477 kA/m at room temperature, depending on their exact compositions [1], while for garnets such as YIG typical values are M_S is 196 kA/m at 0 K and 140 kA/m at room temperatures [25]. Depending on doping, these values can be reduced to near zero, which is relevant for some applications, to about 20% higher than the YIG value.

10.1.1.1.5 Introduction to Relevant Applications of Ferrite Films

Three main relevant applications of ferrite films:

i) MnZn- and NiZn-ferrite films of the spinel category act as flux guides in a number of passive component applications at MHz frequencies.

Owing to their oxidic nature, spinel ferrites have found widespread use in the electronics industry as their high electrical resistivity in comparison to metallic soft-magnetic materials enable applications at frequencies ≥ 10 kHz. Over the years, spinel ferrites such as MnZn- and NiZn ferrites have found widespread use as inductors and transformer cores. The integration of inductors onto ICs for telecommunication applications requires a relatively large surface area that can be reduced significantly if soft-magnetic materials were used as their permeability, μ, would increase the effective inductance dramatically. Moreover, in many instances, transport or concentration of magnetic flux is needed. This can be achieved with a magnetic layer, which has a low reluctance compared to its surroundings, as the magnetic reluctant R is proportional to $1/\mu_r$, with μ_r the relative permeability. It is important to note that ferrite films do not have to exhibit a very high μ to be

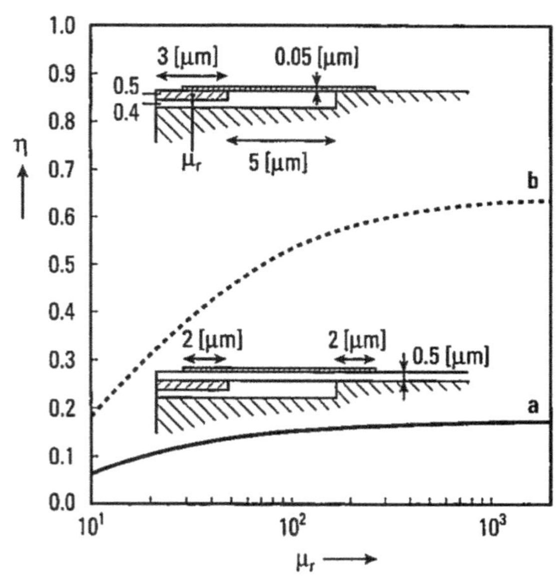

Figure 10.3 Calculated efficiency η of the yoke-type magnetoresistive head (YMRH) versus relative permeability of the flux guide, μ_r, for two conditions. The lower curve (a) corresponds to the situation of a standard YMRH in which the separation oxide is 0.5 μm thick. The upper curve (b) corresponds to the situation in which the flux guide is made of an electrically insulating magnetic material, which permits direct contact between the MR element and the flux guide. Note that even for very low values of $\mu_r \leqslant 10$, the efficiency of this read head is significantly improved by a factor of ≈ 3 (reproduced from [26]).

effective. Figure 10.3 shows this, for the case of using a ferrite film in direct contact with the magnetoresistance element. In the case of a magnetic read head, the sensitivity of the device improves effectively by more than 300% by a mere increase of 1 in μ_r, confirming the permeable material acts as a flux concentrator [26]. Thus, in applications of this nature, it is not the magnetization value, M, of the ferrite layer but the permeability, which can be achieved, that is the crucial parameter to consider.

ii) Spintronic applications: Half-metallic Fe_3O_4 as spin-polarized current injector

Given Fe_3O_4's half-metallic character in which current is only carried by the minority spin [27], using Fe_3O_4 as an electrode material in a tunnel junction has great potential in enhancing the device's spin-transport efficiency.

This can be explained as follows: In magnetic tunnel junctions, the tunnel magnetoresistance (TMR) is defined as the difference in the resistance in the magnetic electrodes between parallel (R_P) and antiparallel (R_{AP}) states divided by the resistance of the antiparallel state [Equation (10.1)]. This definition is adopted in recognizing that the magnetostatic energy of the antiparallel state will be the device zero-field state, and the parallel state will be the high-field state:

$$\text{TMR} = \frac{R_P - R_{AP}}{R_{AP}}\% = \frac{R(H) - R(0)}{R(0)} \tag{10.1}$$

Thus, the TMR will be positive if $R_P > R_{AP}$ and negative when $R_{AP} > R_P$. The TMR can be related to the spin polarization of the electrodes (i.e. P_1 and P_2) through the Julière formula [Equation (10.2)] to be [28]:

$$TMR = \frac{2P_1P_2}{1 - P_1P_2}\% \tag{10.2}$$

In Equation (10.2), the spin polarization P_i of the ith electrode is given by the normalized difference between the densities of the majority spin, $N_{i\uparrow}$, and the minority spin, $N_{i\downarrow}$, at the Fermi level, E_F:

$$P_i = \frac{N_{i\uparrow}(E_F) - N_{i\downarrow}(E_F)}{N_{i\uparrow}(E_F) + N_{i\downarrow}(E_F)} \tag{10.3}$$

The spin polarization of a material is thus positive when the majority spin at the Fermi level is parallel to the bulk magnetization direction and negative when the minority spin at the Fermi level is parallel. Given the theoretical consideration that Fe_3O_4 is a half-metallic ferromagnet [27] with only minority spin polarization at the Fermi level [29], P_i in Equation (10.3) becomes 1, predicting an infinite TMR if two Fe_3O_4 electrodes are employed. The potential to attain a semi-infinite TMR while employing a chemically robust oxide of "simple" composition has motivated many groups to pursue the study of Fe_3O_4 thin-film growth.

As we will discuss later in this chapter for these spintronic applications, in which Fe_3O_4 acts as a spin injector and the spin of the electron needs to be maintained and not flipped by scattering events, high-quality interfaces are the key requirement to prevent spin-flipping scattering and ensure that a spin-current can be injected into an adjacent layer.

iii) YIG for spin-current generation and spin pumping at gigahertz frequencies

A relatively new field of transport electronics, referred to as *magnonics* is based upon long-range communication of spin waves [30–32]. The term *magnonics* falls within the spintronics genus where the quantized spin-wave, i.e. the magnon, couples with injected electron spin-currents. To realize practical magnonic devices, a magnetic medium of high crystal quality and ultralow Gilbert damping is required. Thus far, YIG has shown to be an ideal thin film magnonic media [33–35].

To meet the needs for high-quality YIG films, deposition techniques such as oxygen-assisted PLD and liquid-phase epitaxy (LPE) are of particular utility especially for thin films of insulator-based magnon spintronics [36, 37]. Spin waves manifest as either short-wavelength exchange spin waves or long-wavelength magnetostatic waves. Thus far, most magnon signal processing devices operate with magnetostatic waves that can be excited and detected by inductive antennas.

More recently, attention has focused on exchange waves that allow for nanostructured devices.

Both types of waves have different dispersion characteristics and nonlinear properties and thus offer unique advantages for data processing. To achieve the integration of magnonic devices with electronic circuits efficiently means of exciting magnons by charge currents are required. A promising approach that has been demonstrated is the use of spin-transfer-torque (STT) to realign the magnetization vector or to excite spin precession [38–40]. An alternative approach in achieving spin-polarized electron currents is from the spin Hall effect caused by spin-dependent scattering of electrons in a nonmagnetic metal or semiconductor characterized as having large spin–orbit interaction [41, 42].

Present magnon physics focuses on the development of low-energy magnon excitation and their manipulation and detection in nanometer-scale devices. Such performance and form factors have the greatest potential to impact computational efficiency, functionality, and power. The further development of post-CMOS computer technologies, such as the integration of magnon computing with data storage, magnon logic and memory, have been the subject of great research interest [43–46].

Based on these recent advances, one can imagine the creation of magnonic-based THz communication technologies (i.e. 6G and beyond) [47]. Such breakthroughs would require new low-Gilbert loss ferrites that would enable long-distance magnonic currents to be realized and concomitantly allow decreases in energy consumption. To realize such advances, new magnetic oxides and processing techniques, such as variants of PLD, MBE, CVD, and ALD, must be developed.

In summary, magnonics is a burgeoning field of spintronics that is anticipated to deliver breakthroughs enabling advances in computation, data processing, and communication in the coming years.

10.1.1.1.6 Organization of This Chapter

This chapter begins with a discussion of reported results related to the deposition of Fe_3O_4 by oxygen-assisted MBE. This section builds on and extends an earlier review [48]. An additional reason for this pursuit is that the relative simplicity of the Fe_3O_4 ferrite structure, in combination with a well-controlled deposition technique, provides the basis for understanding the behavior for growth dynamics of increasingly more complex oxides. For these more complex systems, such as MnZn-ferrite, YIG, and hexaferrites, different depositions methods such as PLD, are more suitable. Finally, we discuss deposition techniques such as sputtering and spin-spraying of ferrite layers, which, by their deposition rate and relative costs, are better suited for industrial- and commercial-scale deposition of ferrite layers.

10.2 MBE Growth of Ferrites

10.2.1 MBE Growth of Iron Oxides Films Notably Fe_3O_4

As a natural consequence of advances made in growing high-temperature superconductors during the 1980s, attention naturally turned to the growth of ferrites. The advances in MBE growth of the high-T_c superconducting cuprates, which are composed of copper oxide planes, led several groups to consider applying oxygen-assisted MBE to the growth of ferrite layers and multilayers. However, as ferrites are mixed-valence oxide systems, this raises many challenges. First, the properties of these magnetic oxides are intimately related to cation distribution and stoichiometry, as this influences magnetization and anisotropy [1]. Second, one can envision that these effects are particularly pronounced at interfaces and surfaces giving rise to complications in addition to new effects in thin layers and multilayers comprising mixed-valence oxides such as ferrites.

Table 10.1 presents an overview of various groups that have worked on oxygen-assisted MBE of iron oxides, notably Fe_3O_4, over the last three decades. A previous review by Chambers on the MBE growth of metal oxides in general, focused more on growth-related aspects and appeared in 2000 [49]. Table 10.1 lists typical growth conditions used for Fe_3O_4 deposition, as well as key features of the employed MBE system with respect to previous studies. Typically, in the practice of MBE, layers are deposited in an ultra- high-vacuum chamber, typically having base pressures in the range of low 10^{-10} mbar, minimizing chamber contaminants and allowing the film growth of very pure materials.

Additionally, the use of effusion from Knudsen cells ensures deposition is performed with atomic species arriving at the substrate with very low energies of only a few eV/atom. Consequently, high-quality layers can be produced having sharp interfaces of limited interdiffusion and low surface roughness. In principle, this allows the construction of new artificial materials, i.e. heterostructures, by depositing ideally one layer at a time. Figure 10.4 provides a schematic diagram of the oxygen-assisted MBE system used by the Lind group at Florida State University in Tallahassee [50].

Pioneering work on the deposition of iron oxides layers by MBE was performed by Bando and coworkers [51]. They were first to report the properties of oxygen-assisted MBE-grown Fe_3O_4 films using a NaCl rock salt crystal as substrate [51]. In this initial work, the Fe_3O_4 stabilization phase field, as defined by substrate temperature and deposition rate, was established (see Figure 10.5). Subsequently, this group also reported the growth of Fe_3O_4-based multilayers, through the growth of various Fe_3O_4 and

Table 10.1 MBE growth conditions for Fe_3O_4.

Year	Oxygen pressure (mbar)	T_{dep} (°C)	Grow rate (Å/s)	Substrates	System feature	Ref.
1980	$5–7 \times 10^{-4}$	250–300	5.0–20	NaCl (001)	—	[51]
1992	$2–4 \times 10^{-5}$	240	2.0–2.5	MgO (001)	Oxygen plasma generator to increase oxidizing power	[50]
1994	1.5×10^{-6}	302	0.6	MgO (001), $SrTiO_3$ (001), $MgAl_2O_4$ (001) Al_2O_3 (00001)	Ring-shaped oxygen doser close to the substrate	[52]
1997	—	—	—	—	—	[50]
1999	—	—	—	—	Use of NO_2 as a more powerful oxidizing agent	[53]
2000	—	≤ 300	0.8	MgO (001)	(Similar to system Lind and coworkers [50])	[54–56]
2001	6×10^{-7}	450	0.033	Al_2O_3 (00001)	Use of *atomic* oxygen-assisted growth through the use plasma source of which 5% of the resulting flux is atomic oxygen	[57]
2008	—	—	—	—	—	[58]
2010	1.3×10^{-5}	250	0.3	MgO (001)	80 W electron cyclotron resonance (ECR) plasma source to generate free oxygen radicals	[59]
2011	4×10^{-7}	230	0.033	MgO(001)	*In situ* ultraviolet photoelectron spectroscopy (UPS) and X-ray photoelectron spectroscopy (XPS)	[60]
2014	1.0×10^{-6} 3×10^{-7}	250		MgO (001), $SrTiO_3$ (001), $MgAl_2O_4$ (001)	*In situ* XPS; *in situ* soft X-ray absorption spectroscopy (XAS)	[61]
2011/2016	1×10^{-4} 5×10^{-6}	250		MgO (001) with a Fe cover layer MgO (001)	*In situ* low-energy electron diffraction (LEED) and XPS. Postdeposition annealing of γ-Fe_2O_3 to Fe_3O_4 at 360 °C	[62, 63]
2014	5×10^{-7}	350	0.018	NiO (001)	*In situ* XPS	[64]

(a)

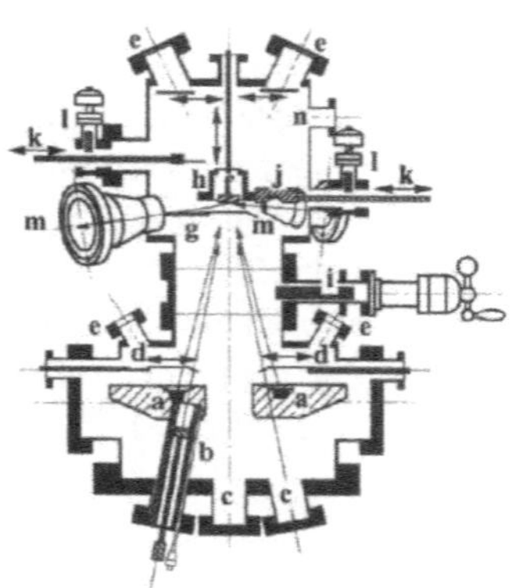

(b)

Figure 10.4 Examples of an oxygen-assisted ultra-high-vacuum MBE deposition system: (A) The system of the Lind group at the Florida State University in Tallahassee (USA) used as an example. This system operates at a typical base pressure of 1×10^{-9} mbar. Others achieve low 10^{-10} mbar base pressure. In this system, the deposition sources are in the lower chamber facing upward and two 15 kW electron gun sources (a), an electron cyclotron resonance (ECR) source magnetically focused gas plasma source (b) with gas inlet and four Knudsen effusion sources [not shown (c)], with all sources pneumatically shuttered (d) and source charge visible from shuttered viewports (e). The substrate holder (f) is shuttered (g) and placed in a heating shroud (h) in the upper chamber. The chambers are joined by an 8-inch inner diameter isolation valve (i). Both quartz crystal deposition rate monitor head (j) and substrate holder (f) may be retracted (k) and gated from the main chamber for replacement. Grazing incidence reflection high-energy electron diffraction (RHEED) diffraction (m) and quadrupole residual gas analysis (n) are also available in the upper chamber. (Reproduced with permission from [50]. (©1992 by the American Physical Society) B) A more recent system in use at CEA Gif-sur Yvette, France. Note the further extension with growth analysis techniques such as XPS. Reproduced with permission from [65].

CoO layers, again using NaCl as the substrate [66]. The advantage of a highly ionic substrate is that the substrate can be readily dissolved in water so as not to interfere with subsequent transmission Mössbauer, transmission electron microscopy (TEM), and magnetization measurements. In the 1990s, other groups led by Lind and Wolf pursued oxidic MBE growth to refine the processing of Fe_3O_4-based layers and multilayers [50, 52, 67]. These groups moved away from NaCl substrates, which with its lattice parameter of 5.62 Å, has rather large lattice mismatch of 34% relative to Fe_3O_4, $a_0 = 4.198$ Å (8.396 Å/2), see the preceding section and Figure 10.2a, to the use of MgO (001). In contrast, the lattice parameter of MgO is 4.212 Å, providing a lattice mismatch of only 0.3% with the oxygen template of Fe_3O_4, thus being better suited for high-quality epitaxial growth. Additionally, these groups demonstrated improved control of the oxidation of Fe, which was evaporated from a Knudsen cell, by using either an oxygen plasma generator [50] or a ring-shaped oxygen doser in close proximity to the substrate [52].

The theme of improving the control of oxidation of the evaporated Fe flux was further elaborated by Hibma and coworkers [53], who pursued MBE growth studies using NO_2 as a more aggressive oxidizing agent such that, in addition to Fe_3O_4 and hematite, i.e. α-Fe_2O_3, other iron oxides of maghemite (γ-Fe_2O_3) and wustite (FeO) were stabilized. Also, Gota et al. contributed to this development by using a source to generate atomic oxygen rather than oxygen radicals commonly yielded by a plasma generator [57, 70]. A further interesting aspect of this work was that these authors extended studies of Fe_3O_4 to the growth of α-Fe_2O_3 (0001) [70] as well as (111) oriented Fe_3O_4 [57] on α-Al_2O_3 (0001) substrates.

Subsequent research groups used MBE systems that enabled the study of the growth of Fe_3O_4 layers by *in situ* spectroscopies, including techniques such as X-ray photoelectron spectroscopy (XPS), soft X-ray absorption spectroscopy (XAS), and ultraviolet photoelectron spectroscopy (UPS) to characterize layers in terms of stoichiometry, bonding valence, local bonding symmetries, and overall crystal quality [61, 64, 71].

From these MBE studies (see Table 10.1), the following conclusions can be derived on optimal deposition conditions for Fe_3O_4. First, the deposition temperature, T_{dep}, should be chosen to be relatively low. This not only follows from careful studies performed by Bando and coworkers, who explored combinations of deposition rate and substrate temperature, T_S (see

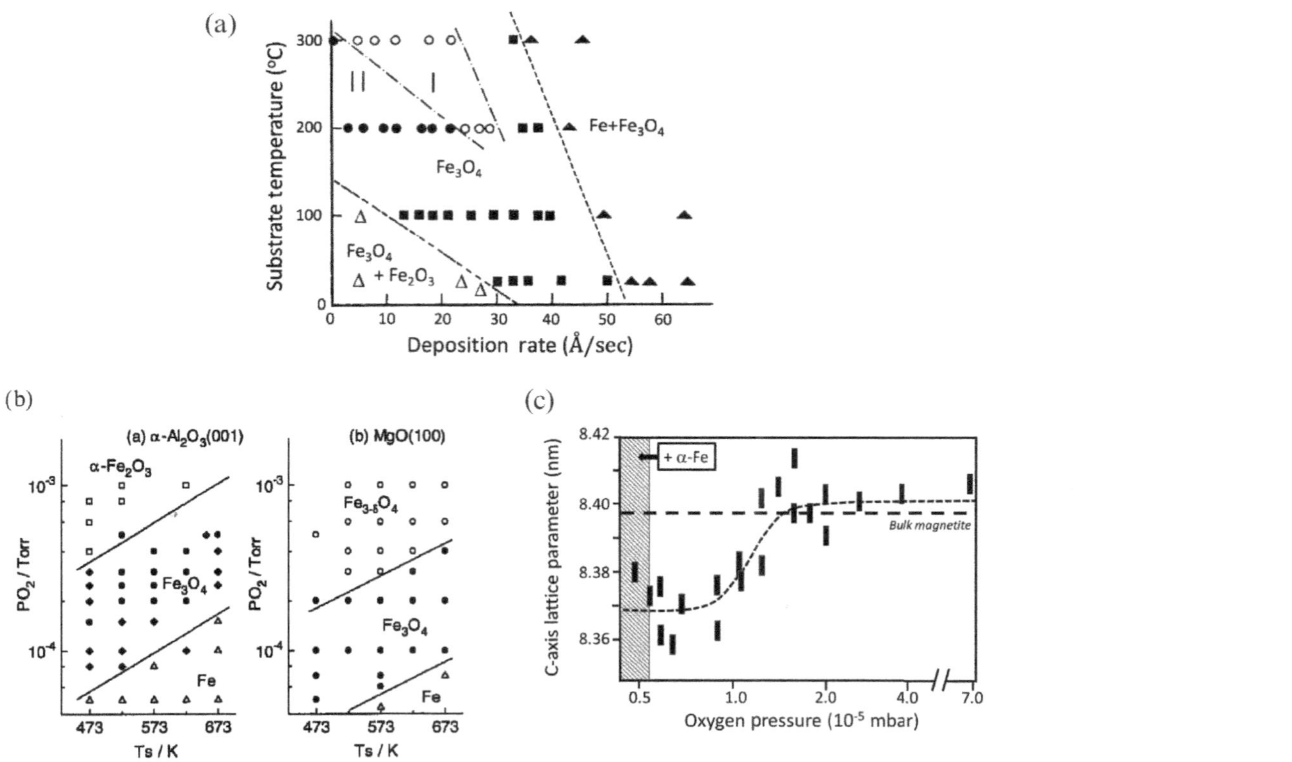

Figure 10.5 (a) Formation range of iron oxides as a function of deposition rate and substrate temperature on NaCl (001) under an oxygen pressure of 6.7 mbar. The films obtained in the regions I (o) and II (●) were considered to be epitaxial Fe_3O_4 films of types I and II, respectively. In film of type I, the anion lattice of Fe_3O_4 (001) is parallel to that the (001) surface of the substrate, whereas for type II films, other anion surfaces such as (110) or (111) are parallel to the (001) substrate surface. The films indicated with (■) were nonoriented Fe_3O_4 films [68]. (b) Formation ranges of Fe, Fe_3O_4, and α-Fe_2O_3 or $Fe_{3-\delta}O_4$ as a function of substrate temperature (T_S) and oxygen partial pressure (pO_2) on α-Al_2O_3 (0001) and MgO (1 0 0), respectively (data taken from [69]). (c) Influence of the oxygen pressure on the magnetite *c*-axis lattice parameter for 1000 Å films grown heteroepitaxially on $SrTiO_3$ (1 0 0). Symbol size represents experimental uncertainty. In the range of 0.6–1.5 × 10^{-5} mbar black featureless films with the bulk Fe_3O_4 where grown. At higher oxygen pressure, red-brown insulating Fe_2O_3 film resulted (figure adapted from [52]).

Figure 10.5a) and oxygen partial oxygen pressure, pO_2 (see Figure 10.5b), to determine optimal pure Fe_3O_4 layers [68, 69]. A further important consideration is that a low deposition temperature of $T_S \sim 250\ °C$ limits intermixing of the Fe_3O_4 layers with the underlying MgO substrate that could give rise to formation of a $Mg_xFe_{2-y}O_{4-\delta}$ ferrite phase at the interface [72, 73]. From Table 10.1, one can further see that the pO_2 should be the order of 3–15 × 10^{-7} mbar. In part, this was due to the use of various improvements to oxidizing conditions through the use of an oxygen plasma generator [50], administering oxygen close to the substrate [52], or the use as NO_2 as an aggressive oxidizing source [53]. With respect to this optimum range of pO_2, the pO_2 used by the Lind group was an order of magnitude higher (see Table 10.1). This can be understood from the fact that in their studies to maintain stability of the Fe and O^+ plasma used in the deposition, the pressure had to be maintained in the 2–4 × 10^{-5} range.

The growth rates employed, which are typically ≤ 1.0 Å/s (see Table 10.1), have also been shown to be very important. This corresponds to a growth rate on the order of 5 nm/min so that the upper limit of what can reasonably be deposited is a layer in the range of 100 nm [61].

10.2.2 Structural Characterization of MBE-Grown Fe_3O_4 Films

The growth of Fe_3O_4 on MgO (001) substrates, or NiO (001), which has an oxygen lattice mismatch of only 0.51% with respect to Fe_3O_4 [64], can lead to excellent *structural* properties of epitaxial Fe_3O_4 layers or Fe_3O_4-based multilayers, as have been amply demonstrated in the literature.

Nearly all groups active in this area of research employ *in situ* RHEED to monitor layer-by-layer growth of films as deduced from RHEED intensity oscillations. Additionally, ever since introduced by the Hibma and Chambers group in the study of the growth of Fe_3O_4 films [53, 74], low-energy electron diffraction (LEED) has also been employed by many groups [60, 61, 75] to verify *in situ* crystalline structure of films. Typically, the presence of sharp RHEED streaks and Kikuchi lines, as well as intense and sharp LEED spots, indicate the presence of well-ordered crystalline surface structure of Fe_3O_4 layers grown under near ideal growth conditions.

Figure 10.6 is illustrative of a Fe_3O_4 film grown on MgO (001). As the growth is fully coherent between the in-plane oxygen lattices of Fe_3O_4 and MgO, while the spinel lattice parameter $a = 8.396$ Å is nearly twice that of MgO, an extra set of RHEED streaks and LEED spots appear at half the lattice spacing of MgO [61]. The occurrence of the half-order diffraction in the zeroth Laue zone, rods and spots in the RHEED and LEED patterns are indicative of the well-known ($\sqrt{2} \times \sqrt{2}$)R45^0 or p(1 × 1) surface reconstruction that is characteristic of single-crystalline Fe_3O_4 [62]. In Figure 10.6, these patterns, illustrated in (c) and (g), correspond to a growth pO_2 of 1.0 × 10^{-6}. At a higher pO_2 of 1.0 × 10^{-5}, patterns become broader, indicating increased disorder. While at a pO_2 of 5 × 10^{-8}, RHEED and LEED do not show characteristic crystalline (surface) structure of Fe_3O_4 [61]. Although these patterns are helpful in assessing crystal quality during growth, LEED patterns alone cannot distinguish between formation of Fe_3O_4 or γ-Fe_2O_3, as their crystal structure and lattice parameters are too similar [64].

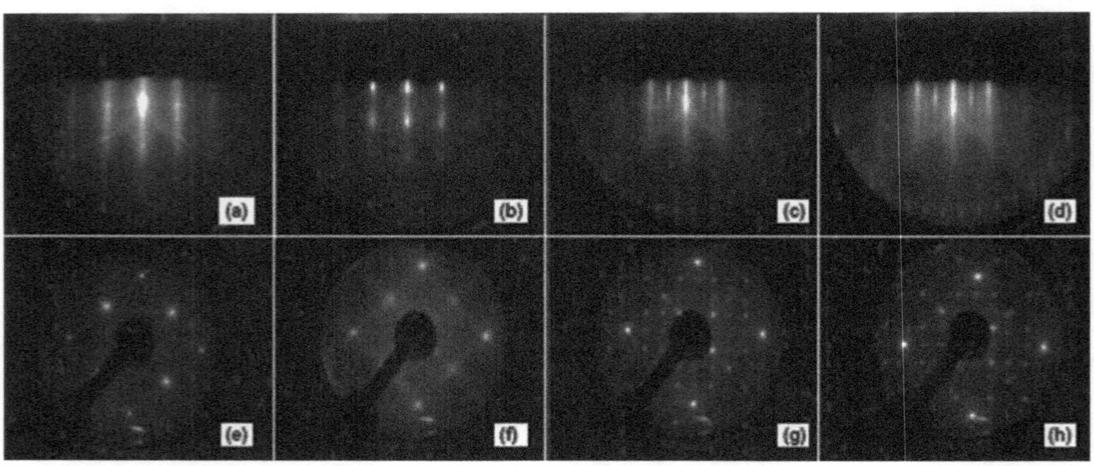

Figure 10.6 (Top row) Reflection high-energy electron diffraction (RHEED) and (bottom row) low-energy electron diffraction (LEED) diffraction patterns of (a) and (e) a clean MgO (001) substrate, and various 40-nm-thick Fe_3O_4 films grown at different at oxygen partial pressure, pO_2 = 5 × 10^{-8} mbar (b) and (f); 1.0 × 10^{-6} mbar (c) and (g); 1.0 × 10^{-5} mbar (d) and (h) by the Tjeng group (figure reproduced from [75]).

Another method to examine of the quality of growth is by *ex situ* high-resolution X-ray diffraction (XRD) characterization. The work performed by Bloemen et al. using high-resolution XRD of (Fe_3O_4/CoO) multilayers is an illustrative example [76]. Figure 10.7 shows a high-resolution XRD pattern of a (36 Å Fe_3O_4/53 Å $CoO)_{x20}$ multilayer grown on MgO (001). From the simulated XRD spectrum of Figure 10.5, it was found that the roughness of the films, included in the simulation through discrete layer thickness fluctuations, amounted to just 1 Å for CoO and Fe_3O_4 [76]. This, and similar XRD patterns, showed single-crystalline superlattices or multilayers could be grown with rocking curve widths of ≈ 0.20° and roughness or interface widths of 2 ± 1 Å [77]. Such high-quality superlattices have been instrumental in elucidating coupling effects in Fe_3O_4-based multilayers [77], a topic that has been reviewed elsewhere [48]. Moreover, the structural quality or sharpness of interfaces is important for spintronic application for which this topic has been studied by others, who find a roughness of 5–7 Å for NiO grown on Fe_3O_4 [64].

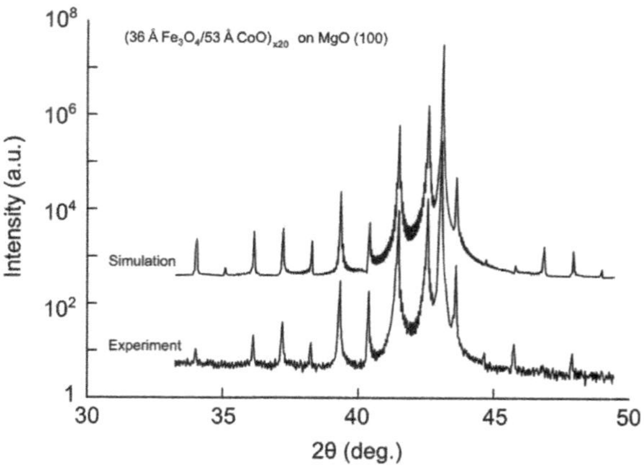

Figure 10.7 X-ray diffraction $\theta-2\theta$ scan for a (36 Å Fe_3O_4/53 Å $CoO)_{x20}$ multilayer grown on an MgO (001) substrate by MBE. The top curve shows the simulated spectrum, from which it follows that the layer thickness variation is less than 1 Å for the CoO and Fe_3O_4 layers (figure reproduced from [76]).

Also, high-resolution transmission electron microcopy (HRTEM) has been used to show that very high-quality layers and multilayers can be grown. This high quality is due to the very close lattice match between the oxygen lattice of MgO, NiO, CoO, and $MgAl_2O_4$ substrates and Fe_3O_4 [52, 78, 79] or $CoFe_2O_4$ [80] films or others. These HRTEM studies show that the interface between the MgO (001) substrate and the Fe_3O_4 layer is structurally coherent, free of an interface phase, and largely free of crystal defects. Even at the step edges on the surface of the MgO (001) substrates, the growth continues epitaxially [79]. Figure 10.8 illustrates this point for a [111] oriented 140 Å Fe_3O_4/10 Å CoO bilayer on α-Al_2O_3 (0001). Even for growth along the (111) plane, which, in principle, is a polar surface, high-quality epitaxial growth was observed. Despite the very thin CoO layer, only ~10 Å, no island formation was detected, and coherent epitaxial growth was realized [78]. Moussy and coworkers found the epitaxial growth of $CoFe_2O_4$ on $MgAl_2O_4$ resulted in a clear out-of-plane deformation while an in-plane deformation was absent, even with a lattice mismatch of 3.86% [80, 81].

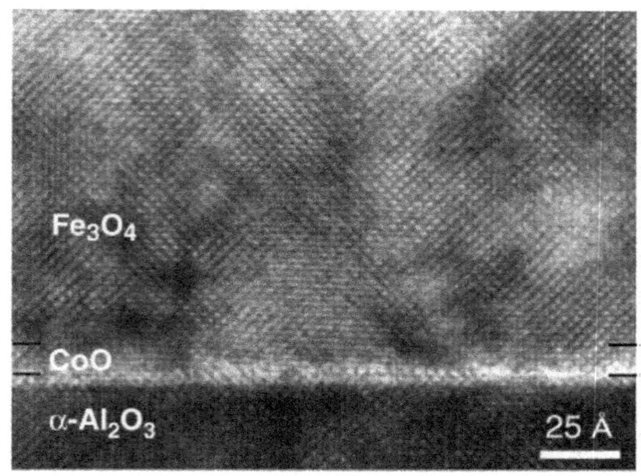

Figure 10.8 High-resolution transmission electron microscopy (HRTEM) image of an oxygen-assisted MBE-grown [111] oriented 140 Å Fe_3O_4/10 Å CoO bilayer on α-Al_2O_3 (0001) (Figure reproduced from [78]).

To summarize, from in-depth structural characterization, be it through *in situ* techniques such as LEED and RHEED or *ex situ* techniques such as XRD and HRTEM, there is solid evidence that high crystal quality and stoichiometric accurate Fe_3O_4 layers can be deposited on a variety of functional substrates that provide a close lattice match to the oxygen atomic template of Fe_3O_4 allowing for epitaxial growth.

10.2.3 Characterization of Stoichiometry of MBE-Grown Fe_3O_4 Films

As it is well established that for ferrites, properties such as magnetization and anisotropy are closely related to cation distribution and stoichiometry [1], the study of stoichiometry of Fe_3O_4 has been an important topic of investigation. In particular, Fujii and Hibma and coworkers have extensively studied this topic using conversion electron Mössbauer spectroscopy (CEMS) [53, 82]. In those studies, a probe layer of 0.42 nm (i.e. two atomic layers) of active ^{57}Fe isotope was grown at various positions within a Fe_3O_4 layer, either in the middle or at the interface of the Fe_3O_4 layer, in order to amplify the sensitivity of CEMS to these regions. Using Mössbauer spectroscopy, the local stoichiometry can be probed by comparison to the well-known Mössbauer spectrum of Fe_3O_4, e.g. the two sextets corresponding to the Fe cations residing on either the tetrahedral A-site (Fe^{3+}) or the octahedral B-site (Fe^{2+} and Fe^{3+}) [82]. From the Mössbauer spectra, the stoichiometry through measured isomer shifts, hyperfine fields and quadrupole splittings are compared to similar values of known Fe_3O_4 bulk values. From a comparison to calculated spectral areas (i.e. the calculated areas under the aforementioned sextets for tetrahedral and octahedral sites), the Fe_3O_4 film stoichiometry, including assessment of the vacancy parameter δ of $Fe_{3-\delta}O_4$ can be determined with a precision of ~0.01 [83]. Some aspects of these studies will be discussed in greater detail in Section 10.2.4.

An alternative approach to assessing stoichiometry of Fe_3O_4 layers is the use of spectroscopic methods. A method that has gained popularity by several research groups, is the use of XPS [60, 61, 84]. In XPS, the Fe 2p core level spectra are recorded and a distinction between α-Fe_2O_3-, FeO-, or Fe_3O_4-like iron oxide structures is discerned by the presence of satellite peaks [61]. However, distinguishing between off-stoichiometric $Fe_{3-\delta}O_4$ and γ-Fe_2O_3 remains challenging.

Magneto-optical (MO) Kerr spectroscopy offers an efficient method to analyze Fe_3O_4-like layers to assess stoichiometry, as within the optical frequency band, many intervalence and intersublattice charge transfer transitions occur [85], through which a sensitive measurement of stoichiometry is possible with an accuracy in γ of ±0.02 [86]. Moreover, as the light penetration changes in optical wavelengths range from 25 to 150 nm, this method is well suited to analyze deposited layers of these thicknesses, which is a typical range for the deposited layers under study. Additionally, MO Kerr spectroscopy allows for some degree of depth sensitivity of stoichiometry [86]. The MO Kerr method has also been directly compared to Mössbauer spectroscopy results on the same films and found to yield consistent results [87].

Finally, the Verwey transition can also be used to assess stoichiometry. The Verwey transition in Fe_3O_4 is interesting in itself as a first-order metal–insulator transition in bulk magnetite [88]. Moreover, it has been well established that the bulk transition, which occurs at 121 K, is a function of stoichiometry [89]. Studying the nature of the metal–insulator Verwey transition in thin films is of interest for its value both in understanding thin Fe_3O_4 film magnetism and in determining how growth conditions impact the Verwey transition, T_V [61]. However, as a method to characterize the stoichiometry of Fe_3O_4 layers, it is fraught with difficulty. For instance, for identically grown Fe_3O_4 layers on $MgAl_2O_4$ substrates, it was found that only for relatively thick layers of ~360 nm, T_V was 122 ± 1 K, consistent with the bulk behavior. While for layers thinner than 115 nm, the Verwey transition significantly broadened and reduced to values below 109 K [90] in accordance with result by others [61]. This suggests, in addition to stoichiometric effects, epitaxial strain may influence T_V [90]. The latter was convincingly demonstrated in experiments in which tensile strain was exerted in MBE-grown Fe_3O_4 layers by deposition on spinel-structured $Co_{2-x-y}Fe_xMn_y TiO_4$ substrates. These substrates have a lattice mismatch to Fe_3O_4 of between +0.66% and +1.1%. It was found that for identical compositions, T_V could be enhanced to above the bulk Fe_3O_4 value merely by adjusting strain [91]. Moreover, the effect of differences in thermal expansion coefficients between substrate and the epilayer must also be considered as an influential variable [90, 92], as early on difference in thermal expansion coefficients giving rise to tensile strain have been shown to influence T_V of magnetite films [92]. Thus, only for comparatively thick films greater than 300 nm, grown on substrates, which exert limited strain, T_V appears as a sharp transition and approach values very near the bulk T_V of stoichiometric Fe_3O_4 [61, 90, 91, 93]. The explanation for this resides with the relationship between film thickness and the size of antiphase domains (APDs; see Section 10.2.4), D, which increases proportional to $\sqrt{t}$, where t is the film thickness [94]. For thinner films, the disorder arising from the more substantial contribution of antiphase boundaries (APBs; see Section 10.2.4) relative to the small domain size contributes to a pronounced shift in T_V [61, 91, 94]. The latter has been elegantly demonstrated in an experiment in which the resistivity, ρ, as a function of T, was measured within a single AP domain as well as across a single APB. No T_V transition was measured across the APB. However, a pronounced T_V transition was detected, even higher than similar thickness films, when measured within the single AP domain [95].

In conclusion, assessment of T_V is a powerful means to illuminate stoichiometry of oxygen-assisted MBE-grown Fe_3O_4 layers; yet, significant care needs to be exercised when interpreting T_V in an absolute sense, as discussed in this section. Knowledge of T_V, when used in conjunction with other spectroscopic methods, make it possible to grow high-crystal-quality stoichiometric layers or multilayers with very close to nominal stoichiometries.

10.2.4 Magnetization of Thin Fe_3O_4 Films and the Influence of Dead Layers and APBs

An important topic to consider is the magnetization of ultrathin Fe_3O_4 films. This topic has garnered significant research interest over the last two decades, most notably on the Fe_3O_4/MgO system. This system will, therefore, be taken as a model from which key aspects of interface effects on magnetic properties will be discussed. A significant fraction of this work has been performed on Fe_3O_4/MgO heterostructures, with the aim of enhancing the magnetic contribution from ultrathin (≤ 5 nm) interface layers.

An understanding of the magnetic structure of thin Fe_3O_4 films, and especially of the interface regions, is crucial for applications in which ferrite films play a role, such as when embedded in multilayer structures to yield all-oxide tunnel junctions or when used as spin filter layers in multilayered structures. The interest in these applications arises from the possibility of spin-polarized transport in ferrite films, in particular Fe_3O_4, owing to the half-metallic nature of Fe_3O_4 as predicted by first-principles calculations, as discussed in greater detail in Section 10.1.1.1. The scientific interest in interface magnetism of Fe_3O_4 layers stems first from the discovery of magnetic interface anisotropy favoring a perpendicular orientation in oxide systems [96] and second from the anomalous behavior of the saturation magnetization of thin Fe_3O_4 layers [96, 97]. We begin our discussion with the latter phenomenon.

10.2.4.1 Reduced Magnetization and the "Dead Layer" Effect at the Interface

A number of studies have found that the saturation magnetization, M_S, of Fe_3O_4 layers decreases with Fe_3O_4 layer thickness to values below the bulk saturation magnetization of 497 kA/m for Fe_3O_4. This effect was initially reported by Parkin et al. for Fe_3O_4 layers deposited on silicon (111) crystals [98]. Further, studies in 1994 by Margulies et al. [97, 99] showed that despite the application of a magnetic field of 7 T, the magnetization of Fe_3O_4 layers deposited by PLD on MgO, as well as those on other substrates such as $MgAl_2O_4$ and Al_2O_3, did not experience saturation. Subsequent experiments performed on MBE-grown Fe_3O_4 layers deposited at 500 K also found anomalous magnetic behavior [76, 96]. The key results of these, and related studies on very well-defined Fe_3O_4/MgO layers, are discussed here.

Van der Heijden et al. studied a series of Fe_3O_4/MgO multilayers grown by oxygen-assisted MBE in a UHV multichamber MBE system [52]. Fe_3O_4 was deposited via electron gun evaporation from Fe targets in an O_2 ambient atmosphere at 2.8×10^{-5} mbar at 500 K [96]. Superconducting Quantum Interference Device (SQUID) magnetometry was used to investigate a series of Fe_3O_4/MgO multilayers on MgO (100), with Fe_3O_4 layer thicknesses varying between 1 and 50 nm and the MgO interlayer thickness chosen such that interlayer coupling effects were minimal [100]. The resulting room temperature magnetic moment, M, normalized to Fe_3O_4 layer thickness, t, is shown in Figure 10.9a as a function of the Fe_3O_4 layer thickness, t_{Fe3O4}. These data show that for $t_{Fe3O4} \geq 30$ nm values for the magnetic moment are equivalent to the bulk moment of Fe_3O_4 at 300 K of 497 kA/m within the experimental accuracy (dashed line in Figure 10.9a) [1]. However, at layer thicknesses below 30 nm the saturation moment, *even* in these high-crystal-quality samples, falls markedly below the bulk value. Following earlier studies [98], the behavior of the Fe_3O_4 layer was modeled using the so-called "dead layer" effect, t_d. This "dead layer" at the interface of the Fe_3O_4 layer contributes no magnetic moment to the overall magnetization. Thus, the overall magnetization of the Fe_3O_4 layer should be given by:

$$M = M_S^{\text{bulk}} - 2\frac{M_I}{t}, \tag{10.4}$$

where M_S^{bulk} is the bulk magnetization, M_I the interface magnetization, and M_I/M_S is the thickness of the magnetic "dead layer" at the interface, t_d. Factor 2 is incorporated as each Fe_3O_4 layer has two interfaces with MgO. The analysis of the data becomes more straightforward by dividing by M_S and multiplying both sides of Equation (10.4) by t yielding:

$$t_M = t_{M_S^{bulk}} - 2t_d. \tag{10.5}$$

In Figure 10.9b, the data have been plotted using this format, and one can see they are well described by this equation for thickness values ranging from 1 to 50 nm. The inset to Figure 10.9b shows the same data over the range to 10 nm. A positive x-axis intercept is found to be 1.4 ± 0.6 nm, yielding a "dead layer" thickness, t_d, of 0.7 ± 0.3 nm. The larger dead layer thicknesses reported by others of 2.5 nm [98] may have resulted from nonoptimal growth

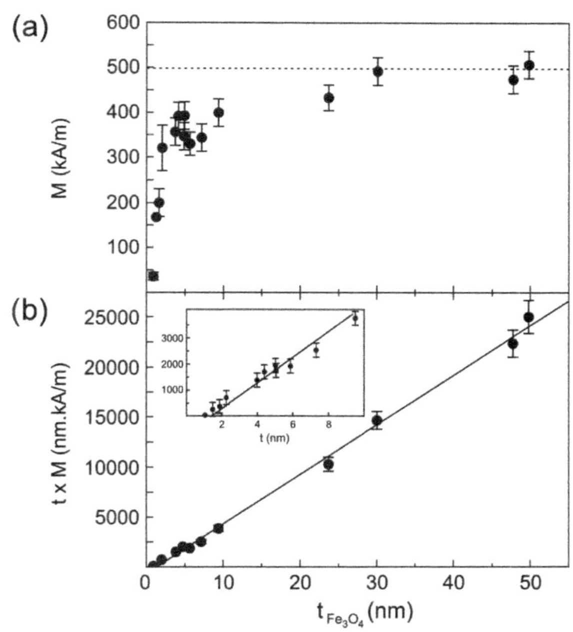

Figure 10.9 Study of the thickness dependence of the magnetization M in (ultra)thin Fe3O4 layers. (a) The magnetic moment normalized to the total Fe_3O_4 volume of the samples at 300 K. (b) *M* by multiplied the Fe_3O_4 layer thickness, t_{Fe3O4}. The solid line is a linear fit of the data. The inset is a blowup for small Fe_3O_4 layer thicknesses (Figure reproduced from [96]).

conditions employed, as they relied on sputtering to prepare the Fe_3O_4 layers. In sputtering the energy of surface adatoms are up to an order of magnitude larger than that of MBE. As the lattice constant of Fe_3O_4 is 8.38 Å [1], the result illustrated in Figure 10.9 might suggest that the magnetization of the Fe_3O_4 layer is disturbed over just the first lattice period and thus validates the very high-quality growth of Fe_3O_4 layers by MBE. Within the study of ferrites, "dead" surface and interface layers have been invoked successfully to account for anomalous saturation magnetization of ultrafine particles due to spin canting at the surface [101] and to explain the dependence of magnetic permeability of bulk ferrites on grain boundary densities [102, 103]. However, one must bear in mind that a magnetically "dead" layer with $M_S = 0$ at the Fe_3O_4/MgO interface is an oversimplification. More likely, the magnetization profile is not abrupt but rather experience some spatial decay. Moreover, the aforementioned description provides limited insight into the exact nature of the magnetic state, giving rise to the observed absence of magnetization at Fe_3O_4 interfaces.

10.2.4.2 Anisotropy and Interface Anisotropy of Thin Fe_3O_4 Layers

The anisotropy in ultrathin MBE-grown Fe_3O_4 films was first systematically investigated by van der Heijden et al. [104]. Later studies by various groups, using either MBE growth [61, 105, 106] or other methods such as PLD [107], confirmed many of their findings. The key findings were again highly film thickness dependent. Specifically, for large layer thicknesses (i.e. $t \geq 30$ nm), the magnitude and temperature dependence of magnetic anisotropy resembles the behavior of bulk Fe_3O_4 [104], while for lower thicknesses ($t < 30$ nm), magnetocrystalline and uniaxial out-of-plane anisotropies decrease monotonically with decreasing Fe_3O_4 layer thickness, ultimately vanishing below 5 nm (see [104, Figure 10.11]). Consequently, the discussion here will focus on what occurs for ultrathin layers due to contributions from interface anisotropy.

The initial study of interface anisotropy of magnetic thin films was motivated to a large part by the relevance of this topic to the development of magneto-optic recording media, which requires a preferred axis of magnetization aligned perpendicular to the surface of the film plane. The study of perpendicular anisotropy had been nearly exclusively limited to metallic layers and multilayers. In the late 1980s, a number of studies investigated the properties of sputtered ferrite films without focusing on obtaining perpendicular anisotropy [108]. However, progress on controlled MBE growth of magnetic oxides allowed more detailed studies of the surface contribution to magnetic anisotropy that gave rise, in some cases, to strong perpendicular anisotropy.

Studies performed on metallic multilayer systems established that out-of-plane anisotropy, K, of thin magnetic layers is composed of two principal contributions, a volume contribution, K_V, and an interface or surface contribution, K_S. These works further showed that anisotropy can be described phenomenologically by [109]:

$$K = K_V + 2\frac{K_S}{t}, \tag{10.6}$$

where t is the layer thickness. The volume anisotropy, K_V, is composed of other terms including shape anisotropy, magnetocrystalline anisotropy, and magnetoelastic anisotropy due to strained coherent growth. For thin layers, the shape anisotropy, favoring in-plane preferential magnetization, can be overcome by surface anisotropy, which gives rise to a preferred magnetization contribution out of the film plane. This effect has been amply demonstrated for metallic systems [110], yet has been investigated for only one oxide system to date, Fe_3O_4 grown on MgO substrates [96]. As discussed in the preceding section, there is no measurable magnetic moment from the interface layer of Fe_3O_4 on MgO. Hence, Equation (10.6) must be modified to account for this "dead layer" effect by correcting the apparent thickness by t_d. In practice, the relevant equation for K becomes [96]:

$$K = K_V + 2\frac{K_S}{(t - 2t_d)}. \tag{10.7}$$

To investigate interface anisotropy of Fe_3O_4, van der Heijden et al. grew a wedge-shaped sample on MgO (1 0 0) [96]. The first sample in a series was composed of 0–40 nm Fe_3O_4/2 nm MgO/3 nm NiO. The MgO component of the capping layer ensured that a symmetrical situation was attained. The additional 3 nm of NiO was needed to prevent a reaction between MgO and H_2O or CO_2 during measurements. To be able to detect any influence of this capping layer, only half of the wedged sample had the capping layer. Figure 10.10 illustrates schematically the layered structure under study. With a wedge-shaped sample, all anisotropy measurements of various layer thicknesses can be performed on a single sample grown in a single growth run without breaking vacuum [111]. As the slope of these wedges is just 3 nm Fe_3O_4/mm, one can easily measure the hysteresis loop at well-defined Fe_3O_4 layer thicknesses using the magneto-optical Kerr effect (MOKE).

To determine the out-of-plane anisotropy, K, van der Heijden et al. [96] measured hysteresis loops along their wedge-shaped sample along both longitudinal and polar MOKE geometries with a magnetic field aligned parallel and perpendicular to the film plane, respectively. Figure 10.11a shows the thickness dependence of uniaxial out-of-plane anisotropy field, H_U, determined from the measured perpendicular saturation field, H_S. Results of longitudinal MOKE experiments show that in-plane saturation magnetization is sufficiently small for Fe_3O_4 layer thicknesses in excess of 5 nm (less than 10 kA/m) compared to the perpendicular saturation field, H_S, such that it can be neglected [112]. Consequently, K can be directly obtained from H_S for Fe_3O_4 thicknesses in excess of 5 nm using [110]:

$$K = -\frac{1}{2}\mu_0 M_S H_S. \tag{10.8}$$

where μ_0 is the vacuum permeability. Below 5 nm, the in-plane saturation is no longer negligible and must be incorporated when calculating K. Figure 10.12 shows the resulting thickness dependence of K multiplied by $(t - 2t_d)$ versus the effective magnetic Fe_3O_4 layer thickness $(t - 2t_d)$ [96]. For two points on this curve, at a Fe_3O_4 layer thickness of 9.3 and 23.7 nm, the value of K was deduced from SQUID magnetometry for corresponding Fe_3O_4/MgO multilayers. In Figure 10.12, open symbols represent data collected from the symmetric wedge, while filled symbols represent data collected from the wedge section that did not have a cap layer. The absence of a cap layer enables one to obtain data at lower Fe_3O_4 thicknesses as less light is absorbed by the cap layer. From this figure, one can immediately see that phenomenologically Equations (10.6) and (10.7) describe the data of this oxide system. The slope of the fit in Figure 10.12 provides values for the volume anisotropy K_V of -0.123 ± 0.010 and -0.129 ± 0.010 MJ/m^3 for the wedged sample with and without a capping layer, respectively.

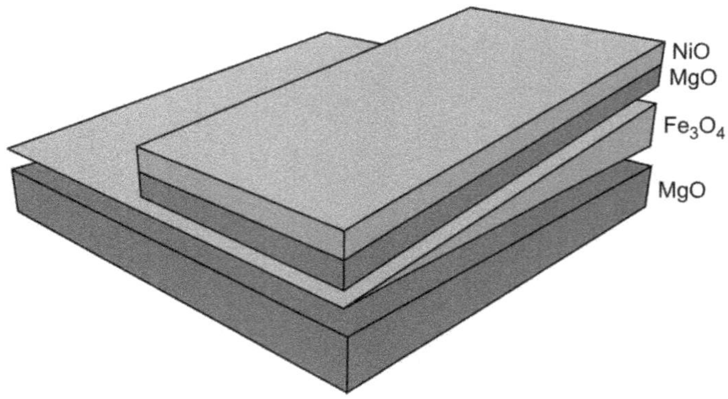

Figure 10.10 Schematic illustration of a wedge-shaped sample grown for the study of the magnetic anisotropy of a Fe_3O_4 film using the magneto-optical Kerr effect (MOKE). Only half of the sample had a MgO/NiO capping layer to be able to examine the effect of the capping layer (see text).

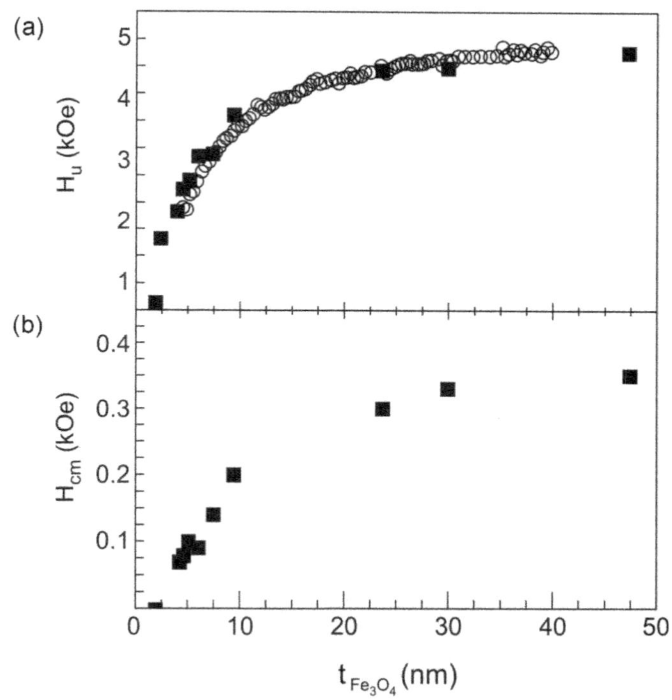

Figure 10.11 (a) The uniaxial out-of-plane field, H_U, and (b) the cubic magnetocrystalline anisotropy field, H_{CM}, as a function of the Fe_3O_4 layer thickness, t_{Fe3O4}. Data were obtained from FMR experiments at room temperature for Fe_3O_4 grown on MgO (1 0 0) (filled squares), and for H_U also from MOKE experiments for a 0–40 nm Fe_3O_4 wedge samples (O) (figure reproduced from [104]) (©1998 with permission from Elsevier).

These values were found to be in accordance with various terms contributing to K_V. Specifically, the shape anisotropy for the system studied is calculated to be $\frac{1}{2}\mu_0 M_S^2 = 0.155$ MJ/m^3, the crystalline anisotropy at $K_1/4 = 0.003$ MJ/m^3 is expectedly small, and the pseudomorphic growth of Fe_3O_4 on MgO (1 0 0) results in a magnetoelastic anisotropy contribution of about 0.027 MJ/m^3. The summation of these values yield a volume anisotropy K_V that is in excellent agreement with experimental data irrespective of the capping layer of -0.125 MJ/m^3 [96]. Nevertheless, the situation concerning K_S is less clear. The fit to

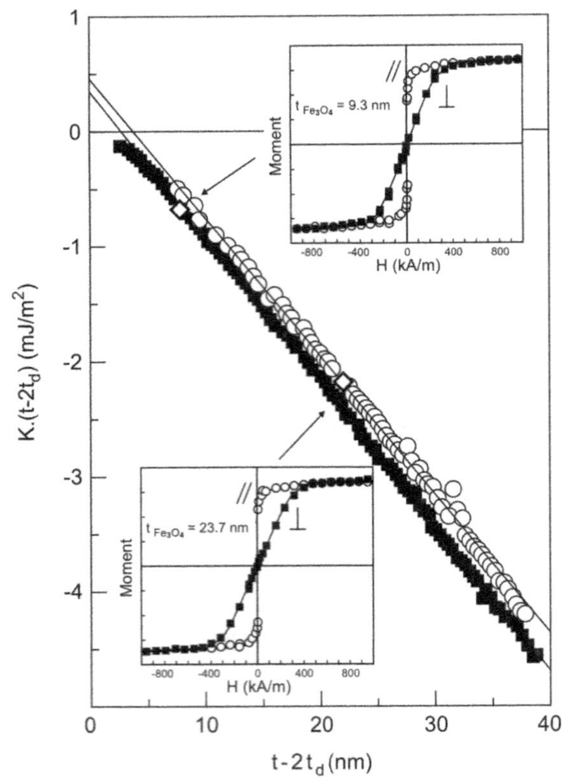

Figure 10.12 Study of the thickness dependence of the out-of-plane anisotropy multiplied by the thickness for ultrathin Fe_3O_4 layers, using the corrected layer thickness $t - 2t_d$ (see text). The data were obtained using MOKE from a wedge-shaped sample grown on MgO(001) (see Figure 10.10). Data with capping layer (O) and without (■) follow the same trend. For two (Fe_3O_4/MgO) multilayers with Fe_3O_4 thickness of 23.7 and 9.3 nm, the out-of-plane anisotropy was also determined from hysteresis loops obtained by SQUID magnetometry, with the field applied along the <1 0 0> in-plane axes and perpendicular to the film plane (figure reproduced from [96]).

the data (see Figure 10.12) suggests that at a Fe_3O_4 layer thickness ($t - 2t_d$) of 3.5 nm, or a total layer thickness of 5 nm, the surface anisotropy, favoring perpendicular anisotropy, eclipses the volume anisotropy. Thus, at thicknesses below 5 nm, the system should exhibit positive anisotropy favoring perpendicular magnetization. However, experimentally this could not be observed. The signal intensities of the MOKE experiment were insufficient to obtain reliable data below 5 and 10 nm for the noncapped and capped wedge-shaped samples, respectively. However, a SQUID magnetometry experiment on a (1.6 nm Fe_3O_4/2.5 nm MgO)$_{x20}$ multilayer capped with 3 nm NiO showed that hysteresis loops along the <1 0 0> in-plane axis and the film normal coincide. FMR experiments confirmed that the out-of-plane, as well as in-plane, anisotropy values *vanish* below a Fe_3O_4 layer thickness of 5 nm. This result led the authors to conclude that the "dead layer" effect that they had observed precludes a meaningful interpretation of the observed interface contribution to K_S in terms of a Néel-type mechanism [96].

10.2.4.3 The Interface Structure: APBs

A key experiment that proved to elucidate the nature of magnetism of the Fe_3O_4 layer interface was the CEMS experiments performed by Voogt et al. [113]. As previously described in Section 10.2.3, these authors successfully grew a series of Fe_3O_4/MgO multilayers by MBE in which 0.42 nm (i.e. two monolayers) of ^{57}Fe Mössbauer probe layers were incorporated at both the Fe_3O_4 interfaces with the MgO layer and within the bulk of the Fe_3O_4 layer. In their series of samples, the Fe_3O_4 layer thickness was, respectively, 1.8, 3.5, and 5.0 nm, and the MgO interlayer thickness was chosen to be 2.0 nm to prevent pinhole coupling between the Fe_3O_4 layers of the multilayer [100]. At first glance, the most surprising result was that the two hyperfine split sextets, indicative of long-range ferromagnetic order in Fe_3O_4, decrease with decreasing Fe_3O_4 layer thickness regardless of the location (depth) of the probe layer (see Figure 10.13 [113]). The breakdown of long-range order for Fe_3O_4 layer thicknesses less than 5 nm is clearly *not* an interface effect. The Mössbauer data clearly show a loss of long-range order as the two sextets peaks gradually diminish with decreasing layer thickness until they appear to coalesce into one peak for the 1.8 nm sample. Thus, this effect occurs throughout the entire layer and not just at the interface.

The two sextets measured for the multilayer, in which the Fe_3O_4 thickness is 5.3 nm, correspond to isolated Fe^{3+} and $Fe^{2.5+}$ ions on octahedral sites. This result demonstrated that the structure of the Fe_3O_4 film was spinel-like in its entirety, and the presence of other iron oxide phases at the interface, such as $Fe_{1-x}O$ or γ-Fe_2O_3, can be ruled out. Moreover, the formation of a mixed $MgFe_2O_4$ ferrite at the interface in these high-quality MBE samples can also be excluded as $MgFe_2O_4$, with its high resistivity, does not support hopping conduction along octahedral sites that would contribute to a $Fe^{2.5+}$ signal in the Mössbauer spectrum [113]. Drawing on a comparison to the CEMS studies of thicker Fe_3O_4 layers [82], Mössbauer data indicate that hyperfine fields of both Fe^{3+} and $Fe^{2.5+}$ contributions become smaller as the Fe_3O_4 thickness is reduced. Even though these films are ferrimagnetic, the magnetic moments of the Fe ions progressively fluctuate faster as films become thinner. Ultimately, this led to the collapse of the sextet spectrum in samples of thickness ~3.5 nm, while a transition to a fully paramagnetic state occurs at a layer thickness of ~1.8 nm. Consequently, from these Mössbauer experiments, there is no evidence of any nonmagnetic or "dead layer" at the Fe_3O_4 interface.

Independent of each other, two groups attributed the anomalous behavior of the magnetization of thin Fe_3O_4 layers to APBs [113, 114]. APBs develop because Fe_3O_4 has a lower crystal symmetry and a larger unit cell than the underlying MgO substrate as well as the MgO interlayer in the case of multilayers. Consequently, Fe_3O_4 islands nucleate on MgO substrates

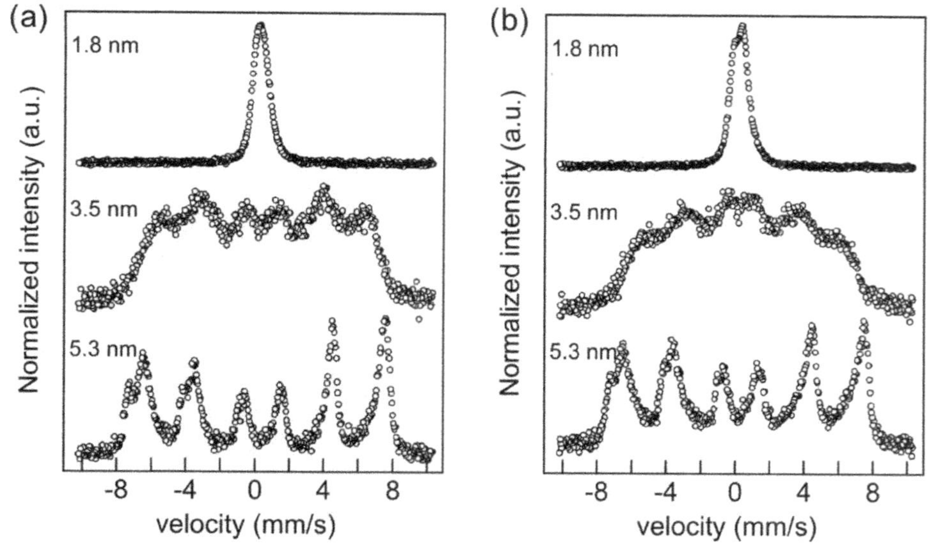

Figure 10.13 Room temperature conversion electron Mössbauer spectra of (1 0 0) Fe_3O_4/MgO multilayers as a function of the Fe_3O_4 thickness (indicated earlier each spectrum) and the position of the Mössbauer active ^{57}Fe probe layers: (a) in the center of the Fe_3O_4 layers and (b) at the Fe_3O_4/MgO interface (figure reproduced with permission from [113]) (©1998 by the American Physical Society).

Figure 10.14 Dark-field TEM images of a 12-nm-thick Fe_3O_4 film. The image in (a) is as grown and while these in (b) and (c) were taken after post-annealing at 300 °C in an oxygen background pressure of 10^{-6} mbar for 1 and 2 h, respectively. The scale bar is 20 nm for all images (figure reproduced from [116]) (©2003 by the American Physical Society).

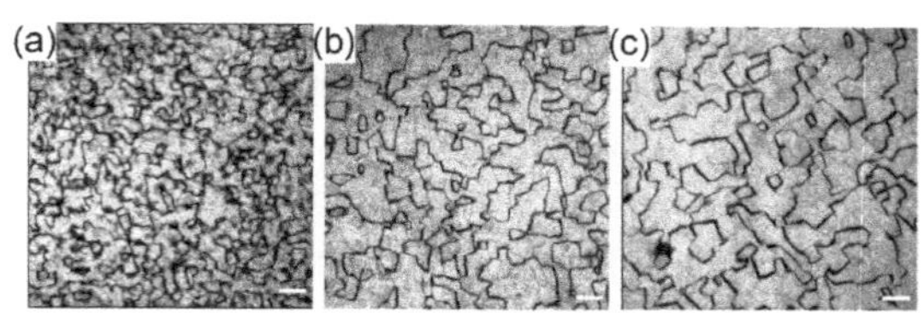

Figure 10.15 Examples of high-resolution imaging of the antiphase boundary itself. (a) Result of a STM study on Fe_3O_4 deposited on MgO (1 0 0) [117] (© 1997 with permission form Elsevier). (b) Result of a scanning transmission electron microscopy (STEM) study of $NiFe_2O_4$ on $MgAl_2O_4$ viewed along the [1 −1 0] direction [118]. For details, see text (Reproduced with permission from [118]).

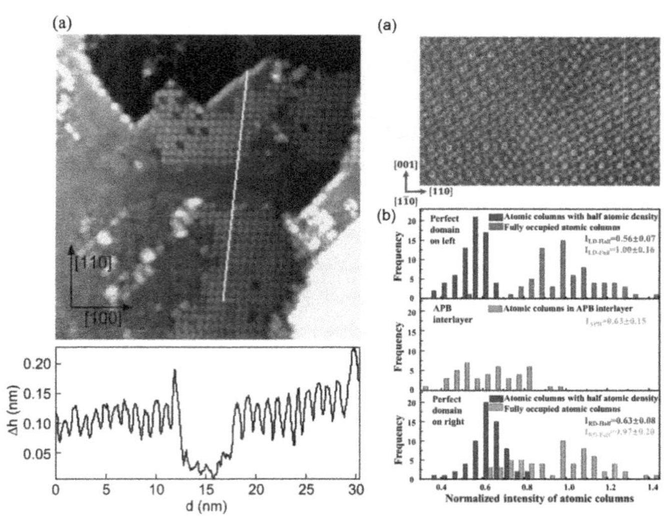

in eight unique ways before coalescing to larger structures. This gives rise to copious lattice defects at the interfaces where different Fe_3O_4 islands meet, since, the Fe_3O_4 layer will have a continuous O_2-sublattice coincident with a fragmented, discontinuous cation sublattice [113, 114]. Such APBs have been observed directly by TEM (see Figure 10.14) [114–116], as well as by scanning tunneling microscopy (STM) studies (see Figure 10.15) [117].

Figure 10.14 shows an example of TEM images of APBs in an MBE-grown Fe_3O_4 layer under lower resolution thereby offering a macroscopic view allowing one to determine the density of APBs [115, 116]. For a similar MBE-grown layer, the APB density, or boundary length per unit area, had been determined to be $6.8 \times 10^{-3}\ nm^{-1}$, which is equivalent to an average APD size of 300 nm^2 [115]. Thus, the average APD size in such MBE-grown Fe_3O_4 layers is more than two orders of magnitude larger than the APD size of 27.5 nm^2 found in sputtered layers [97, 114], indicating a strong dependence on growth conditions. The effect of post-annealing seen in Figure 10.14b and c corroborates this.

It is interesting to gain further microscopic insight into the structure of and the magnetic coupling in ABPs. Figure 10.15a shows an STM image of a $33 \times 33\ nm^2$ area of the MBE-grown Fe_3O_4 surface shows a regular network of brightly colored elongated features aligned along the <110> direction, which indicates a $p(1 \times 1)$ surface reconstruction. This reconstruction was interpreted as a clustering of tetrahedral ions, which have dangling bonds rotated by 90° from one atomic plane to the next [117]. These 90° rotations were observed between two adjacent terraces that are separated by a single step of one atomic layer (0.2 nm). The STM image also shows a disordered region between two Fe_3O_4 islands nucleated with a different cationic orientation relative to one another. Those islands appear to be separated by a disordered region of around 5 nm, which was taken to be an APB. Figure 10.15b shows a high-angle annular dark field (HAADF) scanning TEM (STEM) image of a APB, marked by orange arrows in this image obtained in a recent study [118]. This ABP is in a $NiFe_2O_4$ film deposited on a $Mg_2Al_2O_4$ substrate by reactive RF magnetron sputtering. Interestingly, these authors found that this particular ABP had a rock salt structure (unlike the disordered region found in the STM study). From statistical analysis of the octahedral site occupancy in the columns in the $NiFe_2O_4$ domains left and right of the APB the authors (see the lower half of Figure 10.15b), the authors found that these were either 100% occupied or approximately 60% occupied, consistent with the normal B-site occupancy of Fe^{3+} and Ni^{2+} cations on octahedral sites in $NiFe_2O_4$. Comparing this to the site occupancy in the ABP (middle part lower half of Figure 10.15b), the authors conclude that the octahedral sites in this rock salt ABP were only half occupied. Additional first-principle density functional theory calculations suggest that this particular APB leads to antiferromagnetic coupling and a significant reduction in the magnetic moment in the APB [118].

Although both the San Diego [114] and Groningen [113] groups proposed that APBs play an essential role in explaining the anomalous magnetization behavior of ultrathin Fe_3O_4 layers, models proposed to justify their conclusion differ markedly. In addition to the CEMS studies on Fe_3O_4/MgO multilayers, Voogt et al. performed SQUID studies at low temperature. These authors observed that the remnant magnetization at 5 K attained after cooling in a 10 kOe (=796 kA/m) field decreased rapidly with increasing temperature until it leveled off at what they defined as the blocking temperature [114]. They also determined that this blocking temperature decreased with film thickness with values of 250, 140, and 40 K for 5.3-, 3.5-, and 1.8-nm-thick layers, respectively. This magnetic behavior was attributed to superparamagnetism caused by magnetic domains within the Fe_3O_4 layer having dimensions of approximately $10 \times 10\ nm^2$. It was proposed that these domains originate from a new 180° Fe–O–Fe interaction across the APBs that is not present within Fe_3O_4. They surmised that these 180° Fe–O–Fe interactions are quite strong because a similar Fe–O–Fe superexchange interaction found in orthoferrites have a superexchange constant J of −25 K [119, 120]. As this new interaction at the APB is comparable to the strong superexchange interaction between Fe ions on tetrahedral (A) and octahedral (B) sites within Fe_3O_4 (J_{AB} interaction of −23.4 K [1]), these authors assumed that the A–B coupling across each APB cancels the 180° Fe–O–Fe interaction. This results in an average net coupling strength of $\sim 3 \times 10^{19}\ m^{-2}$ for the antiferromagnetic superexchange interaction between neighboring domains within the Fe_3O_4 layer. This gives a barrier height $W \approx 3 \times 10$ eV per 10 nm of APB for the fluctuation of the domain due to antiferromagnetic coupling across the APB. Comparing this calculation to experimental data, these authors concluded the influence of the superexchange barriers across the APBs must be largely suppressed due to frustration among the interdomain interactions. When Fe_3O_4 islands nucleate on the MgO substrate at random points, junctions between domains that are common to three domains will be in the majority. Each of these triple junctions have competing interdomain interactions in which superexchange barriers effectively cancel each other out and enable the domain to fluctuate freely. In this situation, single-crystalline, epitaxial Fe_3O_4 films become superparamagnetic.

In contrast, the San Diego group did not invoke superparamagnetism to explain how APBs account for magnetism in thin Fe_3O_4 films [114]. These authors concluded that new 180° Fe–O–Fe interactions occur across the APBs. Additionally, these authors also assumed that the number of A–A sublattice interactions increases along with the A-O-A angle relative to 180°. As a consequence, the interaction term also increases [114]. This assumption of an increase in A–A sublattice interactions may be incorrect as it relies on models of the APB in which the octahedral and tetrahedral ions are in close proximity. If one assumes that there must be charge neutrality along the APBs, this is not likely to occur. Nevertheless, Margulies et al. arrived at the same conclusion as Voogt et al., that is, an altered exchange interaction across the APB causes the coupling between two adjacent domains to be antiferromagnetic as 180° Fe–O–Fe coupling will dominate A–B coupling [114]. These authors used this model to explain their observation that thick sputtered Fe_3O_4 films did not saturate, even in extremely high fields up to 70 kOe (= 5572 kA/m). We note, however, that subsequent studies demonstrated that high-quality Fe_3O_4 films grown by MBE could be saturated (see Figure 10.9) presumably due to a lower density of APBs.

The accepted explanation of how APBs and APDs influence the observed magnetic properties of Fe_3O_4 films comes largely from the work of Hibma et al. From resistivity measurements [94], these authors first found that the APD size and thus APB density strongly increases with decreasing film thickness. Specifically, APD size D was found to depend on film thickness, t, as [116]:

$$D \propto \sqrt{t}. \tag{10.9}$$

This effect also explained the thickness dependence of magnetization and anisotropy (see Sections 10.2.4.1 and 10.2.4.2 for further discussion). Subsequently, these authors showed that the APD size is not fixed but could be altered by annealing the Fe_3O_4 films after removing the MgO substrate by dissolution [116]. Figure 10.14a shows a dark-field (TEM) image of the as-grown 12 nm Fe_3O_4 film. Figure 10.14b and c shows the effect of post-annealing the film at 300 °C for 1 and 2 h, respectively, in an oxygen atmosphere of 10^{-6} mbar. Clearly, the APB density and the APD size change. This result is important because the APD size was previously thought to be fixed by the initial growth conditions [115]. Eerenstein et al. showed that the activation energy for the diffusive motion of an APB is quite low at 26 ± 5 kJ/mol (250 meV) [116] primarily because the oxygen sublattice is continuous at the APB, while the cation sublattice is discontinuous. As these cations can migrate to unoccupied interstices of the Fe_3O_4 lattice (see Figure 10.2a) and the rapid exchange of electrons between Fe^{2+} and Fe^{3+} above the Verwey temperature [88] prevents the buildup of long-range electrostatic interactions, this process does not require high energy and can occur at growth temperatures that are typical for Fe_3O_4 films. Consequently, longer deposition times required for the growth of thicker films give rise to lower APB densities. These authors also showed that for MBE-based growth, APD size does not depend on oxygen pressure. Given that the critical nuclei for films consist of both Fe and O ions and that Fe ions diffuse faster, island formation depends on the density of the slower O anions. Hence, conflicting magnetic properties from one research group to another can be attributed to differences in growth chambers, leading to subtle changes in growth conditions that impact magnetic properties, particularly susceptible to growth conditions. This model applies to both Fe_3O_4 films grown by MBE and sputtering but not to pulsed-laser-deposited films as for the latter the atomic species are deposited at roughly the same time. In a subsequent paper, Eerenstein et al. [121], Mössbauer investigations indicated that the magnetic fluctuation of the larger APDs is most likely blocked, while the smaller domains have magnetic moments that are superparamagnetic. A more refined analysis showed that only 50% of the coupling across APBs gives rise to antiferromagnetic coupling [122]. The authors thus could quantitatively explain the earlier results by Voogt et al. [113] on the blocking temperature dependence of remnant magnetization.

APBs were thus directly observed both in sputtered [114] and in MBE-grown films [113, 115], and analyses of magnetization revealed that 180° Fe–O–Fe exchange occurs giving rise to antiferromagnetic coupling between domains among nucleated Fe_3O_4 islands. To gain more insight into the relevant model [113, 114], an experiment was performed to see if the predicted superparamagnetic state [113] could be verified by observation. To explore this hypothesis, hysteresis loops of a $(Fe_3O_4/MgO)_{x50}$ multilayer with a 1 nm Fe_3O_4 and 2.2 nm MgO period was measured by SQUID magnetometry at various temperatures [123] and showed saturation magnetization values decrease with increasing temperature. A key observation is that the temperature dependence of the hysteresis loop cannot be described by the Langevin function, which is expected for superparamagnetic behavior. It is noteworthy that in measuring these hysteresis loops, a correction was made to the signal to account for the diamagnetic contribution of the MgO substrate. However, it was shown that it is not possible to adjust this linear correction such that a Langevin temperature dependence could be obtained. Thus, the proposed model based on superparamagnetic behavior of thin Fe_3O_4 layers induced by the 180° Fe–O–Fe coupling across APBs does not seem to hold true if magnetization measurements are considered.

There appears to be a contradiction between magnetization measurements that find evidence of a "dead layer" and Mössbauer data that see evidence for motional narrowing of the Mössbauer spectra for films less than ~2.4 nm in thickness [113, 122]. These differences may be reconciled by considering that these measurement techniques detect magnetic moments (i.e. magnetization) at different time scales. Specifically, the Mössbauer timescale of 10^{-8} s is much shorter than the timescale of SQUID measurements of ≈ 1 s. Consequently, fluctuations of spins within the measurement time of Mössbauer gives rise to motional narrowing, or "superparamagnetism," while these spin fluctuations are not detectable in conventional DC magnetic measurements that are sensitive to the missing magnetization. The layer thickness dependence can thus be understood in that APD size D depends on the film thickness, i.e. $D \propto \sqrt{t}$ (see Equation 10.9), and magnetic fluctuations of spins in larger AP domains are blocked, or exceed, the Mössbauer measurement time. Consequently, the *relative* contribution of APBs to the total Fe_3O_4 layer thickness decreases as the Fe_3O_4 layer becomes thicker, and layers become easier to magnetize, as shown in Figure 10.9a [96]. A possible interpretation of the strictly linear dependence of anisotropy in Figures 10.9b and 10.12 is that the *absolute* contribution (i.e. amount) of the spins *in* the APBs is more or less constant giving the effect of a (small) fraction of ions without moments (i.e. "dead") that do not contribute to the total magnetization signal. The absence of the so-called "dead layer" *at the Fe_3O_4 interface* would be great importance for oxide tunnel junctions where a Fe_3O_4 layer is employed. The application of a conventional Fe_3O_4 layer with the expected "dead layer" at the Fe_3O_4 interface in such a device would severely limit the spin polarization available for tunneling, thereby reducing the attainable TMR ratio. This will be discussed in detail in Section 10.2.5.

In a recent experiment, Tjeng and coworkers managed to shed light on the growth Fe_3O_4 films and thus on the nature of the Fe_3O_4/MgO interface by using XAS [75]. These authors grew by MBE a series of Fe_3O_4 layers on MgO (0 0 1) between 0.67 and 8 monolayers thick and recorded the Fe L_3 spectra of these films at room temperature. From comparisons with simulated XAS spectra, the relative contributions of Fe^{3+} on tetrahedral-coordinated A sites, the Fe^{2+} and Fe^{3+} ions on octahedral-coordinated B-sites were determined as a function of Fe_3O_4 layer thickness. The resulting data are given as a function of increasing layer thickness in Figure 10.16. The striking aspect of these data is the strong decrease and complete absence of Fe^{3+} on A sites in very thin layers. This absence of Fe^{3+} ions on A sites is compensated by an equivalent surplus of Fe^{3+} ions on B-sites, thus maintaining the overall stoichiometry of the Fe_3O_4 films at a cation/anion ratio of 0.75 at each thickness, as illustrated in Figure 10.16. This is further testament to the high-quality growth of Fe_3O_4 films that the authors achieved.

The data illustrated in Figure 10.16 support the interpretation that the presence of Fe^{3+} ions on A sites gives rise to the polar nature of the Fe_3O_4 (0 0 1) surface. Hence, Chang et al. proposed the model shown in Figure 10.17 to explain the epitaxial growth of ultrathin magnetite layers of Fe_3O_4 on MgO (0 0 1) [75]: The first monolayer that is formed when depositing Fe_3O_4 ultrathin films is a charge neutral rock salt FeO layer with 25% Fe vacancies. This yields a charge neutral or nonpolar layer. When depositing the second monolayer, again a nonpolar layer, similar to the first layer, is formed. Next, both deposited monolayers provide each the Fe^{3+} ions to the intermediary space between both layers to generate A-site Fe^{3+} ions (see Figure 10.17). As this intermediary layer is positioned between two layers from which the Fe^{3+} ions originate, charge neutrality is maintained. The same mechanism is active in the deposition of subsequent monolayers.

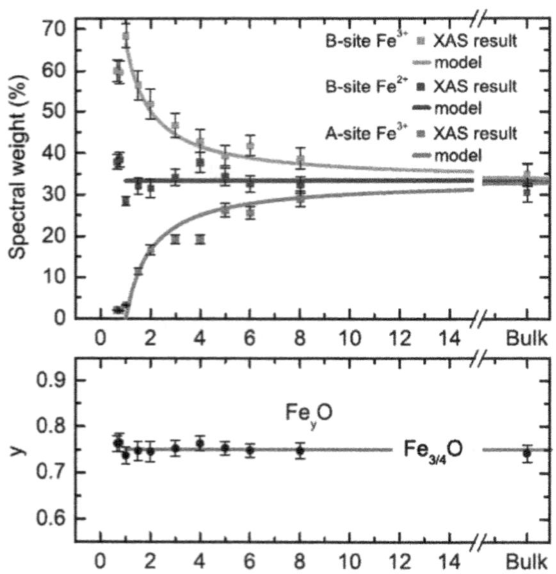

Figure 10.16 The extracted spectral weight from X-ray absorption spectroscopy (XAS) for each of the three Fe cations occurring in Fe_3O_4 versus the film thickness (Fe^{3+}_B is top, Fe^{2+}_B is middle, and Fe^{3+}_A is bottom). The Fe content (y) of the Fe_yO film derived from the average valence is shown in the bottom panel. Error bars reflect the deviation of the fits to the experimental data (figure taken from [75]).

The elegant nature of this growth model is that it is consistent with the experimental observations of Figure 10.16. Additionally, this model avoids the polar catastrophe which would occur, if one were to consider a pure layer-by-layer deposition of the Fe_3O_4 spinel structure. The diffusion of Fe^{3+} ions during film growth in this model cleverly addresses the required population of both cation lattice sites within the magnetite structure [75]. This new model of the Fe_3O_4 growth on MgO (0 0 1) supports the proposition of a "dead layer" at the Fe_3O_4/MgO interface, as the reduced amount of A-site ions present at the interface leads to changed superexchange interactions at the interface. Interestingly, also a recent time-resolved high-energy X-ray-diffraction (tr-HEXRD) study of the MBE growth of ultrathin Fe_3O_4 films on $SrTiO_3$ (0 0 1) and MgO (0 0 1) found that the growth starts with a rock salt $Fe_{1-\delta}O$ phase. This occurs during the first approximately 1 nm of the growth, after which only the characteristic Fe_3O_4 reflections emerge [124]. Yet, as soon as the deposition is stopped, time-resolved hard X-ray photoelectron spectroscopy (tr-HAXPES) suggest that this rock salt layer oxides to one with the characteristic Fe_3O_4 spectrum [124]. Contrary to Chang et al who studied a thickness series of

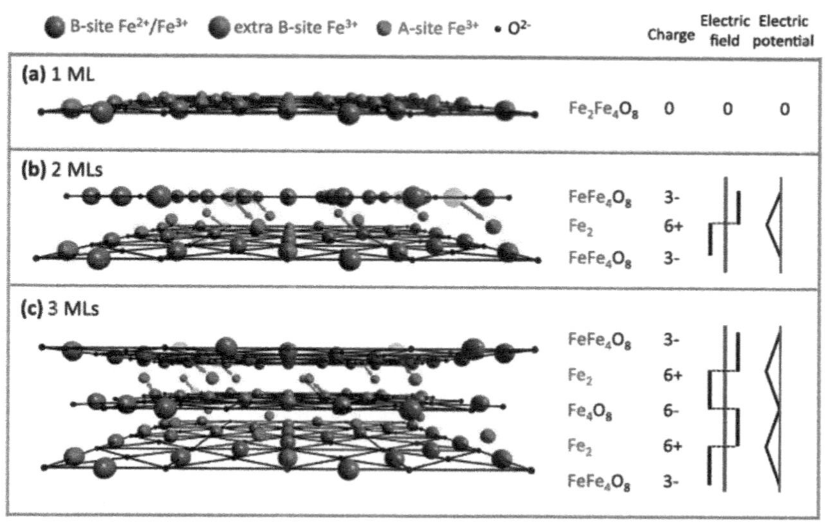

Figure 10.17 Model for the growth process of polar Fe_3O_4 (001) thin films: (a) A single monolayer, (b) two monolayers, and (c) three monolayers (figure reproduced from [75]).

Fe_3O_4/MgO samples [75], Pohlmann et al. studied the Fe_3O_4 formation during growth [124]. Hence, these authors conclude that their data suggest that the $Fe_{1-\delta}O$ phase may be a transient growth phenomenon that exclusively occurs during the growth process but may disappear during completion of the film [124]. A follow-on magnetic characterization by these authors of a 24 nm MBE-grown Fe_3O_4 film on MgO(001) by x-ray resonant magnetic reflectivity (XRMR) and x-ray magnetic circular dichroism (XMCD) suggest that the magnetic order is stable for all three cations ($Fe^{3+}{}_A$, $Fe^{2+}{}_B$, $Fe^{3+}{}_B$) with no interface layers [125]. This is a remarkable results since the surface cation should experience an imbalanced superexchange interaction owing to the nonmagnetic MgO(001) substrate. Note that in contrast to the formation of a rock salt layer at the interface, the APB approach would result in the occurrence of superparamagnetic behavior at the APB which runs *perpendicular* to the substrate and is *not located at the and parallel to interface.*

Summarizing, ferrite layers can be grown with bulk magnetic properties provided their thicknesses exceed 30 nm, as then the effects of reduced dimensions (discussed in this section) do not play a significant role in influencing such properties. Also, at these thicknesses, magnetic anisotropies in ferrite films, based on the results of extensive studies of magnetite, should resemble those of bulk ferrite.

10.2.5 Applications of Thin XFe_2O_4 Films in Spintronics

In this section, we discuss the application of thin ferrites films of composition XFe_2O_4, where X = Fe or Co, Ni and Mn in spintronics. Spintronics is a relatively new branch of electronics in which the spin of an electron plays a key role in transport and device gain and logic functions. For progress in spintronics, initially semiconductor spin devices using GaAs and Si attracted considerable attention due to ease in fabrication of novel spin devices that had potential to impact commercial markets based on existing semiconductor technology (see for instance [126]–[128]). Silicon was especially regarded as a promising platform. As the initially studied metallic magnetic layers (Fe and Co) have a better lattice match to the zinc-blende structure of GaAs and ZnSe, (which leads to better quality magnetic layers), interest shifted to these layers, notably GaAs [126].

In this section, we focus on the two most relevant applications of ferrite layers in the field of spin electronics:

i) Tunnel junctions using Fe_3O_4 as an electrode, which should, in principle, afford an infinite TMR (see Section 10.1.2)

ii) Tunnel junctions using ferrite, e.g. XFe_2O_4 (X = Co, Ni, Mn), layers as spin filters.

10.2.5.1 Tunnel Junctions Using Fe_3O_4 as an Electrode

The results obtained thus far on TMR tunnel junctions having at least one Fe_3O_4 electrode are presented in Table 10.2. In Table 10.2, the substrate, tunnel junction structure, TMR achieved at room and if applicable low temperature, growth method employed, key aspect of this particular TMR study, and source reference are listed.

Table 10.2 Overview of the tunneling magnetoresistance (TMR) for Fe_3O_4-based magnetic tunnel junctions, Note that over time the TMR is found to be negative, as is so be expected, since the conduction band comprises the minority spin.

Year	Substrate	Structure (in brackets the layer thicknesses in nm)	TMR (%)	Dep. method	Key aspect	Ref.
1998	MgO(1 0 0)/ $CoCr_2O_4$ (80 nm)	Fe_3O_4 (40)/MgO (4)/ Fe_3O_4(40)	+0.5 (RT) +1.5 (150 K)	PLD		[129]
1999	Glass	Co (15)/Al_2O_3(2)/ Fe_3O_4 (2)/Al	+13 (RT) +43 (4.2 K)	sputtering	Fe_3O_4 top electrode	[130]
2000	MgO(1 0 0)	Fe_3O_4 (356)/MgO (6)/Fe_3O_4(29)	+ 0.3 (200 K)	MBE	Low switching field≈200 kA/m	[131]
2002	$SrTiO_3$	$L_{0.7}S_{0.3}MO_3$ (80)/ $CoCr_2O_4$ (6)/Fe_3O_4 (60)	-2 (RT) -25 (60 K)	PLD	Fe_3O_4 top electrode. Found expected neg. TMR	[132]
2002	MgO(110)/ TiN(300 nm)	Fe_3O_4 (50)/AlO_x (1.2)/CoFe (10)	+13 (RT) +16 (110 K)	RF sputtering	Suggest change of TMR around T_V	[133]
2003	MgO(110)/10 nm V/60 nm Ru	Fe_3O_4 (20)/AlO_x (1.1)/CoFe (2)/NiFe (4)/Ru(50)	+14 (RT)	Ion beam deposition (IBD)		[134]
2005	Si/SiO_2/Ta (10)	Fe_3O_4 (20)/AlO_x (2)/ CoFe (15)/Ta(10)	+ 11 (RT)	DC and RF sputtering	Fe_3O_4 bottom better than top electrode. Ar sputtering improves Fe_3O_4 interface higher TMR	[135]
2005	Si/SiO_2/Ta(3)/ Cu(50)	NiFe(10)/AlO_x(2)/ Fe_3O_4 (3.5)/Fe(5)/ Ta(3)	-1.2 (RT) -2.4 (110 K)	RF/DC sputtering	Fe_3O_4 formed by oxidation of Fe. Neg TMR on complete oxidation	[136]
2007	α-Al_2O_3 leading to (111) growth	Fe_3O_4 (25)/γ-Al_2O_3 (2)/Co (15)	+2.5 (RT) Barely T dependent	MBE	Spin polarization at Fe_3O_4/γ-Al_2O_3 measured: -40%	[137]
2007	α-Al_2O_3 leading to (111) growth	Fe_3O_4 (60)/MgO (6)/ CoFe (5)/IrMn(10)/ Ta(5)	-14 (RT)	PLD		[138]
2008	MgO (1 0 0)	Fe (12)/Fe_3O_4 (7)/ MgO (2)/Co (10)	-22 (80 K)	UHV sputtering	Fe underlayer to improve growth	[58]
2008	α-Al_2O_3 leading to (111) growth	Fe_3O_4 (60)/ MgO+Al_2O_3 (6)/ CoFe (5)/IrMn(10)/ Ta(5)	-26 (RT)	PLD	Found a clear relation between resistance area (RA) and TMR (see Figure 10.8)	[139]
2011		Fe_3O_4 (.)/AlO_x(2.5)/ Co (.)	+20 (350 K)	PLD	Use of an improved ring-shaped geometry	[140]
2014	MgO(110)	NiO(5)/Fe_3O_4 (60)/ Al_2O_3 (2.4)/Fe(5)/Co (10)	-12 (RT)	MBE		[141]
2015	MgO(1 0 0)	Fe_3O_4 (22)/ MgO(1.5)/CoFeB (4)	-12 (RT)	MBE Fe_3O_4+sputtering (rest)	Annealing Fe_3O_4 at 240–260 °C changes TMR sign from neg. to pos.	[142]
2021	MgO(1 0 0)/ MgO(20)	NiO(5)/Fe_3O_4 (60)/ MgO (2.5)/Fe(10)/ AU (30)	-10 (RT) -50 to 60 (80 120 K)		For the best grown Fe_3O_4 layers the TMR seems correlated with the T_V	[143]

The first studies using Fe_3O_4 as TMR electrodes were reported near the end of the 1990s. In these studies, either a Fe_3O_4/MgO/Fe_3O_4 structure [129, 131] or a structure in which only a single Fe_3O_4 electrode was combined with a metallic counter electrode, e.g. glass/Co/Al_2O_3/Fe_3O_4/Al structure [130] was used. Results were quite disappointing, resulting in either a very low TMR of $\leq + 0.5\%$ [129, 131], or when a large TMR of 13% was found, it was positive, whereas in view of the minority spin-based current in Fe_3O_4, a negative TMR was expected. Also, in subsequent experimental work, positive TMR was observed (see Table 10.2). This led to substantial discussion in the literature concerning if the Fe_3O_4 layer electron polarization should be considered fully spin-polarized and whether a negative or positive TMR should be anticipated.

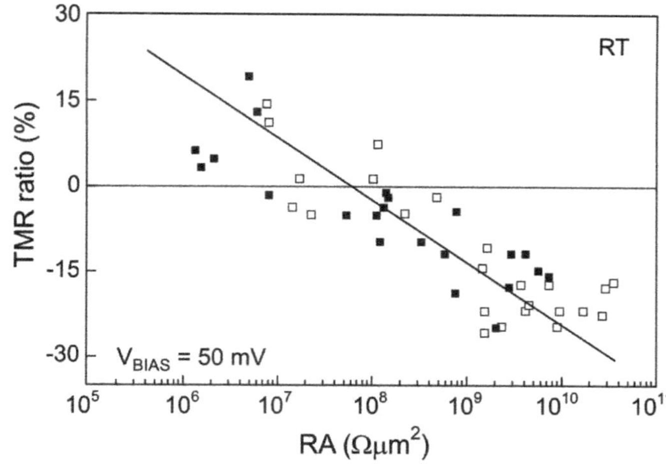

Figure 10.18 Relation between the resistance area (RA) product and the observed room temperature tunneling resistance ratio (TMR) for 60 nm Fe_3O_4/5 nm MgO/1 nm Al_2O_3/CoFe magnetic tunnel junctions grown on sapphire (0001) substrates. Note that for high values of RA only negative TMR values are observed, reaching a maximum value of −26%. The different open and closed symbols refer to samples from different lots (figure reproduced with permission from [139]) (© 2008 by the American Institute of Physics).

The riddle of the observed positive TMR was resolved through experiments by Kado et al. [138, 139]. The work of these authors highlighted the importance of structural quality in fabricated tunnel junctions using Fe_3O_4 electrodes. The authors produced 60 nm Fe_3O_4/6 nm MgO/CoFe/IrMn/Ta junctions grown on α-Al_2O_3, in which the Fe_3O_4 and MgO were deposited by PLD and metal layers by sputtering. These authors found a clear relation between the observed room temperature TMR and the resistance area (RA) product (see Figure 10.18). Tunnel junctions with fewer imperfections that give rise to shunt paths through a tunnel barrier rather than true spin-tunneling currents should exhibit a higher RA. As Figure 10.18 shows, for tunnel junctions with a high RA, the TMR invariably is negative. Hence, the negative TMR for these junctions should be considered intrinsic behavior, as these are tunnel junctions with the least imperfections. These experiments revealed that electron transport through defects in the tunneling barrier, or at the interface within the tunnel barrier, may result in a reversal of TMR polarization [138]. The performance of their tunnel junctions was further improved by adding an amorphous Al_2O_3 layer of 10 Å to the tunnel barrier. This reduced the interfacial roughness from 7 Å, for the Fe_3O_4/60 Å MgO stack, to 3 Å for the Fe_3O_4/50 Å MgO/10 Å Al_2O_3 and produced a record −26% TMR at room temperature.

Examining Table 10.2 shows disappointingly that this is still the finest TMR result as of the writing of this text. What also can be gleaned from comparison between results obtained by various groups (see Table 10.2) is the best results, negative TMRs $\geq -12\%$ are obtained by either using PLD or oxygen-assisted MBE and that both the oxides have been grown in the same growth run to warrant an pristine Fe_3O_4/MgO interface free from contaminants. Nagahama et al. managed to achieve −12% TMR using an Al_2O_3 tunnel barrier directly deposited on Fe_3O_4 by oxygen-assisted MBE in a uninterrupted growth run [141, 143]. Also, the results by Marnitz et al., in which Ar bombardment of the Fe_3O_4 surface before MgO deposition improved the TMR of their tunnel devices from −1% to − 12%, was attributed to a cleaner and smoother Fe_3O_4 surface, leading to a superior interface with the MgO [142]. These results may derive from minimizing the so-called "dead layers" and avoiding the polar catastrophe in epitaxy, as discussed in Section 10.2.4. Table 10.2 also shows that using sputtering to produce high-quality Fe_3O_4 and a high-quality Fe_3O_4/MgO interface remains a significant challenge to which the work by Greullet et al. provides an exception [58]. Note, these authors deposited their samples on the (111) surface of α-Al_2O_3 substrates.

Thus far, the highest room temperature TMR ratios have been achieved in tunnel junctions using:

i) A Fe_3O_4 electrode, with the characteristics:

- Having a low mean roughness, R_a, of ≈ 0.3 nm. Negative TMR is correlated with a mean roughness of electrode/tunnel barrier (Fe_3O_4/MgO-Al_2O_3) stacks (~3 Å) [127, 128] Greullet et al. reported an R_a of 2.3 Å for a Fe_3O_4 layer increasing to 3 Å after an MgO capping layer was deposited [58].
- Use of a thin 10 Å Al_2O_3 layer reduces surface roughness by offering conformal coverage and provide a better interface for the deposition of the metallic counter electrode that is free of pinholes.

ii) A thin, ideally a 20 Å, MgO tunnel barrier as the tunnel current decreases exponentially with barrier thickness
- Magnetic coupling studies of two Fe_3O_4 layers on an MgO interlayer grown by MBE have shown that for $t \geq 1.2$ nm the coupling is due to weak Néel orange peel coupling so that the magnetic electrodes can be considered decoupled. Below this value, the coupling strength increases by a factor of 20 due to pinhole coupling [100]. Later studies of MBE-grown Fe_3O_4/MgO/Fe_3O_4 trilayers, weak antiferromagnetic coupling instead of pinhole coupling at a MgO thickness of 0.8 nm, which the authors attributed to the use of better quality, e.g. unpolished substrates [59].
- For MgO tunnel barriers with $t \geq 20$ Å, the electrode layers should be decoupled.
- A further argument for use of thin, preferably 20 Å, comes from a study of the growth of ultrathin layers of MgO on Fe_3O_4 [63]. Based on the observed Mg yield in XPS studies, these authors found that initially MgO grows in a layer-by-layer manner on Fe_3O_4/MgO (001). Depending on the substrate temperature, after growth of this initial 2- to 3-nm-thick laminar wetting layer, the MgO films begin to roughen due to island growth. As local thickness variations influence tunneling efficiency due to scattering processes, an MgO tunnel barrier should be limited to ~20 Å.

iii) A Fe_3O_4 electrode of $t \geq 30$ nm:
- From MBE-grown Fe_3O_4 layers discussed earlier, for $t \geq 30$ nm good-quality Fe_3O_4 layers exhibit bulk magnetization values [96]. Also, the magnitude and temperature dependence of the anisotropy energy of MBE-grown Fe_3O_4 layers resemble the behavior of bulk Fe_3O_4 when $t \geq 30$ nm [104, 105].
- Authors showing negative TMR when using Fe_3O_4 electrodes used a film thickness of 60 nm [139, 141] or 22 and 77 nm [142]. The work of Greulliet et al. provides an exception who used a 7 nm Fe_3O_4 layer deposited on a 11 nm Fe underlayer [58]. From Table 10.2, the largest negative TMR values were achieved for thicker Fe_3O_4 electrode layers.

iv) Using Fe_3O_4 as a bottom electrode, as this allows a reduction in APB density, spin-polarized transport is enhanced as scattering at APBs would be reduced [139]. The reduction in APB density can be achieved through the following means:
- Increasing the Fe_3O_4 layer thickness as the structural (antiphase) domain size D increases (see Section 10.2.4) proportionally [94]
- Post-annealing in an oxygen background pressure (see Section 10.2.4) [116]
- Use of a Fe underlayer on the substrate [58, 144]
- Or, the use of an electric field during growth [145].

Nevertheless, a maximum TMR value of −26%, achieved at room temperature is disappointing given what can be achieved using conventional metallic CoFe-based MgO tunnel junctions[146]. From room temperature TMR results of reference [75], one can calculate spin polarization of CoFe to be between 55% and 70%. If we take as a conservative estimate to be 55% for the CoFe used in the junction studied by Kado et al., this yields a negative TMR of −26% [139]; using Julière's expression [see Equation (10.2)], one calculates the room temperature spin polarization of Fe_3O_4 to be −27%, which is substantially lower than the predicted half-metallic −100%.

This value also must be compared to the direct measurement of Fe_3O_4 spin polarization in deposited films. These values differ substantially. Güntherodt and coworkers consistently reported this value to be −(80 ± 5)% both for (111) Fe_3O_4 films on W single-crystal substrate, obtained by oxidation of a 50 Å Fe-deposited film [147], as well as for thick (111) Fe_3O_4 films grown by oxygen-assisted MBE [148]. Yet, as these Fe_3O_4 film were produced by oxidation of metallic Fe, the thickness of the Fe_3O_4 layer was could not be reliably measured. It is also noteworthy that Huang et al. obtained for a 31.5-nm-thick (1 0 0) Fe_3O_4 film grown by MBE on MgO a value of −55.5% [65]. To shed light on the conflicting data published in the literature, Morton et al. examined both a sputter-deposited 100 nm (1 0 0) Fe_3O_4 layer grown on MgO (001) using a 300 A Fe buffer layer to alleviate epitaxial strain grown and a 100 nm (1 0 0) Fe_3O_4 film grown on MgO (001) by MBE [149]. The authors concluded that no reliable value could be determined as removing contaminating monolayers of adsorbents could induced surface imperfections that reduce the overall measured polarization by approximately 60%! For the as-received samples before cleaning, these authors put the value of the Fe_3O_4 spin polarization between −30% and – 40% [149]. It is this latter conclusion about the pronounced effect of the role of Fe_3O_4 surfaces that should be taken into account when considering the reported anomalously low TMR values. A previous study raised the potential role of the so-called "dead" Fe_3O_4 surface layer (see Section 10.2.4) [96]. Recent work by Tjeng and coworkers confirmed this "dead layer" hypothesis, as this explanation ensures that the so-called epitaxial polar catastrophe is indeed avoided (see Section 10.2.4) [75].

10.2.5.2 Tunnel Junctions Using XFe_2O_4 as Spin Filters

Using ferrite layers such as XFe_2O_4 in which X = Co, Ni, or Mn as a spin filter to give rise to a spin-polarized current to be injected into a detector layer is an elegant concept (see Figure 10.19) [65]. The spin-filtering effect arises because in a magnetic insulator, such as a ferrite layer, the exchange splitting causes the energy levels for spin-up and spin-down electrons to differ. Consequently, the barrier height Φ differs for spin-up and spin-down electrons, giving rise to a higher probability for one of the two spin states to tunnel through the XFe_2O_4 barrier than the other [150]. The elegance of this concept is that the total current density (J) experiences and exponential relationship with respect to thickness and barrier height Φ:

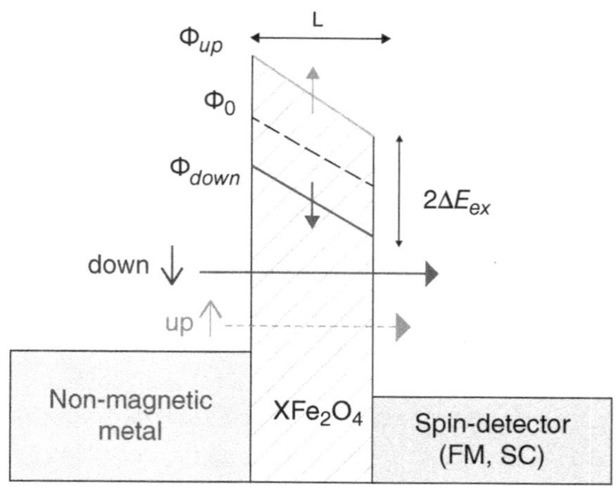

Figure 10.19 Schematic diagram how a spin filter using XFe_2O_4 (X = Co, Ni, or Mn) should operate (figure reproduced from [65]).

$$J \propto \exp(-t\sqrt{\Phi}). \tag{10.10}$$

Hence, even a modest exchange splitting $\Delta E_{exchange}$ will give rise to a significant effect and cause a spin-polarized current to be injected in the detector. Consequently, relatively thin, typically 3–5 nm ferrite layers can be used, which is important as J depends exponentially on t.

A few groups have been working on this approach using ferrite layers as spin filters; the group of Moussy and coworkers is one of them. The results obtained are given in Table 10.3. (For a more detailed account, we refer the reader to the review by J.-B. Moussy [65].) What is striking again is that the room temperature effect observed is limited. In some case, there is no room temperature effect as either the T_C of the detector materials used, i.e. $La_{2/3}Sr_{1/3}MnO_3$, is too low [151] or an Al

Table 10.3 Tunneling magnetoresistance (TMR) and spin polarization results using ferrite filter layers.

Year	Substrate	Filter layer (nm)	Structure (in brackets the layer thicknesses in nm)	TMR (%)	Spin polarization (%)	Deposition ferrite filter layer	Key aspect	Ref.
2006	$SrTiO_3$ (1 0 0)	$NiFe_2O_4$ (NFO)	$La_{0.7}S_{0.3}MO_3$ (35)/NFO/Au	+44 (4 K)	+19 (4 K)	Sputtering	Effect disappears at 140 K	[151]
		(5 nm)	$La_{0.7}S_{0.3}MO_3$ (35)/ $SrTiO_3$(0.8)/NFO/Au	+50 (4 K)	+20 (4 K)			
2007	α-Al_2O_3 (0001)	$CoFe_2O_4$ (CFO) (3 nm)	Pt (20)/CFO(111)/γ-Al_2O_3 (1.5)/Co (10)	-3 (RT) -18 (2 K)	-4 (RT) -25 (2 K)	MBE	Observed the expected neg. spin polarization	[152]
2010	MgO (1 0 0)	$CoFe_2O_4$ (3 nm)	TiN(30)/Cr(2)/Pt(2)/ $CoFe_2O_4$/MgO(1)/Co(5)/ Pt(10)	-3.4 (RT) -30.5(10 K)	4.3 (RT) 44 (10 K)	Thermal oxidation		[153]
2012	α-Al_2O_3 (0001)	$CoFe_2O_4$ 3 nm	Pt(20)/ CFO(111)/γ-Al_2O_3(1.5)/ Co (10)	-6 (RT)	-8 (RT)	MBE	Nanojunctions (area 5×5 nm^2) to probe single CFO domain	[154]
2014	α-Al_2O_3 (0001)	$NiFe_2O_4$ 5 nm	Pt/NFO(111)/γ-Al_2O_3(1.5)/ Al (4.2)		+25 (0.5 K)	MBE	Using a supercon ducting Al electrode	[155]

superconductor detector was used [156]. Generally, the room temperature effect is quite modest at a few percent. Some of the problems encountered in constructing suitable spin filters is that in very thin (3–5 nm) XFe_2O_4 spin filter layers the measured surface resistance (*RA*) as a function of the inverse of temperature, clearly showed a decrease in resistance at low temperatures suggesting a thermally activated transport mechanism due to the presence of defects in the magnetic tunnel barrier [151, 156]. Typically, these defects are oxygen vacancies, giving rise to defect states in the band gap that act to lower the effective barrier height seen by tunneling electrons and thus yielding a stronger *R*(*T*) dependence [156]. These, as well as the APBs known to be present in these ultrathin XFe_2O_4 layers, are invoked to explain the lower-than-expected spin-filtering effect [155].

A further, more fundamental problem is when using an XFe_2O_4 spin filter layer in direct contact with a ferromagnetic (FM) layer that they magnetically couple and *do not switch independently*. To remedy this problem an extra "magnetically" insulating barrier layer is inserted that induces spin scattering. For this purpose, $SrTiO_3$ [151] or γ-Al_2O_3 [154–156] was employed.

We note, as we did for the Fe_3O_4-based spin-tunneling devices, that a possible "dead surface layer" effect may exist and play a dominant role. Such "dead layers" are not just relevant for Fe_3O_4 layers as reports on nanoparticle systems have amply demonstrated [101]. Given the very thin $NiFe_2O_4$ and $CoFe_2O_4$ layers thus far used in spin filtering, such a "dead layer" would contribute significantly in degrading spin polarization and reducing the net spin polarization also in these systems. APBs typically occur orthogonal to the interface, and this forms a possible *local* shunt for defect-mediated electron transport. A magnetic "dead layer" at the XFe_2O_4/FM or XFe_2O_4/γ-Al_2O_3 over the entire interface of the XFe_2O_4 layer is bound to have a very pronounced effect. This could well explain the low spin polarization thus far observed. Not surprisingly, in a study of the electronic and magnetic structure of epitaxial $CoFe_2O_4/Al_2O_3$/Si heterostructures, in which the $CoFe_2O_4$ layer thickness was varied between 1.4 and 11 nm, XAS and XMCD found evidence for a magnetic dead layer of 1.4 nm at the $CoFe_2O_4/Al_2O_3$ interface [157]. This situation in $CoFe_2O_4$ is identical to the situation discussed in Section 10.2.4 for Fe_3O_4 [96]. The conclusions concerning the problem in spin-polarized tunneling is thus identical to the conclusions drawn as to the cause for low TMR in Fe_3O_4-based spin tunnel junctions (see Section 10.2.5.1), namely that the way ferrite layers grow to avoid the "polar catastrophe" [75] causes an interface layer that mixes up the spin polarization, even in high-quality growth methods such as MBE.

10.2.6 Summary and Outlook

The study of MBE-grown thin ferrite layers, mostly Fe_3O_4, has helped reveal the underlying physical properties and phenomena of ferrite layers. It has been demonstrated that stoichiometric thin films can be grown that have magnetization values comparable to those of their bulk equivalents. Yet, when depositing films on higher symmetry substrates, such as MgO (1 0 0), APBs occur. The concentration of APBs is a function of growth conditions and can be improved by postdeposition annealing of the films [116]. However, due to the nature of the MBE growth process, A sites at the interfaces are not filled to the bulk distribution [75]. This changes the superexchange interaction at the interface and may in effect lead to a so-called "dead layer" effect at the Fe_3O_4 interface that causes a preponderance of spin scattering and a reduction of the potentially 100% spin-polarized injection current. Rather than the theoretically predicted infinite spin-tunneling effect, the maximum seen in devices thus far is −26% [139]. Consequently, application of ferrite thin films as spin filters in spintronic devices has been limited.

In other applications of thin ferrite layers, the permeability μ is an important parameter. It has been shown that through MBE a $\approx$ 1-μm-thick Fe_3O_4 film can be grown at a reasonable deposition temperature, T_{dep}, for device integration purposes on glass substrates yielding a $\mu \approx 10$ [26]. Yet, for practical purposes the growth rate of MBE at 6 nm/min is much too low to make this a realistic option in industrial scale device fabrication. Hence, in such applications, it is better to use the more complex ferrites such as MnZn- and NiZn-ferrite, which have intrinsically a higher μ than magnetite [1]. To grow these more complex ferrites under practical conditions, PLD is a more suitable technique. One of the reasons is that PLD generally offers the advantage of congruent ablation, that is, the stoichiometry of the deposited film is equal to that of the target. PLD deposition of ferrite layers will be examined in Section 10.3.

10.3 PLD of Ferrite Films

10.3.1 Introduction

PLD belongs to the physical vapor deposition (PVD) family of process technologies used to deposit thin films and coatings on solid surfaces. PLD makes use of a laser as the source of concentrated energy in order to ablate a solid target within a vacuum environment allowing for congruent (i.e. stoichiometric) vapor transfer to a substrate under a variety of temperature and gas pressure conditions.

The ablation of materials by laser in vacuum was first employed for the deposition of thin films soon after the first practical lasers were developed in the early 1960s. In 1965, Howard and Turner for the first time used a ruby laser to ablate a number of materials and empirically demonstrated the successful condensation of films on substrates [157]. Importantly, it was concluded that most materials cannot be deposited under the same conditions; they would require individual studies to determine optimized conditions. The next milestone occurred in the early 1970s with the emergence of Q-switched lasers [159] and efficient harmonic generators [160] capable of producing much shorter optical pulses and wavelengths, thus providing greater flexibility in laser fluences and growth dynamics.

The first documented attempts to deposit ferrite films by PLD date to the late 1980s [4, 161]. Joshi et al. deposited iron oxide and zinc ferrite films and studied the influence of process parameters on film properties. Ogale and Nawathey deposited zinc ferrite in 1989 [5, 6, 162, 163], and Tanaka et al. deposited Ni–Zn-ferrite films in 1991 [7, 164]. Kidoh et al. [8, 165] synthesized YIG films by laser ablation in 1991. Dorsey et al. [9, 166] and Carosella et al. [10, 167] deposited hexaferrite films in 1992.

In these early developmental years, PLD was considered suitable only for small-area deposition in research studies. Scholarship to address reproducibility, large-area deposition and higher production yields began in earnest in the 1990s. Reliable and reproducible large-area deposition, including blackbody and substrate heaters, laser beam rastering, plasma particle size filters, intelligent windows, multitarget carousels, and *in situ* diagnostic tools, among other advances were also developed at this time.

As of the writing of this chapter, a number of companies (e.g. Neocera and PVD Products, USA; DCA Instruments and Picodeon Ltd. Oy, Finland; Pascal Co., Ltd, Japan; Surface, Germany; TSST and SolMateS, The Netherlands) offer production-oriented turnkey PLD systems for large-area deposition, in some cases up to 10 m^2 continuous flow tape deposition. These companies were first to produce excimer laser-based systems. However, in more recent years, PLD systems have employed femto- or pico-second Nd:YAG lasers. In comparison, excimers offer higher energy densities, while solid-state lasers provide substantial cost of ownership savings.

10.3.2 Development of Excimer and Solid-State Lasers

The development of PLD as an emerging thin film processing technology has paralleled developments in laser technology, in particular, those of excimer and solid-state lasers.

10.3.2.1 Excimer Lasers

Although *excimer*, in this context, refers to a class of gas lasers, the term derives from excited dimers that are short-lived dimeric or heterodimeric molecules formed from two species, one of which must be an element having a complete outer shell of electrons, e.g. one of the noble gases. In such cases, electron bonding by ionic or covalent means are only possible if that atom is in an excited state. The lifetime of an excimer is on the order of nanoseconds.

Excimer laser development began with the early observation of spectral line narrowing in liquid xenon (176 nm) by Basov et al. [168]. Shortly thereafter, Koehler et al. [169] and Ault et al. [170] presented improved demonstrations of stimulated emission using high-pressure xenon gas. Strong stimulated emission was observed as the laser's spectral line narrowed from a continuum of 15 nm to just 0.25 nm, and the intensity increased a thousand-fold. The laser's estimated output was ~1 J [171]. Soon after, noble gas halides, e.g. XeBr, KrF, ArF, and XeCl, were developed by several groups, including: Avco Everett Research Laboratory, Sandia Laboratory, Northrop Research and Technology Center, US Naval Research Laboratory, and Los Alamos National Laboratory [172].

10.3.2.2 Solid-State Lasers

Nd:YAG (neodymium-doped yttrium–aluminum–garnet; Nd:$Y_3Al_5O_{12}$) is a crystal that is used as a lasing medium for solid-state lasers. The dopant, triply ionized neodymium Nd^{3+}, typically replaces a small fraction (~1%) of yttrium ions in the host crystal of the yttrium–aluminum–garnet (YAG), as the two ions are of similar size. It is the neodymium ion that provides the lasing activity in the crystal. Laser operation of Nd:YAG was first demonstrated by Geusic et al. at Bell Laboratories in 1964 [173]. Nd:YAG lasers typically emit light with an IR wavelength of 1064 nm [174]. However, there are also transitions at 946, 1120, 1320, and 1440 nm. Nd:YAG lasers operate in both pulsed and continuous modes. Nd:YAG lasers are typically operated in a Q-switching mode where an optical switch embedded in the cavity triggers when maximum inversions occur in the concentration of neodymium ions giving rise to megawatt output powers as pulses with widths of 10 to 25 ns. The high-intensity pulses may be efficiently frequency doubled to generate laser light at 532 nm, or higher harmonics at 355, 266, and 213 nm [175].

10.3.3 PLD Processes

The interaction of the laser beam and target results in three primarily ablation mechanisms: (i) thermal, (ii) electronic, and (iii) hydrodynamic. The energy of the laser pulses is immediately absorbed by the target material and transformed into thermal, electronic excitations, and hydrodynamic energy. The local temperature at the target surface when exposed to the laser pulse may exceed 5000 K in a few nanoseconds. As a result of the excessive heating rates, congruent ablation of the target's elements occurs. The resulting plasma plume consists of ions, neutral atoms, and splashed particulates. The mechanisms that result in the particulate content of the PLD plume are exfoliation, surface ablation, and shock-wave recoil pressure expulsion (see Figure 10.20) are addressed in [176] and discussed in detail later in this section.

The second stage of the PLD kinetic process is plume expansion. In this stage, ions and atoms leave the target with energies in excess of 100 eV, but eventually thermalize due to collisions with atoms, ions, and molecules in the background gas and plasma. Therefore, the energy of a given particle depends on its mass and scattering length. The resulting element-dependent angular distribution of particle energies lead to nonuniform deposition and limits the uniform deposition area. Excessive energies of arriving particles may also cause element-dependent preferential resputtering from the growing film surface.

It is challenging to grow complex oxide films such as garnets, spinels, and hexaferrites by conventional PVD techniques. The desired outcome of large-area, stoichiometric, and high-crystal-quality films can be achieved only under a narrow range of process parameters due to the challenges associated with stoichiometric transfer of target elements to the substrate having significantly different atomic masses and atomic binding energies.

Alternatively, PLD inherently allows for the deposition of a wide variety of materials, including oxides, nitrides, and carbides, using a wide range of background gases and pressures. Compared to alternative physical vapor approaches, PLD offers a number of advantages including stoichiometric transfer of target materials to films on a variety of substrate materials including reactive deposition with rates as high as 100 nm/min. PLD is the preferred PVD technique for the growth of multielement oxide films of which ferrites are one group [177].

A typical PLD system consists of a vacuum chamber and pulsed excimer gas or solid-state laser (the latter is typically of a Nd:YAG design) as key components. A set of lenses and mirrors is used to focus and, in some cases, raster the laser beam on the target. The vacuum chamber is typically equipped with a turbomolecular pump and scroll pump to provide adequate and dynamic control of the working gas pressure. In most instances, a standard mass flow controller (MFC) enables precise and reproducible gas pressure conditions during deposition. Conventional PLD chambers accommodate a variety of substrate shapes and sizes and usually employ IR lamps or resistive blocks to uniformly heat substrates to temperatures

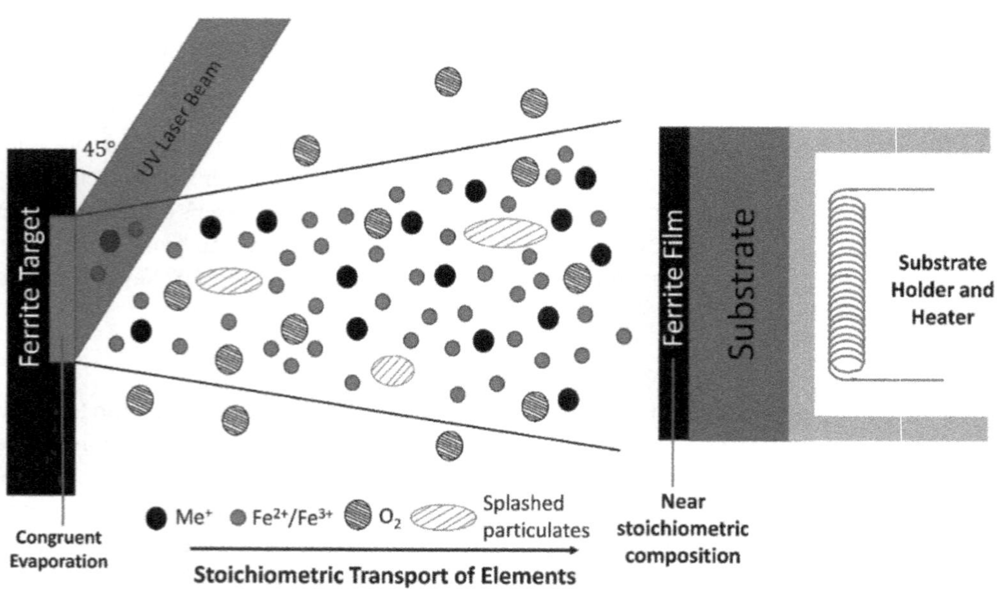

Figure 10.20 Schematic illustration of PLD laser plume processes.

up to and beyond 900 °C. Some PLD systems use liquid cooling of the vacuum chamber to reduce thermal load. Pulsed UV lasers enable the PLD process as nearly all target materials, including high-bandgap oxides, are susceptible to ablation by incident high-energy photons. Laser types determine ablation characteristics. The laser fluence stability in time and space domains is of the greatest importance as the ablation process is acutely sensitive to fluctuations in energy density. As noted earlier, most manufacturers offer excimer and Nd:YAG-based PLD systems. Solid-state lasers (see Figure 10.21) have a lower cost of ownership and also produce ill-suited Gaussian beam profiles. Moreover, they suffer from poor pulse-to-pulse laser fluence stability. Optical properties of excimer lasers are better suited for the PLD process. These gas lasers generate flat-top beam pulses, and their fluence remain stable over millions of pulses but at a high cost of ownership.

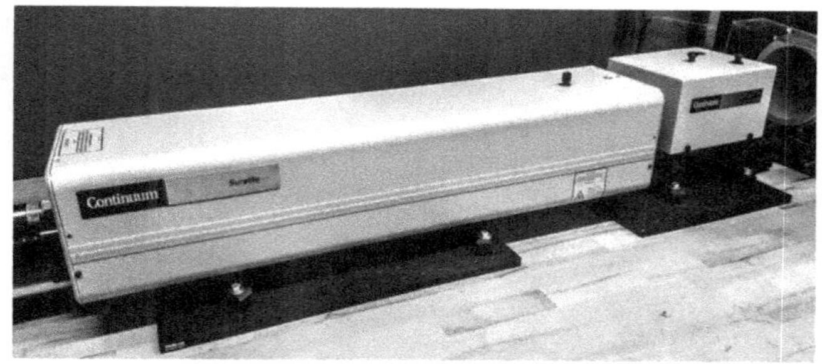

Figure 10.21 Surrelite solid-state Nd:YAG laser produced by Continuum and used in PLD systems in place at Northeastern University (Boston, MA, USA).

Figure 10.22 shows Neocera Pioneer 180 PLD system equipped with a load-lock rapid sample transfer capability for the rapid throughput of 100 mm diameter samples. The load-locked chamber is equipped with a gate valve that isolates it from the main chamber. A magnetically coupled linear transfer arm is used to insert and retract substrates to and from the main chamber. A COMPexPro 110 excimer laser is used in tandem with an optical train consisting of mirrors and lenses mounted on a liner stage allowing for beam focusing and rastering (see Figure 10.23a). A laser energy closed loop allows for precise control of fluence throughout the growth process.

Figure 10.22 Excimer laser-based PLD system manufactured by Neocera in place at Northeastern University (Boston, MA, USA). (A) Laser safety curtains, (B) laser gas exhaust cabinet, (C) excimer laser, (D) optical train, (E) vacuum and power control modules, (F) water chiller, (G) processing chamber, (H) load-lock substrate exchange chamber, (I) low vacuum backing pumps, and (J) process gas cylinders [178].

While most challenges to PLD are common to all PVD techniques, e.g. heating of transparent substrates to high temperatures, few challenges are unique to PLD. For example, the coating of the laser entrance window with ablated flux poses a difficult problem after repeated deposition cycles. The window increasingly becomes coated with ablated flux from the target(s) and attenuates laser beam energy and alters the laser-target energetics compromising reproducibility and quality. Further, the laser increasingly interacts with the window material leading to premature deterioration of its optical properties. Laser interaction with optical windows was investigated first by Greer [179] at Raytheon during the early 1990s, leading to what is today the Intelligent Window (PVD Products, Inc.) [180] (see Figure 10.23b). Regardless of the exact manifestation, the Intelligent Window typically represents a sacrificial fused silica disk placed inside the vacuum chamber exactly behind the laser entrance window that comprises the vacuum/atmosphere interface. The silica disk is shielded from the rest of the chamber with only a small off-center aperture exposed to the incident laser beam. As the aperture becomes coated by ablated materials to the extent that laser energy entering the chamber is attenuated by ~10%, the silica disk is rotated exposing a fresh spot. Depending on the size and configuration of the Intelligent Window, the process is repeated up to 50 times before the fused silica disk must be cleaned or replaced.

The system shown in Figure 10.22 is equipped with a KrF (266 nm) excimer laser generating 10 ns pulses. The laser pulse repetition rate for ferrite film growth is typically ~10 Hz. In the configuration shown in Figure 10.22, laser pulses are incident upon a target held at an angle of 45° aligned parallel the substrate/heater block positioned a distance of 70 mm.

(a)

(b)

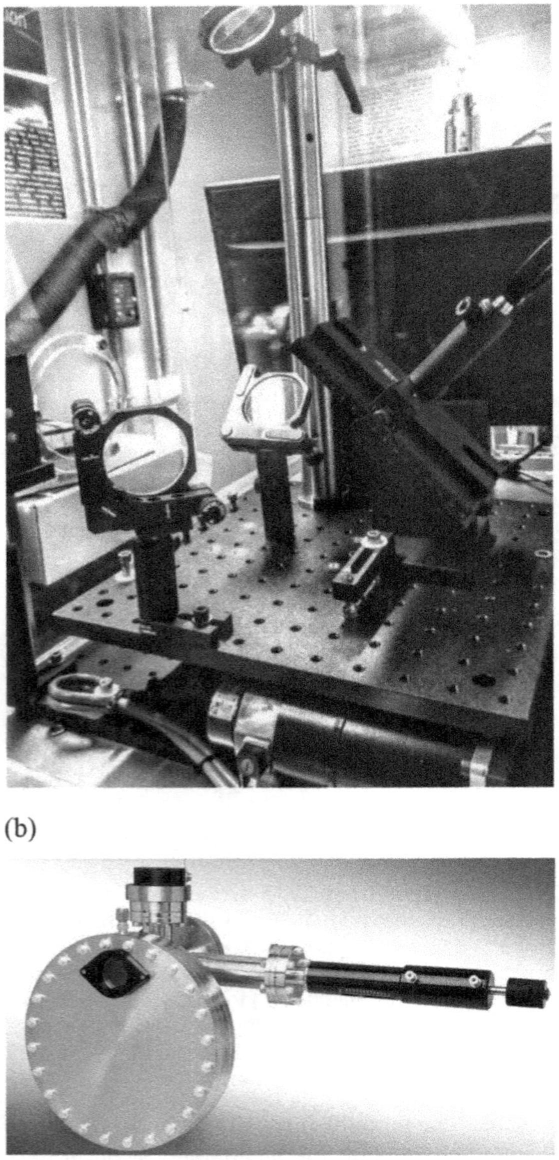

Figure 10.23 (a) Optical train mounted on a liner stage for laser beam focusing and rastering. (b) Intelligent Window (IW) offered by PVD Products, Inc. IW includes a fused silica disk behind the laser entrance window shielded from the rest of the chamber with a small off-center aperture exposed to the incident beam. As the aperture becomes coated by ablated materials the silica disk rotates exposing a fresh spot and the process begins anew [161].

The target sites in a carousel (see Figure 10.24) that allows for rotation and rastering relative to the incident laser beam in order to ensure homogeneous and uniform growth. The laser output power is controlled by varying the voltage output of its power supply, achieving a power density at the target surface of 4–5 J/cm^2. This range of power densities has been shown to allow accurate stoichiometric transfer from the target to the substrate [181].

10.3.4 Ferrite Growth by PLD

10.3.4.1 PLD of Iron Oxide Films

Iron oxide films have been the subject of many studies, primarily due to their interesting and useful magnetic properties. These films exhibit several crystal structures and compositions owing to the different valence states of iron and the penchant for the formation of polyhedral and polyhedra-linking order. These structures include wustite (FeO_x), which has a rocksalt phase, magnetite (Fe_3O_4) and maghemite (γ-Fe_2O_3), which both have the cubic spinel phase, and hematite (α-Fe_2O_3), which has the corundum phase.

Table 10.4 presents a brief listing of reports marking the historical development of PLD iron oxide films. Growth conditions and properties of PLD Fe-oxide films have been reported on substrates of $SrTiO_3$ (STO), $LaAlO_3$, MgO, ZrO:Y, glass, $A1_2O_3$, mica, Kapton, and Si substrates. The most common substrate for epitaxial deposition of every phase of iron oxide is sapphire (0001) (i.e. single-crystal $A1_2O_3$). The (0001) surface of $A1_2O_3$ allows for the deposition of the (111)-oriented cubic films under proper experimental conditions.

MgO (001) substrates conveniently lend themselves for the deposition of the rocksalt phase, or either of the two spinel forms of iron oxide. MgO and spinel phases have a lattice mismatch of 0.4% that enables the epitaxial growth of films with a cube-on-cube stacking. For example, maghemite films on MgO exhibit a high crystal quality with low remanence and high saturation fields.

The earliest published data of iron oxide films deposited by pulsed laser evaporation date back to Joshi et al. and Ogale et al. [161–163] who deposited FeO and Fe_3O_4 films on $A1_2O_3$ substrates These authors showed that the stoichiometry of as-deposited films can be controlled by varying the partial oxygen pressure during growth. Furthermore, γ-Fe_2O_3 films were obtained by post-annealing PLD Fe_3O_4 films. Note, an unusually high (10–12 J/cm^2) laser fluence was employed in that study. The first high-quality (100) oriented Fe_3O_4 films were demonstrated on $LaAlO_3$, $SrTiO_3$, and MgO (100) in 1995 [182].

By circa 2000, a laser fluence of 1.5 J/cm^2 became standard for the deposition of iron oxide films. This value is a few times lower than the fluence typically used for deposition of other ferrite phases. It also became increasingly clear that as-deposited films were mostly iron-deficient. In 2002, Tepper and Ross [183] demonstrated stoichiometric transfer of γ-Fe_2O_3 between target and substrate *in vacuo* as opposed to a partial pressure oxygen atmosphere. They later successfully reactively deposited γ-Fe_2O_3 using a metallic iron target [184].

By the mid-2000s, a consensus formed in the research community that iron oxide films deposited by PLD under low oxygen pressures or vacuum were close to the nominal γ-Fe_2O_3 stoichiometry and exhibited a saturation magnetization typical of bulk maghemite. Additionally, most authors used a substrate temperature of 500 °C to produce the highest quality films [182–185].

Table 10.4 presents a history of published articles briefly detailing the PLD conditions for growth of iron oxide film systems (FeO, α-Fe_2O_3, γ-Fe_2O_3, and Fe_3O_4). The table is segmented with the transition to magnetite films for spintronics applications. Magnetite films, i.e. Fe_3O_4, have been discussed in detail in the MBE section of this chapter as being half-metals. Half-metals possess only one occupied spin density of states (either spin-up or spin-down) at the Fermi energy. As such, in half-metals conduction occurs by 100% spin-polarized electrons. This affords spintronic devices, such as GMR spin valves or MTJs, in theory with infinite gain. The quest for PLD growth of half-metallic magnetite began in 1997 by Gong et al. These authors found, similar to studies on bulk crystals, magnetoresistance exhibits a peak around the Verwey transition, with values peaking to ~32% at 60 K under a 4 T field [186].

From that point forward, PLD processing of magnetite films followed a process, including ~2 J/cm^2 of irradiance and a repetition rate of ~5 to 10 Hz. The choice of well-matched lattice substrates was found equally valuable in reducing stacking fault defects in attaining high-quality films. Substrate temperatures are typically elevated (e.g. 500–1000 °C), but positive results were also obtained for a wide range of combinations of substrate temperatures and working gas pressures.

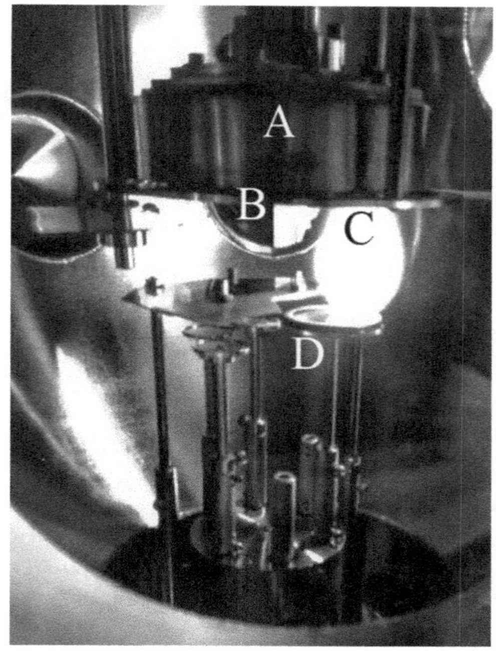

Figure 10.24 PLD target–substrate carousel assembly: (A) heater, (B) 100 mm rotating substrate, (C) plume, and (D) rotating ferrite target.

Table 10.4 PLD conditions and results from early studies of iron oxide film systems (FeO, α-Fe_2O_3, γ-Fe_2O_3, and Fe_3O_4).

Year	Oxygen pressure (Torr) unless denoted	T_{sub} (°C)	Laser parameters	Substrates	Key results	Ref.
1988	5×10^{-7} to 10^{-4} Torr	200	694 nm, 10–12 J/cm^2	$A1_2O_3$	Stoichiometry of pulsed-laser deposition (PLD) films depends on the oxygen partial pressure. Phases of FeO and Fe_3O_4 were stabilized. γ-Fe_2O_3 phase was stabilized by post-annealing of Fe_3O_4 films.	[161]
1988	10^{-4} Torr	200–450	694 nm, 10–12 J/cm^2	$A1_2O_3$ Sapphire	Pulsed ruby laser deposition used to grow iron oxide and zinc ferrite films. The stoichiometry of synthesized iron oxide films depends on the oxygen partial pressure during deposition. Zinc ferrite films were found to be zinc enriched.	[163]
1995	10^{-4} Torr	505	unknown	$SrTiO_3$ (1 0 0), $LaAlO_3$ (1 0 0), MgO (1 0 0), ZrO:Y (1 0 0), glass, $A1_2O_3$, Mica, Kapton	PLD of iron oxide, nickel oxide and cobalt oxide films investigated on various substrates. High-quality (1 0 0) oriented growth: Fe_3O_4 and NiO (1 0 0) $LaAlO_3$, $SrTiO_3$, MgO; Co_3O_4 on $SrTiO_3$ and MgO. (111) oriented growth: Fe_3O_4, NiO, and Co_3O_4 films on cubic zirconia (YSZ). Growth on glass slides result in both (1 1 1) and (1 0 0) orientations.	[182]
1999	15 mTorr	500	KrF excimer, 248 nm, 200 mJ/pulse	MgO (1 0 0)	Under fixed experimental conditions, the structure of PLD iron oxide films depends on the deposition time and thickness. Shorter deposition times (i.e. thinner films) result in spinel-structured Fe-oxide films. Longer deposition times result in corundum-structured Fe-oxide films.	[183]

(Continued)

Table 10.4 (Continued)

Year	Oxygen pressure (Torr) unless denoted	T_{sub} (°C)	Laser parameters	Substrates	Key results	Ref.
2002, 2004	15–50 mTorr	500	KrF excimer, 248 nm, 1.5 J/cm^2	MgO (1 0 0)	PLD in vacuo results in stoichiometric ferrimagnetic γ-Fe_2O_3 films with saturation magnetization values of 371–627 kA/m. PLD in O_2 atmospheres result in iron deficiency and antiferromagnetic α-Fe_2O_3 films with saturation magnetization values of 40–610 kA/m. At least 680 nm of the metastable γ-Fe_2O_3 can be deposited on MgO (1 0 0) by PLD.	[183, 184]
2002	0.5–50 mTorr	390	KrF excimer, 248 nm, 250–500 mJ/pulse	Oxidized Si (1 0 0), Si (111), $LaAlO_3$ (1 0 0)	The phase and orientation of chromium oxide (CrO_x) and iron oxide (FeO_x) films depend on laser fluence, deposition temperature, and oxygen pressure. Films grown at lower laser energies or lower growth temperatures do not exhibit strong XRD peaks.	[201]
2003	UHV – 1 mTorr	RT – 500	KrF excimer, 248 nm, 15–50 Hz, 300–550 mJ/pulse, 1–1.5 J/cm^2	MgO (001), Si (001)	Iron oxide films of iron-deficient stoichiometry were synthesized by (PLD) from a metallic Fe target. The saturation magnetization was ~400 kA/m. Properties were most similar to γ-Fe_2O_3 films grown from oxide targets.	[202]
Growth of magnetite films for spintronic applications						
1997	Unknown	350	Nd:YAG laser, 355 nm, 330 mJ/pulse, 2 J/cm^2	MgO (001)	The magnetoresistance exhibits a peak around the Verwey transition (T_V) as has been reported previously for single crystals. The MR increases monotonically below 100 K with decreasing temperature with values as high as 32% observed at 60 K under a 4-T field. The enhanced low-temperature MR is attributed to a magnetic-field-dependent activation energy for electron hopping transport.	[186]
2003	1.9 mTorr (Ar)	1150	KrF excimer laser, 248 nm, 2–5 J/cm^2, 2–10 Hz	Si(001)	Magnetite epitaxial thin films on Si (001) substrates using a double buffer layer system consisting of TiN and MgO were grown by PLD. The entire growth process was monitored directly by RHEED allowing the determination of surface structure at each process step. By X-ray diffraction and atomic force microscopy (AFM), good crystalline quality of the multilayers and the low roughness values of interfaces verified.	[203]
2005	0.35 mTorr	200	KrF excimer, 248 nm, 5 Hz, 2 J/cm^2	Si (1 0 0)	Stoichiometric magnetite films deposited by PLD at oxygen partial pressures of 0.35 mTorr exhibit saturation magnetization of bulk maghemite.	[187]
2006	3.7 mTorr 0.6 mTorr	210	KrF excimer, 248 nm, 5 Hz, 2 J/cm^2	Si (1 0 0)	Thin films of nearly stoichiometric single-phase polycrystalline magnetite, $Fe_{2.99}O_4$, have been obtained at 210 °C and a working pressure of 0.6 mTorr, with a high-field magnetization of ~490 emu/cm^3 and Verwey transition temperature of 112 K, close to the values reported in the literature for bulk magnetite.	[187]

Table 10.4 (Continued)

Year	Oxygen pressure (Torr) unless denoted	T_{sub} (°C)	Laser parameters	Substrates	Key results	Ref.
2006	7.5×10^{-9} (O_2) Torr	450	KrF excimer, 248 nm, 10 Hz, 2 J/cm^2	α-Al_2O_3 (0001)	A correlation between magneto-transport properties of epitaxial Fe_3O_4 thin films on the order of 5–50 nm in thickness and the APBs in these films. A systematic increase in resistivity and attenuation of the Verwey transition with decreasing film thickness was reported. Trends in transport and magneto-transport properties are attributed to an increase in APB density with decreasing film thickness, and more specifically to additional antiferromagnetic couplings introduced at the APBs, strongly suggesting that the mechanism of magneto-transport is predominantly governed by the spin-polarized transport across APBs.	[204]
2007	10^{-8} Torr	400	KrF excimer, 248 nm, 10 Hz	MgO (001)	High-quality epitaxial Fe_3O_4 films on MgO substrates focusing on magnetoresistance and Hall effect. The magnetoresistance is small as expected for epitaxial films and can be explained by the presence of antiphase boundaries (APBs) with antiferromagnetic correlations. The Hall resistivity is dominated by the anomalous Hall effect, which is proportional to the magnetization.	[192]
2008	UHV – 1 mTorr	150	KrF excimer, 248-nm, 5 Hz, 1 J/cm^2	MgO (001)	Fe_3O_4 films were synthesized on MgO (001) substrates by PLD. Structural and magnetic properties investigated by magnetoresistance measurement and grazing incidence X-ray diffraction.	[193]
2008	2.7 mTorr (Ar)	590	KrF excimer, 248 nm, 2 J/cm^2, 2 Hz	ZnO (0001), Al_2O_3 (0001)	Oriented Fe_3O_4 (111) thin films and Fe_3O_4/ZnO heterostructures on ZnO (0001) and Al_2O_3 (0001) substrates by PLD. High-quality crystalline films with mosaic spread as small as 0.03°, sharp interfaces, and rms surface roughness values of 0.3 nm were achieved. Magnetization measurements show clear ferromagnetic behavior of the magnetite layers with a saturation magnetization of 3.2 μB /f.u. at 300 K.	[199]
2009	2×10^{-6} Torr	350–550	KrF excimer, 248 nm, 10 Hz, 2 J/cm^2	Si (111)	Optimized growth of single-phase-oriented Fe_3O_4 films on Si (111) substrates. XPS data reveal single-phase Fe_3O_4 growth for films grown at 450 and 550 °C. Resistivity data indicates Verwey transition at ~120 K revealing accurate stoichiometry.	[205]
2011	Unknown	100–700	KrF excimer, 248 nm, 2 Hz, 2 J/cm^2	(0001) α-Al_2O_3	Single-phase Fe_3O_4 films were successfully grown on sapphire substrates at a substrate temperature of 500 ° C. The saturation magnetic moment of the single-phase Fe_3O_4 film is 499 emu/cm^2, which is in good agreement with values reported for bulk magnetite, suggesting the Fe_3O_4 film is of high crystal quality without APBs.	[206]

(Continued)

Table 10.4 (Continued)

Year	Oxygen pressure (Torr) unless denoted	T_{sub} (°C)	Laser parameters	Substrates	Key results	Ref.
2012	2×10^{-3} mTorr	350	2.0 J/cm^2, 2 Hz	STO (001)	Fe_3O_4/BFO heterostructures studied for exchange bias (EB) effects between Fe_3O_4 and BFO. EB effects have been observed with a blocking temperature of ~200 K. H_{eb} decreases with increasing temperature or thickness of Fe_3O_4 and does not change after field annealing. Antiferromagnetic superexchange interaction between BFO and Fe_3O_4 can induce uncompensated interface magnetic spins, resulting in the EB effect in heterostructures. Because of half-metallic properties of Fe_3O_4, field-effect devices may enable reversible electrical control of magnetism at higher temperature.	[195]
2013	unknown	507	Nd:YAG laser, 1064 nm, Q-switched, 10 Hz, 4 J/cm^2	$SrTiO_3$ (1 0 0)-doped with 0.1% Nb	Magnetic easy axis directions of stoichiometric magnetite films grown on $SrTiO_3$:Nb by infrared PLD were examined. Spin-polarized, low-energy electron microscopy reveal that individual magnetic domains are magnetized along the in-plane [100] film directions. Magneto-optical Kerr effect measurements show that the maxima of remanence and coercivity are also along in-plane [100] film directions.	[196]
2013	10^{-5} to 0.75 mTorr	600	KrF excimer, 248 nm, 2 Hz, 2 J/cm^2	Sapphire	PLD of iron oxide films on sapphire substrates at 600 °C. The optimum oxygen gas pressure that allows for single-phase magnetite films with high crystal quality are determined to be in a narrow range centered at ~0.0075 mTorr.	[197]
2015	1.5×10^{-6} Torr	600	KrF excimer, 248 nm, 10 Hz, and 300 mJ/pulse	Si (1 0 0)	Inverse MR effect was observed in Fe_3O_4 films with strong (111) texture deposited at high temperatures. The anomalous positive MR behavior and a conduction mechanism is based on the spin filter effect. The observed inverse MR effect extends the understanding of spin-dependent transport in Fe_3O_4.	[207]
2015	1.5×10^{-6} Torr (O_2)	350	Nd:YAG laser, 355 nm, 10 Hz, 1 J/cm^2	$SrTiO_3$ (001)	Films grown by PLD present an expansion of the last atomic layer of the substrate and rough layer surface. However, the obtained structural domain size of 10 nm is identical for the layers grown by both evaporation techniques corresponding to the size of the coincidence lattice. The magnetic and transport properties inherent to the Fe_3O_4/STO heterostructures were found not related to the lattice strain but to the formation of antiphase boundary (APB) defects.	[198]
2017	7.5 mTorr	700	2.0 J/cm^2, 4–5 Hz	(001) Nb-STO single crystal	An interface magneto-elastic ME coefficient of 0.72 Oe-cm/kV at room temperature in Fe_3O_4/$BaTiO_3$/$SrTiO_3$ heterostructures induced by polarization reversal of the $BaTiO_3$ layer was measured. The experimental data was qualitatively explained using a model based on relative magnetization change proportional to the displacement of the $BaTiO_3$ layer due to the electric polarization, indicating the mechanism of ME coupling that atomic displacement at the interface in $BaTiO_3$ layer affect the magnetization at the Fe_3O_4/$BaTiO_3$ interface.	[208]

Table 10.4 (Continued)

Year	Oxygen pressure (Torr) unless denoted	T_{sub} (°C)	Laser parameters	Substrates	Key results	Ref.
2017	1.5×10^{-6} Torr	600–750–950	10 Hz, 300 mJ/pulse	Si (1 0 0)	A systematic study on the evolution of APB densities in Fe_3O_4 films, which were prepared by PLD and post-annealing at different temperatures, was performed. By measuring the electron–phonon coupling parameter and using Allen's formula, the films' APB densities, which showed a decreasing tendency with increasing annealing temperature, was determined. Consequently, a 50% increase of spin polarization in Fe_3O_4 films was achieved, and a 110% increase of the magnetoresistance ratio was found in 900 °C annealed Fe_3O_4 films compared to the as-grown sample.	[209]
2018	15 mTorr and 1.5×10^{-6} Torr	650, 450	KrF excimer, 248 nm, 1 J/cm^2, 5 Hz and 2 J/cm^2, 4 Hz	Au(3 nm)/ Fe_3O_4(6 nm)/ STO (100 nm)/ Pt(111)–TiO_2–SiO_2 –Si	The charge-screening effect influences magnetization and induces the temperature of the Verwey phase transition shifting ~13 K. It suggests that the Verwey phase transition has certain correlations with electron distribution and change in the minority spin electron concentration.	[210]
2019	<10^{-8} Torr, 1×10^{-5} Torr (N_2)	950	KrF excimer, 248 nm, 2.5–20 Hz	MgO (001)	Epitaxial thin films of nitrogen-substituted Fe_3O_4 ($Fe_3O_{4-y}N_y$) were synthesized on MgO (001) substrates by using nitrogen-plasma-assisted PLD. The obtained $Fe_3O_{4-y}N_y$ thin films showed ferrimagnetic behavior at room temperature. The carrier density of the films was decreased by nitrogen substitution up to $y = 0.4$. These properties were similar to those of A-site (tetrahedral site)-substituted Fe_3O_4, such as $Zn_xFe_{3-x}O_4$ or $Mn_xFe_{3-x}O_4$, previously reported. Density functional theory–based first-principles calculation predicted that $Fe_3O_{4-y}N_y$ is half-metallic.	[211]

Paramês et al. showed in 2005 that pure phase magnetite can be grown by PLD at higher oxygen pressures of ≥0.6 mTorr [187]. These authors further demonstrated that a background mixture of oxygen and argon improved the Fe:O ratio and quality of the film surface morphology [188].

While Eerenstein and Hibma [116] and Ramos et al. [170] established fundamental relationships between the film thickness of APB density in MBE and plasma-assisted MBE-grown Fe_3O_4 films, respectively, several groups have reported on the interrelationships of PLD growth dynamics, film properties, and APB density. Among these, Kubo et al. [189] explored the role of magnetic field (up to 2 kG) on film properties of epitaxial $NiFe_2O_4$ films during deposition. These films were deposited on $Y_{0.15}Zr_{0.85}O_{1.93}$ buffered Si (001) substrates. The application of magnetic field during deposition was shown to improve magnetic properties with the saturation magnetization of thin films deposited at 500 and 600 °C increasing 25%. TEM imaging revealed that APB density decreased by the application of a magnetic field during deposition [189]. Also Datta et al. studied films of $NiFe_2O_4$ deposited on a range of substrates using PLD to reveal that the simultaneous nucleation of cation-disordered sites during nonequilibrium growth exhibited APDs irrespective of substrate symmetry [190]. Cobalt ferrite films grown on a multiplicity of substrates allowing for tensile and compressive stress states were extensive studied by the Coey Trinity College Group [191]. Among their findings was that the role of APB density in suppressing magnetization and approach to saturation is related to lattice mismatch and low-temperature deposition. Both Data et al. [190] and Eskandari et al. [191] found that for growth on isostructural spinel substrates, the APB density appears to decrease with decreasing lattice mismatch. HRTEM revealed that the interchange of equivalent tetrahedral cation positions in the host oxygen sublattice as a possible mechanism leading to the formation of APBs [190].

In regards to electrical transport, De Teresa et al. employed a working gas of CO_2 to produce epitaxial Fe_3O_4 and measured the magnetoresistance and Hall effect properties to establish a bulk-like value of the Verwey temperature in ultrathin films [192]. PLD under ultrahigh vacuum conditions were shown by several research groups to provide conditions for single-phase magnetite films with very high-crystal-quality and bulk-like magnetic properties including sharp and bulk-like values of the Verwey transition [193–198].

Nielsen et al. demonstrated the first all-oxide ferromagnet/semiconductor heterostructures that offer substantial advantages for spintronic applications [199]. These heterostructures, i.e. (111) oriented Fe_3O_4/ZnO heterostructures on ZnO (0001) and Al_2O_3 (0001) substrates deposited by PLD, provide clear ferromagnetic behavior of magnetite layers with a room temperature saturation magnetization of 3.2 μ_B/f.u. Results further demonstrate that the Fe_3O_4/ZnO system is a promising candidate as *semiconductor* spintronic devices where spin-polarized charge carriers must be introduced within a conventional semiconductor. Injection of spin-polarized carriers from ferromagnetic electrodes into semiconductor materials have been investigated by Schmidt et al. who found the large conductivity mismatch between 3d ferromagnets and semiconductors prevent efficient spin injection [194]. Hanbicki et al. showed that this problem can be circumvented by the introduction of Schottky or tunnel barriers at the ferromagnet/semiconductor interface [200].

Magnetite films have also been grown on, and incorporated in, ferroelectrically active heterostructures based on STO [196, 198, 210] and $BaTiO_3$ (BTO) [208]. Charge-screening effects were found to influence magnetization and induce the Verwey phase transition shifting ~13 K. Liu et al. suggested that the Verwey phase transition correlates strongly with the minority spin electron concentration [209].

Epitaxial films of nitrogen-substituted Fe_3O_4 (i.e. $Fe_3O_{4-y}N_y$) were synthesized on MgO (001) substrates by nitrogen-plasma-assisted PLD with the carrier density decreasing with nitrogen substitution. The resulting $Fe_3O_{4-y}N_y$ films showed ferrimagnetic behavior above room temperature. Density functional theory–based first-principles calculation predicted that $Fe_3O_{4-y}N_y$ is half-metallic [211].

10.3.4.2 PLD of Garnet Films

Ferrite films of the garnet crystal structure can be prepared by means of LPE, reactive ion beam sputtering, RF-magnetron sputtering, MBE, and PLD, among other soft chemical techniques. While LPE allows for single-crystalline thick films and offers very high crystal quality, PLD is preferred for the deposition of thin ($\leq$ 1 μm) and ultrathin (<50 nm) garnet films because of its flexibility in the rapid and ready deposition of a wide range of compositions with excellent thickness control.

PLD garnets have been the most studied of the ferrite systems. The general formula of garnets is $C_3A_2D_3O_{12}$, where for natural garnets this becomes $C_3A_2(SiO_4)_3$, where C = Fe^{2+}, Mg^{2+}, Mn^{2+}, and Ca^{2+} and A = Al^{3+}, Fe^{3+}, Cr^{3+}, and V^{3+}. For the general formula, O denotes oxygen and the C, A, and D represent cations on dodecahedral, octahedral, and tetrahedral sites, respectively. Garnets can be easily designed for specific applications by substituting cations. YIG, having the stoichiometry of $Y_3Fe_5O_{12}$, is the prototypical garnet.

The optimal pulsed laser processing conditions for the deposition of garnet films include a laser fluence in the range of 3–3.5 J/cm^2, dynamic oxygen pressure from 5 to 50 mTorr, and substrate temperatures from 750 to 950 °C. A lower temperature can be used if the substrate has a limited thermal budget.

Epitaxial growth of garnet films via PLD has only been demonstrated on other garnet substrates such as gadolinium–gallium–garnet (GGG) and YAG. As-deposited films on nongarnet substrates benefit from post-annealing in air or oxygen.

Substrates of GGG, doped GGG, and YAG are popular choices for the PLD growth of garnets. The most common substrate is GGG with a (111) crystal cut orientation. However, garnet films have also been successfully grown on quartz, Corning glass, single-crystal silicon, Al_2O_3, MgO, silicon nitride, and metal-coated semiconductors or insular substrates, as well as metal substrates.

Research performed on PLD-grown rare earth–doped iron garnet (RE-IG) films began in the early 1990s and was primarily motivated by MO memory device applications. Transparency in the near-infrared band and Faraday rotation (FR) angle were the principal figures of merit. The substitution of the trivalent Bi cation has been shown to enhance the FR and improve the signal-to-noise ratio in YIG films. Few publications reported on RF technology–driven studies of nonreciprocal devices, delay lines, filters, resonators, etc. An important figure of merit for those materials, components, and devices is the FMR linewidth that reflects RF loss properties.

The "Dot-com" bubble of the mid-1990s acted to shift the focus from RF to waveguide optical isolators and integrated magnetostatic waves–magneto-optical (MSW–MO) devices. Their performance depended on a large FR. FR magnitude can be enhanced by substitution of light rare earth ions, such as Ce^{3+}or Nd^{3+}, within the garnet unit cell. Therefore, many publications from this era focus on PLD of rare earth–doped YIG films.

Table 10.5 summarizes the historical development of PLD of garnet films between 1990 and 2020.

Table 10.5 Pulsed-laser deposition conditions and key results from early studies of garnet films systems.

Year	Oxygen pressure (Torr) unless denoted	T_{sub} (°C)	Laser parameters	Substrate	Key results	Ref.
1991	0.2–1	200–580	ArF excimer, 193 nm, 5 Hz, 3.5 J/cm^2	GGG (111)	PLD of bismuth-substituted yttrium–iron-garnet (Bi:YIG) films on gadolinium–gallium-garnet (GGG) substrate reported. A single garnet phase was achieved above 490 °C. Confirmed stoichiometric transfer from target to substrate. The saturation magnetization was 1500 G at room temperature.	[165]
1993	50 mTorr –1 Torr	700–850	KrF excimer 248 nm, 3 J/cm^2	GGG (111)	PLD of YIG films on (111) GGG substrates reported. The effects of process parameters on the structural composition and magnetic properties investigated. The properties of PLD YIG are compared to LPE YIG. It demonstrated that narrow FMR linewidths can be achieved at low oxygen pressures and high substrate temperatures. PLD allows deposition of YIG films suitable for microwave applications.	[212, 213]
1995	100–700 mTorr	500–800	KrF excimer 248 nm, 3.3 J/cm^2, 10 Hz	GGG (111)	PLD of alternating ferrimagnetic YIG and bismuth–iron–garnet (BIG) thin film multilayers on GGG. Demonstrated increased saturation magnetization as compared to homogeneous monolayer films.	[214]
1995	1 mTorr – 1 Torr	450–875	Excimer laser 248 nm, 3 J/cm^2, 5–10 Hz	Si, Corning glass, MgO, quartz	PLD of YIG on several substrates under varying processing parameters. All films were found to be polycrystalline. Deposition temperatures below 800 °C resulted in reduced magnetizations. Microwave and magnetic properties improved after post-annealing. YIG films on glass substrates exhibited an FMR linewidth of 55 Oe.	[216]
1999	10–100 mTorr	790	Nd-YAG, 355 nm,90 mJ/pulse, 10 Hz	GGG (111)	PLD Ce-substituted YIG films on (111) GGG. Confirmed epitaxial growth along the [111] axis. Determined the saturation Faraday rotation (FR) values $\lambda = 633$ nm are found to be dependent on the oxygen pressure, with the maximum value of 0.7°/μm for the films grown at 50 mTorr.	[225]
2000	52–130 mTorr	766–875	XeCl excimer laser, 1–3 J/cm^2	S-GGG (111)	PLD growth of YIG and Ce-substituted YIG films up to 2 μm thick. Cracks were found in films over 1 μm thick. Measured saturation magnetization was near to bulk YIG values. Cerium substitution increases the lattice parameter and reduces interface strain.	[226]
2001	10 mTorr	550	KrF excimer 248 nm, 2–3 J/cm^2, 10 Hz	(111) – $(NdGd)_3(ScGa)_5O_{12}$ (NGSGG) And (001) – $Gd_3(ScGa)_5O_{12}$ (GSGG)	Epitaxial $Bi_3Fe_5O_{12}$ (BIG) films are single phase, exclusively (111) and (001) oriented, with less than 0.4° and 0.06° of the full width at half maximum of the rocking curve of main texture Bragg reflection (111) for NGSGG and (001) for GSGG substrate, respectively. The films are strongly in-plane textured with cube-on-cube film-to-substrate epitaxial relationship. Vibrating sample magnetometry (VSM) and ferromagnetic resonance (FMR) measurements revealed in-plane magnetization in BIG/GSGG (001) film, while the BIG/NGSGG (111) film was found to have perpendicular magnetization. For BIG (001) and (111) films the FR at 635 nm was −7.8 and −6.7°/μm; the constant of uniaxial magnetic anisotropy was $K_u = -8.70 \times 10^4$ and $+1.16 \times 10^4$ erg/cm^3; the constant of cubic magnetic anisotropy was $K_1 = -3.6 \times 10^3$ and -7.14×10^3 erg/cm^3.	[224]

(Continued)

Table 10.5 (Continued)

Year	Oxygen pressure (Torr) unless denoted	T_{sub} (°C)	Laser parameters	Substrate	Key results	Ref.
2001	75 and 50 mTorr	750–900	ArF excimer, 193 nm, 4 J/cm^2, 20 Hz	GGG (111)	Cerium-doped YIG (Ce:YIG, $Ce_{0.6}Y_{2.4}Fe_5O_{12}$) films were deposited by PLD for magnetic field sensor applications. Crystalline Ce:YIG films were obtained by depositing at 900 °C substrate temperature and in 10 mTorr argon or an oxygen ambient gas. The FR angle was very large at 4.2 °/μm at 430 nm. The charge state of cerium was important for determining high FR. The lower temperature or higher ambient gas pressure controlled the cerium charge state to Ce^{4+}.	[227]
2002	125–175 mTorr	420–660	ArF excimer, 193 nm, 1 J/cm^2, 10 Hz	GGG (111)	PLD of Bi-doped YIG ($BixY_3\_xFe_5O_{12}$, Bi:YIG) thin films on GGG for electric current sensor applications. PLD process parameters varied to determine optimal growth conditions for high-quality epitaxial Bi:YIG. Epitaxial growth is possible at substrate temperatures of ~500 °C and oxygen pressure between 125 and 175 mTorr. Measured optical properties of synthesized epitaxial films and determined the maximum magneto-optic sensitivity coefficient.	[217]
2001	5–200 mTorr	500–750	Nd:YAG, 180 mJ/pulse, 3–9 J/cm^2, 10 Hz	Amorphous quartz	PLD of thin and ultrathin (<50 nm thick) polycrystalline (YIG) on quartz substrates. Determined that saturation magnetization decreases with the film thickness. The elementary cell volume and the Curie temperature were similar to bulk YIG. Found a change from parallel to perpendicular magnetization in films under 140 Å.	[218]
2001	5–200 mTorr	400	Nd-YAG, 355 nm, 180 mJ/pulse, 3–9 J/cm^2, 10 Hz	Amorphous quartz	Growth of the polycrystalline films with the structural, magnetic, and MO properties corresponding to those of bulk YIG occurred at the substrate temperatures above 400°C and a partial oxygen pressure of 30 mTorr. The appearance of lattice distortion and of perpendicular anisotropy due to thermally induced strains together with the presence of oxygen vacancies lead to structural modifications and anomalous magnetic properties compared to bulk YIG. All the results indicate that the films thicker than 200 Å consist of two layers. One layer near the substrate has a distorted structure and low magnetization. The other layer is nearly perfect in a crystallographic sense. Their relative thicknesses determine the magnetic and resonance behavior of the film.	[219]
2003	30 mTorr (pO_2)	660	Nd:YAG, 355 nm, 180 mJ/pulse 8 J/cm^2, 10 Hz	YAG (111); YAG (110)	Single-crystalline epitaxial YIG films were grown on YAG substrates by PLD. A "bee-waist" shape of the hysteresis loop was observed for ultrathin YIG/YAG films at room temperature. This effect can be explained by a magnetic exchange coupling between two magnetic phases, an interfacial hard phase and a soft phase. The soft phase is pure YIG. The hard phase, $Y_3Fe_{5-x}Al_xO_{12}$, with low magnetization, is formed as a consequence of a slight interdiffusion of Fe and Al at the film/substrate interface. The maximal interface exchange energy for (111) oriented samples is around 0.17 erg/cm^2 at 5 K.	[232]

Table 10.5 (Continued)

Year	Oxygen pressure (Torr) unless denoted	T_{sub} (°C)	Laser parameters	Substrate	Key results	Ref.
2004	1 mTorr	600–850	Nd:YAG, 355 nm, 10 Hz	Si (1 0 0), GGG (111)	PLD of YIG on GGG (1 1 1) and Si (1 0 0) substrates. Textured YIG synthesized on GGG at substrate temperature of 750 °C. Films synthesized on Si substrates *between* 600 and 750°C contain a $YFeO_3$ impurity phase.	[215]
2005	5–400 mTorr (pO_2)	840	Nd:YAG, 355 nm, 3 J/cm^2	Fused silica	PLD of controlled off-stoichiometric YIG thin films. 30 mTorr O_2 pressure allowed for deposition of stoichiometric YIG films. Oxygen-deficient films resulted below this pressure and iron- and yttrium-deficient films above this pressure. Iron and yttrium deficiency resulted in lattice parameter contraction, a 10% increase in Curie temperature, and a 20% increase in saturation magnetization.	[220]
2005	20 to >100 mTorr	850	KrF excimer, 248 nm, 50 mJ/pulse, 4.0 J/cm^2, 20 Hz	GGG	PLD of Ce-substituted YIG (Ce:YIG), Ce,Ga-doped Ce,Ga:YIG, and Ce,Gd-doped semidrying on GGG in oxygen and argon atmospheres. A lower oxygen pressure of 20 mTorr and higher argon pressures in excess of 100 mTorr resulted in crystalline films. The oxidation dynamics in the ablation plume was visualized by laser-induced fluorescence and determined to have a close correlation with crystallinity.	[221]
2007	5–400 mTorr	840	3 J/cm^2 (Asserted to be YIG ablation threshold)	Fused silica	Magnetic properties and superexchange interaction can be controlled by varying oxygen stoichiometry of PLD-grown YIG films. Suggested that the formation of, respectively, Fe^{4+} and Fe^{2+}, in the films with excess oxygen and vacancies can lead to a significant increase in Curie temperature and saturation magnetization.	[222]
2008	50–500 mTorr	570	KrF excimer, 248 nm, 2.5 J/cm^2, 50 Hz	GGG (001)	PLD of bismuth–iron–garnet (BIG) films on GGG. Varying the Bi/Fe ratio of the target and oxygen pressure during deposition allowed the control of film stoichiometry. Investigated the influence of the stoichiometry on structural, magnetic, and MO properties. Determined that the Curie temperature increases with lattice parameter contraction. Suggested bismuth deficiencies affect the MO response of the tetrahedral Fe sublattice, while iron deficiencies affect the MO response of the octahedral sublattice.	[223]
2008	120 mTorr	RT – 850	Nd:YAG, 355 nm, 2.5 J/cm^2, 10 Hz	Fused quartz	PLD of YIG thin films on fused quartz substrates. Varied substrate temperature between room temperature (RT) and 850°C. As-deposited films were amorphous and nonmagnetic at RT. Post-annealing films deposited at RT resulted in X-ray peaks and magnetic order. Post-annealing films deposited between 500 and 850 °C resulted in higher quality and higher saturation magnetization. Saturation magnetization in the very best films reached 68% of the bulk YIG.	[231]

(Continued)

Table 10.5 (Continued)

Year	Oxygen pressure (Torr) unless denoted	T_{sub} (°C)	Laser parameters	Substrate	Key results	Ref.
2009 2010	25 mTorr	750	KrF excimer, 248 nm, 3–4 J/cm^2, 20 Hz	GGG (111)	PLD of epitaxial Fe-deficient YIG films with FMR linewidths of only 0.9 Oe and a high uniaxial anisotropy of 880 Oe. The films are demonstrated to be feasible for magnetostatic waves applications. The origin of a high noncubic magnetic anisotropy investigated. Molecular field theory was used to evaluate saturation magnetization, cubic magnetocrystalline, and uniaxial anisotropies.	[233–235]
2012	0.1	650	KrF excimer, 248 nm, 1.7 J/cm^2, 12 Hz	HEAN/, u/HEAN, HEAN – high entropy alloy nitrides (Lactates)N	PLD of high-quality (several hundred nanometers) YIG films on a sandwich structure made of a thick Cu layer and two thin HEAN cladding layers. HEAN served as a barrier to prevent Cu diffusion and oxidation during PLD growth. Sputtering was employed to fabricate the Cu and HEAN layers. YIG films were smooth to nanometer scales and (111) oriented. An FMR linewidth of 1.1 Oe was measured at 9.45 GHz.	[236]
2012	0.1	790 850	KrF excimer, 248 nm	GGG (111)	PLD of nanometer-thickness YIG films. A (111) orientation detected in films between 5 and 35 nm in thickness. Nanometer-thick films exhibit broader FMR linewidth as compared to thicker YIG films. A linewidth of 6 Oe was measured in 10 nm films. The surface roughness increases the FMR linewidth.	[237]
2012	10^{-6}	100, 550	KrF excimer, 248 nm, 1.3 J/cm^2, 10 Hz	Si-on-insulator and Si	PLD of thin polycrystalline cerium-substituted YIG (Ce:YIG) films on a YIG seed layer on Si. Determined the thermal budget of the YIG/Ce:YIG growth process. Measured optical and MO properties of the synthesized films in the near-IR region.	[228]
2013	7 mTorr	1025–1325	KrF excimer, 248 nm, 3 J/cm^2, 20 Hz	YAG (1 0 0)	PLD of high-quality epitaxial micrometer-thick YIG on YAG substrates by PLD. Investigated the influence of process parameters on composition, crystallinity, and optical transmission and FMR. A substrate temperature of 1325 K and oxygen pressure of 7 mTorr allowed the lowest FMR linewidths.	[238]
2014	20 mTorr	650	KrF excimer, 248 nm, 400 mJ/pulse, 10 Hz	GGG (1 0 0)	PLD of single-phase, epitaxial YIG films on GGG (1 0 0) substrates. Surface roughness was in the range of 0.14 to 0.2 nm. Film thicknesses between 17 and 200 nm allowed for low coercivity.	[34]
2014	5 mTorr	640	KrF excimer, 248 nm, 2–10 Hz	Si_3N_4 on Si (001)	PLD of cerium-substituted YIG (Ce:YIG) films on YIG seed layers on silicon nitride on Si (001). Determined the optimal process conditions for PLD garnet films on silicon nitride. Achieved near-bulk magnetic and MO properties in 160-nm-thick Ce:YIG.	[229]
2016	5 mTorr	615	KrF excimer, 248 nm, 10 Hz	GGG (1 0 0) (110) (111)	PLD of 80-nm-thick Ce:YIG films on GGG with (1 0 0), (110), and (111) orientations. The films exhibit bulk-like structural and magnetic properties. Complete permittivity tensor of the films in the ultraviolet, visible and near-infrared spectral region was measured. The films showed the highest IR FR reported for thin Ce:YIG films – the MO figure of merit.	[230]

Table 10.5 (Continued)

Year	Oxygen pressure (Torr) unless denoted	T_{sub} (°C)	Laser parameters	Substrate	Key results	Ref.
2016	5 mTorr	650	Nd:YAG, 355 nm, 2 J/cm^2	GGG (111)	PLD of thin (10–84 nm) epitaxial YIG on GGG (111) substrates under varying experimental conditions. AFM revealed flat and smooth surface morphologies on micrometer and nanometer scales. X-ray diffraction (XRD) revealed the presence of a rhombohedral distortion (~1%) along the <111> direction. Additional magnetic anisotropy was detected and attributed to a strain-induced magnetocrystalline mechanism. Demonstrated surface spin-wave propagation in in-plane biased films.	[239]
Growth of garnet films for magnonic applications						
2014	60 mTorr	200–650	KrF excimer, 248 nm	Porous Si	Two-dimensional YIG magnonic crystal is deposited on the porous silicon substrate by PLD. It was shown that the YIG magnonic crystals are composed of corn-shaped nanoparticles with excellent soft magnetic properties: 90% of bulk YIG saturation magnetization (M_s), coercivity (H_c) of 30 Oe, and a p-p FMR linewidth of ~96 Oe were obtained.	[240]
2016	19 mTorr	Unknown	KrF Excimer, 248 nm, 2.5 J/cm^2, 5 Hz	GGG (111)	Recrystallization of amorphous YIG by annealing in oxygen atmosphere. Findings show that well below the melting temperature the material transforms into a fully epitaxial layer with exceptional quality, both structural and magnetic. In FMR ultralow damping and extremely narrow linewidth were observed. For a 56-nm-thick layer a damping constant of $\alpha = (6.15 \pm 1.50) \times 10^{-5}$ is found and the linewidth at 9.6 GHz is as small as 1.30 ± 0.05 Oe, which are the lowest values for PLD-grown thin films reported to date. Even for a 20-nm-thick layer a damping constant of $\alpha = (7.35 \pm 1.40) \times 10^{-5}$ is found, which is the lowest value for ultrathin films published so far. The FMR linewidth in this case is 3.49 ± 0.10 Oe at 9.6 GHz.	[241]
2018 2020	97 mTorr	800	1.8 J/cm^2, 2 Hz	GGG (111)	Experimental demonstration of low-loss spin-wave manipulation in magnonic crystals of physically separated nanometer-thick YIG stripes. An enhanced transmission of spin waves was realized in allowed minibands by filling the gaps between YIG stripes with CoFeB. Magnonic crystals exhibit tunable bandgaps of 50–200 MHz with nearly complete suppression of the spin-wave signal. The integration of strong ferromagnets in nanometer-thick YIG-based magnonic crystals provides effective spin-wave manipulation and low-loss propagation.	[242, 243]
2018	97 mTorr	800	1.8 J/cm^2, 2 Hz	GGG (111)	An investigation of propagating spin waves in nanometer-thick YIG films. Broadband spin-wave spectroscopy with integrated coplanar waveguides (CPWs) and microstrip antennas on top of continuous and patterned YIG films was employed to characterize spin waves with wave vectors up to 10 rad/μm. From spin-wave transmission spectra, the Gilbert damping constant, spin-wave dispersion relation, group velocity, relaxation time, and decay length were derived and their dependence on magnetic bias field strength and angle was systematically gauged. For a 40-nm-thick YIG film, a damping constant of 3.5×10^{-4} and a maximum decay length of 1.2 mm was measured.	[244]

The first published report by Kidoh et al. [165] detailed the PLD growth of bismuth-substituted YIG (Bi:YIG) on GGG substrates. Stoichiometric transfer between the target and the substrate was reported. The substrate temperature was varied from 200 to 580 °C, with a single phase of garnet achieved at temperatures above 490 °C. The upper limit of this temperature range is limited by the inefficient absorption of heat by semitransparent substrates. A laser fluence of 3.5 J/cm^2 was shown to be optimal for achieving superior crystal quality and RF properties. The conditions employed by Kidoh et al. have been confirmed by others in the community as optimal.

In 1993, Dorsey et al. systematically investigated the influence of PLD process parameters on the structure, composition, and magnetic properties of YIG films grown on GGG (111) substrates [212, 213]. They found narrow FMR linewidths of < 1 Oe were achieved only at concomitant low oxygen pressures and high substrate temperatures.

YIG films have been deposited on several non-GGG substrates [214, 215]. The best films grown on glass substrates had FMR linewidths of ~55 Oe, which is nearly two orders of magnitude higher than that of epitaxial YIG on GGG.

Stoichiometric YIG and GGG have lattice constants of 1.2373 and 1.2383 nm, respectively, providing a mismatch of less than 0.1%. This slight mismatch can be further reduced by additives to contract that the lattice parameter of GGG or expand that of the YIG. Such near-perfect lattice matching, and close values of thermal expansion coefficients, minimize the strain induced in the growing film, reducing the concentration of stacking faults and other defects thus optimizing RF performance.

Several studies have focused on the influence of the type and pressure of the PLD working gas upon the resulting crystallographic and magnetic properties of garnet films [212, 213, 216–219]. In these studies, oxygen pressure employed during growth was varied between 1 and 1000 mTorr. Dumont et al. demonstrated 30 mTorr of oxygen partial pressure resulted in stoichiometric films showing magnetic and structural properties of bulk YIG. They provide an extension to the YIG phase diagram where nonstoichiometric YIG films grown at $P_{O_2} < P_{stoich}$ are shown to have increased T_c (up to 10%) and $4\pi M_s$ (up to 20%) in films experiencing lattice parameter compression [220]. Generally, higher oxygen partial working gas pressures induce iron and yttrium deficiencies, resulting in crystal lattice distortions and increased losses from magnetocrystalline anisotropy and magnon scattering, the latter manifesting as an increase in the FMR linewidth. However, epitaxial Fe-deficient YIG films, having FMR linewidths of < 1 Oe with high uniaxial anisotropy fields of ~880 Oe, were demonstrated to be suitable for magnetostatic wave device applications [220–223]. Alternatively, excess oxygen and related vacancies can lead to significant increases in Curie temperature and saturation magnetization [223].

Motivated by advances in magneto-optic thin film device technologies, studies of stoichiometric and off-stoichiometric PLD film growth were extended to Bi-substituted [165, 214, 217, 223, 224] and rare earth–doped garnet films [215, 221, 223–230]. Kim et al. deposited films of Ce-YIG on GGG and determined the saturation FR values for $\lambda = 633$ nm to be dependent on oxygen pressure during growth with the maximum value of 0.7°/μm for the films grown at 50 mTorr [225].

Ibrahim et al. explored PLD growth of YIG thin films on 2 mol% magnesium-substituted GGG substrates (S-GGG). S-GGG was employed to provide an improved matching of lattice constants to Ce:YIG films in order to reduce strain and improve the RF performance. A principal finding was those films with thickness greater than ~1 μm, nonetheless, experienced cracking [226].

Stoichiometry variations also influence the Curie temperature T_C by tuning the (Fe)-O-[Fe] geometry: T_C was found to increase when the lattice parameter decreased. The Néel two-sublattice magnetization model was applied to the thermal magnetic properties and FR providing energies of −48, −29, and 84 K, for exchange integrals J_{AA}, J_{AB}, and J_{BB}, respectively [231].

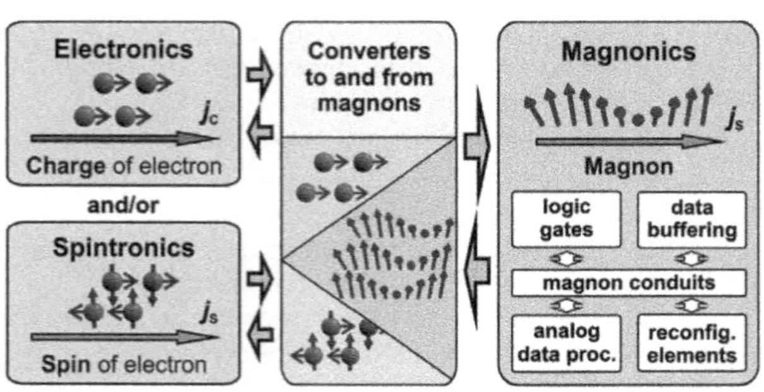

Figure 10.25 The concept of magnon spintronics: information coded into charge- or spin-currents is converted into magnon flows, processed within the magnonic system and converted back. The wave nature of the magnon flow is used for fast processing of complex data inputs (after figure 6 in Stamps et al. [248]).

The burgeoning field of *magnonics* has been a recent driver for PLD garnet films [30, 31, 245]. Magnonics, a subfield of spintronics, first introduced *circa* 2005, employs spin-currents consisting of magnons; the quantum mechanical equivalent particles of magnetic spin waves. As shown in Figure 10.25, similar to electric and spin-currents, magnon currents can be used to communicate and process information [245–248]. As magnons are free of effective mass, they do not suffer from Joule heating and Ohmic losses that plague conventional electronics. Further, magnonic structures, in contrast to conventional electron-based modalities, benefit in performance upon scaling. Finally, due to the wave nature of spin waves and their nonlinear properties, *far more complex data processing concepts* can be conceptualized and realized [248].

A schematic of a transistor based on a magnonic crystal designed on a YIG film with an array of parallel grooves etched into its surface is depicted in Figure 10.26a. Magnons are injected into the transistor's source and detected at its drain using a microstrip antenna. The magnons that control the source-to-drain current are injected directly into the YIG crystal's gate using an antenna of similar design.

Distinct magnon localization, as illustrated by the formation of a standing-wave-like intensity profile (see Figure 10.26b), enhances the nonlinear effects. A magnon transistor allows for the suppression of source-to-drain magnon current via the injection of gate magnons into the gate region (see Figure 10.26c). When magnons from the transistor's source enter the gate region, they are attenuated by nonlinear four-magnon scattering events. The secondary magnons generated due to scattering have large wavevectors and small group velocities that prohibit their flow out of the gate region. The time evolution of the signal at the transistor's drain carried by 50-ns-long magnon packets injected into the transistor's source at the time $t = 0$ ns is shown in Figure 10.26d. In Figure 10.26e, magnon transmission characteristics through the magnonic crystal as a function of the magnon frequency is depicted. The frequency of the magnons injected into the gate is chosen to be in the center of the first bandgap, ensuring localization of magnons inside the crystal [249].

Magnonics demands high crystal quality, low loss factors, and ultrathin-film magnetic materials [30, 245]. As YIG and its variants are well known to have the lowest damping coefficients among popular RF materials, it is a prime candidate to serve this role [22]. However, damping values previously achieved in bulk single crystals and thick films must first be realized in nanometer-thick films. A low-FMR linewidth (ΔH_{FMR}) ~6 Oe was measured in YIG thin films grown on GGG (111) substrates [237], as compared to more typical linewidths of less than 1 Oe for micron-thick films [219]. A typical surface

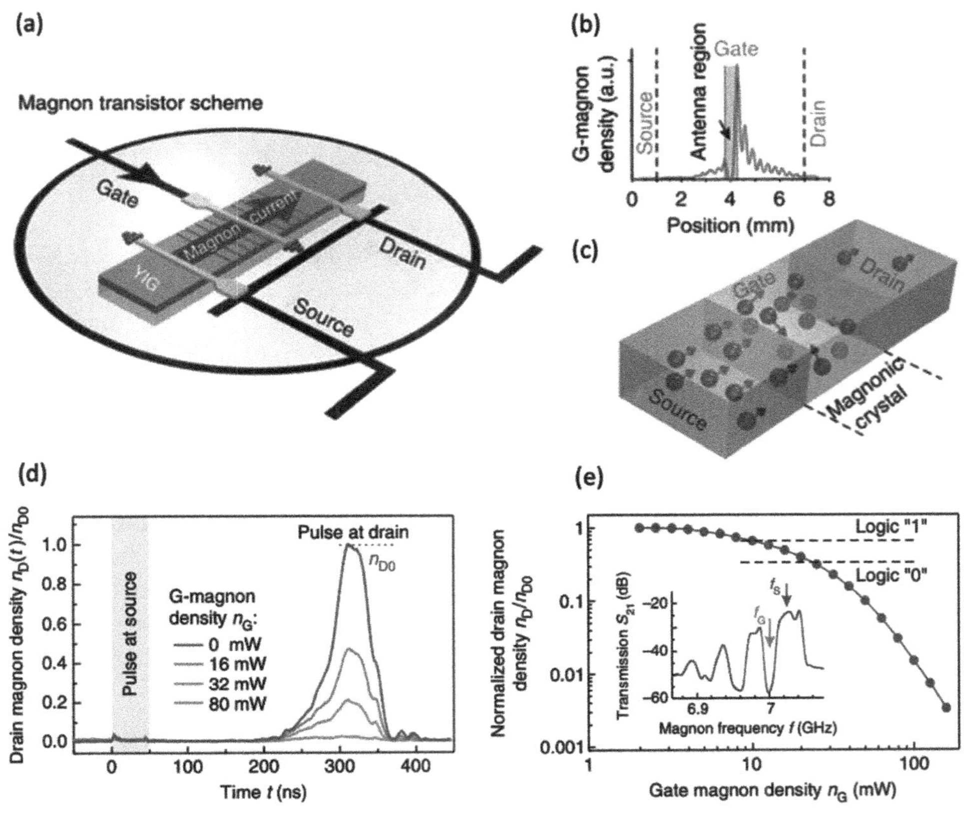

Figure 10.26 Schematic of the magnon transistor. (a) The transistor is based on a magnonic crystal designed in the form of a YIG film with an array of parallel grooves on its surface. (b) Strong magnon localization and the formation of a standing-wave-like intensity profile are visible. (c) A magnon transistor allows for the suppression of the source-to-drain magnon current via the injection of gate magnons into the gate region. (d) The time evolution of the signal at the transistor's drain carried by 50-ns magnon packets injected into the transistor's source at time $t = 0$. (e) The signal decreases with increasing power. The drain magnon density n_D normalized to the original value n_{D_0} is presented as a function of the gate magnon density n_G on a logarithmic scale. The inset to (e) shows the magnon transmission characteristic through the crystal as a function of the magnon frequency. The frequency of the magnons injected into the gate, $f_G = 7$ GHz, is chosen to be in the center of the first bandgap ensuring localization (adapted from [228]).

roughness for ultrathin films ranges from 0.14 to 0.2 nm, as measured by AFM [34]. The surface roughness has been linked to increased coercivity and FMR linewidth.

An innovative approach was demonstrated by Hauser et al. [241] who recrystallized amorphous YIG films produced by PLD at ambient temperatures by postdeposition annealing in an oxygen atmosphere. Upon annealing at temperatures well below melting, the amorphous material transforms into an epitaxial layer with exceptional quality both in structure and magnetic properties. FMR illustrated ultralow damping and extremely narrow linewidths. For a 56-nm-thick film a damping constant of $\alpha = (6.15 \pm 1.50) \times 10^{-5}$ was measured and a linewidth at 9.6 GHz was as low as 1.30 ± 0.05 Oe. These values remain unmatched at the time of the writing of this chapter.

Magnonic metamaterials with periodic structures made of YIG and other ferromagnetic materials have been shown to have promise in providing unusual and valuable constitutive properties with ultralow magnetic damping [242–244].

Several research groups have produced low-loss YIG-based magnonic crystals with large tunable bandgaps. These magnonic crystals are materials where enhanced transmission of spin waves was realized in allowed minibands by filling gaps between YIG stripes with a metallic glass of CoFeB [242, 243]. Magnonic crystals exhibit tunable bandgaps of 50–200 MHz, with nearly complete suppression of the spin-wave signal. The integration of strong ferromagnets in nanometer-thick, YIG-based magnonic crystals provides effective spin-wave manipulation and low-loss propagation [244].

An integrated coplanar waveguide (CPW), with a microstrip antenna atop of a continuous and patterned YIG film, was employed to characterize spin waves with wave vectors up to 10 rad/μm [244]. From spin-wave transmission spectra, the Gilbert damping constant, spin-wave dispersion relation, group velocity, relaxation time, and decay length were derived and their dependence on magnetic bias field strength and angle studied. For a 40-nm-thick YIG film, a damping constant of 3.5×10^{-4} and a maximum decay length of 1.2 mm was measured [244].

10.3.4.3 PLD of Spinel Films

Development of spinel ferrite films evolved in response to technological needs ranging from magnetic memory to integrated inductors to spintronics. For example, spinel coatings and thin films were deemed useful for high-speed digital tape and magnetic recording. On the other hand, communication systems have been continuously shifting toward higher frequencies and smaller and lighter form factors. These latter developments called for integrated thin film inductors, transformers, and nonreciprocal passive microwave circuits.

Other thin-film processes for spinel films had been explored before the application of PLD. These techniques included sol–gel, RF and DC sputtering, and spin-spray plating, among others. Yet, these processes did not yield spinel structures of sufficiently high crystal quality and magnetic properties for many applications. By contrast, PLD produces the desired spinel structure and chemistry directly from high-temperature deposition, congruently transferring target stoichiometry to the substrate. Postdeposition heat treatments are often required to improve the properties of the product.

To date, ferrite spinel films have been demonstrated on fused quartz, Si (1 0 0), MgO (1 0 0) and (011), Si (1 0 0), $ZnAl_2O_4$, Al_2O_3(1 0 0) (sapphire), $SrTiO_3$ (1 0 0), and glass substrates. Spinel targets of nominal compositions are typically fabricated by conventional ceramic methods.

Table 10.6 presents a summary of select published results on the PLD of spinel ferrite films. The first documented PLD of spinel ferrite films dates to 1994 [250]. It's worth noting that while most authors reported stoichiometric transfer between the target and the substrate, some reported Zn-deficient films and correspondingly enriched targets in order to obtain nominal compositions [251]. Nonstoichiometric transfer may be attributed to the use of lower laser fluence such as that employed by Dorsey et al. (2–2.8 J/cm^2) [251]. The laser fluences used for deposition of high-quality spinel ferrites exceed 3 J/cm^2. These values fall between the fluences used in the deposition of high-quality iron oxides and hexaferrites. As a general rule, efficient ablation and stoichiometric transfer of high-*Z* elements require even higher laser fluences.

Due to the overwhelming number of publications in the field of PLD processing of spinel films, we must restrict ourselves to referring to early seminal articles in terms of specific spinel systems beginning with MnZn-ferrite films, proceeding to NiZn ferrites, Co ferrites, and Li ferrites. Many excellent papers will regrettably go without proper recognition.

10.3.4.3.1 Pulsed-Laser-Deposited Mn- and MnZn-Ferrite Films

MnZn ferrites are among the most attractive ferrites for applications up to a few hundred kilohertz. They possess very high permeabilities and magnetization with moderate losses due to their conductivities that approach the level of semiconductors. As mentioned earlier, PLD of MnZn-ferrite films was first demonstrated by the NRL team led by Williams in 1995 [250]. Dorsey et al., also from NRL, soon followed with work that explored a larger swath of the $MnO–ZnO–Fe_2O_3$ ternary

Table 10.6 Pulsed laser deposition conditions and key results from early studies of spinel films.

Year	Oxygen pressure (Torr) unless denoted	T_{sub} (°C)	Laser parameters	Substrate	Key results	Ref.
Growth of Mn- and MnZn-ferrite films						
1994	30 mTorr	300 400 600 800	KrF excimer, 248 nm, 2 J/cm^2, 150 mJ/pulse, 10 Hz	MgO (1 0 0), MgO (011)	Pulsed-laser deposition (PLD) of MnZn-ferrite films on MgO substrates. Uniaxial anisotropy, ferromagnetic resonance (FMR) linewidth, and coercive force depend on microstructure. The saturation magnetization and magnetocrystalline anisotropy depend on film composition. Overall, magnetic properties are similar to bulk MnZn ferrites.	[250]
1995	30 mTorr	850	KrF excimer, 248 nm, 150 mJ/pulse, 2 J/cm^2, 10 Hz	MgO (1 0 0)	PLD of MnZn-ferrite films on single-crystal MgO (1 0 0) substrates. Film properties investigated as a function of composition. Confirmed nonstoichiometric transfer from target to substrate. Synthesized films are typically 40% deficient in Zn with excess of Mn and Fe. Magnetocrystalline anisotropy field strongly depends on composition.	[251]
1996	0.11 Torr –0.15 Torr	800–850	XeCl excimer, 308 nm, 65–75 mJ/pulse, 21 Hz	High-silicon content glass	PLD of MnZn-ferrite films on high-silicon content glass. XRD analysis confirms stoichiometric ablation. Coercive fields are larger than those found in bulk materials, but saturation magnetization (660 emu/cm^3) was comparable.	[252]
1996	1 mTorr 1% O_2 and 99% N_2	600	KrF excimer, 3 J/cm^2, 10 Hz,	STO and $MgAl_2O_4$	Provided a comprehensive study of MnZn-ferrite and Co-ferrite systems grown on a variety of buffered substrates laying the groundwork for a processing – property – performance paradigm. (Buffers: $CoCr_2O_4$, $CuMn_2O_4$, $FeGa_2O_4$, and $NiMn_2O_4$)	[253]
1999	30 mTorr	600	KrF excimer, 248 nm, 150 mJ/pulse, 2 J/cm^2, 10 Hz	MgO (1 0 0)	For the MnZn-ferrite, the linewidth was 49 Oe for perpendicular FMR, increased with the angle between the film normal and the external field to a maximum 80 Oe at 35° and then dropped to 65 Oe at parallel FMR. Linewidths were larger than predicted by Landau–Lifshitz damping theory, which indicated extrinsic contributions to loss. Two-magnon scattering was used to model these linewidth differences.	[257]
1996	300 mTorr	700	XeCl excimer, 308 nm, 3 J/cm^2, 5 Hz	Sapphire	PLD of MnZn-ferrite films on high-silicon-content glass confirmed stoichiometric transfer from target to substrate. Established conditions for congruent compositional transfer from target to substrate providing high-quality MnZn-ferrite film's displaying open hysteresis loops with coercivity, saturation magnetization, and remanent magnetization values comparable to bulk Mn–Zn-ferrite	[258]
2000	300 mTorr	RT	XeCl excimer, 308 nm, 150–300 mJ/pulse, 5–30 Hz	Mo, Al, and $ZnAl_2O_4$	PLD of MnZn-ferrite films on high-silicon-content glass. Confirmed stoichiometric transfer from target to substrate. Saturation and remanent magnetization values are comparable to bulk MnZn-ferrite. Crystalline ferrite substrates allow for the deposition of crystalline MnZn-ferrite thin films at low temperatures with a reduced coercive field by ~25% compared to glass substrates.	[259]

(Continued)

Table 10.6 (Continued)

Year	Oxygen pressure (Torr) unless denoted	T_{sub} (°C)	Laser parameters	Substrate	Key results	Ref.
2001	10^{-2} to 10^{-6} Torr	RT –650	Nd-YAG, 266 nm, 2 J/cm^2, 10 Hz	Sapphire	Films deposited at low oxygen pressures of 1×10^{-4} to 1×10^{-6} Torr crystallized in spinel phase even at room temperature. The structural qualities of these films assessed by X-ray linewidth and surface smoothness were extremely high. MnZn-ferrite film M_s was ~80% (270 emu/cm^3) of that of the bulk.	[260]
2005	1–50 mTorr	700	KrF excimer, 248 nm, laser shots on MnO and Fe_2O_3 targets varied from 1:10 to 6:10 and 4:8, 400 mJ/pulse	MgO (1 0 0) and (111)	Oxygen pressure dependence of magnetic anisotropy in ATLAD manganese ferrite films was investigated and magnetic anisotropy fields are shown to exceed 5 kOe when films were processed at oxygen pressures below 5 mTorr. Further, it is shown that the magnetically preferred direction of H_a can be aligned either along the film plane ($pO_2 < 8$ mTorr or perpendicular to it $pO_2 > 8$ mTorr). The authors speculated that this distortion was a consequence of the local strain caused by cation disorder and is responsible for the induced uniaxial perpendicular anisotropy and spin reorientation via magnetocrystalline anisotropy.	[261]
2008	1–50 mTorr	700	KrF excimer, 248 nm, laser shots on MnO and Fe_2O_3 targets from 1:10 to 6:10 and 4:8, 400 mJ/pulse	MgO (1 0 0) and (111)	Applied diffraction anomalous fine structure (DAFS) to obtain element-specific and site-specific information on site occupancy, local structure, and valency in ATLAD Mn-ferrite films. These properties were introduced to molecular field theory to reproduce thermomagnetization curves and determine the superexchange energy, Néel temperature, and spin canting angle.	[263]
Growth of Ni- and NiZn-ferrite films						
1996	1–100 mTorr	450	355 nm, 0.3–3 J/cm^2	Glass, Quartz, MgO	Demonstrated PLD-congruent compositional transfer from target to film for growth of popular ferrites, including Ni and NiZn ferrites. Low substrate temperatures led to amorphous films that required post-annealing to crystallize into ferrites.	[264]
1996	0.11–0.15 mTorr	800–850	XeCI excimer, 308 mn, 65–75 mJ/pulse, 21 Hz	Fused quartz	The grain size of the crystals is 0.3 to 0.6 pm. VSM measurements show coercive fields of 800 Oe, which are larger than those found for the bulk, in direct relationship to the shape anisotropy originating in the film microstructure. Saturation magnetization values of 660 emu/cm^3 are comparable with the bulk material.	[265]
1997	50–200 mTorr	400–700	KrF excimer, 248 nm, 25 Hz, 2.8 J/cm^2	MgO (1 0 0)	PLD of thick NiZn-ferrite ($Ni_{0.6}Zn_{0.4}Fe_2O_4$) films (10–12 μm) on single-crystal MgO (1 0 0). Film properties investigated as a function of substrate temperature and O_2 working gas pressure. Films are Zn-deficient and require Zn-rich targets. T_s of 700 °C yields high crystalline quality and near-bulk saturation magnetization (5000 G). Low O_2 pressures result in significant compressive stress. It is speculated that oxygen defects are responsible for growth-induced stress.	[266]

Table 10.6 (Continued)

Year	Oxygen pressure (Torr) unless denoted	T_{sub} (°C)	Laser parameters	Substrate	Key results	Ref.
2005	1–200 mTorr	700	KrF excimer, 248 nm, 65–75 mJ/pulse, 10 Hz	MgO (1 0 0), Si (1 0 0), Al_2O_3 (1 0 0), $SrTiO_3$ (1 0 0), glass	PLD of NiZn-ferrite films under varying experimental conditions on a variety of substrates to determine the most favorable conditions for smooth surfaces and attractive magnetic properties. It was determined that films simultaneously include crystalline islands and an amorphous phase. The coercive field is larger than those in bulk materials, but the saturation magnetization was comparable.	[267]
2002	0.0097–97 mTorr	650–800	KrF excimer, 248 nm, 2 J/cm^2, 7 Hz	Si (1 0 0)	This investigation shows that H_c and M_r are strongly influenced by the crystallinity of the films, and M_s, M_r, and H_c depend more on the composition. It turns out as conditions for obtaining optimal M_r and H_c are a substrate temperature of 80°C and O_2 pressure of 9.7 mTorr.	[268]
2004	30–75 mTorr	650	KrF excimer, 200 mJ/pulse, 5 Hz	Si (111)	NiZn-ferrite films fabricated on Si (111) and glass substrates. XRD indicated (400) orientation on Si substrates. The thickness and growth of the films are investigated at different deposition temperatures and pressures. Saturation magnetization and coercivity are found to be strongly dependent upon oxygen pressure, which suggests that cation distribution, defects, and stress are responsible for magnetic properties. Post-annealing improves magnetic properties, especially for films fabricated at low deposition temperatures.	[285]
2007	5 mTorr	700–900	KrF excimer, 248 nm, 400 mJ/pulse, 20 Hz	MgO (1 0 0) MgO (1 0 0)	Spinel nickel ferrite films were deposited by PLD and showed a reduction in $4M_S$ values of 47%, and 27% in films deposited at 700 and 900 °C, respectively. Films deposited at 900 °C were shown to have an increase in magnetization of 25% after post-annealing at 1000°C in air, and the FMR linewidth was reduced from 1.5 to 0.34 kOe. Films experienced tensile stress originating from thermal expansion mismatches between the film and MgO substrates.	[269]
2009	5 mTorr	550–900	KrF excimer, 248 nm	Sapphire (0001)	Authors studied the topology of twin boundaries in films grown under different conditions. For conventional growth, most twin boundaries were lateral in character. In the films involving solid-state reactions, the twin boundaries were found to be mixed: shorter lateral segments and longer {111} segments. The twin boundary segments on the {111} planes are thought to be formed at spinel/Fe_2O_3 interfaces.	[270]12

(Continued)

Table 10.6 (Continued)

Year	Oxygen pressure (Torr) unless denoted	T_{sub} (°C)	Laser parameters	Substrate	Key results	Ref.
2012	Unknown	400	KrF excimer, 248 nm, 550 mJ/pulse, 15 Hz	Si (1 0 0)	The role of oxygen vacancies on electronic structure in nonstoichiometric Ni-ferrite films is studied. X-ray photoelectron spectroscopy shows significant amounts of oxygen vacancies and enhanced cationic inversion for thin films versus the bulk. Films show spin glass behavior, which is contrary to the ferrimagnetic response of the bulk. Films exhibit spin freezing above room temperature at low fields (0.1 T) and shifts to lower temperatures (250 K) in the presence of a larger applied field (3 T). A large exchange bias of 170 Oe at a significantly higher temperature (50 K) is measured in the cooling field of 3 T.	[271]
Growth of Co-ferrite films						
1996	30 mTorr	200–800	KrF excimer, 248 nm, 2.5 J/cm^2, 5 Hz	MgO (1 0 0)	The phase, orientation, and microstructure of as-deposited films were investigated as a function of substrate temperature of 200–800 °C at a constant oxygen deposition pressure of 30 mTorr on MgO (1 0 0). As-deposited films were found to be of single phase, well textured, and matching the stoichiometry of the target. At 800°C, $4\pi M_s$ was measured to be 5370 G that is approximately the accepted bulk value for cobalt ferrite. PLD cobalt ferrite films grown at substrate temperatures of 600 and 800 °C exhibited a uniaxial magnetic anisotropy with an easy direction normal to the film plane. Films grown at 200 and 400 °C exhibited uniaxial magnetic anisotropy but along the planar direction.	[272]
2003	50 mTorr	250–800	KrF excimer, 248 nm, 1.4 J/cm^2, 3 Hz	MgO (1 0 0)	At 300 °C, a strong dependence of magnetic anisotropy on thickness is observed, indicating the dominant role of stress. At 800 °C, magnetic properties are less sensitive to film thickness, suggesting that magnetocrystalline is the only prevailing form of anisotropy.	[273]
2010	120 mTorr	RT	Nd-YAG laser, 355 nm, 10 Hz	Fused quartz	Co-ferrite thin films prepared by PLD at room temperature on fused quartz substrate, as a function of annealing temperature (TA) and annealing time. Spontaneous magnetization value, $4\pi M_s$, of the quenched films was observed to be higher than the furnace-cooled films in all the cases. Magnetization values even higher than the Co-ferrite bulk value of 5300 G were observed in the quenched films annealed at and above 650 °C. In the case of furnace-cooled films, magnetization decreased with the increase in annealing time, while it remained nearly constant in quenched films. Coercivity values of the quenched films were always found to be lower than those of the furnace-cooled films. The observed magnetic properties in the films were attributed to nonequilibrium cation inversion and average grain size.	[276]

Table 10.6 (Continued)

Year	Oxygen pressure (Torr) unless denoted	T_{sub} (°C)	Laser parameters	Substrate	Key results	Ref.
2010	10 mTorr	450	KrF excimer, 248 nm, 2 J/cm^2, 10 Hz	MgO (1 0 0) STO (1 0 0)	Cobalt ferrite films deposited on MgO and STO. Epitaxial films on MgO (1 0 0) with a lattice mismatch of 0.35% showed out-of-plane anisotropy, whereas films on STO (1 0 0) with a lattice mismatch of 7.4% displayed in-plane anisotropy. Calculated stress anisotropy confirmed the change in anisotropy relates to the lattice mismatch. Low-field switching was observed in the *M–H* loops and becomes prominent at lower temperature as evidenced from RF transverse susceptibility measurements and is correlated with the film–substrate interfacial structure and chemistry.	[277]
Growth of Li- and LiZn-ferrite films						
1994	30–375 mTorr	750	Nd-YAG, 1064 nm, 5 J/cm^2, 10 Hz	MgO (1 0 0)	Researchers explored the role of oxygen partial pressure in determining structural and magnetic properties of PLD LiZn-ferrite films. Films grown at lower oxygen pressures displayed cubic anisotropy with properties close to bulk values. The film grown at the lowest pressure had the highest saturation magnetization (4.5 kG) and the narrowest FMR linewidth (190 Oe at 9.5 GHz). The film grown at the highest pressure displayed in-plane uniaxial anisotropy and a relatively low magnetization (3.3 kG).	[280]
1995	75 mTorr	660	KrF excimer, 248 nm, 2 J/cm^2, 30 Hz	Al_2O_3 (0001)	Undoped Li-ferrite films of composition $Li_{0.5}Fe_{2.5}O_4$ on (0001) Al_2O_3 substrates are studied. SEM showed film surfaces with submicron scale roughness and free from particulates and pinholes. XRD revealed (111) epitaxial film growth and magnetization measurements showed an in-plane anisotropy 57 J/T/kg, a coercivity $\mu_0 H_c$ of 55 mT. FR of 0.2/μm across the visible spectrum range and 0.5/μm near the absorption edge at 2.4 eV.	[281]
1995	150 mTorr	750	Nd-YAG, 1064 nm, 5 J/cm^2, 10 Hz	MgO (1 0 0)	Lithium ferrite, pure and Mn and Zn co-substituted, was grown by XeCl excimer and Nd-YAG lasers. Electrical conductivity was shown to be thermally activated with an activation energy for as-grown films of 0.46 eV and after annealing in an oxygen gas environment, i.e. oxidized, of 0.80 eV. Estimated resistivities at room temperature for as-grown films is estimated at $\rho_{as-grown} = 1\times10^4 \frac{\Omega}{cm}$ and after an oxygen anneal at $\rho_{oxidized} = 4\times10^6 \frac{\Omega}{cm}$. Such treatments do not affect the FMR linewidth that was measured to be as low as 200 Oe (at 9.5 GHz).	[282]
2005	400 mTorr	800°C	KrF excimer, 248 nm, 1.7 J/cm^2	MgO (1 0 0)	Zn doping was employed to mediate the mismatch between the Li-ferrite film and the MgO (1 0 0) substrate. XRD patterns indicated a pure ferrite phase with the normal axis deviations below 0.05°. Process conditions were optimized to provide single-phase epitaxial films with half power FMR linewidths as low as 57 Oe at 9.5 GHz.	[284]

phase diagram [251]. Welch et al. explored deposition conditions that led to bulk-like magnetization values [252]. While Suzuki et al. provided a comprehensive study of MnZn-ferrite and Co-ferrite systems grown on a variety of buffered substrates (buffers layers: $CoCr_2O_4$, $CuMn_2O_4$, $FeGa_2O_4$, and $NiMn_2O_4$) laying the groundwork for an improved understanding of the processing–property–performance paradigm [253]. Films are typically grown on lattice-matched substrates of MgO and sapphire at substrate temperatures ranging from 800 to 900 °C in oxygen-rich working pressures of ~30 to 300 mTorr.

Harris et al. studied cation inversion as a function of process conditions in MnZn-ferrite films by applying extended X-ray absorption fine structure (EXAFS) [254] – a technique they refined for the specific use of measuring cation distribution and element-specific valency in thin film ferrites [255, 256]. Harris et al. [254] established that lasing energy and process conditions directly led to cation inversion that determined magnetic and RF properties of PLD ferrite films. This is particularly true for Mn and Fe cations that exist in a multiplicity of valence states and have approximately the same ionic radii allowing them to occupy various sights without giving rise to detrimental strain fields.

Srivastava et al., working with PLD film growers Ramesh, Dorsey, and Chrisey, and FMR experts Patton, Kabos, and Wittenauer, studied gyromagnetic loss mechanisms in PLD garnet and spinel films [257]. This team found MnZn-ferrite films to have FMR linewidths as low as 49 Oe for perpendicular alignments that increased with the angle between the film normal and the external field to a broad maximum of 80 Oe at 35° and to drop to 65 Oe for the parallel alignment. (In reference to NiZn-ferrite films, the linewidth was 310 Oe at perpendicular FMR, increased with the field angle to a broad maximum of 1530 Oe at 45° and then dropped to 960 Oe at parallel FMR.) The linewidths were larger than predicted by the Landau–Lifshitz equation and showed angle dependences that indicated extrinsic contributions. Two-magnon scattering was used to model these differences and showed qualitative agreement with predictions.

Atanasov et al., like others, established conditions for congruent compositional transfer from target to substrate providing high-quality MnZn-ferrite film's with coercivity, saturation magnetization, and remanent magnetization values comparable to bulk MnZn-ferrite [258]. Whereas most employ high substrate temperatures, Nakano et al. [259] and Kiyomura and Gomi [260] explored room temperature deposition on $ZnAl_2O_4$ and sapphire and glass substrates, respectively. They found that use of crystalline substrates was an effective way to obtain crystallized MnZn-ferrite thin films at room temperature. The use of crystalline substrates reduced the coercive force by approximately 25% compared to that of films processed on glass.

In 2000, researchers at Northeastern University (NEU) developed a variant of PLD that employs multiple simple metal oxide targets in a computer-controlled stepping motor carousel to sequentially ablate individual atomic (or blocks) to construct complex ferrites. This technique has been referred to as alternating target laser ablation deposition, or ATLAD, and the films as being atomically engineered. NEU teams have published several studies demonstrating the unique ability to tailor dopants to specific cation sites to control magnetic properties. Zuo et al. showed that a spinel $MnFe_2O_4$ film can be processed using ATLAD that displays bulk properties [261]. In a later study, they demonstrated large induced magnetic anisotropy (H_A), exceeding 5 kOe, when processed at oxygen pressures below 5 mTorr. It was shown that the magnetically preferred direction of H_A can be aligned either along the film plane, i.e. $pO_2 < 8$ mTorr, or perpendicular to it, i.e. $pO_2 > 8$ mTorr. The ability to induce perpendicular magnetic anisotropy in soft spinel ferrites allows for new applications such as isolators and circulators near or above X-band (i.e. 8–12 GHz) frequencies [262]. In these studies, EXAFS measurements provided a direct measure of Mn inversion as δ, defined as $(Mn_{1-\delta}Fe_\delta)^{tet}[Mn_\delta Fe_{2-\delta}]^{oct}O_4$, increased from 0.30 at $pO_2 = 1$ mTorr to > 0.50 at $pO_2 = 20$ mTorr. Another interesting trend seen in the EXAFS results was the increase in the oxygen displacement parameter (commonly denoted by u). An increase in u signals a local distortion of the tetrahedral sublattice. This often leads to a reduction in the unit cell space group symmetry. The authors speculated that this distortion was a consequence of the local strain caused by cation disorder and is responsible for the induced uniaxial perpendicular anisotropy and spin reorientation via magnetocrystalline anisotropy. This also suggest Jahn–Teller distortions from Mn^{3+} ions cannot be ruled out as a contributing mechanism [262].

Yang et al. of NEU, teamed with Haskel and Kravtsov of Argon National Laboratories, to apply diffraction anomalous fine structure (DAFS) [263] to obtain element-specific and site-specific information on cation distribution, local structure, and valency in Mn-ferrite films (see Figure 10.27). DAFS combines the chemical specificity of X-ray absorption fine structure (XAFS; see Figure 10.25b) with the site sensitivity of Bragg diffraction (see Figure 10.25a). Near-edge XAFS, sometimes referred to as NEXAFS or X-ray absorption near-edge spectroscopy (XANES), can be applied in a similar fashion to obtain element-specific valence information (see Figure 10.25c and d). Yang et al. introduced these properties to molecular field theory to reproduce thermomagnetization curves and determine the superexchange energy, Néel temperature, and spin canting angles. Molecular field simulations predicted a room temperature magnetization that was 11% larger (i.e. 4694 G) than the measured 4218 G. This result was consistent with an Mn^{3+} ion–induced Jahn–Teller distortion, whose presence was evidenced by a relatively large EXAFS-determined Debye–Waller factor (DWF) for Mn_B–O bonds. They conjectured that this structural distortion increased the superexchange interaction, which contributed to an increase in Néel temperature and room temperature magnetization [263].

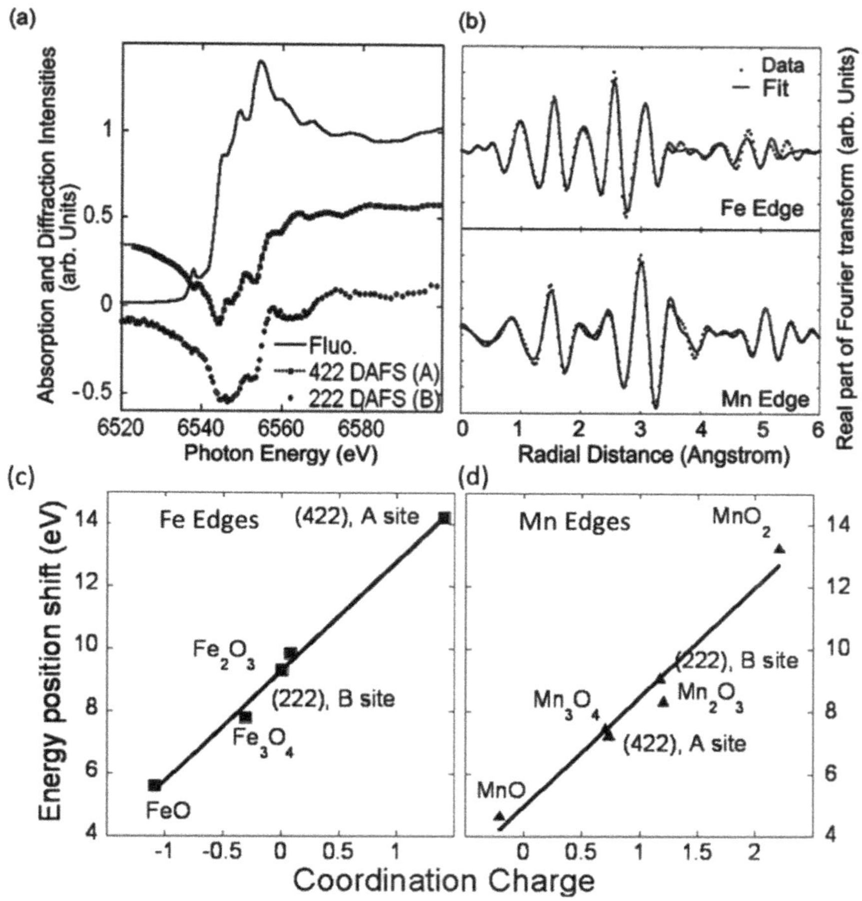

Figure 10.27 (a) Mn K edge DAFS for the (422) and (222) Bragg reflections, together with EXAFS data. The scans are vertically shifted to allow comparison. (b) Fourier-transformed Mn and Fe EXAFS plotted with best fit data. Energy shift plotted as a function of coordination charge for various (c) Fe and (d) Mn oxide standards with site-specific values for Mn-ferrite films (denoted by (*hkl*), A-/B-site) (adapted from [263]).

10.3.4.3.2 Pulsed-Laser-Deposited Ni- and NiZn-Ferrite Films

In contrast to MnZn ferrites, NiZn ferrites have three to six orders of magnitude higher electrical resistivity providing them less conduction losses thus allowing them to operate at far higher frequencies, for example, up to and beyond 1 GHz. However, they also provide lower magnetization and less permeability. Early reports included a 1996 study by Guyot et al. who demonstrated growth of popular ferrites, including Ni and NiZn ferrites [264]. Although not a detailed study, they demonstrated PLD-congruent compositional transfer from target to film. Low substrate temperatures led to amorphous films that required post-annealing to crystallize into ferrites. Soon after, Welch et al. [265] found grain size of thin film crystals ranged from 300 to 600 nm. VSM measurements showed coercive fields of 800 Oe, which are larger than those found for the bulk, reflecting the shape anisotropy from the films' microstructure. Saturation magnetization values of 660 emu/cm^3 are comparable with the bulk material.

In 1997, Dorsey et al. reported on stress-induced magnetic anisotropy in oriented NiZn-ferrite films as thick as ~10 to 12 μm [266]. The morphology, phase, orientation, strain, and magnetic properties of the as-deposited films were investigated as a function of substrate temperature 400–700 °C and oxygen background pressure ranging from 50 to 200 mTorr. PLD NiZn-ferrite films were found to be ~45% deficient in Zn, independent of substrate temperature or pressure. However, Zn-rich targets compensated for the Zn deficiency. NiZn-ferrite films grown at 700 °C exhibited the highest degree of crystalline quality and nearly bulk saturation magnetization values, i.e. 5000 G. At low O_2 pressures, i.e. 0.75 mTorr, films are under compressive stress that was attributed to the oxygen deficiency during deposition. Caltun subsequently investigated the morphological, structural, and magnetic properties of PLD NiZn-ferrite films grown on various crystalline and amorphous substrates at different substrate temperature and different gas pressures in order to establish optimal magnetic properties [267]. As-deposited thin films are composite structures with crystalline islands suspended in an amorphous matrix.

After annealing, the amorphous composite reflects a (311) texture, with the orientation influenced by the deposition conditions, substrate type, and thermal treatment. The film composition was found to be that of the target. The coercive field, due to the crystallite shape anisotropy, was larger than the bulk while magnetization values were nearly the same.

The investigation by Mizukami et al. [268] showed that H_c and M_r are strongly influenced by the crystallinity of the films, and M_s, M_r, and H_c depend more on the composition for films grown on single-crystal Si (1 0 0). Important findings include a linear trend of decreasing Zn content in films with a logarithmic increase in pressure. This is much more profound in samples grown at higher substrate temperatures. Coercivity also is shown to reduce as substrate temperature increases, reflecting grain growth and the reduction in grain boundaries that are highly effective as domain wall pinning sites. Xi et al. investigated NiZn-ferrite films deposited on Si (111). XRD indicated a (400) texture when grown on Si substrates [269]. The thickness and growth of films were investigated at different deposition temperatures and pressures. Saturation magnetization and coercivity were found to be strongly dependent upon oxygen pressure, suggesting cation inversion, defects, and stress play an important role in determining the magnetic properties.

Chinnasamy et al., motivated by first-principles calculations suggesting that occupation of Ni^{2+} ions on the tetrahedral sites would lead to an enhancement in exchange integrals and subsequently the Néel temperature and magnetization, explored the effects of PLD growth temperature on magnetic and RF properties [269]. Magnetic studies showed a magnetization value of 2.7 kG at 300 K. Films deposited at 900 °C were shown to experience a 25% increase in magnetization after post-annealing at 1000 °C in air with the FMR linewidth reduced from 1.5 to 0.34 kOe. The negative uniaxial magnetic anisotropy field (i.e. perpendicular magnetic anisotropy) were measured to be −2 kOe due to a spin reorientation brought on by induced tensile stress.

In 2009, Johnson et al. studied the topology of twin boundaries in films grown under different conditions, including conventional PLD and the formation of layers (similar to ATLAD) followed by solid-state reaction to form a pure phase. For conventional growth, twin boundaries were lateral in character, while in films grown through solid-state reaction, twin boundaries were found to be a mix of shorter lateral segments with longer {111} segments. The twin boundary segments on the {111} planes are thought to be formed at spinel/Fe_2O_3 interfaces [270].

Jaffari et al. studied the role of oxygen vacancies on electronic structure in nonstoichiometric Ni-ferrite films [271]. XPS showed oxygen vacancies and enhanced cationic inversion in films versus bulk. Films also showed spin glass behavior, which is absent in the bulk. Films exhibit spin freezing above room temperature at low fields (0.1 T) and shifts to lower temperatures (250 K) in the presence of large applied fields (3 T). A large exchange bias of 170 Oe is measured at higher temperatures (50 K) in cooling fields of up to 3 T. The role of oxygen vacancies was to induce spin canting by destabilizing super exchange interaction.

10.3.4.3.3 Pulsed-Laser-Deposited Co-Ferrite Films

Co-ferrite is unique among the spinels in that Co^{2+} retains unquenched angular momentum providing it with substantial single-ion anisotropy that gives rise to significant large and negative magnetostriction, rare among oxide materials. Early studies of cobalt ferrite, $CoFe_2O_4$, were reported by Dorsey et al. [272]. The phase, orientation, and microstructure of as-deposited films were investigated as a function of substrate temperature of 200–800 °C at a constant oxygen deposition pressure of 30 mTorr on MgO (1 0 0) substrates. The as-deposited films were found to be single phase, strongly textured, and approximately matching the stoichiometry of the target, but with a cubic lattice constant dependent upon the substrate temperature, indicating that the films were strained. The greatest effect of the substrate temperature was on the magnetic properties of the as-deposited films. At 800 °C, $4\pi M_s$ was measured to be 5370 G that is approximately the accepted bulk value for cobalt ferrite. PLD cobalt ferrite films grown at substrate temperatures of 600 and 800°C exhibited a uniaxial magnetic anisotropy with an easy direction normal to the film plane. Films grown at 200 and 400 °C exhibited uniaxial magnetic anisotropy but along the planar direction. Also in 1996, Suzuki et al. deposited Co-ferrite films on $SrTiO_3$ and $MgAl_2O_4$ substrates buffered by $CoCr_2O_4$ [253]. In that study, saturation magnetization values of 400 emu/cm^2 were measured. Hysteresis loops with the magnetic field applied along an easy (0 0 1) direction are square. An estimate of the crystal anisotropy field of $H_A = 25$ kOe and thus $K_1 = 5 \times 10^6$ ergs/cm^3. This value is of the same order as that of the anisotropy constant $K_1 = 4.3 \times 10^6$ ergs/cm^3 obtained from torque curves of (110) of the films, and the value $K_1 = 3 \times 10^6$ ergs/cm^3 for bulk Co-ferrite. Lisfi and Williams [273] added important observations relating film stress to substrate temperature, film thickness, magnetic anisotropy, and magnetic domain structure. Terzzoli et al. expanded upon such processing studies to realize Co-ferrite films with perpendicular anisotropy at room temperature having coercive fields as high as 5.1 kOe [274]. Yin et al. similarly engineered the induced strain and a nanostructured microstructure to obtain a room temperature perpendicular coercivity of 12.5 kOe [275], making Co-ferrite processed under such conditions a possible candidate for magneto-optic recording applications.

Sahoo et al. investigated Co-ferrite thin films prepared by PLD at room temperature on fused quartz substrate as a function of annealing temperature (T_A) and annealing time [276]. One set of films was quenched immediately after annealing, while the other set was furnace cooled. Saturation magnetization values, $4\pi M_s$, of the quenched films was observed to be higher than the furnace-cooled films in all cases. Magnetization increases monotonically to $T_A = 650$ °C and plateaus. Magnetization values are even higher than the Co-ferrite bulk value of 5300 G in quenched films annealed at and above 650 °C. In the case of furnace-cooled films, magnetization decreased with the increase in annealing time, while it remained nearly constant in quenched films. Coercivity values of the quenched films were always found to be lower than those of the furnace-cooled films. The observed magnetic properties in the films were attributed to nonequilibrium cation inversion and average grain size.

Dhakal reports the observation of contrasting magnetic behavior in cobalt ferrite thin films deposited on MgO and STO [277]. Epitaxial films on MgO (1 0 0) with a lattice mismatch of 0.35% showed out-of-plane anisotropy, whereas the films on STO (1 0 0) with a lattice mismatch of 7.4% displayed in-plane anisotropy. Calculated stress anisotropy confirmed that the change in anisotropy derives from the lattice mismatch. A low-field switching characteristic was observed in the *M–H* loops that becomes prominent at lower temperatures and is evident in RF transverse susceptibility measurements.

10.3.4.3.4 Pulsed-Laser-Deposited Li- and LiZn-Ferrite Films

Lithium ferrite is an attractive and potentially useful material for microwave applications because of its relatively high magnetization, low gyromagnetic losses, and its flexibility to accept cation substitutions such as Zn, Ni, and others. Thin-film microwave applications of lithium ferrites include planar microwave devices that require low microwave loss and low-FMR linewidths [278, 279]. The first report of PLD processing of Li ferrites was by Oliver et al. who explored the role of oxygen partial pressure in determining structural and magnetic properties of LiZn-ferrite films [280]. XRD measurements showed the films grew epitaxially on the (1 0 0) MgO substrates. Films grown at lower oxygen pressures displayed cubic anisotropy with properties close to bulk values, with the film grown at the lowest pressure having the highest saturation magnetization and the narrowest FMR linewidth. The film grown at the highest pressure displayed in-plane uniaxial anisotropy and a relatively low magnetization.

In 1995, Masterson, et al. studied undoped Li-ferrite films $Li_{0.5}Fe_{2.5}O_4$ on (0001) Al_2O_3 substrates. Scanning electron micrographs showed film surfaces with submicron scale roughness and free from particulate contamination and pinholes. XRD revealed (111) epitaxial film growth, and magnetization measurements showed an in-plane anisotropy 57 J/T/kg; a coercivity $\mu_0 H_c$ of 55 mT; a squareness factor, Sq, of 0.58; and FR of 0.2/μm across the visible spectrum range and 0.5/μm near the absorption edge at 2.4 eV [281].

Epitaxial films of lithium ferrite, pure and Mn and Zn co-substituted, were grown by XeCl excimer and Nd-YAG lasers by Balestrino et al. [282] The use of the excimer laser was shown to produce films with superior surface morphology free of particulates. All films showed congruent compositional transfer from target to film. Electrical conductivity was shown to be of a thermally activated hopping mechanism, with an activation energy for as-grown films of 0.46 eV and after annealing in an oxygen gas environment of 0.80 eV. An estimate of room temperature resistivities gives $\rho_{\text{as-grown}} = 1\times10^4 \frac{\Omega}{\text{cm}}$ and $\rho_{\text{oxidized}} = 4\times10^6 \frac{\Omega}{\text{cm}}$. Such thermal treatments did not affect the FMR linewidth that was measured to be as low as 200 Oe at 9.5 GHz [123].

Highly (333) textured Li-ferrite films were synthesized on the *c*-plane sapphire substrates by Cadieu et al. [283] The power absorption derivative peak-to-peak linewidth was measured to be as low as 335 Oe for the in-plane static field configuration. The magnetization versus field measurements from the in-plane FMR configuration yielded a saturation flux density, $4\pi M_s$, of 3.6–4.0 kG. These values are in the same range as bulk Li ferrites. Song et al. employed Zn-doped Li ferrite to mediate the mismatch between the film and MgO (1 0 0) substrate. XRD patterns indicated a pure ferrite phase with local normal axis deviations below 0.05°. Preparation conditions were optimized to provide single-phase epitaxial films with half power ferromagnetic resonance linewidths of as low as 57 Oe at 9.5 GHz. The data yielded a saturation induction $4\pi M_s$ value of 2.73 kG, about the same as for a bulk material with cubic anisotropy [284].

10.3.4.4 PLD of Hexaferrite Films

Hexaferrites are a classification of ferrites of hexagonal symmetry, larger unit cells, and increased structural and magnetic complexity in comparison to the cubic garnets and spinels. Illustrated in the ternary phase diagram of Figure 10.28 are the six well-known hexaferrite variants designated M, Y, X, W, Z, and U.

Table 10.7 provides a listing of these phases, together with the lesser known 18H phase, molecular formula, and block sequence with several of their important magnetic properties. A visualization of the hexaferrite crystal structure is provided by the repetitive stacking of six basic subunits, i.e. the S-block ($[Me_2Fe_4O_8]^{0\pm}$ or $[Fe_6O_8]^{2+}$), R-block ($[BaFe_6O_{11}]^{2-}$),

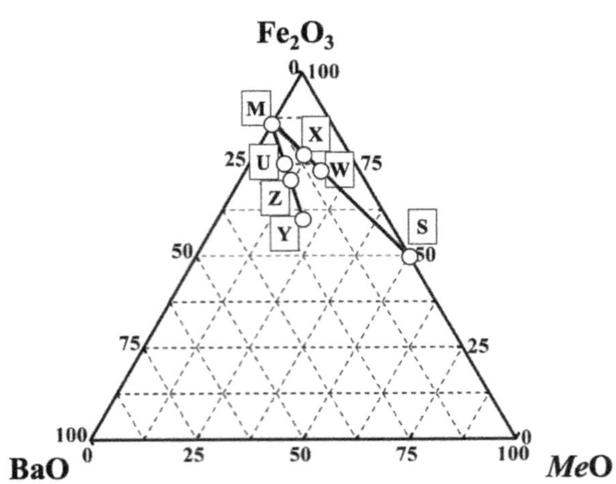

Figure 10.28 Ternary phase diagram of BaO–Fe_2O_3–MeO illustrating six principal microwave hexaferrite phases M, Y, Z, X, W, and U as well as the cubic spinel phase (S).

Table 10.7 Magnetic properties of bulk hexaferrites.[a]

Ferrite type	Molecular formula	Block structure	Magnetic easy alignment	$4\pi M_s^{RT}$ (G)	μ_B	T_C (K)	H_A^θ / H_A^φ (Oe)	K_1 ($\times10^5$)(J/m^3)
M	**$MeFe_{12}O_{19}$**	**SRS*R***						
BaM	$BaFe_{12}O_{19}$	SRS*R*	Uniaxial	4770	20	725	17,000	3.3
SrM	$SrFe_{12}O_{19}$	SRS*R*	Uniaxial	3730	9.8	735	17,600	3.5
Y	**$Ba_2Me_2Fe_{12}O_{22}$**	**3(ST)**						
Co2Y	$Ba_2Co_2Fe_{12}O_{22}$	3(ST)	Planar	2325	14.4	613	28 000/155	-2.6
Zn2Y	$Ba_2Zn_2Fe_{12}O_{22}$	3(ST)	Planar	2850	18.4	403	9000/1	n.a.
W	**$BaMe_2Fe_{16}O_{27}$**	**SSRS*S*R***						
Ni2W	$BaNi_2Fe_{16}O_{27}$	SSRS*S*R*	Conical	4145	n.a.	793	12 700	n.a.
Co2W	$BaCo_2Fe_{16}O_{27}$	SSRS*S*R*	Conical	3980	n.a.	763	21 200	-3.5
Z	**$Ba_3Me_2Fe_{24}O_{41}$**	**STSRS*T*S*R***						
Co2Z	$Ba_3Co_2Fe_{24}O_{41}$	STSRS*T*S*R*	Planar	3390	210.8	683	13 000/16	-1.8
Zn2Z	$Ba_3Zn_2Fe_{24}O_{41}$	STSRS*T*S*R*	Planar	3900	38.4	633	n.a.	n.a.
X	**$Ba_2Me_2Fe_{28}O_{46}$**	**3(SRS*S*R*)**						
Co2X	$Ba_2Co_2Fe_{28}O_{46}$	3(SRS*S*R*)	Conical	2688	46.3	740	9490	n.a
Zn2X	$Ba_2Zn_2Fe_{28}O_{46}$	3(SRS*S*R*)	Conical	5812	55	705	n.a.	n.a.
U	**$Ba_4Me_2Fe_{36}O_{60}$**	**SRS*R*S*T**						
Co2U	$Ba_4Co_2Fe_{36}O_{60}$	SRS*R*S*T	Planar	3624	n.a.	707	n.a.	n.a.
(ZnCo)2U	$Ba_4(Zn_{1.6}Co_{0.4})Fe_{36}O_{60}$	SRS*R*S*T	Planar	3405	n.a.	673	7530	1.03
18H	**$Ba_5Me_2Ti_3Fe_{12}O_{31}$**	**$2[\frac{1}{2}(ST)B\frac{1}{2}(TS)]^{\dagger}$**						
Zn18H [131, 132]	$Ba_5Zn_2Ti_3Fe_{12}O_{31}$	$2[\frac{1}{2}(ST)B\frac{1}{2}(TS)]$	Planar	950	14.1	310	3500	-1.0
Co18H [133]	$Ba_5Co_2Ti_3Fe_{12}O_{31}$	$2\left[\frac{1}{2}(ST)B\frac{1}{2}(TS)\right]$	Planar	1256	9.6	495	40 000	-0.1
(MgZn)18H [134]	$Ba_5Mg_1Zn_1Ti_3Fe_{12}O_{31}$	$2\left[\frac{1}{2}(ST)B\frac{1}{2}(TS)\right]$	Planar	1230	12.5	357	1300	-0.4

† B denotes for hexagonal BTO block.
n.a.: Not available at the time of publication.
a) Data derived in part from [13, 22, 286, 287].

and T-block ($[Ba_2Fe_8O_{14}]^{0\pm}$) and the 180° rotation of these subunits around the crystallographic *c*-axis, conventionally denoted as S*, R*, and T*. Figure 10.29 presents filling and molecular bonding models of the S, R, and T blocks.

Among the most popular of the microwave hexaferrites are those derived from the M-type, or magnetoplumbite crystal structure of space group *Pb3/mmc*. The most popular of the M-type hexaferrites, BaM (i.e. $BaFe_{12}O_{19}$), is prototypical of the structure that consists of 10 oxygen layers in which 4 successive layers each containing four anions followed by the fifth layer that contains three anions and a Ba cation. The BaM unit cell is visualized as the stacking of subunit blocks SRS*R* in which every unit cell consists of two formular units. (see Figure 10.30).

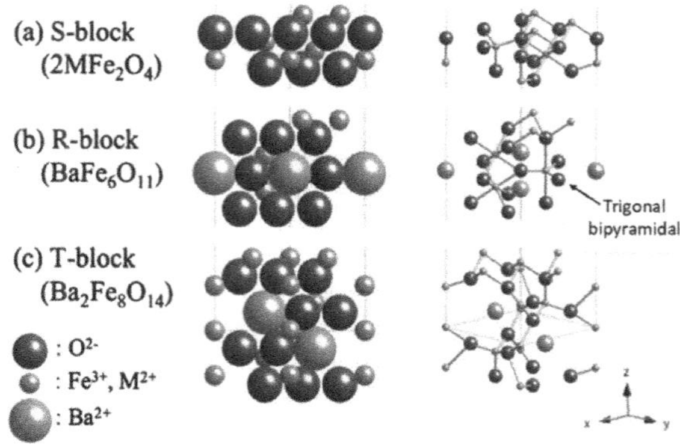

Figure 10.29 Constituent blocks in the Ba-M hexagonal ferrites: (a) S-block $[Me_2Fe_4O_8]^{0\pm}$, (b) R-block ($[BaFe_6O_{11}]^{2-}$), and (c) T-block ($[Ba_2Fe_8O_{14}]^{0\pm}$). Center column is a space filling model, the right column is a molecular bonding model (adapted from [288]).

The Ba^{2+} cation is often, in part or in whole, replaced by a Sr^{2+} cation serving to increase hard magnetic properties for motor applications and/or select microwave device applications. One property, which is of particular value to RF design engineers, is a strong uniaxial anisotropy that aligns its easy magnetization direction along the crystallographic *c*-axis. The Ba(Sr)-M-type magnets have uniaxial magnetic anisotropy fields, H_A, of 17 000 Oe (BaM) and 17 600 Oe (SrM). These large values of magnetic anisotropy field derive predominantly from magnetocrystalline origins.

The high magnetic anisotropy field of the M-type can be further adjusted by the substitution of the Fe cation with other transition metal cations, allowing for the tuning of resonance frequencies in accordance with:

$$f_{FMR} = \frac{\gamma}{2\pi}\sqrt{H_1 H_2}, \tag{10.11}$$

where γ is the electron gyromagnetic ratio (i.e., γ_e) in SI units is $1.76086 \times 10^{11}/(s{\cdot}T)$ $\left(\text{or popularly as } \frac{\gamma_e}{2\pi} = 28\,024.95\frac{\text{MHz}}{\text{T}}\right)$ and H_1 and H_2 are defined in the table within Figure 10.31 for spinel and garnets as well as M-type and Y- and Z-type hexaferrites. Trends in resonant frequency (i.e. f_{FMR}) as a function of H_A and bias field (H) are depicted. This figure makes clear the impressive variation in resonant frequency as a function of applied fields and magnetic anisotropy fields for various compositionally modified hexaferrites. The zero-field resonance values, represented by the *y*-intercept of each curve, extends from UHF to beyond W-band with appropriate structural symmetry and cation doping.

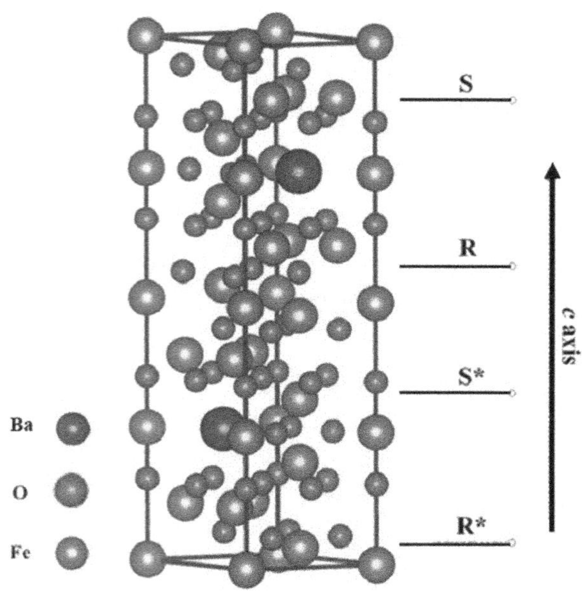

Figure 10.30 Barium magnetoplumbite (BaM) unit cell with atoms identified and structural block sequences (SRS*R*) presented on the right-hand side (adapted from [288]).

This design flexibility makes Ba(Sr)M a choice material for many monolithic microwave IC (MMIC) devices and applications. Each of the other hexaferrite structures Y, U, W, X, and Z has different stacking sequences, oxygen stoichiometry, and markedly different magnetic properties. These other phases do not possess *uniaxial* anisotropy but rather *planar* anisotropy that is highly valued for antenna substrates, filters, phase shifters, etc. [22], or *conical* anisotropy that has found essential in the emerging field of magnetoelectrics (MEs) [289].

Early PLD studies of hexaferrite films were driven by the needs of microwave and millimeter-wave devices, and MMICs. Because hexaferrite phases have large and complex unit cells, fabrication as thin films by PLD are nontrivial. Thus far, PLD of hexaferrites has been dominated by reports of the least complex of the hexaferrites, the M-types. The focus on M-type barium and strontium ferrites is also motivated by their high magnetocrystalline uniaxial anisotropy fields, high

	H_1	H_2
Spinel and garnet	$H+4\pi M$	H
M-type	$H+H_A-4\pi M$	$H+H_A-4\pi M$
Y- and Z-type	$H+H_\theta-4\pi M$	H

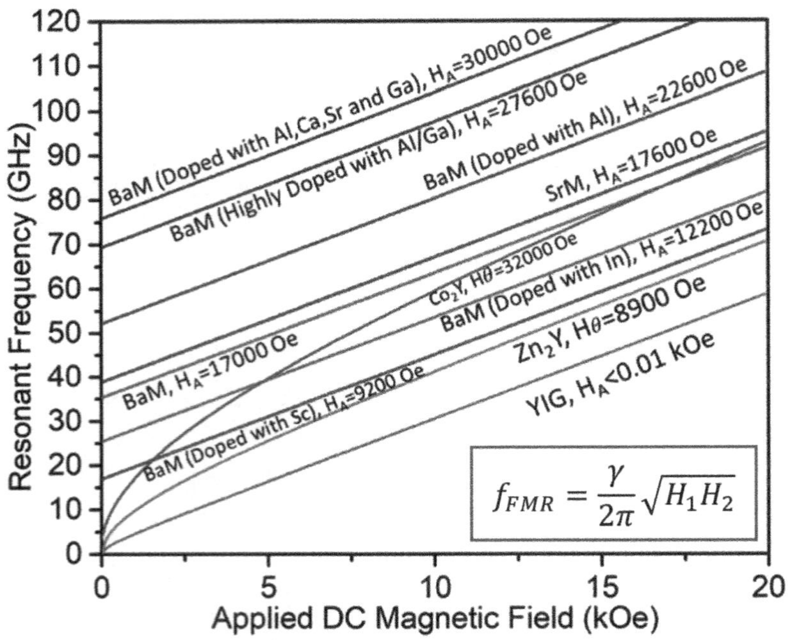

Figure 10.31 Trends in resonant frequency (i.e. f_{FMR}) as a function of H_A, H_θ (with H_θ is the in-plane anisotropy field intrinsic to ferroxplana ferrite systems in the basal plane) and bias field (H). The y-intercept is the zero-field f_{FMR} that defines the operational frequency bands.

coercivities (~0.5 T), and high magnetic retention ($M_r > 0.75M_S$). Epitaxial BaM and SrM films were shown to be amenable to the c-axis texture normal to the substrate plane on (0001) sapphire or with the c-axis parallel to the substrate plane on (1 0 0) sapphire. PLD films deposited on (0001) sapphire substrates experience a 7% lattice mismatch between nominal $BaFe_{12}O_{19}$ and the sapphire crystal, with sapphire being smaller. MgO substrates are also commonly used and provide a reduced lattice mismatch. The growth of crystallographic textured hexaferrite films was also demonstrated on WBG semiconductor substrates such as 6H silicon carbide (SiC) (0001) or gallium nitride (GaN), and narrow bandgap semiconductor substrates such as SiO_2 (thermal oxide) on Si. Table 10.8 presents a historical development of pulsed lased deposited hexaferrite films.

As was shown for garnets and spinels, earlies studies were performed by the NEU and NRL groups. Dorsey et al. [166] and Carosella et al. [167] used similar process conditions and produced high-quality BaM films on sapphire (0001) having bulk-like magnetic properties with FMR linewidth values of ~250 to 300 Oe for as-deposited films, reducing to less than 100 Oe after heat treatments (measured at 50 GHz). Dorsey et al. [290], followed with growth studies of BaM on the c-plane (0001) and m-plane, both (1010) and (1100), single crystal sapphire substrates. Principal findings included bulk-like magnetic properties with retention of more than 90% and linewidths of 450 Oe; properties suitable for many passive nonreciprocal microwave devices. In an explorative study of PLD growth of BaM on ZnO-buffered amorphous substrates at 750–800 °C and 300 mTorr led to the growth of c-axis oriented films with saturation magnetization, magnetic anisotropy field, and ferrimagnetic resonance linewidths in agreement with bulk values [291].

Optical properties of PLD BaM and Co-Ti-doped BaM films were explored by Masterson et al. [309], and Atkinson et al. [292], and Gerber et al. [310], respectively. The dispersion of the optical and magneto-optical activity of Co-Ti-doped BaM is marked by two distinct peaks at wavelengths of 380 and 710 nm. Overall, the optical properties of PLD films reproduce the valued properties of their bulk counterparts.

Table 10.8 Pulsed laser deposition conditions and key results from early studies of hexaferrite films.

Year	Oxygen pressure (Torr) unless denoted	T_{sub} (°C)	Laser parameters	Substrate	Key results	Ref.
Growth of M-type hexaferrite films						
1992	10–400 mTorr	600 935	KrF excimer, 248 nm, 300 mJ/pulse, 2 J/cm^2	Sapphire (0001)	PLD of thin $BaFe_{12}O_{19}$ (BaM) films on (0001) sapphire substrates. Analysis confirms the structure, composition, and magnetic properties were similar to films prepared by other methods, for example, liquid-phase epitaxy ($4\pi M_s = 4400$ G). After an anneal, the FMR linewidth at 50 GHz was 88 Oe.	[166]
1992	10–400 mTorr	600–935	KrF excimer, 248 nm, 300 mJ/pulse, 2 J/cm^2, 10 Hz	Sapphire (0001)	PLD of barium hexaferrite ($BaFe_{12}O_{19}$) films on *c*-plane sapphire. Films were single phase, crystalline, and *c*-textured. Upon post-annealing, the FMR linewidth is reduced by a factor of 3. The derivative of FMR linewidth after post-annealing was measured to be 66 Oe at 58 GHz. The first recorded spin-wave spectra from BaM films are reported.	[167]
1994	50–1000 mTorr	920	KrF excimer, 248 nm, 2.5 J/cm^2, 20 Hz	Sapphire *c*-plane (0001) *m*-plane, both (1010) and (1100)	PLD thick barium hexaferrite films grown on *c*-plane and *m*-plane sapphire substrates. Crystallographically textured thick BaM films up to 15 and 20 μm in thickness were deposited on *c*-plane and *m*-plane sapphire substrates, respectively, before the onset of delamination. FMR linewidth and $4\pi M_s$ were measured to be 200 Oe and 4200 G on *c*-plane and 450 Oe and 4000 G on *m*-plane sapphire, respectively.	[290]
1994	75 mTorr	840	KrF excimer, 248 nm, 200 mJ/pulse, 2 J/cm^2, 20 Hz	Sapphire (0001)	PLD of Co-Ti-doped barium hexaferrite films on (0001) sapphire. Ellipsometry, reflectance photometry, and Kerr polarimetry performed. Kerr rotation and ellipticity of the films in agreement with predictions and mirror bulk properties.	[292]
1995	40–400 mTorr	>850	XeCl excimer 308 nm, 70 mJ/pulse, 20 Hz	Sapphire, silicon, glass	PLD films of barium hexaferrite films on sapphire, silicon, and amorphous substrates. Crystal structure and magnetic properties depend on substrate temperature and oxygen pressure. Barium hexaferrite films on amorphous substrates were weakly polycrystalline. Crystalline orientation of films grown on sapphire were determined by the orientation of the substrate.	[293]
1999	10–100 mTorr	900	KrF excimer, 248 nm, 4 J/cm^2, 50 Hz	*c*-plane sapphire	PLD of $BaSc_xFe_{12-x}O_{19}$ ($x = 0.4$) on *c*-plane sapphire substrates at oxygen pressures in the range of 10–100 mTorr. All films confirmed to be single-phase *c*-axis oriented hexaferrites with the easy magnetization axis normal to the film plane. The saturation magnetization was measured to be ~3.8 kG. The mean uniaxial anisotropy field was measured to be 10.6 kOe.	[294]
2013	5–10 mTorr	850–950	KrF excimer, 248 nm, 400 mJ/pulse	MgO (111)	PLD of Zr- and Zn-doped Ba-hexaferrite films on (111) MgO substrates. Zn and Zr divalent ions reduce the magnetic anisotropy field by 46%, to 7.5 kOe. FMR linewidth was measured at 9.55 GHz to be 120 Oe.	[295]

(Continued)

Table 10.8 (Continued)

Year	Oxygen pressure (Torr) unless denoted	T_{sub} (°C)	Laser parameters	Substrate	Key results	Ref.
2000	20 mTorr	900–925	KrF excimer, 248 nm, 4–5 J/cm^2, 50 Hz	MgO (111)	PLD of highly oriented $BaFe_{12}O_{19}$ (BaM) films on MgO (111). As opposed to sapphire substrates, MgO (111) reduces the in-plane biaxial compressive stress, cracking, or delamination and allows for BaM films up to 28 μm in thickness.	[296]
2002	100 mTorr	600–800	KrF excimer, 248 nm, 2.5 J/cm^2, 15 Hz	Sapphire (0001)	Strontium hexaferrite films were grown by PLD on (0 0 0 1) single-crystal sapphire substrates with postdeposition laser annealing under external magnetic fields. The films deposited at higher substrate temperature were *c*-axis textured. High retention was obtained for films deposited at lower temperatures. Laser modification of ferrite films produced smooth film surfaces and higher coercive forces along the direction of the applied magnetic field.	[297]
2003	300 mTorr	910	KrF excimer, 248 nm, 1.7 J/cm^2	Sapphire (0001)	PLD of hexagonal M-type barium ferrite on (0001) sapphire. Determined optimal conditions for single-phase films with narrow FMR linewidth. $4\pi M_s$, H_A, and FMR linewidths were measured to be 4.2 kG, 16.4 kOe, and 27 Oe (at 60 GHz), respectively.	[298]
2008	300 mTorr	925	KrF excimer, 248 nm, 10 J/cm^2	Sapphire (0001)	Alternating target laser ablation deposition (ATLAD) was employed to selectively place Mn cations in specific interstitial sites of the S-block. As a result of this site selection, the saturation magnetization increased 12–22%, and the Néel temperature increased by 40–60 K compared to bulk materials. Elemental site occupation was confirmed experimentally by EXAFS and its impact predicted by theory.	[299]
2012	100 mTorr	700	KrF excimer, 248 nm, 260 mJ/pulse, 3 J/cm^2, 5 Hz	Silicon (1 0 0)	PLD of strontium hexaferrite ($SrFe_{12}O_{19}$) films on Pt (111) coated Si (1 0 0). Magnetic properties investigated as a function of film thickness in the range between 50 and 700 nm. Confirmed by XRD that films were *c*-textured. Coercivity decreases with increasing thickness due to grain growth and relaxation.	[300]
2020	200 mTorr	920	Nd:YAG laser, 266 nm, 10 Hz, 1 - 5.7 J/cm^2	Sapphire (0001)	PLD of barium hexaferrite on *c*-plane sapphire (0001) substrates to evaluate the effects of laser fluence on films composition, structure, and magnetic properties. The fluence of the Nd:YAG laser was systemically adjusted between 1 and 5.7 J/cm^2. The Fe-to-Ba ratio increased with laser fluence. Insufficient laser energies resulted in the deposition of secondary phases, such as barium monoferrite ($BaFe_2O_4$) and magnetite (Fe_3O_4). Laser fluences above 5 J/cm^2 promoted secondary sputtering of the films by plasma species that degraded film quality, structure, and magnetic properties. A laser fluence of 4.8 J/cm^2 was found to be optimal.	[301]

Table 10.8 (Continued)

Year	Oxygen pressure (Torr) unless denoted	T_{sub} (°C)	Laser parameters	Substrate	Key results	Ref.
2014	50 mTorr	900	KrF excimer, 248 nm, 200 mJ/pulse, 10 Hz	(0001) sapphire	U-type hexaferrite films ($Ba_4Ni_{1.4}Co_{0.6}Fe_{36}O_{60}$) are grown on (0001) sapphire substrates with high retention and low-FMR linewidth (200 Oe at X-band). Anisotropy field measurements indicated an H_A of 8 kOe and a saturation magnetization ($4\pi M_s$) of 3.6 kG. An optimal postdeposition annealing of films resulted in a strong (0 0 *n*) crystallographic texture and a high hysteresis loop squareness (M_r/M_s ~92%) leading to self-properties.	[302]
Integration of hexaferrite films with semiconductor heterostructures						
2006	1–400 mTorr	925	KrF excimer, 248 nm, 350 mJ/pulse	6H- silicon carbide (0001)	PLD of barium hexaferrite films on single-crystal 6H silicon carbide substrates. Atomic force microscopy (AFM) and XRD revealed films to be crystalline and *c*-textured. The saturation magnetization, $4\pi M_s$, was comparable to bulk values (~4320 G) when the oxygen pressure was 20 mTorr during deposition.	[303]
2006	300 mTorr	925	KrF excimer, 248 nm, 200 mJ/pulse	6H silicon carbide (0001)	PLD of barium hexaferrite films on MgO (111) nucleation layers on 6H-SiC (0001) substrates. XRD reveals strong (0 0 2*n*) crystallographic alignment. $4\pi M_s$ was comparable to bulk values (4320 G). FMR linewidths were measured to be 500 Oe at 55 GHz.	[304]
2007	20 mTorr	915	KrF excimer, 248 nm, 250 mJ/pulse	6H silicon carbide (0001)	PLD of barium hexaferrite (BaM) films on 10 nm MgO (111) coated 6H silicon carbide (0001) substrates. The MgO layers were produced by molecular beam epitaxy and reduced lattice mismatch and interdiffusion between film and substrate. XRD and pole figures revealed epitaxial growth and strong crystallographic alignment. Post-annealing allowed for a perpendicular magnetic anisotropy field of 16 900 Oe, a saturation magnetization $4\pi M_s$ of 4.4 kG, and an FMR linewidth of 96 Oe at 53 GHz.	[305]
2010	20 mTorr	900	KrF Excimer, 248 nm, 250 mJ/pulse	GaN/sapphire	Growth of high-quality BaM on GaN/Al_2O_3 by PLD. XRD showed (0 0 2*n*) crystallographic alignment with pole figure analyses confirming epitaxial growth. High-resolution transmission electron microscopy images revealed magnetoplumbite unit cells stacked with limited interfacial mixing. Saturation magnetization, $4\pi M_s$, was measured for as-grown films to be 4100 kG, with a perpendicular magnetic anisotropy field of 16.0 kOe. FMR linewidth, the peak-to-peak power absorption derivative at 53 GHz, was 86 Oe.	[306]

(Continued)

Table 10.8 (Continued)

Year	Oxygen pressure (Torr) unless denoted	T_{sub} (°C)	Laser parameters	Substrate	Key results	Ref.
2021	200 mTorr	920	KrF excimer, 248 nm, 10 Hz	MgO/AlN/SiC (6H)	PLD BaM films on industrial compatible wide bandgap (WBG) semiconductor heterostructures suitable for operation in Ka-band circulators. They demonstrated successful PLD growth of BaM films on WBG semiconductor heterostructures that showed device quality performance in structure, epitaxy, and magnetic properties for BaM/MgO/AlN/SiC(6H) heterostructures. Film properties include bulk-like values of magnetic anisotropy field, H_a~16.5 kOe, and saturation magnetization, $4\pi M_s$ ~4.2 kG. Only heterostructures where SiC substrates having Si-polar surfaces showed superior properties.	[307]
2020	200 mTorr	930	Nd:YAG, 266 nm, 10 Hz, power density at the target intentionally varied	SiO_2/Si	PLD of highly textured $BaFe_{12}O_{19}$ magnetoplumbite thin film on amorphous thermal oxide wafers (i.e. SiO_2/Si). opening an avenue for the monolithic integration of ferrites onto silicon-based platforms compatible with semiconductor systems and processing. The films and interfaces were characterized by vibration sample magnetometry, scanning and transmission electron microscopy, and energy-dispersive X-ray spectroscopy.	[308]

In 1995, Welch et al. [293] conducted a systematic study of the effects of PLD processing conditions on the crystal structure and magnetic properties of BaM films deposited on amorphous Vycor glass, silicon, and sapphire substrates. Film crystallization was observed in films deposited at 570 °C and improved with increases in substrate temperature. A working gas pressure of at least 40 mTorr of oxygen was necessary to completely oxidize the films to nominal stoichiometry. More recent studies further defined optimal processing conditions including oxygen pressures ~400 mTorr and substrate temperatures 900–925 °C.

Oliver et al. [294] and Shi et al. [311] demonstrated the growth of $BaSc_xFe_{12-x}O_{19}$ on *c*-plane sapphire under comparable substrate temperatures and oxygen pressures. Magnetometry measurements showed that all films had easy axes (*c*-axis) normal to the film plane, with a mean saturation magnetization value of 3.8 kG and a uniaxial anisotropy field value of ~10 kOe in accordance with Figures 10.32 (a–d). FMR linewidths varied from 300 to nearly 3000 Oe with larger values corresponding to thicker films suggest the randomness of the c-axis distribution in thick films is the main cause of the increase in FMR linewidth. Hu et al. [295], deposited films of Zr- and Zn-doped Ba-hexaferrite films on (111) MgO substrates. Zn and Zr divalent ions reduce magnetic anisotropy field by 46%, to 7.5 kOe. FMR linewidth was measured at 9.55 GHz to be 120 Oe. These dopants provide a cost-effective alternative to Sc-doped hexaferrites.

Most MMIC devices require thick ferrite films greater than 10 μm. In 2000, Oliver et al. [296] deposited oriented films of $BaFe_{12}O_{19}$ onto MgO (111) substrates. In contrast to epitaxial $BaFe_{12}O_{19}$ films grown on Al_2O_3 (001) substrates, these films experience an in-plane biaxial compressive stress and do not crack or delaminate to thicknesses of at least 28 μm. Further studies exploring low oxygen pressures revealed the ability to realize linewidths of 200 Oe at 52 GHz [312].

Song et al. processed among the highest quality barium hexaferrite films on sapphire substrates by PLD. The key to high-quality deposition is the use of a very low laser pulse rate during the initial stages of deposition and a very slow temperature ramp rate return to room temperature. The half-power FMR main mode linewidth for the best film was 27 Oe at 60.3 GHz matching that of bulk single-crystal BaM platelets. The spectra also exhibited distinct weaker resonances well above the main mode and a sequence of standing spin-wave resonances at lower fields [298].

(a)

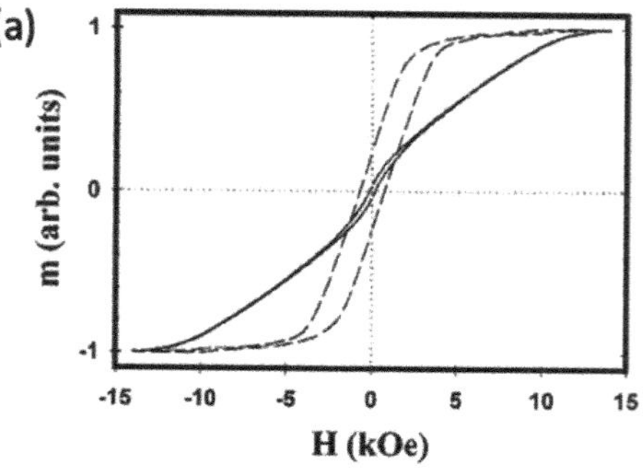

(c)

Thickness (μm)	x	FWHM (008)	H_A (Oe)	$4\pi M_s$ (G)	ΔH (Oe)	H_c (Oe)
5.2	0.6	(Poly)	9200	3850	1720–2740	517
2.3	0.6	0.857	9210	2950	1580–1930	715
2.5	0.3	0.944	12 940	3350	890–1100	631
2.2	0.4	…	11 520	3690	310	281
Bulk	…	…	11 250	3450	80	7.0

(b)

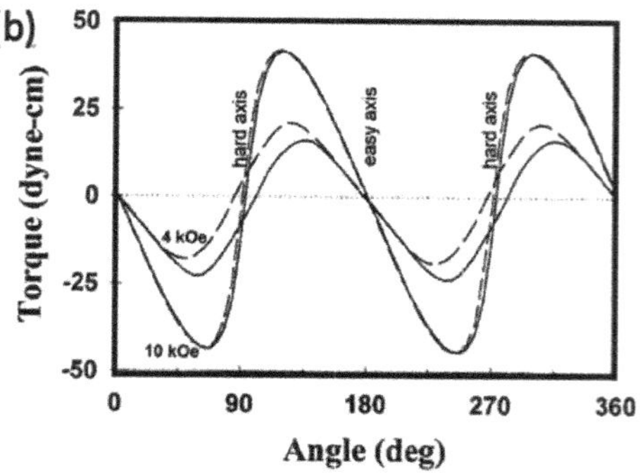

(d)

P (mTorr O_2)	x	$d_{(00n)}$ (Å)	(008) FWHM (deg)	$4\pi M_s$ (kG)	H_c (kOe)	H_A (kOe)
10	0.4	23.286	0.671	3.8	0.72	11.0
20	0.3	23.287	0.816	4.0	0.92	10.1
50	0.4	23.252	0.638	3.9	0.60	11.4
100	0.3	23.288	0.576	3.5	0.52	9.9

Figure 10.32 (a) Magnetization curves for the 10 mTorr film are shown when H lies along the film normal, i.e. easy axis (dashed lines), and in the film plane (solid lines). (b) Magnetic torque vs rotation angle results are presented for a 100 mTorr film for H=4 and H=10 kOe for both cw (solid lines) and ccw (dashed lines) rotations. The rotation is about an axis in the film plane. (c) Data adapted from ref. [294] where x is $BaSc_xFe_{12-x}O_{19}$ and P is the O_2 pressure during deposition, d is the atomic layer distance and (008) full width at half maximum (FWHM) were determined from θ–2θ X-ray diffraction measurements (not shown). Listed magnetic properties were extracted from vibrating sample ($4\pi M_S$, H_C) and torque magnetometry (H_A), and ferromagnetic resonance spectroscopy measurements (ΔH) (the data presented in (d) were taken from Shi et al. [311]).

As previously described as ATLAD, Geiler et al. placed Mn cations in specific interstitial sites of $BaFe_{12}O_{19}$ thin films as opposed to being distributed throughout the unit cell as in conventional PLD processing. As a result of site selection, the saturation magnetization increased 12–22%, and the Néel temperature increased by 40–60 K compared to bulk materials. The increases in the magnetic moment and Néel temperature was related to the differences in specific interstitial site occupation by Mn ions in the S-block of the unit cell. This was confirmed by experiments using EXAFS [299]. In a similar fashion, the magnetocrystalline anisotropy of hexagonal ferrites could be affected by channeling proper cations to the R-block of the unit cell. This approach may usher in the design and processing of a new generation of ferrite materials and also easily extend to other oxide, sulfide, or nitride compounds as well as other well-ordered materials.

Yu et al. performed an impressive systematic study of the impact of laser conditions upon the structural and magnetic properties of PLD BaM films grown on the *c*-plane sapphire (0001) substrates. The fluence of the Nd:YAG laser was systemically adjusted between 1 and 5.7 J/cm^2. The Fe-to-Ba ratio increased with laser fluence. Insufficient laser energies resulted in the deposition of secondary phases, such as barium monoferrite ($BaFe_2O_4$) and magnetite (Fe_3O_4). Laser fluences above 5 J/cm^2 promoted secondary sputtering of the films by plasma species that degraded film quality, structure, and magnetic properties. A laser fluence of 4.8 J/cm^2 was found to be optimal [301].

In a rare demonstration, Su et al. [302], successfully deposited U-type hexaferrite films ($Ba_4Ni_{1.4}Co_{0.6}Fe_{36}O_{60}$) on (0001) sapphire substrates with high retention (M_r/M_s ~92%) and low-FMR linewidths (200 Oe at X-band). Anisotropy field measurements indicated an H_A of 8 kOe and a saturation magnetization ($4\pi M_s$) of 3.6 kG. An optimal postdeposition annealing of films results in a strong (0 0 *n*) crystallographic texture and a high hysteresis loop squareness, leading to self-bias properties.

10.3.4.4.1 Integration of Hexaferrite Films with Semiconductor Heterostructures

Although microwave ICs (MICs) have been under development by RF engineers for several decades, the ability to produce high-crystal-quality ferrite films only now allows for first time realization of true MICs. Ferrite films are now available by industrial OEMs on select substrates of 100 mm diameter or greater, at high process yields, and with reproducible and excellent RF properties using industrial PLD reactors.

Depending upon specific needs of devices and circuits, MICs may require thin or thick films of garnet, spinel, and hexaferrite phases to be integrated with select substrates and heterostructures allowing for component and system integration.

Integrated System-on-Wafer (SoW) solutions come in different modalities, including (i) wafer-level heterogeneous integration of ferrite devices with active semiconductor systems and subsystems, (ii) integration of ferrite pucks/discs into wafer-etched pits allowing for their use as wafer-integrated substrates, and (iii) wafer bond of ferrites onto a wide range of substrates including semiconductors and WBG semiconductors.

Clearly, one path that allows for ease of SoW development is direct ferrite deposition onto industrial substrates and heterostructures that are compatible with high frequencies and high power at ever-decreasing form factors, leading to increased power densities and thermal management challenges. Specifically, such heterostructures include latest generation WBG semiconductors such as GaN, SiC, AlN, and others, including ultimately diamond.

Achieving realistic MICs entails first and foremost an understanding of the chemical, structural and magnetic nature of interfaces and interfacial regions. Also, as modern MICs demand an adaptation to higher frequencies, i.e. circa 2020, this is Ka-band, the choice of ferrite is beyond the garnet and spinel phases and based on magnetocrystalline energy arguments must focus on hexagonal ferrites.

Interfaces introduce chemical and structural roughness arising largely from mismatches in atomic lattices and thermal expansion coefficients. Such mismatches give rise to strain fields confined to the interface region that ultimately manifest as point, line, and three-dimensional defects. Such defects provide perturbations to magnetic, chemical and structural continua giving rise to magnonic-dependent energy dissipation mechanisms that lead to the deterioration of device performance as insertion loss, reflection loss, and other performance metrics.

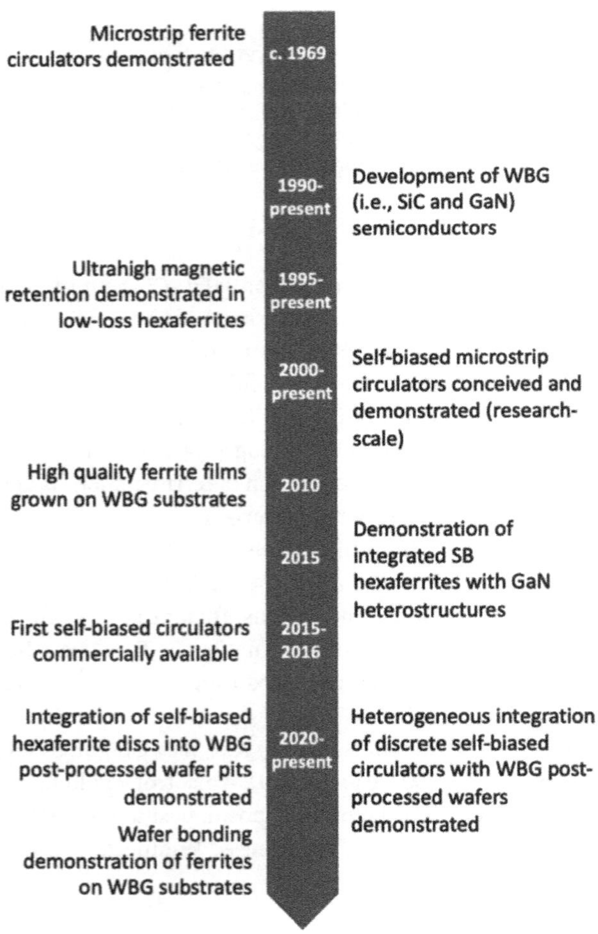

Figure 10.33 Timeline identifying key developments in the integration of ferrite materials with semiconductor substrate-enabling MIC-based system-on-wafer (SoW) solutions.

Figure 10.33 provides a timeline reflecting some of the key developments of WBG semiconductors, self-biased circulators, and ferrite integration with WBG semiconductor substrates and heterostructures, leading to ultimately the demonstration of integrated self-biased circulators with WBG-based systems-on-wafer. The foundation for this technology includes the development of microstrip RF devices circa 1969. As previously mentioned, not until the development of SiC and GaN WBGs, with Ka-band technologies driving the need for hexaferrites, was the table set integration.

Efforts to integrate hexaferrites and semiconductor platforms begin in earnest in the 1990s. WBG semiconductors, such as SiC or GaN, offered sufficient temperature stability and improved crystal structure and lattice matching that did not exist in prior semiconductor systems such as Si and GaAs. At that time, the semiconductor substrates of choice were silicon and gallium arsenide. Silicon and gallium arsenide failed in that the high temperatures required for processing of low-loss microwave ferrite materials resulted in either alloying with the silicon substrate or chemical disassociation of the GaAs substrate. In both cases, the films' microwave performances degrade.

Yet, a nucleation layer grown on the substrate surface was still required to grow crystallographically textured barium hexaferrite films. MgO and Pt layers were explored but offered limited thermal budgets. Today, there are new semiconductor material options, which include the WBG SiC and GaN. These materials are ideal candidates for hexaferrite–semiconductor integration because of their compatibility with high-power, high-frequency device operation. Furthermore, 6H-SiC has the same hexagonal crystal structure as BaM with a lattice mismatch of 4.38% along (0001) plane, providing a suitable substrate for hexaferrite films.

Chen et al. demonstrated barium hexaferrite films with $4\pi M_s$ of 4400 G and H_c of 220 Oe grown by PLD on single-crystal 6H silicon carbide substrates buffered with a 10-nm-thick layer of MBE-grown MgO (111) [303–305]. The BaM films were characterized as having basal planes parallel to the substrate plane. After annealing, the best films had an FMR derivative power linewidth of 96 Oe, which is of sufficiently high quality to be used in microwave device applications. These same authors demonstrated an important process capability in the high-rate reactive ion etching of column arrays having diameters of 1–4 μm with sharp vertical walls, etched at rates as high as 75 nm/min, using an optimized sulfur hexafluoride and methyl trifluoride (SF_6:CHF_3, 3:1) gas mixture. Lateral features as small as 43 nm were fabricated and imaged [313].

An essential advance was the demonstration of the growth of high-quality BaM on GaN/Al_2O_3 by PLD by Chen et al. [306]. XRD showed (0 0 $2n$) crystallographic alignment with pole figure analyses confirming epitaxial growth. HRTEM images revealed magnetoplumbite unit cells stacked with limited interfacial mixing. Saturation magnetization, $4\pi M_s$, was measured for as-grown films to be 4100 kG, with a perpendicular magnetic anisotropy field of 16.0 kOe. FMR linewidth, the peak-to-peak power absorption derivative at 53 GHz, was 86 Oe. These properties will prove enabling for the integration of low-loss BaM ferrite microwave passive devices with active semiconductor circuit elements in SoW architectures.

Yu et al. recently performed a systematic study of PLD BaM films on industrial compatible WBG semiconductor heterostructures suitable for operation in Ka-band circulators. These authors demonstrated successful PLD growth of BaM films on WBG semiconductor heterostructures, i.e. BaM/MgO/AlN/SiC(6H), that showed device quality performance in structure, epitaxy, and magnetic properties. Film properties include bulk-like values of the magnetic anisotropy field, H_a~16.5 kOe, and saturation magnetization, $4\pi M_s$ ~4.2 kG. Only heterostructures where SiC substrates having Si-polar surfaces showed superior RF properties. Figure 10.34a is illustrative in that it depicts a cross-section TEM image of a BaM/MgO/AlN/SiC(X) heterostructure having five distinct regions: (i) the AlN capping layer (50 nm) on the Si-polar SiC crystal substrate, (ii) an amorphous-like diffusion region of Al_2O_3 (45 nm) that forms between AlN and MgO, (iii) a crystalline nucleation layer of MgO (32 nm) with (111) crystallographic orientation, (iv) a diffusion region between the MgO and BaM films forming the spinel $MgFe_2O_4$ (40 nm), and (v) a highly crystalline thick film of BaM. Panel (b) is an expanded HR-TEM view of the BaM film structure showing a near-perfect arrangement of parallel atomic planes defining the length of M-blocks to be 1.186 nm. From these data, BaM lattice parameters of $a = b = 0.601$ nm and $c = 2.366$ nm were deduced. Based on these calculations, a and c lattice parameters experience a 2% dilation. However, these values reflect the atomic structure nearest the interface region where strain is most pronounced. In this region, the transition from the $MgFe_2O_4$ spinel layer to the hexaferrite phase includes a large mismatch in lattice parameters, leading to the formation of stacking faults. The accumulation of copious stacking faults in the immediate region above the interface of the spinel interdiffusion layer and the BaM film is clearly visible in panel (a). The stacking fault region extends to about 30 nm above the interface. Beyond this region, stacking faults are no longer detected in large numbers. These atomic characteristics for BaM film growth suggest that the in-plane strain is relieved as the BaM grows thicker, a common observation in heteroepitaxial growth. Alternatively, the out-of-plane strain remains unchanged. This behavior has been attributed by others to the presence of oxygen deficiencies in the BaM film. Panels (c)–(e) are hysteresis loops for heterostructures whose comparison proves illustrative in relating sample structure and chemistry to magnetic properties. In panel (c) are the hysteresis loops for the BaM/MgO/AlN/SiC(X) sample that show clear magnetic anisotropy with an anisotropy field (H_a) of ≈16.5 kOe and a saturation magnetization ($4\pi M_s$) of 4.2 kOe. For the BaM/MgO/SiC(X) sample (see Figure 10.34b), loops collected for both in-plane and out-of-plane directions nearly overlap, indicating very little magnetic anisotropy. Finally, panel (c) illustrates hysteresis loops collected from the BaM/MgO/GaN/sapphire(X) heterostructure in which the saturation magnetization is near the bulk value. However, electron microscopy reveals a nonuniform film surface with grain outgrowth that induces surface pinning of domain walls resulting in a high coercivity (H_c) of >2 kOe. The high coercivity positively impacts squareness, Sq = M_r/M_s, which reaches 0.70 but degrades ΔH_{FMR} by >60% over the BaM/MgO/GaN/SiC(X) heterostructure value of 340 Oe. These results define a path for integration of magneto-dielectric materials on WBG heterostructures for self-biased devices essential to implementing millimeter-wave *massive-MIMO* array and the enormous potential it offers to 5G technologies [307].

Kulik et al. produced high-quality BaM films on thermal oxide wafers (i.e. SiO_2/Si) – a low-cost heterostructure used in CMOS, MMIC and MEMS applications. Results show that crystallographically textured BaM films deposited on SiO_2/Si

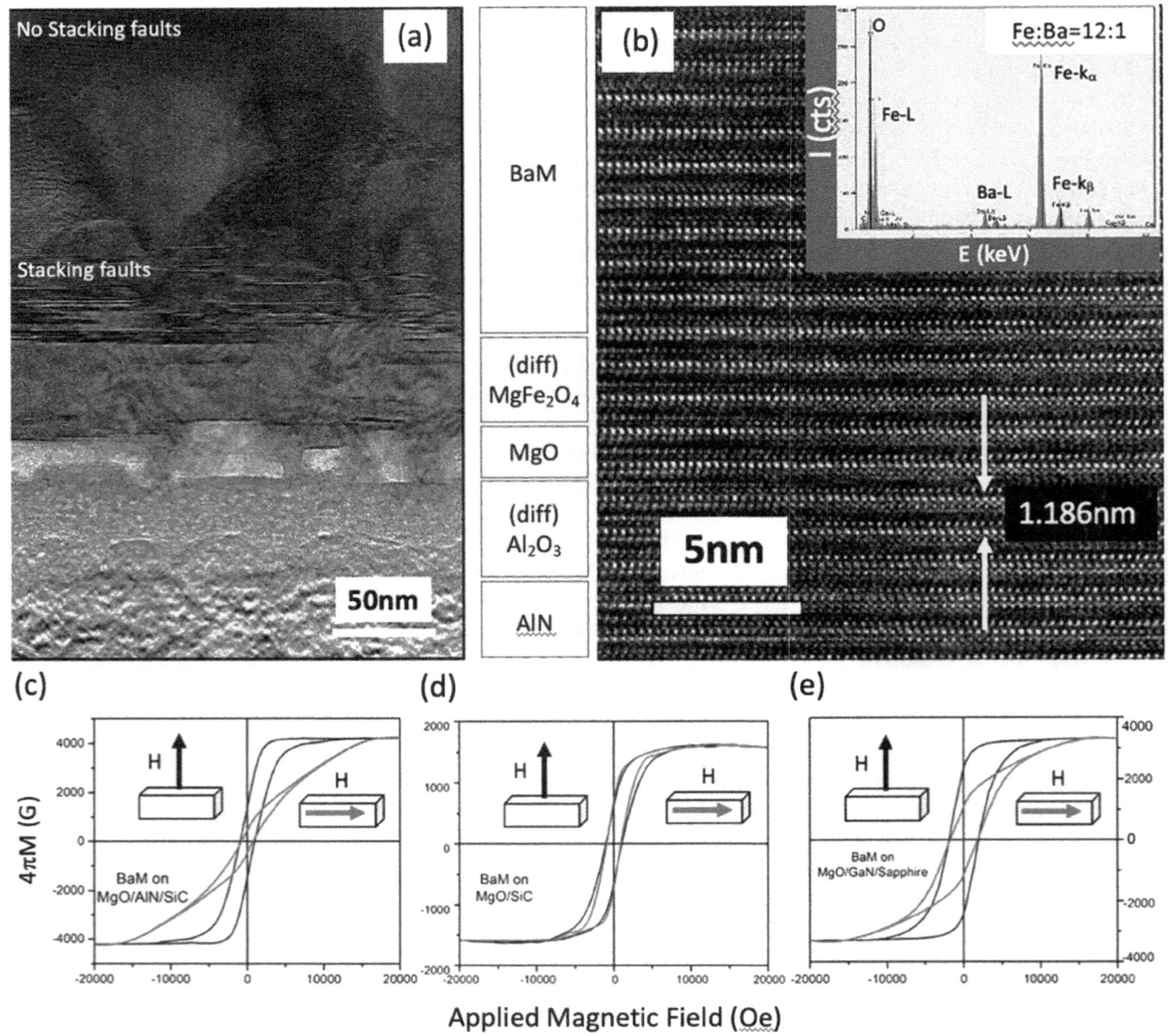

Figure 10.34 (a) Cross-section of transmission electron microscopy image collected in brightfield mode revealing distinct regions of the BaM/MgO/AlN/SiC(X) heterostructure as denoted on the right-hand side. Stacking faults are clearly visible in the first 30 nm of BaM film growth on the $MgFe_2O_4$ diffusion region. (Note: "diff" indicates a diffusion region.) (b) Expanded view of the BaM film cross-section revealing parallel stacking of M-blocks with individual dimension 1.186 nm. Inset to this panel is an energy-dispersive X-ray spectrum confirming the nominal stoichiometric ratio of Fe:Ba = 12:1. Hysteresis loops, as sample magnetization ($4\pi M$, G) versus applied magnetic field (H, Oe), where the applied field is scanned from 0 Oe to ±20 kOe (beyond the magnetic saturation field of ≈17 kOe), are presented for three heterostructure samples. Inset graphics define the orientation of the applied magnetic field relative to the substrate axes. (c) Hysteresis loops collected from the BaM/MgO/AlN/SiC(X) sample, (d) similar data collected for a heterostructure without an AlN capping layer, and (e) loops collected from BaM/MgO/GaN/sapphire(X) (adapted from [293]).

possess similar magnetic and structural properties to those deposited on sapphire despite the fact the SiO_2 layer is amorphous and does not offer atomic lattice matching. Furthermore, an investigation of deposition time and film thickness upon texturing and magnetic properties showed laser fluence and deposition time allowed for enhanced density and texturing. Additionally, to better understand the hexaferrite film growth, TEM and EDXS were conducted and revealed an extensive ~100 nm interdiffusion layer between the $BaFe_{12}O_{19}$ and the SiO_2/Si substrate [308].

10.3.5 Processing Dynamics of PLD of Ferrites

High-quality hexaferrite films can be epitaxially grown on sapphire, MgO, and WBG IV. substrates including GaN and SiC. Crystalline texture, in this case, is determined by atomic lattice matching to the substrate. Highly crystalline and textured hexaferrite films can also be grown on amorphous substrates such as fused quartz and SiO_2/Si. The easy magnetic axis, i.e. the crystallographic *c*-axis, is aligned normal to the substrate plane. As compared to other ferrite systems, PLD hexaferrites generally require higher substrate temperatures, oxygen pressures, and laser fluences in order to attain congruent chemical transfer and obtain superior RF properties.

10.3.5.1 Laser Fluence

Laser fluence, in this context, refers to the energy density incident upon the target surface during PLD growth. The relationships between laser fluence and processing are presented along the top row of Figure 10.35. Much of what we know of the role of the impact of laser fluence upon PLD processing is assembled piecemeal from 10 to 100 s of studies over the past three decades of published results. However, the 2020 publication by Yu et al. [301] indeed focused on this singular relationship investigating stoichiometry, phase, and texture evolution of PLD-grown BaM films as a function of laser process parameters. First, as the laser fluence increases, the ablation of all target constituents, including high-*Z* elements, increases as does the average energy of all plasma species. With the increase in plasma energy also comes an increase in deposition rate and the detrimental impact of secondary sputtering from the growing film.

Film structure and chemistry are affected positively by congruent compositional transfer. With the corresponding increase in specie energy, the film improves in density, crystallinity, and texture. Of course, continued increases in fluence overcompensates this effect and ultimately leads to deleterious properties.

Laser fluence increases congruent transfer, improves film density, crystallinity, and texture, improving the overall film quality. Ultimately, films are judged by their magnetic properties. In response to improved film density, crystallinity, and texture, the saturation magnetization approaches bulk values, the coercivity decreases, and as expected as does RF losses and retention. Although these trends are indisputable, they often experience optimization values.

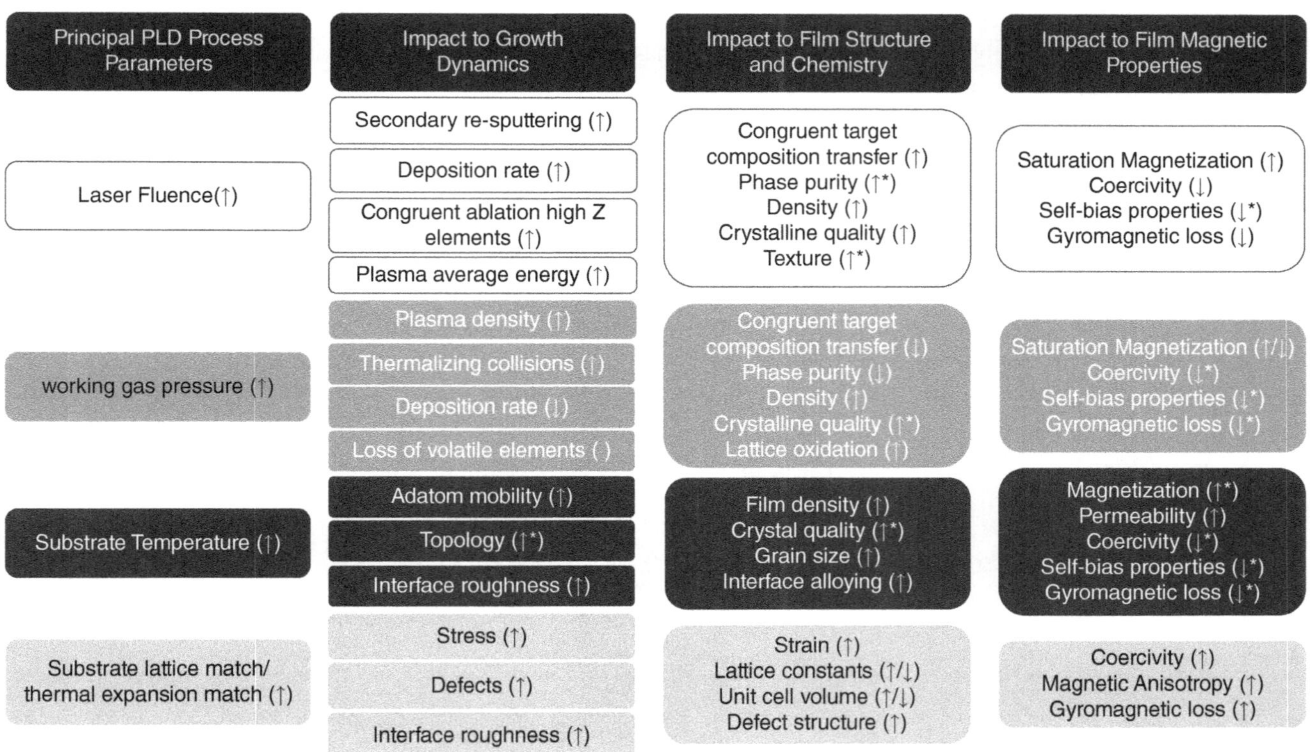

Figure 10.35 Interrelationships between PLD processing parameters and growth dynamics, film chemistry and structure, and performance. (* denotes that upon excessive increase trend may reverse.)

10.3.5.2 Working Gas Pressure

Working gas pressure is a common variable in nearly every PLD paper on ferrite films. When the working gas pressure increases, so does the plasma density, which increases the frequency of thermalizing collisions. As a result, the deposition rate decreases, and highly volatile elements are lost to the plasma and never make it to the substrate. As the working gas pressure increases, the film quality in all measure increases, including the lattice oxidation. As such, the magnetic properties of the film improve in magnetization with lower coercivity, gyromagnetic loss, and reduced self-biased.

10.3.5.3 Substrate Temperature

The manipulation of substrate temperature serves to adjust the kinetic energy of arriving adatoms to the surface of the growing film. The increase in adatom energy leads to additional hopping frequency and hence a lower density of defects and an enhanced topology. This also leads to a smoother chemical and structure interface until the adatom energy increases to the point of enhanced interface diffusion.

10.3.5.4 Substrate Lattice/Thermal Expansion Matching

Substrate lattice and thermal expansion coefficient matching is very important to the templating of the growing film in the manifestation of stress and strain. As stress is a field and strain is a displacement, strain is a more important factor in determining defects in the film and has more impact on magnetic properties. Alternatively, defects directly give rise to coercivity, anisotropy fields, and gyromagnetic loss.

The correlations between film properties and device performance are presented in Table 10.9. In this table, correlations linking Figure 10.35 and RF devices, such as isolators and inductors, M_s relates to isolation (↑) and resonance frequency (↓). Insertion loss is positively related to gyromagnetic loss, and magnetic anisotropy is positively related to operational frequency. Those experiencing negative correlations include electrical resistivity with eddy current losses, core loss with inductor efficiency, and saturation magnetization with resonance frequency.

10.3.6 Outlook: PLD of Ferrite Films

PLD is a highly versatile tool for processing high-crystal-quality complex oxides, nitrides, carbides, and other nonmetallic films with a high degree of reproducibility.

As yet, PLD remains largely a research instrument that has not yet fully transitioned to industrial scale production of engineered materials and components.

Table 10.9 Correlations between film properties and RF device performance.

Film properties	Correlation	RF device performance
Saturation magnetization (M_s)	↑	Isolation[a]
Saturation magnetization (M_s)	↓	Operating frequency[a,d]
Remanence magnetization (M_r)	↑	Self-bias[b]
Gyromagnetic loss (ΔH_{FMR})	↑	Insertion loss[a]
Magnetic anisotropy field (H_A)	↑	Operating frequency[a]
Permittivity (ε), tanδ_ε	↑	Bandwidth[a]
Coercivity (H_c)	↑	Hysteretic losses[c]
Permeability (μ), tanδ_μ	↑	Inductance[c] energy storage efficiency[c]
Electrical resistivity (ρ)	↓	Conduction losses, eddy current losses[c]
Core loss	↓	Inductor efficiency[c]
Snoek's product/limit	↑	Cutoff frequency[c], inductor efficiency[c]

Note:

a) Device being considered is an isolator/circulator.
b) Material property valuable for monolithic integrated circuits.
c) Device being considered is an inductor/convertor.
d) For hexaferrites only.

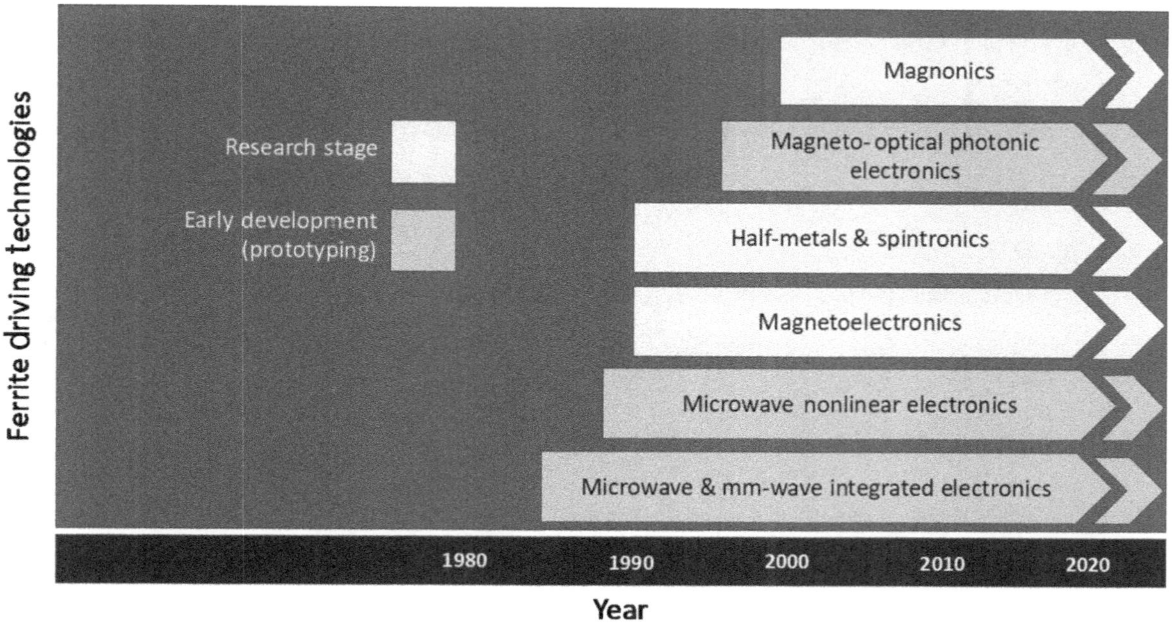

Figure 10.36 Ferrite driving technologies over time.

However, flexibility in working gas pressure, substrate temperature, and laser fluence allows for a wide range of crystalline structures, compositions, and thicknesses. Manufacturers are now offering high-throughput systems equipped with *in situ* real-time growth characterization tools, load-lock systems, and high-power lasers that allow for 100 mm diameter (or larger) wafers. Clearly, this is a precursor to the next step in industrial processing. Emerging ferrite-based technologies that are driving the move toward industrial processing include those that rely upon high-crystal-quality and ultralow magnetic loss materials such as spintronics, magnonics, and microwave- and millimeter-wave electronics, among others.

Figure 10.36 depicts the present and emerging technologies that are dependent upon PLD ferrite films. They include MO electronics, microwave nonlinear electronics, and microwave and millimeter-wave electronics. These provide the foundation of 5G and 6G communication technologies that are expanding hand-held electronics to Ka-band and eventually to quasi-optical bands. Emerging electronics, such as magnetoelectronic, spintronics, and magnonics await. These new technologies are in fact far ahead of our times and imaginations. They include new modalities in logic and depend upon new quantized states of charge carriers providing novel logic systems that provide unimagined potential for computational power.

10.4 Sputter-deposited Ferrite Films

10.4.1 Basic Principles of Sputter Deposition

Sputter deposition is a key processing tool used widely in semiconductor, photovoltaics and magnetic storage industries, among many others, for the deposition of metals, oxides, and semiconductor thin films and heterostructures. Sputtering requires the stabilization of a plasma, typically consisting of an ionized noble gas such as Ar^+, within a vacuum chamber between an electrically biased cathode (i.e. target) and anode (i.e. substrate). Figures 10.37a and 10.37b depict the interaction between the plasma and target giving rise to a "sputtered" flux of neutral atoms. Figure 10.38 is a simplified schematic of a sputtering chamber and its principal components. As shown, free electrons flow from the target in the plasma, collide with neutral Ar gas atoms (i.e. Ar^o), stripping electrons off the atoms and creating Ar^+ ions. Although the plasma consists of a wide variety of neutral and charged particles, singly charged Ar^+ ions are its principal constituent (see Figures 10.37a and 10.38). Ar^+ ions are accelerated to the negatively charged target (i.e. cathode). Upon collision, an exchange of kinetic energy leads to the ejection of neutral atomic particles from the target material (i.e. T^+ and T^o of Figures 10.37a and 10.37b, and 10.38). Collectively, this is referred to as "sputtering" (see Figure 10.3b). The ejected neutrals contain several eVs of

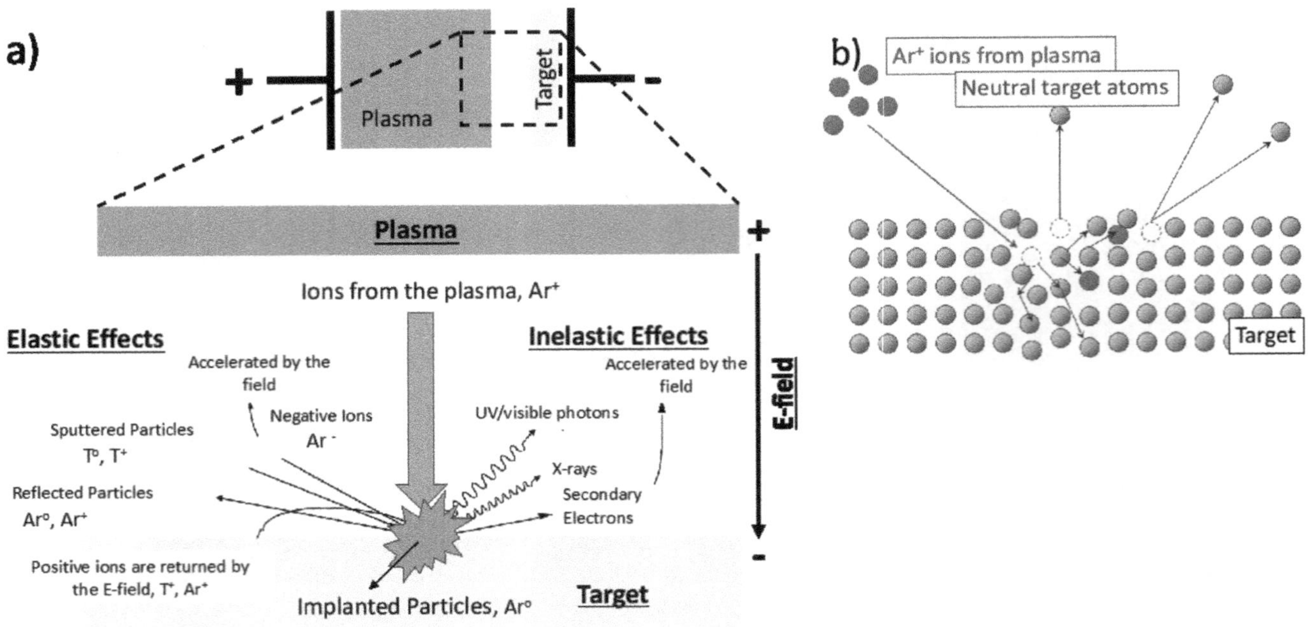

Figure 10.37 a) Schematic depiction of plasma-target interaction. b) Expanded schematic view of Ar ions impinging upon a target and creating a neutral flux of target atoms (adapted from Refs. [314, 315]).

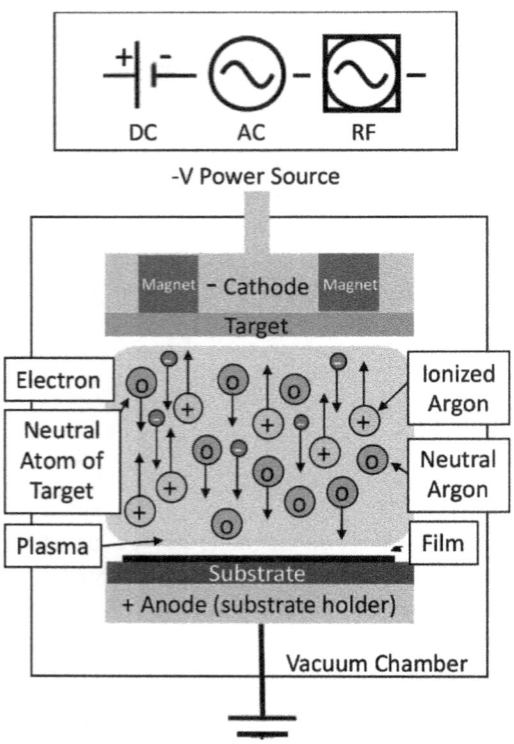

Figure 10.38 Simplified schematic of a sputtering chamber and essential components.

kinetic energy. The sputtered particles traverse the vacuum and deposit as a film of the target composition upon the substrate. Their kinetic energy is converted to adatom mobility and translate to atomic jumps on the surface of the growing film.

Several variants of sputtering are in wide use, including ion beam and ion-assisted sputtering, reactive sputtering, and magnetron sputtering.

Ion beam and ion-assist sputtering involves methods to increase the kinetic energy of the arriving sputtered species at the substrate thus allowing for greater adatom mobility and higher crystal quality of the deposited films (i.e. fewer crystal defects). Alternatively, reactive sputtering involves the growth of a sputtered film in a reactive gas environment that allows for the growth of high-quality oxide, nitride or carbide films and coatings. The plasma consists of charged particles, as such it responds sensitively to magnetic fields. Magnetron sputtering uses magnets placed behind the target (i.e. cathode, see Figures 10.37 and 10.38) to confine a higher concentration of electrons impinging upon the target, allowing for higher deposition rates to be realized [316]. This is very important for industrial processing where thicker coatings and higher throughputs are essential.

The most common magnetron sputter targets are circular (see Figure 10.39) or rectangular in shape. Figure 10.39b is a photograph of the original circular magnetron source called the "Planar Plasmatron" [318] that provided 5 kW of discharge power for a 500 V discharge voltage resulting in a sputter rate for a Cu target of 0.7 g/min.

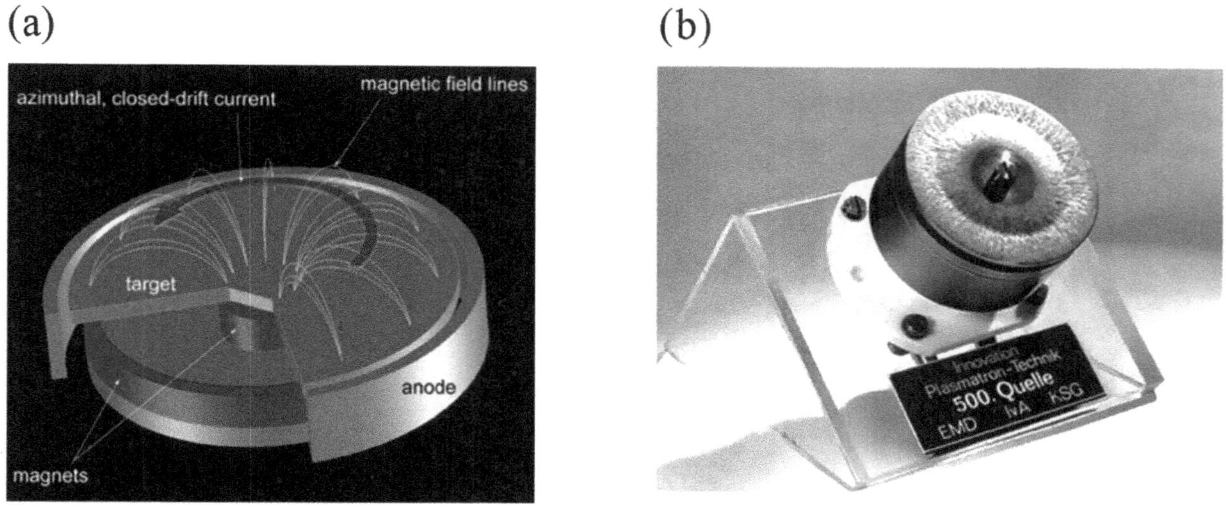

Figure 10.39 a) Illustration of a planar magnetron similar to the one used to produce the current examples in this publication [317]. b) The first "Planar Plasmatron" PPS-5 (reproduced from [318]).

For its day, this was quite an advance in technology. Rectangular magnetrons are commonly used in large scale "in-line" industrial systems where substrates often move past targets on carriers for high-throughput processing. Circular target magnetrons are more commonly found in research scale systems whereas the rectangular geometries are common to industrial process environments.

DC sputtering is used extensively in the deposition of metals. It is used commonly for gold sputter coatings of jewelry, watches and other decorative finishes, non-reflective coatings on glass and optical components, as well as for metalizing plastics. It is also commonly employed in semiconductor processing. During DC sputtering of non-conducting materials, a space charge builds near the target surface that suppresses sputtering. To overcome this limitation, RF sputtering has been developed. RF sputtering alternates the electrical charge of the drive field so as to prevent a charge buildup around the target that allows the generation of a high-density plasma that results in a high degree of ionization of the coating material in the plasma. A principal value of RF sputtering is its ability to efficiently sputter metals or insulating target materials.

10.4.1.1 Commercial Sputtering Systems

Sputter deposition systems come in a great variety of designs to maximize utility and cost efficiency depending upon user needs. Needs vary from basic research, which such as the AJA International Inc., Model ATC 2200 UHV – 9 Position Cassette Auto-Loading System (see Figure 10.40a), provides flexibility with multiple target carrousels and substrate load locks. Such systems can be readily designed for simplicity and low cost for new university professors on limited budgets, or research labs of major companies whose budgets can accommodate complexity and advanced multifunctionality. Other systems can be designed to include different PVD modalities such as MBE, ion-plating, ion beam assist deposition, etc. or incorporate *in situ* characterization tools [319]. Alternatively, for high-throughput manufacturing for thin film photovoltaics, semiconductor, or disk drive memory products (among others), instruments such as the Singulus Technologies production lines have been designed and implemented (see Figure 10.41). The Singulus Technologies systems, promoted under the brand name Vistaris, was developed to meet the requirements of the photovoltaics' industry [319].

10.4.1.2 Sputtering: A Brief History

The term PVD was originally termed by Powell et al. in 1966 [320]. However, PVD processes were invented earlier and associated with the development of vacuum technology. The first person to stabilize a glow discharge was Faraday [321]. However, it was Grove in 1852 [322], while studying glow discharges, who first realized what we now know to be

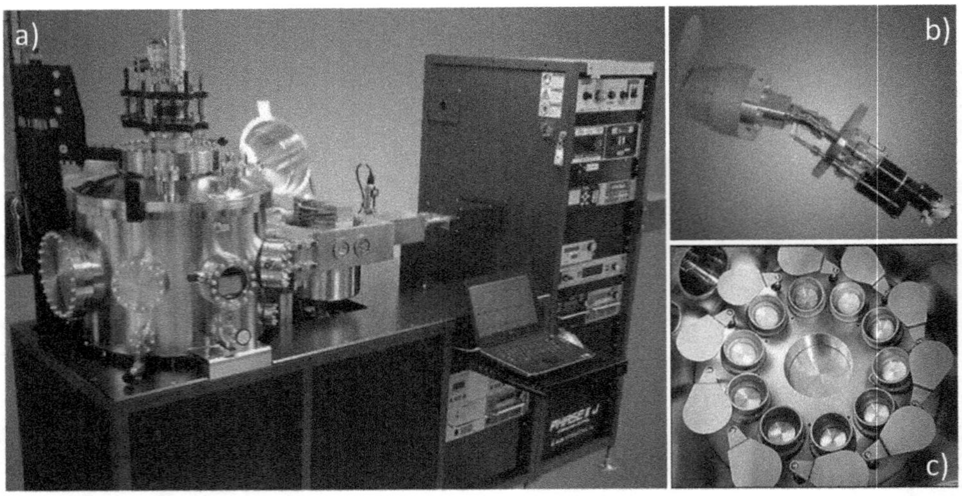

Figure 10.40 a) AJA International Inc. Model ATC 2200 UHV – 9 Position Cassette Auto-Loading System. b) Angled sputter source adjustable from outside the chamber for maximum flexibility. c) Example of an AJA multisource/shutter assembly that provides a variety of target choices without breaking vacuum to replace targets (used with permission from AJA International Inc.).

Figure 10.41 In-line solar cell process line by Vistaris Sputtering Systems [319].

"sputtering." Grove used a tip of a wire as the target (cathode) and sputtered a thin film onto a highly polished silver surface (anode) in close proximity to the wire *in vacuo*. In 1877, Wright used an "electrical deposition apparatus" to create mirrors in a similar manner [323]. In a patent conflict raised by Edison, Wright's deposition was determined by the US Patent and Trade Office to be "arc evaporation" rather than "sputtering." Edison successfully argued that his invention was a continuous arc whereas Wright's was a pulsed arc [324]. Edison is therefore the first person to be credited with the commercial use of sputtering.

In the late 1930s, Penning developed a method to confine electrons near the target surface using a combination of electric and magnetic fields. As such, the first magnetron was created and credited to Penning. This combination of electric and magnetic fields increased the ionization of the plasma near the target surface and was named the "Penning Discharge."

Penning used this to sputter from the inside of a cylinder [325]. This was an important development in the history of sputtering and is the foundation of many modern-day magnetron designs. Such a combination of electric and magnetic fields allowed sputtering to be performed at lower pressures and lower voltages and provided the higher deposition rates required for commercialization. Variations of the Penning magnetron have subsequently been developed, such as the post cathode magnetron by Thornton and Penfold (c. 1970s) [326] and the Peeples, Cuthrell and Mattox's magnetron (c. 1980s) [327]. In 1968, Clarke developed a sputtering source using a magnetic tunnel on the inside of a cylindrical surface. This source became known as the "sputter gun" [328]. Various magnetron configurations, including the planar magnetron, were patented by Corbani [329]. Chapin also developed a planar magnetron source in 1979 and is credited with being the inventor of the planar magnetron source [330]. Major advantages of magnetron sputtering sources include high-rate, large-area, long-life, low-temperature sources that are capable of operating at lower gas pressure and offer higher sputtering rates than nonmagnetic sputtering sources. With these superior characteristics, magnetron sputtering became the most widespread PVD coating technique.

The term reactive sputtering was introduced by Veszi in 1953 while working to develop CdO solar cells [331]. Reactive sputter deposition of tantalum nitride as thin film resistors was an early application [332]. However, it was not until the mid-1970s that reactively sputter-deposited hard coatings on tools [333] began to be developed and they became commercially available.

10.4.1.3 Structure Zone Diagrams

We next focus on the dynamics of the growing film at the surface of the substrate in order to express the growth condition–structure relationships as a structure zone diagram (SZD). An SZD is a convenient way of illustrating microstructural features of relatively thick (i.e. >100 nm) polycrystalline films with respect to sputter process conditions. Although process parameters affecting film growth are many, and their interrelationships complex, we reproduce here an SZD that employs a generalized homologous temperature (T*), the normalized kinetic energy flux (E*), and the net film thickness (t*) to correlate and communicate the effective film growth dynamics.

In 1974, Thornton published an SZD [334] for sputtered films patterned after the Movchan and Demchishin diagram for evaporated coatings [335]. This diagram illustrates the relationship between the coating morphology, the deposition temperature and the pressure in the sputtering chamber. Of course, the sputtering pressure determines the flux and energy of the high-energy neutrals from the sputtering cathode, so the diagram reflects the degree that the depositing material is bombarded by energetic particles during deposition. In 1984, Messier, Giri, and Roy further refined the SZD [336]. The SZD of Figure 10.42, was developed by Anders [337] and examines the microstructure of growing films deposited by magnetron sputtering.

Hultman and Sundgren [338] and Barna and Adamik [339] provided great insight into the microstructure of films in terms of flux energy and substrate temperature. It was Thornton who introduced an SZD for sputtered films in terms of homologous temperature and working gas pressure [333]. Thornton's SZD, and variants thereof, have been reproduced elsewhere [340–342].

Anders proposed changes to the Thornton SZD, including: (I) replace the linear T_h axis with a generalized temperature, T*, which includes the homologous temperature plus a temperature shift caused by the energy of particles arriving at the substrate surface; (II) replace the linear pressure axis with a logarithmic axis for normalized energy, E*, describing heating effects caused by the kinetic energy of bombarding particles; and, (III) replace the z-axis with a net film thickness, t*, which allows one to maintain an illustration of film structure while indicating thickness reduction by densification and sputtering (and resputtering) [343].

For the remainder of this section, we focus on the use of sputtering and sputtering variants in the processing of ferrite films and their applications. We include garnet, spinel and hexaferrite films.

10.4.2 Sputtering of Garnet Films

In 1967, Bobeck led a team at Bell Labs that developed bubble memory. This memory modality was similar to the moving-domain twistor concept [344], but used a single block of magnetic material. The employment of orthoferrites, and later the much-preferred garnets, allowed the writing of magnetized patches that, when a magnetic field was applied to the entire material, the patches would shrink into microscopic *bubbles* (see Figure 10.43). These bubbles were much smaller than magnetic domains, the written bit of the day, and therefore supported higher areal storage densities.

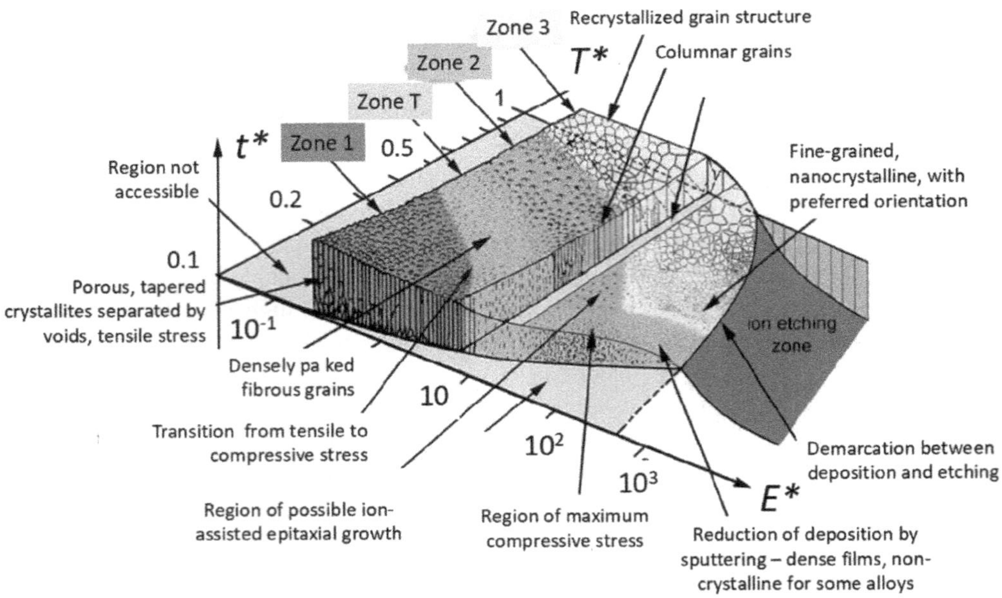

Figure 10.42 The structure zone model of thin film morphology as determined by deposition process variables (adapted from [336]).

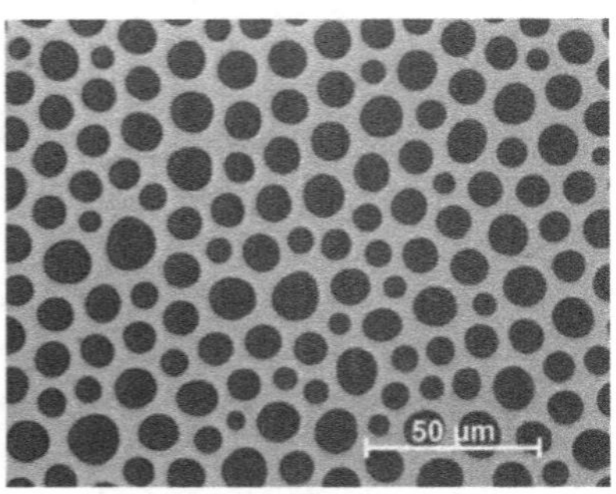

Figure 10.43 Bubble domains visualized by using CMOS-MagView [345].

The bubbles, and their arrangements, remain unchanged when power is disconnected and therefore bubble memory is an effective nonvolatile memory. Other advantages of bubble memory devices include no moving parts, e.g. fields that drive the bubbles along the vector are generated electronically, whereas media of that time, i.e. tapes and disk drives, required mechanical movement of read-write coils.

Among various garnet film growth techniques, sputter deposition has had the highest repeatability and excellent uniformity and as such has become very useful for applications ranging from magneto-optics to spintronics. Table 10.10 presents a comprehensive historical account of the evolution of sputtered garnets in which the year, operating pressure, substrate temperature, growth rate, substrate and key findings are recorded. The first garnet films were sputtered by Sawatzky and Kay from 1967 to 1971 [346–349]. The methods included different variants of sputter deposition applied to Gd-IG (i.e. gadolinium-iron-garnet) films on YAG and GGG (gadolinium-gallium-garnet) substrates motivated largely by use as bubble memory recording media. Cuomo et al. [350] soon followed with the deposition of gadolinium–gallium-doped GG-YIG on GGG substrates. Compared with pure YIG films, bismuth-substituted Bi-YIG films find greater utility in nonreciprocal devices as Bi enhances the Faraday effect and decreases optical losses [351–353]. In the 1980s and 1990s, YIG films were further investigated for their large MO activity and ultralow propagation losses, the latter affecting device and system efficiency.

Motivated by MO properties and technologies, in 1985, Gomi et al. [351] successfully prepared Bi-substituted garnet films on glass substrates. Such substrates are naturally more attractive for industrial application from a supply chain and cost standpoint. Film thicknesses were around 1~1.5 μm. In addition to Bi:YIG, Ce:YIG films were studied during that period [354–356] as they offer intrinsic benefits to advancing magneto-photonic crystal-based devices. For example, Shintaku et al. reported low propagation losses of 9.7 dB/cm in the TE mode (and 5.8 dB/cm in the TM mode) at $\lambda = 1.55$ μm [355], while Gomi et al. enhanced the Faraday effect in both the IR and UV regions [356].

In 2000, Bi-YIG films were grown by reactive ion beam sputtering on $Gd_3ScGa_5O_{12}$ (GSGG) to achieve larger MO effects [357]. In this variant of RF sputtering, a less expensive all metal target is employed and sputtered into a working gas enriched with a reactive gas component such as oxygen to transform metallic neutrals to oxide clusters as they impinge upon the surface resulting in high-quality pure phase oxide films. At the beginning of 21st century, researchers studied

Table 10.10 Sputter deposition conditions and film properties from early studies of garnet systems.

Year	Oxygen pressure (mT)	T_{dep} (°C)	Growth rate (Å/s)	Substrates	Key results	Ref.
1968	20 mT O_2	0–500	8.5 nm/min	NaCl	Gd-IG films grown by three approaches: RF reactive gas sputtering, DC reactive gas sputtering, DC inert gas sputtering.	[347]
1969	20 mT O_2	0–>700	unknown	YAG, glass	Gd-IG films grown by parallel-plate-two electrode-RF-sputtering system.	[348]
1971	unknown	40–>700	unknown	GGG, YAG, MgO,	Magneto-optic memory devices explored. Studies lay the foundation for bubble memory technology.	[349]
1972	25 mT O_2	450	5.6 nm/min	GGG	Epitaxial films of gadolinium-gallium-substituted YIG (Gd: Ga: YIG) films (i.e. $Gd_{0.3}Y_{2.7}Fe_{3.8}Ga_{1.2}O_{12}$), bubble domains observed	[350]
1985	Ar:O_2 = 9:1, total 52 mT	RT-500	6 nm/min	Glass and fused quartz	$Bi_2YFe_{3.8}Al_{1.2}O_{12}$ and $Bi_2GdFe_{3.8}Al_{1.2}O_{12}$ films were RF-sputtered and show large FR, i.e. 1.5×10^4 deg/cm for Bi,Al:YIG film and 1×10^4 deg/cm for Bi,Al:Gd-IG film at $\lambda = 633$ nm, owing to the high Bi substitution.	[351]
2004	0.3 mT	300–700	unknown	GGG, Si	Dual ion beam assisted sputtering of Bi-YIG films followed by RTA.	[352]
1988	(Ar) (P unknown)	480–530	6.0 nm/min	GGG, Nd-GG	Ce:YIG epitaxial films deposited by RF diode sputtering.	[353]
2001	3 mT (3.3–30% O_2:Ar)	RT-500	unknown	GGG, Si, soda lime glass	YIG, Ce-YIG, RF magnetron.	[354]
1997	6 mT; Ar+O_2 (O_2: 0.28 vol.-%)	550	2.1 nm/min	G(Sc)GG	YIG and Ce:YIG films deposited by RF diode sputtering. The lowest reported (to date) propagation losses of 9.7 dB/cm in the TE mode and 5.8 dB/cm in the TM mode obtained for film thicknesses of 0.5 um at $\lambda = 1.55$ μm.	[355]
1991	50 mT (H_2/Ar = 0–5%)	500	6.0 nm/min	GGG, Nd-GG	Ce:YIG, RF diode sputtering, the Ce substitution prominently enhanced the Faraday effect not only in the IR region at $h\nu = 1.4$ eV but also in the UV region at $h\nu = 3.1$ eV.	[356]
2000	0.11 mT	480	2.1 nm/min	$Gd_3ScGa_5O_{12}$ (GSGG)	BIG, reactive ion beam sputtering (RIBS) techniques. Saturation magnetization $4\pi M_s$ and FR at 635 nm were found to be 1400 G and 27.8 deg/mm in PLD-BIG, and 1200 G and 26.9 deg/mm in RIBS-BIG	[357]
2001	3–5 mT Ar+4.7%O_2	700–750	5.0 nm/min	GGG	YIG, magnetron sputter amorphous followed by annealing (SPE)	[358]
2005	O_2/Ar+O_2 =0, 1, 5, 10% DGF	RT	10 nm/min	TOS, GGG	YIG, RF magnetron, high-quality films with H_c~1 Oe, FMR ΔH~4 Oe and without film cracking.	[359]
2007	Ar/O_2, 1:3–3:1, 5.2 mT	RT	2.0 nm/min	Si	Bi-YIG, RF magnetron sputtering, followed by RTA at 600–750 °C.	[360]
2005	Ar/O_2 Differential pressure method	RT	4.78–10.85 nm/min	MgO, quartz	YIG, reactive radio-frequency sputtering using a partial pressure differential to increase sputtering rates. O_2 gas was supplied directly to the substrates using a ring configuration. Films demonstrated refractive indices of ~2.1 and out-of-plane FRs up to 0.2°/μm at 633 nm.	[361]

(Continued)

Table 10.10 (Continued)

Year	Oxygen pressure (mT)	T_{dep} (°C)	Growth rate (Å/s)	Substrates	Key results	Ref.
2004	22.5 mT	RT	unknown	GGG, quartz, zirconi	YIG, RF magnetron. A maximum FR is observed to be 1900°/cm at the wavelength of 594 nm for a YIG thin film formed on quartz substrate	[362]
2004	10 mT, 0–20% O_2+Ar	700–750	unknown	GGG	YIG, sputtering followed by SPE	[361]
2013	11.5 mTorr with 0.15% O_2/Ar	750	0.33 nm/min	GGG	YIG with Pt and W underlayers using off-axis sputtering. FMR spin pumping. $\Delta H = 7.4$ Oe and bulk-like magnetization.	[363]
2014	Ar, 10 mT	RT	unknown	GGG	YIG, RF sputtering. FMR $\Delta H \sim 7$ Oe, $\alpha = 1 \times 10^{-3}$, bulk-like magnetization.	[35]
2014	Ar; 10–40 mT	RT	unknown	GGG	YIG, RF sputtering, $\alpha = 8.58 \times 10^{-5}$, FMR $\Delta H \sim 7.4$ Oe, bulk-like magnetization.	[364]
2014	Ar, 9.7 mT	RT	2.7 nm/min	GGG	YIG/Pt, RF magnetron. Spin-mixing conductance of 2.0×10^{18} m^{-2} was demonstrated with $\alpha = 7.0 \times 10^{-4}$ with FMR $\Delta H \sim 4$ Oe and bulk-like magnetization.	[365]
2013	Ar, 9.7 mT	RT	2.7 nm/min	GGG	Pt/NM/YIG & YIG/Pt, RF magnetron. mV-level inverse spin Hall effect (ISHE) voltage, V_{ISHE} demonstrated.	[366]

TOS: thermally oxidized silicon
GGG: gadolinium-gallium-garnet

low-temperature sputtering techniques followed by post-annealing processes, also called solid phase epitaxy (SPE). This approach was effective in increasing deposition rates to obtain thick (several nm) films [358]. Once again, trends embraced by industry aim to improve production yield and lower processing costs. In 2005, slightly iron-rich amorphous YIG films, deposited on both Si and GGG substrates, were grown by RF sputtering and crystallized by annealing [359]. This technique achieved 10 nm/min deposition rates. Several other studies reported the positive impact of post-annealing under various conditions, including various atmospheres, temperature, rapid thermal anneals and dwell time [360–362, 367]. Post-annealing was found to impart the necessary kinetic energy to ions to increase their mobility within the crystalline lattice leading to the lowering of defect densities and further reduction of losses while concomitantly improving component efficiencies.

Due to the high demand for emerging technologies, e.g. spintronics, nanoscale YIG films with low damping coefficients have attracted the attention of many researchers during the past 10 years (2010–2020). YIG films having such properties have spurred a new field of physics in insulator-based magnon spintronics [36, 37]. Further, the magnon spectrum has been shown to extend to THz frequencies making it useful in future communications technologies [368, 369]. For example, the first YIG magnonic Brillouin zone appears ~7 THz [10, 370, 371]. The Hillebrand Group has provided a detailed review of magnon spintronics [245].

In 2013, Wang et al. deposited 100–200 nm YIG films by off-axis sputtering that maintained stoichiometry and demonstrated high phase purity [363]. In 2014, Liu et al. [35] deposited nm-thick YIG on GGG and found FMR linewidths decreased to ~7 Oe as film thickness increased from several nanometers to about 100 nm. In the same year, Chang et al. [364] deposited ultrathin YIG films with a very low damping coefficient (i.e. $\alpha = 8.6 \times 10^{-5}$) by sputtering at room temperature followed by heat treatment in an oxygen gas atmosphere (see Figure 10.44). In that paper, high-quality YIG films were achieved from the optimization of Ar pressure during deposition and minimization of contamination during the fabrication process. Other excellent sub-micrometer YIG films were produced having compatible properties [365, 366]. In all these efforts, the principal driving factors were reduction in processing costs and low losses. The latter is closely related to the crystal quality of the garnet films that is controlled in large part by processing conditions that allow the tailoring of adatom mobility, oxidation, and growth kinetics.

Figure 10.45 provides and illustrative map of the development of sputtering deposition technologies and corresponding advances in garnet applications in bubble memory, MO devices, nonlinear microwave devices, spintronics, and magnonics.

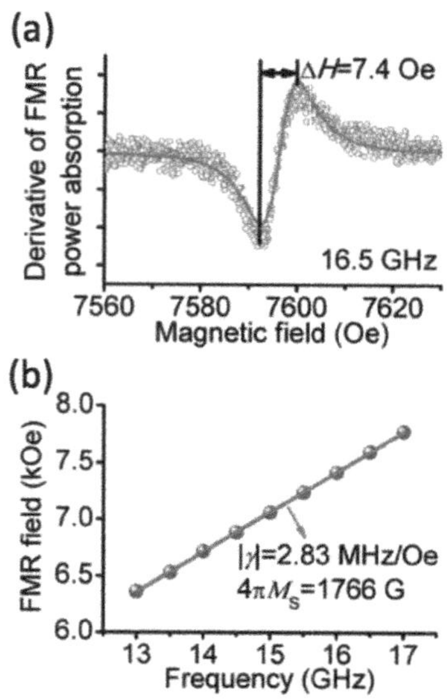

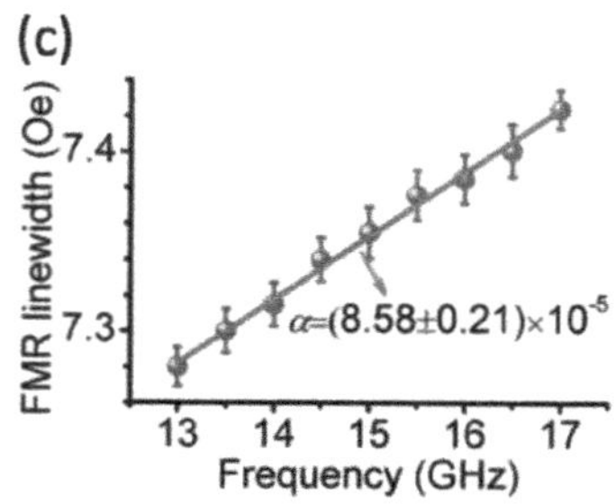

Ar pressure (mTorr)	Film thickness (nm)	Surface roughness (nm)	$4\pi M_s$ (G)	α ($\times 10^{-5}$)
10	39.7	0.10±0.01	1550	20.3±3.2
20	23.4	0.09±0.01	1720	12.5±0.4
25	22.3	0.13±0.05	1770	8.58±0.21
30	19.7	0.10±0.01	1790	28.1±0.5
40	19.5	0.23±0.02	1720	30.7±2.3

Figure 10.44 FMR data obtained from YIG film deposited at 25 mT Ar gas pressure. The FMR data was measured with a static magnetic field applied perpendicular to the plane of the film. Inset table indicates the effect of Ar gas pressure on the properties of YIG thin films (adapted from [365]).

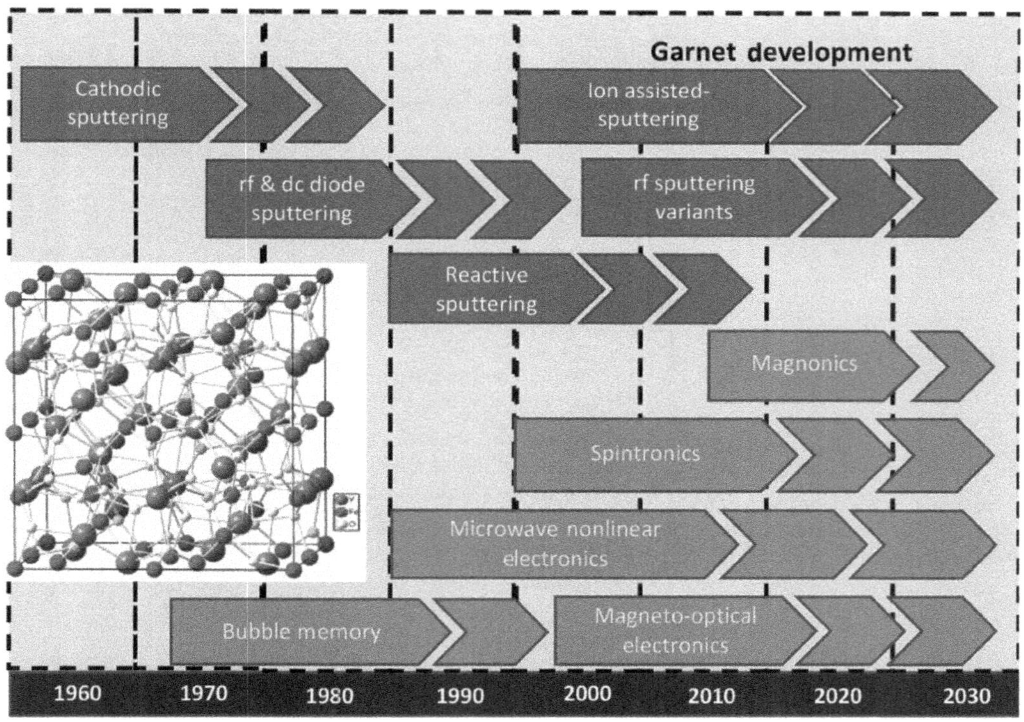

Figure 10.45 Illustration mapping the approximate timeline of sputter-deposition technologies and garnet applications.

10.4.3 Sputtering of Spinel Ferrite Films

Table 10.11 presents a historical development of sputtered ferrite films. Along this timeline one sees a transition from early efforts motivated by magnetic memory to integrated inductors to spintronics.

The earliest published data of sputtered spinel films dates to Westwood et al. [372] who reported single-crystal thin films of Ni-ferrite deposited on (1 0 0) MgO in 1967 by glow discharge sputtering from a nickel ferrite ceramic (i.e. cathode).

Table 10.11 Sputter deposition conditions and film properties from studies of spinel ferrite systems.

Year	Oxygen pressure (mT)	T_{dep} (°C)	Growth rate (Å/s)	Substrates	Key results	Ref.
1967	Ar:O_2 80 mT	<500	5 Å/min	MgO	$NiFe_2O_4$, cathodic sputtering, structural and magnetic characterization support single-crystal growth.	[372]
1975	2.5–20 mT	RT	20–34 Å/min	Al metal	Hematite, RF reactive sputtering. Superior recording performance is attributed to: large residual magnetic flux density leading to high readout amplitudes and large coercive force (>500 Oe) and small thickness leading to high recording density.	[373, 374]
1976	Ar:O_2 (10% O_2) 15–25 mT	12–250	3.0–14.0 nm/min (substrate dependent)	Al metal, glass	Fe metal, hematite, magnetite, RF and DC diode sputtering. Motivated by recording technology, SQ~0.6, Hc = 20–600 Oe.	[375]
1979	Ar:O_2 (10% O_2) 15–25 mT	12–250	3.0–14.0 nm/min (substrate dependent)	Al-Mg alloy; surface is alumite formed by anodic oxidation	α-FeO films are formed by reactive sputtering of an Fe target containing a small amount of Ti and Co in an Ar-O_2 atmosphere. The α-Fe_2O_3 film reduced to Fe_3O_4 at high temperature. Finally, the Fe_3O_4 is converted to γ-Fe_2O_3 by an oxidizing heat treatment. The coercive force (Hc) and the residual magnetization (Br) are 700 Oe and 2,500 G, respectively. A significant improvement of the squareness ratio (Br/Bs), 0.8, is attained for the γ-Fe_2O_3 phase.	[376]
2006	52 mT	600	unknown	(1 0 0) Si	NiZn ferrites films deposited by RF sputtering technique were studied by the magneto-optical Kerr effect. The Jiles–Atherton Model was used to fit the experimental hysteresis loops. The *k*-pinning parameter of the model increases by increasing film thicknesses.	[378]
2009	Ar vol. 75% and O_2 vol. 25% at 5 mT	RT	unknown	Si	Multilayered oxide films were deposited by reactive RF sputtering and crystallized to 1 μm thick ferrite ($Ni_{0.5}Zn_{0.3}Cu_{0.2}Fe_2O_4$) by heat treatment at 800 C for 1 h. This ferrite film was used for the fabrication of planar inductors having Q = 50 at 2.3 MHz.	[380]
2009	3.2 – 6.4 mTorr, O_2/Ar from 0 to 20%	300	unknown	Si	NiZn-ferrite films are deposited by RF sputtering. O_2 gas enrichment increases resistivity of the ferrite and reduces eddy current losses.	[381]
1998	2.2 mT, Ar	300–600	0.2 μm/hr	TOS, Zn-ferrite	$Mn_{0.5}Zn_{0.42}Fe_{2.03}O_4$ films, RF diode sputtering. Grains of ~15 um were obtained by deposition on polycrystalline Zn-ferrite substrates where permeability of 100 was obtained.	[382]
2013	3.7 mT	300	unknown	MgO	Perpendicular magnetic anisotropy (PMA) of cobalt ferrite $Co_xFe_{3-x}O_4$ (x = 0.75 and 1.0) epitaxial thin films grown by a reactive magnetron sputtering. Motivated by spintronics. Large PMA recorded making films suitable for spin filter applications.	[384]
2002	13 mT Ar	573	unknown	MgO	Motivated by spintronics, single-crystal 50 nm films of Fe_3O_4 with saturation magnetization, resistivity, and Verwey point were 438 emu/cm^3, 10,000 μΩ-cm, and ~110 K. These values are comparable to those of the Fe_3O_4 bulk.	[385]

TOS: thermally oxidized silicon

Those films had all the characteristics of Ni ferrites except for the anticipated equilibrium cation distribution. Ni-ferrite is a classic inverted spinel having Ni cations residing on octahedral B-sites. Henceforth, high quench rates, intrinsic to sputter deposition were found to displace cations from their equilibrium sites. This has been well studied and is now understood to result from the ballistic collisions between high-energy particles of the RF plasma impinging upon the growing film surface.

However, wide-scale research into the magnetic properties of sputtered spinel films began in earnest during the 1970s motivated largely by demands of the burgeoning magnetic recording industry for ever-increasing areal storage densities.

During the mid-1970s, researchers found that reactive RF magnetron sputtering was effective in producing high-quality magnetic oxide films including ferrites. Sputtered ferrite films were reported from 1976 to 1979 by Inagaki et al. [373, 374], Heller [375], and Hattori et al. [376] for magnetic memory storage applications. Advances in processing led to high coercivities and squareness ratios and substantial improvement in a large signal-to-noise ratio [377].

MOKE measurements were performed by Calle et al. [378] on NiZn-ferrite films deposited on (1 0 0)-Si substrates by RF sputtering. The Jiles–Atherton model was used to fit the hysteresis loops determining that *k*-pinning increases with increasing film thickness. Coercivity showed a strong dependence upon film thickness with spin reversal governed by interaction of domain walls with impurities. NiZn-ferrite films demonstrate a favorable intrinsic higher magnetization value over YIG-based systems.

The 1980s and 1990s brought a plethora of activity in research into sputtered spinel ferrite films. The technology pull continued to derive from the magnetic recording industry, but also from a new source, electronic applications that were entering into a new paradigm that included thin film topologies related to integrated thin film devices such as inductors, isolators, phase shifters and filters, among others, as well as associated RF-microstrip circuitry toward MICs [379]. Circa 2000, integrated on-chip inductors became a driving technology for thin film sputter-deposited spinel films. In this application, high permeability, low-loss tangents, high magnetization and compatible processing methodologies are desirable. The incorporation of thin film ferrites of moderate to high permeability as on-chip inductors was shown to reduce component size by several 100 times [380, 381]. Gillies et al. of Philips Research, reported the structure and soft magnetic properties of MnZn-ferrite films sputter-deposited on thermally oxidized Si substrates. At high deposition temperatures, i.e. >600°C, the highest value of μ obtained was ~30. MnZn-ferrite films with larger grain sizes, obtained by deposition on polycrystalline Zn-ferrite substrates, led to $\mu \sim 100$ [382]. Results were explained by the *nonmagnetic grain boundary model* [102, 383] shown to be applicable to explain the permeability in polycrystalline ferrites by van der Zaag et al [103]. These films are among the most attractive options for integrated thin film inductors [382]. Nevertheless, the wide-scale incorporation of magnetic materials in VLSI and other technologies mentioned remains a challenge due to industries' trepidation from possible fabrication line contamination, elevated temperature processing, and increased production costs associated with the increase in lithographic layers.

Figure 10.46 provides and illustrative map of the development of sputtering deposition technologies and corresponding advances in spinel applications in magnetic memory, integrated inductors, sensors, microwave devices, and half-metals and spintronics.

The most commonly studied ferrite is Fe_3O_4, which has an inverse spinel structure and a magnetic moment of 4.1 μ_B/f.u [386]. The Curie temperature of Fe_3O_4 is ~850 K and the metal–insulator transition point, i.e. the Verwey transition temperature is 121 K, making Fe_3O_4 conducting at room temperature. According to first-principles band structure calculations, Fe_3O_4 half-metallicity has been predicted [27]. Experimentally, evidence for near half-metallicity of Fe_3O_4 was found by spin-resolved photoelectron spectroscopy by Güntherodt and coworkers who reported a spin polarization of up to ~80% [147].

However, in the study of poly- or nanocrystalline Fe_3O_4 films, the magnetoresistance has been measured to be strongly dependent upon coercivity and temperature but far lower in magnitude than predicted. For example, at low temperature, current is measured to flow in nanocrystalline Fe_3O_4 films deposited in Si (1 0 0) by PLD, having a grain size of 50 nm, and gives rise to a negative MR of ~ −12% at fields of 90 kOe [387] This result is similar to other polycrystalline Fe_3O_4 films: Spin-dependent tunneling through adjacent 8–9 nm particles in pressed compacts has led to a negative magnetoresistance, 28.6% at 200 K and 24.5% at 300 K in a 70 kOe field [388]. While in polycrystalline Fe_3O_4 films prepared by reactive sputtering, prepared and measured at room temperature, the MR reaches −7.4% under a 50 kOe magnetic field [389]. Consequently, in these polycrystalline films resulting from sputtering, an MR effect is only observed at very high fields, which is unattractive for MR-sensor applications. The issues encountered when using epitaxial Fe_3O_4 films are discussed in Section 10.2.5.1.

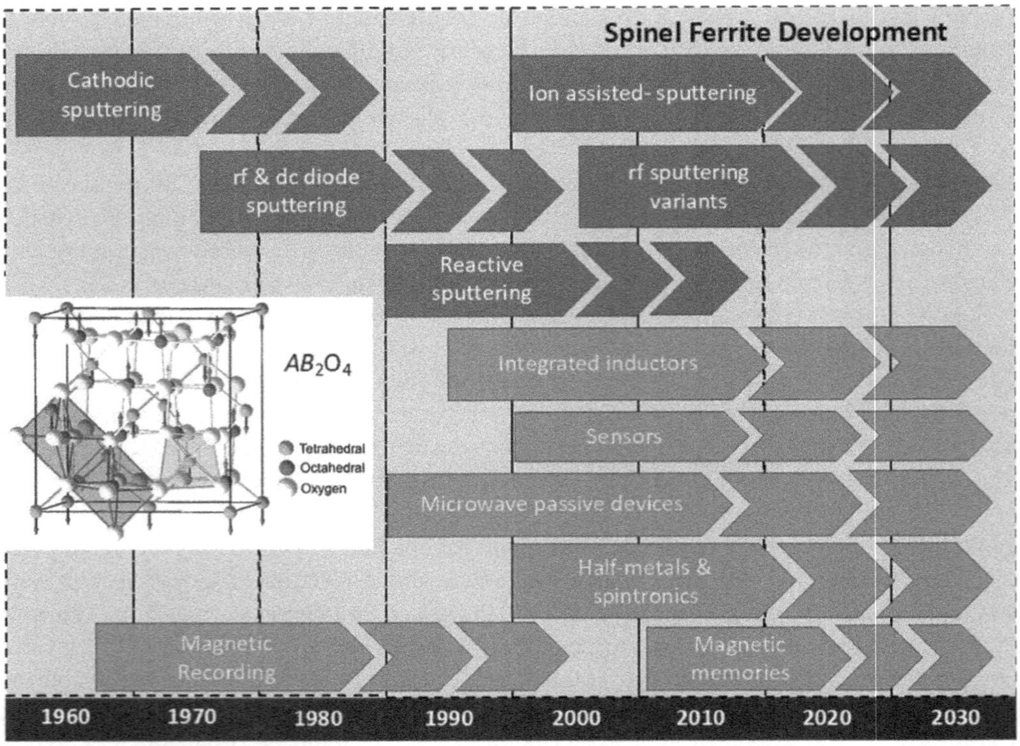

Figure 10.46 Illustration mapping the approximate timeline of sputter-deposition technologies and spinel ferrite applications.

By replacing one of the Fe ions with a cation such as Mn, Co, Ni etc., a mixed ferrite can be formed [1, 22]. Although these materials in bulk form are insulators, Pénicaud et al. [390] predicted half-metallicity in mixed spinel ferrites. In particular, $NiFe_2O_4$ shows a bandgap in the majority band, indicating that this ferrite can become an insulator or a semi-half-metal. Half-metallic ferrites are potential candidates as spin filters because of their magnetic insulating properties and high T_C. A spin filter consists of a magnetic insulator positioned between a nonmagnetic metal and a ferromagnetic metal. Due to exchange splitting of the energy levels in the conduction band of the magnetic insulator, the barrier energy for spin-up electrons differs from that of spin-down electrons, leading to a large difference in tunneling probabilities between the two spin orientations. Therefore, a spin-polarized current is generated, and results in a semi-infinite MR and device gain [391].

Spinel ferrites are typically soft magnets, however, due to unquenched angular momentum of the divalent Co cation, Co ferrites can manifest high magnetic anisotropy energy and substantial magnetostriction. In 2013, Niizeki et al. [392] produced $Co_{1-x}Fe_{3-x}O_4$ (CFO, x = 0.75 and 1.0) epitaxial thin films by RF reactive sputtering deposition on (001) MgO. These films were shown to have strong perpendicular magnetic anisotropy (PMA) and a saturation magnetization value comparable to bulk CFO. The effective PMA coefficients, K_u^{eff}, were estimated at 9.0 and 9.7 × 10^6 erg/cm^3 for x = 0.75 and 1.0, respectively. The intrinsic K_u^{eff}, extrapolated using Miyajima plots, reached values as high as 14.7 × 10^6 erg/cm^3. These films have been proposed as possible spin filters.

The role of ferrites in spintronics is explored in great detail in the MBE section of this chapter. It is covered here as an attempt to provide a comprehensive view of sputtered ferrite films.

10.4.4 Sputtering of Hexaferrite Films

As the early 1980s, sputtering has been successfully employed for depositing hexagonal ferrite films [393, 394]. These early studies were motivated by the development of high-density longitudinal and perpendicular magnetic storage [377, 393–397]. More recently, significant progress has been made in the deposition of high-quality hexagonal ferrites on a variety of substrates. Variants of the hexaferrite family of structures have been deposited by either DC sputtering [398, 399] or RF sputtering [400–404]. Much work has been carried out in the study of sputter deposition of the M-type barium ferrites (BaM).

Sui et al. of Carnegie Mellon University, established sputtering protocols for the growth of BaM films for high-density longitudinal and perpendicular recording media [377, 396]. The processing steps involved the deposition of amorphous films under DC magnetic fields followed by a crystallization heat treatment to produce the final media. A paradox arose, in that intensive characterization by XRD and TEM of each sample set revealed that precursor amorphous films appeared identical and unremarkable. In 1996, Harris and Snyder then of NRL, working with Sui and Kryder, employed polarization-dependent extended X-ray-absorption fine structure (PD-EXAFS) to investigate the anisotropic local structure around both Fe and Ba cations along the films' in-plane and out-of-plane directions [397]. PD-EXAFS data of Figure 10.47 clearly illustrate both significant short-range structure around the cations, as well as a significant degree of local structural anisotropy.

Principal conclusions of that study include that the amorphous films in question had a nonconventional oxide glass structure that was unusual because it was based on Fe as the network forming cation; the Fe-O polyhedral, which made up the network, were predominantly octahedra; octahedra outnumbering tetrahedra by a factor of four. More important for practical reasons, the network was anisotropic, with a local order that was more random in one direction and more like crystalline $BaFe_{12}O_{19}$ in the other. It was this anisotropy in local structure that apparently determined the final crystalline texture and the resulting magnetic anisotropy that formed upon annealing and determined the films' value as longitudinal media [397].

Because of very high magnetocrystalline anisotropy fields and remanent magnetization values, BaM films are also promising candidates for μ-wave and millimeter-wave technologies, and in particular, as key materials in nonreciprocal devices. Other types of hexagonal ferrite films, such as M-type strontium ferrite (SrM) films [399, 402, 405], W-type [403] and Z-type [395] ferrite films, have also been successfully sputter deposited as pure phase films. Most of the hexagonal ferrite films have been deposited on SiO_2/Si [377, 394–397, 399, 400, 402, 404–408] and Al_2O_3 [399, 409] substrates. Hylton et al. [399] and others studied BaM films on various other substrates, including: α-AlO/Si_3N_4 [404] and 6H-SiC [408], Pt [405, 407, 408], Pd [408], and alloys of Pd–Pt [408], that have been employed as underlayers to control grain size and morphology. In conventional sputtering, the source power, working gas pressure (Ar + O_2), and O_2 partial pressure are all key parameters in determining optimal sputtering conditions to attain desirable film properties. As can be seen in Table 10.12 substrate temperature is also an important factor, as is postdeposition annealing conditions, to induce or complete crystallization and improve crystal quality. Most hexagonal ferrite films have their *c*-axis orientation normal to the film plane, while in rare cases the *c*-axis aligns in the film plane often made possible by intentional epitaxy using single-crystal substrates and/or nucleation layers [22].

Perpendicular *c*-axis crystal texture and magnetic easy axis orientation allow for the breaking of time reversal symmetry essential for the operation of nonreciprocal RF and millimeter-wave devices such as isolators and circulators [409, 410].

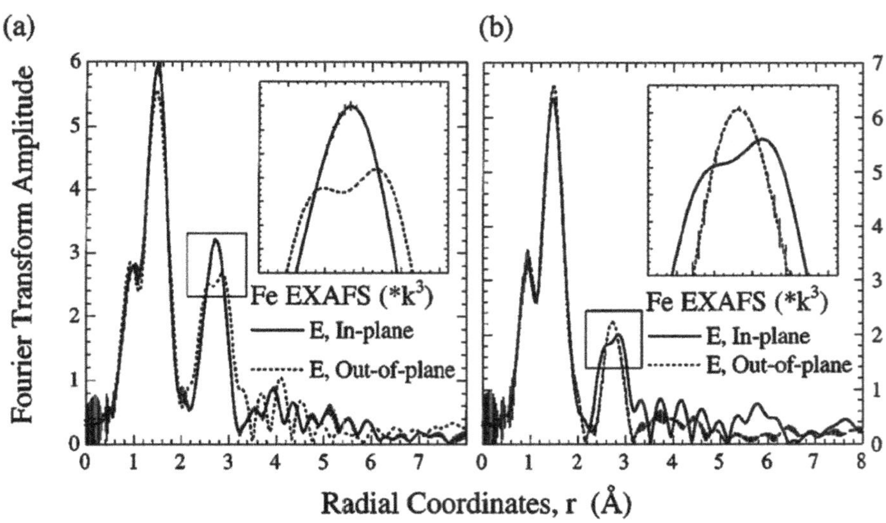

Figure 10.47 Fourier-transformed Fe EXAFS data for amorphous films that crystallize with (a) in-plane magnetic anisotropy and (b) perpendicular magnetic anisotropy. Dashed curves represent structure perpendicular to the film plane. Solid curves represent in-plane structure. Inset panels show the nearest Fe-Fe correlation with expanded scales. Note representative error bars on the in-plane data in panel (a) and out-of-plane data in panel (b). (The small peak at about r ~1 Å is an artifact caused by residual background curvature. It does not correspond to any atomic neighbors and should be ignored.) Used with permission from [396].

Table 10.12 Sputter deposition conditions and film properties from early studies of hexaferrite systems.

Year	Oxygen pressure (mT)	T_{dep} (°C)	Growth rate (Å/s)	Substrates	Key results	Ref.
1981	0.3 mT (O_2) 60 mT ($Ar+O_2$)	300–650	0.16 μm/h	Quartz	$BaFe_{12}O_{19}$, M-type hexaferrite, by DC diode sputtering. A new perpendicular magnetic recording film system is proposed. Its saturation magnetization M_s, coercive force Hci, and magnetic anisotropy constant K_u were 400 emu/cc, 680 Oe, and 1.67×10^6 erg/cc, respectively.	[393]
1982	0.3 mT (O_2) 2 or 60 mT ($Ar+O_2$)	400–650	0.15 μm/h	Fused silica, Si, TOS, sapphire	$BaFe_{12}O_{19}$, M-type, DC diode sputtering, and RF-facing target sputtering. Perpendicular magnetic recording film properties included magnetic anisotropy constant of 3.2×10^6 erg/cc, saturation magnetization of 380 emu/cc, and coercive force of 1.3 kOe.	[394]
1984	2–20 mT	RT - 900	1.0–3.0 nm/min	TOS	Co_2Z ($Ba_3Co_2Fe_{24}O_{41-x}$) films deposited by DC TFT sputtering for longitudinal recording media.	[395]
1986	Total 3.7 mTorr; partial pressure pO_2 was 0.01–3.7 mTorr	620	Unknown	TOS	Magnetoplumbite type of hexagonal Ba-ferrite films were deposited by RF diode sputtering. Films had the *c*-axis well oriented perpendicularly to the film plane. Perpendicular coercivity was ~1.2 kOe and M_s~280 emu/cc. D_{50} of I_{10} and 190 kfrpi for $\Delta\theta_{50}$ of 4.8° and 2.8°, respectively, were attained.	[400]
1988	Ar 3–4 mTorr (small amount of O_2)	<200 °C	0.4–0.5 μm/h	Sapphire	Epitaxial films of barium hexaferrite ($BaFe_{12}O_{19}$) have been prepared on basal-oriented sapphire substrates by RF sputter deposition of an amorphous ferrite film and crystallization of the film by annealing symmetric FMR peaks as narrow as 41 Oe at 60 GHz have been obtained.	[401]
1992	Ar-10% O_2 at 8 mTorr	RT	1.2 nm/min	Sapphire	$BaFe_{12}O_{19}$, M-type, in-plane thin film memory media on single-crystal sapphire substrates were deposited by both DC and RF reactive magnetron sputtering. The easy axis loop is square and shows a coercivity $H_c = 2.5$ kOe, $H_A = 16$–17 kOe, with an anisotropy constant $K_1 = M_s H_A/2 = 2.4 \times 10^6$ erg/cm^3.	[398]
1994	5–10% O_2; 2–8 mT (Ar)	550–700	8.5–11 A/min	Silica, quartz, glass, TOS, Si, Al_2O_3, MgO, Y_2O_3-ZrO_2, $SrTiO_3$, $LaAlO_3$	Ba-ferrite thin films were deposited by reactive magnetron sputtering on a variety of substrates in order to examine their potential as longitudinal magnetic recording media.	[399]
2000	2–8 mT O_2 (0–0.5 mT)	RT	10–40 Å/min	TOS	RF sputtering of La–Zn substituted hexagonal Sr hexaferrite by RF sputtering as a recording media.	[402]
2007	1.5 mT with 20% O_2	450	0.14 nm/min	Si	Growth of W-type hexaferrite by RF sputtering on Si substrates.	[403]
1993	0.5 mT w/0.025 mT O_2 Ar balance	580	5 nm/min	TOS, amorphous Si, Si_3N_4, and SiO/Cr underlayer on carbon substrates	Ba-hexaferrite films were grown using DC facing target sputtering. Underlayer films of Si_3N_4 and SiO_2 were effective in preventing diffusion of carbon atoms into BaFeO films from carbon substrates during deposition. Studies were motivated by high-density magnetic recording technologies.	[404]
1994	2 mT Ar with 0.03 mT O_2	480–550	unknown	TOS	Pd-doped Ba-M-type films grown by DC magnetron sputtering. It was found that the substitution of Pb facilitates the crystallization and improves the crystallinity for hexagonal M phase. All films prepared above 460 °C exhibit *c*-axis orientation with *c*-axis dispersion smaller than 1°.	[406]

Table 10.12 (Continued)

Year	Oxygen pressure (mT)	T_{dep} (°C)	Growth rate (Å/s)	Substrates	Key results	Ref.
1994	2–8 mT with 5–10% O_2	550–700	8.5–11 A/min	TOS, fused silica, or fused quartz glass, and Corning 96% silica sheet glass	Ba-hex are deposited by RF sputtering as polycrystalline films with large in-plane remanence for longitudinal recording.	[399]
2000	2.2–10 mT, 0.7 mT O_2	RT + RTA	unknown	Pt/TOS	RF diode sputtering, 10 wt.% barium-rich nonstoichiometric barium ferrite thin films were deposited in a mixture of Ar and O_2 on the Pt underlayer. Sq~95%.	[407]
2004	Ar and O_2 gases at 2 mTorr	550	Unknown	Si/α-AlO/Al metal; SiO_2/α-AlO/Si	Ba-ferrite films with a small addition of Al deposited by facing target sputtering at 550 °C onto α-AlO underlayer on SiO_2/Si wafer. It is found that a small addition of Al produces a high *c*-axis orientation with a *c*-axis dispersion angle of 2.5 °.	[413]
2005	2 mT, Ar+O_2	550–700	Unknown	Pt/TOS; Pd/TOS; Pt+Pd/TOS;	Ba-hexaferrite films deposited by facing target sputtering. It was found that Pt–Pd alloy underlayer was effective in achieving *c*-axis-orienting barium ferrite films with small grain size (~2 nm).	[408]
2010	Total gas pressure was 1.5 mT	650	0.3 μm/h	Sapphire	Isotropic Ba-hexaferrite films deposited by RF sputter deposition, intrinsic coercivity of films reached ~11.5 kOe at room temperature.	[414]
2003	Total gas pressure 0.2 to 0.003 mT of Ar and Ar+Kr	420–550	Unknown	TOS; glass ceramic disks	BaM and SrM films were deposited on Pt-coated TOS for *c*-axis perpendicularly oriented films using facing targets sputtering. Remanent magnetization, M_r, of SrM/Pt films were more stable than BaM/Pt films.	[405]
2017	Ar-7%O_2 (working gas pressure unknown)	RT	6.3–7.25 nm/min	6H-SiC	Room temperature Ba-hexaferrite films grown on wide bandgap SiC using DC magnetron sputtering with the *c*-axis aligned randomly in-plane for microwave passive device applications.	[410]
2004	Ar pressure set at 0.3 to 7.5 mT	400	Unknown	Alumina, Al_2O_3	Ba-hexaferrite films were deposited by RF sputtering for microwave isolators. Transmission coefficients show a nonreciprocal effect of 8.3 dB/cm at 50 GHz.	[409]
2013	Ar gas pressure 0.8 Pa	500–600	Unknown	Pt-coated Si wafers	Ba-hexaferrite films were deposited by RF sputtering for microwave circulators/isolators. Hysteresis curves also indicate that the BaM thin film exhibits nice self-biased property with 4pMr of 3803.04 Gs, squareness ratio (M_r/M_s) of 0.96, and coercivity of 1767.3 Oe.	[411]

TOS: thermally oxidized silicon

In rare cases, when the remanent magnetization, sometimes referred to as retention, is > 90% films are referred to as self-biased [407]. Retention can be increased by an increasing both crystal texture and coercivity. Li et al., of the Zhang Group of UESTC, demonstrated RF magnetron sputtered BaM thin films on Pt-coated Si wafer substrates having hysteresis curves (see Figure 10.48a) exhibiting self-biased properties with $4\pi M_r$ of 3803 G, high squareness ratios (M_r/M_s) approaching 90%, and coercivities of ~1800 Oe (no FMR linewidths were provided in this reference) [411]. Figure 10.48b depicts the topology for the same sample whose hysteresis data are depicted in panel (a), measured using atomic force microscopy (AFM). In panel (b), one sees a high-density film with copious grains having a narrow size distribution with diameters of ~0.6 to 0.8 μm, which are very near the single domain size. Such self-biased films can be considered for applications in self-biased passive devices that allow for the design and processing of two-dimensional passive devices and their integration into MICs [412].

Figure 10.49 provides an illustrative map of the development of sputtering deposition technologies and corresponding advances in hexaferrite applications in magnetic memory as both longitudinal and perpendicular modalities; microwave devices, including the self-bias variety; and as new electromagnetic interference shielding and absorbers.

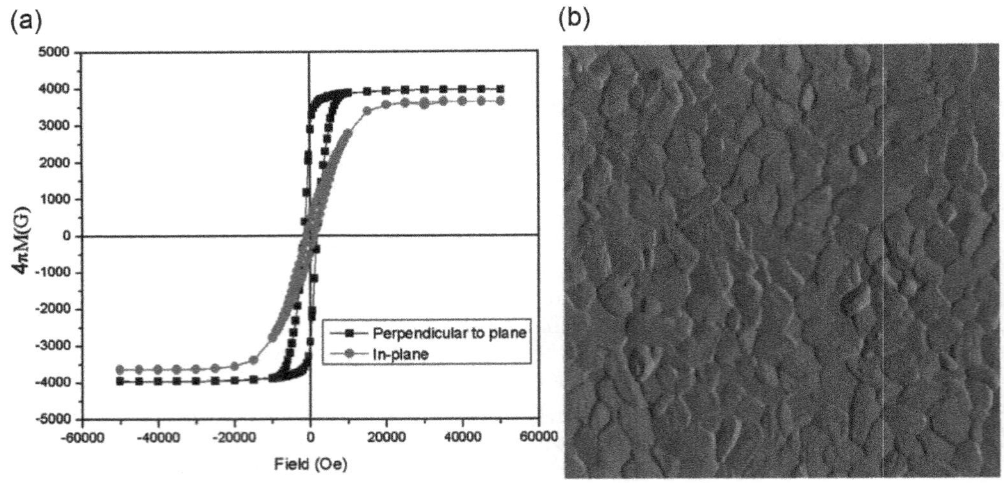

Figure 10.48 (a) Hysteresis loops of a BaM film deposited at 600 °C where the applied magnetic field is aligned along the perpendicular to the sample plane and the in-plane direction. (b) Atomic force microscope image of surface topology of the same sample as presented in (a).

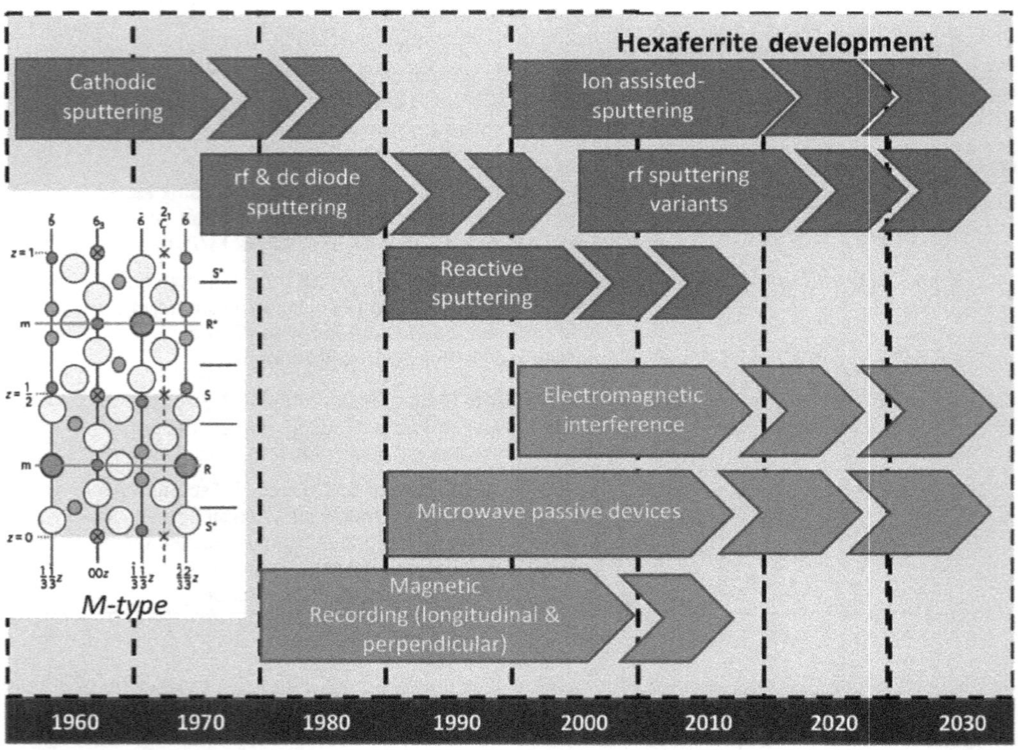

Figure 10.49 Illustration mapping the approximate timeline of sputter-deposition technologies and spinel ferrite applications.

10.5 Atomic Layer Deposition

10.5.1 Introduction

ALD is a thin-film deposition technique capable of producing uniform, high-quality, pinhole-free, conformal, and homogenous films, including ferrites, from gas phase at low temperature. The concept of ALD was established in the early 1950s by Alekskovskii and experimentally demonstrated by Alekskovskii and Kol'tsov as a molecular layering (ML) technique in the 1960s [415]. It was independently developed and patented as an advanced thin-film technology called atomic layer epitaxy (ALE) in late 1970 by Suntola et al. [416, 417]. In early 2000, the name ALE was changed to ALD [418]. A few years later, the term ALE was recoined to refer to atomic layer etching, the "reverse" of ALD.

ALD is unique for its low-temperature (100–400 °C), low-vacuum-deposition conditions, excellent conformality at the nanometer-scale level, and a high degree of composition control and thickness control at the Ångstrom level. ALD processes have been reported to produce ferrite films of thicknesses up to 70 nm [419, 420]. The fabricated ferrite thin films are soft-magnetic at room temperature without further thermal treatment, which is advantageous to thermally sensitive materials and underlaying structures during integration. Precursors utilized are typically simple and widely commercially available organometallics, either alkyl-based such as diethyl zinc ($Zn(C_2H_5)_2$) or cyclopentadienyl (C_5H_5-, abbreviated as Cp)-based metallocenes (Cp_2M). A typical ALD gas-handling system is shown in Figure 10.50a schematically and consists of three main components: (i) a chamber with an internal volume; (ii) a fixture assembly to hold a substrate within the internal volume of the chamber; and (iii) gas injection ports, that is, at least one precursor gas arrangement for the precursor gas, and one inactive gas dispersion arrangement for the inactive gas for introduction of gases into the internal volume. A picture of a typical ALD system is shown in Figure 10.50b.

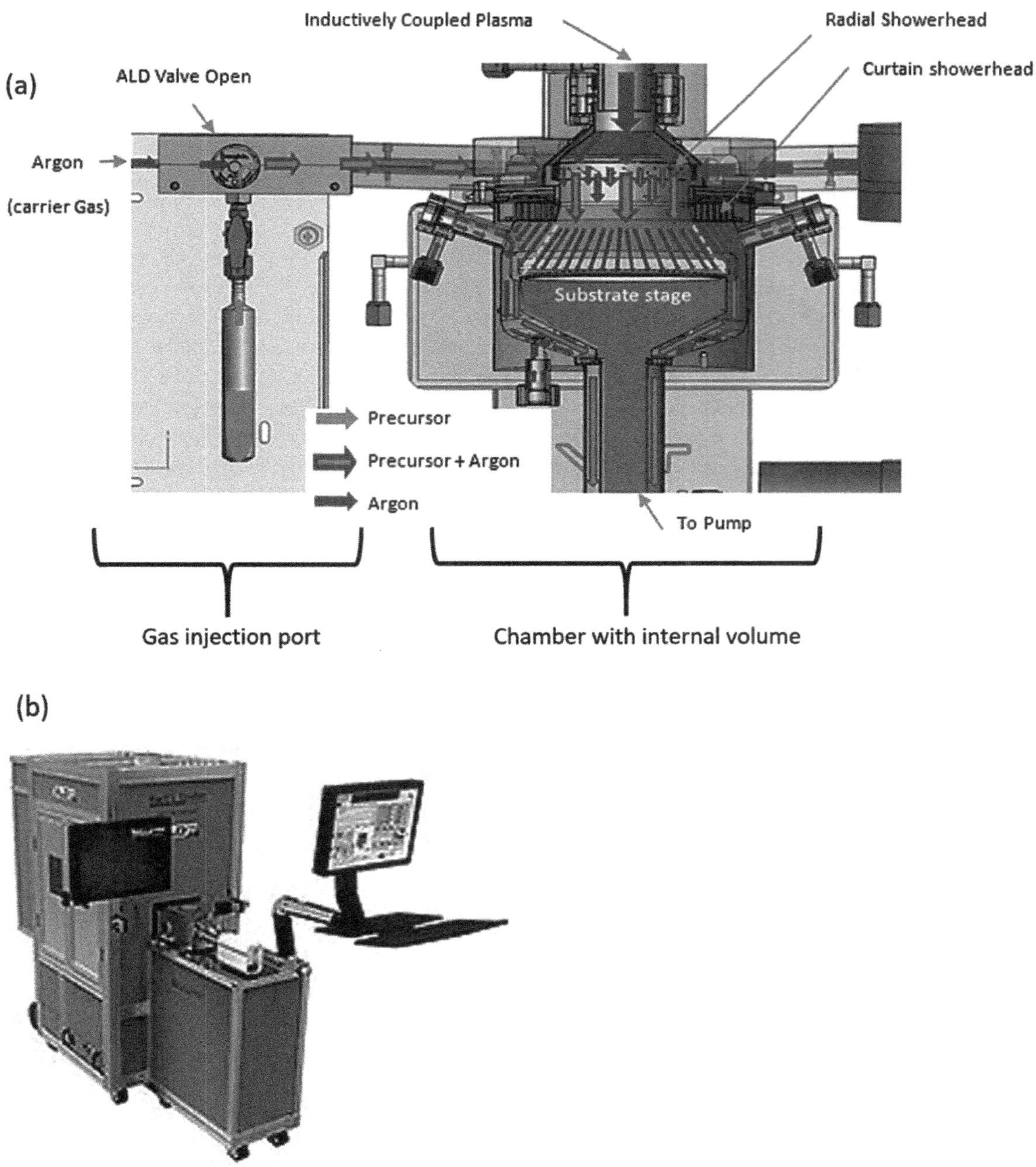

Figure 10.50 (a) Schematic gas-handling layout, and (b) an ALD system (shown is the ALD-150LX system of Kurt Lesker with perpendicular gas inlet design [421]).

In general, film formation in ALD is based on the reaction between the ligands of the precursor vapor exchanging with functional chemical groups present at the surface of a substrate in a self-limiting process regime. This is done by exposing the substrate in an alternating manner to a precursor and to a so-called co-reactant, thereby creating a *layer-by-layer* deposition mode of the materials on the substrate. Each precursor reaction with the substrate is separated by a purge step to remove the unreacted precursor and any by-products. This sequence of reactions and purges constitute a cycle which determines the thickness of the as-deposited ferrite film. In the absence of film nucleation delay, the thickness of the deposited ferrite linearly depends on the number of cycles. For ternary oxides of the type AB_2O_4, one conducts supercycling with one ALD supercycle consisting of n Cp_2Fe/O_3 cycles followed by m Cp_2M/O_3 cycles (M = Co, Ni). These supercycles are repeated until the desired thickness is reached. A schematic diagram of an ALD process is shown in Figure 10.51. The growth rate per cycle (GPC) is typically 0.2 to 3.2 Å, depending on the processing conditions (i.e. temperature, ratio of precursors, etc.) employed [419].

Although ALD has matured considerably in its development of many types of materials (metals, nitrides, etc.) over the last several decades [422–424], it has only been used for spinel ferrite deposition in recent times. This is due to the challenges in achieving a practical ALD process for ferrites that allowed the use of commercially available precursor materials, with fast enough growth rates (0.2 to 3.2 Å per cycle) and sufficiently air-stable ferrite films that demonstrated room temperature ferrimagnetism without further thermal treatment. In the next section, we will discuss ferrite growth through ALD processing.

10.5.2 Ferrite Growth by ALD

ALD processing of ferrites has been limited to spinel ferrites only. In 2000, the Bachmann group [419] was the first to demonstrate an ALD process that can be used to develop air-stable ferrimagnetic oxides of $NiFe_2O_4$ and $CoFe_2O_4$ with growth rates of 0.4 and 0.3 Å/cycle, respectively, using commercially available precursors. They demonstrated the combination of two well-behaved binary oxide ALD processes of Fe_2O_3 and either NiO or Co_3O_4 (i.e. homoleptic cyclopentadienyl complexes of Fe(II), Co(II), and Ni(II) (metallocenes, Cp_2M) combined with ozone into a supercycle mode to allow for the ferrite growth based on ALD reactions to produce the air-stable ferrimagnetic oxides – $NiFe_2O_4$ and $CoFe_2O_4$. They reported growth rates of 0.4 and 0.3 Å/cycle, with thicknesses of 45 and 30 nm for $CoFe_2O_4$ and $NiFe_2O_4$, respectively, after 1000 to 1200 cycles, and demonstrated highly controllable stoichiometry. Energy-dispersive X-ray (EDX) spectroscopy analysis indicated that the atomic ratio between cations, Fe:M, was approximately directly proportional to the corresponding pulse ratios (slopes of ~0.55 for $Co_xFe_{3-x}O_4$ and ~0.19 for $Ni_xFe_{3-x}O_4$), with the M:Fe atomic ratio in the deposited film being

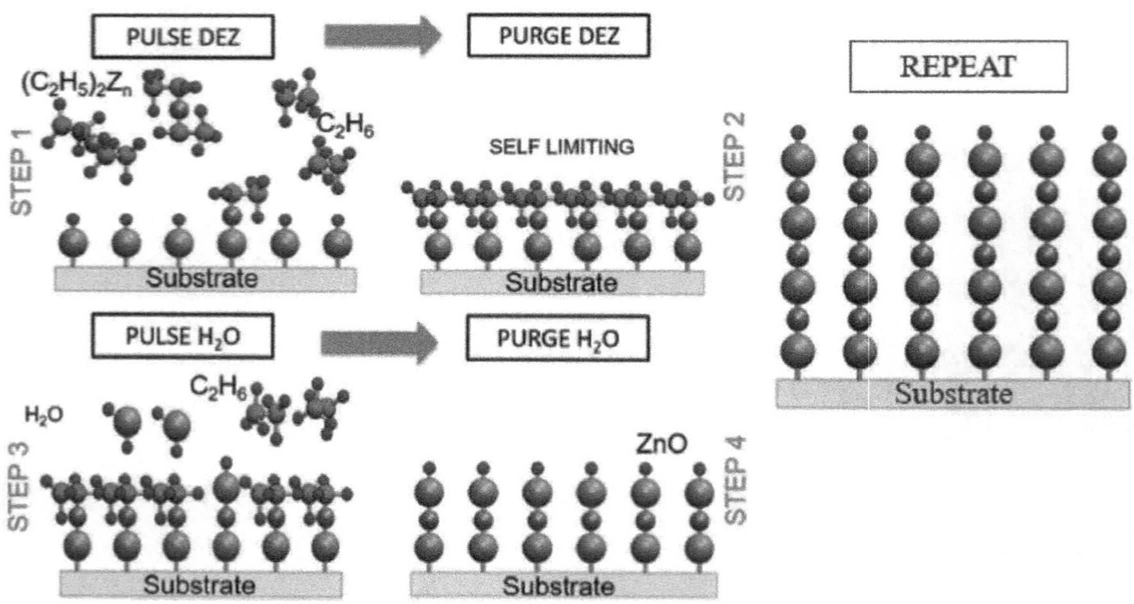

Figure 10.51 Schematics of an ALD step process for thin film deposition from [421]. Excess products are removed with inert carrier gas or pumped off before every next step.

larger than the pulse ratio for both Cp_2Co:Cp_2Fe and Cp_2Ni:Cp_2Fe. This finding, which is caused by the antibonding character of the d electrons in excess of 6, is consistent with the reactivity trend Ni(II,d^8) >Co(II,d^7) >Fe(II,d^6), known for metallocenes [425].

The XRD spectra in Figure 10.50 show the $Co_xFe_{3\text{-}x}O_4$ films as polycrystalline with spinel phase. A noticeable peak shift to lower angles was observed with iron concentration increase, indicating lattice expansion: from 8.24 Å (Fe: Co = 0.58) to 8.34 Å (Fe: Co = 1.2), then 8.38 Å (Fe:Co ratio = 1.7), and finally 8.42 Å (Fe:Co = 2), when the (311) reflection is used for calculating the lattice parameter. On the other hand, XRD patterns of the as-prepared $Ni_xFe_{3\text{-}x}O_4$ films (except for $NiFe_2O_4$) revealed weak, partially crystalline ferrite phases. The $NiFe_2O_4$ film demonstrated stronger signals with pure cubic spinel phase with a slight shift in peak angle compared to $NiFe_2O_4$ standard. Annealing the sample at 700 °C sharpened the XRD peaks with respect to the as-prepared sample, indicating microcrystal growth and shifted them to the positions reported for the $NiFe_2O_4$ standard. Unfortunately, for the annealed film no crystal size, which is so relevant for the permeability value [103, 383], was reported.

These authors did report that the films as deposited were smooth, with grain sizes of 30 and 13 nm for the $CoFe_2O_4$ and $NiFe_2O_4$, respectively. Larger crystallites were formed upon annealing of the initially smooth film under Ar for 1 h at 700 °C.

As shown in Figure 10.53, these authors found that the as-deposited $NiFe_2O_4$ film displayed soft-magnetic properties and weak magnetic signal, requiring postdeposition annealing at 700 °C to enhance the magnetic signal. On the contrary, the

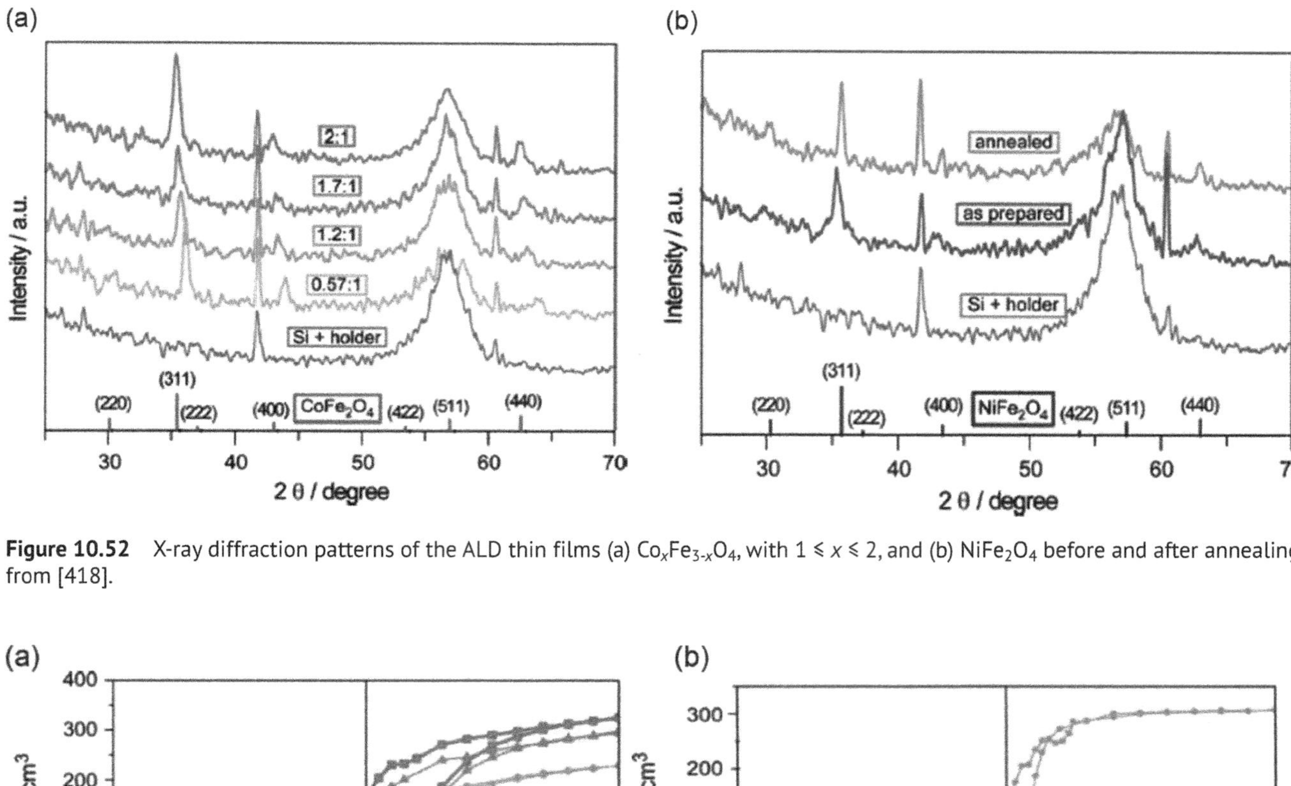

Figure 10.52 X-ray diffraction patterns of the ALD thin films (a) $Co_xFe_{3\text{-}x}O_4$, with $1 \leqslant x \leqslant 2$, and (b) $NiFe_2O_4$ before and after annealing from [418].

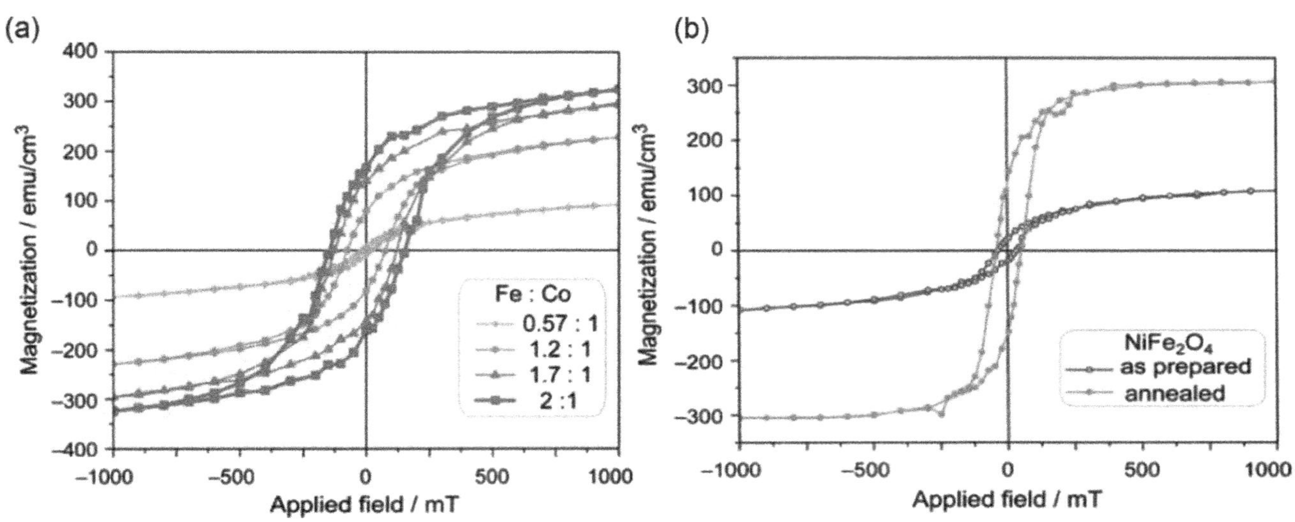

Figure 10.53 Room temperature in-plane magnetic hysteresis loops of the ALD thin films (a) $Co_xFe_{3\text{-}x}O_4$, with $1 \leqslant x \leqslant 2$, and (b) $NiFe_2O_4$ before and after annealing from [419].

Figure 10.54 Cross-sectional SEM image of the ALD iron oxide from $Fe(2,4\text{-}C_7H_{11})_2$ and H_2O_2 on silica from [426].

as-deposited $CoFe_2O_4$ displayed higher magnetization similar to those of bulk materials and showed preferential magnetic orientation parallel to the film surface, suggesting a dominant shape anisotropy. This because in this case, the effect of the large Co^{2+} single-ion magnetocrystalline anisotropy in the $CoFe_2O_4$ film is less significant, as the film thickness and grain size of the fabricated polycrystalline films were small (30 nm).

In 2013, Riha et al. [426] were able to precisely control the oxidation state and crystalline phase of thin-film iron oxides by ALD, utilizing a novel iron precursor, bis(2,4methylpentadienyl) iron [426]. Their approach enabled the selection whether $\alpha\text{-}Fe_2O_3$ to Fe_3O_4 films would be deposited based on using either ozone or H_2O_2 as oxidant, which changes the oxidation state of the films, respectively. These authors demonstrated conformal deposition on a porous high-area silica structure, as shown in SEM image in Figure 10.54, with an average thickness of 43 ± 2 nm. In addition, they showed the ability to tune the film composition, either by the choice of oxygen source used during the deposition or by postdeposition annealing as seen in the grazing incidence XRD (GIXRD) and Raman spectroscopy results in Figure 10.53. Only a single broad peak was observed for the as-deposited thin film made with H_2O_2, which can be indexed to Fe_3O_4 (minor peaks are from quartz substrate). In contrast to H_2O_2-derived films, peaks corresponding to $\alpha\text{-}Fe_2O_3$ were observed in the O_3-dosed films. Annealing the film made with H_2O_2 in O_2 for 30 min at 500 °C led to $\alpha\text{-}Fe_2O_3$ and the presence of some δ-FeOOH even after annealing.

These authors reported that the choice of the complementary oxygen co-reactant under otherwise identical conditions delineates these two distinct film compositions; pointing to the fact that distinct surface reaction mechanisms occur in the two methods [426]. They found that with H_2O_2, ligand protonation may be involved in each deposition cycle concurrent with metal oxidation. However, there is no growth of iron oxide, Fe_2O_3, when H_2O alone is used as the oxygen source, as only two-thirds of the iron atoms are oxidized from Fe^{2+} to Fe^{3+} in Fe_3O_4. Though they could not determine the mechanism that gives rise to distinct Fe^{2+} and Fe^{3+} centers, they reported that O_2 and O_3 are typically used as "proton-free" oxidants, which are believed to completely combust ligand frameworks. Furthermore, under subsaturating O_3 dosing, they observed Fe_3O_4 retention of Fe^{2+}, indicating a more complex mechanism than pure ligand combustion in the O_3 process, as well. They concluded that both processes show self-limiting surface reactions and growth rates of at least 0.6 Å/cycle. Though film deposition occurred at 150 °C, they reported the lowest temperature tested for ALD of iron oxides (65 °C).

Zierold et al. in 2014 reported that magnetite (Fe_3O_4) of high quality could be obtained by post-annealing in flowing argon–hydrogen (95%/5%) gas the ALD processed (using ferrocene and ozone) ferrite thin films [427]. A growth rate of 0.22 Å/cycle with thickness of 60 nm was reported for the films. A complete reduction of $\alpha\text{-}Fe_2O_3$ to Fe_3O_4 was confirmed post-annealing at 400 °C for 1 h in flowing argon–hydrogen (95%/5%) gas. They discovered that the temperature window of the reduction process of mixed iron oxides phases to Fe_3O_4 was about 60 °C. They observed the Verwey transition as well as the isotropic point in temperature-dependent magnetic and electrical measurements to determine stoichiometry and low impurity content in the thin films, thereby proving that the ALD process indeed can achieve high-quality ferrite thin films. They also reported on the isotropic point at which the magnetocrystalline anisotropy of magnetite disappears. They concluded that the properties of the Fe_3O_4 thin films grown by the lower cost ALD with a subsequent thermal reduction procedure were comparable with those grown by MBE [427].

A year later in 2015, Yijun Zhang et al. reported on high-quality superparamagnetic and ferrimagnetic Fe_3O_4 thin films fabricated using ALD technique [428]. They were able to precisely control the morphology and phase of the films by tuning the growth condition and post-annealing process as seen from XRD data in Figure 10.54a. They developed polycrystalline Fe_3O_4 thin films with an average grain size of 7 nm, resulting in a superparamagnetic behavior with a blocking temperature of 210 K (see Figure 10.56b) by post-annealing the as-grown ALD $\alpha\text{-}Fe_2O_3$ film in H_2/Ar atmosphere at 400 °C. The Fe_3O_4 films exhibited ferrimagnetic ordering and distinct magnetic shape anisotropy.

Most recently, in 2021, Márquez et al. demonstrated that controlling the thickness of Fe_3O_4 thin films allow for adjusting their magnetic properties, coercivity and remanence [429]. They showed coercivities between 0.3 and 0.7 kOe and M/M_s between 0.55 and 0.9, with varying thickness from 4 to 16 nm of the synthesized films, as shown in Figure 10.55. In Table 10.13 below, this progression of ALD of ferrites has been collected.

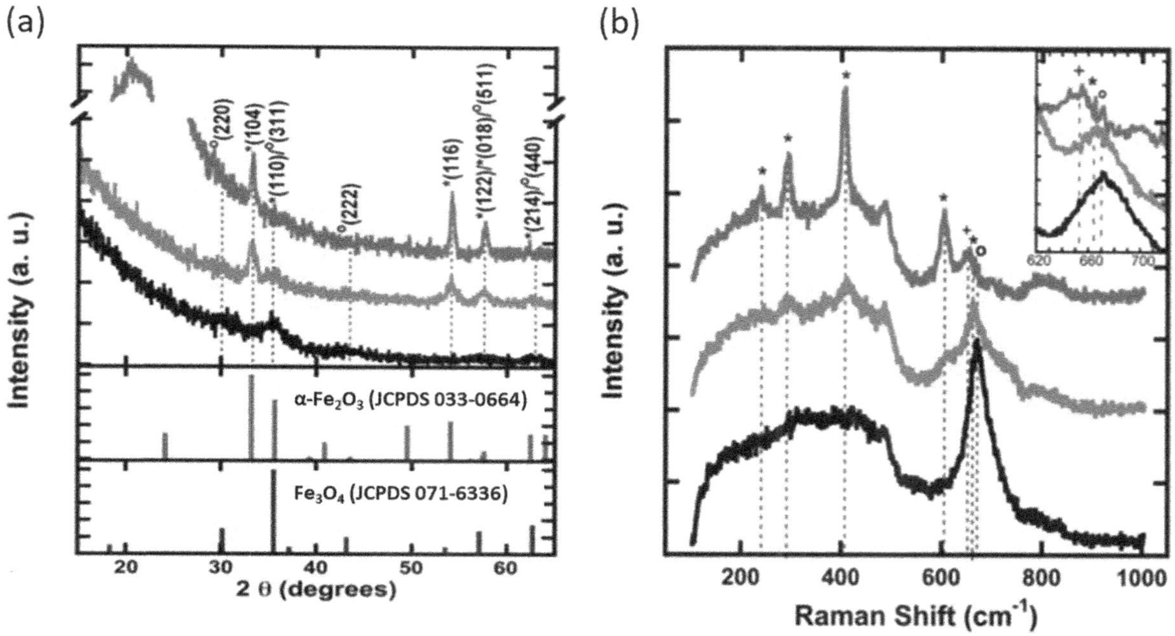

Figure 10.55 (a) GIXRD patterns and (b)Raman spectra of the as-deposited iron oxide films made with H_2O_2 (black, lower curve), 15 s O_3 (light grey, middle curve), and H_2O_2 films annealed in O_2 (dark grey, top curve). (*) α- Fe_2O_3; (O) Fe_3O_4; (+) δ-FeOOH from [426].

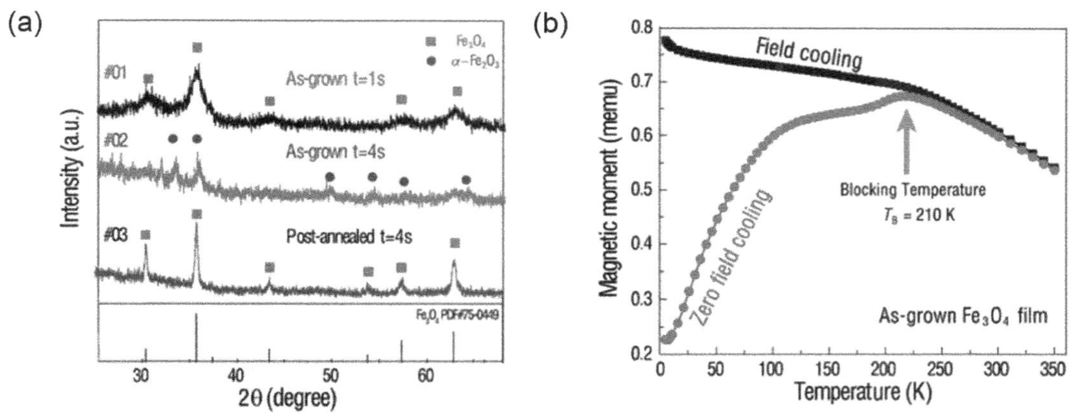

Figure 10.56 (a) X-ray diffraction patterns and (b) magnetization as a function of temperature for as-grown ALD iron oxide thin films from [428]. Sample is cooled with a zero field and a field of 1000 Oe, respectively, before the magnetization measurements.

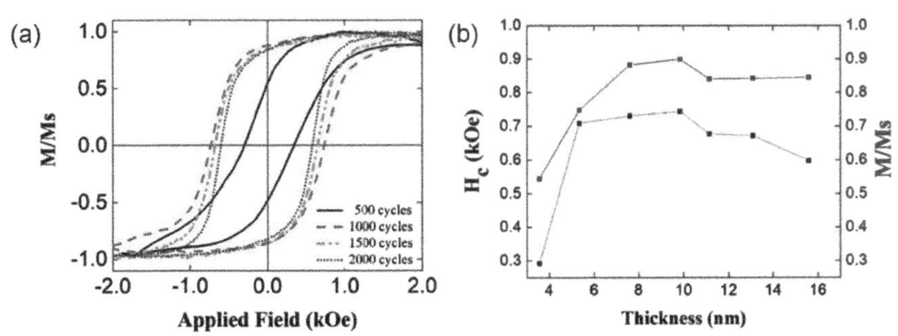

Figure 10.57 (a) Hysteresis loops and (b) coercivity (lower curve) and reduced remanence (top curve) as a function of thickness for Fe_3O_4 thin films from [429].

Table 10.13 Major milestones in ALD of ferrites.

Year	Type of ferrite	Substrate	*t* (nm)	Atm.	T_{dep} (°C)	Growth rate (Å/cycle)	Key result	Ref.
2010	$Co_xFe_{3-x}O_4$ $Ni_xFe_{3-x}O_4$	Si (1 0 0)	45 30	Ozone as oxygen source	250 200	0.4 0.3	First-reported ALD reaction of air-stable room temperature ferro/ferrimagnetic materials using commonly available precursors. [Binary oxide ALD processes (Fe_2O_3 and either NiO or Co_3O_4)]	[419]
2013	α-Fe_2O_3 Fe_3O_4	Si(001), fused quartz, and fluorine-doped tin oxide coated glass	40.7 37.6	Ozone H_2O_2	65 and 120	0.4 (at 65 °C), 0.6 (at 120 °C) 0.5 (at 65 °C), 0.9 (at 120 °C)	Reported the lowest temperature ALD of iron oxides (65 °C). Demonstrate control of α-Fe_2O_3 or Fe_3O_4 films deposition using either ozone or H_2O_2 as oxidant, respectively (ALD process utilizing novel iron precursor, bis(2,4-methylpentadienyl)iron).	[426]
2014	Fe_3O_4	2 or 100 nm layer of SiO_2 on Si (1 0 0)	40–70	10 wt.% ozone in oxygen	200 and 220 in stop-/exposure mode	0.22	Demonstrated high-quality Fe_3O_4 thin films comparable to those grown by MBE. Post-annealed in flowing argon–hydrogen (95/5) gas (Ferrocene/ozone ALD process).	[427]
2015	Fe_3O_4	Si (1 0 0)	—	O_2	400	—	Demonstrated superparamagnetic and ferrimagnetic Fe_3O_4, showing ferrimagnetic ordering and distinct magnetic shape anisotropy [ferrocene and O_2 (post-annealed in 5% H_2/95% Ar at 400 °C for 3 h to reduce the Fe_2O_3 phase to Fe_3O_4)].	[428]
2021	Fe_3O_4	Si (1 0 0) substrate with 208.8 nm of thermally grown SiO2	4–16	N_2	200	0.08	Showed that controlling the thickness of Fe_3O_4 thin films allow for adjusting their magnetic properties, coercivity and remanence [ferrocene/ozone ALD process (post-annealed in 4% H_2/96% Ar at 430 °C for 4 h to reduce Fe_2O_3 phase to Fe_3O_4)].	[429]

They reported that the SEM images of the films showed that they are homogenous at the macroscopic level, though measurements by magneto-optic Kerr effect indicated that the samples should have holes responsible for pinning the domain walls (which would explain the varying coercivity). The authors concluded that the increase in coercivity is due to the structure of holes that appear in the films due to the thermal reduction process to which the film is subjected.

10.5.3 Limitation and Issues

Although ALD is promising for high-quality, conformal, and homogenous ferrite thin film development, there are still limitations associated with its use. For instance, ALD has a long processing time compared to similar processes (e.g. PVD) and thus lower throughput process capability. However, recent developments in ALD technology in general are focusing to overcome this limitation. Finally, the grain size of the ferrites, which are formed using ALD are very small and typically in nanometer range (≤30 nm), which would result in very low permeability values. Although this may be useful in some applications to help flux guiding using oxidic, e.g. nonconduction layers [26], it is not suitable for most integrated inductor or passive component application as the permeability is too small. In general, the fact that ALD can be used to grow MBE-type quality Fe_3O_4 or $NiFe_2O_4$, layers deposited at modest deposition temperature may pave the way for spintronic applications that have been discussed in Section 10.2.5.

10.5.4 Outlook

In this section, we discussed ALD processing of ferrites. ALD processes have been reported to produce ferrite films of thicknesses up to 70 nm [419, 420]. The fabricated ferrite thin films are soft-magnetic at room temperature without further thermal treatment, which is advantageous to thermally sensitive materials. Although ALD has very much matured in its development of many types of materials (metals, nitrides etc.) over the last several decades, it is still a little challenging to use for the synthesis of ferrites. Not only is the growth of magnetic nanostructures by ALD difficult due to their complex binary or ternary chemical reactions, but also the evolution of magnetic ordering, achieving a practical ALD process for ferrites that allowed the use of commercially available precursor materials, with fast enough growth rates (0.2 to 3.2 Å per cycle) and sufficiently air-stable ferrite films that demonstrate room temperature ferrimagnetism without further thermal treatment is challenging. As such, it is important to note that so far, ferrite development by ALD is limited to spinel ferrites perhaps due to their simpler structure compared to hexaferrites and garnets.

10.6 Spin-Spray Plating

10.6.1 Introduction

Spin-spray plating was developed in 1983 by Masanori Abe et al. [430] as a means for polycrystalline spinel ferrite film development. Unlike other conventional processes, spin-spray plating is an electroless type of plating that involves low temperature (70–100 °C), direct deposition of polycrystalline spinel ferrite films from an aqueous solution onto a substrate. It requires no postdeposition heat treatments of the fabricated films. As it is a low-temperature process, it is compatible with a wide range of substrates including metals as well as non-heat-resistant materials like organic materials, plastics, semiconductor substrates, and polymer sheets [431–435]. It is also a rapid growth process with a 40–100 nm/min film growth rate and can produce ferrite films with thicknesses >10 μm ± 10% roughness. Though this technique was developed primarily for spinel ferrite deposition, it has been extended to the deposition of a variety of oxide thin films such as ZnO [436, 437].

Typically, the spin-spray plating process consists of two solutions: reaction/precursor (Fe^{2+} and M^{n+} (M = metal ions such as Fe, Co, Ni, Zn, Mn, etc.) salts, usually chlorides) and oxidizing solution [mixture of a pH buffer (an acetate, e.g. CH_3COONa, CH_3COOK, CH_3COONH_4) and an oxidant ($NaNO_2$)], simultaneously sprayed unto a substrate on a heating spinning table in the presence of nitrogen gas. A schematic layout for the spin-spray plating process is shown in Figure 10.58.

With spin-spray plating, the formation of the ferrite film is facilitated by hydrolysis as well as the oxidation of Fe^{2+} to Fe^{3+}, as shown in Figure 10.59.

Initial absorption of the Fe^{2+} and M^{n+} ions by the substrate is brought about by the presence of OH^- group on its surface. Some of the initially absorbed Fe^{2+} are then oxidized to Fe^{3+} by the oxidizing agent to form the ferrite layer. The absorption and oxidation process are repeated to produce thicker ferrite films. The ferrite film continues to thicken so long as the required ions and the oxidizing reagents exist in the solution. As such, spin-spray plating can be used for the fabrication of thick dense ferrite films.

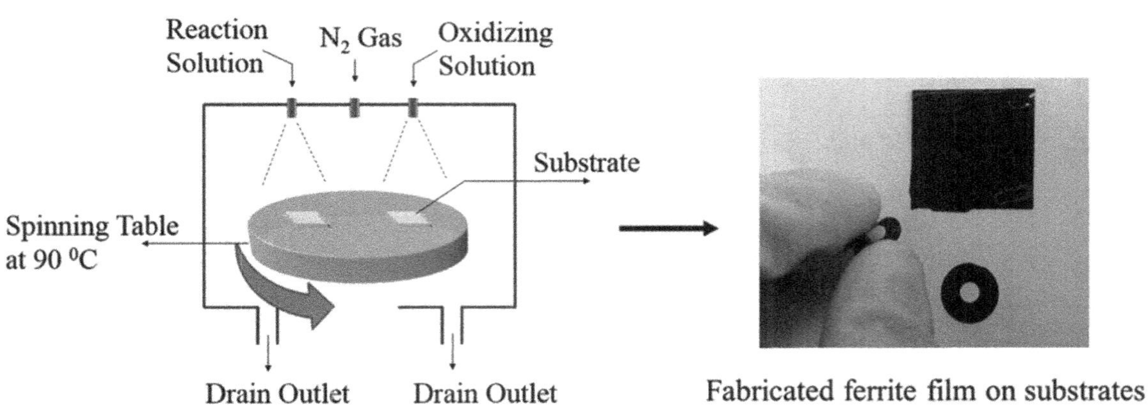

Figure 10.58 Schematics of spin-spray setup with pictures of fabricated films on glass and polyimide substrates.

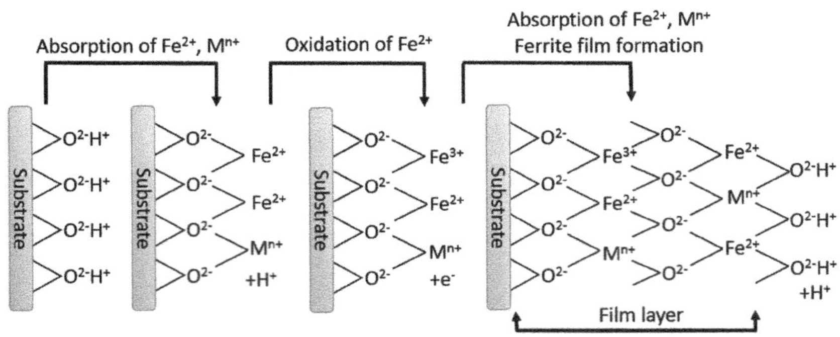

Figure 10.59 Principle of the spin-spray process.

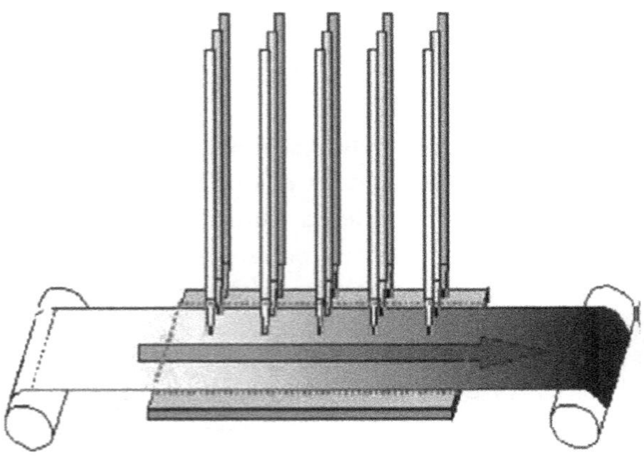

Figure 10.60 Multiple nozzles conveyor belt system to cover large surface areas from [439].

The quality of the ferrite film strongly depends on the adhesive force between the film and the substrate. Poor adhesion will lead to peeling of the film, which leads to degradation of film properties. As described earlier, OH^- group is critical for strong adhesion of the ferrite film to the substrate surface because the Fe^{2+} and M^{n+} ions are chemically bonded onto the substrate surface by the OH^- group. Therefore, substrates without OH^- group must undergo pretreatments before spin-spray plating to improve film adhesion. Common surface treatments include immersing the substrate in a seed-layer bath consisting of diluted HCl and Fe^{2+} to introduce the OH^- group and the initial Fe^{2+} ion onto the surface of the substrate for easy absorption of the ferrite ions by the surface during film fabrication.

Over the years, the spin-spray technique has evolved. In 2006, the one solution method was introduced by Miyasaka et al. for $Fe_3O_{4+\delta}$ thin film fabrication [438]]. This process involved the use of only one aqueous solution, which composed of dext ran+$FeCl_2$+CH_3COOK+$NaNO_2$. Dextran $(C_6H_{12}O_6)_n$, $n = 1100–1700$, was used to suppress iron hydroxides from precipitating in the solution. Though single-phased $Fe_3O_{4+\delta}$ thin film with spinel structure was successfully fabricated using the "one-liquid" spin-spray process, it has yet to be extended to other compositions of spinel ferrites. A modified version of spin-spray plating that allows for the ferrite plating of rolls of polymer materials was discussed by Harris et al. in a review article [439]. This version includes multiple nozzles to cover larger substrate areas, as shown in Figure 10.60.

Though initially developed for spinel ferrite fabrication, the spin-spray technique has been shown to develop other types of oxides. In 2010, Wagata et al. [440] developed ZnO rod arrays and dense films using the spin-spray technique and investigated the post-annealing effect on the UV luminescence of the microstructures. Since then, there have been numerous investigations of ZnO and ZnO-based microstructures using the spin-spray plating method [436, 437, 441].

In the next section, we will take a closer look at the properties of common types of spinel ferrites developed by spin-spray plating.

10.6.2 Properties of Spin-Sprayed Spinel Ferrites

Spin-spray-plated thin ferrites of various compositions (NiZn, NiZnCo, NiCo, MnZn, etc.) have been reported [442–447]. They show good microwave magnetic properties even at gigahertz frequencies and have been used as wave absorbers and to improve performance in different microwave devices such as antennas, inductors and filters operating in the gigahertz range [448–454]. These films were reported to exhibit much higher FMR frequency and higher permeabilities at frequencies >100 MHz than their bulk counterparts, as shown in Figure 10.61, exceeding the Snoek's limit for bulk ferrites.

Snoek's law is a fundamental relation that governs the magnetic properties of ferromagnetic compounds and sets a bound on the permeability and cutoff frequencies. In its general form it can be stated as [1]:

$$(\mu'_{DC} - 1)f_0 \propto 4\pi M_s \tag{10.12}$$

μ'_{DC} is the initial relative permeability measured at DC; f_0 is the cutoff frequency of the ferromagnetic material, signifying the location of FMR in the frequency domain; and $4\pi M_s$ is the saturation magnetization. In general, the saturation magnetization is constant for a given family of magnetic compounds, which sets an upper bound on the initial permeability and cutoff frequency (Snoek's limit). In the case of spin-spray-plated ferrites, the thin film geometry, which induces anisotropy fields, as well as different compositions (which is used to control anisotropy), can be used to manipulate the cutoff frequency. A reduction of initial permeability is noted with increasing cutoff frequency, which is expected from the mutual dependence of initial permeability and cutoff frequency described in Snoek's product.

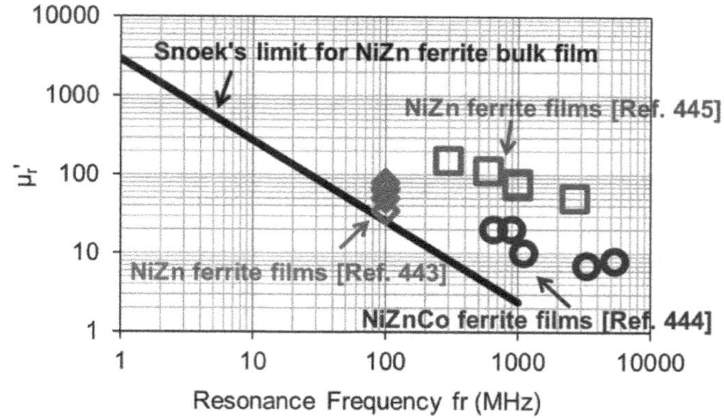

Figure 10.61 Relative permeability (μ_r) versus magnetic resonance frequency of some spin-sprayed ferrites showing applicable frequency range up to 5 GHz.

Initial permeability, μ_i, and FMR frequency of ferrites depend not only on their composition but also on their microstructure. The mechanism behind μ_i of spin-spray-plated ferrites was investigated van der Zaag et al. [103, 445]. They reported that the relatively high values of μ_i obtained by spin-spray plating are due to the low deposition temperature, which limits the adverse magnetostrictive effects due to differences in thermal expansion coefficients between the substrate and the film, in addition to having a reasonable grain size. They were able to improve μ_i of spin-spray-plated NiZn-ferrite by >100% by annealing at 500 °C for 30 min in air and concluded that annealing leads to quenching of the orbital moment of Fe^{2+} ions on the octahedral site by oxidation, which gives rise to reduction in the spin–orbit coupling contribution to the anisotropy, K, of a ferrite. As $\mu_i \propto 1/K$, μ_i increases. Previously, van der Zaag et al. have shown using neutron depolarization on polycrystalline MnZn ferrites that a transition in the intragranular domain structure from one to two domains occurs at grain size ~4 μm [103]. They concluded that NMGB model [102, 383] can account for the entire grain size dependence of the initial permeability of the low-anisotropy ferrites, irrespective of the domain state, and that in wet chemically prepared ferrites such as spin-spray-plated ferrites, the spinel structure persists within 1 nm of the grain boundary [104]. These findings suggest that, although the NMGB model considers "nonmagnetic" grain boundaries, in practice, the grain boundary may be hard magnetic, particularly for MnZn ferrites [103].

Various other authors have investigated the influence of growth parameters of the spin-spray technique upon the microstructure and properties of the developed spinel ferrites [441, 442, 446, 447]. Recently, Wang et al. [446] found that properties of the ferrite film depend on the pH values and concentration of the precursor and oxidizing solutions. They reported that by varying the pH value of precursor and oxidizer, the microstructure of NiZn-ferrite thin film can be manipulated, further bringing varying magnetic/microwave properties. They found out that though the pH of the solutions did not have a significant effect on the grain size of the plated ferrite, the magnetization, permeability, coercivity, and squareness could be controlled by the pH and concentration of the solutions. This again could be due to the reduction of the Fe^{2+} content which leads to a reduction in K as the grain size was not significantly affected but the resistivity varied as well. They achieved a high permeability μ_r of 200 at 0.5 GHz with low-loss magnetic tangent of 0.027 at 3–5 Gigahertz range with a precursor pH value of 4.6 and oxidizer pH value of 9.6.

Commonly used spin-spray-plated ferrites and their magnetic properties are shown in Table 10.14.

Table 10.14 Composition, saturation magnetization M_s, relative permeability μ', magnetic loss tangent tan δ_m, and cutoff frequency of common spin-spray-plated ferrites.

Ferrite composition	M_s (emu/cc)	$4\pi M_s$ (G)	tan δ_m (at 1–1.5 GHz)	μ' (at 1–2.5 GHz)	Cutoff frequency (GHz)
NiZn	358–557	4499–7000	0.027–0.1	>10	1–2.8
NiZnCo	340–390	4273–4900	0.05–0.14	>7	0.65–3.3
NiCo	420	5278	0.04–0.1	>8	Up to 5.4

TOS: thermally oxidized silicon

Other attractive properties of spin-spray-plated ferrites include surface resistivity >1×10^5 Ω/cm^2, stability to temperatures greater than 260 °C (i.e. the lead-free soldering temperature [444]), and the ability to withstand standard bend test with no peel-off after one million cycles of bending at a 3 mm radius of curvature [455].

Here, we will focus on the most used spin-spray-plated ferrites; NiZn and NiZnCo-ferrite.

10.6.2.1 Nickel Zinc (NiZn) Ferrite Thin Films

Spin-spray-plated polycrystalline NiZn-ferrite thin films with various compositions have been fabricated and studied over the years [432, 442, 444, 445, 453]. These fabricated films typically are found to have pure polycrystalline, well-defined spinel phase with no obvious preferential orientations. They exhibit coercivity as low as 15 Oe, FMR frequency as high as 1.2 GHz, and large saturation magnetization (M_s) greater than 400 emu/cc. They were magnetically isotropic in the in-plane (easy and hard magnetic axis) direction, with no appreciable difference in the *M–H* loops but exhibited magnetic anisotropy in the in-plane and out-of-plane *M–H* loops.

NiZn-ferrite noise suppressors absorb noise currents by magnetic losses as they radiate, making them very good thin film electromagnetic noise absorber, which is useful for integrated circuits and microwave devices working in the gigahertz range [434, 435]. In fact, NiZn-ferrite films of 3 μm in thickness, suppress gigahertz noise more effectively than commercialized noise suppressors. The sheet-type noise suppressors, made by a roll-spray ferrite plating process, have surface resistivity >6×10^4 Ω·cm^2, are stable at temperatures greater than 260 °C (i.e. the lead-free soldering temperature) [432, 433, 444], and have the ability to withstand the standard bend test (JIS C5016: no peel-off occurred after one million cycles of bending at a 3 mm radius of curvature) [455].

10.6.2.2 Nickel Zinc Cobalt (NiZnCo) Ferrite Thin Films

Magnetically isotropic NiZnCo-ferrite thin films with low magnetic losses even at 900 MHz have been reported. NiZnCo-ferrite thin films can be used as a shielding material against electromagnetic interference for RFID (900 MHz or 2.45 GHz). Spin-sprayed NiZnCo-ferrite films have been shown to have natural resonance frequencies from 3 to 5 GHz and values of μ' from 5 to 8 at frequencies up to a few gigahertz. This is due to the large Co^{2+} single-ion anisotropy, which reduces μ_i' but increases FMR frequency. These films are promising for magnetic field shielding for 900 MHz RFID applications [443].

10.6.3 Multiferroic Composites

Multiferroic materials have gained a lot of interest due to their capability of efficient energy transfer between electric and magnetic energy, which makes for potential applications in multifunctional devices. These materials have two or more constituent phases of ferri/ferromagnetic phase and ferroelectric phase that can cross interact with each other to create a ME coupling/effect. ME effect is defined as a dielectric polarization variation as a response to an applied magnetic field, or an induced magnetization by an external electric field, which may arise directly between the phases, or indirectly via strain/stress. This direct interaction between ferri/ferromagnetism and ferroelectricity enables *E*-field control of magnetism, which is of great fundamental and technical importance in magnetic and spintronics device applications or M-field control of polarization.

Spin-spray-plated ferrites provide great opportunities for novel multiferroic composite devices. As new chemical bonds are formed during spin-spray process, strong adhesion and ME coupling between the spin-spray-plated ferrite film and a ferroelectric substrate can be realized. A good lattice match between the ferroelectric material and spinel ferrites also leads to excellent wetting between the two materials, forming strong interface bonding that is important for achieving strong ME coupling.

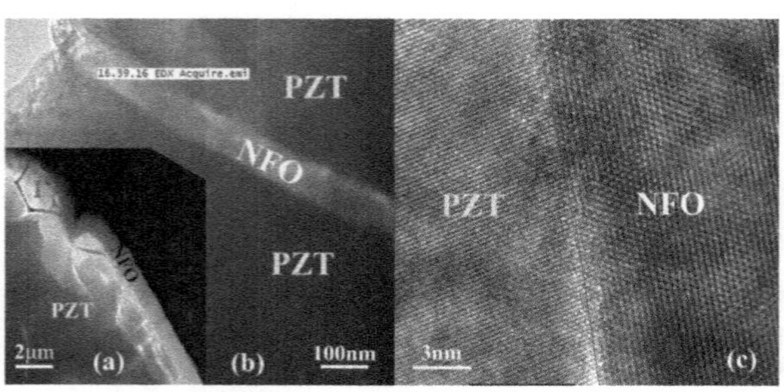

Figure 10.62 (a) Cross-sectional high-angle annular dark field (HAADF) image of NFO/PZT composite, (b) high-resolution transmission microscope (HRTEM) image showing NFO growth in PZT, and (c) enlarged HRTEM showing NFO/PZT interface from [28].

Liu et al. [456] investigated the ME interactions in horizontal heterostructure multiferroic composite of spin-spray-plated $Ni_{0.23}Fe_{2.77}O_4$ (NFO) on commercially available ferroelectric (PZT) substrates. These authors found that though there was tight interface and a strong adhesion between the two phases, there was no obvious lattice matching, even though the lattice parameter of the spinel NFO (a = 8.33 Å) and the perovskite PZT (a = 4.03 Å) phases are well matched with the NFO lattice parameter being very close to two times of PZT as seen in Figure 10.62.

Even without the obvious lattice match, strong adhesion observed in the spin-spray-deposited NFO/PZT multiferroic composite, as well as the low synthesis temperature, still provide a viable route to the synthesis of novel high-quality multiferroic materials by spin-spray deposition. A giant microwave ME interaction was report by Liu et al. [457] in Fe_3O_4/PZN-PT and Fe_3O_4/PMN-PT, which resulted in an electrostatically tunable FMR field range of 860 and 600 Oe, respectively, when the external magnetic fields were applied along the in-plane and the external electric fields were applied across the thickness of substrates starting from 0 to 8 kV/cm. This corresponded to a record-high ME coupling coefficient of 108 and 67 Oe·cm/kV, respectively, as seen in Figure 10.63. Depending on whether the electrostatic field–induced in-plane magnetic anisotropy, H_{eff}, is parallel or perpendicular to the applied external magnetic field, the FMR field can be tuned by an electric field.

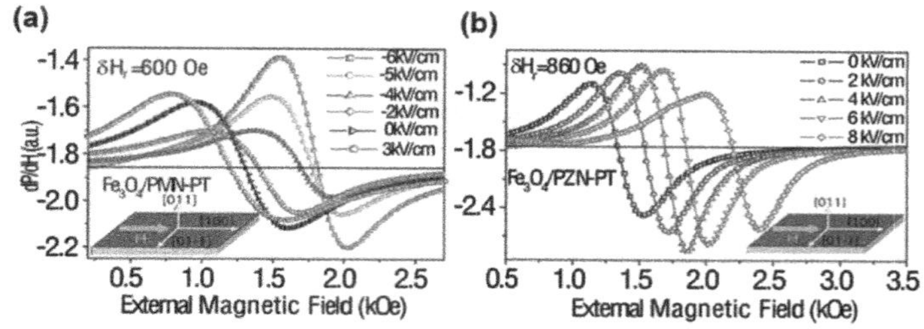

Figure 10.63 FMR absorption spectra of the spin-sprayed Fe_3O_4/PZN-PT and Fe_3O_4/PMN-PT composites subjected to electric fields from [457].

This constitutes a simple but effective approach for achieving twice the tunable FMR frequency range, leading to new opportunities of electrostatically tunable magnetic device design with large effective magnetic field tunability. Spin-spray-plated multiferroic composites have been reported to display electric field dependence of the magnetic hysteresis loops, as shown in Figure 10.64. Due to the strong adhesion between the ferrite and ferroelectric/piezoelectric material, strain on a magnetic material by a mechanical stress alters the magnetization and changes the shape of hysteresis loops. Therefore, magnetization manipulation can be realized.

As then, various authors have reported on spin-spray-plated multifunctional and ME heterostructures [439, 458–460]. Hong et al. [441] investigated the properties of spin-spray-plated Fe_3O_4/ZnO thin films, reporting high optical transmittance above 85% and low electrical resistivity in the range of $\sim 10^{-2}$ Ω cm after UV irradiation, while Zhou et al. [459] demonstrated voltage-induced effective FMR field of 14 Oe in spin-spray-plated Fe_3O_4/ZnO ME heterostructures. Liu et al. [461] developed NiZn thin films with different compositions on enhanced ultrasonic system and significant improvements in microstructure and magnetic properties of NiZn films were observed. They reported formation of triangular grain morphology as a result of anisotropic growth based on thin-film growth mechanism and a preferred (222) anisotropic growth, which they deduced as the consequence of a uniform, fast-rate deposition of high-quality ferrite thin film by the spin-spray method. They also observed a transition of magnetization mechanism along the increment of iron concentration, which resulted in a larger permeability at high frequency. Table 10.15 shows the progression of spin-spray ferrites.

Figure 10.64 Magnetic hysteresis loops and remanence behavior with external magnetic fields along the (a) compressive stress direction (d31 = d32 < 0) of Fe_3O_4/PZT, (b) compressive stress direction [01−1] (d31) of Fe_3O_4/PMN-PT, and (c) tensile stress direction [100] (d32) of Fe_3O_4/PMN-PT from [457].

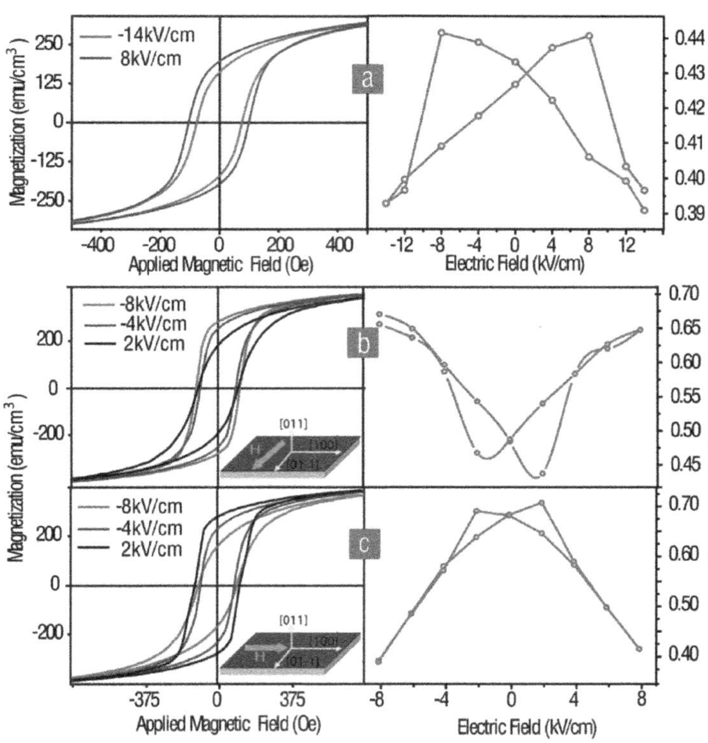

Table 10.15 Major milestones in spin-spray plating of ferrites.

Year	Type of ferrite	Intermediate layer	Substrates	Film thickness (μm)	T_{system} (°C)	Grow rate (Å/min)	Key results	Ref.
1983	$Co_{0.5}Fe_{2.5}O_4$	None	Cu	0.05	70	30–100	First ever reported spin-spray technique on ferrites. Air oxidation by passing air bubbles through solution. Crystalline spinel film formed <80 °C, no preferred orientation.	[430]
	$Co_{0.7}Fe_{2.3}O_4$	Cu thin film	PET	0.16		30–100		
	$Co_{0.4}Fe_{2.6}O_4$	Fe_3O_4	Stainless steel	0.16–0.8		30–100		
	Fe_3O_4	Fe_3O_4		1	65	300–1000		
1987	$Ni_{0.28}Zn_{0.25}Fe_{2.47}O_4$	None, SiO_2	Glass	0.8	70		$NaNO_2$ as an oxidizing agent and CH_3COONH_4 buffer to stabilize the pH. Ferrite plated on a circulator. Circulator behavior observed at 4–4.4 GHz, though very weak with 0.6 dB isolation due to the thin layer of film compared to the substrate.	[433]
	Fe_3O_4	None, SiO_2	GaAs	0.8				
1988	$Co_xFe_{3-x}O_4$ ($x = 0.43 \pm 0.03$)	None	Glass	0.4–0.8	55–90	200–300	$NaNO_2$ as an oxidizing agent and CH_3COONH_4 buffer to stabilize the pH. Films plated at 90 °C on glass substrates exhibited preferential orientation of the (1 0 0) and (1 1 1) planes parallel to the surface, respectively. Result showed deposition rate increased with increase in oxidant concentration and also increased with increase in deposition temperature.	[447]
	Fe_3O_4		α-Al_2O_3 (0 0 1) and $MgAl_2O_4$ (1 0 0) crystal faces	0.4–0.8				
2003	$Ni_{0.2}Zn_{0.2}Fe_{2.6}O_4$	None	Glass PCB	3	90		Ferrite exhibited magnetic and reflection losses suitable for noise suppressors operated up to 3 GHz.	[434]
2004	$Ni_{(0.2\ to\ 0.3)}Zn_xFe_{(2.74\ to\ 2.81)-x}O_4$	None		0.4–3	90		Results strongly suggests that the lower frequency resonance commonly observed in these ferrite films are due to induced anisotropy by the semi-hard residual domain structure plus the intrinsic crystalline anisotropy associated with the spinel structure.	[443]
2006	$Fe_3O_{4+\delta}$	None	Glass and polyimide sheet	1.6	90	520	"One-liquid" spin-spraying method (spin-sprayed ferrite plating method in which the oxidizing and reaction solutions are combined into one)	[438]

Table 10.15 (Continued)

Year	Type of ferrite	Intermediate layer	Substrates	Film thickness (μm)	T_{system} (°C)	Grow rate (Å/min)	Key results	Ref.
2009	Fe_3O_4	None	Ferroelectric PZT, PMN-PT, and PZN-PT	0.9	90	300	Demonstrated a new concept to significantly enhance the microwave resonance field-tunable range up to 1000 Oe by switching the external magnetic field direction with respect to the crystallographic orientation.	[457]
2014	Fe_3O_4	ZnO	Glass	1.4	90		Demonstrated voltage-induced effective FMR field of 14 Oe in spin-spray-plated Fe3O4/ZnO magnetoelectric (ME) heterostructures.	[459]
2015	$Ni_{0.27}Zn_{0.1}Fe_{2.63}O_4$	None	Glass	0.7	90	400	High permeability $\mu_r > 200$ at 0.5 GHz with low-loss tan δ_m (0.027 at 3–5 GHz) was achieved. These results provide viable route for gigahertz range microwave applications.	[446]
2016	Fe_3O_4	ZnO	Glass	1.4	95		Reported high optical transmittance above 85%, and low electrical resistivity in the range of ~10−2 Ω cm after UV irradiation	[441]
2017	$Ni_{0.27}Zn_{0.1}Fe_{2.63}O_4$	Polyimide layer, SiO_2	Silicon wafer	2.1	95	400	Ferrite-based inductor demonstrated stable inductance within a wide frequency ranged from 700 MHz to 6 GHz, with a peak value of quality factor = 23, providing promising applications in RFICs	[452]
2018	$Ni_{0.27}Zn_{0.06}Fe_{2.67}O_4$	None	Rogers TMM 10i printed circuits board	10	90		Ferrite-based isolator with tunable central frequency from 5.63 to 17.57 GHz provides viable route for different microwave components and radar system	[451]
2021	$Ni_{2.09-0.7x}Zn_{0.91-0.3x}Fe_xO_4$	None	Glass		95	300–500	Spin-spray deposition system based on Sono-Tek ultrasonic atomizing solution. Observed a transition of magnetization mechanism along the increment of iron concentration, which resulted in a larger permeability at a high frequency	[461]

10.6.4 Cation Distribution

The cation distribution of spin-spray ferrite was investigated by Harris et al. using conversion electron, EXAFS on spin-spray NiZn-ferrite [$Ni_{0.15}Zn_yFe_{2.85-y}O_4$ ($y = 0.16, 0.23, 0.40, 0.60$)] thin films with thickness 0.5 μm [255, 256]. They measured the X-ray absorption encompassing the K edges of Fe (7111 eV), Ni (8333 eV), and Zn (9659 eV) and then performed data collection through total electron-yield detection scheme on beamline X23B at the National Synchrotron Light Source. Data analysis was carried out using Fourier transform analysis. From these data, the relative occupancy of the cations on the octahedral and tetrahedral sites were determined. They also collected data from standards of Fe_2O_3 and Fe_3O_4 as control in identifying the main features of the Fourier-transformed EXAFS data. Vibrating sample magnetometer (VSM) measurements were also used to compare with the structural results attained from the EXAFS. They found that Ni ions occupied the octahedral sites, while the Zn ions occupied the tetrahedral sites. Fe cations occupy both sites in a ratio determined by the Zn/Fe ratio. The valence distribution of Fe ions on these sites is consistent with that of an inverted spinel. The saturation magnetization varies with increasing the Zn/Fe ratio, first increasing due to the imbalance on the Fe spin lattice and later decreasing due to the disruption of the A–B superexchange. This is in accordance with the findings of Hastings and Corliss [462] who found $ZnFe_2O_4$ particles to be a normal spinel, and $NiFe_2O_4$ particles to be inverse spinel and then when Zn is added to $NiFe_2O_4$ at the expense of Ni, the Zn^{+2} ions preferentially occupy the A sites displacing some of the Fe ions to the B-sites.

10.6.5 Limitations and Issues

So far, the spin-spray plating process can only be used at low temperature, forming oxides such as spinel ferrites. Hexaferrites and garnets cannot be developed using this process due to their more complex structures and the higher temperatures needed.

Though the spin-spray process can be used to produce thicker films need for many microwave applications, the higher FMR linewidths than those processed by PLD, in combination with losses attributed to contaminants from the chemical processing make these films unsuitable for such applications.

10.6.6 Outlook

The spin-spray deposition process is a low-temperature (70–95 °C) process that enables rapid growth rates (300–1000 Å/min) of thick (up to 10 μm) dense films on substrates such as non-heat-resistant materials as plastics, GaAs integrated circuits, and organic materials as biomaterials (e.g. enzymes) having very low heat resistance. The spin-sprayed ferrite films have good magnetic properties comparable to bulk samples, high hydrophilicity, and a strong biocompatibility. The magnetic/microwave properties of these plated films are ideal for gigahertz range microwave application, for example, inductors, antennas, and electromagnetic interference suppression. However, despite the advances and tremendous success in the development of the spin-spray plating process so far, only low-temperature forming oxides such as spinel ferrites can be deposited using this process. Hexaferrites and garnets cannot be deposited, using this process, owing to their more complex structures that require higher temperature deposition. Also, although the spin-spray process can be used to produce thicker films needed for many microwave applications, their larger FMR linewidths (than those processed by PLD), with losses attributed to contaminants from the chemical processing, make these films unsuitable for such applications.

10.7 Conclusions

We have given an overview of the various deposition techniques used for ferrite layer deposition and the progress made using these techniques in depositing ferrite thin for a number of key applications.

The basic understanding of what is feasible has been obtained using MBE studies. Studies with MBE have been largely confined to the simplest of the spin ferrites, mostly centered on Fe_3O_4 and XFe_2O_4 (with X = Ni or Co). These studies have revealed that as long as the layer thickness ≥ 30 nm, ferrite films with bulk-like properties, such as magnetization and anisotropy can be deposited [27]. For layer thickness < 30 nm, the effect of the interfaces comes into play [96], although this can be modeled by a "dead" magnetic interface layer, Tjeng and coworkers have clearly shown that this is the result of the growth [75]. To avoid the creation of a nonstable polar surface during growth of the first layer of Fe_3O_4 deposition proceeds by creating first a charge-neutral FeO layer from which later proper Fe_3O_4 is formed (Section 10.2.4.3). Second, these fundamental studies of the MBE growth of Fe_3O_4 have highlighted the complicating roles APBs play, as Fe_3O_4 may nucleate at various sites and grow to a single layer with a coherent oxygen lattice, but with an incoherent cation ion lattice. The latter two phenomena contribute to impeding the success of Fe_3O_4-based spin tunnel junction for spintronic applications. The growth mode in which first an FeO layer is created, leads to an imbalance between A and B cation sites at the interface. This imbalance at the interface likely contributes to the mixing of spin states and the deterioration of the theoretical 100% spin polarization anticipated for the current passing from the half-metallic Fe_3O_4 layer [27].

Although they play a different role, the presence of APBs also does not help. As APBs grow *orthogonal* to the tunneling interface, they may not necessarily cause spin scattering, yet they may contribute to a nontunneling shunt current that will limit the tunneling magnetoresistance (TMR) effect that can be achieved to 26% at room temperature (Section 10.2.5) [139]. This is makes using Fe_3O_4-based tunnel junction a challenging approach to pursue for practical applications, as metallic-based junctions achieve already far larger room temperature TMR effects [146].

As PLD provides congruent transfer of target material under ablation to a substrate, PLD is, in principle, ideally suited to grow the (complex) magnetic oxides in which stoichiometry plays a key role in their magnetic properties. In addition, on the proper substrate, PLD can provide epitaxial grow, albeit at a somewhat higher deposition temperature of 500–900 °C (see Table 10.16). This implies that growth of some complex layer stacks or devices require further consideration to avoid layer intermixing, for instance, by using NiO buffer layers to avoid migration of Mg^+ ions from underlying layers into deposited ferrite layers.

Sputtering is the routinely used deposition method for thin film deposition in practical applications. Sputtering of ferrite films has been shown to be feasible and hold promise for ferrite film applications: The deposition of YIG films seems an option for magnon spintronic devices [245] and of hexaferrite films, notably BaM, for RF- and millimeter-wave devices. Even soft ferrite MnZn-ferrite films, achieving a μ of $\approx$ 30 have been demonstrated using sputtering [382]. The latter would be ideal for improving the flux transport and thereby the efficiency of magnetic devices (see Fig. 10.3) [26]. Perhaps the deposition temperature of up to 600 ^{0}C still poses a problem, so that the combination with postdeposition anneal processes should be investigated to obtain the same performance of these films at lower deposition temperature.

ALD has been shown to be a promising thin film deposition process for ferrites. However, achieving a practical ALD process for ferrites is still challenging due to the low growth rates (0.2 to 3.2 Å per cycle) of the films and difficulties involved in developing air-stable ferrite films that demonstrate room temperature ferrimagnetism without further thermal treatment. Also, it is important to note that thus far, ferrite development by ALD is limited to spinel ferrites likely due to

Table 10.16 Conclusions: Ferrite deposition methods and their applications. The table gives the typical condition under which good-quality ferrite layers can be deposited.

Deposition method	Deposition temperature (°C)	Deposition Area (typically feasible)	Deposition rate (nm/min)	Typical layer thickness	Typical application
MBE	250–300	150 mm wafers	0.1–5	1–100 nm	• Fundamental studies • Spintronics
PLD	500–1000	150 mm wafers	1–10	0.1–10 μm	• Spintronics • Magnonics • Microwave and millimeter-wave electronics
Sputtering	RT-650	150 mm wafers	10–300	1 nm–1 mm	• Magnetic recording media • Supporting metal structures in TMR, Spintronics, Magnonics, and other
ALD	200–400	150 mm wafers	0.01–0.1	5–100 nm	• Supporting structures in Spintronics, Magnonics, and other
Spin-spray	RT-90	Continuous flow sheets	20–100	1–100 μm	• Inductors • Antennas • EMI-suppression

their simpler structure compared to hexaferrites and garnets. However, with the increased interest in this processing technique, ALD will quickly become an important technique in the fabrication of functional ferrite thin film.

With its low processing temperature advantage (70–95 °C) and rapid growth rates (300–1000 Å/min), the spin-spray process is an excellent candidate for thick (up to 10 μm), highly dense polycrystalline ferrite films. This technique has already been extended to the deposition of ferrites onto rolls of plastic sheet, showing the capability of thick ferrite film deposition on larger substrates. The magnetic/microwave properties of these plated films are ideal for gigahertz range microwave application, for example, inductors, antennas, spintronics, and electromagnetic interference suppression. Despite the advances and tremendous success in the development of the spin-spray plating process so far, only low-temperature forming oxides such as spinel ferrites can be deposited using this process. Hexaferrites and garnets cannot be developed using this process due to their more complex structures and higher temperatures needs.

The general conclusions of this chapter with respect to the suitability of the various ferrite deposition methods for applications of ferrite layers are summarized in Table 10.16. Overall, it seems that future efforts using thin ferrite films are best directed at using PLD to grown layers for magnonics and microwave of millimeter electronics. As in the latter two applications, the interface effect causing spin scattering does not impede the application such as in spintronics.

Acknowledgment

The authors would like to thank Prof. Dr. F. Roozeboom (TU Twente, The Netherland) for the critical reading of parts of the manuscript (Groningen/Eindhoven/Boston, November 2021).

References

1 Smit, J. and Wijn, H.P.J. (1959). *Ferrites*. Eindhoven: Philips Technical Library.

2 Snelling, E.C. (1988). *Soft Ferrites, Properties and Applications*, 2e. London: Butterworth & Co. Ltd.

3 Baibich, M.N. et al. (November 1988). Giant magnetoresistance of (001)Fe/(001)Cr magnetic superlattices. *Physical Review Letters* 61 (21): 2472–2475. doi: 10.1103/PhysRevLett.61.2472.

4 Bednorz, J.G. and Müller, K.A. (June 1986). Possible high Tc superconductivity in the Ba—La—Cu—O system. *Zeitschrift für Physik B Condensed Matter* 64 (2): 189–193. doi: 10.1007/BF01303701.

5 Shukla, P., Chernatynskiy, A., Nino, J.C., Sinnott, S.B., and Phillpot, S.R. (January 2011). Effect of inversion on thermoelastic and thermal transport properties of $MgAl_2O_4$ spinel by atomistic simulation. *Journal of Materials Science* 46 (1): 55–62. doi: 10.1007/s10853-010-4795-7.

6 Mitra, A. (2017). Structural and magnetic properties of YIG thin films and interfacial origin of magnetisation suppression. University of Leeds.

7 Tanwar, K., Gyan, D.S., Gupta, P., Pandey, S., OmParkash, O., and Kumar, D. (2018). Investigation of crystal structure, microstructure and low temperature magnetic behavior of Ce^{4+} and Zn^{2+} co-doped barium hexaferrites ($BaFe_{12}O_{19}$). *RSC Advances* 8 (35): 19600–19609. doi: 10.1039/C8RA02455C.

8 Yoder, H.S. and Keith, L.M. (1951). Complete substitution of aluminum for silicon: The system $3MnO{\cdot}Al_2O_3{\cdot}3SiO_2$—$3Y_2O_3{\cdot}5Al_2O_3$. *American Mineralogist* 36 (7–8): 519–533.

9 Bertaut, F. and Forrat, F. (1956). Structure des ferrites ferrimagnetiques des terres rares. *Comptes Rendus de l'Académie des Sciences* 242: 382.

10 Cherepanov, V., Kolokolov, I., and L'vov, V. (July 1993). The saga of YIG: Spectra, thermodynamics, interaction and relaxation of magnons in a complex magnet. *Physics Reports* 229 (3): 81–144. doi: 10.1016/0370-1573(93)90107-O.

11 Goulon, J. et al. (June 2012). Site-selective couplings in x-ray-detected magnetic resonance spectra of rare-earth-substituted yttrium iron garnets. *New Journal of Physics* 14 (6): 063001. doi: 10.1088/1367-2630/14/6/063001.

12 LeCraw, R.C., Spencer, E.G., and Porter, C.S. (June 1958). Ferromagnetic resonance line width in yttrium iron garnet single crystals. *Physical Review* 110 (6): 1311–1313. doi: 10.1103/PhysRev.110.1311.

13 Pullar, R.C. (September 2012). Hexagonal ferrites: A review of the synthesis, properties and applications of hexaferrite ceramics. *Progress in Materials Science* 57 (7): 1191–1334. doi: 10.1016/j.pmatsci.2012.04.001.

14 Néel, L. (1948). Proprietes magnetiques des ferrites – Ferrimagnetism et antiferromagnetisme. *Annals of Physics* 3 (2): 137–198.

15 Kramers, H.A. (January 1934). L'interaction Entre les Atomes Magnétogènes dans un Cristal Paramagnétique. *Physica* 1 (1–6): 182–192. doi: 10.1016/S0031-8914(34)90023-9.

16 Anderson, P.W. (July 1950). Antiferromagnetism. Theory of superexchange interaction. *Physical Review* 79 (2): 350–356. doi: 10.1103/PhysRev.79.350.

17 Goodenough, J.B. (October 1955). Theory of the role of covalence in the perovskite-type manganites [La, M(II)]MnO_3. *Physical Review* 100 (2): 564–573. doi: 10.1103/PhysRev.100.564.

18 Goodenough, J.B. (August 1958). An interpretation of the magnetic properties of the perovskite-type mixed crystals $La_{1-x}Sr_xCoO_{3-\lambda}$. *Journal of Physics and Chemistry of Solids* 6 (2–3): 287–297. doi: 10.1016/0022-3697(58)90107-0.

19 Kanamori, J. (July 1959). Superexchange interaction and symmetry properties of electron orbitals. *Journal of Physics and Chemistry of Solids* 10 (2–3): 87–98. doi: 10.1016/0022-3697(59)90061-7.

20 Kanamori, J. (May 1960). Crystal distortion in magnetic compounds. *Journal of Applied Physics* 31 (5): S14–S23. doi: 10.1063/1.1984590.

21 Goodenough, J. (2008). Goodenough-Kanamori rule. *Scholarpedia* 3 (10): 7382. doi: 10.4249/scholarpedia.7382.

22 Harris, V.G. (March 2012). Modern microwave ferrites. *IEEE Transactions on Magnetics* 48 (3): 1075–1104. doi: 10.1109/TMAG.2011.2180732.

23 Heeger, A.J. and Houston, T.W. (August 1964). Nuclear magnetic resonance in ferrimagnetic $MnFe_2O_4$. *Physical Review* 135 (3A): A661–A670. doi: 10.1103/PhysRev.135.A661.

24 Wegener, W., Scheerlinck, D., Legrand, E., Hautecler, S., and Brabers, V. (July 1974). Inelastic neutron scattering study of acoustical magnons in $MnFe_2O_4$. *Solid State Communications* 15 (2): 345–349. doi: 10.1016/0038-1098(74)90773-X.

25 Anderson, E.E. (June 1964). Molecular field model and the magnetization of YIG. *Physical Review* 134 (6A): A1581–A1585. doi: 10.1103/PhysRev.134.A1581.

26 van der Zaag, P.J., Ruigrok, J.J.M., and Gillies, M.F. (January 1998). New options in thin film recording heads through ferrite layers. *Philips Journal of Research* 51 (1): Art. no. 1. doi: 10.1016/S0165-5817(97)00002-8.

27 de Groot, R.A., Mueller, F.M., van Engen, P.G., and Buschow, K.H.J. (June 1983). New class of materials: Half-metallic ferromagnets. *Physical Review Letters* 50 (25): Art. no. 25. doi: 10.1103/PhysRevLett.50.2024.

28 Julliere, M. (September 1975). Tunneling between ferromagnetic films. *Physics Letters A* 54 (3): 225–226. doi: 10.1016/0375-9601(75)90174-7.

29 Yanase, A. and Siratori, K. (January 1984). Band structure in the high temperature phase of Fe_3O_4. *Journal of the Physical Society of Japan* 53 (1): 312–317. doi: 10.1143/JPSJ.53.312.

30 Serga, A.A., Chumak, A.V., and Hillebrands, B. (July 2010). YIG magnonics. *Journal of Physics D: Applied Physics* 43 (26): 264002. doi: 10.1088/0022-3727/43/26/264002.

31 Kruglyak, V.V., Demokritov, S.O., and Grundler, D. (July 2010). Magnonics. *Journal of Physics D: Applied Physics* 43 (26): 264001. doi: 10.1088/0022-3727/43/26/264001.

32 Lenk, B., Ulrichs, H., Garbs, F., and Münzenberg, M. (October 2011). The building blocks of magnonics. *Physics Reports* 507 (4–5): 107–136. doi: 10.1016/j.physrep.2011.06.003.

33 d'Allivy Kelly, O. et al. (August 2013). Inverse spin Hall effect in nanometer-thick yttrium iron garnet/Pt system. *Applied Physics Letters* 103 (8): 082408. doi: 10.1063/1.4819157.

34 Onbasli, M.C. et al. (October 2014). Pulsed laser deposition of epitaxial yttrium iron garnet films with low Gilbert damping and bulk-like magnetization. *APL Materials* 2 (10): 106102. doi: 10.1063/1.4896936.

35 Liu, T. et al. (May 2014). Ferromagnetic resonance of sputtered yttrium iron garnet nanometer films. *Journal of Applied Physics* 115 (17): 17A501. doi: 10.1063/1.4852135.

36 Pirro, P. et al. (January 2014). Spin-wave excitation and propagation in microstructured waveguides of yttrium iron garnet/Pt bilayers. *Applied Physics Letters* 104 (1): 012402. doi: 10.1063/1.4861343.

37 Hahn, C. et al. (April 2014). Measurement of the intrinsic damping constant in individual nanodisks of Y3Fe5O12 and Y3Fe5O12|Pt. *Applied Physics Letters* 104 (15): 152410. doi: 10.1063/1.4871516.

38 Slonczewski, J.C. (June 1996). Current-driven excitation of magnetic multilayers. *Journal of Magnetism and Magnetic Materials* 159 (1–2): L1–L7. doi: 10.1016/0304-8853(96)00062-5.

39 Berger, L. (October 1996). Emission of spin waves by a magnetic multilayer traversed by a current. *Physical Review B* 54 (13): 9353–9358. doi: 10.1103/PhysRevB.54.9353.

40 Slavin, A. and Tiberkevich, V. (April 2009). Nonlinear auto-oscillator theory of microwave generation by spin-polarized current. *IEEE Transactions on Magnetics* 45 (4): 1875–1918. doi: 10.1109/TMAG.2008.2009935.

41 Dyakonov, M.I. and Perel, V.I. (July 1971). Current-induced spin orientation of electrons in semiconductors. *Physics Letters A* 35 (6): 459–460. doi: 10.1016/0375-9601(71)90196-4.

42 Hirsch, J.E. (August 1999). Spin Hall effect. *Physical Review Letters* 83 (9): 1834–1837. doi: 10.1103/PhysRevLett.83.1834.

43 Fert, A., Cros, V., and Sampaio, J. (March 2013). Skyrmions on the track. *Nature Nanotechnology* 8: 152–156.

44 Urazuka, Y., Imamura, K., Oyabu, S., Tanaka, T., and Matsuyama, K. (November 2014). Successive logic-in-memory operation in spin wave-based devices with domain wall data coding scheme. *IEEE Transactions on Magnetics* 50 (11): 1–3. doi: 10.1109/TMAG.2014.2320759.

45 Schütte, C. and Garst, M. (September 2014). Magnon-skyrmion scattering in chiral magnets. *Physical Review B* 90 (9): 094423. doi: 10.1103/PhysRevB.90.094423.

46 Dutta, S. et al. (September 2015). Non-volatile clocked spin wave interconnect for beyond-CMOS nanomagnet pipelines. *Scientific Reports* 5 (1): 9861. doi: 10.1038/srep09861.

47 Frey, P. et al. (December 2020). Reflection-less width-modulated magnonic crystal. *Communications Physics* 3 (1): 17. doi: 10.1038/s42005-020-0281-y.

48 van der Zaag, P.J. and Borchers, J.A. (2010). Antiferromagnetic-ferromagnetic oxide multilayers: Fe_3O_4-based systems as a model. In: *Magnetic Properties of Antiferromagnetic Oxide Materials* (ed. L. Duò, M. Finazzi, and F. Ciccacci), 239–300. Weinheim, Germany: Wiley-VCH Verlag GmbH & Co. KGaA. doi: 10.1002/9783527630370.ch7.

49 Chambers, S.A. (August 2000). Epitaxial growth and properties of thin film oxides. *Surface Science Reports* 39 (5–6): 105–180. doi: 10.1016/S0167-5729(00)00005-4.

50 Lind, D.M., Berry, S.D., Chern, G., Mathias, H., and Testardi, L.R. (January 1992). Growth and structural characterization of Fe_3O_4 and NiO thin films and superlattices grown by oxygen-plasma-assisted molecular-beam epitaxy. *Physical Review B* 45 (4): 1838–1850. doi: 10.1103/PhysRevB.45.1838.

51 Shigematsu, T., Ushigome, H., Bando, Y., and Takada, T. (December 1980). Epitaxial growth of magnetite on rock salt by reactive evaporation. *Journal of Crystal Growth* 50 (4): 801–806. doi: 10.1016/0022-0248(80)90141-4.

52 Wolf, R.M., De Veirman, A.E.M., van der Sluis, P., van der Zaag, P.J., and aan de Stegge, J.B.F. (1994). MBE growth of CoO and Fe_3O_4 films and CoO/Fe_3O_4 multilayers. *MRS Proceedings* 341: 23. doi: 10.1557/PROC-341-23.

53 Voogt, F.C., Fujii, T., Smulders, P.J.M., Niesen, L., James, M.A., and Hibma, T. (October 1999). NO_2 -assisted molecular-beam epitaxy of Fe_3O_4, $Fe_{3-\delta}O_4$, and $\gamma-Fe_2O_3$ thin films on MgO(100). *Physical Review B* 60 (15): 11193–11206. doi: 10.1103/PhysRevB.60.11193.

54 Chern, G. et al. (February 2000). Structural and magnetic characterization of Fe_3O_4/Mn_3O_4 superlattices. *Journal of Magnetism and Magnetic Materials* 209 (1–3): 138–141. doi: 10.1016/S0304-8853(99)00668-X.

55 Chern, G., Horng, L., Shieh, W.K., and Wu, T.C. (February 2001). Antiparallel state, compensation point, and magnetic phase diagram of Fe_3O4/Mn_3O_4 superlattices. *Physical Review B* 63 (9): 094421. doi: 10.1103/PhysRevB.63.094421.

56 Horng, L., Chern, G., Chen, M.C., Kang, P.C., and Lee, D.S. (April 2004). Magnetic anisotropic properties in Fe_3O_4 and $CoFe_2O_4$ ferrite epitaxy thin films. *Journal of Magnetism and Magnetic Materials* 270 (3): 389–396. doi: 10.1016/j.jmmm.2003.09.005.

57 Gota, S., Moussy, J.-B., Henriot, M., Guittet, M.-J., and Gautier-Soyer, M. (June 2001). Atomic-oxygen-assisted MBE growth of Fe_3O_4 (111) on α-Al_2O_3 (0001). *Surface Science* 482–485: 809–816. doi: 10.1016/S0039-6028(00)01066-9.

58 Greullet, F. et al. (February 2008). Large inverse magnetoresistance in fully epitaxial Fe/Fe_3O_4/MgO/Co magnetic tunnel junctions. *Applied Physics Letters* 92 (5): 053508. doi: 10.1063/1.2841812.

59 Wu, H.-C., Arora, S.K., Mryasov, O.N., and Shvets, I.V. (May 2008). Antiferromagnetic interlayer exchange coupling between Fe_3O_4 layers across a nonmagnetic MgO dielectric layer. *Applied Physics Letters* 92 (18): Art. no. 18. doi: 10.1063/1.2919081.

60 Moyer, J.A., Vaz, C.A.F., Negusse, E., Arena, D.A., and Henrich, V.E. (January 2011). Controlling the electronic structure of $Co_{1-x}Fe_{2+x}O_4$ thin films through iron doping. *Physical Review B* 83 (3): 035121. doi: 10.1103/PhysRevB.83.035121.

61 Liu, X.H., Rata, A.D., Chang, C.F., Komarek, A.C., and Tjeng, L.H. (September 2014). Verwey transition in Fe_3O_4 thin films: Influence of oxygen stoichiometry and substrate-induced microstructure. *Physical Review B* 90 (12): 125142. doi: 10.1103/PhysRevB.90.125142.

62 Bertram, F., Deiter, C., Pflaum, K., Suendorf, M., Otte, C., and Wollschläger, J. (November 2011). In-situ x-ray diffraction studies on post-deposition vacuum-annealing of ultra-thin iron oxide films. *Journal of Applied Physics* 110 (10): 102208. doi: 10.1063/1.3661655.

63 Nordmann, T., Kuschel, O., and Wollschläger, J. (September 2016). Epitaxial growth of ultrathin MgO layers on Fe_3O_4(001) films. *Applied Surface Science* 381: 28–31. doi: 10.1016/j.apsusc.2016.02.133.

64 Gálvez, S., Rubio-Zuazo, J., Salas-Colera, E., Muñoz-Noval, A., and Castro, G.R. (December 2014). Sharp chemical interface in epitaxial Fe_3O_4 thin films. *Applied Physics Letters* 105 (24): 241603. doi: 10.1063/1.4904459.

65 Moussy, J.-B. (April 2013). From epitaxial growth of ferrite thin films to spin-polarized tunnelling. *Journal of Physics D: Applied Physics* 46 (14): 143001. doi: 10.1088/0022-3727/46/14/143001.

66 Terashima, T. and Bando, Y. (September 1987). Formation and magnetic properties of artificial superlattice of CoO-Fe_3O_4. *Thin Solid Films* 152 (3): 455–463. doi: 10.1016/0040-6090(87)90261-6.

67 Lind, D.M., Berry, S.D., Chern, G., Mathias, H., and Testardi, L.R. (November 1991). Characterization of the structural and magnetic ordering of Fe_3O_4/NiO superlattices grown by oxygen-plasma-assisted molecular-beam epitaxy. *Journal of Applied Physics* 70 (10): 6218–6220. doi: 10.1063/1.350000.

68 Bando, Y. and Terashima, T. (October 1987). Structure and magnetic properties of artificial superlattice and multilayered films of magnetic oxides. *IEEE Translation Journal on Magnetics in Japan* 2 (10): 948–958. doi: 10.1109/TJMJ.1987.4549651.

69 Fujii, T., Takano, M., Katano, R., Bando, Y., and Isozumi, Y. (October 1989). Preparation and characterization of (111)-oriented Fe_3O_4 films deposited on sapphire. *Journal of Applied Physics* 66 (7): 3168–3172. doi: 10.1063/1.344154.

70 Gota, S., Guiot, E., Henriot, M., and Gautier-Soyer, M. (November 1999). Atomic-oxygen-assisted MBE growth of $\alpha-Fe_2O_3$ on $\alpha-Al_2O_3$ (0001): Metastable FeO(111)-like phase at subnanometer thicknesses. *Physical Review B* 60 (20): 14387–14395. doi: 10.1103/PhysRevB.60.14387.

71 Wang, H.-Q., Gao, W., Altman, E.I., and Henrich, V.E. (July 2004). Studies of the electronic structure at the Fe_3O_4–NiO interface. *Journal of Vacuum Science and Technology A: Vacuum, Surfaces and Films* 22 (4): 1675–1681. doi: 10.1116/1.1763900.

72 Thevuthasan, S., McCready, D.E., Jiang, W., McDaniel, E.D., Yi, S.I., and Chambers, S.A. (1999). Rutherford backscattering and channeling studies of Al and Mg diffusion in iron oxide thin films. In: *AIP Conference Proceedings*, Denton, Texas (USA), 508–511. doi: 10.1063/1.59180.

73 Lee, D.S., Wang, J.S., Modak, D.K., Liu, Y.S., Chang, C.L., and Chern, G. (May 2007). The magnetic properties of strained and relaxed $Fe_{3-x}Mg_xO_4$ ferrite films on MgO(001) and $SrTiO_3$(001) by molecular beam epitaxy. *Journal of Applied Physics* 101 (9): 09M523. doi: 10.1063/1.2712826.

74 Kim, Y.J., Gao, Y., and Chambers, S.A. (February 1997). Selective growth and characterization of pure, epitaxial α-Fe_2O_3(0001) and Fe_3O_4(001) films by plasma-assisted molecular beam epitaxy. *Surface Science* 371 (2–3): 358–370. doi: 10.1016/S0039-6028(96)00999-5.

75 Chang, C.F. et al. (October 2016). Dynamic atomic reconstruction: How Fe_3O_4 thin films evade polar catastrophe for epitaxy. *Physical Review X* 6 (4): 041011. doi: 10.1103/PhysRevX.6.041011.

76 Bloemen, P.J.H. et al. (1996). Magnetic and structural properties of MBE-grown oxidic multilayers. *Materials Research Society Symposium Proceedings* 401: 485–499.

77 Ijiri, Y., Borchers, J.A., Erwin, R.W., Lee, S.-H., van der Zaag, P.J., and Wolf, R.M. (January 1998). Perpendicular coupling in exchange-biased Fe_3O_4/CoO superlattices. *Physical Review Letters* 80 (3): 608–611. doi: 10.1103/PhysRevLett.80.608.

78 van der Zaag, P.J. et al. (June 2000). Difference between blocking and Néel temperatures in the exchange biased Fe_3O_4/CoO system. *Physical Review Letters* 84 (26): Art. no. 26. doi: 10.1103/PhysRevLett.84.6102.

79 Wu, H.-C. et al. (September 2014). Magnetic and transport properties of epitaxial stepped Fe_3O_4 (100) thin films. *Applied Physics Letters* 105 (13): 132408. doi: 10.1063/1.4897001.

80 Matzen, S. et al. (August 2011). Restoration of bulk magnetic properties by strain engineering in epitaxial $CoFe_2O_4$ (001) ultrathin films. *Applied Physics Letters* 99 (5): 052514. doi: 10.1063/1.3622307.

81 Gatel, C., Warot-Fonrose, B., Matzen, S., and Moussy, J.-B. (August 2013). Magnetism of $CoFe_2O_4$ ultrathin films on $MgAl_2O_4$ driven by epitaxial strain. *Applied Physics Letters* 103 (9): 092405. doi: 10.1063/1.4819178.

82 Fujii, T., Takano, M., Katano, R., Isozumi, Y., and Bando, Y. (February 1994). Surface and interface properties of epitaxial Fe_3O_4 films studied by Mössbauer spectroscopy. *Journal of Magnetism and Magnetic Materials* 130 (1–3): 267–274. doi: 10.1016/0304-8853(94)90683-1.

83 Voogt, F.C., Hibma, T., Zhang, G.L., Hoefman, M., and Niesen, L. (July 1995). Growth and characterization of non-stoichiometric magnetite $Fe_{3-\delta}O_4$ thin films. *Surface Science* 331–333: 1508–1514. doi: 10.1016/0039-6028(95)00347-9.

84 Wang, H.-Q., Altman, E.I., and Henrich, V.E. (February 2008). Interfacial properties between CoO (100) and Fe_3O_4 (100). *Physical Review B* 77 (8): 085313. doi: 10.1103/PhysRevB.77.085313.

85 Fontijn, W.F.J., van der Zaag, P.J., Devillers, M.A.C., Brabers, V.A.M., and Metselaar, R. (September 1997). Optical and magneto-optical polar Kerr spectra of Fe_3O_4 and Mg^{2+} – or Al^{3+} -substituted Fe_3O_4. *Physical Review B* 56 (9): 5432–5442. doi: 10.1103/PhysRevB.56.5432.

86 Fontijn, W.F.J., Wolf, R.M., Metselaar, R., and van der Zaag, P.J. (January 1997). Investigation of the stoichiometry of MBE-grown Fe_3O_4 layers by magneto-optical Kerr spectroscopy. *Thin Solid Films* 292 (1–2): 270–276. doi: 10.1016/S0040-6090(96)08910-9.

87 Fontijn, W.F.J., van der Heijden, P.A.A., Voogt, F.C., Hibma, T., and van der Zaag, P.J. (January 1997). Comparison of a stoichiometric analysis of $Fe_{3-\delta}O_4$ layers by magneto-optical Kerr spectroscopy with Mössbauer results. *Journal of Magnetism and Magnetic Materials* 165 (1–3): 401–404. doi: 10.1016/S0304-8853(96)00569-0.

88 Verwey, E.J.W. (August 1939). Electronic conduction of magnetite (Fe_3O_4) and its transition point at low temperatures. *Nature* 144 (3642): 327–328. doi: 10.1038/144327b0.

89 Shepherd, J.P., Koenitzer, J.W., Aragón, R., Spałek, J., and Honig, J.M. (April 1991). Heat capacity and entropy of nonstoichiometric magnetite $Fe_{3(1-\delta)}O_4$: The thermodynamic nature of the Verwey transition. *Physical Review B* 43 (10): 8461–8471. doi: 10.1103/PhysRevB.43.8461.

90 van der Zaag, P.J. et al. (1996). A study of the magneto-optical Kerr spectra of bulk and ultrathin Fe_3O_4. *Journal of Applied Physics* 79 (8): 5936. doi: 10.1063/1.362112.

91 Liu, X.H., Liu, W., and Zhang, Z.D. (2017). Extremely low coercivity in Fe_3O_4 thin film grown on Mg_2TiO_4 (001). *RSC Advances* 7 (69): 43648–43654. doi: 10.1039/C7RA08916C.

92 Feng, J.S.-Y. and Nicolet, M.-A. (September 1975). Substrate effects on the Verwey transition of thin magnetite films. *Journal of Physics and Chemistry of Solids* 36 (9): 835–838. doi: 10.1016/0022-3697(75)90156-0.

93 Lee, A.K.H. et al. (October 2010). Magnetic force microscopy of epitaxial magnetite films through the Verwey transition. *Applied Physics Letters* 97 (16): 162502. doi: 10.1063/1.3502606.

94 Eerenstein, W., Palstra, T.T.M., Hibma, T., and Celotto, S. (November 2002). Origin of the increased resistivity in epitaxial Fe_3O_4 films. *Physical Review B* 66 (20): 201101. doi: 10.1103/PhysRevB.66.201101.

95 Wu, H.-C., Abid, M., Chun, B.S., Ramos, R., Mryasov, O.N., and Shvets, I.V. (April 2010). Probing one antiferromagnetic antiphase boundary and single magnetite domain using nanogap contacts. *Nano Letters* 10 (4): 1132–1136. doi: 10.1021/nl902973p.

96 van der Heijden, P.A.A. et al. (July 1996). Magnetic interface anisotropy of MBE-grown ultra-thin (001) Fe_3O_4 layers. *Journal of Magnetism and Magnetic Materials* 159 (3): Art. no. 3. doi: 10.1016/0304-8853(96)00376-9.

97 Margulies, D.T. et al. (April 1996). Anomalous moment and anisotropy behavior in Fe_3O_4 films. *Physical Review B* 53 (14): 9175–9187. doi: 10.1103/PhysRevB.53.9175.

98 Parkin, S.S.P., Sigsbee, R., Felici, R., and Felcher, G.P. (March 1986). Observation of magnetic dead layers at the surface of iron oxide films. *Applied Physics Letters* 48 (9): 604–606. doi: 10.1063/1.96481.

99 Margulies, D.T., Parker, F.T., and Berkowitz, A.E. (May 1994). Magnetic anomalies in single crystal Fe_3O_4 thin films. *Journal of Applied Physics* 75 (10): 6097–6099. doi: 10.1063/1.355472.

100 van der Heijden, P.A.A. et al. (May 1997). Interlayer coupling between Fe_3O_4 layers separated by an insulating nonmagnetic MgO layer. *Physical Review B* 55 (17): Art. no. 17. doi: 10.1103/PhysRevB.55.11569.

101 Coey, J.M.D. (October 1971). Noncollinear spin arrangement in ultrafine ferrimagnetic crystallites. *Physical Review Letters* 27 (17): 1140–1142. doi: 10.1103/PhysRevLett.27.1140.

102 Visser, E.G. and Johnson, M.T. (October 1991). A novel interpretation of the complex permeability in polycrystalline ferrites. *Journal of Magnetism and Magnetic Materials* 101 (1–3): 143–147. doi: 10.1016/0304-8853(91)90707-H.

103 van der Zaag, P.J. et al. (September 1993). The initial permeability of polycrystalline MnZn ferrites: The influence of domain and microstructure. *Journal of Applied Physics* 74 (6): Art. no. 6. doi: 10.1063/1.354454.

104 van der Heijden, P.A.A., van Opstal, M.G., Swüste, C.H.W., Bloemen, P.H.J., Gaines, J.M., and de Jonge, W.J.M. (February 1998). A ferromagnetic resonance study on ultra-thin Fe_3O_4 layers grown on (001)MgO. *Journal of Magnetism and Magnetic Materials* 182 (1–2): Art. no. 1–2. doi: 10.1016/S0304-8853(97)01017-2.

105 McGuigan, L., Barklie, R.C., Sofin, R.G.S., Arora, S.K., and Shvets, I.V. (May 2008). In-plane magnetic anisotropies in Fe_3O_4 films on vicinal MgO(100). *Physical Review B* 77 (17): 174424. doi: 10.1103/PhysRevB.77.174424.

106 Schemme, T., Krampf, A., Bertram, F., Kuschel, T., Kuepper, K., and Wollschläger, J. (September 2015). Modifying magnetic properties of ultra-thin magnetite films by growth on Fe pre-covered MgO(001). *Journal of Applied Physics* 118 (11): 113904. doi: 10.1063/1.4930998.

107 Brandlmaier, A. et al. (March 2008). In situ manipulation of magnetic anisotropy in magnetite thin films. *Physical Review B* 77 (10): 104445. doi: 10.1103/PhysRevB.77.104445.

108 Martens, J. and Voermans, A. (September 1984). Cobalt ferrite thin films for magneto-optical recording. *IEEE Transactions on Magnetics* 20 (5): 1007–1012. doi: 10.1109/TMAG.1984.1063167.

109 den Broeder, F.J.A., Hoving, W., and Bloemen, P.J.H. (February 1991). Magnetic anisotropy of multilayers. *Journal of Magnetism and Magnetic Materials* 93: 562–570. doi: 10.1016/0304-8853(91)90404-X.

110 Johnson, M.T., Bloemen, P.J.H., den Broeder, F.J.A., and de Vries, J.J. (November 1996). Magnetic anisotropy in metallic multilayers. *Reports on Progress in Physics* 59 (11): 1409–1458. doi: 10.1088/0034-4885/59/11/002.

111 Purcell, S.T. et al. (August 1991). Oscillations with a period of two Cr monolayers in the antiferromagnetic exchange coupling in a (001) Fe/Cr/Fe sandwich structure. *Physical Review Letters* 67 (7): 903–906. doi: 10.1103/PhysRevLett.67.903.

112 van der Heijden, P.A.A. et al. (1995). Magnetic properties of epitaxial MBE grown thin Fe_3O_4 films on MgO(100). *Materials Research Society Symposium Proceedings* 384: 27–32.

113 Voogt, F.C., Palstra, T.T.M., Niesen, L., Rogojanu, O.C., James, M.A., and Hibma, T. (April 1998). Superparamagnetic behavior of structural domains in epitaxial ultrathin magnetite films. *Physical Review B* 57 (14): R8107–R8110. doi: 10.1103/PhysRevB.57.R8107.

114 Margulies, D.T. et al. (December 1997). Origin of the anomalous magnetic behavior in single crystal Fe_3O_4 films. *Physical Review Letters* 79 (25): 5162–5165. doi: 10.1103/PhysRevLett.79.5162.

115 Hibma, T. et al. (April 1999). Anti-phase domains and magnetism in epitaxial magnetite layers. *Journal of Applied Physics* 85 (8): 5291–5293. doi: 10.1063/1.369857.

116 Eerenstein, W., Palstra, T.T.M., Hibma, T., and Celotto, S. (July 2003). Diffusive motion of antiphase domain boundaries in Fe_3O_4 films. *Physical Review B* 68 (1): 014428. doi: 10.1103/PhysRevB.68.014428.

117 Gaines, J.M. et al. (February 1997). An STM study of Fe_3O_4(100) grown by molecular beam epitaxy. *Surface Science* 373 (1): 85–94. doi: 10.1016/S0039-6028(96)01145-4.

118 Li, Z. et al. (March 2021). Atomic structure and electron magnetic circular dichroism of individual rock salt structure antiphase boundaries in spinel ferrites. *Advanced Functional Materials*: 2008306. doi: 10.1002/adfm.202008306.

119 Motida, K. and Miyahara, S. (May 1970). On the 90° exchange interaction between cations (Cr^{3+}, Mn^{2+}, Fe^{3+} and Ni^{2+}) in oxides. *Journal of the Physical Society of Japan* 28 (5): 1188–1196. doi: 10.1143/JPSJ.28.1188.

120 Boekema, C., van der Woude, F., and Sawatzky, G.A. (1972). Covalency effects and hyperfine interactions in the rare earth orthoferrites. *International Journal of Magnetics* 3 (4): 341–348.

121 Eerenstein, W., Hibma, T., and Celotto, S. (November 2004). Mechanism for superparamagnetic behavior in epitaxial Fe_3O_4 films. *Physical Review B* 70 (18): 184404. doi: 10.1103/PhysRevB.70.184404.

122 Celotto, S., Eerenstein, W., and Hibma, T. (November 2003). Characterization of anti-phase boundaries in epitaxial magnetite films. *European Physical Journal B: Condensed Matter* 36 (2): 271–279. doi: 10.1140/epjb/e2003-00344-7.

123 van der Heijden, P.A.A. (1998). Magnetic properties of oxide-based thin films and multilayers. Eindhoven: Eindhoven University of Technology.

124 Pohlmann, T. et al. (January 2022). Time-resolved diffraction and photoelectron spectroscopy investigation of the reactive molecular beam epitaxy of Fe_3O_4 ultrathin films. *Phys. Review B* 105: 045412Phys Review B. doi.org/10.1103/PhysRevB.105.045412.

125 Pohlmann, T., Bertram, F., Thien, J., et al. (June 2021). Structural and magnetic investigation of the interfaces of Fe_3O_4 / MgO(001) with and without NiO interlayer. *Physical Review B* 105: 235436.

126 Prinz, G.A. (November 1990). Hybrid ferromagnetic-semiconductor structure. *Science* 250 (4984): 1092–1097. doi: 10.1126/science.250.4984.1092.

127 Fiederling, R. et al. (December 1999). Injection and detection of a spin-polarized current in a light-emitting diode. *Nature* 402 (6763): 787–790. doi: 10.1038/45502.

128 Grundler, D. (April 2002). Spintronics. *Physics World* 15 (4): 39–43. doi: 10.1088/2058-7058/15/4/38.

129 Li, X.W., Gupta, A., Xiao, G., Qian, W., and Dravid, V.P. (November 1998). Fabrication and properties of heteroepitaxial magnetite (Fe_3O_4) tunnel junctions. *Applied Physics Letters* 73 (22): 3282–3284. doi: 10.1063/1.122745.

130 Seneor, P., Fert, A., Maurice, J.-L., Montaigne, F., Petroff, F., and Vaurès, A. (June 1999). Large magnetoresistance in tunnel junctions with an iron oxide electrode. *Applied Physics Letters* 74 (26): 4017–4019. doi: 10.1063/1.123246.

131 van der Zaag, P.J. et al. (March 2000). On the construction of an Fe_3O_4-based all-oxide spin valve. *Journal of Magnetism and Magnetic Materials* 211 (1–3): Art. no. 1–3. doi: 10.1016/S0304-8853(99)00751-9.

132 Hu, G. and Suzuki, Y. (December 2002). Negative spin polarization of Fe_3O_4 in magnetite/manganite-based junctions. *Physical Review Letters* 89 (27): 276601. doi: 10.1103/PhysRevLett.89.276601.

133 Matsuda, H. et al. (April 2002). Fabrication and magnetoresistance properties of spin-dependent tunnel junctions using an epitaxial Fe_3O_4 film. *Japanese Journal of Applied Physics* 41 (Part 2, 4A): L387–L390. doi: 10.1143/JJAP.41.L387.

134 Aoshima, K. and Wang, S.X. (May 2003). Fe_3O_4 and its magnetic tunneling junctions grown by ion beam deposition. *Journal of Applied Physics* 93 (10): 7954–7956. doi: 10.1063/1.1558633.

135 Yoon, K.S., Koo, J.H., Do, Y.H., Kim, K.W., Kim, C.O., and Hong, J.P. (January 2005). Performance of $Fe_3O_4/AlO_x/CoFe$ magnetic tunnel junctions based on half-metallic Fe_3O_4 electrodes. *Journal of Magnetism and Magnetic Materials* 285 (1–2): 125–129. doi: 10.1016/j.jmmm.2004.07.025.

136 Park, C., Zhu, J.-G., Peng, Y., Laughlin, D.E., and White, R.M. (October 2005). Inverse magnetoresistance in magnetic tunnel junction with an Fe_3O_4 electrode. *IEEE Transactions on Magnetics* 41 (10): 2691–2693. doi: 10.1109/TMAG.2005.855294.

137 Bataille, A.M. et al. (September 2007). On the spin polarization at the Fe_3O_4/gamma-Al_2O_3 interface probed by spin-resolved photoemission and spin-dependent tunneling. *Journal of Magnetism and Magnetic Materials* 316 (2): e963–e965. doi: 10.1016/j.jmmm.2007.03.156.

138 Kado, T., Saito, H., and Ando, K. (May 2007). Room-temperature magnetoresistance in magnetic tunnel junctions with Fe_3O_4 electrode. *Journal of Applied Physics* 101 (9): 09J511. doi: 10.1063/1.2713045.

139 Kado, T. (March 2008). Large room-temperature inverse magnetoresistance in tunnel junctions with a Fe_3O_4 electrode. *Applied Physics Letters* 92 (9): 092502. doi: 10.1063/1.2890852.

140 Opel, M. et al. (February 2011). Novel multifunctional materials based on oxide thin films and artificial heteroepitaxial multilayers. *Physica Status Solidi A* 208 (2): 232–251. doi: 10.1002/pssa.201026403.

141 Nagahama, T. et al. (September 2014). Magnetic properties of epitaxial Fe_3O_4 films with various crystal orientations and tunnel magnetoresistance effect at room temperature. *Applied Physics Letters* 105 (10): 102410. doi: 10.1063/1.4894575.

142 Marnitz, L. et al. (April 2015). Sign change in the tunnel magnetoresistance of Fe_3O_4/MgO/Co-Fe-B magnetic tunnel junctions depending on the annealing temperature and the interface treatment. *AIP Advances* 5 (4): 047103. doi: 10.1063/1.4917018.

143 Yasui, S., Honda, S., Okabayashi, J., Yanase, T., Shimada, T., and Nagahama, T. (March 2021). Large inverse tunnel magnetoresistance in magnetic tunnel junctions with an Fe_3O_4 electrode. *Physical Review Applied* 15 (3): 034042. doi: 10.1103/PhysRevApplied.15.034042.

144 Anisimov, A.V., Goikhman, A.Y., Kupriyanova, G.S., Nevolin, V.N., Popov, A.P., and Rodionova, V.V. (June 2012). Change in the magnetic properties of polycrystalline thin-film magnetite upon introduction of an iron sublayer. *Physics of the Solid State* 54 (6): 1153–1159. doi: 10.1134/S1063783412060030.

145 Kumar, A. et al. (May 2018). Effect of in situ electric-field-assisted growth on antiphase boundaries in epitaxial Fe_3O_4 thin films on MgO. *Physical Review Materials* 2 (5): 054407. doi: 10.1103/PhysRevMaterials.2.054407.

146 Parkin, S.S.P. et al. (December 2004). Giant tunnelling magnetoresistance at room temperature with MgO (100) tunnel barriers. *Nature Materials* 3 (12): Art. no. 12. doi: 10.1038/nmat1256.

147 Dedkov, Y.S., Rüdiger, U., and Güntherodt, G. (January 2002). Evidence for the half-metallic ferromagnetic state of Fe_3O_4 by spin-resolved photoelectron spectroscopy. *Physical Review B* 65 (6): 064417. doi: 10.1103/PhysRevB.65.064417.

148 Fonin, M., Dedkov, Y.S., Pentcheva, R., Rüdiger, U., and Güntherodt, G. (August 2007). Magnetite: A search for the half-metallic state. *Journal of Physics: Condensed Matter* 19 (31): 315217. doi: 10.1088/0953-8984/19/31/315217.

149 Morton, S.A., Waddill, G.D., Kim, S., Schuller, I.K., Chambers, S.A., and Tobin, J.G. (August 2002). Spin-resolved photoelectron spectroscopy of Fe_3O_4. *Surface Science* 513 (3): L451–L457. doi: 10.1016/S0039-6028(02)01824-1.

150 Moodera, J.S., Santos, T.S., and Nagahama, T. (April 2007). The phenomena of spin-filter tunnelling. *Journal of Physics: Condensed Matter* 19 (16): 165202. doi: 10.1088/0953-8984/19/16/165202.

151 Lüders, U. et al. (February 2006). Spin filtering through ferrimagnetic $NiFe_2O_4$ tunnel barriers. *Applied Physics Letters* 88 (8): 082505. doi: 10.1063/1.2172647.

152 Ramos, A.V. et al. (September 2007). Room temperature spin filtering in epitaxial cobalt-ferrite tunnel barriers. *Applied Physics Letters* 91 (12): 122107. doi: 10.1063/1.2787880.

153 Takahashi, Y.K., Kasai, S., Furubayashi, T., Mitani, S., Inomata, K., and Hono, K. (February 2010). High spin-filter efficiency in a Co ferrite fabricated by a thermal oxidation. *Applied Physics Letters* 96 (7): 072512. doi: 10.1063/1.3318297.

154 Matzen, S., Moussy, J.-B., Mattana, R., Bouzehouane, K., Deranlot, C., and Petroff, F. (July 2012). Nanomagnetism of cobalt ferrite-based spin filters probed by spin-polarized tunneling. *Applied Physics Letters* 101 (4): 042409. doi: 10.1063/1.4738790.

155 Matzen, S. et al. (May 2014). Structure, magnetic ordering, and spin filtering efficiency of $NiFe_2O_4$ (111) ultrathin films. *Applied Physics Letters* 104 (18): 182404. doi: 10.1063/1.4871733.

156 Ramos, A.V., Santos, T.S., Miao, G.X., Guittet, M.-J., Moussy, J.-B., and Moodera, J.S. (November 2008). Influence of oxidation on the spin-filtering properties of $CoFe_2O_4$ and the resultant spin polarization. *Physical Review B* 78 (18): 180402. doi: 10.1103/PhysRevB.78.180402.

157 Wakabayashi, Y.K. et al. (September 2017). Electronic structure and magnetic properties of magnetically dead layers in epitaxial $CoFe_2O_4/Al_2O_3/Si$ (111) films studied by x-ray magnetic circular dichroism. *Physical Review B* 96 (10): 104410. doi: 10.1103/PhysRevB.96.104410.

158 Smith, H.M. and Turner, A.F. (January 1965). Vacuum deposited thin films using a ruby laser. *Applied Optics* 4 (1): 147. doi: 10.1364/AO.4.000147.

159 Letokhov, V.S. (1969). Generation of ultrashort light pulses in a laser with a nonlinear absorber. *Soviet Physics JETP* 28: 562.

160 Smith, R. (April 1970). Theory of intracavity optical second-harmonic generation. *IEEE Journal of Quantum Electronics* 6 (4): 215–223. doi: 10.1109/JQE.1970.1076440.

161 Joshi, S. et al. (November 1988). Pulsed laser deposition of iron oxide and ferrite films. *Journal of Applied Physics* 64 (10): 5647–5649. doi: 10.1063/1.342258.

162 Ogale, S.B. and Nawathey, R. (February 1989). Deposition of zinc ferrite ($Zn_xFe_{3-x}O_4$) films by pulsed laser evaporation process. *Journal of Applied Physics* 65 (3): 1367–1369. doi: 10.1063/1.343008.

163 Ogale, S.B. et al. (October 1988). Deposition of iron oxide thin films by pulsed laser evaporation. *Applied Physics Letters* 53 (14): 1320–1322. doi: 10.1063/1.100427.

164 Tanaka, K., Omata, Y., Nishikawa, Y., Yoshida, Y., and Nakamura, K. (November 1991). Formation of Ni-Zn-ferrite films using laser ablation. *IEEE Translation Journal on Magnetics in Japan* 6 (11): 1001–1006. doi: 10.1109/TJMJ.1991.4565294.

165 Kidoh, H., Morimoto, A., and Shimizu, T. (July 1991). Synthesis of ferromagnetic Bi-substituted yttrium iron garnet films by laser ablation. *Applied Physics Letters* 59 (2): 237–239. doi: 10.1063/1.105977.

166 Dorsey, P. et al. (September 1992). Oriented barium hexaferrite thin films prepared by pulsed laser deposition. *IEEE Transactions on Magnetics* 28 (5): 3216–3218. doi: 10.1109/20.179763.

167 Carosella, C.A. et al. (May 1992). Pulsed laser deposition of epitaxial $BaFe_{12}O_{19}$ thin films. *Journal of Applied Physics* 71 (10): 5107–5110. doi: 10.1063/1.350614.

168 Basov, N.G., Danilychev, V.A., and Popov, Y.M. (January 1971). Stimulated emission in the vacuum ultraviolet region. *Soviet Journal of Quantum Electronics* 1 (1): 18–22. doi: 10.1070/qe1971v001n01abeh003011.

169 Koehler, H.A., Ferderber, L.J., Redhead, D.L., and Ebert, P.J. (September 1972). Stimulated VUV emission in high-pressure xenon excited by high-current relativistic electron beams. *Applied Physics Letters* 21 (5): 198–200. doi: 10.1063/1.1654342.

170 Ault, E., Bhaumik, M., Hughes, W., Jensen, R., Robinson, C., Kolb, A., and Shannon, J. (1973). Xe laser operation at 1730 Å. *Journal of the Optical Society of America. 63* 7: 907.

171 Ault, E. et al. (October 1973). Xenon molecular laser in the vacuum ultraviolet. *IEEE Journal of Quantum Electronics* 9 (10): 1031–1032. doi: 10.1109/JQE.1973.1077396.

172 Basting, D., Pippert, K.D., and Stamm, U. (February. 2002). History and future prospects of excimer lasers. Singapore, 25. doi: 10.1117/12.456812.

173 Geusic, J.E., Marcos, H.M., and Van Uitert, L.G. (1964). Laser oscillations in Nd-doped yttrium aluminum, yttrium gallium and gadolinium garnets. *Applied Physics Letters* 4 (10): 182–184.

174 Yariv, A. (1989). *Quantum Electronics*, 3e. New York, NY: Wiley.

175 Koechner, W. (1965). *Solid-state Laser Engineering*. Springer-Verlag.

176 Kuppusami, P. and Raghunathan, V.S. (April 2006). Status of pulsed laser deposition: Challenges and opportunities. *Surface Engineering* 22 (2): 81–83. doi: 10.1179/174329406X98502.

177 Chrisey, D.B. and Hubler, G.K. (eds.) (1994). *Pulsed Laser Deposition of Thin Films*. New York, NY: J. Wiley.

178 http://neocera.com/products/large-area-pulsed-laser-deposition-systems (accessed 27 August 2021).

179 Greer, J. (June, 2021). Private communication.

180 http://www.pvdproducts.com/pulsed-laser-deposition-systems/pld-components/pld-intelligent-window (accessed 3 June 2021).

181 http://www.pvdproducts.com/pulsed-laser-deposition-systems/pld-components/pld-intelligent-window (accessed 27 August 2021).

182 Kennedy, R.J. (November 1995). The growth of iron oxide, nickel oxide and cobalt oxide thin films by laser ablation from metal targets. *IEEE Transactions on Magnetics* 31 (6): 3829–3831. doi: 10.1109/20.489786.

183 Tepper, T. and Ross, C.A. (April 2002). Pulsed laser deposition of iron oxide films. *Journal of Applied Physics* 91 (7): 4453–4456. doi: 10.1063/1.1456248.

184 Tepper, T., Ross, C.A., and Dionne, G.F. (May 2004). Microstructure and optical properties of pulsed-laser-deposited iron oxide films. *IEEE Transactions on Magnetics* 40 (3): 1685–1690. doi: 10.1109/TMAG.2004.827215.

185 Johnson, M.T., Michael, J.R., Gilliss, S.R., and Carter, C.B. (December 1999). Iron oxide on (001) MgO. *Philosophical Magazine A* 79 (12): 2887–2898. doi: 10.1080/01418619908212031.

186 Gong, G.Q., Gupta, A., Xiao, G., Qian, W., and Dravid, V.P. (September 1997). Magnetoresistance and magnetic properties of epitaxial magnetite thin films. *Physical Review B* 56 (9): 5096–5099. doi: 10.1103/PhysRevB.56.5096.

187 Paramês, M.L., Mariano, J., Rogalski, M.S., Popovici, N., and Conde, O. (April 2005). UV pulsed laser deposition of magnetite thin films. *Materials Science and Engineering: B* 118 (1–3): 246–249. doi: 10.1016/j.mseb.2004.12.037.

188 Paramês, M.L. et al. (April 2006). PLD of Fe_3O_4 thin films: Influence of background gas on surface morphology and magnetic properties. *Applied Surface Science* 252 (13): 4610–4614. doi: 10.1016/j.apsusc.2005.07.090.

189 Kubo, T., Sakamoto, N., Shinozaki, K., Suzuki, H., and Wakiya, N. (July 2011). Magnetic properties of epitaxial $NiFe_2O_4$ thin films prepared using Dynamic Aurora PLD in a magnetic field. *Key Engineering Materials* 485: 221–224. doi: 10.4028/www.scientific.net/KEM.485.221.

190 Datta, R., Kanuri, S., Karthik, S.V., Mazumdar, D., Ma, J.X., and Gupta, A. (August 2010). Formation of antiphase domains in $NiFe_2O_4$ thin films deposited on different substrates. *Applied Physics Letters* 97 (7): 071907. doi: 10.1063/1.3481365.

191 Eskandari, F., Porter, S.B., Venkatesan, M., Kameli, P., Rode, K., and Coey, J.M.D. (December 2017). Magnetization and anisotropy of cobalt ferrite thin films. *Physical Review Materials* 1 (7): 074413. doi: 10.1103/PhysRevMaterials.1.074413.

192 De Teresa, J.M. et al. (May 2007). Magnetotransport properties of Fe_3O_4 thin films for applications in spin electronics. *Microelectronic Engineering* 84 (5–8): 1660–1664. doi: 10.1016/j.mee.2007.01.120.

193 Kado, T. (February 2008). Structural and magnetic properties of magnetite-containing epitaxial iron oxide films grown on MgO(001) substrates. *Journal of Applied Physics* 103 (4): 043902. doi: 10.1063/1.2840118.

194 Schmidt, G., Ferrand, D., Molenkamp, L.W., Filip, A.T., and van Wees, B.J. (August 2000). Fundamental obstacle for electrical spin injection from a ferromagnetic metal into a diffusive semiconductor. *Physical Review B* 62 (8): R4790–R4793. doi: 10.1103/PhysRevB.62.R4790.

195 Qu, T.L., Zhao, Y.G., Yu, P., Zhao, H.C., Zhang, S., and Yang, L.F. (June 2012). Exchange bias effects in epitaxial Fe_3O_4/$BiFeO_3$ heterostructures. *Applied Physics Letters* 100 (24): 242410. doi: 10.1063/1.4729408.

196 Monti, M. et al. (December 2013). Room temperature in-plane ⟨100⟩ magnetic easy axis for Fe_3O_4/$SrTiO_3$(001):Nb grown by infrared pulsed laser deposition. *Journal of Applied Physics* 114 (22): 223902. doi: 10.1063/1.4837656.

197 Guo, Q. et al. (March 2013). Effects of oxygen gas pressure on properties of iron oxide films grown by pulsed laser deposition. *Journal of Alloys and Compounds* 552: 1–5. doi: 10.1016/j.jallcom.2012.10.088.

198 Rubio-Zuazo, J., Onandia, L., Salas-Colera, E., Muñoz-Noval, A., and Castro, G.R. (January 2015). Incommensurate growth of thin and ultrathin films of single-phase Fe_3O_4 (001) on $SrTiO_3$ (001). *Journal of Physical Chemistry C* 119 (2): 1108–1112. doi: 10.1021/jp510615j.

199 Nielsen, A. et al. (October 2008). All oxide ferromagnet/semiconductor epitaxial heterostructures. *Applied Physics Letters* 93 (16): 162510. doi: 10.1063/1.2998576.

200 Hanbicki, A.T. et al. (June 2003). Analysis of the transport process providing spin injection through an Fe/AlGaAs Schottky barrier. *Applied Physics Letters* 82 (23): 4092–4094. doi: 10.1063/1.1580631.

201 Shima, M., Tepper, T., and Ross, C.A. (2002). Magnetic properties of chromium oxide and iron oxide films produced by pulsed laser deposition. *Journal of Applied Physics* 91 (10): 7920. doi: 10.1063/1.1451881.

202 Ilievski, F., Tepper, T., and Ross, C.A. (September 2003). Optical and magnetic properties of γ-iron oxide made by reactive pulsed laser deposition. *IEEE Transactions on Magnetics* 39 (5): 3172–3174. doi: 10.1109/TMAG.2003.816047.

203 Reisinger, D. et al. (August 2003). Epitaxy of Fe_3O_4 on Si(001) by pulsed laser deposition using a TiN/MgO buffer layer. *Journal of Applied Physics* 94 (3): 1857–1863. doi: 10.1063/1.1587885.

204 Ramos, A.V. et al. (November 2006). Magnetotransport properties of Fe_3O_4 epitaxial thin films: Thickness effects driven by antiphase boundaries. *Journal of Applied Physics* 100 (10): 103902. doi: 10.1063/1.2386927.

205 Tiwari, S., Choudhary, R.J., and Phase, D.M. (April 2009). Effect of growth temperature on the structural and transport properties of magnetite thin films prepared by pulse laser deposition on single crystal Si substrate. *Thin Solid Films* 517 (11): 3253–3256. doi: 10.1016/j.tsf.2008.11.074.

206 Ding, J. et al. (December 2011). Growth and characterization of Fe_3O_4 films. *Materials Research Bulletin* 46 (12): 2212–2216. doi: 10.1016/j.materresbull.2011.09.008.

207 Liu, E. et al. (November 2015). Inverse magnetoresistance in textured Fe_3O_4 film. *Journal of Alloys and Compounds* 649: 239–243. doi: 10.1016/j.jallcom.2015.07.108.

208 Ji, H., Wang, Y.G., and Li, Y. (November 2017). Electric modulation of magnetization at the $Fe_3O_4/BaTiO_3$ interface. *Journal of Magnetism and Magnetic Materials* 442: 242–246. doi: 10.1016/j.jmmm.2017.05.091.

209 Liu, E. et al. (April 2017). Increasing spin polarization in Fe_3O_4 films by engineering antiphase boundary densities. *Applied Physics Letters* 110 (14): 142402. doi: 10.1063/1.4979586.

210 Ji, H., Wang, Y.G., and Li, Y. (March 2018). Charge screening-controlled Verwey phase transition in $Fe_3O_4/SrTiO_3$ heterostructure. *Journal of Physics: Condensed Matter* 30 (11): 11LT01. doi: 10.1088/1361-648X/aaae37.

211 Fujiwara, S., Kurauchi, Y., Hirose, Y., Harayama, I., Sekiba, D., and Hasegawa, T. (April 2019). Modification of electrical and magnetic properties of Fe_3O_4 epitaxial thin films by nitrogen substitution for oxygen. *ACS Applied Electronic Materials* 1 (4): 595–599. doi: 10.1021/acsaelm.9b00051.

212 Dorsey, P.C., Bushnell, S.E., Seed, R.G., and Vittoria, C. (July 1993). Epitaxial yttrium iron garnet films grown by pulsed laser deposition. *Journal of Applied Physics* 74 (2): 1242–1246. doi: 10.1063/1.354927.

213 Dorsey, P.C., Bushnell, S.E., Seed, R.G., and Vittoria, C. (November 1993). Magnetic and structural characteristics of pulsed laser deposited epitaxial $Y_3Fe_5O_{12}$ films. *IEEE Transactions on Magnetics* 29 (6): 3069–3071. doi: 10.1109/20.281118.

214 Simion, B.M., Thomas, G., Ramesh, R., Keramidas, V.G., and Pfeffer, R.L. (February 1995). Growth and characterization of ($Y_3Fe_5O_{12}$ –$Bi_3Fe_5O_{12}$) heterostructures by pulsed laser deposition. *Applied Physics Letters* 66 (7): 830–832. doi: 10.1063/1.113436.

215 Kumar, N., Misra, D.S., Venkataramani, N., Prasad, S., and Krishnan, R. (May 2004). Magnetic properties of pulsed laser ablated YIG thin films on different substrates. *Journal of Magnetism and Magnetic Materials* 272–276: E899–E900. doi: 10.1016/j.jmmm.2003.12.223.

216 Karim, R., Oliver, S.A., and Vittoria, C. (November 1995). Laser ablation deposition of YIG films on semiconductor and amorphous substrates. *IEEE Transactions on Magnetics* 31 (6): 3485–3487. doi: 10.1109/20.489544.

217 Hayashi, H. et al. (September 2002). Fabrication of Bi-doped YIG optical thin film for electric current sensor by pulsed laser deposition. *Applied Surface Science* 197–198: 463–466. doi: 10.1016/S0169-4332(02)00364-1.

218 Popova, E. et al. (September 2001). Perpendicular magnetic anisotropy in ultrathin yttrium iron garnet films prepared by pulsed laser deposition technique. *Journal of Vacuum Science and Technology A: Vacuum, Surfaces and Films* 19 (5): 2567–2570. doi: 10.1116/1.1392395.

219 Popova, E. et al. (August 2001). Structure and magnetic properties of yttrium–iron–garnet thin films prepared by laser deposition. *Journal of Applied Physics* 90 (3): 1422–1428. doi: 10.1063/1.1379344.

220 Dumont, Y. et al. (May 2005). Superexchange and iron valence control by off-stoichiometry in yttrium iron garnet thin films grown by pulsed laser deposition. *Journal of Applied Physics* 97 (10): 10G108. doi: 10.1063/1.1860991.

221 Nakata, Y., Okada, T., Maeda, M., Higuchi, S., and Ueda, K. (February 2006). Effect of oxidation dynamics on the film characteristics of Ce:YIG thin films deposited by pulsed-laser deposition. *Optics and Lasers in Engineering* 44 (2): 147–154. doi: 10.1016/j.optlaseng.2005.03.003.

222 Dumont, Y. et al. (September 2007). Tuning magnetic properties with off-stoichiometry in oxide thin films: An experiment with yttrium iron garnet as a model system. *Physical Review B* 76 (10): 104413. doi: 10.1103/PhysRevB.76.104413.

223 Vertruyen, B. et al. (September 2008). Curie temperature, exchange integrals, and magneto-optical properties in off-stoichiometric bismuth iron garnet epitaxial films. *Physical Review B* 78 (9): 094429. doi: 10.1103/PhysRevB.78.094429.

224 Jalali-Roudsar, A.A., Denysenkov, V.P., Khartsev, S.I., Grishin, A.M., Adachi, N., and Okuda, T. (July 2001). Microwave and magneto-optic properties of pulsed laser deposited bismuth iron garnet films. *IEEE Transactions on Magnetics* 37 (4): 2454–2456. doi: 10.1109/20.951201.

225 Kim, H., Grishin, A.M., Rao, K.V., Yu, S.C., Sbiaa, R., and Le Gall, H. (September 1999). Ce-substituted YIG films grown by pulsed laser deposition for magneto-optic waveguide devices. *IEEE Transactions on Magnetics* 35 (5): 3163–3165. doi: 10.1109/20.801115.

226 Ibrahim, N.B., Edwards, C., and Palmer, S.B. (October 2000). Pulsed laser ablation deposition of yttrium iron garnet and cerium-substituted YIG films. *Journal of Magnetism and Magnetic Materials* 220 (2–3): 183–194. doi: 10.1016/S0304-8853(00)00331-0.

227 Higuchi, S. et al. (July 2001). Fabrications of cerium-substituted YIG thin films for magnetic field sensor by pulsed-laser deposition. *IEEE Transactions on Magnetics* 37 (4): 2451–2453. doi: 10.1109/20.951200.

228 Goto, T., Onbaşlò, M.C., and Ross, C.A. (December 2012). Magneto-optical properties of cerium substituted yttrium iron garnet films with reduced thermal budget for monolithic photonic integrated circuits. *Optics Express* 20 (27): 28507. doi: 10.1364/OE.20.028507.

229 Onbasli, M.C., Goto, T., Sun, X., Huynh, N., and Ross, C.A. (October 2014). Integration of bulk-quality thin film magneto-optical cerium-doped yttrium iron garnet on silicon nitride photonic substrates. *Optics Express* 22 (21): 25183. doi: 10.1364/OE.22.025183.

230 Onbasli, M.C. et al. (July 2016). Optical and magneto-optical behavior of Cerium Yttrium Iron Garnet thin films at wavelengths of 200–1770 nm. *Scientific Reports* 6 (1): 23640. doi: 10.1038/srep23640.

231 Kumar, N., Prasad, S., Misra, D.S., Venkataramani, N., Bohra, M., and Krishnan, R. (September 2008). The influence of substrate temperature and annealing on the properties of pulsed laser-deposited YIG films on fused quartz substrate. *Journal of Magnetism and Magnetic Materials* 320 (18): 2233–2236. doi: 10.1016/j.jmmm.2008.04.112.

232 Popova, E. et al. (January 2003). Exchange coupling in ultrathin epitaxial yttrium iron garnet films. *European Physical Journal B: Condensed Matter* 31 (1): 69–74. doi: 10.1140/epjb/e2003-00010-2.

233 Manuilov, S.A., Fors, R., Khartsev, S.I., and Grishin, A.M. (February 2009). Submicron $Y_3Fe_5O_{12}$ film magnetostatic wave band pass filters. *Journal of Applied Physics* 105 (3): 033917. doi: 10.1063/1.3075816.

234 Manuilov, S.A., Khartsev, S.I., and Grishin, A.M. (December 2009). Pulsed laser deposited $Y_3Fe_5O_{12}$ films: Nature of magnetic anisotropy I. *Journal of Applied Physics* 106 (12): 123917. doi: 10.1063/1.3272731.

235 Manuilov, S.A. and Grishin, A.M. (July 2010). Pulsed laser deposited $Y_3Fe_5O_{12}$ films: Nature of magnetic anisotropy II. *Journal of Applied Physics* 108 (1): 013902. doi: 10.1063/1.3446840.

236 Sun, Y., Song, -Y.-Y., and Wu, M. (August 2012). Growth and ferromagnetic resonance of yttrium iron garnet thin films on metals. *Applied Physics Letters* 101 (8): 082405. doi: 10.1063/1.4747465.

237 Sun, Y. et al. (October 2012). Growth and ferromagnetic resonance properties of nanometer-thick yttrium iron garnet films. *Applied Physics Letters* 101 (15): 152405. doi: 10.1063/1.4759039.

238 Sposito, A., May-Smith, T.C., Stenning, G.B.G., de Groot, P.A.J., and Eason, R.W. (May 2013). Pulsed laser deposition of high-quality μm-thick YIG films on YAG. *Optical Materials Express* 3 (5): 624. doi: 10.1364/OME.3.000624.

239 Sokolov, N.S. et al. (January 2016). Thin yttrium iron garnet films grown by pulsed laser deposition: Crystal structure, static, and dynamic magnetic properties. *Journal of Applied Physics* 119 (2): 023903. doi: 10.1063/1.4939678.

240 Zheng, H. et al. (May 2014). Preparation of two-dimensional yttrium iron garnet magnonic crystal on porous silicon substrate. *Materials Letters* 123: 181–183. doi: 10.1016/j.matlet.2014.03.006.

241 Hauser, C. et al. (August 2016). Yttrium iron garnet thin films with very low damping obtained by recrystallization of amorphous material. *Scientific Reports* 6 (1): 20827. doi: 10.1038/srep20827.

242 Qin, H., Both, G.-J., Hämäläinen, S.J., Yao, L., and van Dijken, S. (December 2018). Low-loss YIG-based magnonic crystals with large tunable bandgaps. *Nature Communications* 9 (1): 5445. doi: 10.1038/s41467-018-07893-5.

243 Qin, H. and van Dijken, S. (May 2020). Nanometer-thick YIG-based magnonic crystals: Bandgap dependence on groove depth, lattice constant, and film thickness. *Applied Physics Letters* 116 (20): 202403. doi: 10.1063/5.0009807.

244 Qin, H., Hämäläinen, S.J., Arjas, K., Witteveen, J., and van Dijken, S. (December 2018). Propagating spin waves in nanometer-thick yttrium iron garnet films: Dependence on wave vector, magnetic field strength, and angle. *Physical Review B* 98 (22): 224422. doi: 10.1103/PhysRevB.98.224422.

245 Chumak, A.V., Vasyuchka, V.I., Serga, A.A., and Hillebrands, B. (June 2015). Magnon spintronics. *Nature Physics* 11 (6): 453–461. doi: 10.1038/nphys3347.

246 Khitun, A., Bao, M., and Wang, K.L. (July 2010). Magnonic logic circuits. *Journal of Physics D: Applied Physics* 43 (26): 264005. doi: 10.1088/0022-3727/43/26/264005.

247 Chumak, A.V., Serga, A.A., and Hillebrands, B. (June 2017). Magnonic crystals for data processing. *Journal of Physics D: Applied Physics* 50 (24): 244001. doi: 10.1088/1361-6463/aa6a65.

248 Stamps, R.L. et al. (August 2014). The 2014 magnetism roadmap. *Journal of Physics D: Applied Physics* 47 (33): 333001. doi: 10.1088/0022-3727/47/33/333001.

249 Chumak, A.V., Serga, A.A., and Hillebrands, B. (December 2014). Magnon transistor for all-magnon data processing. *Nature Communications* 5 (1): 4700. doi: 10.1038/ncomms5700.

250 Williams, C.M., Chrisey, D.B., Lubitz, P., Grabowski, K.S., and Cotell, C.M. (February 1994). The magnetic and structural properties of pulsed laser deposited epitaxial MnZn–ferrite films. *Journal of Applied Physics* 75 (3): 1676–1680. doi: 10.1063/1.356353.

251 Dorsey, P.C., Lubitz, P., Harris, V.G., Chrisey, D.B., and Horwitz, J.S. (November 1995). Magnetic properties of pulsed laser deposited MnZn-ferrite films as a function of composition. *IEEE Transactions on Magnetics* 31 (6): 3455–3457. doi: 10.1109/20.489534.

252 Welch, R.G., Neamtu, J., Rogalski, M.S., and Palmer, S.B. (December 1996). Polycrystalline MnZn ferrite films prepared by pulsed laser deposition. *Materials Letters* 29 (4–6): 199–203. doi: 10.1016/S0167-577X(96)00146-2.

253 Suzuki, Y. et al. (January 1996). Structure and magnetic properties of epitaxial spinel ferrite thin films. *Applied Physics Letters* 68 (5): 714–716. doi: 10.1063/1.116601.

254 Harris, V.G. et al. (1998). Non-equilibrium inversion of MnZn-ferrite powders and films. *Journal of the Magnetics Society of Japan* 22 (S_1_ISFA_97): S1_157–161. doi: 10.3379/jmsjmag.22.S1_157.

255 Harris, V.G. et al. (November 1995). Direct measurement of octahedral and tetrahedral site environments in NiZn-ferrites. *IEEE Transactions on Magnetics* 31 (6): 3473–3475. doi: 10.1109/20.489540.

256 Harris, V.G., Koon, N.C., Williams, C.M., Zhang, Q., Abe, M., and Kirkland, J.P. (April 1996). Cation distribution in NiZn-ferrite films via extended x-ray absorption fine structure. *Applied Physics Letters* 68 (15): 2082–2084. doi: 10.1063/1.115591.

257 Srivastava, A.K. et al. (June 1999). Angle dependence of the ferromagnetic resonance linewidth and two magnon losses in pulsed laser deposited films of yttrium iron garnet, MnZn ferrite, and NiZn ferrite. *Journal of Applied Physics* 85 (11): 7838–7848. doi: 10.1063/1.370595.

258 Atanasov, P.A., Koleva, M.E., Tomov, R.I., and Krastev, V.I. (1999). Pulsed laser deposition of Mn-Zn ferrite thin films. *Journal of Materials Science: Materials in Electronics* 10 (4): 295–298. doi: 10.1023/A:1008924802224.

259 Nakano, M., Tomohara, K., Song, J.M., Fukunaga, H., and Matsuo, Y. (May 2000). Mn–Zn ferrite thin films fabricated on crystalline substrates using laser ablation technique. *Journal of Applied Physics* 87 (9): 6217–6219. doi: 10.1063/1.372659.

260 Kiyomura, T. and Gomi, M. (January 2001). Structural and magnetic properties of spinel ferrite epitaxial films pulsed-laser-deposited at low temperature. *Japanese Journal of Applied Physics* 40 (Part 1, 1): 118–122. doi: 10.1143/JJAP.40.118.

261 Zuo, X., Yang, A., Yoon, S.-D., Christodoulides, J.A., Harris, V.G., and Vittoria, C. (May 2005). Magnetic properties of manganese ferrite films grown at atomic scale. *Journal of Applied Physics* 97 (10): 10G103. doi: 10.1063/1.1849072.

262 Zuo, X., Yang, A., Yoon, S.-D., Christodoulides, J.A., Harris, V.G., and Vittoria, C. (October 2005). Large induced magnetic anisotropy in manganese spinel ferrite films. *Applied Physics Letters* 87 (15): 152505. doi: 10.1063/1.2084341.

263 Yang, A. et al. (August 2008). Element- and site-specific oxidation state and cation distribution in manganese ferrite films by diffraction anomalous fine structure. *Applied Physics Letters* 93 (5): 052504. doi: 10.1063/1.2969406.

264 Guyot, M., Lisfi, A., Krishnan, R., Porte, M., Rougier, P., and Cagan, V. (April 1996). Ferrimagnetic thin films prepared by pulsed laser deposition. *Applied Surface Science* 96–98: 802–806. doi: 10.1016/0169-4332(95)00586-2.

265 Welch, R.G., Neamtu, J., Rogalski, M.S., and Palmer, S.B. (February 1996). Pulsed laser deposition of polycrystalline NiZn ferrite films. *Solid State Communications* 97 (5): 355–359. doi: 10.1016/0038-1098(95)00662-1.

266 Dorsey, P.C., Rappoli, B.J., Grabowski, K.S., Lubitz, P., Chrisey, D.B., and Horwitz, J.S. (May 1997). Stress-induced magnetic anisotropy in thick oriented NiZn–ferrite films on (100) MgO substrates. *Journal of Applied Physics* 81 (10): 6884–6891. doi: 10.1063/1.365249.

267 Caltun, O.F. (2005). Pulsed laser deposition of Ni-Zn ferrite thin films. *Journal of Optoelectronics and Advanced Materials* 7 (2): 739–744.

268 Mizukami, S., Shimizu, K., Wakiya, N., Shinozaki, K., and Mizutani, N. (2002). Special issue ceramics integration. effect of composition and crystallinity of epitaxial Ni-Zn ferrite films by pulsed laser deposition on their magnetic properties. *Journal of the Ceramic Society of Japan* 110 (1281): 428–431. doi: 10.2109/jcersj.110.428.

269 Chinnasamy, C.N., Yoon, S.D., Yang, A., Baraskar, A., Vittoria, C., and Harris, V.G. (May 2007). Effect of growth temperature on the magnetic, microwave, and cation inversion properties on NiFe2O4 thin films deposited by pulsed laser ablation deposition. *Journal of Applied Physics* 101 (9): 09M517. doi: 10.1063/1.2714204.

270 Johnson, M.T., Kotula, P.G., and Carter, C.B. (November 1999). Growth of nickel ferrite thin films using pulsed-laser deposition. *Journal of Crystal Growth* 206 (4): 299–307. doi: 10.1016/S0022-0248(99)00342-5.

271 Jaffari, G.H., Rumaiz, A.K., Woicik, J.C., and Shah, S.I. (May 2012). Influence of oxygen vacancies on the electronic structure and magnetic properties of $NiFe_2O_4$ thin films. *Journal of Applied Physics* 111 (9): 093906. doi: 10.1063/1.4704690.

272 Dorsey, P.C., Lubitz, P., Chrisey, D.B., and Horwitz, J.S. (1996). $CoFe_2O_4$ thin films grown on (100) MgO substrates using pulsed laser deposition. *Journal of Applied Physics* 79 (8): 6338. doi: 10.1063/1.361991.

273 Lisfi, A. and Williams, C.M. (May 2003). Magnetic anisotropy and domain structure in epitaxial $CoFe_2O_4$ thin films. *Journal of Applied Physics* 93 (10): 8143–8145. doi: 10.1063/1.1541651.

274 Terzzoli, M.C., Duhalde, S., Jacobo, S., Steren, L., and Moina, C. (April 2004). High perpendicular coercive field of $CoFe_2O_4$ thin films deposited by PLD. *Journal of Alloys and Compounds* 369 (1–2): 209–212. doi: 10.1016/j.jallcom.2003.09.086.

275 Yin, J., Ding, J., Liu, B., Miao, X., and Chen, J. (April 2006). Nanocrystalline Co-ferrite films with high perpendicular coercivity. *Applied Physics Letters* 88 (16): 162502. doi: 10.1063/1.2196472.

276 Sahoo, S.C., Venkataramani, N., Prasad, S., Bohra, M., and Krishnan, R. (March 2010). Stability of nonthermodynamic equilibrium cation distribution frozen during pulsed laser deposition of Co-ferrite thin films. *Applied Physics A* 98 (4): 889–894. doi: 10.1007/s00339-009-5471-0.

277 Dhakal, T. et al. (March 2010). Magnetic anisotropy and field switching in cobalt ferrite thin films deposited by pulsed laser ablation. *Journal of Applied Physics* 107 (5): 053914. doi: 10.1063/1.3327424.

278 Patton, C.E., Edmondson, C.A., and Liu, Y.H. (March 1982). Magnetic properties of lithium zinc ferrite. *Journal of Applied Physics* 53 (3): 2431–2433. doi: 10.1063/1.330835.

279 Patton, C.E. and Liu, Y. (November 1983). Localised canting models for substituted magnetic oxides. *Journal of Physics C: Solid State Physics* 16 (31): 5995–6010. doi: 10.1088/0022-3719/16/31/014.

280 Oliver, S.A. et al. (November 1994). Magnetic and structural properties of laser deposited lithium ferrite films. *IEEE Transactions on Magnetics* 30 (6): 4933–4935. doi: 10.1109/20.334270.

281 Masterson, H.J., Lunney, J.G., Ravinder, D., and Coey, J.M.D. (February 1995). Magnetic properties of (111) $Li_{0.5}Fe_{2.5}O_4$ films. *Journal of Magnetism and Magnetic Materials* 140–144: 2081–2082. doi: 10.1016/0304-8853(94)00646-6.

282 Balestrino, G. et al. (December 1995). Lithium ferrite epitaxial films grown by pulsed laser deposition: Structural and electrical transport properties. *Solid State Communications* 96 (12): 997–1002. doi: 10.1016/0038-1098(95)00256-1.

283 Cadieu, F.J. et al. (April 1997). Static magnetic and microwave properties of Li-ferrite films prepared by pulsed laser deposition. *Journal of Applied Physics* 81 (8): 4801–4803. doi: 10.1063/1.365468.

284 Song, -Y.-Y., Grinolds, M.S., Krivosik, P., and Patton, C.E. (May 2005). Pulsed laser-deposited single-crystal LiZn-ferrite films with low microwave loss. *Journal of Applied Physics* 97 (10): 103516. doi: 10.1063/1.1897833.

285 Xi, X.W., Chen, Y.J., and Zhang, X.Y. (July 2004). Growth and magnetic properties of soft ferrite films by pulsed laser deposition. *Vacuum* 75 (2): 161–167. doi: 10.1016/j.vacuum.2004.02.001.

286 Vassiliev, A., Nicolas, J., and Hildebrand, M. (1961). Sur les proprieties des grenats mixtes d'yttrium—gadolinium. *Comptes Rendus de l'Académie des Sciences* 252 (17): 2529.

287 Wilson, A.J.C. (May 1966). Handbook of microwave ferrite materials edited by W. H. von Aulock. *Acta Crystallographica* 20 (5): 702–702. doi: 10.1107/S0365110X66004596.

288 Kamishima, K., Hosaka, N., Kakizaki, K., and Hiratsuka, N. (January 2011). Crystallographic and magnetic properties of Cu_2X, Co_2X, and Ni_2X hexaferrites. *Journal of Applied Physics* 109 (1): 013904. doi: 10.1063/1.3527933.

289 Kimura, T. (March 2012). Magnetoelectric hexaferrites. *Annual Review of Condensed Matter Physics* 3 (1): 93–110. doi: 10.1146/annurev-conmatphys-020911-125101.

290 Dorsey, P.C., Chrisey, D.B., Horwitz, J.S., Lubitz, P., and Auyeung, R.C.Y. (November 1994). Oriented barium hexaferrite thick films grown on c-plane and m-plane sapphire substrates. *IEEE Transactions on Magnetics* 30 (6): 4512–4517. doi: 10.1109/20.334133.

291 Dorsey, P.C. and Vittoria, C. (October 1994). Oriented barium hexaferrite films grown on amorphous substrates. *Journal of Magnetism and Magnetic Materials* 137 (1–2): 89–97. doi: 10.1016/0304-8853(94)90191-0.

292 Atkinson, R., Papakonstantinou, P., Salter, I.W., and Gerber, R. (November 1994). Optical and magneto-optical properties of Co-Ti-substituted barium hexaferrite single crystals and thin films produced by laser ablation deposition. *Journal of Magnetism and Magnetic Materials* 138 (1–2): 222–231. doi: 10.1016/0304-8853(94)90420-0.

293 Welch, R.G., Jackson, T.J., and Palmer, S.B. (November 1995). Barium hexaferrite films prepared by pulsed laser deposition. *IEEE Transactions on Magnetics* 31 (6): 2752–2754. doi: 10.1109/20.490140.

294 Oliver, S.A., Chen, M.L., Vittoria, C., and Lubitz, P. (April 1999). Properties of pulsed laser deposited scandium-doped barium hexaferrite films. *Journal of Applied Physics* 85 (8): 4630–4632. doi: 10.1063/1.370430.

295 Hu, B. et al. (July 2013). Magnetocrystalline anisotropy and FMR linewidth of Zr and Zn-doped Ba-hexaferrite films grown on MgO (111). *IEEE Transactions on Magnetics* 49 (7): 4234–4237. doi: 10.1109/TMAG.2013.2243122.

296 Oliver, S.A., Yoon, S.D., Kozulin, I., Chen, M.L., and Vittoria, C. (June 2000). Growth and characterization of thick oriented barium hexaferrite films on MgO (111) substrates. *Applied Physics Letters* 76 (24): 3612–3614. doi: 10.1063/1.126723.

297 Koleva, M.E. et al. (January 2002). Simultaneous laser-magnetic field treatment of $SrFe_{12}O_{19}$ thin films grown by pulsed laser deposition. *Applied Surface Science* 186 (1–4): 463–468. doi: 10.1016/S0169-4332(01)00737-1.

298 Song, -Y.-Y., Kalarickal, S., and Patton, C.E. (2003). Optimized pulsed laser deposited barium ferrite thin films with narrow ferromagnetic resonance linewidths. *Journal of Applied Physics* 94 (8): 5103. doi: 10.1063/1.1608475.

299 Geiler, A.L. et al. (August 2008). Atomic scale design and control of cation distribution in hexagonal ferrites. *Physical Review Letters* 101 (6): 067201. doi: 10.1103/PhysRevLett.101.067201.

300 Masoudpanah, S.M., Seyyed Ebrahimi, S.A., and Ong, C.K. (August 2012). Magnetic properties of strontium hexaferrite films prepared by pulsed laser deposition. *Journal of Magnetism and Magnetic Materials* 324 (17): 2654–2658. doi: 10.1016/j.jmmm.2012.03.040.

301 Yu, C., Sokolov, A.S., Kulik, P., and Harris, V.G. (January 2020). Stoichiometry, phase, and texture evolution in PLD-grown hexagonal barium ferrite films as a function of laser process parameters. *Journal of Alloys and Compounds* 814: 152301. doi: 10.1016/j.jallcom.2019.152301.

302 Su, Z., Bennett, S., Hu, B., Chen, Y., and Harris, V.G. (May 2014). Magnetic and microwave properties of U-type hexaferrite films with high remanence and low ferromagnetic resonance linewidth. *Journal of Applied Physics* 115 (17): 17A504. doi: 10.1063/1.4854935.

303 Chen, Z., Yang, A., Yoon, S.D., Ziemer, K., Vittoria, C., and Harris, V.G. (June 2006). Growth of Ba-hexaferrite films on single crystal 6-H SiC. *Journal of Magnetism and Magnetic Materials* 301 (1): 166–170. doi: 10.1016/j.jmmm.2005.06.030.

304 Chen, Z. et al. (October 2006). Structure and magnetism of Ba-hexaferrite films grown on single crystal 6-H SiC with graduated interfacial MgO buffer layers. *IEEE Transactions on Magnetics* 42 (10): 2855–2857. doi: 10.1109/TMAG.2006.879883.

305 Chen, Z. et al. (October 2007). Epitaxial growth of M-type Ba-hexaferrite films on MgO (111)||SiC (0001) with low ferromagnetic resonance linewidths. *Applied Physics Letters* 91 (18): 182505. doi: 10.1063/1.2794011.

306 Chen, Z. et al. (June 2010). Structure, magnetic, and microwave properties of thick Ba-hexaferrite films epitaxially grown on GaN/Al_2O_3 substrates. *Applied Physics Letters* 96 (24): 242502. doi: 10.1063/1.3446867.

307 Yu, C. et al. (August 2021). Interface-engineered barium magnetoplumbite–wide-bandgap semiconductor integration enabling 5G system-on-wafer solutions for full-duplexing phased arrays. *Applied Physics Letters* 119 (5): 051906. doi: 10.1063/5.0058784.

308 Kulik, P., Yu, C., Sokolov, A., Liang, W., and Harris, V.G. (November 2020). $BaFe_{12}O_{19}$ magnetoplumbite films grown on SiO_2/Si substrates for widescale magnetic film semiconductor systems integration. *Scripta Materialia* 188: 190–194. doi: 10.1016/j.scriptamat.2020.07.041.

309 Masterson, H.J., Lunney, J.G., Coey, J.M.D., Atkinson, R., Salter, I.W., and Papakonstantinou, P. (April 1993). Thin films of barium ferrite with perpendicular magnetic anisotropy produced by laser ablation deposition. *Journal of Applied Physics* 73 (8): 3917–3921. doi: 10.1063/1.352880.

310 Gerber, R., Atkinson, R., and Šimša, Z. (November 1997). Magnetism and magneto-optics of hexaferrite layers. *Journal of Magnetism and Magnetic Materials* 175 (1–2): 79–89. doi: 10.1016/S0304-8853(97)00151-0.

311 Shi, P., Yoon, S.D., Zuo, X., Kozulin, I., Oliver, S.A., and Vittoria, C. (May 2000). Microwave properties of pulsed laser deposited Sc-doped barium hexaferrite films. *Journal of Applied Physics* 87 (9): 4981–4983. doi: 10.1063/1.373222.

312 Oliver, S.A., Chen, M.L., Kozulin, I., and Vittoria, C. (May 2000). Structure and magnetic properties of barium hexaferrite films deposited at low oxygen pressures. *Journal of Magnetism and Magnetic Materials* 213 (3): 326–334. doi: 10.1016/S0304-8853(00)00004-4.

313 Chen, Z., Yang, A., Xie, C., Yang, Q., Vittoria, C., and Harris, V.G. (March 2009). High-rate reactive ion etching of barium hexaferrite films using optimal CHF_3/SF_6 gas mixtures. *Applied Physics Letters* 94 (11): 112505. doi: 10.1063/1.3099884.

314 Nastasi, M., Mayer, J.W., and Hirvonen, J.K. (1996). *Ion-Solid Interactions: Fundamentals and Applications*. Cambridge, UK: Cambridge University Press.

315 K.J. Lesker Co. (2010). Practical process tips. *Lesker Tech* 7 (1).

316 Sarakinos, K., Alami, J., and Konstantinidis, S. (February 2010). High power pulsed magnetron sputtering: A review on scientific and engineering state of the art. *Surface and Coatings Technology* 204 (11): 1661–1684. doi: 10.1016/j.surfcoat.2009.11.013.

317 Anders, A. (July 2011). Discharge physics of high power impulse magnetron sputtering. *Surface and Coatings Technology* 205: S1–S9. doi: 10.1016/j.surfcoat.2011.03.081.

318 Seyfert, U., Heisig, U., Teschner, G., and Strümpfel, J. (fall 2015). 40 years of industrial magnetron sputtering in Europe. *SVC Bulletin*: 22–26.

319 https://www.singulus.com/products/vistaris/ (accessed 20 September 2022).

320 Powell, C.F., Oxley, J.H., Blocher, J.M., and Klerer, J. (1966). Vapor deposition. *Journal of the Electrochemical Society* 113 (10): 266C. doi: 10.1149/1.2423765.

321 Faraday, M. (December 1843). Experimental researches in electricity, Thirteenth series. *Philosophical Transactions of the Royal Society of London* 4: 54–56. doi: 10.1098/rspl.1837.0020.

322 Grove, W.R. (December 1852). VII. On the electro-chemical polarity of gases. *Philosophical Transactions of the Royal Society of London* 142: 87–101. doi: 10.1098/rstl.1852.0008.

323 Wright, A.W. (September 1877). On a new process for the electrical deposition of metals, and for constructing metal-covered glass specula. *American Journal of Science* s3–14 (81): 169–178. doi: 10.2475/ajs.s3-14.81.169.

324 Edison, T.A. (18 October 1892). Process of duplicating phonograms. U.S. patent 484,582.

325 Penning, F.M. (7 February 1939). Coating by cathode disintegration. U.S. patent 2,146,025.

326 Thornton, J.A. and Penfold, A.S. (1978). Cylindrical magnetron sputtering. In: *Thin Film Processes* (ed. J.L. Vossen and W. Kern), 75–113. New York, NY: Academic.

327 Peeples, C.R., Cuthrell, R.E., and Mattox, D.M. (November 1988). Post cathode magnetron sputter deposition inside a long, small-diameter closed-end tube. *Journal of Vacuum Science and Technology A: Vacuum, Surfaces and Films* 6 (6): 3159–3160. doi: 10.1116/1.575049.

328 Clarke, P.J. (29 November 1977). Sputtering apparatus and method. U.S. patent 4,060,470.

329 Corbani, J.F. (April 15, 1975). Cathode sputtering apparatus. U.S. patent 3,878,085.

330 Chapin, J.S. (August 28, 1979). Sputtering process and apparatus. U.S. patent 4,166,018.

331 Veszi, G.A. (April 1953). The modern single-layer selenium photo-electric cell. *Journal of the British Institution of Radio Engineers* 13 (4): 183–189. doi: 10.1049/jbire.1953.0022.

332 Krikorian, E. and Sneed, R.J. (September 1966). Deposition of tantalum, tantalum oxide, and tantalum nitride with controlled electrical characteristics. *Journal of Applied Physics* 37 (10): 3674–3681. doi: 10.1063/1.1707903.

333 Wakefield, G.F. (October 14, 1975). Low temperature metal carbonitride coatings. U.S. Patent 3,912,461.

334 Thornton, J.A. (July 1974). Influence of apparatus geometry and deposition conditions on the structure and topography of thick sputtered coatings. *Journal of Vacuum Science and Technology* 11 (4): 666–670. doi: 10.1116/1.1312732.

335 Movchan, B.A. and Demchishin, A.V. (1969). Structure and properties of thick condensates of nickel, titanium, tungsten, aluminum oxides, and zirconium oxides in vacuum. *Fiz Met Metalloved* 28: 653–660.

336 Messier, R., Giri, A.P., and Roy, R.A. (April 1984). Revised structure zone model for thin film physical structure. *Journal of Vacuum Science and Technology A: Vacuum, Surfaces and Films* 2 (2): 500–503. doi: 10.1116/1.572604.

337 Anders, A. (May 2010). A structure zone diagram including plasma-based deposition and ion etching. *Thin Solid Films* 518 (15): 4087–4090. doi: 10.1016/j.tsf.2009.10.145.

338 Hultman, L. and Sundgren, J.E. (2001). Structure/properties relationship for hard coatings. In: *Handbook of Hard Coatings: Deposition Technologies, Properties and Applications* (ed. R.F. Bunshah), 108–180. Park Ridge, NJ: Noyes Publications.

339 Barna, P.B. and Adamik, M. (April 1998). Fundamental structure forming phenomena of polycrystalline films and the structure zone models. *Thin Solid Films* 317 (1–2): 27–33. doi: 10.1016/S0040-6090(97)00503-8.

340 Greene, J.E. (2010). (Vince did you meant) Thin film nucleation, growth and microstructural evolution: A atomistic view. In: *Handbook of Deposition Technologies for Films and Coatings: Science, Technology, and Applications* (ed. P.M. Martin), 554–620. Amsterdam, The Netherlands: Elsevier.

341 Mattox, D.M. (1998). *Handbook of Physical Vapor Deposition (PVD) Processing*. Park Ridge, NJ: Noyes Publications.

342 Anders, A. (2008). *Cathodic Arcs: From Fractal Spots to Energetic Condensation*, vol. 50. New York, NY: Springer New York. doi: 10.1007/978-0-387-79108-1.

343 Harris, V.G. and Pokhil, T. (July 2001). Selective-resputtering-induced perpendicular magnetic anisotropy in amorphous TbFe films. *Physical Review Letters* 87 (6): 067207. doi: 10.1103/PhysRevLett.87.067207.

344 Bobeck, A.H. (November 1957). A new storage element suitable for large-sized memory arrays — The twistor. *Bell Labs Technical Journal* 36 (6): 1319–1340. doi: 10.1002/j.1538-7305.1957.tb01513.x.

345 https://matesy.de/en/products/magnetic-field-visualization/applications/scientific-research (accessed 25 January 2021).

346 Sawatzky, E. and Kay, E. (September 1967). Magnetic properties of polycrystalline GdIG films. *IEEE Transactions on Magnetics* 3 (3): 530–530. doi: 10.1109/TMAG.1967.1066046.

347 Sawatzky, E. and Kay, E. (September 1968). Preparation of garnet films by sputtering. *Journal of Applied Physics* 39 (10): 4700–4706. doi: 10.1063/1.1655822.

348 Sawatzky, E. and Kay, E. (March 1969). Some magnetic and structural properties of epitaxial garnet films prepared by rf sputtering. *Journal of Applied Physics* 40 (3): 1460–1462. doi: 10.1063/1.1657720.

349 Sawatzky, E. and Kay, E. (January 1971). Magnetic and structural properties of epitaxial and polycrystalline GdIG films prepared by rf sputtering. *Journal of Applied Physics* 42 (1): 367–375. doi: 10.1063/1.1659603.

350 Cuomo, J.J., Sadagopan, V., DeLuca, J., Chaudhari, P., and Rosenberg, R. (December 1972). Growth of uniaxial magnetic garnet films by rf sputtering. *Applied Physics Letters* 21 (12): 581–584. doi: 10.1063/1.1654264.

351 Gomi, M., Tanida, T., and Abe, M. (April 1985). rf sputtering of highly Bi-substituted garnet films on glass substrates for magneto-optic memory. *Journal of Applied Physics* 57 (8): 3888–3890. doi: 10.1063/1.334905.

352 Bandyopadhyay, A.K., Rios, S.E., Fritz, S., Garcia, J., Contreras, J., and Gutierrez, C.J. (July 2004). Ion beam sputter-fabrication of Bi–YIG films for magnetic photonic applications. *IEEE Transactions on Magnetics* 40 (4): 2805–2807. doi: 10.1109/TMAG.2004.834215.

353 Gomi, M., Satoh, K., and Abe, M. (August 1988). Giant Faraday rotation of Ce-substituted YIG films epitaxially grown by RF sputtering. *Japanese Journal of Applied Physics* 27 (Part 2, 8): L1536–L1538. doi: 10.1143/JJAP.27.L1536.

354 Park, M.-B. and Cho, N.-H. (June 2001). Structural and magnetic characteristics of yttrium iron garnet (YIG, Ce:YIG)films prepared by RF magnetron sputter techniques. *Journal of Magnetism and Magnetic Materials* 231 (2–3): 253–264. doi: 10.1016/S0304-8853(01)00068-3.

355 Shintaku, T., Tate, A., and Mino, S. (September 1997). Ce-substituted yttrium iron garnet films prepared on $Gd_3Sc_2Ga_3O_{12}$ garnet substrates by sputter epitaxy. *Applied Physics Letters* 71 (12): 1640–1642. doi: 10.1063/1.120003.

356 Gomi, M., Furuyama, H., and Abe, M. (December 1991). Strong magneto-optical enhancement in highly Ce-substituted iron garnet films prepared by sputtering. *Journal of Applied Physics* 70 (11): 7065–7067. doi: 10.1063/1.349786.

357 Adachi, N., Denysenkov, V.P., Khartsev, S.I., Grishin, A.M., and Okuda, T. (September 2000). Epitaxial $Bi_3Fe_5O_{12}$(001) films grown by pulsed laser deposition and reactive ion beam sputtering techniques. *Journal of Applied Physics* 88 (5): 2734–2739. doi: 10.1063/1.1287227.

358 Jang, P.W. and Kim, J.Y. (July 2001). New growth method of solid phase epitaxy in sputtered YIG films. *IEEE Transactions on Magnetics* 37 (4): 2438–2440. doi: 10.1109/20.951196.

359 Kang, Y.-M. et al. (May 2005). Magnetic properties of YIG ($Y_3Fe_5O_{12}$) thin films prepared by the post annealing of amorphous films deposited by rf-magnetron sputtering. *Journal of Applied Physics* 97 (10): 10A319. doi: 10.1063/1.1855460.

360 Yang, Q., Huaiwu, Z., Yingli, L., and Qiye, W. (September 2007). Effect of post-annealing on the magnetic properties of Bi:YIG film by RF magnetron sputtering on Si substrates. *IEEE Transactions on Magnetics* 43 (9): 3652–3655. doi: 10.1109/TMAG.2007.900979.

361 Sung, S.-Y., Qi, X., and Stadler, B.J.H. (September 2005). Integrating yttrium iron garnet onto nongarnet substrates with faster deposition rates and high reliability. *Applied Physics Letters* 87 (12): 121111. doi: 10.1063/1.2046733.

362 Boudiar, T., Payet-Gervy, B., Blanc-Mignon, M.-F., Rousseau, -J.-J., Le Berre, M., and Joisten, H. (December 2004). Magneto-optical properties of yttrium iron garnet (YIG) thin films elaborated by radio frequency sputtering. *Journal of Magnetism and Magnetic Materials* 284: 77–85. doi: 10.1016/j.jmmm.2004.06.046.

363 Wang, H.L., Du, C.H., Pu, Y., Adur, R., Hammel, P.C., and Yang, F.Y. (September 2013). Large spin pumping from epitaxial $Y_3Fe_5O_{12}$ thin films to Pt and W layers. *Physical Review B* 88 (10): 100406. doi: 10.1103/PhysRevB.88.100406.

364 Chang, H. et al. (2014). Nanometer-thick yttrium iron garnet films with extremely low damping. *IEEE Magnetics Letters* 5: 1–4. doi: 10.1109/LMAG.2014.2350958.

365 Lustikova, J. et al. (October 2014). Spin current generation from sputtered $Y_3Fe_5O_{12}$ films. *Journal of Applied Physics* 116 (15): 153902. doi: 10.1063/1.4898161.

366 Du, C.H. et al. (December 2013). Probing the spin pumping mechanism: Exchange coupling with exponential decay in $Y_3Fe_5O_{12}$/Barrier/Pt heterostructures. *Physical Review Letters* 111 (24): 247202. doi: 10.1103/PhysRevLett.111.247202.

367 Jang, P., Yamamoto, S., and Kuniki, H. (June 2004). Growth of YIG films by solid phase epitaxy and their properties. *Physica Status Solidi A* 201 (8): 1851–1854. doi: 10.1002/pssa.200304546.

368 Owens, J.M., Collins, J.H., and Carter, R.L. (March 1985). System applications of magnetostatic wave devices. *Circuits, Systems, and Signal Processing* 4 (1–2): 317–334. doi: 10.1007/BF01600088.

369 Adam, J.D. (February 1988). Analog signal processing with microwave magnetics. *Proceedings of the IEEE* 76 (2): 159–170. doi: 10.1109/5.4392.

370 Balashov, T., Buczek, P., Sandratskii, L., Ernst, A., and Wulfhekel, W. (October 2014). Magnon dispersion in thin magnetic films. *Journal of Physics: Condensed Matter* 26 (39): 394007. doi: 10.1088/0953-8984/26/39/394007.

371 Chuang, T.-H. et al. (May 2014). Magnetic properties and magnon excitations in Fe(001) films grown on Ir(001). *Physical Review B* 89 (17): 174404. doi: 10.1103/PhysRevB.89.174404.

372 Westwood, W.D., Eastwood, K., Poulsen, R.G., and Sadler, A.G. (March 1967). Single-crystal ferrite films prepared by cathodic sputtering. *Journal of the American Ceramic Society* 50 (3): 119–123. doi: 10.1111/j.1151-2916.1967.tb15059.x.

373 Inagaki, N., Hattori, S., Ishii, Y., and Katsuraki, H. (September 1975). Ferrite thin films for high recording density. *IEEE Transactions on Magnetics* 11 (5): 1191–1193. doi: 10.1109/TMAG.1975.1058852.

374 Inagaki, N., Hattori, S., Ishii, Y., Terada, A., and Katsuraki, H. (November 1976). Ferrite thin films for high recording density. *IEEE Transactions on Magnetics* 12 (6): 785–787. doi: 10.1109/TMAG.1976.1059174.

375 Heller, J. (July 1976). Deposition of magnetite films by reactive sputtering of iron. *IEEE Transactions on Magnetics* 12 (4): 396–400. doi: 10.1109/TMAG.1976.1059043.

376 Hattori, S., Ishii, Y., Shinohara, M., and Nakagawa, T. (November 1979). Magnetic recording characteristics of sputtered γ-Fe_2O_3 thin film disks. *IEEE Transactions on Magnetics* 15 (6): 1549–1551. doi: 10.1109/TMAG.1979.1060465.

377 Sui, X., Scherge, M., Kryder, M.H., Snyder, J.E., Harris, V.G., and Koon, N.C. (March 1996). Barium ferrite thin-film recording media. *Journal of Magnetism and Magnetic Materials* 155 (1–3): 132–139. doi: 10.1016/0304-8853(95)00722-9.

378 Calle, C. et al. (October 2006). Magneto-optical Kerr effect in NiZn ferrite films of variable thickness. *Physica B: Condensed Matter* 384 (1–2): 103–105. doi: 10.1016/j.physb.2006.05.163.

379 Glass, H.L. (February 1988). Ferrite films for microwave and millimeter-wave devices. *Proceedings of the IEEE* 76 (2): 151–158. doi: 10.1109/5.4391.

380 Bae, S. et al. (October 2009). High Q Ni-Zn-Cu ferrite inductor for on-chip power module. *IEEE Transactions on Magnetics* 45 (10): 4773–4776. doi: 10.1109/TMAG.2009.2023856.

381 Kaneko, K., Inoue, N., Furutake, N., Hijioka, K., and Hayashi, Y. (April 2009). Microstructure control of low-loss Ni–Zn ferrite by low-temperature sputtering for on-chip magnetic film. *Japanese Journal of Applied Physics* 48 (4): 04C030. doi: 10.1143/JJAP.48.04C030.

382 Gillies, M.F., Coehoorn, R., van Zon, J.B.A., and Alders, D. (June 1998). Structure and soft magnetic properties of sputter deposited MnZn-ferrite films. *Journal of Applied Physics* 83 (11): 6855–6857. doi: 10.1063/1.367866.

383 Rikukawa, H. (November 1982). Relationship between microstructures and magnetic properties of ferrites containing closed pores. *IEEE Transactions on Magnetics* 18 (6): 1535–1537. doi: 10.1109/TMAG.1982.1062065.

384 Hou, Y.H. et al. (November 2010). Structural, electronic and magnetic properties of partially inverse spinel $CoFe_2O_4$: A first-principles study. *Journal of Physics D: Applied Physics* 43 (44): 445003. doi: 10.1088/0022-3727/43/44/445003.

385 Soeya, S. et al. (February 2002). Development of half-metallic ultrathin Fe_3O_4 films for spin-transport devices. *Applied Physics Letters* 80 (5): 823–825. doi: 10.1063/1.1446995.

386 Swartzendruber, L.J. (November 1991). Properties, units and constants in magnetism. *Journal of Magnetism and Magnetic Materials* 100 (1–3): 573–575. doi: 10.1016/0304-8853(91)90841-W.

387 Tang, J., Wang, K.-Y., and Zhou, W. (June 2001). Magnetic properties of nanocrystalline Fe_3O_4 films. *Journal of Applied Physics* 89 (11): 7690–7692. doi: 10.1063/1.1358350.

388 Liu, K., Zhao, L., Klavins, P., Osterloh, F.E., and Hiramatsu, H. (May 2003). Extrinsic magnetoresistance in magnetite nanoparticles. *Journal of Applied Physics* 93 (10): 7951–7953. doi: 10.1063/1.1556133.

389 Liu, H., Jiang, E.Y., Bai, H.L., Zheng, R.K., Wei, H.L., and Zhang, X.X. (October 2003). Large room-temperature spin-dependent tunneling magnetoresistance in polycrystalline Fe_3O_4 films. *Applied Physics Letters* 83 (17): 3531–3533. doi: 10.1063/1.1622440.

390 Pénicaud, M., Siberchicot, B., Sommers, C.B., and Kübler, J. (January 1992). Calculated electronic band structure and magnetic moments of ferrites. *Journal of Magnetism and Magnetic Materials* 103 (1–2): 212–220. doi: 10.1016/0304-8853(92)90255-M.

391 Hirohata, A. et al. (October 2015). Roadmap for emerging materials for spintronic device applications. *IEEE Transactions on Magnetics* 51 (10): 1–11. doi: 10.1109/TMAG.2015.2457393.

392 Niizeki, T. et al. (October 2013). Extraordinarily large perpendicular magnetic anisotropy in epitaxially strained cobalt-ferrite $Co_xFe_{3-x}O_4$ (001) ($x = 0.75, 1.0$) thin films. *Applied Physics Letters* 103 (16): 162407. doi: 10.1063/1.4824761.

393 Naoe, M., Hasunuma, S., Hoshi, Y., and Yamanaka, S. (November 1981). Preparation of barium ferrite films with perpendicular magnetic anisotropy by DC sputtering. *IEEE Transactions on Magnetics* 17 (6): 3184–3186. doi: 10.1109/TMAG.1981.1061729.

394 Matsuoka, M., Hoshi, Y., Naoe, M., and Yamanaka, S. (November 1982). Formation of Ba-ferrite films with perpendicular magnetization by targets-facing type of sputtering. *IEEE Transactions on Magnetics* 18 (6): 1119–1121. doi: 10.1109/TMAG.1982.1062006.

395 Morisako, A. and Naoe, M. (September 1984). Preparation of double layer hexagonal ferrite films for high density longitudinal magnetic recording media. *IEEE Transactions on Magnetics* 20 (5): 815–817. doi: 10.1109/TMAG.1984.1063366.

396 Sui, X. and Kryder, M.H. (September 1993). Magnetic easy axis randomly in-plane oriented barium hexaferrite thin film media. *Applied Physics Letters* 63 (11): 1582–1584. doi: 10.1063/1.110755.

397 Snyder, J.E., Harris, V.G., Koon, N.C., Sui, X., and Kryder, M.H. (October 1996). Local structure of the amorphous precursor to Ba-hexaferrite thin films: An anisotropic octahedral Fe-O glass network. *Physical Review Letters* 77 (16): 3383–3386. doi: 10.1103/PhysRevLett.77.3383.

398 Hylton, T.L., Parker, M.A., and Howard, J.K. (August 1992). Preparation and magnetic properties of epitaxial barium ferrite thin films on sapphire with in-plane, uniaxial anisotropy. *Applied Physics Letters* 61 (7): 867–869. doi: 10.1063/1.107772.

399 Hylton, T.L., Parker, M.A., Ullah, M., Coffey, K.R., Umphress, R., and Howard, J.K. (May 1994). Ba-ferrite thin-film media for high-density longitudinal recording (invited). *Journal of Applied Physics* 75 (10): 5960–5965. doi: 10.1063/1.356987.

400 Morisako, A., Matsumoto, M., and Naoe, M. (September 1986). Ba-ferrite thin film rigid disk for high density perpendicular magnetic recording. *IEEE Transactions on Magnetics* 22 (5): 1146–1148. doi: 10.1109/TMAG.1986.1064350.

401 Yuan, M.S., Glass, H.L., and Adkins, L.R. (July 1988). Epitaxial barium hexaferrite on sapphire by sputter deposition. *Applied Physics Letters* 53 (4): 340–341. doi: 10.1063/1.100601.

402 Liu, X. et al. (March 2000). La–Zn substituted hexagonal Sr ferrite thin films for high density magnetic recording. *Journal of Applied Physics* 87 (5): 2503–2506. doi: 10.1063/1.372210.

403 Han, M., Lu, H., Lin, C., and Deng, L. (June 2007). Comparisons on the properties of $(Zn_{1-x}Co_x)_2$–W type barium hexaferrite thin films prepared on the Si (100) and (111) substrates. *Thin Solid Films* 515 (18): 7294–7298. doi: 10.1016/j.tsf.2007.03.013.

404 Sin, K., Sivertsen, J.M., Judy, J.H., Hoshi, Y., and Speliotis, D.E. (May 1993). Preparation and characterization of thin films of M-type barium ferrite on microcrystalline carbon substrates with diffusion barrier sublayers. *Journal of Applied Physics* 73 (10): 6689–6691. doi: 10.1063/1.352503.

405 Nakagawa, S., Mizuno, K., and Naoe, M. (May 2003). Low temperature deposition of c-axis oriented Sr-ferrite thin films prepared by facing targets sputtering with mixture gas of Ar and Kr. *Journal of Applied Physics* 93 (10): 8170–8172. doi: 10.1063/1.1555846.

406 Morisako, A., Nakanishi, H., Matsumoto, M., and Naoe, M. (May 1994). Low-temperature deposition of hexagonal ferrite films by sputtering. *Journal of Applied Physics* 75 (10): 5969–5971. doi: 10.1063/1.355528.

407 Zhuang, Z., Rao, M., White, R.M., Laughlin, D.E., and Kryder, M.H. (May 2000). Barium ferrite thin film media with perpendicular c-axis orientation and small grain size. *Journal of Applied Physics* 87 (9): 6370–6372. doi: 10.1063/1.372709.

408 Shams, N.N., Liu, X., Matsumoto, M., and Morisako, A. (April 2005). Manipulation of crystal orientation and microstructure of barium ferrite thin film. *Journal of Magnetism and Magnetic Materials* 290–291: 138–140. doi: 10.1016/j.jmmm.2004.11.166.

409 Capraro, S. et al. (December 2004). Thick barium ferrite films use for passive isolators. *Physica Status Solidi C* 1 (12): 3373–3376. doi: 10.1002/pssc.200405460.

410 Zhang, X., Yue, Z., and Zhang, Y. (August 2017). Structure characterization and magnetic properties of barium hexaferrite films deposited on 6H-SiC with random in-plane orientation. *Ceramics International* 43 (12): 8611–8615. doi: 10.1016/j.ceramint.2017.03.169.

411 Li, Y., Liu, Y., Li, J., Yang, K., and Zhang, H. (May 2013). Magnetic and self-biased properties of highly oriented M-type barium ferrite films on Pt-coated Si substrate by magnetron sputtering. *Journal of Applied Physics* 113 (17): 17B306. doi: 10.1063/1.4798608.

412 Harris, V.G. and Sokolov, A.S. (January 2019). The self-biased circulator: Ferrite materials design and process considerations. *Journal of Superconductivity and Novel Magnetism* 32 (1): 97–108. doi: 10.1007/s10948-018-4928-9.

413 Shams, N.N., Matsumoto, M., and Morisako, A. (2004). Effect of Al-addition on C-axis orientation of Ba-ferrite film deposited on amorphous aluminum oxide (α-AlO) layer. *Transactions of the Magnetics Society of Japan* 4 (2): 43–45. doi: 10.3379/tmjpn2001.4.43.

414 Zhang, W., Peng, B., Zhang, W., Zhou, S., and Schmidt, H. (July 2010). Ultra large coercivity in barium ferrite thin films prepared by magnetron sputtering. *Journal of Magnetism and Magnetic Materials* 322 (13): 1859–1862. doi: 10.1016/j.jmmm.2009.12.041.

415 Malygin, A.A., Drozd, V.E., Malkov, A.A., and Smirnov, V.M. (December 2015). From V. B. Aleskovskii's 'Framework' hypothesis to the method of molecular layering/atomic layer deposition. *Chemical Vapor Deposition* 21 (10–11–12): 216–240. doi: 10.1002/cvde.201502013.

416 Parsons, G.N. et al. (September 2013). History of atomic layer deposition and its relationship with the American Vacuum Society. *Journal of Vacuum Science and Technology A: Vacuum, Surfaces and Films* 31 (5): 050818. doi: 10.1116/1.4816548.

417 Puurunen, R.L. (December 2014). A short history of atomic layer deposition: Tuomo Suntola's atomic layer epitaxy. *Chemical Vapor Deposition* 20 (10-11-12): 332–344. doi: 10.1002/cvde.201402012.

418 George, S.M. (January 2010). Atomic layer deposition: An overview. *Chemical Reviews* 110 (1): 111–131. doi: 10.1021/cr900056b.

419 Chong, Y.T., Yau, E.M.Y., Nielsch, K., and Bachmann, J. (December 2010). Direct atomic layer deposition of ternary ferrites with various magnetic properties. *Chemistry of Materials* 22 (24): 6506–6508. doi: 10.1021/cm102600m.

420 Jalkanen, P. et al. (December 2014). Magnetic properties of polycrystalline bismuth ferrite thin films grown by atomic layer deposition. *Journal of Physical Chemistry Letters* 5 (24): 4319–4323. doi: 10.1021/jz502285f.

421 https://www.lesker.com/newweb/vacuum_systems/deposition_systems_ald_ald150lx.cfm. Last accessed in November 2021.

422 Oviroh, P.O., Akbarzadeh, R., Pan, D., Coetzee, R.A.M., and Jen, T.-C. (December 2019). New development of atomic layer deposition: Processes, methods and applications. *Science and Technology of Advanced Materials* 20 (1): 465–496. doi: 10.1080/14686996.2019.1599694.

423 Langston, M.C. et al. (November 2012). In situ cycle-by-cycle flash annealing of atomic layer deposited materials. *Journal of Physical Chemistry C* 116 (45): 24177–24183. doi: 10.1021/jp308895e.

424 Sheng, J., Lee, J.-H., Choi, W.-H., Hong, T., Kim, M., and Park, J.-S. (November 2018). Review article: Atomic layer deposition for oxide semiconductor thin film transistors: Advances in research and development. *Journal of Vacuum Science and Technology A* 36 (6): 060801. doi: 10.1116/1.5047237.

425 Crabtree, R.H. (2014). *The Organometallic Chemistry of the Transition Metals*, 6e. New York, NY: Wiley.

426 Riha, S.C., Racowski, J.M., Lanci, M.P., Klug, J.A., Hock, A.S., and Martinson, A.B.F. (March 2013). Phase discrimination through oxidant selection in low-temperature atomic layer deposition of crystalline iron oxides. *Langmuir* 29 (10): 3439–3445. doi: 10.1021/la305027k.

427 Zierold, R. et al. (December 2014). Magnetic characterization and electrical field-induced switching of magnetite thin films synthesized by atomic layer deposition and subsequent thermal reduction. *Journal of Physics D: Applied Physics* 47 (48): 485001. doi: 10.1088/0022-3727/47/48/485001.

428 Zhang, Y., Liu, M., Zhang, Y., Chen, X., Ren, W., and Ye, Z.-G. (May 2015). Atomic layer deposition of superparamagnetic and ferrimagnetic magnetite thin films. *Journal of Applied Physics* 117 (17): 17C743. doi: 10.1063/1.4916818.

429 Márquez, P., Alburquenque, D., Celis, F., Freire, R.M., and Escrig, J. (July 2021). Structural, morphological and magnetic properties of iron oxide thin films obtained by atomic layer deposition as a function of their thickness. *Journal of Magnetism and Magnetic Materials* 530: 167914. doi: 10.1016/j.jmmm.2021.167914.

430 Abe, M. and Tamaura, Y. (August 1983). Ferrite-plating in aqueous solution: A new method for preparing magnetic thin film. *Japanese Journal of Applied Physics* 22 (Part 2, 8): L511–L513. doi: 10.1143/JJAP.22.L511.

431 Abe, M. (June 2000). Ferrite plating: A chemical method preparing oxide magnetic films at 24–100°C, and its applications. *Electrochimica Acta* 45 (20): 3337–3343. doi: 10.1016/S0013-4686(00)00403-5.

432 Kondo, K. et al. (2005). Spin-sprayed ferrite films highly permeable in the GHz range, with excellent heat resistance. *Transactions of the Magnetics Society of Japan* 5 (4): 161–164. doi: 10.3379/tmjpn2001.5.161.

433 Abe, M., Itoh, T., Tamaura, Y., Gotoh, Y., and Gomi, M. (September 1987). Ferrite plating on GaAs for microwave monolithic integrated circuit. *IEEE Transactions on Magnetics* 23 (5): 3736–3738. doi: 10.1109/TMAG.1987.1065205.

434 Kondo, K. et al. (May 2003). Conducted noise suppression effect up to 3 GHz by NiZn ferrite film plated at 90 °C directly onto printed circuit board. *Journal of Applied Physics* 93 (10): 7130–7132. doi: 10.1063/1.1555362.

435 Kondo, K. et al. (2004). GHz conducted noise suppression by NiZn ferrite films plated onto polyimide sheets. In *Proceedings of the Ninth International Conference on Ferrites (ICF-9)*, San Francisco, CA, 659–663.

436 Hong, J.S., Wagata, H., Ohashi, N., Katsumata, K., Okada, K., and Matsushita, N. (August 2015). Transparent ZnO films deposited by aqueous solution process under various pH conditions. *Journal of Electronic Materials* 44 (8): 2657–2662. doi: 10.1007/s11664-015-3670-3.

437 Obi, O. et al. (June 2013). Growth behaviors and characteristics of low temperature spin-sprayed ZnO and Al-doped ZnO microstructures. *Journal of Materials Science: Materials in Electronics* 24 (6): 2058–2066. doi: 10.1007/s10854-012-1056-3.

438 Miyasaka, J., Tada, M., Abe, M., and Matsushita, N. (April 2006). $Fe_3O_{4+\delta}$ films prepared by 'one-liquid' spin spray ferrite plating for Gigahertz-range noise suppressors. *Journal of Applied Physics* 99 (8): 08M916. doi: 10.1063/1.2177394.

439 Harris, V.G. et al. (July 2009). Recent advances in processing and applications of microwave ferrites. *Journal of Magnetism and Magnetic Materials* 321 (14): 2035–2047. doi: 10.1016/j.jmmm.2009.01.004.

440 Wagata, H. et al. (August 2010). Single-step fabrication of ZnO rod arrays on a nonseeded glass substrate by a spin-spray technique at 90 °C. *Crystal Growth & Design* 10 (8): 3502–3507. doi: 10.1021/cg100386c.

441 Hong, J.S., Katsumata, K., and Matsushita, N. (May 2016). The morphological properties of heterostructured Fe_3O_4/ZnO film by aqueous solution process. *IEEE Transactions on Magnetics* 52 (5): 1–5. doi: 10.1109/TMAG.2015.2513740.

442 Shimada, Y., Matsushita, N., Abe, M., Kondo, K., Chiba, T., and Yoshida, S. (July 2004). Study on initial permeability of Ni–Zn ferrite films prepared by the spin spray method. *Journal of Magnetism and Magnetic Materials* 278 (1–2): 256–262. doi: 10.1016/j.jmmm.2003.12.1312.

443 Kondo, K., Yoshida, S., Ono, H., and Abe, M. (May 2007). Spin sprayed Ni(–Zn)–Co ferrite films with natural resonance frequency exceeding 3GHz. *Journal of Applied Physics* 101 (9): 09M502. doi: 10.1063/1.2710465.

444 Obi, O. et al. (April 2011). Spin-spray deposited NiZn-Ferrite films exhibiting $\mu_r' > 50$ at GHz range. *Journal of Applied Physics* 109 (7): 07E527. doi: 10.1063/1.3562879.

445 van der Zaag, P.J., Lubitz, P., Kitamoto, Y., and Abe, M. (September 1999). The permeability of plated ferrite films. *IEEE Transactions on Magnetics* 35 (5): 3436–3438. doi: 10.1109/20.800549.

446 Wang, X. et al. (March 2015). Growth behavior and RF/microwave properties of low temperature spin-sprayed NiZn ferrite. *Journal of Materials Science: Materials in Electronics* 26 (3): 1890–1894. doi: 10.1007/s10854-014-2625-4.

447 Goto, Y., Tamaura, Y., Abe, M., and Gomi, M. (February 1988). Improvement in deposition rate and quality of films prepared by 'thin liquid-film' ferrite plating method. *IEEE Translation Journal on Magnetics in Japan* 3 (2): 159–165. doi: 10.1109/TJMJ.1988.4563663.

448 Yang, G.-M., Obi, O., and Sun, N.X. (May 2011). Small global positioning system patch antennas with self-biased NiCo-ferrite films. *Microwave and Optical Technology Letters* 53 (5): 1162–1165. doi: 10.1002/mop.25909.

449 Yang, G.-M., Obi, O., and Sun, N.X. (January 2012). Enhancing ground plane immunity of dipole antennas with spin-spray-deposited lossy ferrite films. *Microwave and Optical Technology Letters* 54 (1): 230–233. doi: 10.1002/mop.26490.

450 Yang, G.-M. et al. (October 2009). Miniaturized antennas and planar bandpass filters with self-biased NiCo-ferrite films. *IEEE Transactions on Magnetics* 45 (10): 4191–4194. doi: 10.1109/TMAG.2009.2023996.

451 Guo, R. et al. (May 2018). An integrated tunable isolator based on NiZn film fabricated by spin-spray plating. *AIP Advances* 8 (5): 056620. doi: 10.1063/1.5006722.

452 Wang, X. et al. (May 2017). A novel NiZn ferrite integrated magnetic solenoid inductor with a high quality factor at 0.7–6 GHz. *AIP Advances* 7 (5): 056606. doi: 10.1063/1.4973283.

453 Chao, L., Sharma, A., Afsar, M.N., Obi, O., Zhou, Z., and Sun, N. (November 2012). Permittivity and permeability measurement of spin-spray deposited Ni-Zn-ferrite thin film sample. *IEEE Transactions on Magnetics* 48 (11): 4085–4088. doi: 10.1109/TMAG.2012.2200665.

454 Xing, Y., Myers, J., Obi, O., Sun, N.X., and Zhuang, Y. (April 2012). Excessive grain boundary conductivity of spin-spray deposited ferrite/non-magnetic multilayer. *Journal of Applied Physics* 111 (7): 07A512. doi: 10.1063/1.3676242.

455 Matsushita, N., Kondo, K., Yoshida, S., Tada, M., Yoshimura, M., and Abe, M. (July 2006). Ni-Zn ferrite films synthesized from aqueous solution usable for sheet-type conducted noise suppressors in GHz range. *Journal of Electroceramics* 16 (4): 557–560. doi: 10.1007/s10832-006-9918-9.

456 Liu, M. et al. (April 2008). Spin-spray deposited multiferroic composite $Ni_{0.23}Fe_{2.77}O_4/Pb(Zr,Ti)O_3$ with strong interface adhesion. *Applied Physics Letters* 92 (15): 152504. doi: 10.1063/1.2911743.

457 Liu, M. et al. (June 2009). Giant electric field tuning of magnetic properties in multiferroic ferrite/ferroelectric heterostructures. *Advanced Functional Materials* 19 (11): 1826–1831. doi: 10.1002/adfm.200801907.

458 Liu, M. et al. (February 2009). Strong magnetoelectric coupling in ferrite/ferroelectric multiferroic heterostructures derived by low temperature spin-spray deposition. *Journal of Physics D: Applied Physics* 42 (4): 045007. doi: 10.1088/0022-3727/42/4/045007.

459 Zhou, Z. et al. (March 2014). Low-temperature spin spray deposited ferrite/piezoelectric thin film magnetoelectric heterostructures with strong magnetoelectric coupling. *Journal of Materials Science: Materials in Electronics* 25 (3): 1188–1192. doi: 10.1007/s10854-014-1707-7.

460 Xing, Y., Myers, J., Obi, O., Sun, N.X., and Zhuang, Y. (March 2012). Scanning microwave microscopy characterization of spin-spray-deposited ferrite/nonmagnetic films. *Journal of Electronic Materials* 41 (3): 530–534. doi: 10.1007/s11664-011-1874-8.

461 Liu, H. et al. (January 2021). Anisotropic growth and magnetic properties of nickel–zinc ferrite thin film by spin spray deposition. *Ceramics International* 47 (1): 1318–1324. doi: 10.1016/j.ceramint.2020.08.253.

462 Hastings, J.M. and Corliss, L.M. (January 1953). Neutron diffraction studies of zinc ferrite and nickel ferrite. *Reviews of Modern Physics* 25 (1): 114–119. doi: 10.1103/RevModPhys.25.114.*(Continued)*

11

Properties and Applications of Single-Crystal Ferrite Films Grown by Liquid-Phase Epitaxy

Huaiwu Zhang[1], Qinghui Yang[1], Qiye Wen[1], Yingli Liu[1], and Vincent G. Harris[2]

[1] *State Key Laboratory of Electronic Thin Film and Integrated Devices, University of Electronic Science and Technology of China, Chengdu, 610054 P. R. China*
[2] *Northeastern University, College of Engineering, 360 Huntington Ave., Boston, MA 02115-5000 USA*

11.1 Introduction

The development of radiofrequency (RF) ferrite devices creates the need for growth of both thin film and thick films of high crystal quality. Thin film development of ferrites has been covered thoroughly in Chapter 10. Thin-film growth techniques adaptable to ferrites include molecular beam epitaxy, pulsed laser deposition, sputtering, spin-spray plating, and a myriad of soft chemical processes.

Liquid-phase epitaxy (LPE) is distinguished from those film growth techniques in that its processes occur close to thermodynamic equilibrium. This distinction affords LPE with its unique advantages, such as excellent thick film material quality and anisotropic and faceted crystal growth. The near-equilibrium growth also accounts for some intrinsic shortcomings of LPE, including precise control of film thickness; difficulty in producing films with thicknesses less than ~0.1 μm; incompatibility with growth of heterostructures due to lattice mismatch; unwanted compositional and doping gradients; and less-than-optimal surface morphologies. LPE is sensitive to small perturbations in temperature, melt composition, wetting, and surface tension, as well as substrate surface specifications such as crystallographic orientation and dislocation density.

Nonetheless, when one requires thick films, i.e. 1 to >100 μm, of near-single crystal quality, LPE is singular in its value and utility.

Specifically, LPE deposits single crystal quality, highly oriented thin or thick films from the liquid phase, typically at rates of 0.1 to >1 μm/min., on lattice-matched substrates. LPE has been shown uniquely valuable in the growth of semiconductor, optical, magnetic, or other functional materials allowing for the development of devices and the advancement of emerging technologies.

The congruent transfer of chemistry from melt to substrate occurs at temperatures well below the melting point of the film material. In practice, the material to be deposited is dissolved in a melt mixture of the targeted composition in combination with a low melting point flux under conditions that are close to the equilibrium between dissolution and deposition.

The growth of the film on the substrate is intentionally slow and measured. The conditions depend very much upon the temperature and concentration of the melt. The growth of the film from the liquid phase can be closely controlled by the careful cooling of the melt. Film compositions may be altered by introducing additives directly to the melt.

To facilitate nucleation, and avoid tensile and compressive stresses in the growing film that may lead to defects, the coefficient of thermal expansion (CTE) of the substrate and the film should be very similar.

In LPE, a supersaturated solution of the material, i.e. melt, to be grown is brought into direct contact with the substrate. If the substrate is single crystalline and the material has a similar lattice constant to that of the substrate, the material nucleates onto the substrate, maintaining high crystalline quality, resulting ultimately in a lattice matched epitaxial film.

11.2 Historical Development of LPE Technique and Reactor Design

LPE was first developed circa 1963 [1] and was adapted and shown to be instrumental in the advance of III–V compound semiconductor devices, including light-emitting diodes (LEDs), photodetectors, solar cells, bipolar transistors, and field effect transistors (FETs) among others [2].

Modern Ferrites: Basic Principles, Processing and Properties, Volume 1, First Edition. Edited by Vincent G. Harris.

Nelson's method [1], representing the earliest manifestation of LPE, was followed by several improvements to instrumentation that resulted in improved surface quality, reproducibility, crystal quality that led to a reduction in defects and increase in wafer yields.

These improvements were required to meet the growing needs for Gunn diodes, FETs, and Schottky barrier diodes [3–6].

From 1963 to 1970, several studies investigated the role of temperature gradients in reactor designs and upon epitaxy and film morphology [3–12].

In the early 1970s, LPE was rapidly developed as a tool for the growth of not only III–V compound semiconductors [13, 14] but also ferrite films in the work of Linares in 1968 [15] using a "rocking-crucible" variant of LPE (see Figure 11.1a–d), also popularized by Shick and Nielson [16].

In this variant of LPE, the melt is shifted from one end of an elongated crucible to the other by gravitational force. The substrate is fixed at one end of the crucible with a screen positioned midway to filter solute fragments (see Figure 11.1c). The tilting of the crucible to and frow represents a single growth cycle with many cycles required to complete a single film deposition.

Other variants of the rocking crucible (sometimes referred to as tilting crucible) are depicted in Figure 11.1a–d, as discussed by Linares [15], Shick and Nielson [16], and Mauk [17]. Still another variant of LPE is the "sliding-crucible" technique in which crucibles containing the melt are passed over windows exposing the melt to the substrate for a predetermined time to allow nucleation and growth. This variant can also be performed in a cyclic fashion (see Figures 11.2a and b).

First-use vertical reactor designs similar to today's reactors are presented in Figure 11.3 [18].

In 1971, this new dipping technique successfully produced uniform-thickness low-defect-density garnet films. The technique is based on the observation that melts having the appropriate flux and garnet composition can be supercooled over a large temperature range and will remain supercooled, permitting the growth at a constant temperature of epitaxial garnet films.

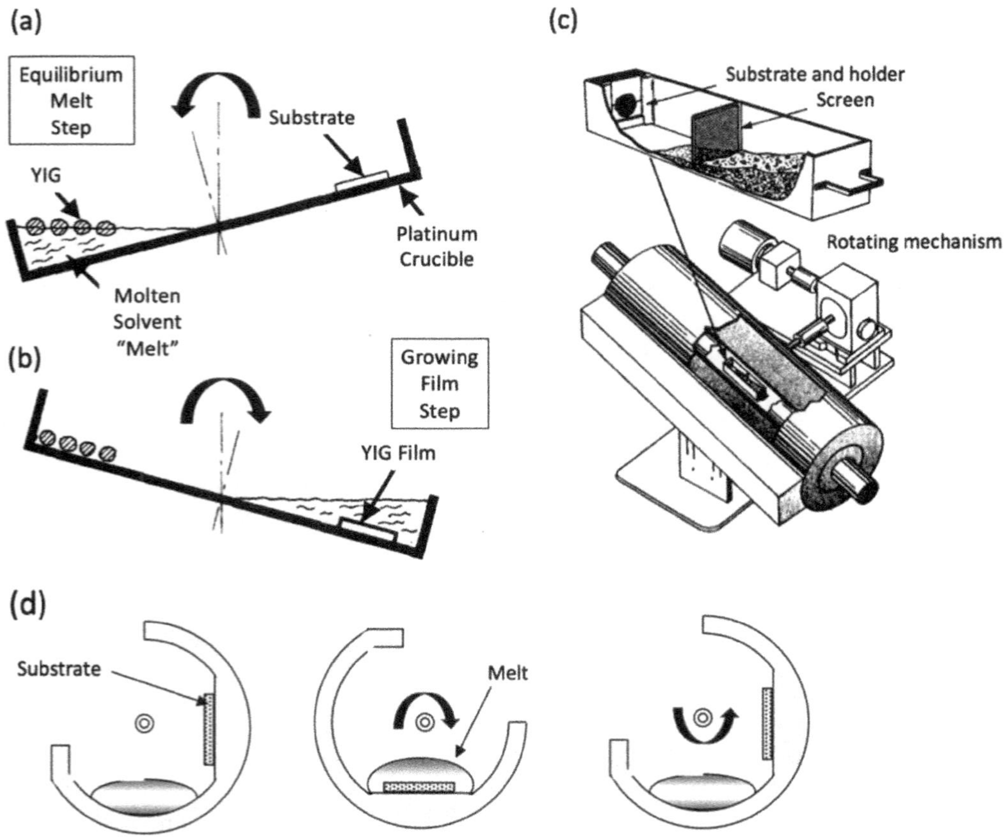

Figure 11.1 Schematic drawings of (a and b) "rocking" or "tilting" crucible LPE methods [15]. (c) Platinum boats with a platinum screen separating the boat into saturating and growing zones and preventing excess solute from drifting to the substrate. The boat is covered as tightly as possible with platinum foil to minimize solvent loss to evaporation [16]. (d) Conceptual drawing of a variant to the rocking LPE method by Nelson as reported in [17].

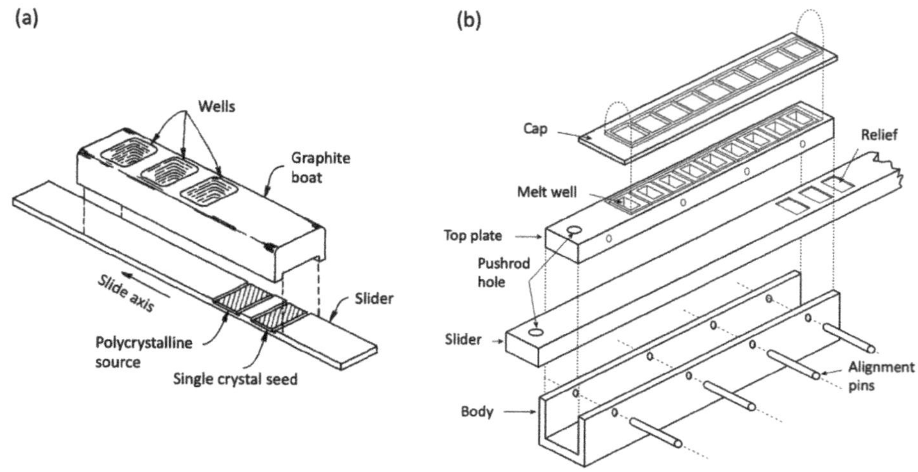

Figure 11.2 Schematic drawings of the "sliding" LPE techniqie [17].

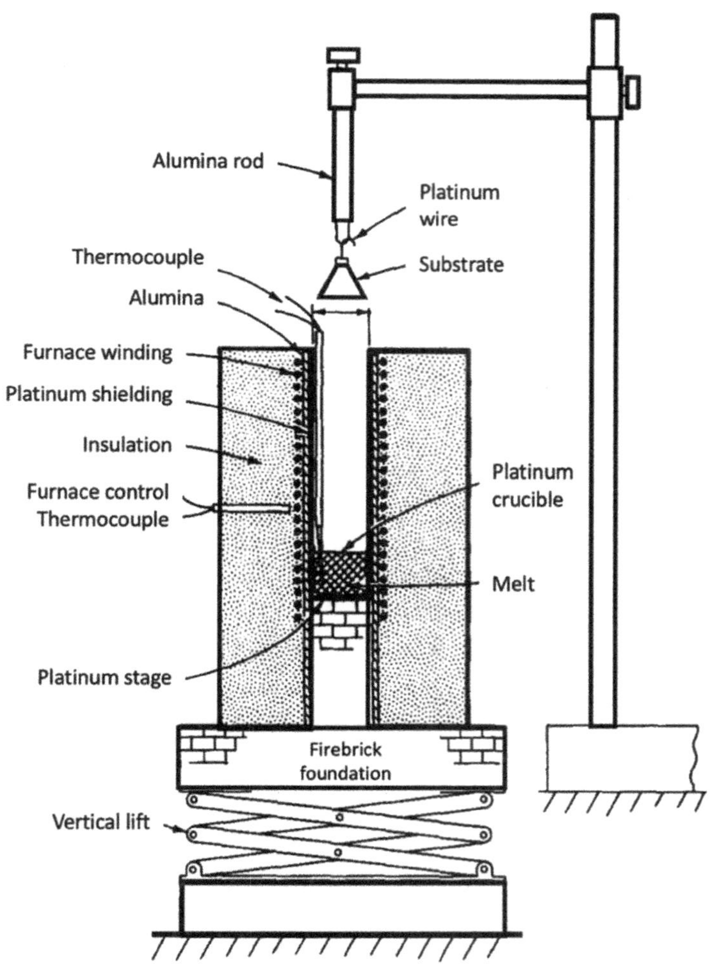

Figure 11.3 Schematic drawing of the vertical LPE reactor used by Levinstein et al. [18].

Levinstein et al. presented an LPE reactor design that mirrors modern vertical reactor designs (see Figure 11.3). This advance demonstrated that flux and garnet compositions could be maintained under supercooled conditions allowing growth at constant temperatures, leading to reproducibility in garnet films having uniform-thickness and low-defect density [18].

In a vertical LPE reactor, single-crystal substrate(s) is dipped into a molten solution containing a solvent flux and solute composition in a ratio close to the nominal value of the targeted film composition in an arrangement such as that depicted in Figure 11.4a–f.

In Figure 11.4, a platinum crucible is depicted to contain the melt and limit reaction and contamination (see specifically Figure 11.4b), while a motor-controlled substrate-holder rod is rotated axially to maintain the crystal substrate relative to the melt. Figure 11.4c–f shows substrate design alternatives that allow for single and multiple substrates with vertical and horizontal alignments relative the crucible and melt [19].

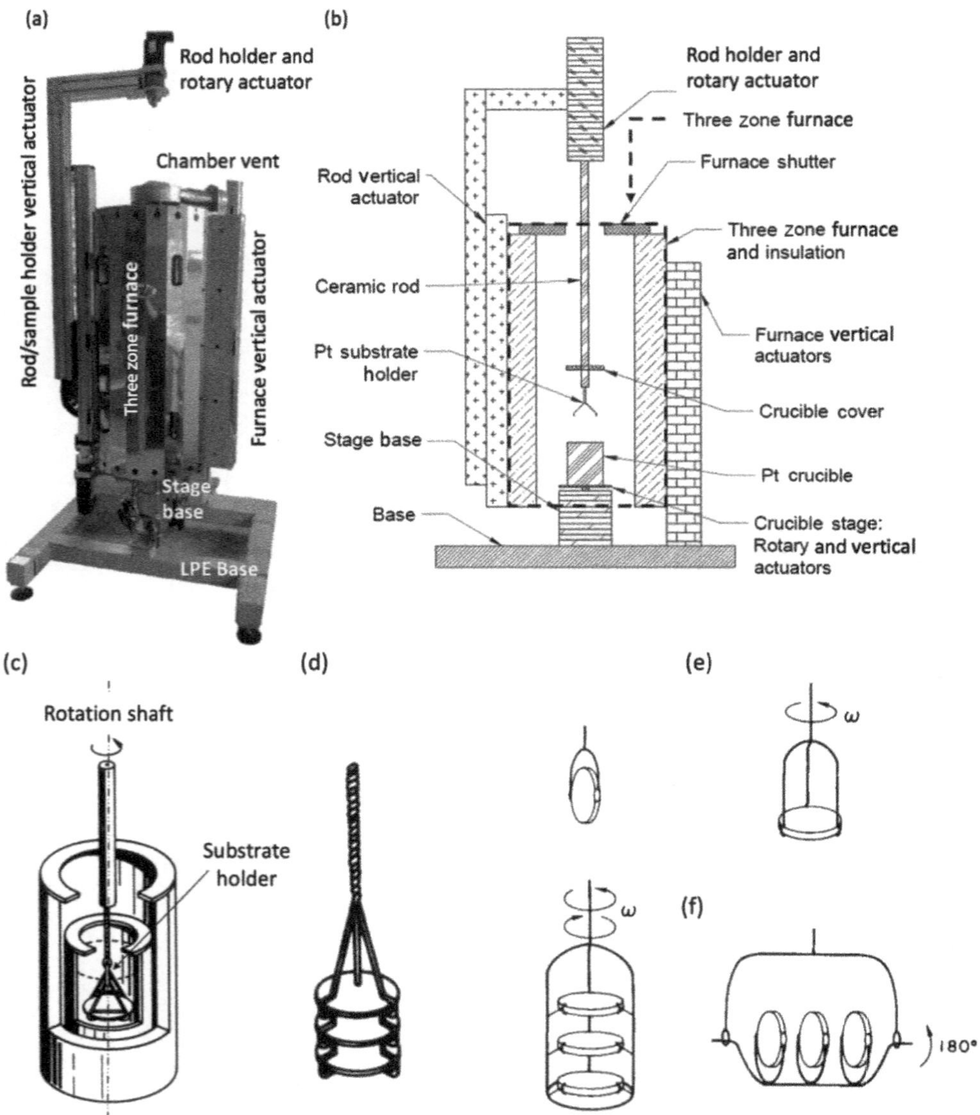

Figure 11.4 (a) Cyberstar's (http://www.cyberstar.fr/index.html) vertical LPE research reactor with its corresponding (b) cross-section schematic with major components labeled (courtesy of Jacob Remz of Metamagnetic Inc., circa 2021). (c)–(f) Schematics of alternative designs of substrate holders, including (c) the double baffle, single-substrate holder; (d) multiple substrate horizontal holder; (e) single-wafer vertical and unidirectional horizontal dipping mode holder; and (f) muliple-wafer oscillatory horizontal and vertical dipping mode sample holder [19].

11.3 Growth of Epitaxial Ferrites by LPE: Process Considerations

In this section, we address LPE growth methodologies and specifically the roles of solvents, substrates, and growth kinetics.

11.3.1 Solvents

Solvents are oxides that are added to the target composition (i.e. the solute) collocated within the crucible to form a homogeneous melt with a much lower melting point than the targeted composition. The principal role of solvents in the LPE process is to lower the melt and growth temperatures of the crystalline film.

As pointed out by Hiske [19], the basic requirements of solvents to achieve this goal include five properties:

i) A low vapor or dissociation pressure and low reactivity with the crucible material at growth temperatures
ii) Low liquidus temperatures and low viscosity
iii) High solubility for crystal constituents and a steep liquidus to permit precise supersaturation conditions
iv) Temperature-invariant distribution coefficients of crystal constituents that are near unity
v) Very low solid solubility of solvent ions in the crystal film

Early successful demonstrations of growth of bulk-YIG single crystals were demonstrated by Nielsen and Dearborn using PbO-based fluxes [39]. This provided a starting point for further development of solvents for the LPE growth of ferrites.

Table 11.1 contains a listing of select Pb-free and Pb-containing, binary and ternary, solvents that have found popular use in LPE growth of garnets, spinels, and hexaferrites. In this table, solvents are listed with examples of film grown with the right-hand-side column listing experimentally proven advantages and disadvantages.

Solvent properties of $BaO–B_2O_3$ were investigated as early as 1962 by Linares [15] to grow $Y_3Fe_5O_{12}$ (YIG) from eutectic compositions of 60.5 BaO–39.5 B_2O_3 and determined the relevant phase equilibria. One disadvantage of the 60.5 BaO–39.5 B_2O_3 eutectic for LPE garnet growth is its relatively high melting point of 915 °C.

In response, lead oxide–boron oxide (i.e. $PbO–B_2O_3$) solvents became popular due to their low volatility and viscosity and the deep eutectics that occur in ternary phase diagrams formed with garnet oxides resulting in dramatic reductions in saturation temperatures, i.e. temperatures at which a crystalline film nucleates upon the substrate being introduced to melt.

The Blank and Nielsen [40] study of 1972 and the Hiskes [19] study of 1974 represent seminal works revealing the role of solvents in LPE film growth and properties. Specifically, Blank and Nielsen introduced ratios, i.e. defined in Figure 11.5a that allowed for ease in correlating solvent and solute compositions with growth kinetics and film properties. These ratios were embraced by other scholars over the next decade as research groups contributed to the understanding of different solvents for LPE growth of new electronic and optical materials systems.

Hiskes [19] performed a comparative study of BaO-based and PbO-based binary and ternary solvents in the LPE growth of mixed rare earth magnetic bubble garnets (i.e. $(SmY)_3(GaFe)_5O_{12}$ and $(GdYbY)_3(GaFe)_5O_{12}$) where growth times, rotation rates and melt temperatures were systematically investigated to determine their impact upon growth rate and film properties.

The experimental film thickness as a function of growth time for $Sm_{0.9}Y_{2.1}Ga_{1.2}Fe_{3.8}O_{12}$ grown at 1000 °C with a rotation rate of 120 rpm is shown in Figure 11.5b. A best-fit line is seen to extrapolate through the origin. The anticipated initial-stage growth transient (defined in terms of a competition between interfacial and diffusion kinetics) for the $PbO–B_2O_3$ solvent, is estimated by Ghez and Giess [41] to be of the order of 1 s, does not appear on the scales shown in Figure 11.5b. The growth rate for films grown using $BaO–B_2O_3–BaF_2$ is seen to be nearly twice that of films grown using the $PbO–B_2O_3$ solvent using identical undercooling and substrate rotation rate. This behavior is attributable to the effective increase in diffusion coefficients from the increased growth temperature (~100 °C).

As illustrated in Figure 11.5c, the solubility of the garnet phase increases as R_1 increases, similar to the behavior reported for the $PbO–B_2O_3$ system. The solubility of the garnet phase in the solvent 43 BaO–41 B_2O_3–16 BaF_2 is far greater than that in the $PbO–B_2O_3$ solvent.

Blank and Nielsen [40] provided part of the pseudo-binary phase diagram, as depicted in Figure 11.5d, in which one observes that T_{Sat} decreases from 1000 °C at 9.7 mole-% to 780 °C for 4.9 mole-% garnet oxides. A pseudo-eutectic occurs near 4.9 mole-% garnet oxides, below which one observes the precipitation of the primary PbO phase.

Elwell et al. [42] added Bi_2O_3 to lower the liquidus further to below 800 °C and grew nickel ferrite films. As related by Johansen et al. [25], Wittekoek et al. [43] reported that LPE-grown Bi-substituted $Y_3Fe_5O_{12}$ films (produced by Robertson et al. [44])

Table 11.1 Commonly used solvents for LPE growth of ferrites and select examples.

Solvents	Crystal film system	Comments
	Pb-free solvents	
Bi_2O_3–BaO	$BaFe_{10.7}Al_{1.3}O_{19}$ and $BaFe_{9.3}A1_{2.7}O_{19}$ [20]	Allows very low melt and substrate temperatures (~750 °C) and is free of lead (Pb) and (F). Disadvantages include the dissolving of platinum from the crucible leading to contamination. Bi_2O_3 leads to excessive flux adhesion and surface damage.
BaO–B_2O_3	YIG [15] $NiFe_2O_4$ [21] Fe_2TiO_4 [22] $Zn_2Ba_2Fe_{12}O_{22}$ [23] $BaFe_{12}O_{19}$ [24]	Advantages include low toxicity; free of lead (Pb) and (F). Disadvantages include phase instability and high melt and substrate temperatures.
Bi_2O_3–BaO–B_2O_3	Bi-YIG [25] $BaFe_{10.7}Al_{1.3}O_{19}$ and $BaFe_{9.3}A1_{2.7}O_{19}$ [20]	Allows very low melt and substrate temperatures (~750 °C) and is free of lead (Pb) and (F). The lower melt temperatures lead to higher levels of Bi contamination. Disadvantages also include the dissolving of platinum from the crucible, leading to contamination. Bi_2O_3 leads to excessive flux adhesion and surface damage.
BaO–BaF_2–B_2O_3	$(Sm_1Y_2)(Ga_{1.2}Fe_{3.8})O_{12}$ [19] $Gd_3(Al,Ga)_5O_{12}$:Ce [26] $Lu_{3-x}Gd_xGa_yA_{15-y}O_{12}$:Ce [27]	Advantages include low volatility and low toxicity. Disadvantages include high toxicity from presence of fluorine (F). Bi_2O_3 leads to excessive flux adhesion and surface damage.
	Pb-containing solvents	
PbO–V_2O_5	$(LuLaSm)_3(GaFe)_5O_{12}$ [28]	Advantages include a reduced solute solubility that reduces barriers to steady-state composition gradients and slower and controlled growth rates.
PbO–B_2O_3	$Y_3(FeGa)_5O_{12}$ and $Sm_1Y_2Ga_{1.2}Fe_{3.8}O_{12}$ [19, 29] $Eu_RY_{3-R}Fe_{5-x}Ga_xO_{12}$ [30] $Y_{2.40}Eu_{0.60}Ga_{1.1}Fe_{3.9}O_{12}$ and $EuEr_2Ga_{0.7}Fe_{4.3}O12$ [29] $NiFe_2O_4$ [21] $(FeMgIn)_2O_4$ [31] $LiFe_2O_4$ [32] $MgFe_2O_4$, $NiFe_2O_4$, $ZnFe_2O_4$, and $Ni_{1-x}Zn_xFe_2O_4$ [33] $Y_{1.5}Gd_{0.8}Eu_{0.7}Fe_{4.3}Al_{0.6}Ga_{0.1}O_{12}$ $Y_{1.0}Gd_{1.0}Er_{1.0}Fe_{4.5}Al_{0.3}Ga_{0.2}O_{12}$, $Y_{0.9}Gd_{1.5}Yb_{0.6}Fe_{4.5}Al_{0.5}O_{12}$, $Y_{1.0}Gd_{1.0}Tm_{1.0}Fe_{4.4}Ga_{0.6}O_{12}$, $Y_{1.1}Gd_{0.8}Tm_{1.1}Fe_{4.4}Ga_{0.6}O_{12}$ and $Y_{1.2}Gd_{0.6}Tm_{1.2}Fe_{4.4}Ga_{0.6}O_{12}$ [34] Various garnets [35]	Advantages include low volatility and viscosity. Disadvantages include high toxicity from the presence of lead (Pb).
PbF_2–B_2O_3	Various garnets including $Y_3(FeGa)_5O_{12}$ [36]	Advantages include phase purity. Disadvantages include high toxicity from the presence of lead (Pb).
PbO–B_2O_3–Bi_2O_3	$(Y_{2.5}Bi_{0.5})(FeGa)_5O_{12}$ [37] $(Y_{2.7}Bi_{0.3})(Fe_{3.8}Ga_{1.2})_5O_{12}$ [38]	Advantages include low volatility and viscosity. Disadvantages include high toxicity from the presence of lead (Pb), erosion of platinum from the crucible, leading to contamination. Bi_2O_3 leads to excessive flux adhesion and surface damage.

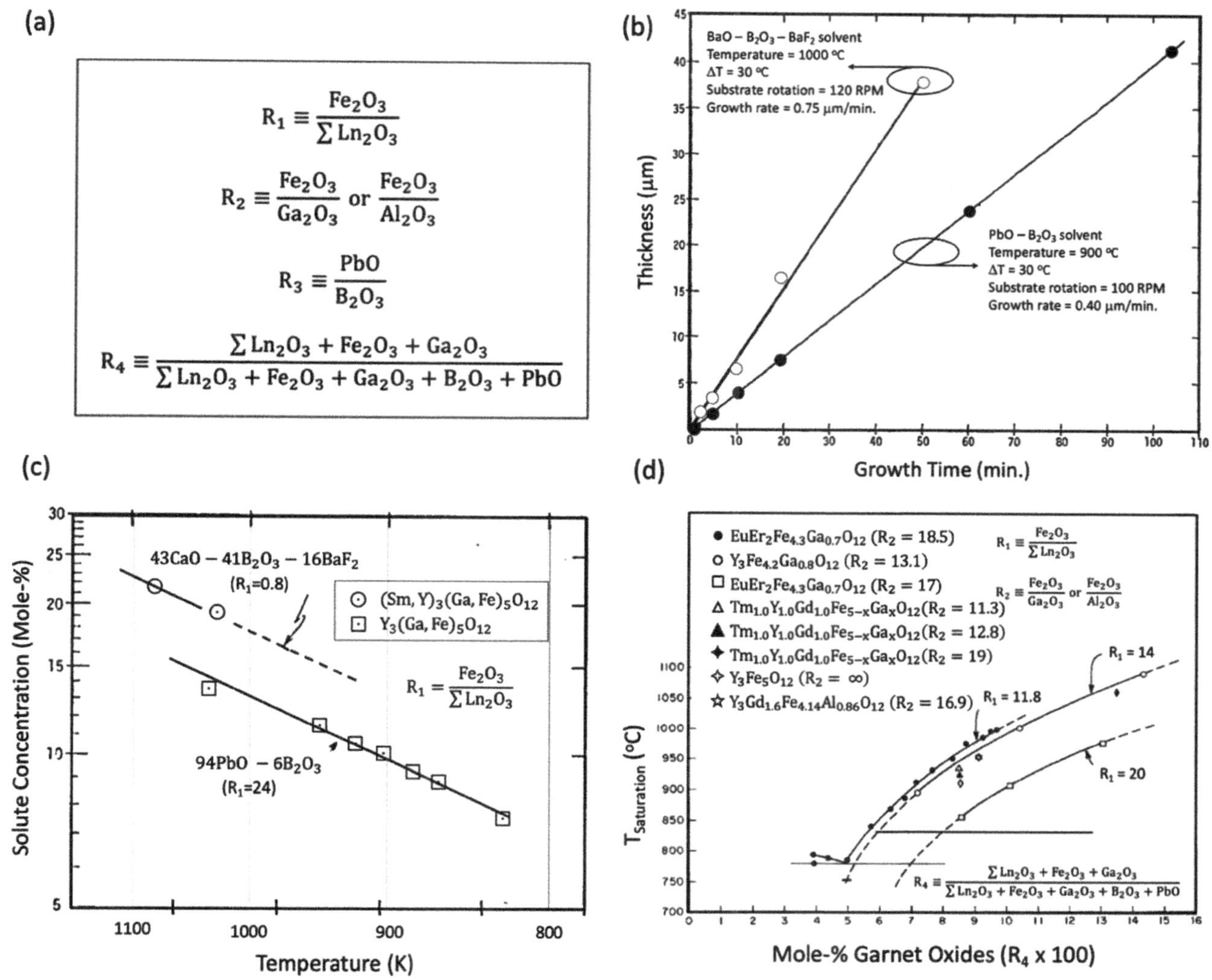

Figure 11.5 (a) Blank and Nielsen ratios allowing for ease in correlating solvent and solute compositions with growth kinetics and film properties. (b) Growth rate for $Sm_{0.3}Y_{2.7}Ga_{1.2}Fe_{3.8}O_{12}$ grown in $BaO-B_2O_3-BaF_2$ and $PbO-B_2O_3$ solvents [19]. (c) Solubility of magnetic garnets in BaO-based and PbO-based solvents [19]. (d) Logarithmic portion of pseudo-binary $PbO-B_2O_3$ garnet oxides [40].

experienced a marked increase in Faraday rotation in the visible and near-infrared bands. Introduction of Bi into LPE garnet films derived from the addition of Bi_2O_3 to the PbO-based flux, which rapidly grew in popularity. This led to a plethora of studies advancing such materials for magneto-optic applications.

It was soon found that the induced low melt temperatures employed during LPE using these solvents resulted in enhanced supersaturation and greater Bi uptake in crystalline films [44, 45]. Correspondingly, the introduction of the larger Bi ion ($r_{ion} = 117$ pm) [46] for the smaller Y ion ($r_{ion} = 104$ pm) [44] caused proportional lattice distortions and subsequent challenges related to cracking [47–49].

The use of Bi_2O_3-rich solvents also led to excessive flux adhesion and surface damage as well as increased erosion of platinum crucibles, all leading to deleterious effects to film properties.

11.3.1 Ternary Solvents

A shortcoming pointed out by Linares [15] in the composition he explored in 1962, i.e. 60.5 BaO–39.5 B_2O_3, was that it lies just at the edge of the garnet phase field. This problem was alleviated by the addition of a third component, i.e. BaF_2, to shift the phase field as well as lower the melting point via stabilization of a ternary eutectic. BaF_2 additions to the $BaO–B_2O_3$ solvent was used to grow $Y_3Fe_5O_{12}$ and several mixed rare earth bubble garnets [50, 51], including films of Ga- or

Al-substituted rare earth iron garnets on (111) GGG substrates [52]. Supercooling and growth rates were comparable to those obtained using $PbO–B_2O_3$ solvents, but the viscosity of the $BaO–BaF_2–B_2O_3$ solvent was much higher than $PbO–B_2O_3$, making it a more attractive solvent [20, 53].

This flux mixture possesses all the properties of an ideal solvent with the exception of its problematic volatility and toxicity. LPE undercooling of more than 100 °C was realized without experiencing secondary crystallization.

The most serious shortcoming of $PbO–B_2O_3$ and other PbO-containing solvents experience is the vaporization of PbO that is both corrosive and toxic.

Finally, during successive films cycles high vapor-pressure melt constituents experience selective evaporation, leading to the depletion of solute concentrations from the melt called *melt nutrient depletion (MND)*. MND limits the number of films produced per crucible load. To compensate for MND, Hewitt et al. [52] lowered the growth temperature and lengthened the growth duration for each film deposition. Alternatively, Stein [54] used the method of replenishing depleted oxides and remelting to extend the lifetime of each melt.

11.3.2 Substrates

In considering the LPE process, a chemically polished and cleaned substrate is affixed to the substrate holder and lowered under axial rotation into the undercooled melt. After a predetermined time, the substrate is removed from the solution, held just above the melt, and the melt remnants are spun free of the substrate surface using a high rotational speed. The substrate surface can be further cleaned by using an etch in a dilute acid mixture (i.e. typically of heated nitric acid, hydrochloric acid, and acetic acid).

In this section, we examine properties of LPE-grown films resulting from the use of a variety of substrates.

Kramer et al. [55] introduced the characteristics that make for an excellent substrate for LPE growth of films; these include:

i) Lattice parameter and thermal expansion coefficient match those of the ferrite film.
ii) Mechanical strength allows for slicing and polishing without fracture and chemical and thermal compatibility with melt constituents limiting dissolution and diffusion.
iii) Provides a phase diagram allowing for large high-quality single-crystal growth.
iv) Is a good electrical insulator and is paramagnetic or diamagnetic.
v) Low dielectric loss at RF frequencies.

These five characteristics provide a stable template for the nucleation and epitaxial growth of a largely stress-free film. Furthermore, if the substrate is free of dislocations and inclusions, there is the likelihood that the film will also grow free of defects provided the lattice constant of the film and substrate are within ~0.15%. Although this seems like an unreasonably rigid constraint upon lattice matching, Table 11.2 lists several substrates that meet this criterion.

For a more perfect lattice match, the thermal expansion difference between film and substrate should also be considered since films are grown at high temperatures but operate near room temperature.

Gadolinium gallium garnet ($Gd_3Ga_5O_{12}$), henceforth GGG, grown by the Czochralski crystal growth method has a lattice parameter (a_0) of 12.383, while high-purity stoichiometric GGG has an a_0 of 12.376. This ~0.05% to 0.06% difference is caused by contaminants introduced during crystal growth. Such contamination is also inherent to high-temperature LPE processing where impurity ions such as Pb^{2+} and Pt^{4+} are introduced by solvents and the erosion of the crucible.

Table 11.2 provides a list of substrates and their properties that have been used for LPE growth of garnet, spinel, and hexaferrite films. The substrate and film lattice constants are presented with the anticipated, or measured, mismatch, and film stress values of tensile or compressive stresses. In the comment column of this table, relevant comments are provided, e.g. high-temperature lattice parameters or key findings. The film stress is estimated here by:

$$\frac{a_{\text{film}} - a_{\text{substrate}}}{a_{\text{substrate}}} \times 100\%$$

where positive values coincide to compressive stresses and negative values coincide to tensile stresses.

For garnets, other garnets of similar crystal symmetry and lattice parameters make logical choices as substrates. The most attractive is the previously mentioned GGG. The GGG mismatch provides an anticipated near perfect match to YIG with a 0.056% film grown in slight tension.

For the LPE growth of spinel films, nonmagnetic spinel substrates have been studied, including Zn, Al, and Ga spinels [21]. Depending upon the chosen ferrite to be grown, the lattice mismatch can be as small as 0.02%. However, the large-scale growth of these exotic substrates is not routine and often prove too costly. MgO, which is widely available and cost-effective in large-scale production, provides a <1% tensile stress in the growth of most ferrite films and in some cases <0.03% mismatch with select spinel films (see Table 11.2), e.g. $Ni_{1-x}Zn_xFe_2O_4$ grown on MgO [33].

Robertson et al. demonstrated that small additions of Ga_2O_3, SnO_2, and Cr_2O_3 to spinel melts could improve lattice matches between $MgFe_2O_4$ films and their MgO substrates. An example of the approach is depicted in Figure 11.6, with the addition of 0.2 wt% SnO_2 to the melt (i.e. 510 g mixture of $PbO+B_2O_3+Fe_2O_3+MgO$) provides a near-zero mismatch [33].

The first attempt to grow hexaferrite films by LPE was reported by Sterns and Glass in 1975 [61]. That report presented growth and properties of hexaferrite Y-type (i.e. Zn_2Y: $Ba_2Zn_2Fe_{12}O_{22}$) films on substrates of M-type hexagonal ferrite crystals having the chemical formula $BaFe_{12}O_{19}$. Since then, most efforts have focused on the growth of M-type barium hexaferrite on a wide variety of substrates (see Table 11.2), including GGG [62], nonmagnetic spinels [55, 63], hexagallates [20, 55, 64], MgO [65, 66], and sapphire [60, 62, 67, 68], among others. In an effort to integrate hexaferrites with wide-bandgap

Table 11.2 Common LPE substrates and their properties for the growth of ferrite films.

Film composition	Film lattice constant (a_f)	Substrate composition	Substrate lattice constant (a_s)	Lattice mismatch %^{RT} (%^{HT}) T: tensile C: compressive	Comments	Ref.
			Garnet			
$Y_3Fe_5O_{12}$	12.376	$Gd_3Ga_5O_{12}$	12.383	−0.057 T	GGG can be reliably grown with few surface defects	[15]
$Y_3Fe_5O_{12}$	12.376	$Gd_3Sc_2Ga_3O_{12}$	12.567	−1.520 T	$a_f = 12.4951$; $a_s = 12.6740$ (950 °C)	[40]
$Y_3Fe_5O_{12}$	12.376	$Nd_3Ga_5O_{12}$	12.509	−1.063 T	$a_f = 12.4951$; $a_s = 12.6155$ (950 °C)	
$Y_3Fe_5O_{12}$	12.376	$Sm_3Ga_5O_{12}$	12.438	−0.498 T	$a_f = 12.4951$; $a_s = 12.5443$ (950 °C)	
$Y_3Fe_5O_{12}$	12.376	$Ho_3Ga_5O_{12}$	12.293	0.675 C	$a_f = 12.4951$; $a_s = 12.3973$ (950 °C)	
$Y_3Fe_5O_{12}$	12.376	$Y_3Al_5O_{12}$	12.008	3.065 C	$a_f = 12.4951$; $a_s = 12.1035$ (950 °C)	
$Tm_{1.0}Y_{1.0}Gd_{1.0}Fe_{3.9}Ga_{0.63}Al_{0.7}O_{12}$	12.357	$Gd_3Sc_2Ga_3O_{12}$	12.567	−1.671 T	$a_f = 12.4757$; $a_s = 12.674$ (950 °C)	
$Y_3Fe_{4.3}Al_{0.07}O_{12}$	12.324	$Nd_3Ga_5O_{12}$	12.509	−1.479 T	$a_f = 12.4426$; $a_s = 12.6155$ (950 °C)	
$Y_3Fe_{4.3}Al_{0.07}O_{12}$	12.324	$Sm_3Ga_5O_{12}$	12.438	−0.917 T	$a_f = 12.4426$; $a_s = 12.5443$ (950 °C)	
$Y_3Fe_{4.3}Al_{0.07}O_{12}$	12.324	$Ho_3Ga_5O_{12}$	12.293	0.252 C	$a_f = 12.4426$; $a_s = 12.3973$ (950 °C)	
$Y_3Fe_{4.3}Al_{0.07}O_{12}$	12.324	$Y_3Al_5O_{12}$	12.008	2.632 C	$a_f = 12.4426$; $a_s = 12.1035$ (950 °C)	
			Spinels			
$NiFe_2O_4$	8.332	$ZnGa_2O_4$	8.334	−0.024 T	Crack-free films with a uniaxial anisotropy energy exceed in demagnetization energy.	[21]
NiFe2-xAl0.68O4	8.23	$ZnGa_2O_4$	8.23–8.28	−0.363 T		[56]

(*Continued*)

Table 11.2 (Continued)

Film composition	Film lattice constant (a_f)	Substrate composition	Substrate lattice constant (a_s)	Lattice mismatch %^{RT} (%^{HT}) T: tensile C: compressive	Comments	Ref.
Li.5Fe2.5O4	8.33	$Mg(In,Ga)_2O_4$	8.381	−0.609 T	Crack due to misfit strain, undercool 5–10 °C; garnet 100 °C, hexaferrite 40 °C;	[32]
$Li_{0.5}Fe_{2.0-x}Al_xO_4$	8.305	$Zn(Ga,Al)_2O_4$	8.310	−0.060 T	$a_f = 8.39$ (1200 °C)	[57]
$Li_{0.5}Fe_{2.0}O_4$	8.330	$Zn(Ga,Al)_2O_4$	8.310	0.241 C	$a_f = 8.39$ (1200 °C)	
$Li_{0.5}Fe_{2.2}Al_{0.03}O_4$	8.305	$Zn(Ga,Al)_2O_4$	8.280	0.302 C	$a_f = 8.39$ (1200 °C)	
$Mn_{0.9}Zn_{0.5}Fe_{1.6}O_4$	8.416	$MnGa_2O_4$	8.440–8.447 w/in error	−0.332 T		[58]
$Mn_{0.3}Zn_{0.3}Fe_{2.4}O_4$	8.450	$MnGa_2O_4$	8.440–8.447 w/in error	0.071 C		
$Mn_xZn_{1-x}Fe_2O_4$	8.4955	Zn_2TiO_4	8.4694	0.308 C	Films show two distinct layers with different lattice parameters	[59]
$MgFe_2O_4$	8.36	MgO	8.424	−0.760 T	($2a_s$)	[33]
$NiFe_2O_4$	8.364	MgO	8.424	−0.712 T	($2a_s$)	
$ZnFe_2O_4$	8.42	MgO	8.424	−0.047 T	($2a_s$)	
			Hexaferrites			
$BaFe_{12}O_{19}$	$a = 5.89$	$CoGa_2O_4$	5.883	0.119 C	Supercooling 9–10 °C; thick 2–18 μm; Permitivity 9.2 tanδ = 0.0023 Congruent melt on substrate	[55]
$BaFe_{12}O_{19}$	$a = b = 5.89$ $c = 23.17$	$KGa_{11}O_{17}$	$a = b = 5.80$ $c = 23.5$	+1.55 C/ −1.40 T	An interdiffusion layer whose thickness depended on the time and temperature of LPE growth. For good-quality layer growth on (MgZr) $SrGa_{12}O_{19}$ substrates, Al was substituted to compensate the size effect of the Bi ion incorporated from the flux.	[55]
$BaFe_{12}O_{19}$	$a = b = 5.89$ $c = 23.17$	$Ba_3(VO_4)_2$	$a = b = 5.78$ $c = 21.34$	+1.90 C/ +8.57 C		
$BaFe_{12}O_{19}$	$a = b = 5.89$ $c = 23.17$	(MgZr) $(SrGa)_{12}O_{19}$	$a = b = 5.82$ $c = 21.34$	+1.20 C/ +8.43 C		
$BaFe_{12}O_{19}$	$a = 5.892$ $c = 23.198$	$Sr(Ga,Mg,Zr)_{12}O_{19}$	$a = 5.823$ $c = 23.073$	+ 1.17 C/ +0.54 C	Supercooling 30–60 °C $BaO–Bi_2O_3–Fe_2O_3$	[20]
$BaFe_{9.3}Al_{2.7}O_{19}$	$a = 5.845$ $c = 22.988$	$Sr(Ga,Mg,Zr)_{12}O_{19}$	$a = 5.823$ $c = 23.073$	+0.38 C/ −0.3 T		
$BaFe_{9.3}Al_{2.7}O_{19}$	$a = 5.801$ $c = 22.884$	$Sr(Ga,Mg,Zr)_{12}O_{19}$	$a = 5.823$ $c = 23.073$	−0.38 T/ −0.82 T		
$BaFe_{12}O_{19}$	23.23 +/−0.03	*m*-plane, $(1\underline{1}00)$ $(10\underline{1}1)$, sapphire (Al_2O_3)	$a = 4.785$ $c = 12.991$		The saturation magnetization $4\pi M_s$ was 4.6 ± 0.1 kG with the uniaxial anisotropy field $H_A = 16.9 \pm 0.5$ kOe	[60]

semiconductor platforms, Chen et al. impressively demonstrated LPE growth on GaN/sapphire [69].

Yoo et al. [20] demonstrated LPE growth of barium hexaferrite and aluminum-substituted barium hexaferrite on nonmagnetic $Sr(Ga,Mg,Zr)_{12}O_{19}$ hexaaferrite substrates using an optimized flux of Bi_2O_3–BaO–B_2O_3. Figure 11.7 reveals optical images of $BaFe_{9.3}Al_{2.7}O_{19}$ at progressively thinner samples and the corresponding microstructure with reduction in grain boundary density, leading to improved properties. In this study, Yoo et al. employed supercoolings of 30 to 60 °C.

Advances that remediate film stress, thus allowing for the growth of high-quality crack-free thick films, included the employment of double-sided growth on substrates that facilitated the cancelling of film stress [65], the use of a high-quality seed crystals deposited by plused laser deposition prior to LPE growth [62, 68, 70], and the use of an MgO buffer layer to limit reactivity and diffusion between the ferrite film and substrate.

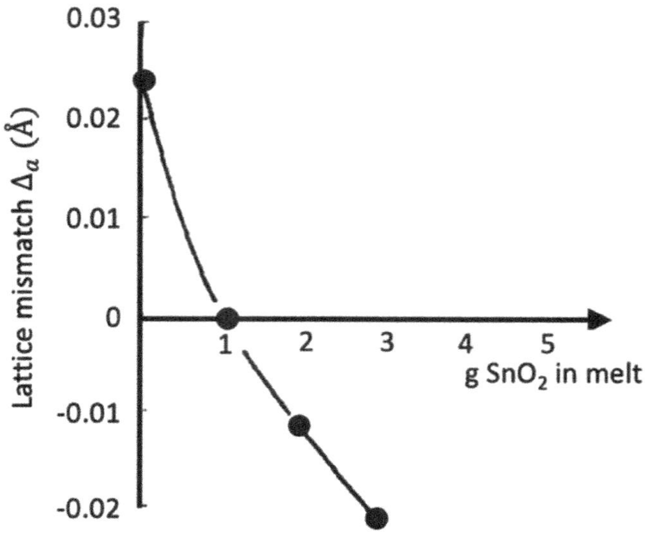

Figure 11.6 Lattice parameter mismatch as a function of SnO_2 concentration in the melt for $MgFe_2O_4$ epilayers on (111) MgO substrates grown at 863± 2°C [33].

Figure 11.8 shows hysteresis loops of BaM films grown on GGG and sapphire substrates by Yoon et al. [62]. A comparison of Figure 11.8a and b reveals anisotropy fields of BaM/GGG of ~16 kOe and BaM/Al_2O_3 of ~25 kOe. The zero-field f_{FMR} is 34 and 52 GHz, respectively. These values are in near exact agreement with calculated values presented in [60] for pure BaM and Al-doped BaM, suggesting that the BaM grown on sapphire experience significant diffusion of Al into the BaM film.

Krasnov et al. demonstrated the growth of 550-μm-thick BaM films on sapphire (Al_2O_3) (0001) substrates without the need for a seed layer with a modified LPE reflow method. The growth rate for the BaM thick films was considerably higher, on the order of 300 μm/h. This method is cost- and time-effective in production. However, the average saturation magnetization value was relatively low ($4\pi M_S$~2050 G) in comparison to other techniques with coercivities of about 100 Oe in the perpendicular plane [71].

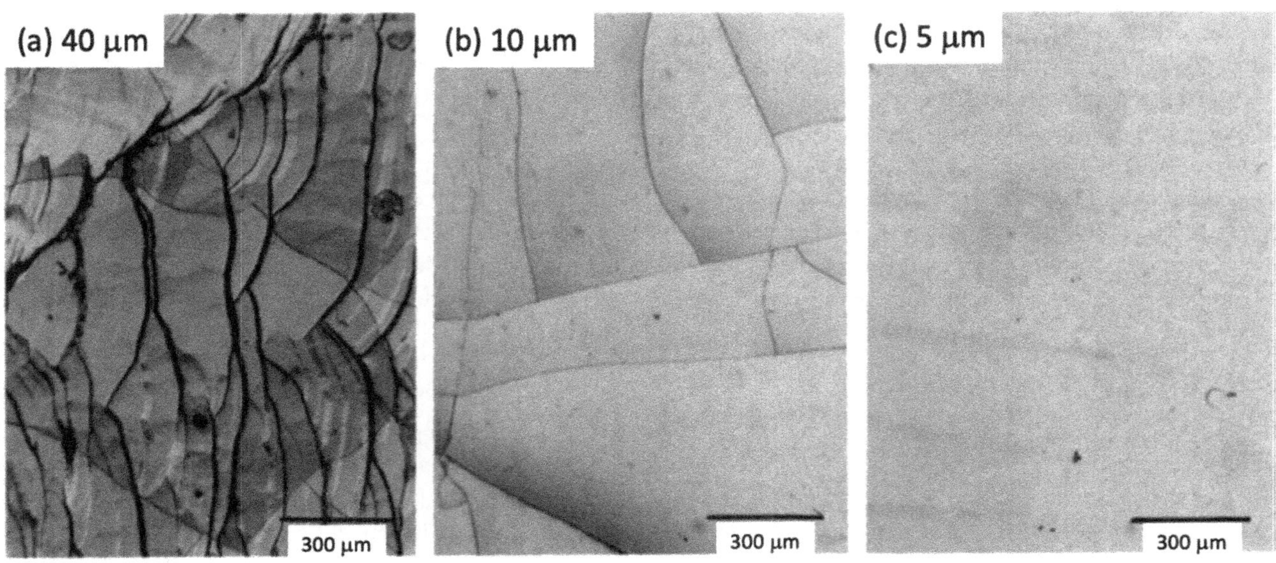

Figure 11.7 Optical microscopes of $BaFe_{9.3}Al_{2.7}O_{19}$ films of different thicknesses: (a) 40 μm, (b) 10 μm, and (c) 5 μm (adapted from [20]).

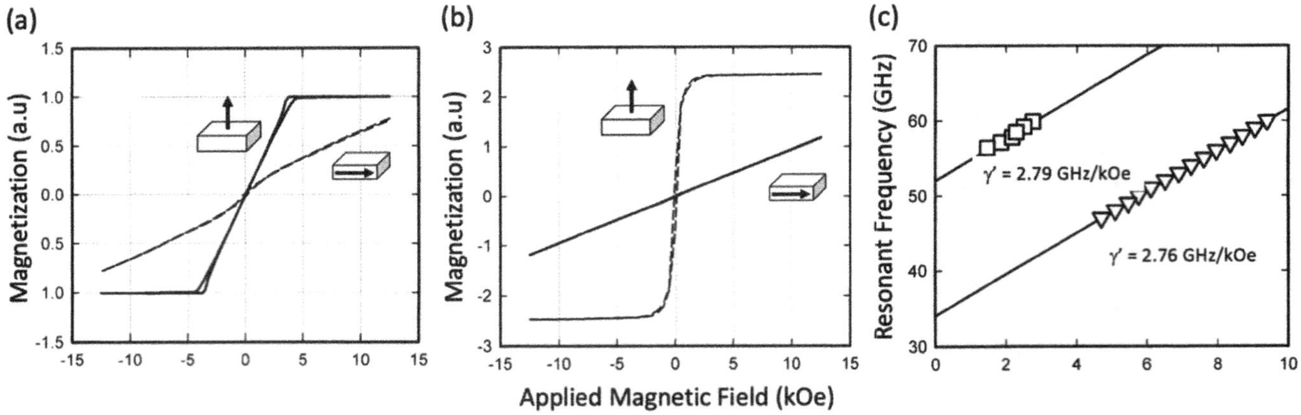

Figure 11.8 Hysteresis loops of (a) 50-μm-thick BaM film on (111) GGG substrates and (b) 80-μm-thick BaM film on (1100) Al_2O_3 substrate. (c) Ferrimagnetic resonance frequency (f_{FMR}) versus resonance magnetic field (square and triangle symbols present those data collected from the 80-μm-thick BaM film grown on (1100) Al_2O_3 and 50-μm-thick BaM film grown on (111) GGG substrate) [60].

Chen et al. [69] demonstrated growth of BaM on GaN/sapphire: A major step in the integration of ferrites with wide bandgap (WBG) heterostructures. Figure 11.9 is a composite figure depicting X-ray pole figures, suggesting a strong degree of epitaxing exists between the BaM film and its underlying substrate in the heterostructure: MgO(100)/GaN(0001)/Al_2O_3(0001). Figure 11.9e–f shows the illustrations of the TEM cross section of this sample with panel (f) showing and expanded view of the interfacial region depicting interfacial diffusion regions marked 2, 3, and 4, corresponding to $MgFe_2O_4$ (2), MgO (3), and MgO/GaN (4). In panel (g) are hysteresis loops corresponding to the out-of-plane and in-plane orientations. The retention of this sample is surprisingly high as the hard axis coercivity is very low, providing potential for self-bias device operation. Alternatively, the hard axis loop shows a "knee" near remanence consistant with a second soft magnetic phase, likely arising from the formation of a spinel phase at interdiffusion region #2. Panel (h) is a plot of the power derivative as a function of applied magnetic field collected at 53 GHz. The linewidth measured as the peak-to-peak power derivative is 86 Oe, which is one of the lowest values reported [69].

11.3.3 Thermodynamics of LPE Growth

11.3.3.1 Transient Cooling of the Melt

By far, the most common method in LPE for inducing crystallization from the melt is the transient cooling of the solution. This is facilitated by advances made in the programmable temperature control of furnaces that can introduce substrates into and out of contact with the melt under precise conditions allowing for near-exact determination of saturation temperatures and supercooling settings.

There are several variants of transient cooling including step cooling, ramp cooling, supersaturation (a combination of step and ramp cooling), transient-mode growth, and two-phase growth. In these methods, the limiting step is often assumed to be diffusion through the melt of the epitaxial layer. Simple growth kinetics, i.e. growth rate or layer thickness as a function of substrate contact time with the melt, are solutions to the one-dimensional diffusion equation. Diffusion coefficients in the liquid phase, including their temperature and composition dependence, are often found as best fits. However, determination of diffusion coefficients may become complicated by interface kinetics [17].

11.3.3.2 Ultrafast Cooling

In transient cooling methods, supercoolings are in the range of 1–10 °C and cooling rates of 0.1–1 °C/min are typical. Much higher supercoolings (several hundred degrees) and cooling rates (10^2–10^3 °C/s) are possible [17, 18, 20, 72].

The introduction of modern furnaces having independently heated zones (on motorized rails) allow for the rapid adjustment of large temperature gradients, cooling rates, and more exact substrate–melt contact times.

Effective supercooling at the melt–substrate interface can be inferred by careful experimental measurement of meltback. These techniques generate uncharacteristically high supersaturation that prove particulary useful in ferrite systems.

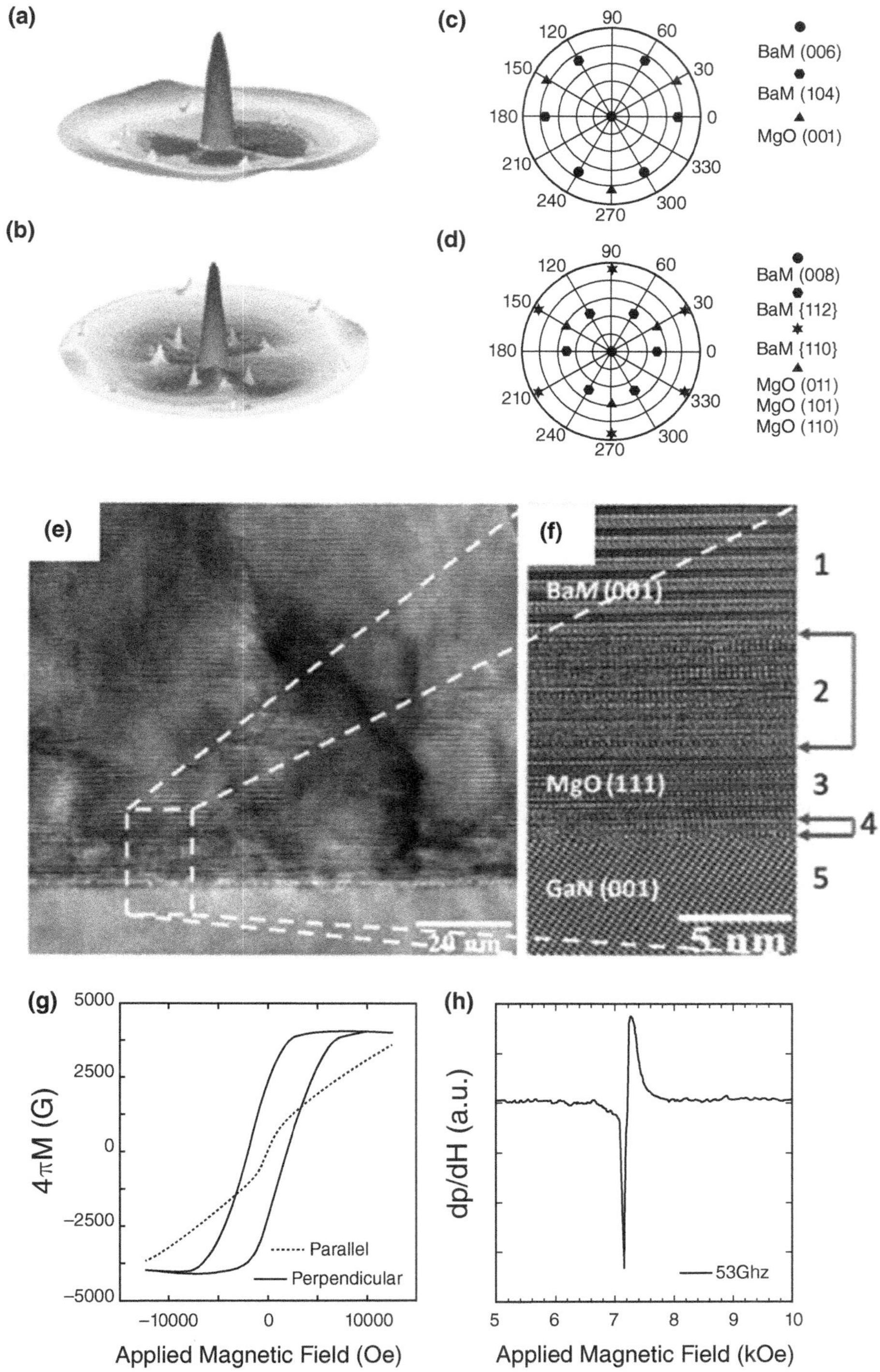

Figure 11.9 Pole figures obtained for (a) (006) and (b) (008) reflections and corresponding two-dimensional projections (c) and (d), with 2θ values fixed at 23.00° and 30.30°, respectively. The single dominant peak in (a) corresponds to $\phi = \xi = 0°$ for the BaM (006) reflection. The weaker peaks in (a) and (c) exhibiting sixfold symmetry correspond to BaM (104) reflections. The minor peaks (solid triangle) exhibiting threefold symmetry in (c) and (d) derive from MgO (100) reflections illustrating the epitaxial nature of the BaM films grown on $MgO/GaN/Al_2O_3$. Transmission electron microscopy images of the BaM film grown on $MgO/GaN/Al_2O_3$ in the cross section. The image (f) is an expanded view of the section denoted in (e). The regions denoted by 2, 3, and 4 signify intermixing. Hysteresis loops (g) were obtained with a maximum applied field of $\pm$13 000 Oe aligned perpendicular solid and parallel dashed to the film plane. (h) Power derivative as a function of applied magnetic field (@ 53 GHz). The linewidth measured as the peak-to-peak power derivative is 86 Oe.

11.3.3.3 Interfacial Energy LPE

The driving force behind crystal growth from an undersaturated liquid derives from the minimization of interfacial energy between melt and substrate. When the substrate–melt interface energy exceeds the heteroepitaxial layer–melt interface energy, a lower state of free energy allows for wetting of the substrate surface. By superheating the melt, the driving force for epitaxy can be further minimized to produce atomically sharp interfaces. As a result, heteroepitaxial layers can be deposited from melts under superheated conditions from 10 to 100 °C [17, 20].

11.4 LPE Growth of Garnets

11.4.1 Structure, Chemistry, and Magnetism of Garnets

The phase diagram for YIG was provided first by van Hook [74] and is adapted here as Figure 11.10 in both binary (a) and ternary (b) forms, respectively.

In Figure 11.10, it is seen that YIG forms in a narrow phase field at ~1555 °C. The structure of garnets is cubic, belonging to the space group *Ia*3*d*, as was first proposed by Menzer [75]. The measured lattice constant for YIG is 12.376+/−0.005 [76]. Bertaut and Forrat in 1956 [77] reported on the magnetic properties of $Y_3Fe_5O_{12}$, YIG.

YIG has a high Verdet constant, which provides a large magneto-optical response, high *Q* factor at microwave frequencies, low absorption in the infrared (up to 600 nm), and small FMR linewidth. As a result, YIG finds use in RF, microwave, optical, and magneto-optical applications, as microwave filters, solid-state lasers, Faraday rotators, and in nonlinear optical applications, among others.

Figure 11.11 is a representation of the YIG unit cell [78]. In the YIG structure, Fe cations occupy two octahedral 16[a] and three tetrahedral 24(d) sites. The yttrium ions are too large to occupy either tetrahedral or octahedral sites and too small to substitute for oxygen anions. They occupy dodecahedron 24{c} sites having eightfold oxygen coordination. There are eight formula units per unit cell. The magnetism of the two Fe ions on the octahedral sites oppose the three on the Fe on tetrahedral sites: The net result is a magnetization, $4\pi M_s$, of ~1740 G (see Figure 11.12) [79].

Magnetism in garnets, and other ferrite structures, arise from superexchange interactions [80–82]. In the ground state, superexchange is a negative exchange interaction that results in the antialignment of cation spins (i.e. ferrimagnetic). This, together with the dilution introduced by the oxygen lattice, is the principal reason that ferrite magnetization is significantly reduced in comparison to the magnetization of *3d* metallic alloys, e.g. Fe, Co, Ni, etc., which are governed by direct exchange

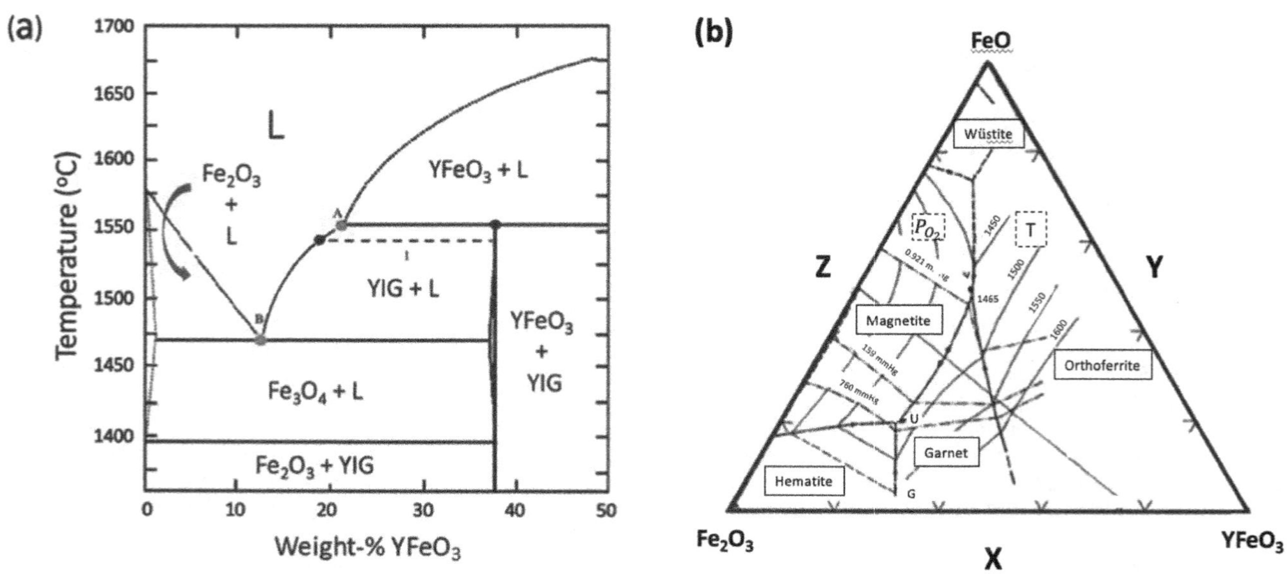

Figure 11.10 (a) Binary Fe_2O_3–$YFeO_3$ phase diagram adapted from [39]. (b) Ternary Fe_2O_3–$YFeO_3$–FeO phase diagram reproduced from [73].

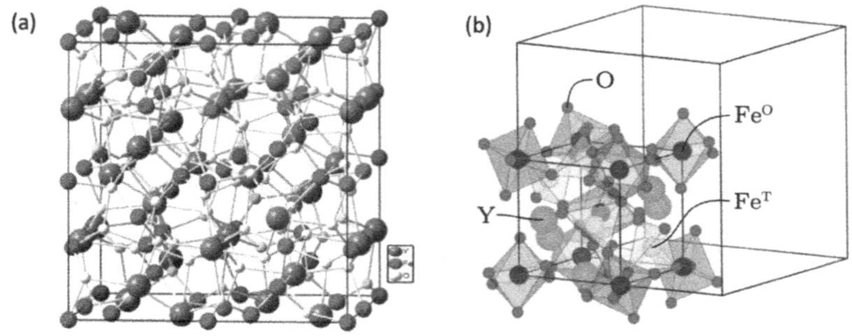

Figure 11.11 (a) A representation of the YIG unit cell [78]. (b) Fraction of the YIG unit cell depicting tetrahedrally coordinated iron (Fe^T) and octahedrally coordinated iron (Fe^O).

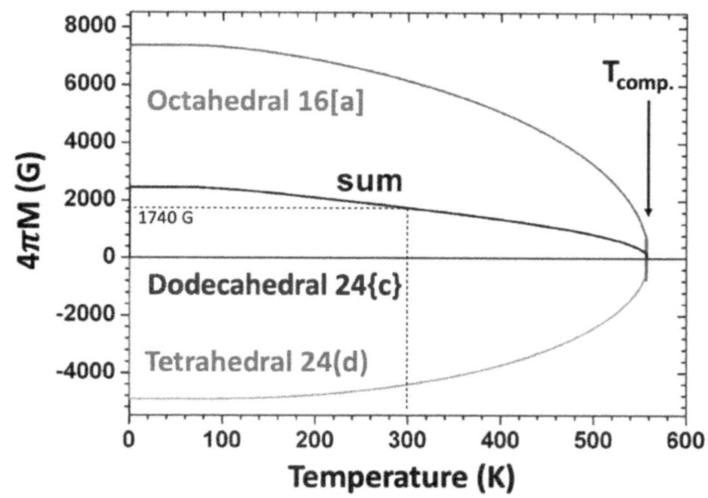

Figure 11.12 Contribution of different lattice magnetization to the net magnetization of YIG (adapted from [79]).

where spins align parallel (i.e. ferromagnetic) to one another. It is also the reason that it is challenging to increase the magnetization of ferrite materials being that there is little flexibility in creating long-range networks of collinear spins of high moment.

The Curie temperature of YIG is ~560 K and varies little among other magnetic garnets.

The saturation magnetization and the magnetic anisotropy field are rather low at $4\pi M_s$ ~ 1740 G and H_A ~ 10–100 Oe, respectively, resulting in a low FMR frequency. For these reasons, YIG is usually biased by a DC magnet field provided by a permanent magnet to shift the FMR frequency and increase operational frequencies from ultrahigh frequency (UHF) to X band. At frequencies beyond X band, the size and weight introduced to devices by such bias magnets often become prohibitively cumbersome.

The properties of YIG can be modified by cation substitution (see Tables 11.3 and 11.4). For example, lanthanum and gallium ions are common dopants in YIG films for microwave applications. Lanthanum substitution for yttrium expands the lattice [87, 88] and gallium substitution [84] for iron on the tetrahedral sites lowers the magnetization to values between 100 and 1740 G.

Table 11.4 lists several garnet compositions with their magnetic and rf properties for single-crystal and polycrystal samples.

In pure YIG, magnetic anisotropy fields range from 10 to 100 Oe, depending upon processing history. Some fraction of this anisotropy field is attributable to secondary crystalline effects that influence the MSW propagation. For co-doped Ga-La YIG, $4\pi M_s$ values less than 400 Oe and anisotropy fields as high as 700 Oe have been measured [88].

In LPE films, the anisotropy field can be modified by flux composition, growth temperature, and growth rate as well as postdeposition heat treatments.

The stress between film and substrate becomes more and more important with increasing film thickness. For this reason, lanthanum ions were introduced on dodecahedron sites to expand the lattice constant due to its larger ionic radius relative to the yttrium ion [87].

Table 11.3 Cation occupation on garnet lattice sites (Wycoff notation) with typical valence.

24{c}	16[a]	24(d)
Y^{3+} Lu^{3+} Tm^{3+} Bi^{3+}	Sc^{3+} In^{3+} Cr^{3+} Fe^{2+}	Ga^{3+} Al^{3+} Si^{4+} V^{5+}
La^{3+} Ce^{3+} Er^{3+} Ca^{2+}	Ni^{2+} Mg^{2+} Ti^{4+} Sn^{4+}	Ge^{4+} Co^{2+} Co^{3+} Fe^{4+}
Ho^{3+} Yb^{3+} Dy^{3+} Tb^{3+}	Zr^{4+} Fe^{4+} Pt^{4+} Rh^{4+}	
Gd^{3+} Eu^{3+} Sm^{3+} Nd^{3+}	Ru^{4+} V^{5+}	
Pr^{3+} Pr^{4+} Pb^{2+} Pb^{4+}		

Table 11.4 Properties of LPE garnets (microwave properties were measured at X band).

Composition	$4\pi M_s$ (G)	T_C (K)	ΔH (Oe)	g (X band)	ρ (Ω·cm)	Ref.
$Y_3Fe_5O_{12}$	1740	560	15 ~0.1 (SX)	2.00	1×10^{10}	[83]
$Y_3Fe_{4.67}Al_{0.33}O_{12}$	1200	500	44	2.02		[84]
$Y_3Fe_{4.5}Ga_{0.5}O_{12}$	1000	490	50	2.01	1×10^{8}	[84]
$Y_3Fe_{4.5}Cr_{0.5}O_{12}$	1800	505	140	2.15		[84]
$Y_2GdFe_5O_{12}$	1200	560	110	2.00		[84, 85]
$Y_2NdFe_5O_{12}$	1930	551	2 (SX)		5×10^{7}	[86]
$Y_2SmFe_5O_{12}$	1710	550–555	100 (SX)	2.00		[86]

Unless denoted, samples are polycrystalline compacts. SX: single crystal

11.4.2 Historical Development of LPE Garnets

11.4.2.1 LPE Garnet Films for Bubble Memory

Table 11.5 presents the history of key developments in the evolution of LPE garnets. Early work by Linares was motivated to develop two-dimensional S-band filters to challenge the market position held by conventional yttrium iron garnet (YIG, $Y_3Fe_5O_{12}$) devices that used single-crystal spheres. Spheres proved difficult to manufacture in large numbers. Linares demonstrated growth rates of ~0.5 μm/min yielding films of 0.5 to 100 μm having ferrimagnetic resonance (FMR) linewidths of less than 1 Oe [15].

In addition to filters, there are other devices that can be realized using LPE YIG films. Noteworthy applications include tunable oscillators, linearly dispersive, and nondispersive delay lines, the latter enabling the development of phased array antennas.

Rapid development of high-quality LPE-grown garnet films continued throughout the 1970s motivated in large part by the emerging bubble memory technology invented at the Bell Labs circa 1970 [99].

Bobeck et al. discovered the formation and precise movement of circular magnetic domains called *bubbles* that appear in YIG films upon magnetization and demagnetization. This discovery led to the establishment of the magneto-optic "bubble" data storage industry. The storage efficacy of the medium depended on the reproducible growth of largely defect-free garnet films: LPE was shown effective for this task.

11.4.2.2 Bubble Memory – Basic Principles

Figure 11.13 provides a schematice illustration that defines the energetic conditions of stable magnetic domains or "bubbles." Specifically, the total energy is defined in terms of contributions from magnetostatic energy, domain wall energy, and applied magnetic field strength. The minimum of this total energy defines a stable bubble radius. As the applied magnetic field increases above this critical value, the bubble domain collapses. This is shown schematically in Figure 11.14a–c and by experiment in (d–f). In panels (d–f), magnetic domains in a YIG sample under polarized light microscopy are shown in a demagnetized state where the shaded regions corresponding to north- or south-oriented spins represent 50% of the same volumes so they sum to a net zero magnetization. (e) When an external magnetic field is applied perpendicular to the specimen, domains magnetized in the opposite direction narrow with a few contracting into *bubbles*. (f) A further increase in

Table 11.5 Historical development of liquid-phase epitaxy deposition conditions and key results of garnet films.

Year	Film composition	Flux	Rate and thickness	Substrates	Key results	Ref.
			LPE Garnet Films			
1968	YIG	BPO_4–BaB_2O_4–WO_3	25–75 /s; $t = 0.5$–100 μm	GGG	Single-crystal, crack-free films, 0.5 to 100 μm thick are grown on substrates of gadolinium gallium garnet (GGG) from molten salt solutions of lead borate, barium borate, and sodium tungstate. Growth rates of 25 to 75 Å/s yielded films with ferrimagnetic resonance linewidths less than 1 Oe.	[15]
1971	$Eu_2Er_1Ga_{0.7}Fe_{4.3}O_{12}$ and $Er_2Er_1Ga_{0.7}Fe_{4.3}O_{12}$	Eu_2O_3, Er_2O_3, Ga_2O_3, Fe_2O_3, PbO, B_2O_3	5–20 μm	(110) and (111) GGG	Magnetically uniaxial epitaxial films of $Eu_2Er_1Ga_{0.7}Fe_{4.3}O_{12}$ and $Er_2Er_1Ga_{0.7}Fe_{4.3}O_{12}$ 5–20 μm thick are grown free of cracks from PbO–B_2O_3 solutions on (110) and (111) GGG substrates, respectively. Bubble propagating T-bar and Y-bar circuits demonstrated.	[89]
1971	$EuEr_2Fe_{4.3}Ga_{0.7}O_{12}$	50/1 weight ratio of PbO to B_2O_3 Bi_2O_3 w/ 10 wt% V_2O_5	$0.6<t<4.85$ μm at rates of −0.485 μm/min	GGG	A new dipping technique is described for producing uniform-thickness low-defect-density magnetic garnet films suitable for bubble-domain devices. The technique is based on the observation that melts having the appropriate flux and garnet composition can be supercooled over a large temperature range and remain supercooled, permitting the growth at constant temperature of epitaxial magnetic garnet films.	[18]
1971	$Eu_2Er_1Ga_{0.7}Fe_{4.3}O_{12}$ and $Er_2Er_1Ga_{0.7}Fe_{4.3}O_{12}$	Eu_2O_3, Er_2O_3, Ga_2O_3, Fe_2O_3, PbO, B_2O_3	5–20 μm	(110) and (111) GGG	Anisotropy magnitudes and orientations are studied under annealing behaviors and effects of strain relief. Observations indicate that noncubic anisotropies in bulk materials and LPE films are similar, and are consistent with a pair-ordered anisotropy (POA) model. Annealing reduces the POA in LPE films.	[90]
1971	$Eu_3Fe_{5-x}Ga_xO_{12}$; $Y_{3-2x}Ca_{2x}V_xFe_{5-x}O_{12}$; $Bi_{0.9}Ca_{2.1}V_{1.05}Fe_{3.95}O_{0.2}$	Unknown	23–25 μm	Unknown	Noncubic, growth-induced anisotropy was measured in $Eu_3Fe_{5-x}Ga_xO_{12}$ films with a single rare earth ion or in $Y_{3-2x}Ca_{2x}V_xFe_{5-x}O_{12}$ films with only diamagnetic ions. The noncubic anisotropy was explained by distortion of the local symmetry around tetrahedral ions. A similar distortion of tetrahedral symmetry by ordering of dodecahedral ions contributes in mixed rare earth garnets.	[91]

(Continued)

Table 11.5 (Continued)

Year	Film composition	Flux	Rate and thickness	Substrates	Key results	Ref.
1972	$Y_3Fe_5O_{12}$ $Y_3Ga_{1-x}Fe_{5-x}O_{12}$ $Y_3Al_{1-x}Fe_{5-x}O_{12}$ $Y_3Gd_{1.6}Al_{0.86}Fe_{4.14}O_{12}$ $EuEr_2Fe_{4.3}Ga_{0.7}O_{12}$	Eu_2O_3, Er_2O_3, Fe_2O_3, Ga_2O_3, PbO, B_2O_3	t<0.1–0.5 μm/min	GGG	Phase relationships in the PbO–B_2O_3–garnet system and its relationship to high quality – low-defect epitaxial films are presented. The effects of growth temperature, lattice mismatch, and growth rate on the magnetic properties of rare earth-doped iron garnets explored.	[40]
1972	Ga:Y(Gd,Eu)IG $Eu_1Er_2Ga_{0.7}Fe_{4.3}O_{12}$	Unknown	Unknown	(110) and (111) GGG	Ga:YIG films with partial replacements of Y by Gd or Eu were grown by LPE on (111) and (100) GGG show annealing effects in air and oxygen between 800 and 1200 °C. The uniaxial anisotropy has a nearly linear variation with film–substrate lattice mismatch, indicating a stress-induced anisotropy.	[92]
1974	$Y_{2.5}Sm_{0.5}Ga1.2Fe_{3.8}O_{12}$	PbO–B_2O_3	10 μm at t = 0.05–0.2 μm/min	GGG	The solute diffusion coefficient in a PbO–B_2O_3 flux system used for the growth of garnet films has been measured and found to be $(5 \pm 1.5) \times 10^{-7}$ cm^2/s in the growth range of 840–887 °C. The growth rate is constant after an initial transient, and is proportional to the square root of the rotation rate.	[35]
1974	$Y_{1.93}Sm_{0.1}Ca_{0.96}Ge_{0.96}Fe_{4.04}O_{12}$ $Y_{2.5}Sm_{1.5}Ga_{1.3}Fe_{3.7}O_{12}$	Unknown	Unknown	GGG	A new method for the suppression of hard bubbles has been developed. The method involves annealing films in an inert atmosphere at temperatures near 1100 °C. The proposed mechanism involves enhanced diffusion of Ga from the substrate into the film due to oxygen vacancies produced by annealing in the inert atmosphere. The diffused Ga replaces Fe on tetrahedral sites.	[93]
1974	$EuYb_2Fe_5O_{12}$	PbO B_2O_3 Fe_2O_3	Unknown	Unknown	$EuYb_2Fe_5O_{12}$ magnetic garnet films are grown by LPE under undercooling conditions. With increasing undercooling the growth rate first rises linearly and then reaches a temperature insensitive plateau. The data are analyzed according to a simple diffusion–reaction model of the deposition process. Data-fitting yields values 1.2×10^{-5} cm^2/s, 74.5 μm, 1.085 eV, and 1.61 eV for the diffusion coefficient, solute boundary layer thickness, heat of solution, and activation energy of the reaction, respectively.	[41]

Table 11.5 (Continued)

Year	Film composition	Flux	Rate and thickness	Substrates	Key results	Ref.
1974	$Y_{3-x}(M1)Fe_{5-y}(M2)_yO_{12}$, where M1 = Gd,Yb, Tm, E, Eu; M2 = Al,Ga	varies	Unknown	Unknown	Compositions of $Y_{0.9}Gd_{1.5}Yb_{0.6}Fe_{4.5}Al_{0.5}O_{12}$, $Y_{1.2}Gd_{0.6}Tm_{1.2}Fe_{4.4}Ga_{0.6}O_{12}$, $Y_{1.1}Gd_{0.8}Tm_{1.1}Fe_{4.4}Ga_{0.6}O_{12}$, $Y_{1.0}Gd_{1.0}Tm_{1.0}Fe_{4.4}Ga_{0.6}O_{12}$, $Y_{1.0}Gd_{1.0}Er_{1.0}Fe_{4.5}Al_{0.3}Ga_{0.2}O_{12}$, $Y_{1.5}Gd_{0.8}Eu_{0.7}Fe_{4.3}Al_{0.6}Ga_{0.1}O_{12}$ were prepared by standard LPE techniques. Over the postdeposition temperature range from 0 to 100 °C, K_u decreases approximately exponentially with temperature. The dynamic behavior of hard bubbles was found to differ among the material compositions.	[34]
1974	YIG	Unknown	1–16 μm	GGG	Optical absorption, derivative absorption, reflectivity, and derivative reflectivity are presented for YIG films. The real and imaginary parts of the dielectric response functions are compared with previously published results for bulk crystals and shown to be substantially larger.	[94]
1976	YIG	$PbO–B_2O_3$	100 μm	GGG	Single-crystal films of YIG up to about 100 μm thick have been grown on (111) GGG by LPE using a $PbO–B_2O_3$-fluxed melt. Although interface instability arises from supersaturation during film growth the film surfaces were found to be smooth.	[95]
1976	Ho-IG	$PbO–B_2O_3–Fe_2O_3$	2 μm	GGG	The incorporation of Pb, from the flux, into $Ho_3Fe_5O_{12}$ garnet films grown by LPE onto GGG substrates (111) oriented support earlier findings that Pb increases significantly with decreasing growth temperature and that Pb increases, to a lesser degree, with increasing growth rate, induced by axial substrate rotation, at a fixed temperature. Also Pt, from the Pt crucibles, increases with Pb but not in sufficient amounts to completely charge compensate Pb.	[53]
1977	$(Y,Sm,Lu,Ca)_3(Fe)_5O_{12}$	PbO/B_2O_3	1.67 +/− 0.46 μm/min.	GGG	The effect of the melt PbO/B_2O_3 ratio on the Curie temperature of LPE films containing lutetium is described along with data showing the flux effect on the bubble collapse field temperature.	[96]

(Continued)

Table 11.5 (Continued)

Year	Film composition	Flux	Rate and thickness	Substrates	Key results	Ref.
1977	CaGe-Substituted $EuTm_2Fe_5O_{12}$	$PbO–B_2O_3–Fe_2O_3$	0.5–3 μm/min	GGG	Measured growth rates for LPE films, produced from melts with four different concentrations of CaGe-substituted $EuTm_2Fe_5O_2$ garnet fit a quantitative model having adjustable variables: diffusivity, D; reactivity, K; viscosity, ν; and concentration, c_e for rotation, ω, over a defined growth temperature range. It is shown that the garnet melt concentration terms are the most critical ones in determining the growth rate, and hence, the activation energy (heat of solution) for the equilibruim concentration.	[97]
1978	SmCaGe- garnet	PbO/B_2O_3	14–17 μm	GGG	Authors have investigated thickness dependence of four film compositions in the SmCaGe–garnet system. In thin films, coercivity is dominated by a high coercivity region near the substrate. As the film thickness increases, the coercivity decreases approaching a "bulk" value. The two regions of the film, with their different properties, produce a linear relationship of coercivity versus reciprocal thickness.	[98]

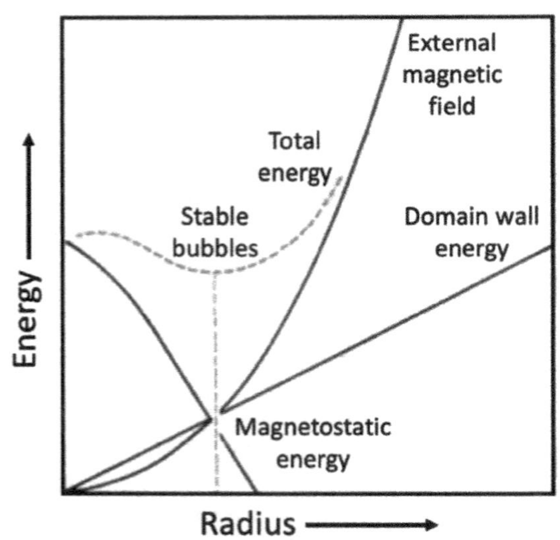

Figure 11.13 The minimum of a total energy curve composed of magnetostatic energy, domain wall energy, and applied magnetic field strength components. The minimum of the total energy curve defines the stable bubble radius. Above this critical magnetic field value, the bubble domain collapses [90].

the external bias field converts the remaining domains to bubbles. A further increase above the critical field would lead to the destabilization of the bubbles and their eventual collapse.

In 1971, seminal research ensued in the development of bubble memory largely spurred by the production of high-quality LPE garnet films.

Research groups of Kurtzig and Hagedorn [90] and Akselrad et al. [91] studied LPE-grown films of bubble-memory compositions (i.e. Er-, Eu-, Ga-co-doped garnets) that elucidated the origins of noncubic magnetic anisotropy. Specifically, results suggested the anisotropy fields could be described using a POA model where chemical ordering of cations on tetraheral sites was responsible for the magnetic anisotropy energy. Post-annealing and quenching are shown to reduce growth-induced anisotropy, presumable by randomizing this chemical ordering, resulting in the enhancement of domain wall mobility and stabilization of bubble circuits. LeCraw et al. [93] demonstrated that brief anneals in an inert gas environment led to diffusion of Ga ions from the GGG substrate into the bubble medium (i.e. garnet film). Ga ions preferentially displace Fe ions from the tetrahedral sublattice, thus suppressing the POA and lessoning the impact of magnetically "hard" bubbles.

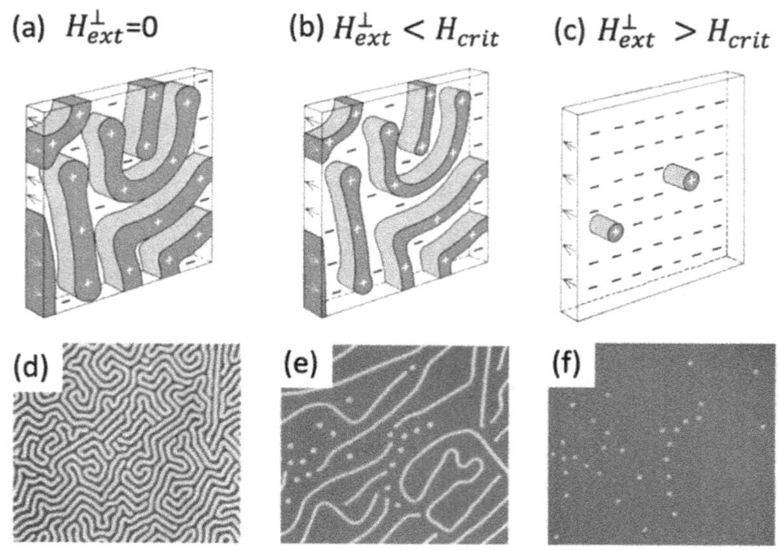

Figure 11.14 (a) An imagined slice of a YIG sample with no external magnetic field applied the domain configuration will reach a minimum energy state where serpentine domains of equal volume nullify magnetization. (b) Upon application of a magnetic field aligned perpendicular to the sample surface, the domains magnetized opposite the applied field direction narrow. (c) As this magnetic field increases to a critical stabilizing value, all the remaining domains convert to bubbles. (d) Magnetic domains in a YIG sample under polarized light microscopy. By adjusting polarizing filters, domains with the same orientation can be made to look either bright or dark. (e) When an external magnetic field is applied perpendicular to the specimen, domains magnetized in the opposite direction narrow with a few contracting into bubbles. (f) A further increase in the external bias field converts the remaining domains to bubbles [100].

Once the processing of stable bubbles was achieved, approriate circuitry was required to realize functional memory devices.

Shick et al. reported on properties of LPE growth of 5- to 20-μm-thick epitaxial $Eu_2Er_1Ga_{0.7}Fe_{4.3}O_{12}$ and $Er_2Eu_1Ga_{0.1}Fe_{4.3}O_{12}$ films on (110) and (111) GGG substrates, respectively, and demonstrated bubble propagating T-bar and Y-bar circuits deposited on these films demonstrated 100-step domain shift registers [89, 101].

Figure 11.15a–c presents a series of photographs illustrating bubble generators, movers and manipulators. In Figure 11.15a, a T-bar bubble mover is demonstrated as an external magnetic field rotates clockwise in the plane of the garnet wafer (see black arrows). As the field rotates, different parts of the Ts and bars, made of Permalloy (a soft magnetic iron nickel alloy), are sequentially polarized + to –, attracting the surface poles of the bubbles as they progress from left to right through tee and bar patterns [100].

Figure 11.15b contains a series of photographs demonstrating a bubble generator. This device creates a new bubble every time an in-plane field makes a rotation. The domain appears as a white crescent projecting from the black square becoming circular after it splits from the generator. Thereafter, it is carried to the right by a modified Y-bar element [100].

Figure 11.15c demonstrates a divide-by-two circuit. This circuit employs an idler that traps every other bubble. In these series of photographs, two bubbles are seen to approach the idler from the left. The first bubble is diverted into the idler (located between the second and the third T-bar from the bottom). Its presence in the idler forces the second bubble to stay in the horizontal track. The passage of the second bubble, in turn, forces the first bubble out of the idler [100].

A schematic diagram of a basic bubble-memory package containing a bubble-memory chip, magnetic field coils, and permanent magnets with control circuitry is presented in Figure 11.15d. A rotating magnetic field created by two mutually perpendicular coils manipulate the bubbles as described in (a–c). Two permanent magnets provide nonvolatility [102].

During the 1970s, researchers also explored the roles of other rare earth ions [28–30, 34], flux compositions [22, 25, 28, 35–37, 51], process modification impact on film quality and defect mitigation [25, 52, 90], and studies to clarify the thermodynamic underpinnings of the LPE film growth [29, 40], all with the goal of enhancing bubble memory technologies.

11.4.2.3 LPE Garnet Films for Magnetostatic Wave Devices

Epitaxially grown YIG films up to 100 μm in thickness is known to have ultralow damping coefficients valuable for MSW devices. Such samples are difficult to obtain reproducibly from bulk crystals where they are subject to fracture upon cutting and polishing.

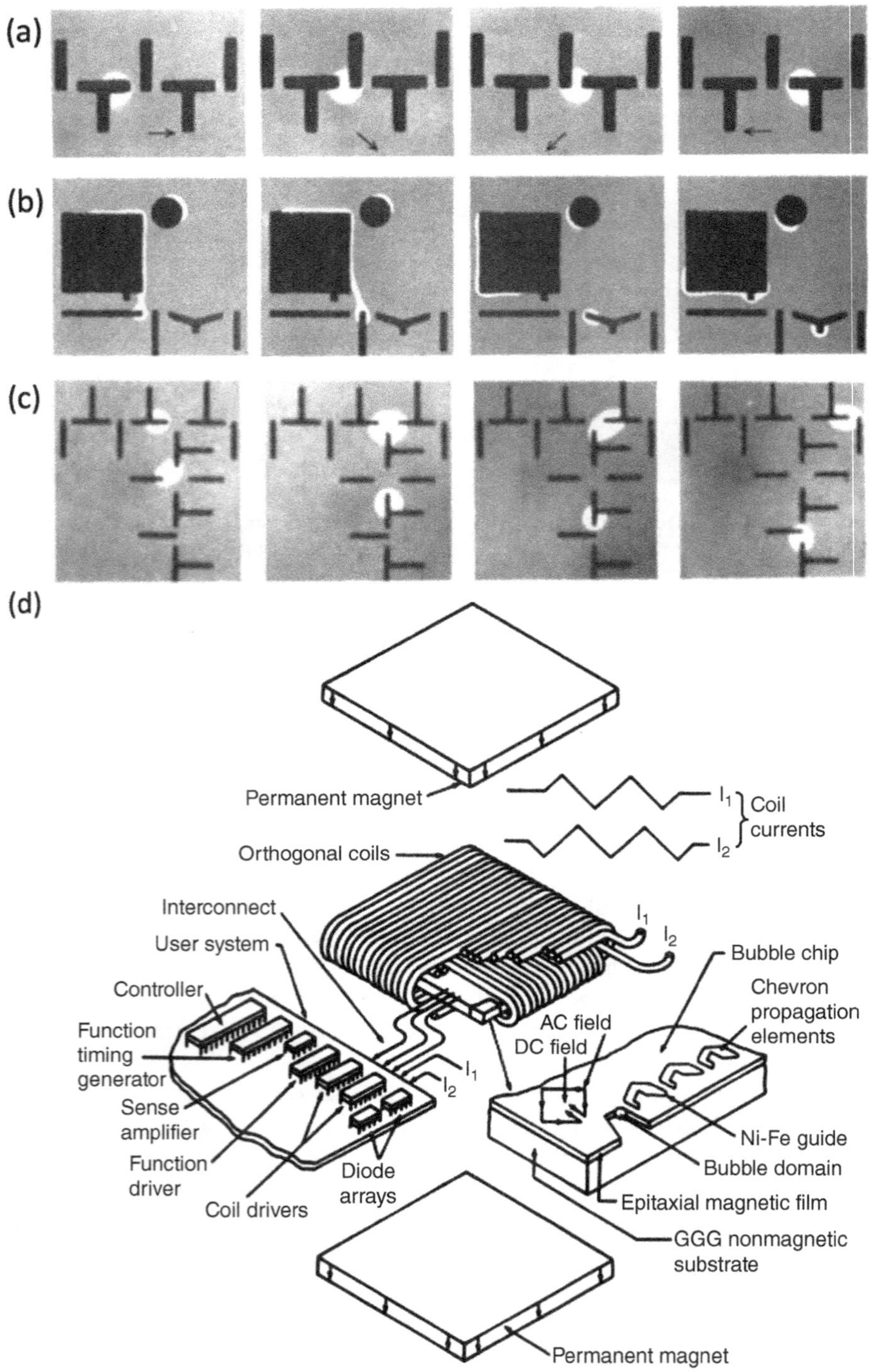

Figure 11.15 (a) Demonstration of a T-bar bubble mover, (b) a bubble generator, and (c) a divide-by-two circuit. (d) Schematic diagram of a basic bubble memory package containing a bubble memory chip, magnetic field coils, and permanent magnets with control circuitry. Two permanent magnets provide nonvolatility [100, 102].

In the LPE growth of YIG films, the ideal substrate has been shown to be GGG. This substrate can be produced largely defect free by the Czochralski crystal growth technique with orientations within 0.1 arcseconds of desired orientations. Additionally, GGG has a dielectric constant of ~10 and can serve to isolate YIG layers if multiple-layer device topologies are desired.

MSW devices have the potential to perform at RF frequencies many of the signal processing functions performed at UHF by more sophisticated and costly SAW techniques.

Circa 1964, Olsson and Yaeger demonstrated the propagation, dispersion, and attenuation of backward-travelling magnetostatic waves (MSWs) in YIG crystals [103]. Damon and van der Vaart followed with demonstrations of spin wave propagation at microwave frequencies in transverse and axially magnetized rods and discs [104].

Baek et al. [105] performed an in-depth study of MSW modes of an LPE-grown YIG film. Those authors found an applied DC magnetic field applied parallel to the film plane; magnetostatic surface wave (MSSW) and magnetostatic backward volume wave (MSBVW) modes were observed in FMR spectra. Figure 11.16a illustrates the derivative power absorption spectra of a YIG film when a DC magnetic field was applied in the direction parallel to the sample surface and the microwave frequency was fixed at X band (i.e. 9.433 GHz). The signals that appear at a field less than 2615 Oe are MSSW modes, while those that appear at higher fields are MSBVW modes. The MSSW modes propagate along the direction perpendicular to the magnetic field, while MSBVW modes propagate along the direction of the magnetic field.

Figure 11.16b and c shows the derivative power absorption of a YIG thin film of 1.65 mm × 1.20 mm, collected with the DC magnetic field parallel to the sample surface as a function of temperature. The intensities of the MSSW modes and the MSBVW modes increase slightly with decreasing temperature to 333 K and then decrease gradually below 333 K. The field widths (intervals between the modes) increase slightly with decreasing temperature (see Figure 11.16d) [105].

From these experiments, it became evident that MSW devices promised low-loss dispersive delay lines that find utility in sensors, oscillators, pulse compressors, and a variety of filter designs operating over broad regions of the RF frequency spectrum.

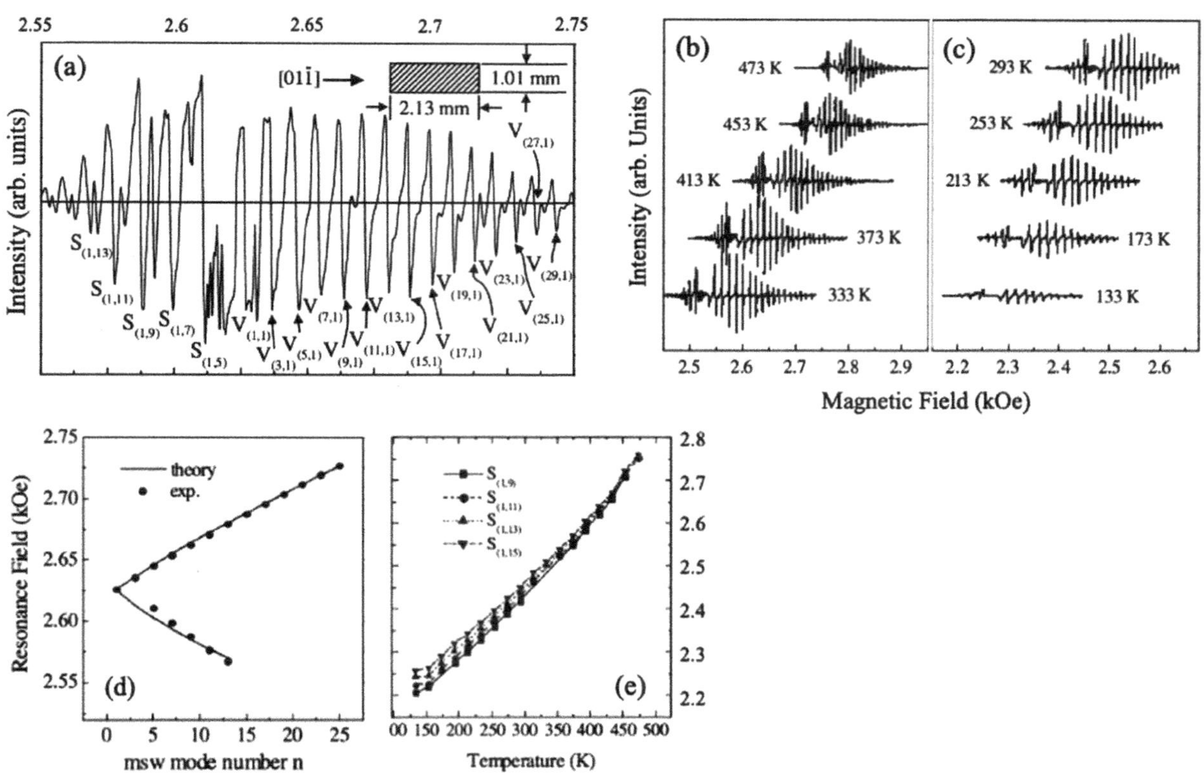

Figure 11.16 (a) The derivative of power absorption of a YIG film when the magnetic field (*H*) is applied parallel to the sample's surface at 9.433 GHz (room temperature). (b and c) Resonance fields of MSW modes as a function of temperature for a YIG film with *H* in the parallel configuration. (d) Measured resonance fields of MSW modes as a function of mode number (*n*) (symbols) and calculated (curve). (e) *E* – the field width of MSW modes as a function of temperature [105].

Due to the progress made in the growth of high quality ($1<t<100$ μm) epitaxial YIG films mostly grown on GGG, together with approximately uniform internal dc magnetic fields (i.e. magnetic anisotropy fields), MSW propagation in these films have led to efficient transduction of EM waves, spatial tapping, and modification of propagation characteristics through the use of layered topologies.

Table 11.6 presents a list of MSW devices, functions, and conditions of their operation provided by Adam et al. [106] and modified here. These with those listed in Table 11.7 compare and contrast the SAW devices of the day with MSW devices enabled by LPE YIG films. MSW devices offer advanced functionality in tunability at gigahertz frequencies with low insertion loss not readily available in SAW devices.

With the advent of thick LPE YIG films, Sykes et al. [108] demonstrated the ability to etch channels into YIG to provide tailored resonances. Figure 11.17a shows a schematic of the cross section of a periodic channeled structure prepared by selective etching of an epitaxial LPE-grown YIG film [108]. Figure 11.17b shows the predicted (top) and measured (bottom)

Table 11.6 Applications of MSW devices based on LPE YIG (adapted in part from [106]).

Magnetic bias	Device	Microwave function
Fixed bias field, H_0	• Nondispersive delay lines • Tapped delay lines • Dispersive delay lines	Phase locking of pulsed oscillators, PSK-matched filter, signal transformation, group delay equalizer, pulse compression, compressive receiver, rate sensors, frequency synthesis switches, sprectral analysis switch, spectral analysis signal correlation, and Fourier transformation
Variable bias field, H_0	• Acoustic delay lines • Variable delays • Meander line couplers • Oscillators	Narrow band tunable bandpass filter, pulse gating target simulation, electronic timing tunable bandpass filter, and MSW multipliers
Variable or fixed, H_0	• Resonant waveguides	Two-cavity etched-groove MSW resonators, guided-wave ring resonators, and one-port and two-port straight-edge MSW resonators
Variable or fixed, H_0	• Directional couplers	Full power transfer between input and the output transducers
Variable or fixed, H_0	• Nonlinear devices	Frequency selective limiters, signal-to-noise enhancers, MSFVW convolvers

Table 11.7 SAW and MSW devices, applications, and advantages and disadvantages (after [107]).

Devices	Application	Advantages	Disadvantages
SAW frequency filter	Satellite channel multiplexing filter	Design-out group delay variations	High insertion loss
SAW oscillator	Local oscillator Drive and analog communications modulator	High operating IF, easy harmonic selection	Poor long-term stability
SAW discriminator	Analog communications demodulator	High operating IF	Moderate operating bandwidth, moderate linearity
SAW fixed delay line	Delay equalization for digital links employing space diversity	Stable and accurate delay	Marginal bandwidth
SAW bit-matched filter	Digital IF and to baseband decoder	Optimized SNR	Data rate limited
MSW group delay equalizer	Equalizer waveguide dispersion millimetric trunk systems	Adjustable delay slope	Delay ripple and delay linearity event
MSW variable delay line	Delay equalization for digital links employing space diversity	Electronically adjustable, GHz operation IF	Moderate insertion losses

insertion loss (dB) properties versus frequency (GHz) at a bias field of 380 Oe through the periodic structure of (a).

Measurements are seen to be in good agreement with theory based on repetitively mismatched transmission lines. This demonstration laid the groundwork for YIG film-based MSW resonant devices such as resonators.

During the 1980s, low-loss MSW propagation in YIG crystals generated great interest from advances made in LPE growth of YIG films.

The advantages of MSW devices at microwave frequencies include low propagation loss and ease of transducer fabrication. Specifically, above 3 GHz MSWs are projected to have lower losses than SAWs built upon lithium niobate.

Due to the inherent dispersion of MSWs, which are slow electromagnetic waves where the magnetic fields dominate the characteristics, group velocities in the range 10–10^3 km/s are obtained for wavenumbers in the range 10–103 cm^{-1} throughout the microwave region. Microsecond-order delays can be realized using single 10-m wide microstrips as transducers at frequencies up to and above 12 GHz.

In contrast, SAWs have equal phase and group velocities in the range 2–6 km/s. The low SAW phase velocity coupled with a small piezoelectric coupling coefficient necessitates photolithographic fabrication of interdigital transducers, having finger widths in the 1-μm range, for efficient transduction at low microwave frequencies (i.e. 1 GHz) [106].

MSWs, similar to SAWs, are accessible for tapping at the film surface, thus allowing construction of magnetic microwave analogues of available SAW VHF signal processing devices. Additionally, the bias field dependence of the MSW dispersion allows the construction of electronically variable delay lines and resonators. Selective etching and planar fabrication provide MSW signal routing techniques and MIC-compatible devices [108].

With an ever-increasing competition for access to the RF spectrum, signal interference is becoming a serious challenge to users. In response, Adam et al. [110] developed frequency selective limiter (FSL) technology.

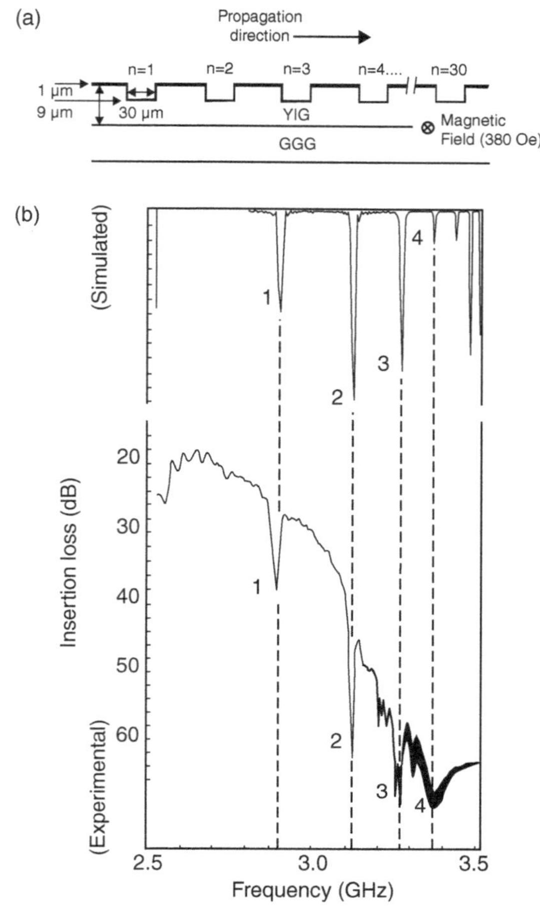

Figure 11.17 (a) Schematic of cross section for the periodic grooved structure prepared by selective etching of an epitaxial YIG film (not drawn to scale). (b) Simulated (top) and experimental (bottom) insertion loss (dB) as a function of frequency (GHz) for MSSW propagation through the periodic structure depicted in (a). The MSSW $k = 0$ mode cutoff is most clearly seen in the simulated data at $f \sim 2.53$ GHz and the transducer cutoff appear near (notch #4) $f \sim 3.37$ GHz [108].

In Figure 11.18a, the response of an FSL in the presence of simultaneous strong (1, 4, and 5) and weak (2 and 3) signals is illustrated [109]. The most common FSL devices use nonlinear excitations of magnetic spin waves in a magnetized ferrite such as LPE-grown YIG. Low power microwave signals couple with a sample of YIG produce a linear, low-loss response. However, above a critical RF magnetic field strength (h_c), the precession of magnetic dipoles in the YIG becomes nonlinear and transfer energy exponentially to the subharmonic spin wave envelop. These "half-frequency" spin waves transfer energy to the crystal lattice where excess power is dissipated as heat. This nonlinear coupling takes place in a bandwidth of the spin wave linewidth. For LPE-grown YIG, the spin wave linewidth is ~ 0.25 Oe, which represents a frequency bandwidth of about 1 MHz [109].

When this signal is passed through an FSL, the FSL's response changes to provide selective attenuation (see Figure 11.18b) of above-threshold signals. Below-threshold signals of interest remain unaffected (see Figure 11.18c).

Many more garnet-based technologies are under development and space limitations preclude a thorough treatment of these.

We next turn our attention to spinel ferrites.

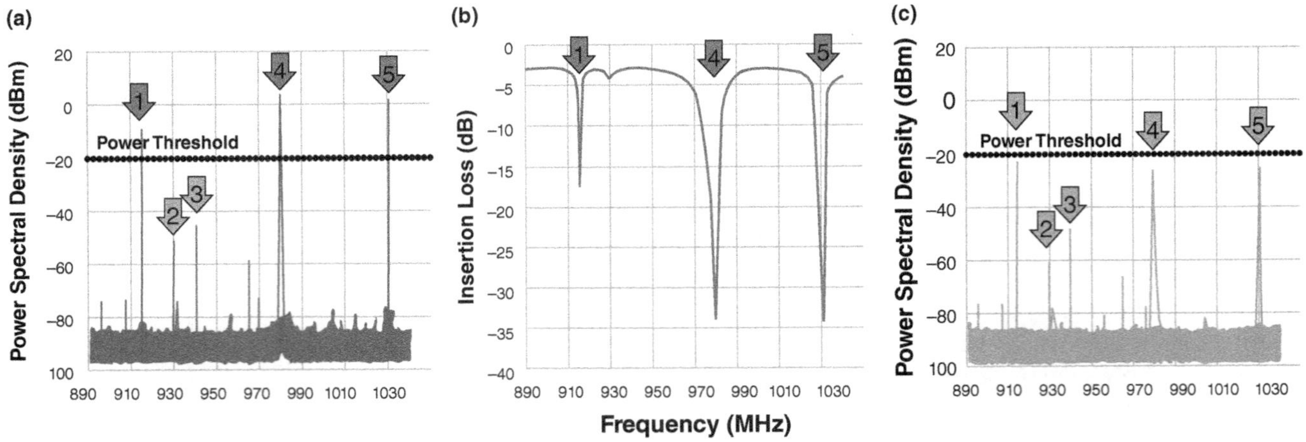

Figure 11.18 (a) Representative spectrum with above-threshold (1, 4, and 5) signals and below-threshold (2 and 3) signals used to demonstrate the FSL concept. The FSL's response function changes to provide selective attenuation (b) of above-threshold signals. Below-threshold signals of interest remain unaffected (c) [109].

11.5 LPE Growth of Spinels

The first systematic study of spinel ferrite compositions and magnetic properties such as manganese, copper, cobalt, magnesium, and zinc was reported by Hilpert in 1909 [112].

Two decades later, Forestier [113] performed careful studies of the preparation of various spinel ferrites and measurements of saturation magnetization and Curie temperatures.

In 1932, Kato and Takei discovered that a solid solution of magnetite and cobalt ferrite displayed high-room-temperature magnetization, suggesting its viability for commercial applications [114]. This led to the development of CuZn ferrites.

Philips researchers in the Netherlands were significant contributors to the development ferrite technology of ferrites. Starting in 1935, Snoek [115] prepared ferrites with the highest permeability and lowest loss factor of that time.

Snoek predicted that MnZn and NiZn ferrites would be very important electronic materials [116]. History would prove him correct.

Remarkable progress in the global electronics industry spurred by communication equipment, consumer electronics, and computers promoted the expansion of the ferrite market [117].

11.5.1 Structure, Chemistry, and Magnetism of Spinels

Figure 11.19a is a depiction of the Fe–O binary phase diagram defining phase fields for wustite, magnetite, and hematite compounds as a function of composition and temperature [117]. The magnetite (Fe_3O_4) phase field is defined for x~27.5 wt.-% O_2 and at a eutectic temperature of 580 °C. Figure 11.19b is the ternary phase diagram for MnZn ferrite formed by ZnO–Fe_2O_3–MnO phases with saturation polarization (J_s), saturation magnetostriction (λ_s), and magnetocrystalline anisotropy constant (K_1) properties superimposed over compositions [111].

The spinel structure, space group *Fd*3*m*, is typically expressed in the form $(A)[B]_2O_4$ (where ( ) indicates divalent cations occupying fourfold tetrahedra lattice sites and [ ] indicates trivalent cations occupying sixfold octahedra lattice sites). $(A)[B]_2O_4$ is a single formula unit that constitutes 1/8 of the spinel unit cell. Cations in this structure have either fourfold or sixfold coordination forming tetrahedra (A) and octahedra [B] sublattices that are arranged in a close-packed arrangement with respect to each other. *A* cations reside on 8 of the 64 available tetrahedral sites, whereas *B* cations reside on 16 of the 32 available octahedral sites. Oxygen anions form a close-packed structure and contribute 32 ions that electrically balance the unit cell (see Figure 11.20) [78].

The most popular and abundant of the magnetic spinels is magnetite, Fe_3O_4, or $(Fe^{3+})[Fe^{3+}Fe^{2+}]_2O_4$, which has 8 trivalent Fe ions on the A sublattice and a mix of divalent (8) and trivalent (8) ions on the B sublattice. This structure and cation distribution is referred to as an *inverse* spinel. The γ form of Fe_2O_3 is also a spinel in which vacancies form preferentially on the B sublattice in the form of $(Fe^{3+})[Fe^{3+}_{1.7}\square_{0.3}]O_4$, where $\square$ denotes vacancies [78].

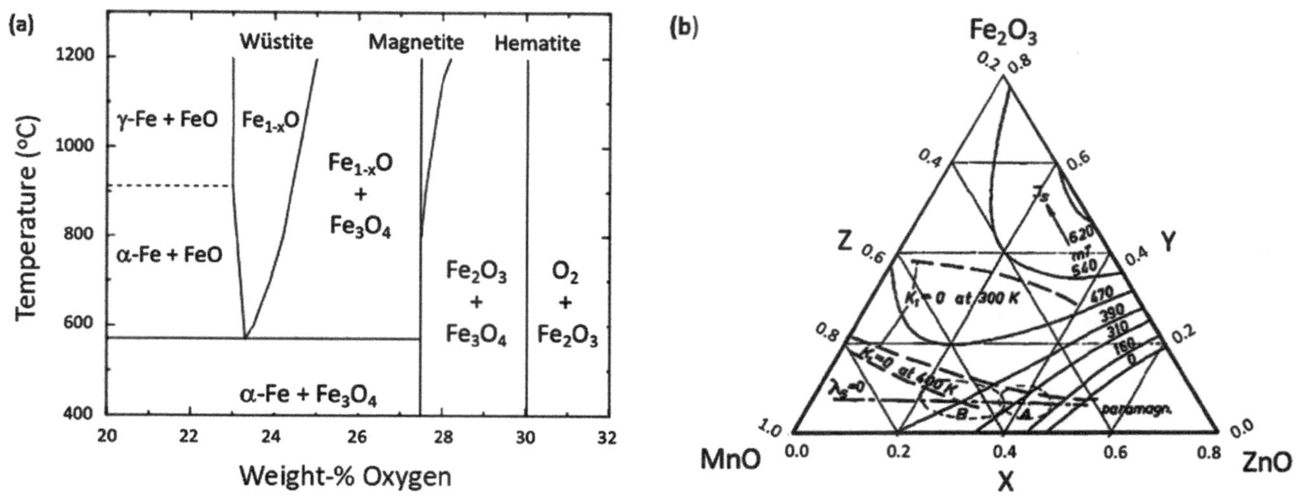

Figure 11.19 (a) Fe–O binary equilibrium phase diagram [111]. (b) Properties of manganese–zinc ferrites (ternary phase diagram formed by $ZnO-Fe_2O_3-MnO$). Anisotropy constant K_1 and magnetostriction constant λ_s, and saturation polarization J_s at 400 K [118].

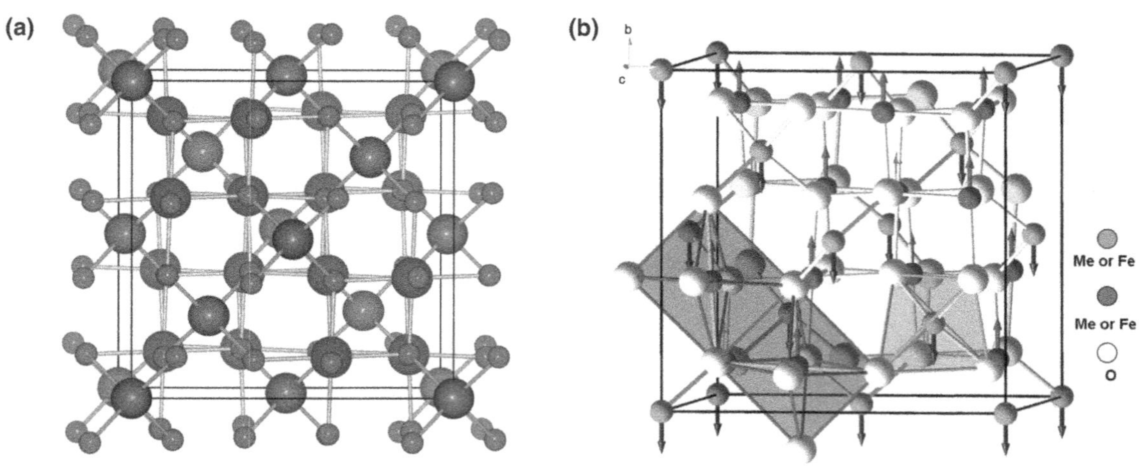

Figure 11.20 A 56-atom AB_2O_4 spinel unit cell with 8 A (smaller circles, lighter in appearance, tetrahedrally coordinated) and 16 B cations (smaller circles, darker in appearance, octahedrally coordinated), with 32 oxygen anions (larger circles, lighter in appearance). Only representative octahedral (B) and tetrahedral (A) polyhedral are shown [78].

In addition to inverse, spinels can be classified as *normal* or *mixed*. The *normal* spinel has only divalent cations residing on the 8 A sites with trivalent ions filling the 16 B sites. There are not many known *normal* spinels. An example is the aforementioned Zn spinel ferrite (i.e. $(Zn^{2+})[Fe^{3+}]_2O_4$) where divalent Zn fill predominantly A sites and trivalent Fe reside on the B sites.

The *mixed* spinel has mixed valence ions occupying both A and B sublattices. Because each species of ion has thermodynamically preferred lattice distributions based on ionic radius, electrostatic energy, and electronic configuration, this class of ferrite commonly results from nonequilibrium processing such as those that involve vapor quenching (e.g. pulsed laser deposition and sputter deposition) and high kinetic energy transfer (e.g. ball milling), among other techniques. Cation species of similar ionic radii that exist in a multiplicity of valence states are susceptible to this type structure. For example, Mn ions with common valences of 2, 3, 4, 6, and 7, and of comparable sizes to Fe cations, allow for such a distribution as $(Mn^{2+,3+}{}_{1-\delta}Fe^{2+,3+}{}_{\delta})[Mn^{2+,3+}{}_{\delta}Fe^{2+,3+}{}_{2-\delta}]O_4$ to exist (where δ represents the cation inversion parameter). The tendency to stabilize defects in such structures adds yet another degree of complexity. As one might imagine, determining such a structure with robust measurement statistics represents a significant challenge to the experimentalist. It is likely that most ferrites produced using today's advanced processing techniques fall into this *mixed* category [78].

Table 11.8 presents some common spinel ferrite materials with their preferred cation distributions, magnetization, Curie temperatures, g factors, and related properties. It is important to realize, as stated earlier, that nonequilibrium processing leads to deviations from the listed equilibrium cation distributions.

Table 11.9 presents a historical evolution of liquid-phase epitaxy deposition of spinel ferrite films. Spinel films of $MgFe_2O_4$, $NiFe_2O_4$, $ZnFe_2O_4$, and $Ni_{1-x}Zn_xFe_2O_4$ were first grown by LPE on MgO from $PbO–B_2O_3$ fluxes by Robertson et al. in 1977 [33] and later Mn and MnZn ferrites [59]. Glass and Liaw followed soon after with the growth of Li ferrite on nonmagnetic spinels and attained ΔH as low as 7.8 Oe at 9.07 GHz [32]. van der Straten et al. grew $Ni(Fe,Al)_2O_4$ films on (111) MgO and on (111) $ZnGa_2O_4$ substrates using a $PbO–B_2O_3–Fe_2O_3$ solvent and correlated magnetostrictive properties with stress-induced uniaxial anisotropy, domain structure, and bubble stability. This report was the first-time magnetic bubbles were studied in LPE-grown spinel ferrites [56]. Plaskett et al. employed reactive gasses in the LPE growth of MnZn-ferrite films and found a strong dependence on gas environments and film stoichiometry and properties [58]. van der Straten et al. followed with studies of the growth of LPE spinel films of Mn-, Ni-, and Al-doped Cu ferrite (on MgO) and $Ni[Fe(Al)]_2O_4$ (on

Table 11.8 Spinel magnetism, Curie temperature, microwave properties, and resistivity.

References		[119]	[120]	[121]	[121]
Designation $(Me_1)^A[Me_2]^B{}_2O_4$	Spinel type	$4\pi M_s$ (G)[13)]	T_c (K)	g (X band)	ρ (Ω·cm)
Magnetite Fe_3O_4 [122]	Inverse	6000	858	2.17	10–2
$(Zn)[Fe]_2O_4$	Normal	Anti-FM	10.5	—	10^6
Jacobsite $(Mn_{0.8}Fe_{0.2})[Mn_{0.2}Fe_{1.8}]O_4$ [86]	Mixed	4800	577	2.06	10^4
$(Fe)[FeCo]_2O_4$	Inverse	5300	793	2.7(2.27) Rt/200c	10^4
Trevorite $(Fe)[FeNi]_2O_4$ [86]	Inverse	3000	860	2.19	10^9
Cuprospinel $(Fe)[FeCu]_2O_4$	Inverse	2227	728	2.05–2	
Magnesioferrite $(Fe)[FeMg]_2O_4$	Inverse	1500	710	2.06.2.20 +450 to −200	10^6
$(Fe)[Li_{0.5}Fe_{1.5}]_2O_4$ [123]	Inverse	3750	893	2.08	10^{-2}
$(Fe_{0.5}Zn_{0.5})[Li_{0.25}Fe_{1.75}]O_4$ [124]	Mixed	3030	600	—	10^8
$(Fe_{1-y-z}Ni_yZn_z)$ $[Ni_{0.5-y-z/2}Zn_zFe_{1.5+y+z/2}]O_4$ [125-n40]	Mixed	2500	875	—	10^{10}
Franklinite $(Fe_{1-y-z}Mn_yZn_z)$ $[Mn_{0.5-y-z/2}Zn_zFe_{1.5+y+z/2}]O_4$ [126-MN30]	Mixed $y = 0.6, z = 0.2$	4300	500	—	10^3

Table 11.9 Historical development of liquid-phase epitaxy deposition conditions and key results of spinel ferrite films.

Year	Film composition	Flux	Rate and thickness	Substrates	Key results	Ref.
			LPE spinel films			
1977	$MgFe_2O_4$, $NiFe_2O_4$, $ZnFe_2O_4$, $Ni_{1-x}Zn_xFe_2O_4$	$PbO–B_2O_3$	1–3.3 μm	MgO	Thin films of $MgFe_2O_4$, $NiFe_2O_4$, $ZnFe_2O_4$, and $Ni_{1-x}Zn_xFe_2O_4$ were grown on MgO substrates from a $PbO–B_2O_3$ flux. Films were examined by interference microscopy and X-ray topography. Substitution of Ga, Sn, and Cr in $MgFe_2O_4$ films for $1.11<x<1.32$; $1.06<x<1.56$ and $0.005<x<0.012$ for Ga, Sn, and Cr, respectively. Doping of flux is shown to minimize the lattice mismatch.	[33]
1978	$(Fe)[Li_{0.5}Fe_{1.5}]_2O_4$	$PbO–B_2O_3$	Unknown	Nonmagnetic spinels	Results on the growth of lithium ferrite by LPE are described. A $PbO–B_2O_3$ flux was used with nonmagnetic spinel substrate crystals. Compared with growth of magnetic garnets, the LPE melt was found to be relatively unstable with respect to undercooling. All of the epitaxial films suffer from cracks; ferrimagnetic resonance linewidths as low as 7.8 Oe at 9.07 GHz have been measured.	[32]
1978	$Ni(Fe,Al)_2O_4$	$PbO–B_2O_3–Fe_2O_3$	Unknown	(111) MgO, (111) $ZnGa_2O_4$	The growth of single-crystal $Ni(Fe,Al)_2O_4$ films on (111) MgO and on (111) $ZnGa_2O_4$ substrates using a $PbO–B_2O_3–Fe_2O_3$ solvent is described. While films grown upon MgO show stress relief at growth temperatures, films grown upon $ZnGa_2O_4$ possess a tensile strain due to elastic deformation. Since λ_{111} for $NiFe_2O_4$ is strongly negative, resulting in a stress-induced uniaxial anisotropy. Stripe domains are observed when a magnetic field is applied perpendicular to the plane of the film. A bubble stability factor q exceeding unity is obtained. For the first time magnetic bubbles are found in LPE grown spinel ferrites.	[56]
1979	ZnFe2O_4 Mn_xZnFeO_4 $(Fe)[Li_{0.5}Fe_{1.5}]_2O_4$	$PbO–B_2O_3–Na_2B_4O_7$, $Na_2P_2O_7$	0.05–0.55 μm/min	$MgAl_2O_4$, MgO, MnZn ferrite, $MnGa_2O_4$	LPE films were grown from $PbO–B_2O_3–Na_2B_4O_7–Na_2P_2O_7$ fluxes on different substrates. A $PbO–B_2O_3$ flux is necessary to have to obtain misorientation of a few degrees from the (111) direction to eliminate hillock growth. Different fluxes result in different misorientation angles.	[59]

(Continued)

Table 11.9 (Continued)

Year	Film composition	Flux	Rate and thickness	Substrates	Key results	Ref.
1980	$Mn_{.9}Zn_{.5}Fe_{1.6}O$, $Mn_{0.3}Zn_{0.3}Fe_{2.4}O_4$	$Na_2B_4O_7$ fluxes under reactive H_2 and CO_2 gas environments	Unknown	$MnGa_2O_4$	Spinel films of MnZn ferrite were grown on $MnGa_2O_4$ substrates under controlled gas mixtures of H_2 and CO_2. Each layer was deposited at 660 °C from a single $Na_2B_4O_7$-based charge. Film composition varied with the growth ambient; CO_2 atmospheres yielded films with a nominal composition of $Mn_{0.9}Zn_{.5}Fe_{1.6}O_4$ and Curie temperatures near 110 °C. As the oxygen partial pressure in the growth ambient was reduced by the addition of H_2 to CO_2, the film composition changed, the Fe content increased while the proportions of Mn and Zn decreased to $Mn_{.3}Zn_{.3}Fe_{2.4}O_4$, which had the Curie temperature of about 430°C.	[58]
1980	Mn-, Ni-, and Al-substituted Cu ferrite	$PbO–B_2O_3$	0.05–0.9 μm/min; $t = 2–9$ μm	(111)-MgO	Single-crystalline Mn-, Ni-, and Al-substituted copper ferrite films are grown from a $PbO–B_2O_3$ flux on (111)-MgO substrates. Solid solutions between copper ferrite and Mn_3O_4, $NiFe_2O_4$, and $CuAl_2O_4$ were obtained. The segregation coefficients for Ni and Al were shown to be linearly dependent on growth temperature. From the observed domain structure and from torque measurements, it was concluded that a positive uniaxial anisotropy is present in Cu-ferrite films.	[127]
1981	$Ni(Fe,Al)_2O_4$	$PbO–B_2O_3$	0.05–0.47 μm/min; t~6 μm	(100), (110), and (111) MgO and on (111) $ZnGa_2O_4$, $Zn(Ga,Al)_2O_4$, and $Zn(Ga,Fe)_2O_4$	Properties of LPE-grown $Ni(Fe,Al)_2O_4$ films on (100), (110), and (111) MgO and on (111) $ZnGa_2O_4$, $Zn(Ga,Al)_2O_4$, and $Zn(Ga,Fe)_2O_4$ substrates are reported. For low film thicknesses and small lattice mismatches, elastically deformed films, showing minimal stress relief, can be grown on spinel substrates. For films grown in tension, positive uniaxial anisotropies are found and stripe domains observed. Crack-free films with a uniaxial anisotropy energy exceeding the demagnetization energy are obtained when the misfit is lower than 0.03 Å and $0.50<x<0.65$. For those films, the stripe domains can be contracted to bubbles.	[21]

substrates of $ZnGa_2O_4$, $Zn(Ga,Al)_2O_4$ and $Zn(Ga,Fe)_2O_4$) and correlated growth characteristics with the films' stress state, cracks, and bubble stability [21, 127].

Recent interests in the LPE growth of spinel films are motivated to extend the frequency range of microwave devices such as FSLs. To date, FMR linewidths of LPE spinels remain prohibitively large.

11.6 LPE Growth of Hexaferrites

Circa 1950, barium hexaferrite ($BaO \bullet 6Fe_2O_3$ or $BaFe_{12}O_{19}$) was discovered somewhat serendipitously at Philips Physics Laboratory when engineers preparing samples of hexagonal lanthanum ferrite for use as a semiconductor material produced barium hexaferrite of the magnetoplumbite structure. Barium hexaferrite, advertised under the trade name *Ferroxdure*, was shown to have both a high coercivity (i.e. ~170 kA/m) and relatively low raw materials cost. Soon after, Philips developed strontium hexaferrite (i.e. $SrO \bullet 6Fe_2O_3$ or $SrFe_{12}O_{19}$), with superior properties to barium hexaferrite for some applications. Barium and strontium hexaferrites have since dominated the market for hard ferrites due their excellent performance and low costs.

Hexaferrites exist in several structural variants, i.e. M, Y, Z, X, U, W, in which the metal and oxygen stoichiometry vary [128]. Due to their highly symmetric structures and low intrinsic magnetocrystalline anisotropy energy, cubic garnets and spinel ferrites tend to have low magnetic anisotropy fields. In comparison, crystal symmetry breaking in the hexaferrites provides them with extremely high magnetic anisotropy fields. As will be discussed in greater detail, the ferrimagnetic resonance (FMR) frequency shifts with changes in magnetic anisotropy field, together with demagnetization field, magnetization field, and applied magnetic field (i.e. bias fields), leading to the operation of hexaferrites at microwave, millimeter-wave, quasi-optical frequencies.

11.6.1 Structure, Chemistry, and Magnetism of Hexaferrites

The M-type barium hexaferrite (also known as Ferroxdure or BaM) has a chemical formula $BaFe_{12}O_{19}$, melting point of 1390 °C, and theoretical density of 5.3 g/cm^3. The symmetry of the barium hexaferrite crystal structure is characterized by the space group (*P63/mmc*) with lattice parameter $a = b = 5.89$ Å, $c = 23.17$ Å, $\beta = \alpha = 90°$, $\gamma = 120°$ [128].

In contrast to garnet and spinel ferrites, the hexaferrites have a hexagonal crystal symmetry, leading to high magnetocrystalline anisotropy energies and magnetic anisotropy fields (H_A) and subsequently high FMR frequencies.

Among the most popular of the microwave hexaferrites are those derived from the Ba M-type (or BaM) hexaferrite, $BaFe_{12}O_{19}$ (see Figure 11.21a). BaM has the magnetoplumbite structure, space group *Pb3/mmc* and consists of 10 oxygen layers in which four successive layers each containing four oxygen anions are followed by a fifth layer containing three anions and a Ba cation. The structure is constructed of spinel (*S*) blocks containing two oxygen layers separated by an *R* block containing the Ba ion. Every other *S* and *R* block are rotated 180° with respect to the other and are denoted as *S** and *R**, respectively. Trivalent Fe ions occupy tetrahedral and octahedral sites (derived from the spinel blocks) as well as one trigonal bipyramidal (TBP) site. The TBP site resides on the Ba-containing layer amidst the trigonal arrangement of oxygen anions [129].

There exist other hexaferrite structures designated as *Y*, *U*, *W*, *X*, and *Z*, each having different structures and oxygen stoichiometry and markedly difference magnetic properties. For example, the *Y*-type structure, $Ba_2Me_2Fe_{12}O_{22}$, consists of the stacking of S and T blocks where the T blocks can be thought of as R blocks in which the inner most atomic layer is repeated. Having defined the *S*, *R*, and *T* blocks, one can describe the *W*-type, $BaMe_2Fe_{16}O_{27}$, as a stacking of *SSRS*S*R**, and the *Z*-type, $Ba_3Me_2Fe_{24}O_{41}$, as *R*STSRS**. The reader is directed to the literature to explore further types of hexaferrite structures. The magnetization of the hexaferrites derives from superexchange interactions [80–82].

A recent development of another less-known hexaferrite phase is the MgZn-18H hexaferrite. The 18H hexaferrites exhibit planar magnetocrystalline anisotropy except for the Cu-18H, which possesses a uniaxial magnetocrystalline anisotropy. This suggests 18H hexaferrites to be promising Co-free microwave materials that are nontoxic and low-cost alternatives to other microwave hexaferrites.

The crystal structure has been determined to be a repetitive stacking of three basic subunits: S-block ($[Me_2Fe_4O_8]^{0\pm}$ or $[Fe_6O_8]^{2+}$), R-block ($[BaFe_6O_{11}]^{2-}$), and T-block ($[Ba_2Fe_8O_{14}]^{0\pm}$). As shown in Figure 11.22, the unit cell of the Mg-18H hexaferrite, $Ba_5Mg_2Ti_3Fe_{12}O_{31}$, contains two formula units, where 18 layers of close-packed positions are occupied by a total of 10 Ba^{2+} and 62 O^{2-} ions. The crystal structure is hexagonal, with the space group *P63/mmc*, which can be viewed as the structure of Y-type hexaferrites with intercalation of a three-layered hexagonal barium titanate block (*h*-BTO) into the middle of the T-block [130].

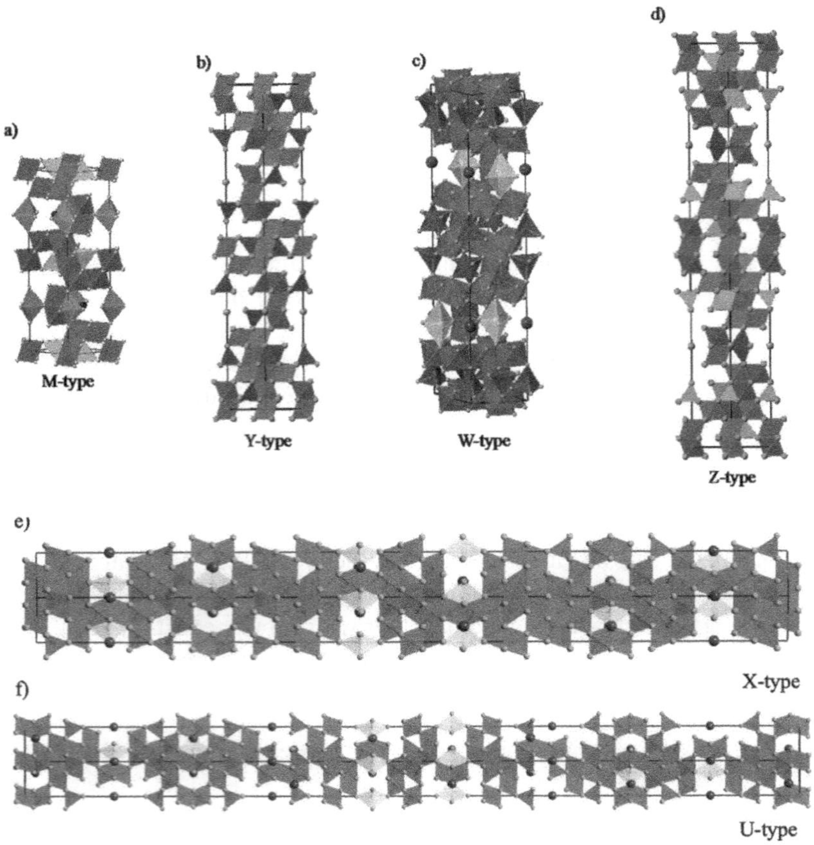

Figure 11.21 Hexaferrite unit cells depicting polyhedra arrangements forming the unit cells for (a) M-type [$(Ba,Me)Fe_{12}O_{19}$], (b) Y-type ($Ba_2Me_2Fe_{12}O_{22}$), (c) W-type ($BaMe_2Fe_{16}O_{27}$), (d) Z-type ($Ba_3Me_2Fe_{24}O_{41}$), (e) X-type ($Ba_2Me_2Fe_{28}O_{46}$), and (f) U-type ($Ba_4Me_2Fe_{36}O_{60}$). The atoms forming the vertices of the polyhedra are oxygen anions, while atoms within the polyhedra cage are iron cations. The atoms positioned apart from the polyhedra are the barium (or Me) cations. Space groups and specific properties are discussed in the text (illustrations provided courtesy of Dr. Su Zhijuan).

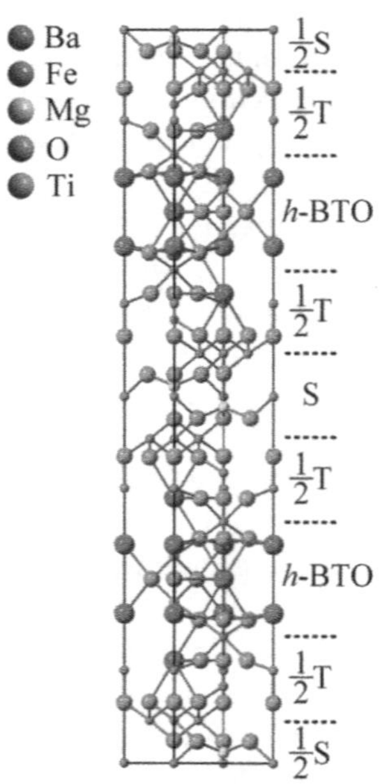

Figure 11.22 Schematic illustration of the Mg 18H hexaferrite unit cell with the stacking subunit blocks denoted [130].

Figure 11.23 is an illustration of (a) the binary phase diagram of $BaO–Fe_2O_3$ [131]. In this figure, ~86 -% of Fe_2O_3 coincides to a well-defined phase field for $BaFe_{12}O_{19}$ up to 1455 °C. Correspondingly, Figure 11.23b is an illustration modified from [132] depicting the ternary phase diagram formed from Fe_2O_3–BaO–MeO and the position of the principal hexaferrite phases and spinel.

As a prototypical hexaferrite, the BaM system is remarkably versatile in that substitution for the Fe cation can drastically reduce or increase H_A, thus shifting the FMR frequency and the operating frequency of related devices. For example, the substitution of Sc or In for Fe reduces H_A, allowing for applications from X, Ku, K, to Ka bands. Substitution of Al and Ga leads to increase in H_A and device applications at frequencies up to and including U, E, and W bands.

In essence, the BaM hexaferrite, and its substitutional systems, allow for device applications from <<10 to >>100 GHz. This has been presented most clearly in the work of Harris and Sokolov [133].

The utility of M-type ferrites stems in part from the alignment of the easy magnetic direction along the crystallographic *c*-axis and the ability to process these materials with a high degree of crystal texture and preferential magnetic anisotropy. For example, the growth of BaM films with the crystallographic *c*-axis aligned perpendicular to the film plane leads to perpendicular magnetic anisotropy: a requirement for conventional circulator and isolator devices and perpendicular magnetic storage media [134].

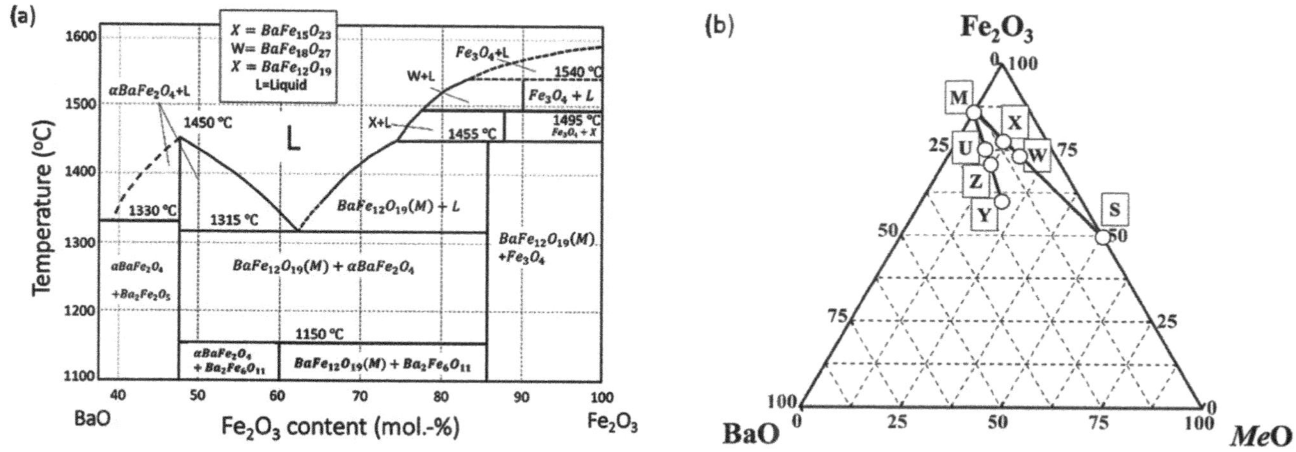

Figure 11.23 (a) Binary phase diagram of Bao–Fe_2O_3 [131]. (b) A modified ternary diagram from [132] depicting the location of principal hexaferrite phases and the spinel.

Unlike the M-type ferrites, *Y*- and *Z*-type ferrites have their magnetic easy axis aligned within the *a–b* or basal plane. Where the M-type hexaferrites can be saturated relatively easily by fields applied along the *c*-axis, the *Y*- and *Z*-type ferrites can be easily saturated by fields aligned along the basal plane. This is of particular value for a conventional phase shifter, delay lines, filter devices, and antenna substrate applications.

In the hexaferrites, R and T blocks break the crystal symmetry, resulting in the hexagonal structure and introduction of large magnetocrystalline anisotropy energies. Also contributing to the magnetic anisotropy are dipolar interactions and single-ion anisotropies. Remarkably, the magnetic anisotropy field in hexaferrites can reach several 10 000 Oe, for example, Al- and Ca-doped BaM, where H_A approaches 30 000 Oe, or ~10 000 times greater than cubic ferrites [133, 135].

The large H_A places the zero-field FMR frequency of undoped BaM near 36 GHz. Thus, while an externally applied magnetic field is still often necessary to saturate the ferrite during device operation, the magnitude of the applied field required to shift FMR to high frequencies is substantially reduced. As such, devices based upon this ferrite can operate at frequencies as high as Ku band for below-resonance operations and beyond the Q band for above resonance operations.

Textured polycrystalline hexaferrites can be produced with permanent magnet properties such that they remain in a magnetized state in the absence of an externally applied magnetic field. In this condition, referred to as self-biased, the high internal bias field required for device operation at high frequencies is achieved without any external magnets. Bias magnets hinder efforts to reduce the size and weight of devices, assemblies, and systems, such as transmit and receive modules in radar and communication systems. Gorbachev et al. [135] have recently demonstrated that textured samples of Al and Ca-doped SrM hexaferrites have H_A values that approach 40 000 Oe and zero-field FMR frequencies ranging from 160 to 250 GHz.

Table 11.10 presents a listing of key magnetic and physical properties of hexaferrite materials.

Table 11.11 lists key historical developments of liquid-phase epitaxy deposition conditions and key results for hexaferrite films. LPE growth of hexaferrites begin in the 1970s with the work of Stearns and Glass. In 1975, they were the first to grow by LPE hexagonal ferrite Zn_2Y, having the chemical formula $Ba_2Zn_2Fe_{12}O_{22}$, grown by the isothermal dipping method using a $PbO–BaO–B_2O_3$ flux. The substrates were M-type hexagonal ferrite, $BaFe_{12}O_{19}$ [138]. In 1976, using a $PbO–BaO–B_2O_3$ flux, they produced Zn-substituted M-type hexagonal ferrite and a heavily Zn-substituted spinel epitaxial films on nonmagnetic spinel substrates. The substrates studied were single crystals of three different compositions: $MgAl_2O_4$, $MgGa_2O_4$, and $Mg(In,Ga)_2O_4$ [61].

Barium M-type hexaferrite films were grown by LPE on Zr- and Mg-substituted strontium hexagallate substrates with the film surfaces exhibiting terrace growth with step heights of ~10 nm. An FMR linewidth of 45 Oe at 53 GHz was measured for a 2.1-μm-thick film [64].

Graetsch *et al.* [139] produced LPE-grown single crystals as solid solution films of $SrZn_{2-x}Co_xFe_{16}O_{27}$ ($SrZn_{2-x}Co_xW$). The magnetocrystalline anisotropy was measured to change from uniaxial to planar upon variation in temperature and cobalt substitution for $x = 0.5$ for 6 K and $x = 0.85$ for 298 K. The sixfold anisotropy in the basal plane of the planar W hexaferrites increases strongly with increasing cobalt content.

Conditions favoring the LPE growth of BaM films were investigated for the Bi_2O_3–BaO–B_2O_3 flux system. The barium hexaferrite–saturated solutions exhibited relatively large supercooling values (~65 °C). However, the $BaFe_{12}O_{19}$ phase field occupies a limited region of the BaO/Fe_2O_3/Bi_2O_3 pseudoternary phase diagram. PbF_2 was added to the melt to extend this region. LPE films of pure barium hexaferrite and aluminum-substituted hexaferrite were successfully grown on Sr(Ga, Mg, $Zr)_{12}O_{19}$ substrates using this flux [20].

The Vittoria-led group published a series of papers on the LPE growth of $BaFe_{12}O_{19}$ films from 2002 to 2004. In these studies, LPE BaM films were investigated for their microwave properties. Growth entailed the precoating of both sides of the substrate with ~0.5 μm BaM nucleation layers by pulsed layer deposition. LPE growth at ~40 μm/h followed on both sides of the substrate in an effort to cancel compressive and tensile stresses. These authors explored this approach using multiple substrate systems, including (111) MgO [66], (111) GGG [62, 68], and *m*-plane sapphire [60, 62]. Ferrimagnetic linewidth measurements on films ~50 μm revealed 27 Oe at 60 GHz for growth on (111) MgO, 68 Oe at 58 GHz for growth on (111) GGG, and 80 Oe at 60 GHz for growth on (1100) m-plane Al_2O_3.

The long-sought goal of the ferrite community has been the integration of ferrite passive devices (e.g. isolators and circulators, phase shifters, filters, etc.) with semiconductor systems on a common semiconductor platform. Such an achievement is the first step to integrated system-on-a-wafer architectures.

Table 11.10 Hexaferrite structure, magnetism, Curie temperature, and anisotropy fields.

Ferrite type	Composition	Space group and block sequence	Magnetic anisotropy type	$4\pi M_s$ (kG)	μ_B	T_c (K)	$H_{A\Theta}$/ (kOe$H_{A\Phi}$/Oe)	K_1 (×10^5) (J/m^3)
Magnetoplumbite (M)								
M	$BaFe_{12}O_{19}$	SRS*R* *Space group: P63/mmc*						
BaM	$BaFe_{12}O_{19}$	SRS*R*	Uniaxial	4.77	20	725	17.0	3.3
SrM	$SrFe_{12}O_{19}$	SRS*R*	Uniaxial	3.73	9.8	735	17.6	3.5
Ferroxplana (Y, Z, U, 18-H)								
Y [136]	$Ba_2Me_2Fe_{12}O_{22}$	3(ST) *Space group: P63/mmc*						
Co2Y	$Ba_2Co_2Fe_{12}O_{22}$	3(ST)	Planar	2.33	10.8	613	28.0/155	−2.6
Zn2Y	$Ba_2Zn_2Fe_{12}O_{22}$	3(ST)	Planar	2.85	18.4	403	9.0 /1	n.a.
Z	$Ba_3Me_2Fe_{24}O_{41}$	STSRS*T*S*R* *Space group: P63/mmc*						
Co2Z	$Ba_3Co_2Fe_{24}O_{41}$	STSRS*T*S*R*	Planar	3.39	210.8	683	13.0/ 16	−1.8
Zn2Z	$Ba_3Zn_2Fe_{24}O_{41}$	STSRS*T*S*R*	Planar	3.90	38.4	633	n.a.	n.a.
U	$Ba_4Me_2Fe_{36}O_{60}$	SRS*R*S*T *Space group: P63/mmc*						
Co2U	$Ba_4Co_2Fe_{36}O_{60}$	SRS*R*S*T	Planar	3.62	n.a.	707	n.a.	n.a.
(ZnCo) 2U	$Ba_4(Zn_{1.6}Co_{0.4})Fe_{36}O_{60}$	SRS*R*S*T	Planar	3.41	n.a.	673	7.53	1.03
18H	$Ba_5Me_2Ti_3Fe_{12}O_{31}$	2[(ST)B(TS)] *Space group: P63/mmc*						

Table 11.10 (Continued)

Ferrite type	Composition	Space group and block sequence	Magnetic anisotropy type	$4\pi M_s$ (kG)	μ_B	T_c (K)	$H_{A\theta}$/ (kOe$H_{A\Phi}$/Oe)	K_1 ($\times 10^5$) (J/m^3)
Zn-18H	$Ba_5Zn_2Ti_3Fe_{12}O_{31}$	2[(ST)B(TS)]	Planar	0.95	14.1	310	3.5	−1.0
Co-18H	$Ba_5Co_2Ti_3Fe_{12}O_{31}$	2[(ST)B(TS)]	Planar	1.26	9.6	495	40.0	−0.1
(MgZn)-18H	$Ba_5Mg_1Zn_1Ti_3Fe_{12}O_{31}$	2[(ST)B(TS)]	Planar	1.23	12.5	357	1.3	−0.4
			Conical hexaferrites (X, W)					
X	$Ba_2\mathit{Me}_2Fe_{28}O_{46}$	3(SRS*S*R*) *Space group: P63/mmc*						
Co2X	$Ba_2Co_2Fe_{28}O_{46}$	3(SRS*S*R*)	Conical	2.69	46.3	740	9.49	n.a.
Zn2X	$Ba_2Zn_2Fe_{28}O_{46}$	3(SRS*S*R*)	Conical	5.81	55	705	n.a.	n.a.
W [137]	$Ba\mathit{Me}_2Fe_{16}O_{27}$	SSRS*S*R* *Space group: P63/mmc*						
Ni2W	$BaNi_2Fe_{16}O_{27}$	SSRS*S*R*	Conical	4.15	n.a.	793	12.7	n.a.
Co2W	$BaCo_2Fe_{16}O_{27}$	SSRS*S*R*	Conical	3.98	n.a.	763	21.2	−3.5

Table 11.11 Historical development of liquid-phase epitaxy deposition conditions and key results of hexaferrite films.

Year	Film composition	Flux	Rate and thickness	Substrates	Key results	Ref.
LPE hexaferrite films						
1975	Zn_2Y, $Ba_2Zn_2Fe_{12}O_{22}$	PbO–BaO–B_2O_3	Unknown	$BaFe_{12}O_{19}$	The first attempt to grow hexaferrite films by LPE were of the hexagonal ferrite Zn_2Y, having the chemical formula $Ba_2Zn_2Fe_{12}O_{22}$, grown by the isothermal dipping method of liquid phase epitaxy using a PbO–BaO–B_2O_3 flux. The substrates were flux-grown M-type hexagonal ferrite crystals having the chemical formula $BaFe_{12}O_{19}$. The films and substrates were distinguishable by the differences in their crystallographic and magnetic properties.	[138]
1976	Zn_2Y	PbO–BaO–B_2O_3		Hexagonal ferrite M substrates $MgAl_2O_4$, $MgGa_2O_4$, and $Mg(In,Ga)_2O_4$	Using a PbO–BaO–B_2O_3 fluxed melt, which had been employed previously to grow hexagonal ferrite Zn_2Y films on hexagonal ferrite M substrates, isothermal dipping liquid phase epitaxy produced epitaxial films on nonmagnetic spinel substrates. The substrates studied were single crystals of three different compositions: $MgAl_2O_4$, $MgGa_2O_4$, and $Mg(In,Ga)_2O_4$. Two different film phases were identified: a lightly Zn-substituted M-type hexagonal ferrite and a heavily Zn-substituted magnetite.	[61]

(Continued)

Table 11.11 (Continued)

Year	Film composition	Flux	Rate and thickness	Substrates	Key results	Ref.
1983	BaM	Unknown	2.1 μm	Zr and Mg substituted Strontium hexagallates	Barium hexaferrite films have been grown by liquid-phase epitaxy on Zr- and Mg-substituted strontium hexagallate substrates. Continuous films of 20 mm diameter have been obtained. The film surface exhibits terrace growth with a step height of about 10 nm. The saturation magnetization and anisotropy field as measured by static and FMR techniques are close to data of bulk crystals. An FMR linewidth of 45 Oe at 53 GHz was found for a 2.1-μm-thick film.	[64]
1984	$SrZn_{2-x}Co_xFe_{16}O_{27}$ W-type	SrO/B_2O_3 flux	Unknown	Unknown	The saturation magnetization and the magnetocrystalline anisotropy were measured in the solid solutions $SrZn_{2-x}Co_xFe_{16}O_{27}$ ($SrZn_{2-x}Co_x$-W) at 298 K and 6 K. Lattice constants and Curie temperatures are given as a function of the composition. The magnetocrystalline anisotropy changes from uniaxial to planar dependent on temperature and cobalt substitution at $x = 0.5$ for 6 K and $x = 0.85$ for 298 K.	[139]
1987	BaM	Bi_2O_3–BaO–B_2O_3 flux	$t = 5$–40	$Sr(Ga, Mg, Zr)_{12}O_{19}$	Conditions favoring the LPE growth of barium hexaferrite films have been investigated using the Bi_2O_3–BaO–B_2O_3 flux system. The barium hexaferrite–saturated solutions exhibit relatively large degrees of supercooling (up to 65 °C). However, the $BaFe_{12}O_{19}$ phase field occupies a limited region of the $BaO/Fe_2O_3/Bi_2O_3$ pseudo ternary phase diagram. LPE films of pure barium hexaferrite and aluminum-substituted hexaferrite were successfully grown on $Sr(Ga, Mg, Zr)_{12}O_{19}$ substrates using this flux system.	[20]
2002	BaM	Unknown	45 μm	(111) MgO	$BaFe_{12}O_{19}$ films were grown on both sides of (111) MgO substrates. The total thickness was ~45 μm. A low ferrimagnetic resonance linewidth of 27 Oe was measured at 60 GHz.	[66]
2003	BaM	Unknown	45 μm/hr	(111) GGG	LPE deposition and characterization of thick $BaFe_{12}O_{19}$ films deposited on (111) oriented single-crystal garnet GGG substrates are reported. The thickness of the films ranged from 45 and 80 μm with growth rates of up to 40 μm/h. The ferrimagnetic resonance linewidth is narrow (~50 Oe) for external fields applied along the *c*-axis.	[68]

Table 11.11 (Continued)

Year	Film composition	Flux	Rate and thickness	Substrates	Key results	Ref.
2003	BaM	Unknown	4 μm/h 25 to 100 μm/h	(111) MgO, (111) GGG, and *m*-plane, sapphire Al_2O_3	The deposition of thick and high-quality films of BaM on (111) gadolinium gallium garnet and *m*-plane, (1<u>1</u>00) or 10<u>1</u>0), sapphire (Al_2O_3) substrates. The total thickness ranged from 50 to 200 μm for a 2 h deposition. The ferrimagnetic resonance (FMR) linewidth (ΔH) was 68 Oe at 58 GHz for the BaM films on (111) GGG and 80 Oe at 59.9 GHz for the *m*-plane sapphire substrates.	[62]
2003	BaM	Unknown	60–200 μm	PLD BaM seed layers were grown onto *m*-plane (1<u>1</u>00) or (10<u>1</u>1) sapphire (Al_2O_3)	The LPE growth of thick barium hexaferrite films on *m*-planes (1<u>1</u>00) or (10<u>1</u>0) sapphire substrate is reported. The total film thickness ranged from 60 to 200 μm. The coercive field ranged from 7 to 80 Oe, and the saturation magnetization ($4\pi M_s$) was 4.42 kG. The ferrimagnetic resonance (FMR) linewidth (ΔH) at 59.9 GHz was 80 Oe.	[60]
2010	BaM	Unknown	4 μm	GaN/Al_2O_3	Thick barium hexaferrite films were epitaxially grown on the *c*-axis-oriented GaN/Al_2O_3 substrates by pulsed laser deposition followed by liquid-phase epitaxy. X-ray diffraction showed (0 0 2*n*) crystallographic alignment with pole figure analyses confirming epitaxial growth. Saturation magnetization, $4\pi M_s$, was measured for as-grown films to be 4.1 kG with a perpendicular magnetic anisotropy field of 16.3 kOe. Ferrimagnetic resonance linewidth, as peak-to-peak power absorption derivative, at 53 GHz was 86 Oe.	[69]

To this end, Chen et al. [69] deposited barium hexaferrite $BaFe_{12}O_{19}$ films, having the magnetoplumbite structure, on the *c*-axis-oriented GaN/Al_2O_3 substrates by pulsed laser deposition followed by LPE. X-ray diffraction showed (0 0 2*n*) crystallographic alignment with pole figure analyses confirming the epitaxial growth. High-resolution transmission electron microscopy images revealed magnetoplumbite unit cells stacked with limited interfacial mixing. Saturation magnetization, $4\pi M_s$, was measured to be 4.1 kG for as-grown films, with a perpendicular magnetic anisotropy field of 16.3 kOe. Ferrimagnetic resonance linewidth, as a peak-to-peak power absorption derivative, was measured at 86 Oe (at 53 GHz) [69]. This marked the first successful demonstration of ferrite growth on WBG substrates and provided the foundational steps necessary to move forward toward RF active–passive systems integration. Yu et al. built upon this work in a study of PLD growth of BaM films on WBG semiconductor heterostructures. BaM films that show device quality performance in structure, epitaxy, and magnetic properties were realized for BaM/MgO/AlN/SiC(X) where interface-engineered BaM – WBG semiconductors are proving enabling for 5G system-on-wafer solutions [140].

11.7 Outlook

Microwave and millimeter wave technologies are the foundation of next-generation commercial electronics, wireless communication, as well as commercial and military sensing and tracking technologies, among others. These technologies require advanced materials that provide broadband, high-power density, volumetric efficiency, and the ability to form higher level systems integration.

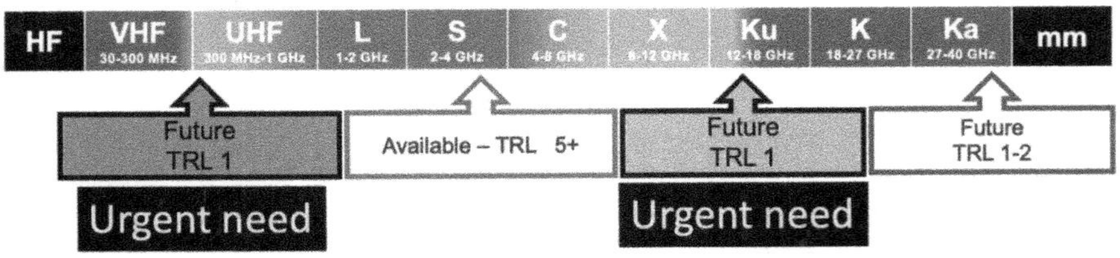

Figure 11.24 State of frequency selective limiter (FSL) technology (circa 2022).

In all instances, these materials must be thin or thick films of high-crystal-quality magnetoceramics of high permeability and permittivity with low losses. LPE is the most promising of film processing techniques that produce such materials.

Within this chapter, we have reviewed the LPE processing conditions of garnets, spinels and hexaferrites films.

With the availability of thick single-crystal YIG films, devices that could only be made previously with spheres, can now be made two dimensional with films. These include linearly dispersive delay lines and FSLs used in electronic countermeasure receivers and nondispersive delay lines used in phased-array antennas. The physical advantages of using planar YIG are the ease with which it can be grown on GGG and the ability to use integrated circuit methods rather than complicated sphere fabrication and mounting techniques.

Other emerging technologies driving the LPE YIG film growth include the fields of spintronics and magnonics [141, 142].

Applications requiring LPE-grown spinel and hexaferrite films are, at this time, few. Figure 11.24 provides a snapshot of FSL technology circa 2022. TRL in this context refers to technology readiness level. A TRL of 5+ indicates that commercial products are available. TRL of 1–2 indicates that the technology is at a research level. Those bands defined as *urgent* represent immediate technology needs. FSL development at frequencies above C band require LPE-grown spinels and hexaferrites. As yet, such materials do not possess the low FMR linewidths to meet these needs.

References

1 Nelson, H. (1963). Epitaxial growth from the liquid state and its application to the fabrication of tunnel and lasar diodes. *RCA Review* 24 (4): 603–615.

2 Stareev, G.D. (1976). A new method for liquid phase epitaxy. *Journal of Crystal Growth* 32 (2): 189–196.

3 Panish, M.B. and Sumski, S. (1969). Ga-Al-As: Phase, thermodynamic and optical properties. *Journal of Physics and Chemistry of Solids* 30 (1): 129–137.

4 Deitch, R.H. (1970). Liquid-phase epitaxial growth of gallium arsenide under transient thermal conditions. *Journal of Crystal Growth* 7 (1): 69–73.

5 Donahue, J.A. and Minden, H.T. (1970). A new technique for liquid phase epitaxy. *Journal of Crystal Growth* 7 (2): 221–226.

6 Vilms, J. and Garrett, J.P. (1972). The growth and properties of LPE GaAs. *Solid-State Electronics* 15 (4): 443–455.

7 Nelson, H. (1974). Liquid-phase epitaxy—its role in crystal growth technology. *Journal of Crystal Growth* 27: 1–5.

8 Longo, J.T., Harris, J.S., Jr., Gertner, E.R., and Chu, J.C. (1972). Improved surface quality of solution grown GaAs and Pb1− xSnxTe epitaxial layers: A new technique. *Journal of Crystal Growth* 15 (2): 107–116.

9 Mattes, B.L. and Route, R.K. (1972). A temperature gradient cell for liquid phase epitaxial growth of GaAs. *Journal of Crystal Growth* 16 (3): 219–222.

10 Minden, H.T. (1970). Constitutional supercooling in GaAs liquid phase epitaxy. *Journal of Crystal Growth* 6 (3): 228–236.

11 Crossley, I. and Small, M.B. (1973). Some observations of the surface morphologies of GaAs layers grown by liquid phase epitaxy. *Journal of Crystal Growth* 19 (3): 160–168.

12 Crossley, I. and Small, M.B. (1971). Computer simulations of liquid phase epitaxy of GaAs in Ga solution. *Journal of Crystal Growth* 11 (2): 157–165.

13 Jain, V.K. and Sharma, S.K. (1970). On the preparation of epitaxial films of III–V compounds. *Solid-State Electronics* 13 (8): 1145–1162.

14 Miki, H. and Otsubo, M. (1971). High purity GaAs crystals grown by liquid phase epitaxy. *Japanese Journal of Applied Physics* 10 (4): 509.

15 Linares, R.C. (1968). Epitaxial growth of narrow linewidth yttrium iron garnet films. *Journal of Crystal Growth* 3: 443–446.

16 Shick, L.K. and Nielsen, J.W. (1971). Liquid-phase homoepitaxial growth of rare-earth orthoferrites. *Journal of Applied Physics* 42 (4): 1554–1556.

17 Mauk, Michael G. (2015). 6 - Liquid-Phase Epitaxy, Editor(s): Thomas F. Kuech, In: *Handbook of Crystal Growth* (Second Edition), 225–316. North-Holland, ISBN 9780444633040.

18 Levinstein, H.J., Licht, S., Landorf, R.W., and Blank, S.L. (1971). Growth of high-quality garnet thin films from supercooled melts. *Applied Physics Letters* 19 (11): 486–488.

19 Hiskes, R. (1974). LPE growth and characterization of magnetic garnets grown in BaO-based solvents. *Journal of Crystal Growth* 27: 287–298.

20 Yoo, K.C., Storrick, R.P., Kramer, W.E., Stewart, A.M., and Hopkins, R.H. (1987). Investigation of growth conditions for the liquid phase epitaxy of hexaferrite films using a Bi2O3-BaO-B2O3 flux. *Journal of Crystal Growth* 85 (3): 389–395.

21 Van der Straten, P.J.M., Bondarenko, V.V., and Metselaar, R. (1981). LPE growth and magnetic anisotropy of Ni (Fe, Al) 2O4 films. *Journal of Crystal Growth* 51 (1): 119–130.

22 Hauptman, Z., Wanklyn, B.M., and Smith, S.H. (1973). Flux growth of Fe2TiO4 from BaO-B2O3 in an iron crucible. *Journal of Materials Science* 8 (12): 1695–1698.

23 Tolksdorf, W. (1973). Herstellung hexagonaler Ferrit-Einkristalle mit y-struktur aus schmelzflüssiger lösung. *Journal of Crystal Growth* 18 (1): 57–60.

24 Lemke, M., Hoppe, W., Tolksdorf, W., and Welz, F. (1979). Magnetically tunable millimetre-wave filter with single-crystal barium ferrite. *IEE Journal on Microwaves, Optics and Acoustics* 3 (6): 253–254.

25 Johansen, T.R., Hewitt, F.G., Torok, E.J., and Fleming, D.L. (1976, May). Lead free bismuth substituted garnet films by LPE. In: *AIP Conference Proceedings* (Vol. 29, No. 1, pp. 580–582). American Institute of Physics.

26 Chewpraditkul, W., Chewpraditkul, W.R., Yawai, N., Wantong, K., Kucera, M., Lucenicova, Z., and Nikl, M. (2018). Scintillation characteristics of GAGG: Ce single-crystalline films grown by liquid phase epitaxy. *IEEE Transactions on Nuclear Science* 65 (8): 2132–2135.

27 Prusa, P., Kucera, M., Mares, J.A., Onderisinova, Z., Hanus, M., Babin, V., ... Nikl, M. (2015). Composition tailoring in Ce-doped multicomponent garnet epitaxial film scintillators. *Crystal Growth & Design* 15 (8): 3715–3723.

28 Berkstresser, G.W. and Blank, S.L. (1981). LPE of LuLaSmGaFe garnet from a PbO-V2O5 flux. *Journal of Applied Physics* 52 (3): 2332–2334.

29 Blank, S.L., Hewitt, B.S., Shick, L.K., and Nielsen, J.W. (1973, January). Kinetics of LPE growth and its influence on magnetic properties. In: *AIP Conference Proceedings* (Vol. 10, No. 1, pp. 256–270). American Institute of Physics.

30 Giess, E.A., Argyle, B.E., Cronemeyer, D.C., Klokholm, E., McGuire, T.R., O'kane, D.F., ... Sadagopan, V. (1972, March). Europium-yttrium iron-gallium garnet films grown by liquid phase epitaxy on gadolinium gallium garnet. In: *AIP Conference Proceedings* (Vol. 5, No. 1, pp. 110–114). American Institute of Physics.

31 Van Der Straten, P.J.M. and Metselaar, R. (1977). Liquid phase epitaxial growth of indium substituted magnesium ferrite films. *Materials Research Bulletin* 12 (7): 707–715.

32 Glass, H.L. and Liaw, J.H.W. (1978). LPE growth of lithium ferrite on spinel substrate crystals. *Materials Research Bulletin* 13 (4): 353–359.

33 Robertson, J.M., Jansen, M., Hoekstra, B., and Bongers, P.F. (1977). Growth of spinel ferrite films by liquid phase epitaxy. *Journal of Crystal Growth* 41 (1): 29–35.

34 Smith, A., Kestigian, M., and Bekebrede, W. (1976). The temperature dependence of bubble-domain properties in ygdeu-, YGdEr-, YGDTM-, and YGdYb-Garnets. *IEEE Transactions on Magnetics* 12 (5): 579–582. doi: 10.1109/tmag.1976.1059089.

35 Knight, S., Hewitt, B.S., Rode, D.L., and Blank, S.L. (1974). Measurement of the diffusion coefficient of rare earth species in PbO-B2O3 flux used for liquid phase epitaxial growth of magnetic garnet films. *Materials Research Bulletin* 9 (7): 895–905.

36 Van Uitert, L.G., Grodkiewicz, W.H., and Dearborn, E.F. (1965). Growth of large optical-quality Yttrium and rare-earth aluminum garnets. *Journal of the American Ceramic Society* 48 (2): 105–108.

37 Chani, V.I. and Balbashov, A.M. (1984). Investigation of the phase diagram of the system PbO-B2O3-Bi2O3-Y2O3-Fe2O3-Ga2O3 for (YBi) 3 (FeGa) 5O12 crystal growth from flux. *Journal of Crystal Growth* 66 (3): 616–620.

38 Ito, S., Mikami, I., Sugita, Y., and Taniguchi, S. (1973). LPE films of bismuth-substituted bubble garnet. *IEEE Transactions on Magnetics* 9 (3): 460–463.

39 Nielsen, J.W. and Dearborn, E.F. (1958). The growth of single crystals of magnetic garnets. *Journal of Physics and Chemistry of Solids* 5 (3): 202–207.

40 Blank, S.L. and Nielsen, J.W. (1972). The growth of magnetic garnets by liquid phase epitaxy. *Journal of Crystal Growth* 17: 302–311.

41 Ghez, R. and Giess, E.A. (1974). The temperature dependence of garnet liquid phase epitaxial growth kinetics. *Journal of Crystal Growth* 27: 221–226.

42 Elwell, D., Morris, A.W., and Neate, B.W. (1972). The flux system BaO/Bi2O3/B2O3. *Journal of Crystal Growth* 16 (1): 67–70.

43 Wittekoek, S., Popma, T.J., Robertson, J.M., and Bongers, P.F. (1975). Magneto-optic spectra and the dielectric tensor elements of bismuth-substituted iron garnets at photon energies between 2.2–5.2 eV. *Physical Review B* 12 (7): 2777.

44 Robertson, J.M., Wittekoek, S., Popma, T.J., and Bongers, P.F. (1973). Preparation and optical properties of single crystal thin films of bismuth substituted iron garnets for magneto-optic applications. *Applied Physics* 2 (5): 219–228.

45 Akselrad, A., Novak, R.E., and Patterson, D.L. (1974). New bi-based garnet films for magnetic-bubble devices with magneto-optic applications. In: *MAGNETISM AND MAGNETIC MATERIALS — 1973: Nineteenth Annual Conference*. doi: 10.1063/1.3141858.

46 Shannon, R.D. (1976). Revised effective ionic radii and systematic studies of interatomic distances in halides and chalcogenides. *Acta Crystallographica Section A: Crystal Physics, Diffraction, Theoretical and General Crystallography* 32 (5): 751–767.

47 Besser, P.J. et al. (1971). "Film/Substrate Matching Requirements for Bubble Domain Formation in CVD Garnet Films," *AIP Conference Proceedings* No. 5, New York, American Institute of Physics, 1972, p. 125.

48 Besser, P.J., Mee, J.E., Elkins, P.E., and Heinz, D.M. (1971). A stress model for heteroepitaxial magnetic oxide films grown by chemical vapor deposition. *Materials Research Bulletin* 6 (11): 1111–1123. doi: 10.1016/0025-5408(71)90046-8.

49 Makino, H., Hibiya, T., and Matsumi, K. (1974). Misfit strain in LPE. In: *MAGNETISM AND MAGNETIC MATERIALS — 1973: Nineteenth Annual Conference*. doi: 10.1063/1.3141746.

50 Suemune, Y. and Inoue, N. (1974). Liquid phase epitaxial growth of garnets in BaO–BaF2–B2O3 solvents. *Japanese Journal of Applied Physics* 13 (1): 204.

51 Hiskes, R., Felmlee, T.L., and Burmeister, R.A. (1972). Growth of rare earth orthoferrites and garnets using BaO-based solvents. *Journal of Electronic Materials* 1 (4): 458–473.

52 Hewitt, B., Pierce, R., Blank, S., and Knight, S. (1973). Technique for controlling the properties of Magnetic Garnet Films. *IEEE Transactions on Magnetics* 9 (3): 366–372. doi: 10.1109/tmag.1973.1067614.

53 Davies, J.E., Giess, E.A., Kuptsis, J.D., and Reuter, W. (1976). Lead in holmium iron garnet films grown from fluxed melts by liquid phase epitaxy. *Journal of Crystal Growth* 36 (2): 191–197.

54 Stein, B.F. (1974, March). Growth of garnet films by liquid phase epitaxy. In: *AIP Conference Proceedings* (Vol. 18, No. 1, pp. 48–62). American Institute of Physics.

55 Kramer, W., Stewart, A., Hopkins, R., Kun, Z., Storrick, R., and Daniel, M. (1986). Investigation of hexagonal ferrite substrate and film growth. *IEEE Transactions on Magnetics* 22 (5): 981–983.

56 van der Straten, P.J.M. and Metselaar, R. (1978). Stress-induced anisotropy in LPE grown Ni (Fe, Al) 2O4 films. *Materials Research Bulletin* 13 (11): 1143–1151.

57 van der Straten, P.J.M. and Metselaar, R. (1980). Liquid phase epitaxial growth of lithium ferrite-aluminate films. *Journal of Crystal Growth* 48 (1): 114–120. doi: 10.1016/0022-0248(80)90200-6.

58 Plaskett, T.S., Shivashankar, S.A., Gilbert, B.L., Olson, B.L., Chien, C.L., and Herman, D.A., Jr. (1982). Synthesis of Mn-Zn ferrite from Na2O-B2O3 flux under CO-CO2 ambients. *Journal of Applied Physics* 53 (3): 2428–2430.

59 Robertson, J., Damen, J., and Algra, H. (1979). Liquid phase epitaxial growth and properties of spinel thin films. *IEEE Transactions on Magnetics* 15 (6): 1870–1872.

60 Yoon, S.D. and Vittoria, C. (2003). Microwave and magnetic properties of barium hexaferrite films having the c-axis in the film plane by liquid phase epitaxy technique. *Journal of Applied Physics* 93 (10): 8597–8599.

61 Stearns, F.S. and Glass, H.L. (1976). Liquid phase epitaxy of hexagonal ferrites and spinel ferrites on non-magnetic spinel substrates. *Materials Research Bulletin* 11 (10): 1319–1325.

62 Yoon, S.D. and Vittoria, C. (2003). Preparation of high-quality hexaferrite thick films by an improved liquid phase epitaxy deposition technique. *IEEE Transactions on Magnetics* 39 (5): 3163–3165.

63 Röschmann, P., Lemke, M., Tolksdorf, W., and Welz, F. (1984). Anisotropy fields and FMR linewidth in single-crystal Al, Ga and Sc substituted hexagonal ferrites with M structure. *Materials Research Bulletin* 19 (3): 385–392.

64 Dötsch, H., Mateika, D., Röschmann, P., and Tolksdorf, W. (1983). Growth and properties of epitaxial barium hexaferrite films. *Materials Research Bulletin* 18 (10): 1209–1216.

65 Glass, H.L. (1988). Ferrite films for microwave and millimeter-wave devices. *Proceedings of the IEEE* 76 (2): 151–158.

66 Wang, S.G., Yoon, S.D., and Vittoria, C. (2002). Microwave and magnetic properties of double-sided hexaferrite films on (111) magnesium oxide substrates. *Journal of Applied Physics* 92 (11): 6728–6732.

67 Yuan, M.S., Glass, H.L., and Adkins, L.R. (1988). Epitaxial barium hexaferrite on Sapphire by sputter deposition. *Applied Physics Letters* 53 (4): 340–341. doi: 10.1063/1.100601.

68 Yoon, S.D. and Vittoria, C. (2004). Thick M-type barium hexaferrite films grown on garnet substrates. *Journal of Applied Physics* 96 (4): 2131–2135.

69 Chen, Z., Yang, A., Mahalingam, K., Averett, K.L., Gao, J., Brown, G.J., ... Harris, V.G. (2010). Structure, magnetic, and microwave properties of thick Ba-hexaferrite films epitaxially grown on GaN/Al 2 O 3 substrates. *Applied Physics Letters* 96 (24): 242502.

70 Song, Y.Y., Das, J., Wang, Z., Tong, W., and Patton, C.E. (2008). In-plane c-axis oriented barium ferrite films with self-bias and low microwave loss. *Applied Physics Letters* 93 (17): 172503.

71 Kranov, Y.A., Abuzir, A., Prakash, T., McIlroy, D.N., and Yeh, W.J. (2006). Barium hexaferrite thick films made by liquid phase epitaxy reflow method. *IEEE Transactions on Magnetics* 42 (10): 3338–3340.

72 Dorfman, V.F., Petrushinina, S.A., and Shupegin, M.L. (1979). Growth kinetics and structure of epitaxial garnet layers as functions of the mismatch of crystal lattice parameters. *Thin Solid Films* 62 (2): 157–164.

73 Jonker, H.D. (1975). Investigation of the phase diagram of the system PbO-B2O3-Fe2O3-Y2O3 for the growth of single crystals of Y3Fe5O12. *Journal of Crystal Growth* 28 (2): 231–239.

74 HOOK, H.J. (1962). Phase relations in the ternary system FE2O3-feo-YFEO3. *Journal of the American Ceramic Society* 45 (4): 162–165. doi: 10.1111/j.1151-2916.1962.tb11112.x.

75 Menzer, G. (1928). Die Kristallstruktur der Granate: Zeitschr.

76 Strocka, B., Holst, P., and Tolksdorf, W. (1978). An empirical formula for the calculation of lattice constants of oxide garnets based on substituted yttrium-and gadolinium-iron garnets. *Philips Journal of Research* 33: 186–202.

77 Bertaut, F. and Forrat, F. (1956). Studying the compounds of rare earth oxides with aluminum and gallium. *CR Acad. Sci. Paris* 243: 1219–1222.

78 Harris, V.G. (2011). Modern microwave ferrites. *IEEE Transactions on Magnetics* 48 (3): 1075–1104.

79 Winkler, G., Hansen, P., and Holst, P. (1972). Variation of magnetic material parameters and lattice-constants of polycrystalline yttrium-iron garnet by incorporation of nonmagnetic ions. *Philips Research Reports* 27 (2): 151.

80 Kramers, H.A. (1934). L'interaction Entre Les Atomes Magnétogènes dans un Cristal Paramagnétique. Physica 1 (1–6): 182–192. doi: 10.1016/s0031-8914(34)90023-9.

81 Anderson, P.W. (1950). Antiferromagnetism. Theory of superexchange interaction. Physical Review 79 (2): 350.

82 Néel, L. (1948). Propriétés magnétiques des ferrites; ferrimagnétisme et antiferromagnétisme. In: Annales de physique (Vol. 12, No. 3, pp. 137–198).

83 Harrison, G.R. and Hodges, L.R., Jr. (1961). Microwave properties of polycrystalline hybrid garnets. *Journal of the American Ceramic Society* 44 (5): 214–220.

84 Vassiliev, A., Nicolas, J., and Hildebrant, M. (1961). In: *Handbook of Ferromagnetic Materials*, Vol. 252, 2529. essay. C.R. Acad. Sci.

85 Goldring, G., Schieber, M., and Vager, Z. (1960). On the incorporation of neodymium in mixed Yttrium-Neodymium-Iron garnets. *Journal of Applied Physics* 31 (11): 2057–2059.

86 von Aulock, W.H., Boxer, A.S., Ollom, J.F., and Rauchmiller, R.F. (1965). Handbook of Microwave Ferrite Materials. Edited by Wilhelm H. Von Aulock. Contributors: Arnold S. Boher. John F. Ollom, Robert F. Rauchmiller. Academic Press.

87 Choi, D.Y. and Chung, S.J. (1998). Annealing behaviors of lattice misfit in YIG and La-doped YIG films grown on GGG substrates by LPE method. Journal of Crystal Growth 191 (4): 754–759.

88 Glass, H.L., Liaw, J.H.W., and Elliott, M.T. (1977). Temperature stabilization of ferrimagnetic resonance field in epitaxial YIG by Ga, La substitution. Materials Research Bulletin 12 (7): 735–740.

89 Shick, L., Nielsen, J., Bobeck, A., Kurtzig, A., Michaelis, P., and Reekstin, J. *Liquid phase epitaxial growth of uniaxial garnet films: Circuit Deposition and bubble propagation, J. Appl. Phys., vol 42, 1277 (1971).*

90 Kurtzig, A. and Hagedorn, F. (1971). Noncubic magnetic anisotropies in bulk and thin-film garnets. IEEE Transactions on Mag*netics* 7 (3): 473–476. doi: 10.1109/tmag.1971.1067073.

91 Akselrad, A. and Callen, H. (1971). Growth-induced noncubic anisotropy arising from the tetrahedral sites in garnets. Applied Physics Letters 19 (11): 464–466.

92 Cronemeyer, D.C., Giess, E.A., Klokholm, E., Argyle, B.E., and Plaskett, T.S. (1972, March). Annealing of LPE garnet films. In: AIP Conference Proceedings (Vol. 5, No. 1, pp. 115–119). American Institute of Physics.

93 LeCraw, R.C., Gyorgy, E.M., and Wolfe, R. (1974). Suppression of hard bubbles in LPE garnet films by inert atmosphere annealing. *Applied Physics Letters* 24 (11): 573–574.

94 Wemple, S.H., Blank, S.L., Seman, J.A., and Biolsi, W.A. (1974). Optical properties of epitaxial iron garnet thin films. *Physical Review B* 9 (5): 2134–2144. doi: 10.1103/physrevb.9.2134.

95 Glass, H.L. (1976). Growth of thick single-crystal layers of yttrium iron garnet by liquid phase epitaxy. *Journal of Crystal Growth* 33 (1): 183–184. doi: 10.1016/0022-0248(76)90099-3.

96 Blank, S., Biolsi, W., and Nielsen, J. (1977). The effect of melt composition on the Curie temperature and flux spin-off from lutetium containing LPE garnet films. *IEEE Transactions on Magnetics* 13 (5): 1095–1097. doi: 10.1109/tmag.1977.1059546.

97 Giess, E.A. and Guerci, C.F. (1977). LPE growth kinetics of CaGe-Substituted EuTm2Fe5O12 garnet films. *Journal of Crystal Growth* 42: 328–333.

98 Moore, E.B., Calhoun, B.A., and Lee, K. (1978). Thickness dependence of coercivity in LPE Garnet Films. *Journal of Applied Physics* 49 (3): 1879–1881. doi: 10.1063/1.324791.

99 Bobeck, A. H., Spencer, E. G., Van Uitert, L. G., Abrahams, S. C., Barns, R. L., Grodkiewicz, W. H., Sherwood, R. C., Schmidt, P. H., Smith, D. H., and Walters, E. M. (1970). Uniaxial magnetic garnets for domain wall "bubble" devices", *Appl. Phys. Lett.* 17, 131 https://doi.org/10.1063/1.1653335.

100 Andrew H. Bobeck and H. E. D. Scovil, MAGNETIC BUBBLES. *Scientific American*, Vol. 224, No. 6 (June 1971), pp. 78-91

101 Nielsen, J.W. (1979). Magnetic bubble materials. *Annual Review of Materials Science* 9 (1): 87–121.

102 Texas Instruments Tib0203 bubble memory. (n.d.). Retrieved February 24, 2022, from http://www.decodesystems.com/tib0203.html.

103 Olson, F.A. and Yaeger, J.R. (1965). Microwave delay techniques using YIG. *IEEE Transactions on Microwave Theory and Techniques* 13 (1): 63–69.

104 Damon, R.W. and van de Vaart, H. (1965). Dispersion of spin waves and magnetoelastic waves in YIG. *Proceesings of the IEEE* 53 (4): 348–354.

105 Baek, J.S., Ha, S.Y., Lim, W.Y., and Lee, S.H. (2004). Magnetostatic wave study of YIG films grown by the LPE method. *Physica Status Solidi (A)* 201 (8): 1806–1809.

106 Adam, J.D. and Collins, J.H. (1976). Microwave magnetostatic delay devices based on epitaxial yttrium iron garnet. *Proceedings of the IEEE* 64 (5): 794–800.

107 Grant, P., Adam, J., and Collins, J. (1974). Surface wave device applications in microwave communication systems. *IEEE Transactions on Communications* 22 (9): 1410–1419.

108 Skyes, C.G., Adam, J.D., and Collins, J.H. (1976). Magnetostatic wave propagation in periodic structures. *Appl. Phys. Left.* 29: 388–391.

109 Gillette, S.M., Geiler, M., Adam, J.D., and Geiler, A.L. (2018, July). Ferrite-based reflective-type frequency selective limiters. In: *2018 IEEE MTT-S International Microwave Workshop Series on Advanced Materials and Processes for RF and THz Applications (IMWS-AMP)* (pp. 1–3). IEEE.

110 Adam, J.D. and Stitzer, S.N. (1993). Frequency selective limiters for high dynamic range microwave receivers. *IEEE Transactions on Microwave Theory and Techniques* 41 (12): 2227–2231.

111 Yu, X.L., Jiang, Z.Y., Yang, D.J., Wei, D.B., Yang, Q., 2012. Precipitation Behavior of Magnetite in Oxide Scale during Cooling of Microalloyed Low Carbon Steel. *Advanced Materials Research*, 572, 249–254. https://doi.org/10.4028/www.scientific.net/amr.572.249

112 Hilpert, S. (1909). Correspondence as to structure and origin in magnetic properties of ferrite and iron oxide. *Berichte der Deutschen Chemischen Gesellschaft* 42: 2248–2261.

113 Forestier, H. (1928). Magnetic transformation of the sesquioxid of iron, of its solid solution and of its ferromagnetic combinations (in Fr.). *Liebigs Annalen der Chemie.* 9: 316.

114 Kato, Y. and Takei, T., 1932. Manufacturing method of ferrite coves. *Jpn.), Japan Pat*, (98), p.844.

115 Snoek, J.L. (1936). Magnetic and electrical properties of the binary systems MO. Fe_2O_3. *Physica* 3 (6): 463–483.

116 T., F. C. (1948). New developments in ferromagnetic materials. *Nature* 161 (4096): 666–666. doi: 10.1038/161666b0.

117 Sugimoto, M. (1999). The past, present, and future of ferrites. *Journal of the American Ceramic Society* 82 (2): 269–280.

118 Konig, U. (1975). Improved manganese-zinc ferrites for power transformers. *IEEE Transactions on Magnetics* 11 (5): 1306–1308.

119 Smit, J. and Wijn, H.P.J. (1959). Ferrites, 149.

120 Smit, J. and Wijn, H.P.J. (1959). Ferrites, 157.

121 Smit, J. and Wijn, H.P.J. (1959). Ferrites, 174.

122 Cornell, R.M. and Schwertmann, U. (2007). *The Iron Oxides: Structure, Properties, Reactions, Occurrences and Uses.* Wiley-VCH. ISBN 978-3-527-60644-3.

123 White, G.O. and Patton, C.E. (1978). Magnetic properties of lithium ferrite microwave materials. *Journal of Magnetism and Magnetic Materials* 9 (4): 299–317. doi: 10.1016/0304-8853(78)90085-9.

124 Assar, S.T., Abosheiasha, H.F., and El Nimr, M.K. (2014). Comparison study of the magnetic permeability and dc conductivity of Co–Ni–Li ferrite nanoparticles and their bulk counterparts. *Journal of Magnetism and Magnetic Materials* 354: 1–6.

125 *High Frequency Ni-Zn Ferrite*. www.magneticsgroup.com. (2019). https://www.magneticsgroup.com/wp-content/uploads/2019/09/N40-ISO-WEB-DATA.pdf. Accessed Aug. 19, 2022.

126 *General Purpose Mn-Zn Ferrite*. https://www.magneticsgroup.com. (2019). https://www.magneticsgroup.com/wp-content/uploads/2019/09/MN30-ISO-WEB-DATA.pdf. Accessed Aug. 19, 2022.

127 Van der Straten, P.J.M. and Metselaar, R. (1980). LPE growth of Mn, Ni-and Al-substituted copper ferrite films. *Journal of Applied Physics* 51 (6): 3236–3240.

128 Pullar, R.C. (2012). Hexagonal ferrites: A review of the synthesis, properties and applications of hexaferrite ceramics. *Progress in Materials Science* 57 (7): 1191–1334.

129 Smit, J. and Wijn, H.P.J. (1959). Ferrites, 204–206.

130 Li, Qifan, Chen, Yajie, Yu, Chengju, Young, Lance, Spector, Jonah, and Harris, Vincent G. (2022). Emerging magnetodielectric materials for 5G communications: 18H hexaferrites. *Acta Materialia*. Volume 231, 2022,117854

131 Goto, Y. and Takada, T. (1960). Phase diagram of the system BaO-Fe2O3. *Journal of the American Ceramic Society* 43 (3): 150–153.

132 Chandel, M., Singh, V.P., Jasrotia, R., Singha, K., and Kumar, R. (2020). A review on structural, electrical and magnetic properties of Y-type hexaferrites synthesized by different techniques for antenna applications and microwave absorbing characteristic materials. *AIMS Materials Science* 7 (3): 244–268.

133 Harris, V.G. and Sokolov, A.S. (2019). The self-biased circulator: Ferrite materials design and process considerations. *Journal of Superconductivity and Novel Magnetism* 32 (1): 97–108.

134 Sui, X., Scherge, M., Kryder, M.H., Snyder, J.E., Harris, V.G., and Koon, N.C. (1996). Barium ferrite thin-film recording media. *Journal of Magnetism and Magnetic Materials* 155 (1–3): 132–139.

135 Gorbachev, E.A., Trusov, L.A., Sleptsova, A.E., Kozlyakova, E.S., Alyabyeva, L.N., Yegiyan, S.R., Prokhorov, A.S., Lebedev, V.A., Roslyakov, I.V., Vasiliev, A.V., and Kazin, P.E. (2020). Hexaferrite materials displaying ultra-high coercivity and sub-terahertz ferromagnetic resonance frequencies. *Materials Today* 32: 13–18. doi: 10.1016/j.mattod.2019.05.020.

136 Smit, J. and Wijn, H.P.J. (1959). Ferrites, 204.

137 Smit, J. and Wijn, H.P.J. (1959). Ferrites, 210.

138 Stearns, F.S. and Glass, H.L. (1975). Liquid phase epitaxy of hexagonal ferrites. *Materials Research Bulletin* 10 (11): 1255–1258.

139 Graetsch, H., Haberey, F., Leckebusch, R., Rosenberg, M., and Sahl, K. (1984). Crystallographic and magnetic investigation on W-type-hexaferrite single crystals in the solid solution series SrZn 2-x Co x Fe 16 O 27. *IEEE Transactions on Magnetics* 20 (3): 495–500.

140 Yu, C., Andalib, P., Sokolov, A., Fitchorova, O., Liang, W., Beam, E., and Harris, V.G. (2021). Interface-engineered barium magnetoplumbite–wide-bandgap semiconductor integration enabling 5G system-on-wafer solutions for full-duplexing phased arrays. *Applied Physics Letters* 119 (5): 051906.

141 Qin, H., Both, G.J., Hämäläinen, S.J., Yao, L., and van Dijken, S. (2018). Low-loss YIG-based magnonic crystals with large tunable bandgaps. *Nature Communications* 9 (1): 1–10.

142 Chumak, A.V. and Serga, A.A. (2015). Magnon spintronics. *Nature Physics* 11: 453.

12

Ferrite-Based Electronic Bandgap Heterostructures and Metamaterials

Martha Pardavi-Horvath

Department of Electrical and Computer Engineering, School of Engineering and Applied Science, The George Washington University, Washington, DC 20052, USA

12.1 Introduction

Metamaterials are man-made media, not found in nature, but built bottom up from components, and designed to have specific properties, unique to the arrangement of the components. Electromagnetic metamaterials are usually associated with engineered double-negative composite media, fabricated into a periodic structure. Thus, metamaterials are imitating nature by crafting artificial periodic lattice structures, similar to atoms arranged in a crystal lattice. The advancement of materials technology made it possible to create one-dimensional (1D), two-dimensional (2D), or three-dimensional (3D) periodic artificial structures with amazing electromagnetic properties, making it possible to greatly extend our fundamental understanding of natural phenomena. Based on this new knowledge, novel devices for a broad range of applications in modern technology can be designed. The interest to metamaterials is motivated not only by their fascinating properties, but the potential applications in information technology, communication, space, security and defense, health and environmental industry, as sensors, new type of laser and LED cavities, superlensing, and cloaking. All this is made possible by controlling electromagnetic wave (EMW) propagation in metamaterials.

The properties of electromagnetic metamaterials originate in the spatial arrangement of the constituents, without scaling limitations of size, spectral range, or dimensionality, in the sense that such periodic arrangements of components can be designed for any frequency, based on the knowledge of the equation of motion of components, i.e. Maxwell's equations with the constituent relationships, and the Landau–Lifshitz equation of the motion of magnetization.

12.1.1 Classification of Electromagnetic Metamaterials

The most important classes of metamaterials, relevant to electromagnetics, are the bandgap structures, i.e. photonic/magnetophotonic crystals (MPCs), and the double-negative (left-handed) media.

Photonic crystals or photonic bandgap (PBG) materials are periodic structures, made of dielectrics, with spatially varying dielectric constants. The periodicity of PBG is on the order of the photonic wavelength [1–3].

The original idea was to design an artificial crystal where the photons behave the same way as electrons in the usual atomic arrangements of crystal lattice, resulting in a band structure, depending on the novel crystal structure and the electrodynamic properties of the nodes. Thus, one can create an optical bandgap, due to the properties of the periodic potential of the arrangements of dielectric components with different dielectric constants. The photonic band structure and bandgaps are very similar to the concept of semiconductor band structure. In the frequency range of the PBG, the Bragg scattering at the dielectric interfaces produces effects analogous to semiconductor forbidden bandgaps, i.e. no light in that frequency range can enter the crystal. If a defect center is created in the photonic crystal then a defect level, i.e. a localized photonic state, emerges in the bandgap, acting as a dopant in the regular semiconductor, making it possible to manipulate light, i.e. the electromagnetic radiation. For optical communication, and information technology PGBs promised a revolutionary future. By the 80s, materials' technology achieved the level to fabricate such structures having a periodicity on the order of 1 μm for optical communication in the crucial 1–1.5 μm optical communication range.

If one of the components or the defects in the PBG is magnetic, a powerful new degree of freedom is added to the control mechanism of the PBG. In this case, the material is a magneto-PBG structure or an MPC [4]. MPCs manifest unique electromagnetic properties, and greatly enhanced magneto-optical effects, due to the slow light in the PBG. Moreover, the propagation of the light (EMWs) can be controlled by a bias magnetic field.

Modern Ferrites: Basic Principles, Processing and Properties, Volume 1, First Edition. Edited by Vincent G. Harris.

The idea of PBGs and MPCs is not limited to the optical spectrum. There is no limitation on the operating frequency. Metamaterials can be built of arbitrary period, from macroscopic down to atomic arrangements, depending on the application wavelength. This means that all these metamaterials can be treated as a single category of electromagnetic bandgap structures. Properties of ferrite-based MPC structures and devices will be described in Sections 12.2.1 and 12.4.

The term metamaterial is frequently used to describe another important class of electromagnetic composite structures, the so-called left-handed, or double-negative, or negative index media (NIM). The idea and the potential consequences of the possibility of finding a material having simultaneously both negative permittivity ($\varepsilon < 0$) and permeability ($\mu < 0$) belong to Veselago, who predicted the existence of the backward waves, having oppositely directed group and phase velocities, anomalous (left-handed) refraction at the interface, and anomalous high transparency by these materials [5]. The formal consequence of both $\varepsilon < 0$ and $\mu < 0$ is that the sign of the refractive index, $n = \pm\sqrt{\varepsilon\mu}$, is chosen as negative. The practical realization of such materials is based on one conducting component (semiconductor, metal, wires, etc.), providing the negative permittivity, while the other component is magnetic (special metallic structure or ferrite) contributing the negative permeability. The negative permittivity is the consequence of the dispersion of $\varepsilon = \varepsilon(\omega)$:

$$\varepsilon(\omega) \approx \varepsilon_0(\omega) - \left[\omega_p^{\,2}/\omega^2\right]$$

where $\omega_p = ne^2/m^*$ is the plasma frequency of the conduction electrons and m^* is the "effective mass" of the electrons. Well below the plasma frequency ($\omega_p \sim 10^{15}$ Hz), but above the static limit, the permittivity is negative, $\varepsilon(\omega) < 0$, a classical "text-book" result [6].

For a double-negative material, the permeability should be negative, $\mu(\omega) < 0$, in the frequency range overlapping with $\varepsilon(\omega) < 0$. Fortunately, for ferro- or ferrimagnetic materials, including ferrites, the dispersion of the magnetic permeability above the ferromagnetic resonance (FMR) frequency provides $\mu(\omega) < 0$, the necessary mechanism for left-handed behavior in the very important microwave frequency range [7].

The field of metamaterials combines concepts from solid-state physics, electrodynamics, and materials science. In the following Section 12.2, a short theoretical background will be given to describe the most important characteristics of metamaterials, followed by a review of the state of art in material and device development.

12.2 Fundamentals of Electrodynamics of Metamaterials

The most important classes of electromagnetic metamaterials, the MPCs and the double-negative media (NIM) represent qualitatively different behavior: MPCs show bandgap behavior for short, optical wavelengths (high frequencies), i.e. $\lambda < a$, while NIM show effectively continuous behavior at microwave frequencies for wavelengths much larger than the periodicity of the structure $\lambda > a$ [8]. The description of the basic concepts and relevant terms are introduced in the following section.

The foundation of the electrodynamics of metamaterials lies in Maxwell's equations, describing the space- and time-dependent behavior of electric and magnetic fields in media, in terms of partial differential equations:

$$\nabla \times \boldsymbol{E}(\boldsymbol{r},t) = -\frac{\partial}{\partial t}\boldsymbol{B}(\boldsymbol{r},t) \tag{12.1}$$

$$\nabla \times \boldsymbol{H}(\boldsymbol{r},t) = \frac{\partial}{\partial t}\boldsymbol{D}(\boldsymbol{r},t) + \boldsymbol{J}(\boldsymbol{r},t) \tag{12.2}$$

$$\nabla \cdot \boldsymbol{D}(\boldsymbol{r},t) = \rho(\boldsymbol{r},t) \tag{12.3}$$

$$\nabla \boldsymbol{B}(\boldsymbol{r},t) = 0 \tag{12.4}$$

All quantities are functions of space ($\boldsymbol{r}$) and time (t), and the spatial and temporal differentiations are denoted as ∇ and $\partial/\partial t$, respectively. The quantities in these equations and their units in the SI system are the following: $\boldsymbol{E}$ is the electric field vector (V/m), $\boldsymbol{H}$ the magnetic field vector (A/m), $\boldsymbol{D}$ the electric flux density (displacement) vector (A·s/m^2), $\boldsymbol{B}$ the magnetic flux density (induction) vector [V·s/m^2 or tesla (T)], $\boldsymbol{J}$ is the current density vector (A/m^2), and ρ is the scalar charge density (A·s/m^2). As all quantities are physical and observable, they are real valued.

To fully describe the interaction of electromagnetic fields, including light, with matter, Maxwell's equations have to be supplemented by the constitutive relations. They relate the flux densities to the fields as $\boldsymbol{D} = \boldsymbol{D}(\boldsymbol{E}, \boldsymbol{H})$ and $\boldsymbol{B} = \boldsymbol{B}(\boldsymbol{E}, \boldsymbol{H})$ through the material properties. In vacuum, the constitutive relations are simple:

$$D(\mathbf{r},t) = \varepsilon_0 \boldsymbol{E}(\boldsymbol{r},t) \text{ and } \boldsymbol{B}(\boldsymbol{r},t) = \mu_0 \boldsymbol{H}(\boldsymbol{r},t) \tag{12.5}$$

with the free-space permittivity $\varepsilon_0 = 8.854 \times 10^{-12}$ V·s/A·m and permeability $\mu_0 = 4\pi \times 10^{-7}$ A·s/V·m, and c is the speed of light in vacuum (c = 299 792 458 m/s).

Polarization of the matter is included as electric polarization $\boldsymbol{P}(\boldsymbol{r},t)$ and magnetization $\boldsymbol{M}(\boldsymbol{r},t)$. The constitutive relations become:

$$\boldsymbol{D}(\boldsymbol{r},t) = \varepsilon_0 \boldsymbol{E}(\boldsymbol{r},t) + \boldsymbol{P}(\boldsymbol{r},t) \text{ and } \boldsymbol{B}(\boldsymbol{r},t) = \mu_0[\boldsymbol{H}(\mathbf{r},t) + \boldsymbol{M}(\boldsymbol{r},t)] \tag{12.6}$$

For harmonic fields of angular frequency ω, Maxwell's equations in the frequency domain will be simplified to:

$$\nabla \times \boldsymbol{E}(\boldsymbol{r}) = -j\omega \boldsymbol{B}(\mathbf{r}) \tag{12.7}$$

$$\nabla \times \boldsymbol{H}(\boldsymbol{r}) = j\omega \boldsymbol{D}(\boldsymbol{r}) + \boldsymbol{J}(\boldsymbol{r}) \tag{12.8}$$

$$\nabla \boldsymbol{D}(\boldsymbol{r}) = \rho(\boldsymbol{r}) \tag{12.9}$$

$$\nabla \boldsymbol{B}(\boldsymbol{r}) = 0 \tag{12.10}$$

Assuming causality (future events do not influence polarizations) and all quantities to be linear (directly proportional to the fields) and local (the flux at point $\boldsymbol{r}$ depends only on the fields at the same point), we introduce the temporal dispersion of these quantities, i.e. they depend on the frequency as:

$$\boldsymbol{P}(\boldsymbol{r},\omega) = \varepsilon_0 \chi_e(\boldsymbol{r},\omega)\boldsymbol{E}(\boldsymbol{r},\omega) \quad \textit{and} \quad \boldsymbol{M}(\boldsymbol{r},\omega) = \chi_m(\boldsymbol{r},\omega)\boldsymbol{H}(\boldsymbol{r},\omega) \tag{12.11}$$

where the electric and magnetic susceptibilities $\chi_{e,m}$ are defined as (R is a response function):

$$\chi_{e,m}(\boldsymbol{r},\omega) = \int_{-\infty}^{\infty} R_{e,m}(\boldsymbol{r},t) e^{j\omega\tau} d\tau \tag{12.12}$$

If there is a coupling to both $\boldsymbol{E}$ and $\boldsymbol{H}$, the constitutive relations are becoming very complicated. However, the knowledge of the constitutive relations is essential for the solution of Maxwell's equations. In the case of nanostructured metamaterials and high power devices, the inclusion of nonlinearities, i.e. higher order terms in the field response, is unavoidable [9].

12.2.1 Magnetophotonic Electromagnetic Bandgap Materials

Metamaterials with artificial periodic modulation of the refractive index, where the electromagnetic wavelength λ is comparable to the characteristic size/periodicity of the structure a, i.e. $\lambda/a \cong 1$, are called electromagnetic bandgap structures, or electromagnetic crystals (EMCs). EMCs with magnetic components have extra degrees of freedom and provide a convenient way for controlling the EM properties of such an EMC. Periodic metamaterials having a magnetic component are called MPCs, and the field of studying and designing of those based on the properties of specific magnetodynamic modes, i.e. spin waves, is called magnonics [4, 10, 11].

As introduced in Section 12.1.1, the periods of these bandgap crystals, or super heterostructures, are comparable with the wavelengths of incident electromagnetic radiation in the microwave or optical range. This fact leads to band structure formation in the normal EMW spectrum, including the presence of bands and forbidden bandgaps, where mode propagation is forbidden for selected, or for any, polarization. The last case corresponds to a complete bandgap. These peculiarities in the EMW propagation through EM crystals can be observed via measurements of transmission spectra. Defect states in the forbidden bandgap can be created by inserting a (different) magnetic component into the periodic MPC or EMC. The resulting structure exhibits unique optical and magneto-optical response in the optical range. There is a strong photon confinement at a magnetic defect what can be utilized to enhance the magneto-optical response and nonlinearities of such materials. It is shown theoretically that the Faraday rotation becomes anomalously large and exhibits extraordinary behavior near the frequencies of the extraordinary optical transmittance through optically thick perforated metal film with holes filled with a magneto-optically active material. This phenomenon is explained as result of strong confinement of the evanescent electromagnetic field within the magnetic material, which occurs due to excitation of the coupled plasmon–polaritons on the opposite surfaces of the film [12]. By applying an external bias magnetic field to an MPC, the mutual orientation between the light wave vector $\boldsymbol{k}$ and the magnetization vector $\boldsymbol{M}$ can be controlled, and it is possible to change the light polarization state from elliptically polarized modes to circularly polarized modes (if $\boldsymbol{k} \parallel \boldsymbol{M}$), and to TE and TM modes (if $\boldsymbol{k} \perp \boldsymbol{M}$, and $\boldsymbol{M}$ is perpendicular to the EMW propagation plane).

Magnetic materials possess a unique electromagnetic property of nonreciprocity, the asymmetry of EM waves propagating in opposite directions in a magnetic medium. Most natural materials are reciprocal, with the exceptions of magnetic materials, first of all ferrites. For example, nonreciprocity enables the separation of the transmitter and the receiver in communication. Ferrites are essential for nonreciprocal devices, as it was shown in Chapter 12 of this book. MPCs have greatly enhanced nonreciprocal magneto-optical and microwave effects.

The influence of magnetic permeability μ on the EMCs constructed from ferrites was considered theoretically in [13]. The dielectric permittivity ε and magnetic permeability μ were taken into account as scalar values. The wave equation for description of the propagation of EMWs through ferrite PCs was presented in the following form:

$$\nabla \times \left[\varepsilon^{-1} \times (\mu^{-1}\boldsymbol{B})\right] = \frac{\omega^2}{c^2}\mathrm{B} \tag{12.13}$$

where $\boldsymbol{B}$ is the magnetic induction and $\nabla \boldsymbol{B} = 0$. Similar equation holds for $\boldsymbol{D}$ vector, ω is the frequency of the plane EMW, and c is the velocity of light in vacuum. In general, for magnetic materials, both the dielectric permittivity and magnetic permeability are tensor quantities. The permeability and permittivity tensors for reciprocal materials are symmetric. The measure of the nonreciprocity is in the asymmetric parts of these tensors. In fact, in modeling and theoretical calculations, this condition is usually ignored.

The band structure was calculated for several cases of 2D and 3D structures. The existence of the bandgap and its dependence on the relative values of ε and μ was shown. In the case where both ε and μ have their maximum values in the same material of the composite structure, the photonic bandgaps tend to disappear. However, in the opposite case, where ε and μ have their maximum values in the different components of the EMC, the gaps become wider.

A more detailed theoretical analysis of the 1D MPC with a scalar model for ε and μ was offered in [14]. It is based on a 1D array of two-component alternating layers with thicknesses d_a and d_b and a period $D = d_a + d_b$, characterized by dielectric permittivities ε_a and ε_b, magnetic permeabilities μ_a and μ_b, respectively. The exact positions for the band structure were found as the solution of the dispersion equation. Generally speaking, for magnetic materials, this is valid only for the transversal magneto-optical configuration, when EMWs can be presented as TE and TM modes. As is mentioned earlier, both the dielectric permittivity and the magnetic permeability are tensor quantities and the off-diagonal components of these tensors are determined by the magnetization vector or the dc magnetic field. The bandgaps in the normal EMW spectra for a 1D PC can then be determined from the solution, as presented in [14].

An electrically controlled nonreciprocal electromagnetic bandgap material is proposed and studied in [15]. The material is a periodic 3D regular lattice of 3 mm periodicity of small magnetized $r = 1$ mm ferrite spheres. The propagation of plane EMW was considered, and an analytical solution for plane wave reflection from a planar interface is also presented. In the proposed material, a new electrically controlled stop band appears for one of the two circularly polarized eigenwaves in a frequency band around the ferrimagnetic resonance frequency. This frequency can be well below the usual lattice bandgap, which allows the realization of rather compact structures. It was shown that for right circular polarizations, an additional bandgap, corresponding to the FMR, is found, while there is no gap for left circular polarizations.

Magnetic materials respond to and interact with electromagnetic radiation in the microwave range via spin waves. Depending on the dominating interaction, those can be associated with longer magnetostatic (dipolar) or the high-wavenumber-exchange-dominated spin waves. Spin waves interact with the EMWs and contribute to the transmission and reflection phenomena, especially in the range of resonance frequencies. The excitation of spin waves at band edges opens new ways of information processing, and a novel type of bandgap structures can be designed based on magnonics [11].

The overview of some typical ferrite based bandgap structures and devices, operating in the microwave range, are presented in Section 12.2.2.

12.2.2 Negative Index Metamaterials (NIM)

In contrast to EM bandgap structures, treated as a periodic lattice with a characteristic wavelength, comparable to the lattice periodicity, negative index NIM (double-negative, left-handed) metamaterials have a periodic modulation of the refractive index, but in this case, the electromagnetic wavelength λ is much larger than the characteristic size/periodicity of the structure a, i.e. $\lambda/a > 1$. As a result, negative index metamaterials (NIM) can be considered as effectively continuous media, i.e. the NIM can be treated as the long wavelength limit of the discrete bandgap materials.

The simplest case of an electromagnetically continuous medium is isotropic, and it is characterized by two scalar quantities, $\varepsilon = \text{const}$ and $\mu = \text{const}$, relating the fields $\boldsymbol{E}$ and $\boldsymbol{H}$, and displacement $\boldsymbol{D}$, $\boldsymbol{B}$ induction, and $\boldsymbol{M}$ magnetization vectors as:

$$\boldsymbol{D} = \varepsilon \boldsymbol{E}, \boldsymbol{B} = \mu \boldsymbol{H} = \mu_0 (H + M) \tag{12.14}$$

The plane wave solution of Maxwell's equations in a homogeneous and source-free ($\boldsymbol{J} = 0$, $\rho = 0$) space can be written for a plane wave as:

$$\frac{d}{dz}\mathbf{E}(z) + \omega^2 \varepsilon\mu \mathbf{E}(z) = 0 \tag{12.15}$$

By defining the wave vector (propagation constant) as $k = \omega\sqrt{(\varepsilon\mu)} = \omega n$, where n is the index of refraction of the medium, the solutions are:

$$\boldsymbol{E}(\boldsymbol{r}) = \boldsymbol{E}_{1,2} e^{\mp jkz} \tag{12.16}$$

These solutions are transverse EMWs, orthogonal to the z-axis. For a lossless medium, μ and ε are real numbers. But if these two parameters have different signs, then k might be either real or pure imaginary! For a real-valued k, the waves in Equation (12.15) undergo a phase shift, and the EMW propagating in the z-direction has a wavelength of:

$$\lambda_{\text{ph}} = \frac{2\pi}{\omega\sqrt{\varepsilon\mu}} = \frac{2\pi c}{\omega n}, \tag{12.17}$$

i.e. the phase wavelength λ_{ph} becomes shorter for high-index materials. The phase velocity v_{ph} is defined as $v_{\text{ph}} = \omega / k$ for a given EMW of frequency ω.

The index of refraction is defined as $n = c/v_{\text{ph}}$, i.e. the ratio of the phase velocity in the medium with respect to the velocity of light, c. In contrast, the group velocity v_g is the velocity of the envelope of the wave packet; this is the rate how the information (energy) is propagating, defined as $v_g = \partial\omega / \partial k$.

For lossy materials, μ and ε are complex numbers, with the imaginary part causing an exponential decay of the amplitude of the EMW, propagating in the medium. For a conductor of conductivity σ, the generalized permittivity is always complex:

$$\varepsilon(\omega) = \varepsilon_0 + \sigma(\omega) / j\omega \tag{12.18}$$

The signs of ε and μ are not restricted by any physical law; they can be positive or negative. For example, in metals, plasma oscillations cause $\varepsilon(\omega) \approx \varepsilon_0(\omega) - [\omega_p^2/\omega^2] < 0$ below the UV range, i.e. at microwave frequencies, and the real part of the metal permittivity has very large negative values. At the same time, the imaginary part of the permittivity, due to the large conductivity of metals according to Equation (12.17), might dominate over the negative real part. The sign of the permeability, μ, may be negative in the region of magnetic resonances, for example as in ferrites above the FMR. Ferrites might exhibit several types of distinct spin wave modes with distinct resonance frequencies, and those might be exploited in order to create multiple band NIMs in low-loss ferrites. As a consequence, metamaterials, with both negative permittivity and permeability in the same frequency range, can be engineered.

The double-negative materials are also called left-handed media, because of the vectors $\boldsymbol{E}$, $\boldsymbol{H}$, and $\boldsymbol{k}$ form a left-handed system, and, as a result, the EMW's energy $\boldsymbol{S} = \boldsymbol{E} \times \boldsymbol{H}$ (Poynting vector) is moving opposite to the phase velocity, like a backward wave shown in Figure 12.1a. Backward waves can propagate in isotropic, double-negative media. At the interface of a usual material and that of a double-negative material, negative refraction takes place. Backward wave means that the direction of propagation of energy (or information, i.e. the Poynting vector $\boldsymbol{S}$) is opposite to the wave vector $\boldsymbol{k}$, i.e. phase velocity, in contrast to "normal" media, where $\boldsymbol{k} || \boldsymbol{S}$, as illustrated in Figure 12.1b. Experimentally observed left-handed behavior is shown for a self-biased Sr-hexaferrite rod metamaterial NIM structure in Figure 12.2.

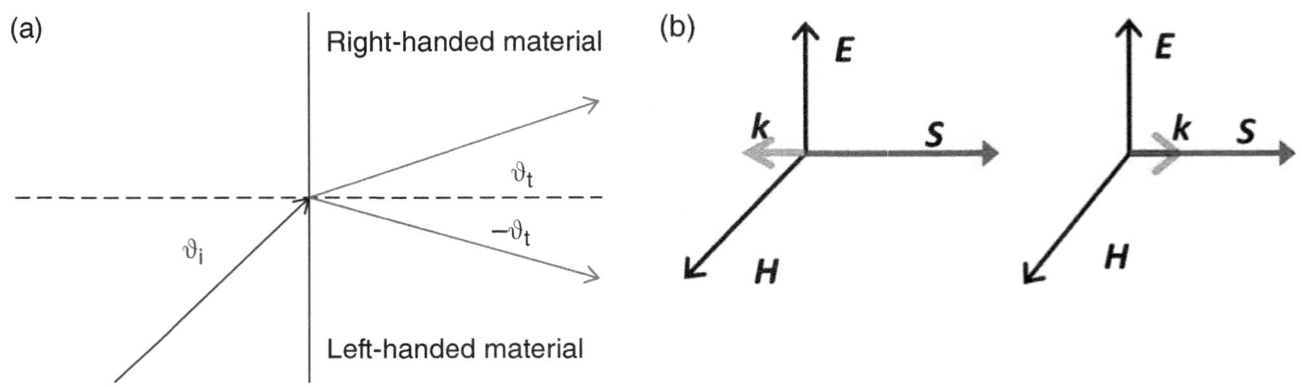

Figure 12.1 (a) Backward propagation of EMWs in a negative index material (NIM). EMW incident from left on the boundary of a normal and a double-negative metamaterial with $\varepsilon < 0$ and $\mu < 0$. Angle of incidence ϑ_i and angle of refraction ϑ_t. (b) Backward wave propagation in isotropic, double-negative medium. The direction of propagation of energy **S** (or information, i.e. the Poynting vector) is opposite to the wave vector $\boldsymbol{k}$, i.e. the phase velocity, in contrast to "normal" media, where $\boldsymbol{k} || \boldsymbol{S}$, shown on the right side.

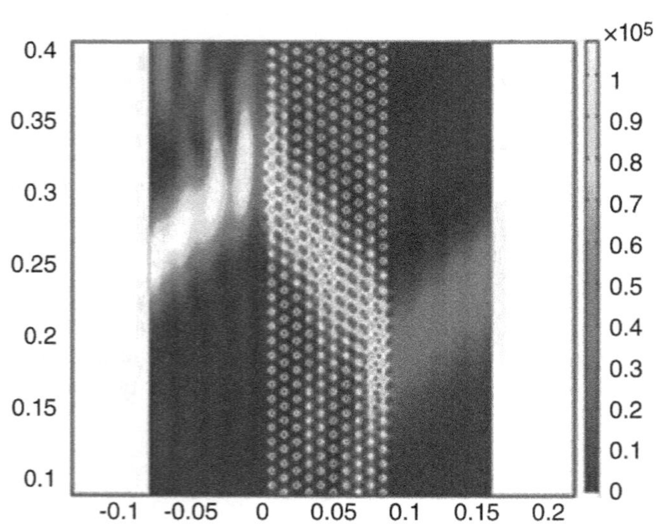

Figure 12.2 Experimentally observed left handed behavior for a self-biased Sr-hexaferrite rod metamaterial NIM structure. Measured distribution of power flow intensity around the block of ferrite rod structure at 11.1 GHz. A Gaussian beam is incident at 45° on the left surface of the ferrite rod block and refracts negatively into the block; then it has another negative refraction on the right surface of the block. Reprinted from [16], with the permission of AIP Publishing.

12.3 Modeling of Metamaterials

To describe the electromagnetic behavior, i.e. the scattering of EMWs as a result of interaction of an incident EMW with a composite medium, the knowledge of the constituent parameters of the medium is necessary. The material parameters for the MPC and NIM are essential in order to solve the nonlinear Maxwell's equations together with the Landau–Lifshitz equation to model and design materials and devices based on these periodic structures.

For composite microwave media, the common approach is to find effective material parameters for the media as a continuum, what is applicable when the characteristic length of the structure is much smaller than the operating wavelength, i.e. $\lambda/a \gg 1$. The classical approach is using the mixing rule based on the Maxwell Garnett formula [17]. Another approach is using the Bruggeman mixing rule [18]. At low-volume fractions, the two methods yield similar results. However, above the percolation limit, Maxwell Garnett predicts much lower permittivity, especially for the imaginary part. In [19], a YIG–semiconductor NIM structure was modeled as an effective continuous medium, using the transmission line approach, and the index of refraction for the composite was determined as:

$$n_{\text{eff}}^2 = \varepsilon_{\text{eff}}\mu_{\text{eff}} = \left(\frac{r\varepsilon_a + \varepsilon_b}{1+r}\right)\left(\frac{r\mu_a + \mu_b}{1+r}\right)$$

where r is the thickness ratio of the semiconductor (a) to ferrite (b) layers. It was shown that the structure is tunable, and the TM mode has a negative index.

There are several problems with modeling of EMW interaction with composite media, including metamaterials. While the unit cell of the media is small, frequently in the nanometer range, and contains only one magnetic inclusion, the general size of the device is much larger and it is in the millimeter range. Mathematical and numerical modeling of a device having two drastically different length scales presents serious difficulties. The method of decomposition of a large device into autonomous blocks, containing the inclusion, solving the EMW diffraction problem for the blocks, and then reconstituting the blocks, is a robust and electrodynamically precise method [20, 21]. This method, as applied to periodic magnetic nanocomposites, takes into account electrodynamically correct nonradiating boundary conditions, often neglected in modeling. Mixing rules do not take into account an important characteristic of the composite media. Although the small inclusion approximation is valid, as $\lambda/a \gg 1$, the periodicity of the magnetic inclusions (spheres, wires, or thin layers) is not commensurate with the wavelength, and as a result, some of the magnetic inclusions are located at zero nodes of the EM fields, some at its maxima. This situation changes with frequency, causing seemingly inconsistent results and problems with experimental verification. The use of autonomous blocks avoids this problem.

There are several other ways to describe the interaction of metamaterials with electromagnetic fields. One way is to find the eigenmodes in the material, after substituting the constitutive equations into Maxwell's equations and applying the appropriate boundary conditions. The superposition of the eigenmodes results in the field distributions in the metamaterial. From there, the propagation constants and reflection and transmission coefficients can be found [22, 23].

Another method to describe metamaterials is by building equivalent circuit models. The magnetic component is modeled as an equivalent *RLC* circuit, and then applying circuit theory, one can obtain the currents in the circuit, and from there, the frequency behavior of the magnetization as $\mu(\omega)$. The R, L, and C parameters depend on the geometry of the metamaterial [24, 25].

In the case of large scale, complicated metamaterial structures, the only way to solve the EMW scattering and propagation problems is by numerical methods, requiring large computational resources. Full-wave electromagnetic modeling is a powerful tool to describe metamaterials, using finite difference time-domain (FTDT), finite element (FE), and finite difference (FD) approaches [26]. Commercial software packages are available to model metamaterials and structures. Usually, the excitation is a harmonic EMW, and the output is either the $\boldsymbol{E}$ and $\boldsymbol{H}$ field distribution in the structures, the eigenmode representations, and/or the S matrix parameters (transmission and reflection coefficients). Simple model calculations and approximate solutions are regularly checked against the results of a full-wave analysis [27–30].

12.4 Magneto-PBG Structures and Devices

Starting from the 70s, PBG materials were in the center of electromagnetics research. These PBGs are 1D, 2D, or 3D periodically ordered structures, made from materials with dissimilar refractive indices. The periods of these superstructures are comparable with the wavelength of the incident electromagnetic radiation in the microwave or optical range. By the end of the 80s, the intensive theoretical research was followed by the investigation of tangible 1D multilayers, 2D arrays and 3D dimensional structures, built mostly on Si/SiO_2 substrates and operating in the technologically crucial near IR optical communication band [1–3].

The critical material parameters of a PBG are both ε and μ. Initially, magnetic materials were not considered in PBGs as $\mu = 1$ in the optical range. However, in the microwave range, ferrites have μ values quite different from 1, and this can be exploited for microwave MPCs. When ferrites are used in microwave devices, they usually operate in the saturated state, as in the microwave range below saturation, they are too lossy. The microwave permeability of the ferrites depends on the saturation magnetization, the microwave frequency, and the external bias magnetic field:

$$\mu = \frac{\left(\omega_{ex} + \omega_m\right)^2 - \omega^2}{\left[\omega_{ex}\left(\omega_{ex} + \omega_m\right) - \omega^2\right.}, \tag{12.19}$$

where $\omega_{ex} = \gamma H_{ex}$ and $\omega_m = 4\pi\gamma M_s$, γ is the gyromagnetic ratio, M_s is the magnetization, and H_{ex} is the dc bias magnetic field. Therefore, ferrites with applied magnetic fields make tunable MPCs possible.

By the 90s, magnetic components were inserted into the PBG structures, resulting in magnetophotonic materials (MPC), offering nonreciprocity, tunability by external magnetic field, and strong nonlinearity due to the Kerr effect, with a potential to build switchable optical circuits. At the same time, the operating frequencies were extended into the microwave range, promising novel and better microwave devices for the emerging immense communication market. By the early 2000s, the field matured to the point of summarization of the theoretical and experimental accomplishments, and it was published in several reviews on MPC [4, 10, 11].

In an early paper on experiments and modeling of a PBG structure, a microstrip line was built with a periodic hole array in a Duroid substrate. The introduction of a defect created a single transmission mode in the PBG. The structure exhibited a PBG between 9.85 and 13.54 GHz. The authors, using techniques from microwave engineering, proposed an interpretation that the bandgap is the result of the behavior of the wave impedance at the interface, which depends on $Z = \sqrt{\mu/\varepsilon}$, rather than the index of refraction $n = \sqrt{\mu\varepsilon}$. According to the authors, the wave impedance plays the essential role by means of the boundary conditions at interfaces, but not through Maxwell's equations, in which the refractive index is important, although the position of bandgaps and the formation of defect modes are related to the refractive index [31]. Figure 12.3 shows a 1D MPC consisting of alternating nanolayers of Bi:YIG and SiO_2 and a periodic PBC structure containing a Bi:YIG defect layer. Figure 12.4 illustrates the optical and magneto-optical spectra of the defect structure in Figure 12.3a [4].

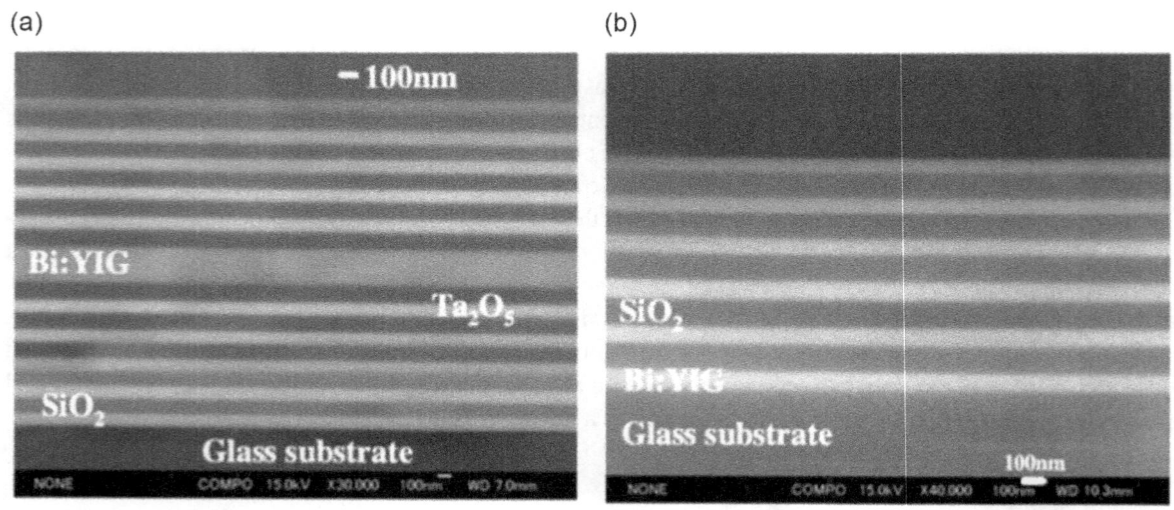

Figure 12.3 (a) Cross-sectional TEM image of a Ta_2O_5 and SiO_2 periodic 1D structure, with a Bi:YIG magnetic defect layer inclusion, (b) $(Bi{:}YIG/SiO_2)_n$ one-dimensional magnetophotonic crystal based on a periodic structure of 98 nm thick bismuth substituted yttrium iron garnet and 147-nm-thick SiO layers on a glass substrate. Reproduced from [4] with the permission of ©IOP Publishing. All rights reserved.

Modeling and device research both intensified from the early 2000s. Several experimental structures were built, measured, and reviewed in [32]. Band structures and possible violations of periodicity (Tamm states, defect modes) and the influence of external magnetic field on the spectral characteristics of MPC are described. Plane wave propagation, transmission spectra, and experiments for several basic structures were demonstrated for 1D multilayered ferrite–dielectric structures, including ferrite or semiconductor defect elements. Theoretical analysis based on Maxwell's equations with proper boundary conditions, the corresponding band structures, and experimental realization are provided for several interesting cases, for example for combined bandgap and metamaterial structures, for the case of two interfacing MPCs, for MPC and semiconductor interface properties are treated in [32].

Device development goes hand in hand with material and structure optimization. An MPC-based microwave switch was proposed in [33]. For this purpose, a frequency-selective multiplexer was implemented combining 1D MPC structures. The tunability of microwave photonic crystals (MPCs) was achieved with ferrites and/or ferroelectric materials.

A theoretical calculation performed for a 2D hexagonal lattice of air rods embedded in a rectangular ferrite slab shows that the MPC is well tunable with an external magnetic field, applied parallel to the air rod axis. The implementation of a microwave switch was also discussed. The photonic band structure was modeled for a bulk magnetically saturated ferrite material having a triangular array of circular holes. The band structure and the bandgap are illustrated in Figure 12.5.

A bias field is applied along the axis of the holes. In this 2D MPC, the ferrite can affect the transverse electric TE mode whose electric field lies in the z-axis but not the transverse magnetic TM mode. The photonic bandgaps of the TE mode shift to lower frequencies, and their widths become narrower compared to the case when the relative magnetic permeability $\mu = 1$. Both the frequency position and the width of the photonic bandgap increase as the bias field H_{ex} increases but the gap never becomes larger than that for $\mu = 1$ [34].

The propagation characteristics of spin waves in 2D MPCs formed by a periodic structure in yttrium iron garnet (YIG) film based on a 2D array of etched holes, was studied by atomic force and magnetic force magnetometry, and the MPCs were measured and analyzed in [35].

An early MPC transmission line approach to form planar microwave devices and circuits based on MPCs was proposed in [36]. Materials used were dielectric substrates, containing nanoscaled ferromagnetic inclusions that can be arranged in a periodic way with specific impedances and phase velocities. They form photonic or periodic bandgap MPC substrates, on which planar guiding structures are deposited. Using the transmission line concept, defect modes in PBG planar devices were studied. A theoretical explanation of the defect mode operation, validated by measurement on a dielectric PBG structure, was proposed. It is then transposed to a magnetic PBG structure based on arrays of nanoscaled ferromagnetic

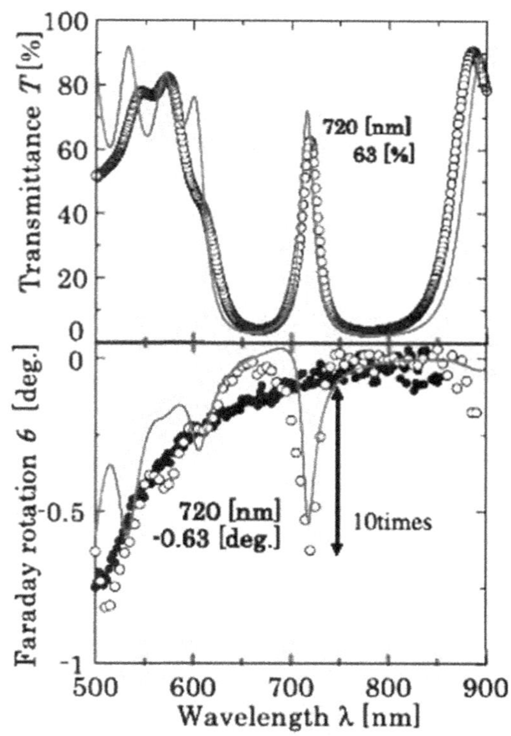

Figure 12.4 Optical and magneto-optical response of the one-dimensional (1D) MPC with the Bi:YIG defect layer in the $(Ta_2O_5/SiO_2)_n$ structure shown in Figure 12.3a. Solid lines are simulated spectra, open symbols are for the 1D MPC cavity of Figure 12.3a, filled symbols are for comparison with a single layer of Bi:YIG. Reproduced from [4], with the permission of ©IOP Publishing. All rights reserved.

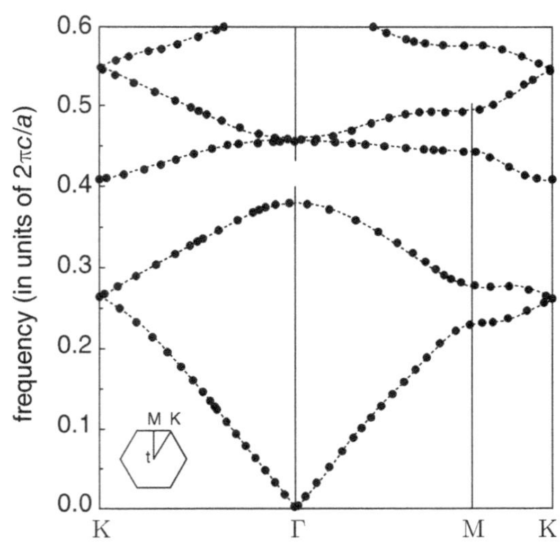

Figure 12.5 The lowest five photonic bands for the TE mode of a triangular lattice of air rods in a ferrite medium with $\mu = 1.0$. The frequency is normalized to $2\pi c/a$, where a is the lattice constant and c is the light velocity. The radius of the air rods is $0.34a$ and the ferrite medium has $\varepsilon = 12.25$. Reprinted with permission from [34]. Copyright (2000) by the American Physical Society.

nanowires. The effect on its frequency response of varying defect length was discussed, in order to design planar sensors and filters.

Magnetically tunable wideband microwave bandstop and bandpass filters using FMR absorption in yttrium iron garnet/gadolinium gallium garnet–gallium arsenide (YIG/GGG-GaAs) layered structures are reported in [37]. The measured characteristics of the bandstop filter, with nonuniform bias magnetic field distribution along the YIG/GGG film, shows a large absorption bandwidth of 1900 MHz at the stopband center frequency of 11.0 GHz, average peak absorption of 13 dB, and an out-of-stopband insertion loss of 1.5 dB. The measured characteristics of the bandpass filter at the passband center frequency of 7.1 GHz demonstrates a 3 dB bandwidth of 1450 MHz, out-of-band rejection of 33 dB, and an insertion loss of 1.5 dB. The measured transmission characteristics of the device are shown in Figure 12.6. A good agreement between the experimental results and the simulation results based on a lumped element equivalent circuit modeling of the YIG filters has been achieved.

A high-efficiency, T-shaped magneto-optical circulator, constructed in a 2D square-lattice photonic crystal, was recently proposed in [38]. In the T-shaped structure, the single-direction light transmission for 90° light bending is achieved by coupling two magneto-optical rods. Aided with a side-coupled cavity, two paths of single-direction 90° light bending are effectively combined, which realizes the nonreciprocal light transmission for two waveguides arranged and linked along a straight line. The optical properties of the system are investigated by FE method. The circulator considered here can be used for isolating light reflections and improving system stabilization in designing photonic crystal integrated circuits.

Omnidirectional total reflection (ODTR) has a wide variety of applications, such as in optical communication, electromagnetic energy transportation, and solar energy engineering. Such a photonic structure made of dielectric and magnetic 1D photonic crystals was proposed in [39]. Simulations by the optical transmission matrix method demonstrate that such structures can achieve ODTR in a wide wavelength range. In such structures, the total reflections of the *s*-polarization part and *p*-polarization part of the incident wave are caused by the 1D dielectric sub-PC and the 1D magnetic sub-PC, respectively.

Omnidirectional complete bandgap behavior can be achieved in 3D MPC structures. From the very early times of PBG research, opals were a favorite structure to build 3D photonic crystals [2]. Opal structures have voids inside the network of 100–200 nm size dielectric SiO_2 spheres, forming the close-packed 3D opal network. These voids can be infiltrated chemically by magnetic materials, first of all ferrites, creating a 3D MPC. The propagation of EMW at microwave frequencies was investigated in SiO_2 opal-based magnetic nanostructures, using rigorous mathematical models solving Maxwell's equations with electrodynamic boundary conditions, complemented by the Landau–Lifshitz equation. The numerical approach is based on the Galerkin's projection method using the decomposition algorithm on autonomous blocks with Floquet channels. The opal structure consists of 250 nm SiO_2 nanospheres, with intersphere voids infiltrated with nanoparticles of either $Ni_{0.7}Zn_{0.3}Fe_2O_4$, with $4\pi M_s = 5$ kG, or $NiFe_2O_4$ with $4\pi M_s = 3.12$ kG. Both the opal matrix and the ferrite are assumed to be lossy through complex dielectric constants. The field dependence of the complex wave number of the fundamental extraordinary mode of the propagating EMWs in the 3D opal-based MPCs was determined for transverse orientation of the bias magnetic field at a frequency of 9.375 GHz, as illustrated in Figure 12.7. The numerical technique shows an excellent agreement when applied to model recent experimental data of waveguide measurements on similar ferrite opals [40].

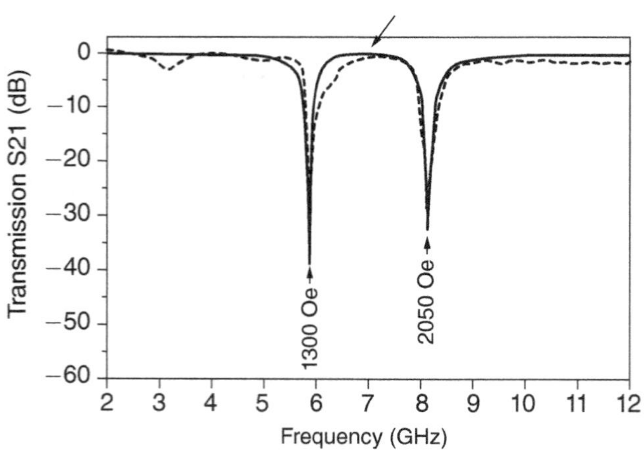

Figure 12.6 Comparison of measured and simulated transmission characteristics of a YIG/GGG-GaAs-based microwave bandpass filter. ©(2005) IEEE. Reprinted with permission from [37].

Observation of magnetic antiresonance phenomenon is reported in similar 3D opal nanocomposite with embedded ferrite particles. At antiresonance, there is a sharp maximum of the reflection coefficient of an EMW. Antiresonance at microwave frequencies of millimeter wave band was observed as a minimum of absorbed power. Measurements were carried out in the frequency range from 26 to 38 GHz for embedded ferrite particles, $Co_{0.5}Zn_{0.5}Fe_2O_4$ and $Ni_{0.5}Zn_{0.5}Fe_2O_4$. The physical nature of antiresonance is discussed in [41].

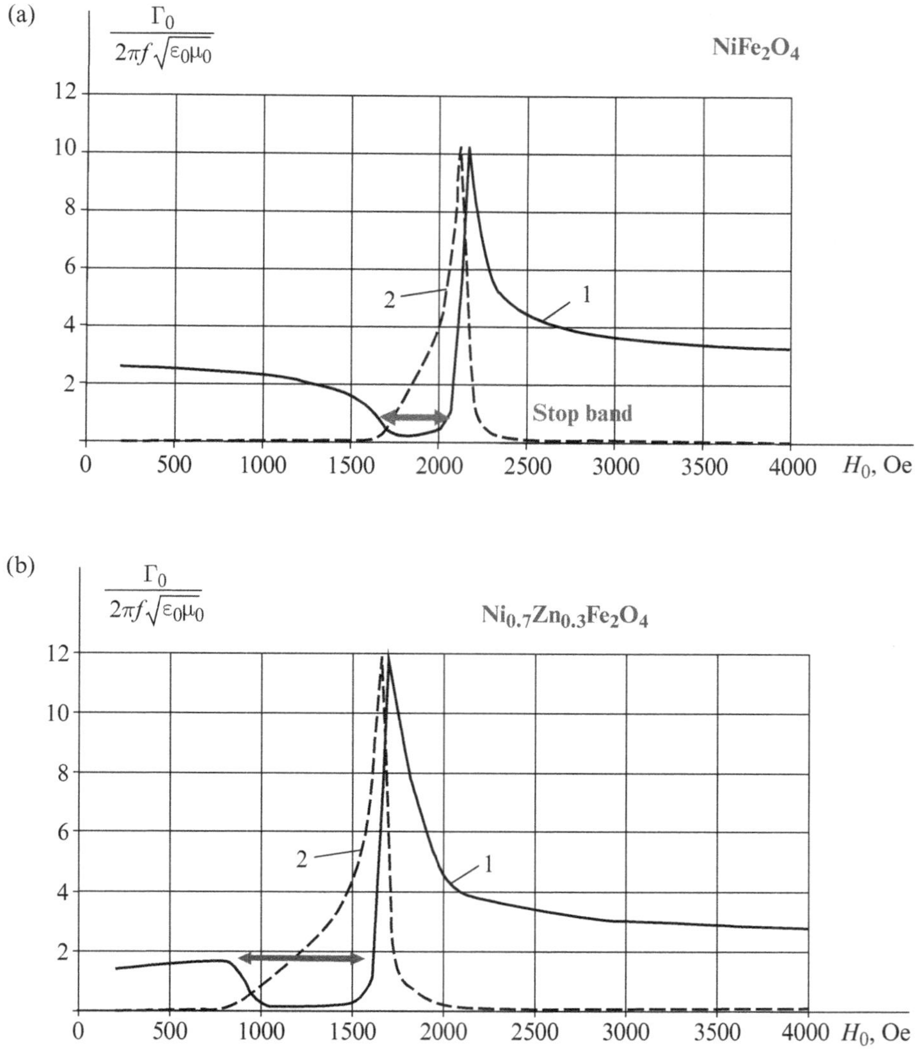

Figure 12.7 Real and imaginary parts of complex wave number Γ_0 of the fundamental mode of quasi-extraordinary EMW in 3D opal-based MPC, depending on transverse bias magnetic field $\boldsymbol{H}_0 = H_0\,\boldsymbol{x}_0$ for magnetic nanoparticles of (a) $NiFe_2O_4$: $4\pi M_s = 3.2$ kG; (b) $Ni_{0.7}Zn_{0.3}Fe_2O_4$: $4\pi M_s = 3.2$ kG; SiO_2 opal nanospheres $r = 125$ nm, $f = 9.375$ GHz, 1—Re Γ_0; 2—Im Γ_0. ©(2011) IEEE. Reprinted with permission from [40].

12.5 Negative Index Ferrite Metamaterial (NIM) Structures and Devices

Following the discovery of double-negative index metamaterials (NIM), first demonstrated on large-size metallic *RLC* structures, soon the attention turned to small microwave structures utilizing ferrites, providing negative permeability $\mu < 0$ above the FMR range, usually in combination with a metallic component offering $\varepsilon < 0$, the negative permittivity, due to plasmonic effects, as described in Section 12.2. NIMs are, in fact, bandpass type structures, where the μ and ε relations are valid in restricted frequency range of the passband. The exact broadband dispersion relations can be obtained from the Bloch–Floquet theory [24].

Double-negative electromagnetic metamaterials are usually constructed by combining a conducting component (thin film, wires, or semiconductor) serving as the source of negative ε and a magnetic subsystem (magnetized ferrite) with $\mu < 0$. The characteristic size of the structure should be much smaller than the order of the operating wavelength of EMW. Microwave applications are a direct consequence of this fact. The transmission of EMW of normal incidence on the NIM, composed of an infinite structure of alternating ferrite and semiconductor layers, is derived in [32], and the results of simulation show the existence of the backward wave and its attenuation into the medium. The authors performed T-junction waveguide microwave experiments in the 22–40 GHz range in bias fields from 0 to 11 kG on a similar structure and found good agreement with the results of modeling, as shown in Figure 12.8.

Theoretical treatment and experimental results were given for a finite length stack of a fine stratified ferrite–semiconductor structure, composed of periods of InSb semiconductor/1SCh4-ferrite in [42]. It was shown that the propagation of EMW can be controlled by external magnetic field. The transitional region between uniform medium and photonic crystal corresponds to the thickness of the ferrite layer $d_1 = 2.0, 2.5, 3.0$, and 3.5 mm. Figure 12.9 shows the measured band spectrum of the ferrite–semiconductor periodic structure with different thicknesses of the ferrite layer, measured at constant magnetic field H_{ex} = 5.24 kOe. The transmission bands are indicated by gray hatching. It is shown that while the thickness of the elements decreases, the transition bands shift to higher frequencies and become broader. In the limit of very small layer thickness, the band structure of the spectrum tends to disappear.

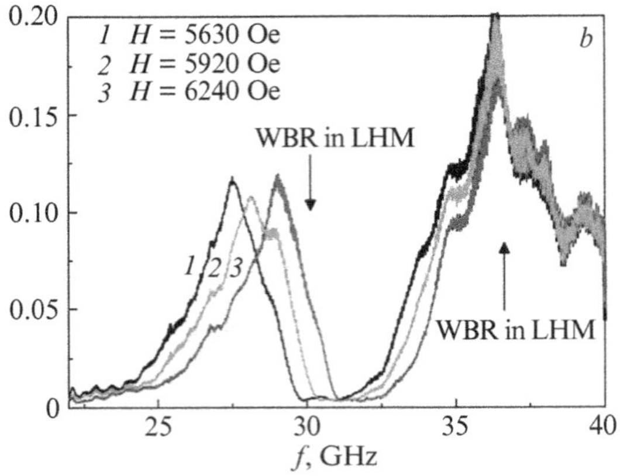

Figure 12.8 Measured bias field dependence of the transmission spectra for a layered ferrite (1SCH4)/ InSb semiconductor 1D MPC structure. The lower frequency peaks are for the left-handed medium (LHM), the peaks on the high-frequency side are the regular Bragg peaks (WBR) for the right-handed region (RHM). Reprinted from [32], with the permission of AIP Publishing.

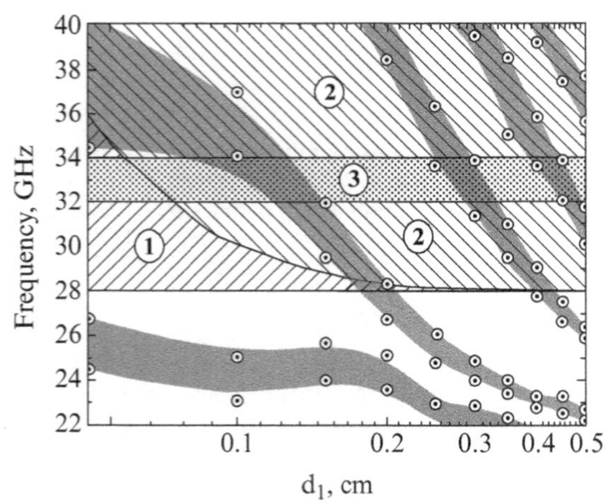

Figure 12.9 Measured and simulated transmission band spectral structure of 10 periods of InSb semiconductor/1SCh4 $NiOFe_2O_3$ ferrite-layered periodic structure versus the thickness of the ferrite layer, measured in a bias magnetic field of H_{ex} = 5.24 kOe. The transmission bands are indicated by gray hatching. The transition regions between uniform medium and photonic crystal are 1 and 3. The blank region is bigyrotropic, and the hatched region 2 is a photonic crystal. ©(2011) IEEE. Reprinted with permission from [42].

Significant efforts were devoted to designing and optimizing various ferrite-based metamaterial structures, e.g. as in [43]. Negative index materials were made of YIG–semiconductor and YIG–metal mesh structures, and it was shown that the losses are reduced for the ferrite-mesh structure. Figure 12.10 demonstrates tunability by magnetic field, shifting the negative index frequency range for a YIG-doped Si metamaterial. Theoretical and experimental methods of verifying the NIM character of a 2D ferrite rods–metallic wires metamaterial are described in [44]. It is possible to construct dual-band NIM metamaterials, as it was demonstrated for a wire – ferrite rod structure, where the dual passbands are due to two distinct FMR frequencies [45].

Surprisingly, not much interest was given to temperature-dependent properties of the metamaterials and devices, in spite of the well-known temperature variations of the underlying properties. Based on temperature dependent measurements, it is suggested that the temperature sensitivity of NIM structures could be used for tuning of NIM devices. Upon increasing the temperature, the passband frequency range shifts to lower frequencies [46].

One of the most important microwave component is the phase shifter. Phase shifters are critical elements for electronically steerable antennas, employed in military and civilian radar and communication systems. A bias magnetic field is applied to change the permeability of the ferrite and thus the refractive index and the phase velocity of the signal are changing; as a result, the phase of the transmission coefficient changes too. Tunable ferrite-based NIM metamaterials opened new ways to design and build efficient phase shifters. Built on YIG slabs and plasmonic copper wires, the microstrip device has a low insertion loss of 5 dB at the center of the transmission band. The YIG component enabled the magnetic field tuning of the negative refractive index in a dynamic frequency band from 7.0 to 11.0 GHz. The insertion phase can be tuned by 45° by continuously varying the bias field from 3.8 to 4.6 kOe at 9.0 GHz [47]. A broadband ferrite tunable phase shifter is proposed and validated by full-wave simulations in [48].

Hexagonal ferrites make it possible to design and fabricate non-biased microwave devices, eliminating the bulky and expensive permanent magnets [49]. A hexagonal ferrite-based tunable NIM was tested in [50]. The structure consists of one slab of single-crystalline scandium-doped barium hexaferrite (Sc-BaM), aligned parallel to two rows of periodic copper wires. The presence of magnetic field-tunable passband was measured, indicating the occurrence of negative index n. The center frequency of the 5 GHz wide passband, having a transmission peak of −13 dB, is shifted linearly from 40.9 to 43.9 GHz by varying the bias field (H_{ex}) from 4.0 to 7.0 kOe. The structure and its measured transmission characteristics are shown in Figure 12.11 [51].

A new type of low loss Sr hexaferrite (FB6H, TDK) was shown to be a good candidate for high-efficient, sub-millimeter range nonreciprocal devices in [52]. A full-band microwave isolator, working over the entire X-band spectrum, was constructed from metal wire–coated ferrite (YIG) disks, and the structure was analyzed taking into account the plasmon modes. The results of S parameter measurements agree with the analysis of the new measurement method [53].

Some composite metamaterials might show right- or left-handed (CRLH) behavior, depending on the frequency range. Due to unique dispersion characteristics, electronic beam steering is possible by varying either the operational frequency or the bias field. This offers a promising opportunity for application of ferrite loaded transmission line NIM metamaterials as a new type of leaky-wave antenna. The structure is a combination of left-handed and righ-thanded metamaterials, with a gapless transition between the two sides, and unique

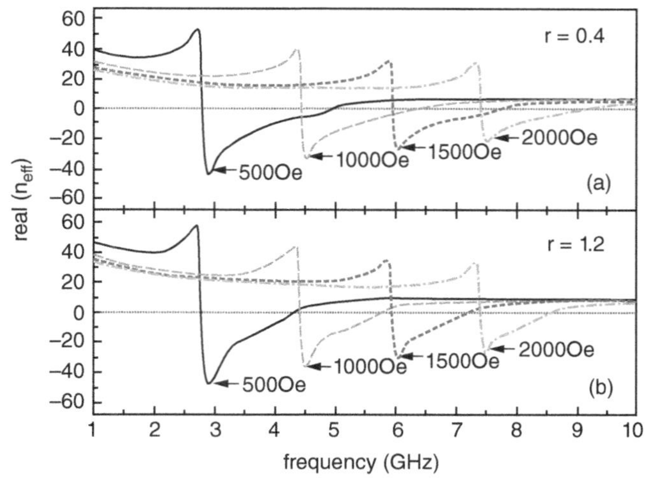

Figure 12.10 Bias field tuning of the frequency range for a negative index ferrite-doped Si metamaterial, for different thickness ratios r of the components: (a) r = 0.4 and (b) r = 1.2. Reprinted from [19], with the permission of AIP Publishing.

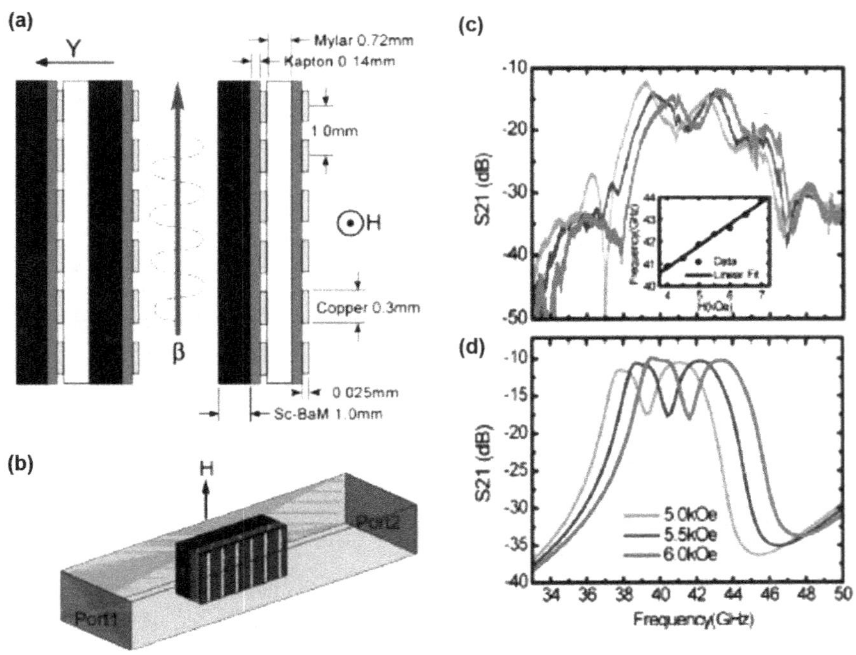

Figure 12.11 (a) Top views of the NIM design, consisting of two Sc-BaM slabs, two rows of copper wires on Kapton substrate, a Mylar spacer, and the simplified NIM consisting of only one Sc-BaM slab; (b) Schematic drawing of the NIM composite mounted in a Q-band rectangular waveguide; (c) measured and (d) simulated S_{21}s of the NIM composite containing a 1.0-mm-thick Sc-BaM slab in bias fields of 5.0, 5.5, and 6.0 kOe, respectively. Inset: Measured center frequency of the NIM passband versus the bias magnetic field. Reproduced from [51], with the permission of © IOP Publishing. All rights reserved.

dispersion characteristics. The ferrite-filled rectangular waveguide is open from one side and biased in an orthogonal direction. It is a low-loss structure with a full-space scanning. The design and analysis of such a proposed structure is described in [54]. A similar antenna, tunable in the 30 GHz band, is described in [55]. Another interesting feature of the CRLH media is the highly directed and efficient radiation from such an antenna. An asymmetrically loaded coplanar configuration is proposed with strong nonreciprocity in the phase constant, which means larger scanning angles for leaky-wave antenna applications [56].

Nonreciprocity is one of the critical parameters of the metamaterial structures. A theoretical analysis of the EMW transmission and dispersion characteristics shows that it is possible to create nonreciprocal transmission even above the FMR frequency [57].

12.6 Future Trends

Both bandgap (MPC) and negative index (NIM)-type metamaterials stimulate the development of novel microwave devices. The manufacturing of electromagnetic metamaterials is becoming a maturing technology. With the progress in flexible device fabrication technologies, new roads open up for miniaturizing of multiscale and multiphysics integrated devices, having functionalized components from diverse classes of natural materials, combined to utilize the new physics, materials science, and microwave engineering of metamaterials.

Ferrites are good candidates to serve as the magnetic component in both MPC and NIM applications. A review of properties and applications, demonstrating the progress of ferrite manufacturing and applications, was published recently in [58]. Soft magnetic ferrites are essential materials for microwave devices up to the sub-millimeter wave range due to their high permeability, moderate to high permittivity, high Q, low conductivity and low losses, nonreciprocity in propagation of EMW, and tunability by bias magnetic fields. Ferrites are high-quality, robust, and stable materials, with well-established technologies of preparation. Low-cost, high-power, lightweight, small-size devices are necessary for application in communication, signal processing, radar, and other information management devices (phase shifters, circulators, isolators, etc.). Ferrites satisfy these requirements. The advantages outweigh their relatively low magnetization and difficulties in integration with high-frequency planar semiconductors. The advantages of microwave devices, built on ferrite – semiconductor or ferrite – metal metamaterial structures prove the potential of the extended functionality of these materials.

References

1 John, S. (1991). Localization of light. *Physics Today* 44 (5): 32–40.

2 Joannopoulos, J.D., Johnson, S.G., Winn, J.N., and Meade, R.D. (2011). *Photonic Crystals: Molding the Flow of Light*. Princeton: Princeton University Press.

3 Joannopoulos, J.D., Villeneuve, P.R., and Fan, S. (1997). Photonic crystals: Putting a new twist on light. *Nature* 386 (6621): 143–149.

4 Inoue, M., Fujikawa, R., Baryshev, A., Khanikaev, A., Lim, P.B., Uchida, H., Aktsipetrov, O., Fedyanin, A., Murzina, T., and Granovsky, A. (2006). Magnetophotonic crystals. *Journal of Physics D: Applied Physics* 39 (8): R151–R161.

5 Veselago, V.G. (1968). The electrodynamics of substances with simultaneously negative values of ϵ and μ. *Soviet Physics Uspekhi* 10: 509–514.

6 https://en.wikipedia.org/wiki/Plasma_oscillation.

7 Soohoo, R.F. (1985). *Microwave Magnetics*. New York: Harper and Row Publishers.

8 European Commission. (2010). *Nanostructured Metamaterials* (ed. A.F. de Baas). Luxembourg: Publication Office of the European Union. doi: 10.2777/54953.

9 Makeeva, G.S., Golovanov, O.A., and Pardavi-Horvath, M. (2006). Numerical analysis of electromagnetic wave instability in nonlinear ferrite structures using bifurcation points of the nonlinear Maxwell's operator. *IEEE Transactions on Magnetics* 42: 3550–3552.

10 Lyubchanskii, I.L., Dadoenkova, N.N., Lyubchanskii, M.I., Shapovalov, E.A., and Rasing, T. (2003). Magnetic photonic crystals. *Journal of Physics D: Applied Physics* 36 (18): R277–R287.

11 Kruglyak, V.V., Demokritov, S.O., and Grundler, D. (2010). Magnonics. *Journal of Physics D: Applied Physics* 43 (26): 264001.

12 Khanikaev, A.B., Baryshev, A.V., Fedyanin, A.A., Granovsky, A.B., and Inoue, M. (2007). Anomalous Faraday effect of a system with extraordinary optical transmittance. *Optics Express* 15 (11): 6612–6622. doi: 10.1364/OE.15.006612.

13 Sigalas, M.M., Soukoulis, C.M., Biswas, R., and Ho, K.M. (1997). Effect of the magnetic permeability on photonic band gaps. *Physical Review B* 56 (3): 959–962.

14 Kim, K., Lee, D.-H., and Lim, H. (2005). Theory of the propagation of coupled waves in arbitrarily inhomogeneous stratified media. *EPL (Europhysics Letters)* 69 (2): 207–213. doi: 10.1209/epl/i2004-10341-0.

15 Belov, P.A., Tretyakov, S.A., and Viitanen, A.J. (2002). Nonreciprocal microwave band-gap structures. *Physical Review E* 66: 016608.

16 Gu, Y., Wu, R.X., Yang, Y., Poo, Y., Chen, P., and Lin, Z. (2013). Self-biased magnetic left-handed material. *Applied Physics Letters* 102 (23): 231914.

17 Garnett, J.-C.M. (1906). Colours in metal glasses, in metallic films, and in metallic solutions, II. *Philosophical Transactions of the Royal Society of London. Series A*203 (359–371): 237–288.

18 Sihvola, A.H. (ed.) (1999). *Electromagnetic Mixing Formulas and Applications*, No. **47**. IET.

19 Zou, D.Y., Jiang, A.M., and Wu, R.X. (2010). Ferromagnetic metamaterial with tunable negative index of refraction. *Journal of Applied Physics* 107 (1): 013507. doi: 10.1063/1.3275857.

20 Makeeva, G.S., Golovanov, O.A., Pardavi-Horvath, M., and Kouzaev, G.A. (2008, July). Decomposition approach to nonlinear diffraction problems of nanoelectromagnetics and nanophotonics using autonomous blocks with Floquet channels. In: *Advanced Applications of Electrical Engineering. Proceedings of the 7th WSEAS International Conference on Application of Electrical Engineering (AEE'08)*, 2–4.

21 Golovanov, O.A. and Makeeva, G.S. (2009). The method of autonomous blocks with magnetic nanoinclusions and Floquet channels applied for simulation of magnetic nanostructures with allowance for the exchange and boundary conditions. *Journal of Communications Technology and Electronics* 54 (12): 1345–1352.

22 Golovanov, O.A. (2006). Solution of applied problems of electromagnetics based on the use of autonomous blocks with virtual Floquet channels. *Journal of Communications Technology and Electronics* 51 (12): 1338–1345.

23 Nikol'skii, V.V. and Lavrova, T.I. (1979). Application of the method of minimum autonomous blocks to the problem of the propagation of radiation in a nonlinear medium. *Radiophysics and Quantum Electronics* 22 (9): 761–768.

24 Caloz, C. and Itoh, T. (2005). *Electromagnetic Metamaterials: Transmission Line Theory and Microwave Applications*. John Wiley & Sons.

25 Bilotti, F., Toscano, A., Vegni, L., Aydin, K., Alici, K.B., and Ozbay, E. (2007). Equivalent-circuit models for the design of metamaterials based on artificial magnetic inclusions. *IEEE Transactions on Microwave Theory and Techniques* 55 (12): 2865–2873.

26 Bergmann, T., Robertsson, J.O., and Holliger, K. (1998). Finite-difference modeling of electromagnetic wave propagation in dispersive and attenuating media. *Geophysics* 63 (3): 856–867.

27 Cui, T.J., Smith, D.R., and Liu, R.P. (2014). *Metamaterials*. New York: Springer.

28 Kern, D.J., Werner, D.H., and Lisovich, M. (2005). Metaferrites: Using electromagnetic bandgap structures to synthesize metamaterial ferrites. *IEEE Transactions on Antennas and Propagation* 53 (4): 1382–1389.

29 Yang, F. and Rahmat-Samii, Y. (2009). *Electromagnetic Band Gap Structures in Antenna Engineering*. Cambridge, UK: Cambridge University Press.

30 Kono, N. and Tsuji, Y. (2004). A novel finite-element method for nonreciprocal magneto-photonic crystal waveguides. *Journal of Lightwave Technology* 22 (7): 1741.

31 Kee, C.-S., Kim, J.-E., Park, H.Y., Kim, S.J., Song, H.C., Kwon, Y.S., Myung, N.H., Shin, S.Y., and Lim, H. (1999). Essential parameter in the formation of photonic band gaps. *Physical Review E* 59 (4): 4695–4698. doi: 10.1103/PhysRevE.59.4695.

32 Tarapov, S.I. and Belozorov, D.P. (2012). Microwaves in dispersive magnetic composite media. *Low Temperature Physics* 38 (7): 603–625.

33 Kee, C.-S., Kim, J.-E., Park, H.Y., and Lim, H. (1999). Roles of wave impedance and refractive index in photonic crystals with magnetic and dielectric properties. *IEEE Transactions on Microwave Theory and Techniques* 47: 2148–2150.

34 Kee, C.-S., Kim, J.-E., Park, H.Y., Park, I., and Lim, H. (2000). Two-dimensional tunable magnetic photonic crystals. *Physical Review B* 61 (23): 15523–15525. doi: 10.1103/PhysRevB.61.15523.

35 Nikitov, S.A., Tsai, C.S., Gulyaev, Y.V., Filimonov, Y.A., Volkov, A.I., Vysotskii, S.L., and Tailhades, P. (2004). YIG thin film-based two-dimensional magnonic and magneto-photonic crystals. *MRS Proceedings* 834, J2.2. doi: 10.1557/PROC-834-J2.2.

36 Saib, A. and Huynen, I. (2005). Transmission lines on periodic bandgap metamaterials: From microwaves to optics applications. *Journal of Optics A: Pure and Applied Optics* 7 (2): S124. doi: 10.1088/1464-4258/7/2/017.

37 Tsai, C.S., Qiu, G., Gao, H., Yang, L.W., Li, G.P., Nikitov, S.A., and Gulyaev, Y. (2005). Tunable wideband microwave band-stop and band-pass filters using YIG/GGG-GaAs layer structures. *IEEE Transactions on Magnetics* 41: 3568–3570. doi: 10.1109/TMAG.2005.855191.

38 Wang, Q., Ouyang, Z., Tao, K., Lin, M., and Ruan, S. (2012). T-shaped optical circulator based on coupled magneto-optical rods and a side-coupled cavity in a square-lattice photonic crystal. *Physics Letters A* 376 (4): 646–649. doi: 10.1016/j.physleta.2011.11.032.

39 Ouyang, Z., Mao, D., Liu, C.P., and Wang, J.C. (2008). Photonic structures based on dielectric and magnetic one-dimensional photonic crystals for wide omnidirectional total reflection. *Journal of the Optical Society of America B-Optical Physics* 25 (3): 297–301. doi: 10.1364/JOSAB.25.000297.

40 Pardavi-Horvath, M., Makeeva, G.S., and Golovanov, O.A. (2011). Interactions of electromagnetic waves with 3-D opal-based magnetophotonic crystals at microwave frequencies. *IEEE Transactions on Magnetics* 47 (2): 341–344.

41 Ustinov, V.V., Rinkevich, A.B., Perov, D.V., Samoilovich, M.I., and Klescheva, S.M. (2012). Anomalous magnetic antiresonance and resonance in ferrite nanoparticles embedded in opal matrix. *Journal of Magnetism and Magnetic Materials* 324 (1): 78–82.

42 Bulgakov, A.A., Girich, A.A., Shramkova, O.V., and Tarapov, S.I. (2013, June). Transmission and reflection of electromagnetic waves from ferrite-semiconductor periodic multilayered structures. In: *2013 International Kharkov Symposium on Physics and Engineering of Microwaves, Millimeter and Submillimeter Waves (MSMW)*, 136–138. IEEE.

43 Dewar, G. (2005). A thin wire array and magnetic host structure with n< 0. *Journal of Applied Physics* 97 (10): 10Q101. doi: 10.1063/1.1846032.

44 Poo, Y., Wu, R.X., He, G.H., Chen, P., Xu, J., and Chen, R.F. (2010). Experimental verification of a tunable left-handed material by bias magnetic fields. *Applied Physics Letters* 96 (16): 161902.

45 Bi, K., Zhou, J., Zhao, H., Liu, X., and Lan, C. (2013). Tunable dual-band negative refractive index in ferrite-based metamaterials. *Optics Express* 21 (9): 10746–10752.

46 Bi, K., Dong, G., Fu, X., and Zhou, J. (2013). Ferrite based metamaterials with thermo-tunable negative refractive index. *Applied Physics Letters* 103 (13): 131915, 131915-4. doi: 10.1063/1.4823598.

47 He, Y., He, P., Harris, V.G., and Vittoria, C. (2006). Role of ferrites in negative index materials. *IEEE Transactions on Magnetics* 42 (10): 2852–2854.

48 Abdalla, M.A. and Hu, Z. (2010, July). Ferrite tunable metamaterial phase shifter. In: *2010 IEEE Antennas and Propagation Society International Symposium*, 1–4. IEEE.

49 Geiler, A., Daigle, A., Wang, J., Chen, Y., Vittoria, C., and Harris, V. (2012). Consequences of magnetic anisotropy in realizing practical microwave hexaferrite devices. *Journal of Magnetism and Magnetic Materials* 324 (21): 3393–3397.

50 He, P., Gao, J., Marinis, C.T., Parimi, P.V., Vittoria, C., and Harris, V.G. (2008). A microstrip tunable negative refractive index metamaterial and phase shifter. *Applied Physics Letters* 93 (19): 193505. doi: 10.1063/1.3025303.

51 He, P., Gao, J., Chen, Y., Parimi, P.V., Vittoria, C., and Harris, V.G. (2009). Q-band tunable negative refractive index metamaterial using Sc-doped BaM hexaferrite. *Journal of Physics D: Applied Physics* 42 (15): 155005. doi: 10.1088/0022-3727/42/15/155005.

52 Yang, B., Donnan, R.S., Dubrovka, R., and Tang, W. (2011). Negative permeability characterization of gyrotropic hexaferrite in the millimeter wave band for engineering of double-negative devices. *Journal of Applied Physics* 109 (10): 104505.

53 Rahman, N., Obol, M., Sharma, A., and Afsar, M.N. (2010). Determination of negative permeability and permittivity of metal strip coated ferrite disks using the transmission and reflection technique. *Journal of Applied Physics* 107 (9): 09A513.

54 Kodera, T. and Caloz, C. (2009). Uniform ferrite-loaded open waveguide structure with CRLH response and its application to a novel backfire-to-endfire leaky-wave antenna. *IEEE Transactions on Microwave Theory and Techniques* 57 (4): 784–795.

55 Simion, S., Marcelli, R., Bartolucci, G., Craciunoiu, F., Lucibello, A., De Angelis, G., Muller, A.A., Bunea, A.C., and Sajin, G. (2010). Composite Right/Left Handed (CRLH) based devices for microwave applications. *Ch.6.* In: *Advanced Microwave and Millimeter Wave Technologies: Semiconductor Devices Circuits and Systems* (ed. M. Mukherjee), 89–112. ISBN: 978-953-307-031-5, InTech. doi: 10.5772/8742.

56 Porokhnyuk, A., Ueda, T., Kado, Y., and Itoh, T. (2014). Phase-constant-nonreciprocal composite right/left-handed metamaterials based on coplanar waveguides. *Journal of Applied Physics* 115 (17): 17E519.

57 Butylkin, V.S., Kraftmakher, G.A., and Mal'Tsev, V.P. (2013). The nonreciprocity of microwave transmission along a bianisotropic-ferrite metastructure. *Journal of Communications Technology and Electronics* 58 (6): 543–550.

58 Harris, V.G. (2012). Modern microwave ferrites. *IEEE Transactions on Magnetics* 48 (3): 1075–1104.

Index

Note: Page numbers followed by "*f*" denotes figure and "*t*" denotes table respectively.

Modern Ferrites: Basic Principles, Processing and Properties, Volume 1, First Edition. Edited by Vincent G. Harris.

d

t